Advances in Experimental Medicine and Biology

Volume 993

Advances in Experimental Medicine and Biology presents multidisciplinary and dynamic findings in the broad fields of experimental medicine and biology. The wide variety in topics it presents offers readers multiple perspectives on a variety of disciplines including neuroscience, microbiology, immunology, biochemistry, biomedical engineering and cancer research. Advances in Experimental Medicine and Biology has been publishing exceptional works in the field for over 30 years and is indexed in Medline, Scopus, EMBASE, BIOSIS, Biological Abstracts, CSA, Biological Sciences and Living Resources (ASFA-1), and Biological Sciences. The series also provides scientists with up to date information on emerging topics and techniques.

More information about this series at http://www.springer.com/series/5584

Klaus Groschner • Wolfgang F. Graier •
Christoph Romanin
Editors

Store-Operated Ca^{2+} Entry (SOCE) Pathways

Emerging Signaling Concepts in Human (Patho)physiology

Second Edition

Springer

Editors
Klaus Groschner
Institute of Biophysics
Medical University of Graz
Graz, Austria

Wolfgang F. Graier
Institute of Molecular Biology and Biochemistry
Medical University of Graz
Graz, Austria

Christoph Romanin
Institute of Biophysics
Johannes Kepler University Linz
Linz, Austria

ISSN 0065-2598 ISSN 2214-8019 (electronic)
Advances in Experimental Medicine and Biology
ISBN 978-3-319-57731-9 ISBN 978-3-319-57732-6 (eBook)
DOI 10.1007/978-3-319-57732-6

Library of Congress Control Number: 2017944273

Printed on acid-free paper

This Springer imprint is published by Springer Nature
The registered company is Springer International Publishing AG
The registered company address is: Gewerbestrasse 11, 6330 Cham, Switzerland

Preface

A brief recapitulation of the overwhelming recent progress in our understanding of Ca^{2+} signaling and (patho)physiological processes linked to or associated with depletion/refilling of the cells' sarcoplasmic/endoplasmic reticulum led us to initiate the compilation of a second edition, an update of this book, less than 5 years after the release of the first edition. We kept the original structure to address the key issues of (1) fundamental mechanistic concepts, (2) cross talk between organelle and cell compartments, and (3) molecular and cellular (patho)physiology of these processes. Each of these sections has not only been significantly updated by amendatory, detailed information complementing the previous topics and chapters but also extended by entirely novel aspects addressed in separate, new chapters. Among these important extensions of the scope are the recently gained information on the molecular structure of the STIM–Orai machinery at the atomic resolution level (Chap. 2), novel insights into the supramolecular domain organization of the STIM–Orai coupling machinery (Chap. 5), the recent developments in STIM–Orai optogenetics (Chap. 7), novel insights into the structure and function of membrane (in particular plasma membrane)-endoplasmic reticulum contact sites (Chaps. 15, 17), as well as recently gained knowledge on the role of SOCE in cancer (Chaps. 30, 31). In turn, a few other aspects were found better suitable for combined synopsis within a single chapter such as integrative aspects of cardiovascular disease and therapy (Chap. 24).

Overall, we hope that this second edition may be inspiring and supportive by its comprehensive and timely information on the SOCE phenomenon for both students and advanced colleagues who focus on SOCE-related aspects of cellular Ca^{2+} homeostasis. Even more, we hope that this updated compilation of current expertise in SOCE signaling will serve as an influential knowledge base to further groundbreaking developments in this steadily growing field of cell biology/pathophysiology.

Graz, Austria	Klaus Groschner
Graz, Austria	Wolfgang F. Graier
Linz, Austria	Christoph Romanin

Acknowledgments

The editors express their sincere thanks and appreciation to all contributors, who repetitively and unswervingly supported this project. Again, we wish to express our particular gratitude to Karin Osibow for her professional and exceptionally dedicated help with all management aspects required for the realization of this book.

Graz, Austria — Klaus Groschner
Graz, Austria — Wolfgang F. Graier
Linz, Austria — Christoph Romanin

Contents

Editors Biography

Klaus Groschner studied pharmacy at the School of Pharmacy, University of Graz, Austria. He completed his PhD studies at the Department of Pharmacodynamics in Graz and received part of his postdoctoral education at the University of Miami, Florida, USA, studying electrophysiology and cell biology. In 1990, he became an assistant professor at the Department of Pharmacology and Toxicology and received his habilitation in pharmacology and toxicology along with promotion to associate professor in 1993. Until 2011, he was head of a research unit for physiology and pharmacology of membrane transport at the Institute of Pharmaceutical Sciences at the University of Graz. Since 2012, he is full professor and chairman at the Institute of Biophysics at the Medical University of Graz.

Wolfgang F. Graier studied pharmacy at the University of Graz, Austria, and received his PhD in pharmacology at the Department of Pharmacology and Toxicology, University of Graz. In 1993–1994, he was a postdoctoral research fellow, analyzing physiology and membrane biophysics at the Dalton Cardiovascular Research Center, University of Missouri, Columbia, USA. In 1994, he became an assistant professor at the Department of Medical Biochemistry, University of Graz. In 1995, he became associate professor (habilitation) in biochemical pharmacology and in physiology in 2001 at the Department of Medical Biochemistry, University of Graz. Since 2009, he is full professor for molecular biology and chairman of the Institute of Molecular Biology and Biochemistry at the Medical University of Graz. Since 2015, he is also head of the Nikon Center of Excellence for Super-Resolution Microscopy: *Cells and Organelles* that is part of BioTechMed, the concerted research platform of the Medical University of Graz, the University of Graz, and the Graz University of Technology. In 2016, he cofounded a spin-off company named Next Generation Fluorescence Imaging or NGFI (www.ngfi.eu).

Christoph Romanin studied chemistry at the Graz University of Technology, Austria, and completed his doctoral studies at the Department of Pharmacology and Toxicology, University of Graz. In 1986, he started with a postdoc position at the Institute of Biophysics of the Johannes Kepler University Linz, Austria, where he became professor in biophysics (habilitation) in 1993 and served as an associate professor since 1997. In 2001, he was a guest researcher at the National Institute on Aging (NIA) of the NIH in Bethesda, USA. Currently, he is vice-chairman of the institute and head of the Ion Channel Group at the Institute of Biophysics in Linz, Austria.

List of Contributors

Eunüs S. Ali Department of Medical Biochemistry, School of Medicine, Flinders University, Adelaide, SA, Australia

Centre for Neuroscience, School of Medicine, Flinders University, Adelaide, SA, Australia

Indu S. Ambudkar Secretory Physiology Section, Molecular Physiology and Therapeutics Branch, National Institute of Dental and Craniofacial Research – NIDCR, National Institutes of Health – NIH, Bethesda, MD, USA

Greg J. Barritt Department of Medical Biochemistry, School of Medicine, Flinders University, Adelaide, SA, Australia

Centre for Neuroscience, School of Medicine, Flinders University, Adelaide, SA, Australia

Fiona Bartoli Inserm UMR S1180, University Paris-Sud, Université Paris-Saclay, Châtenay-Malabry, France

Alexis Bavencoffe Department of Integrative Biology and Pharmacology, McGovern Medical School, The University of Texas Health Science Center at Houston, Houston, TX, USA

Lothar A. Blatter Department of Physiology & Biophysics, Rush University Medical Center, Chicago, IL, USA

Dora Bodnar Epithelial Signaling and Transport Section, Molecular Physiology and Therapeutics Branch, National Institute of Dental and Craniofacial Research – NIDCR, National Institute of Health – NIH, Bethesda, MD, USA

Sunitha Bollimuntha Department of Biomedical Sciences, School of Medicine and Health Sciences, University of North Dakota, Grand Forks, ND, USA

Martin D. Bootman School of Life, Health and Chemical Sciences, The Open University, Milton Keynes, UK

Xiangyu Cai Department of Cellular and Molecular Physiology, The Pennsylvania State University College of Medicine, Hershey, PA, USA

Woo Young Chung Epithelial Signaling and Transport Section, Molecular Physiology and Therapeutics Branch, NIDCR, NIH, Bethesda, MD, USA

Nicolas Demaurex Department of Cell Physiology and Metabolism, University of Geneva, Geneva 4, Switzerland

Isabella Derler Institute of Biophysics, Johannes Kepler University Linz, Linz, Austria

Hayley Dingsdale School of Biosciences, Cardiff University, Cardiff, UK

Petra Eder Department of Internal Medicine I, Comprehensive Heart Failure Center Würzburg, University Hospital Würzburg, Würzburg, Germany

A. Mark Evans Centre for Integrative Physiology, College of Medicine and Veterinary Medicine, University of Edinburgh, Edinburgh, UK

Marc Fahrner Institute of Biophysics, Johannes Kepler University Linz, Linz, Austria

Nicola Fameli Institute of Biophysics, Medical University of Graz, Graz, Austria

Department of Anesthesiology, Pharmacology, and Therapeutics, University of British Columbia, Vancouver, BC, Canada

Qingping Feng Department of Physiology and Pharmacology, Schulich School of Medicine and Dentistry, Western University, London, ON, Canada

Maud Frieden Department of Cell Physiology and Metabolism, University of Geneva, Geneva, Switzerland

Donald L. Gill Department of Cellular and Molecular Physiology, The Pennsylvania State University College of Medicine, Hershey, PA, USA

Wolfgang F. Graier Institute of Molecular Biology and Biochemistry, Medical University of Graz, Graz, Austria

Klaus Groschner Institute of Biophysics, Medical University of Graz, Graz, Austria

Aparna Gudlur Division of Signalling and Gene Expression, La Jolla Institute for Allergy & Immunology, La Jolla, CA, USA

Daniele Guido Department of Cell Physiology and Metabolism, University of Geneva, Geneva 4, Switzerland

Yousang Gwack Department of Physiology, David Geffen School of Medicine at UCLA, Los Angeles, CA, USA

Lee Haynes The Physiological Laboratory, Department of Cellular and Molecular Physiology, The University of Liverpool, Liverpool, UK

Patrick G. Hogan Division of Signalling and Gene Expression, La Jolla Institute for Allergy & Immunology, La Jolla, CA, USA

Jeong Hee Hong Department of Physiology, College of Medicine, Gachon University, Yeonsu-gu, Incheon City, Korea

Yun Huang Center for Epigenetics and Disease Prevention, Institute of Biosciences and Technology, Texas A&M University, Houston, TX, USA

Archana Jha Epithelial Signaling and Transport Section, Molecular Physiology and Therapeutics Branch, National Institute of Dental and Craniofacial Research – NIDCR, National Institute of Health – NIH, Bethesda, MD, USA

Kiho Lee Department of Animal and Poultry Sciences, Virginia Tech, Blacksburg, VA, USA

Natalia A. Loktionova Department of Cellular and Molecular Physiology, The Pennsylvania State University College of Medicine, Hershey, PA, USA

Jose J. Lopez Cell Physiology Research Group, Department of Physiology, University of Extremadura, Cáceres, Spain

Gyorgy Lur Department of Neurobiology, Yale School of Medicine, New Haven, CT, USA

Guolin Ma Center for Translational Cancer Research, Institute of Biosciences and Technology, Texas A&M University Health Science Center, Houston, TX, USA

Zoltan Machaty Department of Animal Sciences, Lilly Hall of Life Sciences, Purdue University, West Lafayette, IN, USA

Roland Malli Institute of Molecular Biology and Biochemistry, Medical University of Graz, Graz, Austria

Rajender K. Motiani Systems Biology Group, CSIR-Institute of Genomics and Integrative Biology, New Delhi, India

Shmuel Muallem Epithelial Signaling and Transport Section, Molecular Physiology and Therapeutics Branch, National Institute of Dental and Craniofacial Research – NIDCR, National Institute of Health – NIH, Bethesda, MD, USA

Martin Muik Institute of Biophysics, Johannes Kepler University Linz, Linz, Austria

Barbara A. Niemeyer Molecular Biophysics, Center for Integrative Physiology and Molecular Medicine, Saarland University, Homburg, Germany

Robert M. Nwokonko Department of Cellular and Molecular Physiology, The Pennsylvania State University College of Medicine, Hershey, PA, USA

Emmanuel Okeke The Physiological Laboratory, Department of Cellular and Molecular Physiology, The University of Liverpool, Liverpool, UK

Hwei Ling Ong Secretory Physiology Section, Molecular Physiology and Therapeutics Branch, National Institute of Dental and Craniofacial Research – NIDCR, National Institutes of Health – NIH, Bethesda, MD, USA

Jonathan Pacheco Instituto de Fisiología Celular, Universidad Nacional Autónoma de México, Ciudad Universitaria, México, DF, México

Biswaranjan Pani Department of Biomedical Sciences, School of Medicine and Health Sciences, University of North Dakota, Grand Forks, ND, USA

Natalia Prevarskaya Laboratory of Excellence, Ion Channels Science and Therapeutics, Université Lille I Sciences et Technologies, Villeneuve d'Ascq, France

Inserm U1003, Equipe Labellisée par la Ligue Nationale Contre le Cancer, Université de Sciences et Technologies de Lille (USTL), Villeneuve d'Ascq, France

James W. Putney The Signal Transduction Laboratory, Department of Health and Human Services, National Institute of Environmental Health Sciences – NIH, Research Triangle Park, NC, USA

Katja Rietdorf School of Life, Health and Chemical Sciences, The Open University, Milton Keynes, UK

Abigaël Ritaine Laboratory of Excellence, Ion Channels Science and Therapeutics, Université Lille I Sciences et Technologies, Villeneuve d'Ascq, France

Inserm U1003, Equipe Labellisée par la Ligue Nationale Contre le Cancer, Université de Sciences et Technologies de Lille (USTL), Villeneuve d'Ascq, France

Christoph Romanin Institute of Biophysics, Johannes Kepler University Linz, Linz, Austria

Juan A. Rosado Cell Physiology Research Group, Department of Physiology, University of Extremadura, Cáceres, Spain

Grigori Y. Rychkov School of Medicine, The University of Adelaide, Adelaide, SA, Australia

South Australian Health and Medical Research Institute, Adelaide, SA, Australia

Jessica Sabourin Inserm UMR S1180, University Paris-Sud, Université Paris-Saclay, Châtenay-Malabry, France

Gines M. Salido Cell Physiology Research Group, Department of Physiology, University of Extremadura, Cáceres, Spain

Sophie Saüc Department of Cell Physiology and Metabolism, University of Geneva, Geneva, Switzerland

Rainer Schindl Institute of Biophysics, Johannes Kepler University Linz, Linz, Austria

George Shapovalov Laboratory of Excellence, Ion Channels Science and Therapeutics, Université Lille I Sciences et Technologies, Villeneuve d'Ascq, France

Inserm U1003, Equipe Labellisée par la Ligue Nationale Contre le Cancer, Université de Sciences et Technologies de Lille (USTL), Villeneuve d'Ascq, France

Niroj Shrestha Institute of Biophysics, Medical University of Graz, Graz, Austria

Brij B. Singh Department of Biomedical Sciences, School of Medicine and Health Sciences, University of North Dakota, Grand Forks, ND, USA

András Spät Laboratory of Molecular Physiology, Hungarian Academy of Sciences, Budapest, Hungary

Department of Physiology, Semmelweis University, Budapest, Hungary

Sonal Srikanth Department of Physiology, David Geffen School of Medicine at UCLA, Los Angeles, CA, USA

Peter B. Stathopulos Department of Physiology and Pharmacology, Schulich School of Medicine and Dentistry, Western University, London, ON, Canada

Zuoming Sun Department of Molecular Immunology, Beckman Research Institute of the City of Hope, Duarte, CA, USA

Gergö Szanda Department of Physiology, Semmelweis University, Budapest, Hungary

Jyoti Tanwar Systems Biology Group, CSIR-Institute of Genomics and Integrative Biology, New Delhi, India

Alexei V. Tepikin The Physiological Laboratory, Department of Cellular and Molecular Physiology, Institute of Translational Medicine, University of Liverpool, Liverpool, UK

Jin-bin Tian Department of Integrative Biology and Pharmacology, McGovern Medical School, The University of Texas Health Science Center at Houston, Houston, TX, USA

Mohamed Trebak Department of Cellular and Molecular Physiology, The Pennsylvania State University College of Medicine, Hershey, PA, USA

Luis Vaca Instituto de Fisiología Celular, Universidad Nacional Autónoma de México, Ciudad Universitaria, México, DF, México

Cornelis van Breemen Department of Anesthesiology, Pharmacology and Therapeutics, The University of British Columbia, Vancouver, BC, Canada

Chunmin Wang Vivere Health, Houston, TX, USA

Youjun Wang Beijing Key Laboratory of Gene Resources and Molecular Development, College of Life Sciences, Beijing Normal University, Beijing, People's Republic of China

Shufan Wen Center for Translational Cancer Research, Institute of Biosciences and Technology, Texas A&M University, Houston, TX, USA

Jin Seok Woo Department of Physiology, David Geffen School of Medicine at UCLA, Los Angeles, CA, USA

Dongki Yang Department of Physiology, College of Medicine, Gachon University, Yeonsu-gu, Incheon City, Korea

Lu Zhang Department of Animal Sciences, Lilly Hall of Life Sciences, Purdue University, West Lafayette, IN, USA

Yandong Zhou Department of Cellular and Molecular Physiology, The Pennsylvania State University College of Medicine, Hershey, PA, USA

Yubin Zhou Center for Translational Cancer Research, Institute of Biosciences and Technology, Texas A&M University Health Science Center, Houston, TX, USA

Department of Medical Physiology, College of Medicine, Texas A&M University Health Science Center, Houston, TX, USA

Jinhui Zhu Department of Physiology and Pharmacology, Schulich School of Medicine and Dentistry, Western University, London, ON, Canada

Michael Xi Zhu Department of Integrative Biology and Pharmacology, McGovern Medical School, The University of Texas Health Science Center at Houston, Houston, TX, USA

Abbreviations

α-SNAP	α-Soluble NSF attachment protein
$\Delta\Psi_m$	Mitochondrial membrane potential
σ1R	Sigma-1 receptor
–/–	Double knockout
+TIP	Plus-end-tracking protein
$[Ca^{2+}]$	Ca^{2+} concentration
$[Ca^{2+}]_{cyt}/[Ca^{2+}]_c$	Cytosolic Ca^{2+} concentration
$[Ca^{2+}]_{ER}$	Free Ca^{2+} concentration of the ER
$[Ca^{2+}]_i$	Intracellular free Ca^{2+} concentration
$[Na^+]_{ns}$	Sodium concentration within the nanospace
2-APB	2-Aminoethoxydiphenyl borate
aa	Amino acid
AA	Arachidonic acid
ABCA	ATP-binding cassette subfamily A member
AC	Adenylyl cyclase
AD	Alzheimer's disease
ADPKD	Autosomal dominant polycystic kidney disease
AM	Atrial myocyte
Ang II	Angiotensin II
ANT	Adenine nucleotide translocase
APC	Adenomatous polyposis coli
APP	Amyloid precursor protein
ARC	Arachidonate-regulated Ca^{2+} channel
Arf6	ADP-ribosylation factor 6
AtCRY2	*Arabidopsis thaliana* cryptochrome 2
ATF6	Activating transcription factor 6
ATP	Adenosine triphosphate
ATXN1	Ataxin 1
BACCS	Blue light-activated Ca^{2+} channel switch
bFGF	Basic fibroblast growth factor
BHQ	2,5-Di-(tert-butyl)-1,4-benzohydroquinone
BKCas	Big conductance Ca^{2+}-activated K^+ channels

BTP2	*N*-[4-[3,5-Bis(trifluoromethyl)-1H-pyrazol-1-yl]phenyl]-4-methyl-1,2,3-thiadiazole-5-carboxamide
Ca^{2+}	Calcium
CAD	Channel-activating domain
CaM	Calmodulin
CaMK	Ca^{2+}-/CaM-dependent protein kinase
cAMP	Cyclic adenosine monophosphate
CaN	Calcineurin
CAPE	Caffeic acid phenethyl ester
CAR	Ca^{2+} accumulation region
CASK	Ca^{2+}-/CaM-dependent serine protein kinase
cat-SOC	Cation store-operated channel
Cav-1	Caveolin-1
CAX	Ca^{2+}/hydrogen exchanger
CC	Coiled coil
CCE	Capacitative Ca^{2+} entry
CCh	Carbachol
CDC42	Cell division control protein 42 homolog
CDI	Ca^{2+}-dependent inactivation
cER	Cortical endoplasmic reticulum
CFP	Cyan fluorescent protein
CGD	Chronic granulomatous disease
cGMP	Cyclic guanosine monophosphate
CIRB	CaM- and IP_3R-binding site
CK1	Casein kinase 1
Cl^-	Chloride
CLEM	Correlative light and electron microscopy
CM	Cardiac myocyte
CnA	Catalytic A subunit of calcineurin
CnB	Calcineurin B
Co-IP	Co-immunoprecipitation
COX	Cyclooxygenase
CPA	Cyclopiazonic acid
CPAE	Calf pulmonary endothelial cell
CRAC	Ca^{2+} release-activated Ca^{2+} channel
CRACR2A	CRAC regulator 2A
CREB	cAMP response element-binding transcription factor
CRMP2	Collapsin response mediator protein-2
cRNA	Complementary RNA
CRYs	Cryptochromes
CSQ	Calsequestrin
CTID	C-terminal inhibitory domain
DAG	Diacylglycerol
DAPC	Dystrophin-associated protein complex
DBD	DNA-binding domain

DC	Dendritic cell
DCX	Doublecortin
DHPR	Dihydropyridine receptor
DKO	Double knockout
dLNs	Draining lymph nodes
DMD	Duchenne muscular dystrophy
DN	Dominant negative
DTS	Dense tubular system
DVF	Divalent-free
DYRK	Dual-specificity tyrosine phosphorylation-regulated kinase
EAE	Experimental autoimmune encephalomyelitis
EB1	End-binding 1 protein
EC	Endothelial cell
ECC	Excitation–contraction coupling
ECCE	Excitation-coupled Ca^{2+} entry
ECM	Extracellular matrix
EDCF	Endothelium-derived contracting factor
EDL	Extensor digitorum longus
EDRF	Endothelium-derived relaxing factor
EE	Early endosome
EGFR	Epidermal growth factor receptor
EGTA	Ethylene glycol-bis(β-aminoethyl ether)-*N*,*N*,*N*′,*N*′-tetraacetic
EMRE	Essential MCU regulator
En	Endosome
eNOS	Endothelial NO synthase (NOS III)
EPAC	Exchange protein directly activated by cAMP
ER	Endoplasmic reticulum
E_{rev}	Reversal potential
ERM	Ezrin–radixin–moesin
ESCRT	Endosomal sorting complex required for transport
E-Syt	Extended synaptotagmin
ET-1	Endothelin 1
ETC	Excitation–transcription coupling
ETON	Extended transmembrane Orai1 N-terminal
EVH1	Ena (Enabled)/VASP (vasodilator-stimulated phosphoprotein) homology 1
FAD	Familial Alzheimer's disease
FAK	Focal adhesion kinase
FCCP	Carbonyl cyanide 4-(trifluoromethoxy)phenylhydrazone
FCDI	Fast Ca^{+2}-dependent inactivation
FF	Diphenylalanine
FFAT	Acidic tract motif
FGF2	Fibroblast growth factor 2
FKBP	FK506-binding protein

FRB	FKBP–rapamycin binding
FRET	Förster/fluorescence resonance energy transfer
GC-A	Guanylyl cyclase-A
GECAs	Genetically encoded Ca^{2+} actuators
GECIs	Genetically encoded Ca^{2+}-sensitive indicators
GEF	Guanine nucleotide exchange factor
GLP-1	Glucagon-like peptide 1
GLUT4	Glucose transporter 4
GPCR	G protein-coupled receptor
GPN	Glycyl-L-phenylalanine-2-naphthylamide
GqPCR	Gq protein-coupled receptor
Grx's	Glutaredoxins
GSH	Glutathione
GSK-3β	Glycogen synthase kinase 3β
GST	Glutathione S-transferase
H_2O_2	Hydrogen peroxide
H2S	Hydrogen sulfide
HAD	HIV-associated dementia
HCMD	High Ca^{2+} microdomain
HDAC	Histone deacetylase
hESC-CM	Human embryonic stem cell-derived cardiomyocyte
HF	Heart failure
HIV	Human immunodeficiency virus
HMGB1	High-mobility group box 1 protein
Hrs	Hepatocyte growth factor-regulated tyrosine kinase substrate
HSG	Human submandibular gland
HUVEC	Human umbilical vein EC
I_{CRAC}	Ca^{2+} release-activated Ca^{2+} current
IκB	Inhibitor of NF-κB
IKK	IκB kinase
IMM	Inner mitochondrial membrane
iNKT	Invariant natural killer T
IP_3	Inositol 1,4,5-trisphosphate
IP_3R	IP_3 receptor
IQGAP	IQ motif-containing GTPase-activating protein
IRE1	Inositol-requiring protein 1
IS	Immunological synapse
Iso	Isoproterenol
I_{SOC}	Store-operated Ca^{2+} current
ITAM	Immunoreceptor tyrosine-based activation motif
JNK	c-Jun N-terminal kinase
JP	Junctophilin
K^+	Potassium
KCa	Ca^{2+}-activated K^+ channel
KD	Knockdown

LB	Lewy body
LCK	Lymphocyte-specific protein tyrosine kinase
LOV	Light–oxygen–voltage sensing
LOXs	Lipoxygenases
LPS	Lipopolysaccharide
LRC	LTC_4-regulated Ca^{2+} channel
LRD	Lipid raft domain
LRRK2	Leucine-rich repeat kinase 2
LTC_4	Leukotriene C_4
LTCCs	L-type Ca^{2+} channels
MAM	Mitochondria-associated membrane
MAPK	Mitogen-activated protein kinase
MAPPER	Membrane-attached peripheral ER
MBP	Myelin basic protein
MCP-1	Monocyte chemoattractant protein-1
MCS	Membrane contact site
MCU	Mitochondrial Ca^{2+} uniporter
MCUP	Mitochondrial Ca^{2+} uniporter complex
MCUR1	Mitochondrial Ca^{2+} uniporter regulator 1
MEF	Mouse embryonic fibroblast
MEF2	Myocyte enhancer factor 2
MEK	Mitogen-activated protein kinase kinase
MG	Myasthenia gravis
MHC	Major histocompatibility complex
MICU	Mitochondrial calcium uptake
mitoNOS	Mitochondria-specific NO synthase
MLC	Myosin light chain
MLCK	Myosin light chain kinase
MMP	Matrix metalloproteinase
MORN	Membrane occupation and recognition nexus
MPF	M-phase promoting factor
MPTP	1-Methyl-4-phenyl-1,2,3,6-tetrahydropyridine
MRs	Mineralocorticoid receptors
MRF	Myogenic regulatory factor
mRFP	Mitochondrially targeted red fluorescent protein
MS	Multiple sclerosis
MT	Microtubule
Mwk mice	Moonwalker mice
Na^+	Sodium
NAADP	Nicotinic acid adenine dinucleotide phosphate
NAD(P)H	Nicotinamide adenine dinucleotide phosphate
NAFLD	Nonalcoholic fatty liver disease
NASH	Nonalcoholic steatohepatitis
NCKX	Na^+/Ca^{2+}–K^+ exchanger
NCX	Na^+/Ca^{2+} exchanger

NCXmito	Mitochondrial Na^+/Ca^{2+} exchanger
NPC1	Niemann–Pick type C protein 1
NFATs	Nuclear factor of activated T cells
NF-κB	Nuclear factor kappa-light-chain-enhancer of activated B cells
NIK	NF-κB-inducible kinase
NIR	Near infrared
$NKA\alpha_{2,3}$	Na^+/K^+-ATPase isoforms $\alpha_{2,3}$
NLS	Nuclear localization sequence
NMR	Nuclear magnetic resonance
nNOS	Neuronal nitric oxide synthase
NO	Nitric oxide
NOX	NADPH oxidase
NRON	Noncoding RNA repressor of NFAT
NSC	Cortical neural stem cell
$O_2^{\bullet-}$	Superoxide anion
OASF	Orai-activating small fragment
OMM	Outer mitochondrial membrane
ORP	Oxysterol-binding protein-related protein
ox-LDL	oxidized low-density lipoprotein
P2Y receptor	Purinergic G protein-coupled receptor
PA	Phosphatidic acid
PAEC	Porcine aortic endothelial cell
PAH	Pulmonary arterial hypertension
PAMs	Plasma membrane-associated membranes
PAR	Protease-activated receptor
PARP	Poly ADP ribose polymerase
PASMC	Pulmonary artery smooth muscle cell
PC	Polycystin
PD	Parkinson's disease
PDAC	Pancreatic ductal adenocarcinoma cell
PDGF	Platelet-derived growth factor
PDZ	PSD95–disc large–zonula occludens protein
PE	Phenylephrine
PERK	RNA-like ER kinase
Ph	Phagosome
PHR	Photolyase homology region
PhyBs	Phytochromes
PI	Phosphoinositide
PI4P	Phosphatidylinositol-4-phosphate
PIP_2	Phosphatidylinositol 4,5-bisphosphate
PKA	Protein kinase A
PKB/Akt	Protein kinase B
PKC	Protein kinase C
PKG	Protein kinase G
PLB	Phospholamban

PLC	Phospholipase C
PLN	Phospholamban
PM	Plasma membrane
PMCA	Plasma membrane Ca^{2+} ATPase
POST	Partner of STIM
PRD	Proline-rich domain
PS	Phosphatidylserine/presenilin
PTP1B	Protein tyrosine phosphatase 1B
RA	Rheumatoid arthritis
RAAS	Renin–angiotensin–aldosterone system
Rab 7	Ras-related protein 7
RasGRP1	Ras guanine nucleotide-releasing protein 1
RBL	Rat basophil leukemia
REG	Regulatory domain
RHD	Rel homology domain
RNAi	RNA interference
ROCs	Receptor-operated channels
ROCE	Receptor-operated Ca^{2+} entry
ROS	Reactive oxygen species
RR	Ruthenium red
RTKs	Receptor tyrosine kinases
RyR	Ryanodine receptor
S/ER	Sarcoplasmic/endoplasmic reticulum
S/P	Serine/proline rich
$S1^{CT}$	Soluble STIM1
S1P	Sphingosine-1-phosphate
SA node	Sinoatrial node
SACs	Stretch-activated cation channels
SACE	Stretch-activated Ca^{2+} entry
SAM	Sterile alpha motif
SARAF	SOC-associated regulatory factor
SCA	Spinocerebellar ataxia
SCD	Sudden cardiac death
$Scgd^{-/-}$ mice	δ-Sarcoglycan deleted mouse model for muscular dystrophy
SCID	Severe combined immune deficiency
SD	Sporadic Alzheimer's disease
SERCA	Sarcoplasmic/endoplasmic reticulum Ca^{2+}-ATPase
SF	Shape factor
SICE	Store-independent Ca^{2+} entry
SLP76	SH2 domain-containing leukocyte protein of 76 kDa
SMC	Smooth muscle cell
SMOCE	Second messenger-operated Ca^{2+} entry
SNAP25	Synaptosome-associated protein 25
SNARE	Soluble *N*-ethylmaleimide-sensitive factor attachment receptor
sNPF	Short neuropeptide F

SOAP	STIM1–Orai1 association pocket
SOAR	STIM1–Orai-activating region
SOC	Store-operated channel
SOCE	Store-operated Ca^{2+} entry
SOCIC	Store-operated Ca^{2+} influx complex
SOD	Superoxide dismutase
SPCA	Secretory pathway Ca^{2+}-ATPase
SPL	Subplasmalemmal
SR	Sarcoplasmic reticulum
SR/ER	Sarcoplasmic/endoplasmic reticulum
SRR	Serine-rich region
STARD	(StAR)-related lipid transfer domain protein
STIM	Stromal interaction molecule
STIMATE	STIM-activating enhancer
Synta66	*N*-(2′,5′-Dimethoxy[1,1′-biphenyl]-4-yl)-3-fluoro-4-pyridinecarboxamide
TAC	Transverse aortic constriction
TAD-C	C-terminal transcription activation domain
TAM	Tubular aggregate myopathy
TBHQ	2,5-Di-(tert-butyl)-1,4-benzohydroquinone
TCR	T-cell receptor
TG	Thapsigargin
TGN	Trans-Golgi network
Th_{eff}	T helper effector cells
TIRF	Total internal reflectance fluorescence
TM	Transmembrane
TMD	Transmembrane domain
TPC	Two-pore domain channel
TPEN	*N,N,N′,N′*-Tetrakis(2-pyridylmethyl)ethylenediamine
TRP	Transient receptor potential
TRPCs	Transient receptor potential canonical family of ion channels
TRPL	TRP-like
Trx	Thioredoxin
TTCC	T-type Ca^{2+} channel
t-tubules	Transverse tubules
UCP	Uncoupling protein
UPR	Unfolded protein response
UVRs	Ultraviolet-B receptors
VAP	Vesicle-associated membrane protein-associated protein
VAPA	Vesicle-associated membrane protein-associated protein A
VDAC	Voltage-dependent anion channel
VEC	Vascular endothelial cell
VEGF	Vascular endothelial growth factor
VEGFR	VEGF receptor
VGCCs	Voltage-gated Ca^{2+} channels

VM	Ventricular myocyte
VOCC	Voltage-operated Ca^{2+} channel
VSMC	Vascular smooth muscle cell
VT	Ventricular tachycardia
WT	Wild type
YFP	Yellow fluorescent protein
YPS	York platelet syndrome
ZAP-70	Z chain-associated protein kinase 70

Part I

SOCE: Fundamental Mechanistic Concepts

Introduction

1

James W. Putney

Abstract

This second edition volume will present an updated, state-of-the art description and analysis of the rapidly expanding field of store-operated Ca^{2+} entry (SOCE). And this first part will deal with the most fundamental mechanistic concepts underlying this process. In this brief introduction, I will try to summarize the historical development of the concept of store-operated Ca^{2+} entry and say a bit about some recent work that speaks to its general function in cell signaling. Much of the material below is taken from the Introduction to the first edition, updated for the second edition.

Keywords

Calcium channels • Orai • STIM1 • Oscillations • Store-operated channels • Mouse models

1.1 SOCE: Historical Development of the Concept

Many would attribute the origins of this concept to my 1986 hypothesis paper in *Cell Calcium* (Putney 1986), but in fact no idea is born in a vacuum, and much of the key elements for this concept developed from much earlier findings. One earlier and fundamental concept is that Ca^{2+} signals can arise in two very general ways: either by influx to the cytoplasm across the plasma membrane or by discharge to the cytoplasm from storage depots within the cell. Although it is now clear that this is a general property of Ca^{2+} signaling pathways, it was the smooth muscle physiologists who first appreciated it, based largely on the differential sensitivity

J.W. Putney (✉)
The Signal Transduction Laboratory, Department of Health and Human Services, National Institute of Environmental Health Sciences – NIH, Research Triangle Park, NC, USA
e-mail: putney@niehs.nih.gov

K. Groschner et al. (eds.), *Store-Operated Ca^{2+} Entry (SOCE) Pathways*, Advances in Experimental Medicine and Biology 993, DOI 10.1007/978-3-319-57732-6_1

of initial and sustained contractures to removal of extracellular Ca^{2+} (Bohr 1963; Van Breemen 1969; Sitrin and Bohr 1971; Van Breemen et al. 1973; Steinsland et al. 1973; reviews: Bohr 1973; Bolton 1979). The finding that this same dual mechanism of Ca^{2+} mobilization applied to other, non-excitable cell types led to the suggestion that it was a general property of Ca^{2+} signaling systems (Putney et al. 1981). Commonly both processes occur and they interact and regulate one another in a variety of ways. One clear example is the heart where the influx of Ca^{2+} is amplified manyfold through a process of Ca^{2+}-induced Ca^{2+} release through ryanodine receptors in the sarcoplasmic reticulum (Meissner 1994). However, the more common mechanism operates in the reverse mode: release of intracellular Ca^{2+} activates Ca^{2+} influx channels in the plasma membrane, which is the process of SOCE, and the focus of this volume.

Early in my career, I attempted to understand Ca^{2+} signaling in exocrine gland cells by monitoring a downstream Ca^{2+}-regulated event, the activation of Ca^{2+}-activated K^+ channels. Membrane permeability to K^+ could be assessed by the rate of efflux of a radioactive K^+ mimic, $^{86}Rb^+$. In a pattern reminiscent of the earlier studies on smooth muscle contraction, activation of any of three surface receptors (muscarinic, α-adrenergic, or substance P) stimulated the rate of $^{86}Rb^+$ efflux, and this response occurred in two phases: a transient phase which did not depend upon extracellular Ca^{2+} and a sustained phase that required extracellular Ca^{2+} (Putney 1976). The nature of the Ca^{2+}-independent phase was not known until in a subsequent study, it was shown to depend upon an internal pool of Ca^{2+} that was released in response to receptor activation (Putney 1977). Thus, these three receptors appeared each capable of releasing stored Ca^{2+} and also activating Ca^{2+} influx through plasma membrane channels. Further, by use of a series of experiments employing serial application of different agonists and antagonists under differing conditions, it was shown that a single pool of intracellular Ca^{2+} and a single population of surface membrane channels were commonly regulated by the three different receptor types. In the absence of extracellular Ca^{2+}, the released Ca^{2+} was lost from the cells, presumably through active extrusion at the plasma membrane, but could be replenished from the outside, apparently through the receptor-activated channels. This was the first suggestion that plasma membrane channels were responsible for maintaining the intracellular stores, although it was not that clear that this could occur completely independently of receptor activation. That receptor activation was not required to refill the stores was shown by experiments of a similar nature carried out with slices of rat lacrimal gland. In the previous study with parotid slices (Putney 1977), refilling was achieved with the following protocol: in the absence of external Ca^{2+}, stores were emptied with agonist for one receptor (muscarinic), then Ca^{2+} was restored to permit influx, and then the agonist was removed pharmacologically by application of a potent receptor antagonist (atropine). Ca^{2+} was then removed, and a second agonist (substance P) was applied which induced a robust transient $^{86}Rb^+$ efflux, indicative of efficient refilling of the pools. In the subsequent study with lacrimal slices (Parod and Putney 1978), a similar protocol employing epinephrine as the first agonist, phentolamine as the antagonist, and carbachol to assess refilling produced essentially the same result as before. However, in this study an alternative sequence was also tested: following depletion of the intracellular store by epinephrine, phentolamine was added before restoration of extracellular Ca^{2+}. In this condition, it was assumed that the receptor-operated channels were inactive. However, a brief (2 min) application of external Ca^{2+}, even after phentolamine, was equally efficient in refilling the intracellular stores. This experiment indicated that refilling could occur efficiently in the absence of receptor activation. Three years later, Casteels and Droogmans (1981) reported a similar phenomenon in smooth muscle. They proposed a privileged route by which Ca^{2+} could enter the intracellular stores, in this case sarcoplasmic reticulum, without traversing the cytoplasm.

In 1983 came the discovery of the Ca^{2+} mobilizing second messenger, inositol 1,4,5-trisphosphate (IP_3) (Berridge 1983; Streb et al. 1983). The initial experiments involved demonstration of IP_3-induced release from intracellular stores, mainly by use of permeable cell models (Streb et al. 1983; Burgess et al. 1984; Biden et al. 1984) but also with microsomal fractions (Prentki et al. 1984; Streb et al. 1984; Ueda et al. 1986). However, when plasma membrane fractions were separated from endoplasmic reticulum, it appeared that IP_3 did not activate release from the plasma membrane vesicles (Streb et al. 1984; Ueda et al. 1986). Yet, injection of IP_3 into sea urchin oocytes produced a full fertilization response, known to require activation of Ca^{2+} influx (Slack et al. 1986). At this point I attempted to put together the prior work on Ca^{2+} entry with the more recent findings with IP_3 to come up with a model that could account for the biphasic nature of Ca^{2+} signaling. This was the often cited 1986 hypothesis paper published in *Cell Calcium* (Putney 1986). I chose this instrument, a hypothesis paper, for publishing this idea because at the time I did not know how to proceed to test it. I used the term "capacitative" to describe a mechanism by which the empty state of the intracellular store signaled to and activated Ca^{2+} channels in the plasma membrane. Closely associated components of the endoplasmic reticulum would then rapidly capture this entering Ca^{2+} resulting in efficient refilling of intracellular stores. I envisioned the process of Ca^{2+} entry as a constant refilling of the stores and constant emptying through the IP_3 receptor, thus the analogy with electrical circuitry whereby current flows through a resistor and capacitor in series. This idea was somewhat similar to the model proposed by Casteels and Droogmans (1981), although I assumed a series arrangement of channels and ER pumps, rather than a direct conduit. Subsequent findings confirmed the basic idea of store Ca^{2+} content regulating influx, but the route of entry was not as envisioned by either myself or Casteels and Droogmans. This was clearly demonstrated in two subsequent studies that provided the initial proofs for the concept of store-operated Ca^{2+} entry.

The first was a report in the *Biochemical Journal* in 1989 in which transient increases in $[Ca^{2+}]_i$ were observed during refilling of intracellular stores, independently of receptor activation (Takemura and Putney 1989). The protocol was similar to the earlier study in lacrimal cells in which stores refilled rapidly following block of the agonist receptor. The major difference was that $[Ca^{2+}]_i$ was more directly monitored by using the newly developed Ca^{2+} indicator, fura-2- (Grynkiewicz et al. 1986). In the earlier study, no increase in $^{86}Rb^+$ efflux was seen during the process of refilling, suggesting a highly protected route to the stores. But with fura-2, one could clearly see a transient rise in $[Ca^{2+}]_i$ during the refilling process. The rise was transient because as the stores refilled, the store-operated process shut down. The second key finding, reported in that same year, served to simplify and redefine the primary experimental paradigm for demonstrating and investigating store-operated Ca^{2+} entry. Thapsigargin is an inflammatory plant toxin that was shown to increase Ca^{2+} in cells (Ali et al. 1985; Thastrup et al. 1987). Jackson et al. demonstrated that thapsigargin released the same pool of Ca^{2+} as did phospholipase C-linked agonists, but without increasing the formation of IP_3 (Jackson et al. 1988). The mechanism for this release was not known at the time, but was later shown to result from inhibition of the sarcoplasmic-endoplasmic reticulum Ca^{2+} ATPase which is responsible for concentrating Ca^{2+} in the intracellular stores (Thastrup et al. 1990). I recognized that this reagent could serve as a test of the store-operated Ca^{2+} entry idea, but I was discouraged by the fact that Jackson et al. did not apparently observe any increased Ca^{2+} entry in the cells they studied. Nonetheless, there was evidence from the earlier studies in platelets that thapsigargin could cause a sustained $[Ca^{2+}]_i$ increase (Thastrup et al. 1987), and very early experiments examining the effects of Ca^{2+} repletion on secretion provided evidence that thapsigargin likely did augment entry (Patkar et al. 1979). When we applied thapsigargin to parotid acinar cells, a robust and sustained elevation in $[Ca^{2+}]_i$ ensued (Takemura et al. 1989). Removal of external Ca^{2+} and its subsequent restoration clearly demonstrated that this sustained elevation was due to increased Ca^{2+} entry across the plasma membrane. Importantly, thapsigargin and a phospholipase C-linked agonist, methacholine, were not additive in increasing Ca^{2+} influx, indicating that thapsigargin had stimulated the same pathway for Ca^{2+} entry as had the more physiological receptor agonist. Two important implications came from this study. First, it provided direct evidence that simply depleting Ca^{2+} stores could quantitatively account for all of the Ca^{2+} entry produced by activation of the phospholipase C-IP_3 signaling pathway. Second, since IP_3 receptors were not activated, it indicated that the pathway to the cytoplasm did not traverse the IP_3-regulated Ca^{2+} pool but that the Ca^{2+} entered the cytoplasm directly through the store-operated channels (Takemura et al. 1989; Muallem et al. 1990). Why did Jackson et al. not see Ca^{2+} entry in their studies with thapsigargin? We now know that the cell line used in that study, NG115-401L, a neuroblastoma/glioma hybrid line, is deficient in store-operated Ca^{2+} entry (Csutora et al. 2008) due to a deficiency in the Ca^{2+} sensor STIM1. And a third important consequence of the 1989 paper on thapsigargin was that it provided perhaps the simplest and most readily understandable evidence for store-operated Ca^{2+} entry. Interest in this phenomenon increased as a new tool for its investigation became available.

In 1992 a major advance occurred when Hoth and Penner published the first recordings of whole-cell current activated by Ca^{2+} store depletion (Hoth and Penner 1992). They called the current I_{CRAC} for Ca^{2+} release-activated Ca^{2+} current. The current was similar to other selective Ca^{2+} currents in being highly selective for Ca^{2+} and thus strongly inwardly rectifying with a positive reversal potential. The current was small, however, in the range of only a few pA/pF. Actually, in the mast cells which Hoth and Penner used in their initial studies, and other hematopoietic cells, such as T-cells, I_{CRAC} seems to be somewhat larger than other cell types where is may be close to threshold levels of detection, <1 pA/pF (e.g., Vig et al. 2006; DeHaven et al. 2007). And even these small currents are generally only seen when Ca^{2+} inactivation of the current is reduced or prevented, by either inclusion of chelators in the patch pipette solution or by holding at relatively positive membrane potentials. Hoth and Penner used the whole-cell patch clamp technique to measure total membrane current; attempts to detect single channels underlying I_{CRAC} were unsuccessful. Subsequently, Zweifach and Lewis (1993) used a noise analysis algorithm to estimate the single-channel conductance of CRAC channels to be ~24 fS, well below the level detectable by conventional electrophysiological techniques. Two clues to the molecular nature of the channels were provided in these early studies: first, I_{CRAC} developed rather slowly following store depletion, with a time constant of the order of 20 s; second, the extremely low estimated single-channel conductance predicted that the molecular nature of the CRAC channel might be very different from other ion channels.

From 1992 to 2005, research on store-operated Ca^{2+} entry followed three general lines: investigations into the properties and regulatory mechanisms for I_{CRAC} and store-operated Ca^{2+} entry, for example, the complex regulation by intracellular and extracellular Ca^{2+}, attempts to delineate the mechanism by which depleted Ca^{2+} stores signaled to plasma membrane Ca^{2+} channels, and attempts to identify the store-operated channels. Whereas the first thread of investigation produced a lot of useful information, the other two sometimes yielded briefly encouraging findings which usually did not survive further scrutiny. Some issues from this earlier work still remain unsettled, for example, the role of the hypothetical signaling molecule, Ca^{2+} influx factor (Csutora et al. 2008), and to some degree the function of transient receptor potential channels in store-operated Ca^{2+} entry (Liao et al. 2008). The latter issue has been substantially clarified since the first edition and is discussed in Chaps. 8 and 9. A number of excellent reviews have been written on the basic nature of SOCE, too numerous to completely list here (e.g., Parekh and Putney 2005; Hogan et al. 2010; Feske 2010; Cahalan et al. 2007; Prakriya and Lewis 2015).

In 2005 the modern, molecular era of store-operated Ca^{2+} entry began with the initial identification of the endoplasmic reticulum Ca^{2+} sensors, STIM1 and STIM2 (Roos et al. 2005; Liou et al. 2005), followed shortly thereafter by the Ca^{2+} channel subunit proteins, Orai1, Orai2, and Orai3 (Feske et al. 2006; Vig et al. 2006; Zhang et al. 2006). Much will be said about these proteins in all three parts of this edition. Briefly, the STIM proteins are endoplasmic reticulum-resident single-pass membrane proteins with a Ca^{2+} binding domain directed to the lumen of the endoplasmic reticulum. Upon release of Ca^{2+} from the endoplasmic reticulum, Ca^{2+} dissociates from the domain resulting in a conformational change, association of STIM proteins into dimers and higher-order oligomers, and accumulation of STIM at endoplasmic reticulum—plasma membrane junctions where STIM interacts with and activates channels composed of Orai subunits (Prakriya and Lewis 2015).

In addition to the canonical pathway briefly outlined above, it has become clear that both STIM and Orai proteins function in other signaling pathways. STIM1 interacts with the microtubule cytoskeleton and is involved in remodeling of the endoplasmic reticulum (Grigoriev et al. 2008; Smyth et al. 2012). In addition to its role in forming classical CRAC channels, Orai1 interacts in a complex with STIM and TRPC proteins to produce a less selective current called I_{SOC} (Cheng et al. 2011), discussed in Chaps. 8 and 9. Orai1 and Orai3 form a heteromeric channel that depends on STIM1, but not on Ca^{2+} store depletion. Rather, these channels are activated by arachidonic acid (Thompson and Shuttleworth 2013) or by its metabolite, leukotriene C4 (Zhang et al. 2013). In the latter case, a specific N-terminal extended form of Orai1 is specifically required (Desai et al. 2015). Finally, Orai1 can also be constitutively activated in calcium-transporting epithelia by an interaction with a Golgi calcium transporter, SPCA2 (Feng et al. 2010).

1.2 The Physiological Function of SOC Channels

The availability of molecular tools to investigate store-operated Ca^{2+} entry has afforded an opportunity to address a long-standing question: what is the physiological function of store-operated Ca^{2+} channels? At the level of the organism, it is clear that this process is very important as mutations in the major players, whether arising by chance in humans or by design in animal models, in all cases produce severe phenotypes characterized by general immune deficiencies as well as problems in musculoskeletal development and problems with ectodermally derived tissues and functions (Feske 2010). The availability of Orai1 and STIM1 knockout mice and more recently mice with FLOXed Orai1 and STIM1 genes has afforded the opportunity to investigate the role of SOCE in a wide variety of additional physiological pathways. In work from my own laboratory, in some cases, the role of SOCE was predictable, for example, where phospholipase C-linked receptors are known to be key. Examples include muscarinic regulation of lacrimal secretion (Xing et al. 2013), calcium-sensing receptor regulation of keratinocyte differentiation (Numaga-Tomita and Putney 2013), regulation of neutrophil chemotaxis by formylated peptides (Steinckwich et al. 2015), and regulation of milk expulsion from mammary glands by oxytocin receptors (Davis et al. 2015). However, in some instances, the upstream pathways are not obvious. Loss of Orai1 significantly impacts bone formation, affecting the functions of both bone-forming osteoblasts and bone-resorbing osteoclasts (Hwang and Putney 2012; Hwang et al. 2012). Male mice lacking Orai1 are sterile and fail to develop mature, functional sperm (Davis et al. 2016). In neither of the latter two examples are the upstream regulators of Orai1 known. And there are also examples of physiological pathways in which Orai1 seems to be unnecessary: to our surprise, Orai1-deficient female mice are fertile (Davis et al. 2015).

But what is the basis for these effects at the cellular level? The basic assumption has been that these channels serve to refill and maintain endoplasmic reticulum Ca^{2+} stores, which are important for proper protein synthesis and protein folding. In terms of signaling, the role of SOCE may serve to maintain the stores that are released by IP_3, thereby activating downstream Ca^{2+}-regulated pathways. However, there is a growing body of evidence suggesting that in many instances, it is the Ca^{2+} entering the cytoplasm through SOC channels that directly couples to downstream pathways (Dupont et al. 2011; Parekh 2011). In this context, the function of IP_3-induced Ca^{2+} release is to lower Ca^{2+} in the endoplasmic reticulum to a level sufficient to activate SOC channels. One of the clearest examples is seen in a study by Di Capite et al. (2009) investigating the regulation of gene expression in the mast cell line, RBL. These cells respond to leukotriene B4 with repetitive Ca^{2+} oscillations and expression of the early gene, *c-fos*. In the absence of external Ca^{2+}, the oscillations quickly ran down, and there was no increase in *c-fos* expression. In the absence of external Ca^{2+}, but with Ca^{2+} extrusion blocked by a high concentration of La^{3+} (van Breemen 1969), the oscillations did not run down and appeared indistinguishable from those observed in the presence of Ca^{2+}. However, in this latter instance, there was no increase in *c-fos* expression. Thus, Ca^{2+} influx through plasma membrane channels appears essential for activating the downstream pathway responsible for the induction of gene expression (see also Parekh 2011). In this case, the global rise in Ca^{2+} produced by each Ca^{2+} spike does not provide the primary Ca^{2+} signal, because this global signal was normal in the high La^{3+} yet gene expression was not increased.

There is an increasing number of examples of downstream responses that have been shown to require SOC entry specifically, albeit not necessarily in the context of Ca^{2+} oscillations (Cooper et al. 1994; Chang et al. 2008; Ng et al. 2008; Kar et al. 2011; Samanta et al. 2014; Willoughby et al. 2012). There are also clear examples of signaling specifically linked to Ca^{2+} release through IP_3 receptors, the perhaps best documented example being the regulation of mitochondrial metabolism (Rizzuto and Pozzan 2006; Nicholls 2005). An emerging concept is that Ca^{2+} signaling generally takes place within highly constrained temporal and spatial parameters. This is not unexpected given the potential deleterious consequences of uncontrolled elevation in global cytoplasmic Ca^{2+} (Orrenius et al. 1989; Caron-Leslie and Cidlowski 1991; Sattler and Tymianski 2000; Tymianski 1996). A challenge for future research will be to define these spatial and temporal parameters and to elucidate the structural and molecular entities responsible for establishing and maintaining them.

References

Ali H, Christensen SB, Foreman JC, Pearce FL, Piotrowski W, Thastrup O (1985) The ability of thapsigargin and thapsigargicin to activate cells involved in the inflammatory response. Br J Pharmacol 85:705–712

Berridge MJ (1983) Rapid accumulation of inositol trisphosphate reveals that agonists hydrolyse polyphosphoinositides instead of phosphatidylinositol. Biochem J 212:849–858

Biden TJ, Prentki M, Irvine RF, Berridge MJ, Wollheim CB (1984) Inositol 1,4,5-trisphosphate mobilizes intracellular Ca2+ from permeabilized insulin-secreting cells. Biochem J 223:467–473

Bohr DF (1963) Vascular smooth muscle: dual effect of calcium. Science 139:597–599
Bohr DF (1973) Vascular smooth muscle updated. Circ Res 32:665–672
Bolton TB (1979) Mechanisms of action of transmitters and other substances on smooth muscle. Physiol Rev 59:606–718
Burgess GM, Godfrey PP, McKinney JS, Berridge MJ, Irvine RF, Putney JW Jr (1984) The second messenger linking receptor activation to internal Ca release in liver. Nature 309(5963):63–66
Cahalan MD, Zhang SL, Yeromin AV, Ohlsen K, Roos J, Stauderman KA (2007) Molecular basis of the CRAC channel. Cell Calcium 42:133–144
Caron-Leslie LM, Cidlowski JA (1991) Similar actions of glucocorticoids and calcium on the regulation of apoptosis in S49 cells. Mol Endocrinol 5:1169–1179
Casteels R, Droogmans G (1981) Exchange characteristics of the noradrenaline-sensitive calcium store in vascular smooth muscle cells of rabbit ear artery. J Physiol (Lond) 317:263–279
Chang WC, Di Capite J, Singaravelu K, Nelson C, Halse V, Parekh AB (2008) Local Ca2+ influx through Ca2+ release-activated Ca2+ (CRAC) channels stimulates production of an intracellular messenger and an intercellular pro-inflammatory signal. J Biol Chem 283:4622–4631
Cheng KT, Liu X, Ong HL, Swaim W, Ambudkar IS (2011) Local Ca(2)+ entry via Orai1 regulates plasma membrane recruitment of TRPC1 and controls cytosolic Ca(2)+ signals required for specific cell functions. PLoS Biol 9:e1001025
Cooper DMF, Yoshimura M, Zhang Y, Chiono M, Mahey R (1994) Capacitative Ca2+ entry regulates Ca2+-sensitive adenylyl cyclases. Biochem J 297:437–440
Csutora P, Peter K, Kilic H, Park KM, Zarayskiy V, Gwozdz T, Bolotina VM (2008) Novel role for STIM1 as a trigger for calcium influx factor production. J Biol Chem 283:14524–14531
Davis FM, Janoshazi A, Janardhan KS, Steinckwich N, D'Agostin DM, Petranka JG, Desai PN, Roberts-Thomson SJ, Bird GS, Tucker DK, Fenton SE, Feske S, Monteith GR, Putney JW Jr (2015) Essential role of Orai1 store-operated calcium channels in lactation. Proc Natl Acad Sci U S A 112:5827–5832
Davis FM, Goulding EH, D'Agostin DM, Janardhan KS, Cummings CA, Bird GS, Eddy EM, Putney JW (2016) Male infertility in mice lacking the store-operated Ca(2+) channel Orai1. Cell Calcium 59:189–197
DeHaven WI, Smyth JT, Boyles RR, Putney JW Jr (2007) Calcium inhibition and calcium potentiation of Orai1, Orai2, and Orai3 calcium release-activated calcium channels. J Biol Chem 282:17548–17556
Desai PN, Zhang X, Wu S, Janoshazi A, Bolimuntha S, Putney JW, Trebak M (2015) Multiple types of calcium channels arising from alternative translation initiation of the Orai1 message. Sci Signal 8:ra74
Di Capite J, Ng SW, Parekh AB (2009) Decoding of cytoplasmic Ca(2+) oscillations through the spatial signature drives gene expression. Curr Biol 19:853–858
Dupont G, Combettes L, Bird GS, Putney JW (2011) Calcium oscillations. Cold Spring Harb Perspect Biol 3(3):a004226
Feng M, Grice DM, Faddy HM, Nguyen N, Leitch S, Wang Y, Muend S, Kenny PA, Sukumar S, Roberts-Thomson SJ, Monteith GR, Rao R (2010) Store-independent activation of Orai1 by SPCA2 in mammary tumors. Cell 143:84–98
Feske S (2010) CRAC channelopathies. Pflugers Arch 460(2):417–435
Feske S, Gwack Y, Prakriya M, Srikanth S, Puppel SH, Tanasa B, Hogan PG, Lewis RS, Daly M, Rao A (2006) A mutation in Orai1 causes immune deficiency by abrogating CRAC channel function. Nature 441:179–185
Grigoriev I, Gouveia SM, van der Vaart B, Demmers J, Smyth JT, Honnappa S, Splinter D, Steinmetz MO, Putney JW Jr, Hoogenraad CC, Akhmanova A (2008) STIM1 is a MT-plus-end-tracking protein involved in remodeling of the ER. Curr Biol 18:177–182
Grynkiewicz G, Poenie M, Tsien RY (1986) A new generation of Ca2+ indicators with greatly improved fluorescence properties. J Biol Chem 260:3440–3450
Hogan PG, Lewis RS, Rao A (2010) Molecular basis of calcium signaling in lymphocytes: STIM and ORAI. Annu Rev Immunol 28:491–533

Hoth M, Penner R (1992) Depletion of intracellular calcium stores activates a calcium current in mast cells. Nature 355:353–355

Hwang SY, Putney JW (2012) Orai1-mediated calcium entry plays a critical role in osteoclast differentiation and function by regulating activation of the transcription factor NFATc1. FASEB J 26:1484–1492

Hwang SY, Foley J, Numaga-Tomita T, Petranka JG, Bird GS, Putney JW Jr (2012) Deletion of Orai1 alters expression of multiple genes during osteoclast and osteoblast maturation. Cell Calcium 52:488–500

Jackson TR, Patterson SI, Thastrup O, Hanley MR (1988) A novel tumour promoter, thapsigargin, transiently increases cytoplasmic free Ca2+ without generation of inositol phosphates in NG115-401L neuronal cells. Biochem J 253:81–86

Kar P, Nelson C, Parekh AB (2011) Selective activation of the transcription factor NFAT1 by calcium microdomains near Ca2+ release-activated Ca2+ (CRAC) channels. J Biol Chem 286:14795–14803

Liao Y, Erxleben C, Abramowitz J, Flockerzi V, Zhu MX, Armstrong DL, Birnbaumer L (2008) Functional interactions among Orai1, TRPCs, and STIM1 suggest a STIM-regulated heteromeric Orai/TRPC model for SOCE/Icrac channels. Proc Natl Acad Sci U S A 105:2895–2900

Liou J, Kim ML, Heo WD, Jones JT, Myers JW, Ferrell JE Jr, Meyer T (2005) STIM is a Ca2+ sensor essential for Ca2+-store-depletion-triggered Ca2+ influx. Curr Biol 15:1235–1241

Meissner G (1994) Ryanodine receptor/Ca2+ release channels and their regulation by endogenous effectors. Annu Rev Physiol 56:485–508

Muallem S, Khademazad M, Sachs G (1990) The route of Ca2+ entry during reloading of the intracellular Ca2+ pool in pancreatic acini. J Biol Chem 265:2011–2016

Ng SW, Di Capite J, Singaravelu K, Parekh AB (2008) Sustained activation of the tyrosine kinase syk by antigen in mast cells requires local Ca2+ influx through Ca2+ release-activated Ca2+ channels. J Biol Chem 283:31348–31355

Nicholls DG (2005) Mitochondria and calcium signaling. Cell Calcium 38:311–317

Numaga-Tomita T, Putney JW (2013) Role of STIM1- and Orai1-mediated Ca2+ entry in Ca2+-induced epidermal keratinocyte differentiation. J Cell Sci 126:605–612

Orrenius S, McConkey DJ, Bellomo G, Nicotera P (1989) Role of Ca2+ in toxic cell killing. Trends Pharmacol Sci 10:281–285

Parekh AB (2011) Decoding cytosolic Ca2+ oscillations. Trends Biochem Sci 36:78–87

Parekh AB, Putney JW Jr (2005) Store-operated calcium channels. Physiol Rev 85:757–810

Parod RJ, Putney JW Jr (1978) The role of calcium in the receptor mediated control of potassium permeability in the rat lacrimal gland. J Physiol Lond 281:371–381

Patkar SA, Rasmussen U, Diamant B (1979) On the mechanism of histamine release induced by thapsigargin from Thapsia garganica L. Agents Actions 9:53–57

Prakriya M, Lewis RS (2015) Store-operated calcium channels. Physiol Rev 95:1383–1436

Prentki M, Biden TJ, Janjic D, Irvine RF, Berridge MJ, Wollheim CB (1984) Rapid mobilization of Ca2+ from rat insulinoma microsomes by inositol-1,4,5-trisphosphate. Nature 309:562–564

Putney JW Jr (1976) Biphasic modulation of potassium release in rat parotid gland by carbachol and phenylephrine. J Pharmacol Exp Ther 198:375–384

Putney JW Jr (1977) Muscarinic, alpha-adrenergic and peptide receptors regulate the same calcium influx sites in the parotid gland. J Physiol (Lond) 268:139–149

Putney JW Jr (1986) A model for receptor-regulated calcium entry. Cell Calcium 7:1–12

Putney JW Jr, Poggioli J, Weiss SJ (1981) Receptor regulation of calcium release and calcium permeability in parotid gland cells. Philos Trans R Soc Lond B 296:37–45

Rizzuto R, Pozzan T (2006) Microdomains of intracellular Ca2+: molecular determinants and functional consequences. Physiol Rev 86:369–408

Roos J, DiGregorio PJ, Yeromin AV, Ohlsen K, Lioudyno M, Zhang S, Safrina O, Kozak JA, Wagner SL, Cahalan MD, Velicelebi G, Stauderman KA (2005) STIM1, an essential and conserved component of store-operated Ca2+ channel function. J Cell Biol 169:435–445

Samanta K, Bakowski D, Parekh AB (2014) Key role for store-operated Ca2+ channels in activating gene expression in human airway bronchial epithelial cells. PLoS One 9:e105586

Sattler R, Tymianski M (2000) Molecular mechanisms of calcium-dependent excitotoxicity. J Mol Med 78:3–13

Sitrin MD, Bohr DF (1971) Ca and Na interaction in vascular smooth muscle contraction. Am J Physiol 220:1124–1128

Slack BE, Bell JE, Benos DJ (1986) Inositol 1,4,5-trisphosphate injection mimics fertilization potentials in sea urchin eggs. Am J Physiol 250:C340–C344

Smyth JT, Beg AM, Wu S, Putney JW Jr, Rusan NM (2012) Phosphoregulation of STIM1 leads to exclusion of the endoplasmic reticulum from the mitotic spindle. Curr Biol 22:1487–1493

Steinckwich N, Myers P, Janardhan KS, Flagler ND, King D, Petranka JG, Putney JW (2015) Role of the store-operated calcium entry protein, STIM1, in neutrophil chemotaxis and infiltration into a murine model of psoriasis-inflamed skin. FASEB J 29:3003–3013

Steinsland OS, Furchgott RF, Kirpekar SM (1973) Biphasic vasoconstriction of the rabbit ear artery. Circ Res 32:49–58

Streb H, Irvine RF, Berridge MJ, Schulz I (1983) Release of Ca2+ from a nonmitochondrial store in pancreatic cells by inositol-1,4,5-trisphosphate. Nature 306:67–68

Streb H, Bayerdorffer E, Haase W, Irvine RF, Schulz I (1984) Effect of inositol-1,4,5-trisphosphate on isolated subcellular fractions of rat pancreas. J Membr Biol 81:241–253

Takemura H, Putney JW Jr (1989) Capacitative calcium entry in parotid acinar cells. Biochem J 258:409–412

Takemura H, Hughes AR, Thastrup O, Putney JW Jr (1989) Activation of calcium entry by the tumor promoter, thapsigargin, in parotid acinar cells. Evidence that an intracellular calcium pool, and not an inositol phosphate, regulates calcium fluxes at the plasma membrane. J Biol Chem 264:12266–12271

Thastrup O, Foder B, Scharff O (1987) The calcium mobilizing tumor promoting agent, thapsigargin elevates the platelet cytoplasmic free calcium concentration to a higher steady state level. A possible mechanism of action for the tumor promotion. Biochem Biophys Res Commun 142:654–660

Thastrup O, Cullen PJ, Drobak BK, Hanley MR, Dawson AP (1990) Thapsigargin, a tumor promoter, discharges intracellular Ca2+ stores by specific inhibition of the endoplasmic reticulum Ca2+-ATPase. Proc Natl Acad Sci U S A 87:2466–2470

Thompson JL, Shuttleworth TJ (2013) Exploring the unique features of the ARC channel, a store-independent Orai channel. Channels (Austin) 7:364–373

Tymianski M (1996) Cytosolic calcium concentrations and cell death in vitro. Adv Neurol 71:85–105

Ueda T, Church SH, Noel MW, Gill DL (1986) Influence of inositol 1,4,5-trisphosphate and guanine nucleotides on intracellular calcium release within the N1E-115 neuronal cell line. J Biol Chem 261:3184–3192

van Breemen C (1969) Blockade of membrane calcium fluxes by lanthanum in relation to vascular smooth muscle contractility. Arch Int Physiol Biochim 77:710–716

van Breemen C, Farinas BR, Casteels R, Gerba P, Wuytack F, Deth R (1973) Factors controlling cytoplasmic Ca 2+ concentration. Philos Trans R Soc Lond B Biol Sci 265:57–71

Vig M, Peinelt C, Beck A, Koomoa DL, Rabah D, Koblan-Huberson M, Kraft S, Turner H, Fleig A, Penner R, Kinet JP (2006) CRACM1 is a plasma membrane protein essential for store-operated Ca2+ entry. Science 312:1220–1223

Willoughby D, Everett KL, Halls ML, Pacheco J, Skroblin P, Vaca L, Klussmann E, Cooper DM (2012) Direct binding between Orai1 and AC8 mediates dynamic interplay between Ca2+ and cAMP signaling. Sci Signal 5:ra29

Xing J, Petranka JG, Davis FM, Desai PN, Putney JW, Bird GS (2013) Role of orai1 and store-operated calcium entry in mouse lacrimal gland signaling and function. J Physiol (Lond) 592:927–939

Zhang SL, Yeromin AV, Zhang XH, Yu Y, Safrina O, Penna A, Roos J, Stauderman KA, Cahalan MD (2006) Genome-wide RNAi screen of Ca2+ influx identifies genes that regulate Ca2+ release-activated Ca2+ channel activity. Proc Natl Acad Sci U S A 103:9357–9362

Zhang X, Gonzalez-Cobos JC, Schindl R, Muik M, Ruhle B, Motiani RK, Bisaillon JM, Zhang W, Fahrner M, Barroso M, Matrougui K, Romanin C, Trebak M (2013) Mechanisms of STIM1 activation of store-independent leukotriene C4-regulated Ca2+ channels. Mol Cell Biol 33:3715–3723

Zweifach A, Lewis RS (1993) Mitogen-regulated Ca2+ current of T lymphocytes is activated by depletion of intracellular Ca2+ stores. Proc Natl Acad Sci U S A 90:6295–6299

The STIM-Orai Pathway: STIM-Orai Structures – Isolated and in Complex

2

Jinhui Zhu, Qingping Feng, and Peter B. Stathopulos

Abstract

Considerable progress has been made elucidating the molecular mechanisms of calcium (Ca^{2+}) sensing by stromal interaction molecules (STIMs) and the basis for Orai channel activity. This chapter focuses on the available high-resolution structural details of STIM and Orai proteins with respect to the regulation of store-operated Ca^{2+} entry (SOCE). Solution structures of the Ca^{2+}-sensing domains of STIM1 and STIM2 are reviewed in detail, crystal structures of cytosolic coiled-coil STIM fragments are discussed, and an overview of the closed *Drosophila melanogaster* Orai hexameric structure is provided. Additionally, we highlight structures of human Orai1 N-terminal and C-terminal domains in complex with calmodulin and human STIM1, respectively. Ultimately, the accessible structural data are discussed in terms of potential mechanisms of action and cohesiveness with functional observations.

Keywords

Stromal interaction molecules • Orai1 • Calmodulin • X-ray crystallography • Solution nuclear magnetic resonance spectroscopy • Store-operated calcium entry • Structural mechanisms

2.1 Background

The calcium ion (Ca^{2+}) is a universal messenger which controls a vast number of cellular processes such as the prolonged regulation of transcription, cell division, and apoptosis, as well as more short-lived secretion and contraction (Berridge et al.

J. Zhu • Q. Feng • P.B. Stathopulos (✉)
Department of Physiology and Pharmacology, Schulich School of Medicine and Dentistry, Western University, London, ON, Canada, N6A 5C1
e-mail: peter.stathopulos@schulich.uwo.ca

K. Groschner et al. (eds.), *Store-Operated Ca^{2+} Entry (SOCE) Pathways*, Advances in Experimental Medicine and Biology 993, DOI 10.1007/978-3-319-57732-6_2

2003; Berridge 2009; Bootman and Lipp 2001). The voltage-independent mode of store-dependent Ca^{2+} mobilization is conserved among eukaryotic cells: agonist-induced stimulation of G-protein-coupled, T-cell, or tyrosine kinase receptors activates phospholipase C β or $\gamma 2$, leading to the hydrolysis of membrane-associated phosphoinositide 4,5-bisphosphate, yielding inositol 1,4,5-trisphosphate (IP_3). IP_3 is a small diffusible second messenger which binds to the IP_3 receptor (IP_3R) on the endoplasmic reticulum (ER) membrane. Binding of IP_3 allosterically opens this Ca^{2+} release channel, thus allowing Ca^{2+} to move down the concentration gradient from the lumen into the cytoplasm. The ER lumen can only transiently supply the cytosol with Ca^{2+} before it is rapidly depleted. In excitable cells, ryanodine receptors (RyRs) play a larger role in sarcoplasmic reticulum (SR) luminal Ca^{2+} depletion than the IP_3Rs. After diminishment of ER/SR-stored Ca^{2+}, highly Ca^{2+}-selective and Ca^{2+}-permeable store-operated channels (SOCs) on the plasma membrane (PM) open, providing sustained Ca^{2+} influx into the cytosol from the virtually inexhaustible extracellular Ca^{2+} supply. Ultimately, the cytosolic influx of Ca^{2+} replenishes the luminal stores via the SR/ER Ca^{2+} ATPase pump. This specific communicative interchange of Ca^{2+} between the ER/SR lumen, cytosol, and extracellular space is termed as store-operated Ca^{2+} entry (SOCE).

Although SOCE was proposed over two decades ago (Putney 1986), the major molecular players were identified and began to be characterized ~20 years following the conceptualization of this model, with the stromal interaction molecules (STIMs) functioning as the ER/SR Ca^{2+} sensors and PM SOC activators (Liou et al. 2005; Roos et al. 2005; Zhang et al. 2005) and the Orai proteins serving as the major PM channel components (Feske et al. 2006; Prakriya et al. 2006; Vig et al. 2006a, b; Yeromin et al. 2006; Zhang et al. 2006). Orai1-composed SOCs are termed Ca^{2+} release-activated Ca^{2+} (CRAC) channels due to the voltage-independent, highly Ca^{2+}-selective, and inward-rectifying currents generated during activity which is distinct from other SOCs. SOCE via CRAC channels is the primary mode of augmenting cytosolic Ca^{2+} in lymphocyte signaling.

Human Orai1 is composed of 301 amino acids and has four transmembrane (TM) segments (Cai 2007; Feske 2007). Both the amino and carboxy termini of this PM protein reside in the cytoplasm, and each has been implicated as a critical accessory region in Orai1 activation via direct and indirect interactions with STIM1 (Derler et al. 2013; Frischauf et al. 2009; McNally et al. 2013; Muik et al. 2008, 2012; Palty et al. 2015; Palty and Isacoff 2016; Park et al. 2009; Yuan et al. 2009; Zheng et al. 2013; Zhou et al. 2010a). Humans encode three homologues (i.e., Orai1, Orai2, Orai3) showing conservation in the predicted TM domains, acidic residues involved in ion permeability and selectivity, as well as basic residue position associated with severe combined immunodeficiency disease (SCID). Transmission electron microscopy images of Orai1 suggest that Orai1 tetramers form a teardrop-shaped structure that extends into the cytosol sufficiently for direct interaction with STIM1 at ER-PM junctions (i.e., ~10 nm) (Maruyama et al. 2009). The TM1 segment constitutes the central pore of Orai channels (McNally et al. 2009, 2012; McNally and Prakriya 2012; Prakriya et al. 2006; Yeromin et al. 2006; Zhang et al. 2011; Zhou et al. 2010b); moreover, published data supports both tetrameric (Demuro et al. 2013; Maruyama et al. 2009; Mignen et al. 2008; Penna et al. 2008; Thompson and Shuttleworth 2013) and hexameric architectures (Balasuriya et al. 2014; Hou et al. 2012; Zhou et al. 2010b).

Upon ER Ca^{2+} store depletion, STIM1 moves from a pervasive ER distribution to distinct punctate aggregates at ER-PM junctions (Liou et al. 2005; Zhang et al. 2005). This cluster of Ca^{2+}-depleted STIM1 facilitates recruitment of Orai1 to the same junctions, establishing sites of open CRAC channels (Luik et al. 2006; Varnai et al. 2007; Wu et al. 2006; Xu et al. 2006). The present chapter focuses on a comparison of the high-resolution structural properties of the Ca^{2+}-sensing regions of human STIM1 and STIM2; further, the work discusses available crystal structure information on the cytosolic coiled-coil (CC) domains of STIM1 as well as *Drosophila melanogaster* Orai. In addition to this structural data of STIM and Orai fragments in isolation, the structural elucidations of the N- and C-terminal regions of Orai1 in complex with calmodulin (CaM) and a human STIM1 CC region, respectively, are reviewed. Collectively, these high-resolution structural details provide important insights into the mechanisms that activate and promote relocalization of STIMs to ER-PM junctions and the basis for STIM coupling with Orai subunits.

2.2 Stromal Interaction Molecule Domain Organization

Human STIM1 is a type I TM protein of 685 amino acids localized on ER/SR membranes or on the PM following posttranslational *N*-glycosylation of Asn131 and Asn171 (Manji et al. 2000; Williams et al. 2001, 2002). Vertebrates encode a second homologue, STIM2, which contains regions highly conserved with STIM1 within both the luminal and cytosolic domains. Unlike STIM1, STIM2 does not appear to localize on the PM, despite conservation of the STIM1 Asn131 residue. The most homologous regions between STIM1 and STIM2 include the EF-hand, sterile α motif (SAM) domain, as well as two cysteines within the luminal domains and three CC domains, a Lys-rich and a Pro/Ser-rich segment within the cytosolic region (Fig. 2.1a). Both proteins encode a single-pass TM region identifiable in hydropathy plots. ER localization of STIM1 is signaled through the first 22 amino acids. STIM2, on the other hand, encodes an additional 87 residues distally toward the N-terminus outside of the homologous ER signal peptide sequences. Early studies proposed that translation initiation of STIM2 occurs at a non-AUG site (Williams et al. 2001); however, more recently it has been suggested that the much longer signal sequence of STIM2, upstream of and including the homologous STIM1 signal peptide (i.e., residues 1–101) (Fig. 2.1b), is necessary for appropriate ER localization of STIM2 proteins (see Sect. 2.3) (Graham et al. 2011). Phosphorylation of various STIM1 Ser and Thr residues within the Ser/Pro-rich region has also been described (Pozo-Guisado et al. 2010; Smyth et al. 2009).

CRAC channel activation by ER-residing STIM1 is a multistep process: first, STIM1 oligomerization on the ER membrane occurs in response to ER luminal Ca^{2+} depletion; second, STIM1 homotypic oligomers translocate to

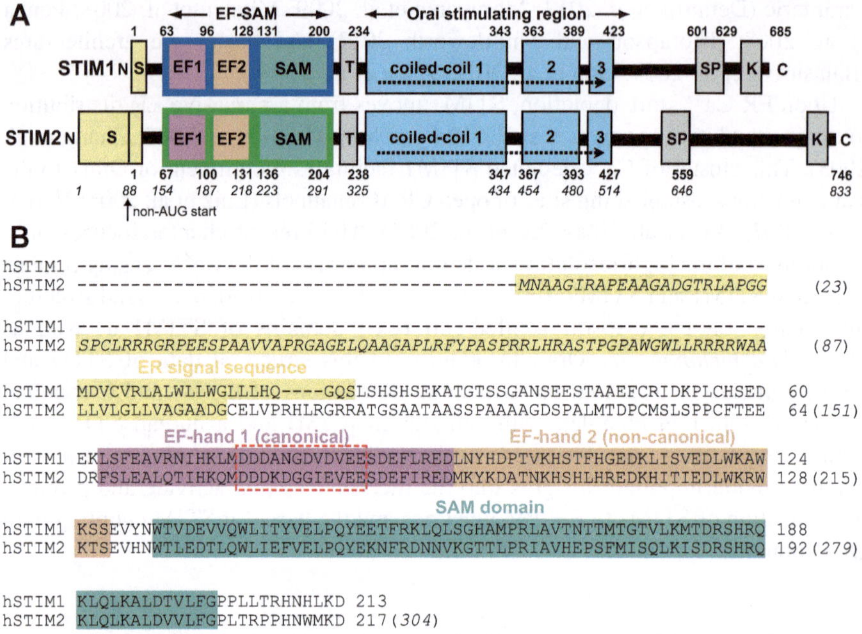

Fig. 2.1 Primary sequence and domain architecture of STIM1 and STIM2. (**a**) *Conserved domains in human STIM1 and STIM2*. Upstream of the signal peptides, S (*yellow*), the luminal domains include the canonical EF-hand, EF1 (*violet*), the noncanonical EF-hand, EF2 (*beige*), and the SAM domain (*green*). A single-pass transmembrane region, T, separates the cytosolic portion that includes three coiled-coil domains (*blue*), a serine/proline-rich region, SP, and a lysine-rich (i.e., polybasic) region, K. Residue boundaries are indicated above and below STIM1 and STIM2, respectively. STIM2 contains a second set of numbering (*italics*) corresponding to the entire STIM2 open reading frame. (**b**) *Sequence conservation between human STIM1 and STIM2 luminal regions*. Amino acids through the domain boundaries are colored as described in (**a**). The canonical EF-hand loop residues involved in Ca^{2+} coordination are bounded by a box (*red, broken lines*). Residues shown in *italics* are upstream of the non-AUG start site for STIM2. Alignment was performed using ClustalW (Larkin et al. 2007)

ER-PM junctions; third, CRAC channels composed of PM Orai1 subunits are recruited and open at these junctions (Liou et al. 2007). The cytosolic regions of STIMs play a role in the oligomerization of Ca^{2+}-depleted STIM1 (Covington et al. 2010; Muik et al. 2009), in targeting STIM1 to the ER-PM junctions and in mediating the interactions between STIM1 and Orai1 pore subunits (Baba et al. 2006; Huang et al. 2006; Li et al. 2007; Liou et al. 2007). Three separate investigations identified cytosolic STIM1 residues 233–450 (i.e., Orai1-activating small fragment, OASF) (Muik et al. 2009), 342–448 (Park et al. 2009) (i.e., CRAC-activating domain, CAD), and 344–443 (Yuan et al. 2009) (i.e., STIM1-Orai1-activating region, SOAR) as critical amino acid stretches through the conserved CC domains for inducing Orai1 channel activation.

The essential role of STIM1 in the activation of CRAC channels is evident from RNA interference studies which demonstrate a significant attenuation in CRAC entry after STIM1 knockdown (Liou et al. 2005; Roos et al. 2005) and from STIM1/Orai1 co-overexpression data which show very large augmentations in ER Ca^{2+} depletion-dependent SOCE (Mercer et al. 2006; Soboloff et al. 2006). Along with a role in SOCE, STIM2 more prominently controls basal Ca^{2+} homeostasis (Bird et al. 2009; Brandman et al. 2007). A fraction of STIM2 is coupled to Orai1 at resting ER Ca^{2+} levels, probably due to a somewhat lower affinity for Ca^{2+} (Brandman et al. 2007; Zheng et al. 2011). Both STIM1 and STIM2 are indispensable in CRAC-induced immune cell activation, though STIM2 knockout ($^{-/-}$) affects SOCE in T cells and fibroblasts to a lesser extent than STIM1$^{-/-}$ (Bird et al. 2009; Oh-hora and Rao 2008). While both homologues are expressed in a variety of cell types, a vital role for STIM2 has been emphasized in neuronal Ca^{2+} signaling (Berna-Erro et al. 2009).

The EF-hand and SAM domains located in the luminal region of STIMs are conserved from roundworms to vertebrates (Collins and Meyer 2011; Stathopulos et al. 2009). In humans, the EF-hand together with the SAM domain (i.e., EF-SAM) exhibits greater than 85% sequence similarity (Fig. 2.1b). Live cell studies have demonstrated that EF-SAM provides the machinery necessary to sense changes in luminal Ca^{2+} levels. Specifically, disruption of Ca^{2+} binding via mutations in the canonical EF-hand causes STIM1 to constitutively form puncta independent of ER Ca^{2+} levels (Liou et al. 2005; Mercer et al. 2006; Roos et al. 2005; Spassova et al. 2006). Furthermore, deleting the SAM domain from STIM1 abrogates the ability of the protein to form inducible puncta (Baba et al. 2006), and exchanging EF-SAM with a rapamycin-inducible oligomerization domain allows control of STIM1 puncta formation via rapamycin treatment, independent of ER Ca^{2+} levels (Luik et al. 2008). Finally, at resting ER Ca^{2+} concentrations, fluorescent protein-tagged STIM1 constructs engineered without the cytosolic domains exhibit very low intermolecular fluorescence resonance energy transfer (FRET) levels (i.e., consistent with non-interacting EF-SAM domains) that are largely augmented upon Ca^{2+} store depletion (Covington et al. 2010). In addition to cell biology work, recombinant expression and isolation of STIM1 and STIM2 EF-SAM domains has been a valuable approach in garnering unambiguous biochemical and structural data on this critical Ca^{2+}-sensing region.

2.3 Folding and Ca^{2+} Sensitivity of Isolated EF-SAM Domains

Throughout nature, EF-hands are found as pairs of helix-loop-helix motifs comprising single domains, and these EF-hand domains are fully capable of self-folding and cation binding. Similarly, SAM domains have been purified and characterized, folding into compact five-helix bundles, often showing a tendency for intermolecular association. The identification of STIMs as important regulators of CRAC entry inspired the recombinant expression, purification, and characterization of these domains, in cis, on the same polypeptide chain, as they occur in STIMs. Remarkably, recombinant STIM1 EF-SAM unfolds through a single, highly cooperative transition using temperature or chemical denaturation (Stathopulos et al. 2006). Similarly, a single and cooperative transition is observed for STIM2 EF-SAM (Zheng et al. 2008), suggesting that the EF-hand and SAM domains of STIMs are codependent in overall stability and folding (Fig. 2.2a). The Ca^{2+}-depleted forms

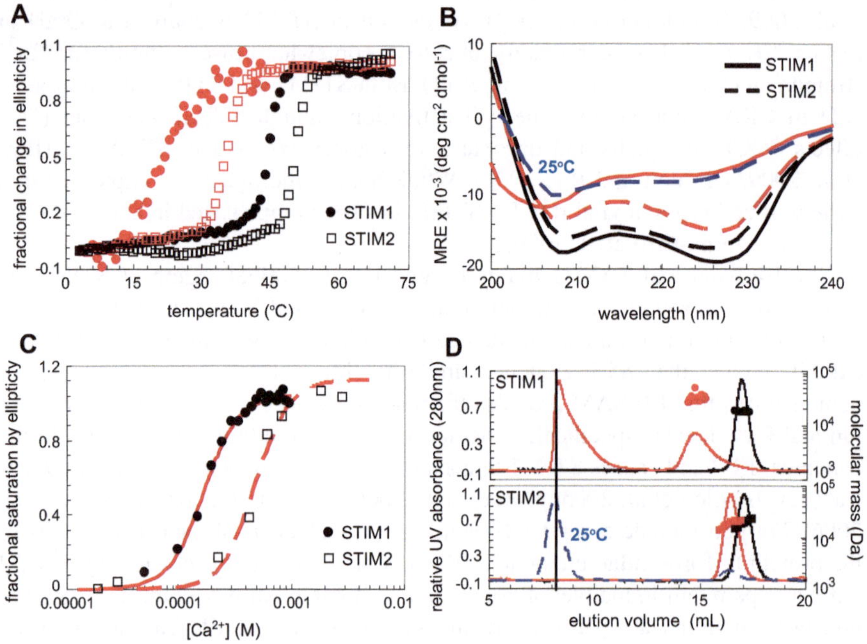

Fig. 2.2 In vitro Ca^{2+}-sensing characteristics of STIM1 and STIM2 EF-SAM. (**a**) *Thermal stability of EF-SAM proteins*. The change in far-UV CD at 225 nm is plotted as a function of temperature. Proteins in the presence of excess Ca^{2+} (i.e., holo) are plotted with *black symbols* and proteins in the absence of Ca^{2+} (i.e., apo) are plotted with *red symbols*. *Circles* and *squares* represent STIM1 and STIM2 EF-SAM, respectively [adapted from Stathopulos et al. (2006) and Zheng et al. (2008)]. (**b**) *Secondary structure of EF-SAM proteins*. The far-UV CD spectra of holo and apo STIM1 are shown as *solid black* and *red lines*, respectively. The CD spectra of holo and apo STIM2 are shown as *broken black* and *red lines*, respectively. Apo STIM2 only exhibits a spectrum resembling apo STIM1 EF-SAM at 25°C, shown as a *broken blue line* [adapted from Stathopulos et al. (2006) and Zheng et al. (2008)]. (**c**) *Ca^{2+} binding to EF-SAM proteins*. Binding to STIM1 EF-SAM monitored by the fractional increase in negative ellipticity at 222 nm is plotted in *circles*. Binding to STIM2 EF-SAM is plotted in *squares*. The Hill equation *fitted lines* are shown in *solid* and *broken red lines* for STIM1 and STIM2, respectively [adapted from Zheng et al. (2011)]. (**d**) *Quaternary structure of EF-SAM proteins*. The STIM1 EF-SAM elution profiles in the presence (*black line*) and absence (*red line*) of Ca^{2+} are shown (*upper panel*). The MALS-determined monomeric molecular weight of holo and the dimeric molecular weight of apo STIM1 EF-SAM are shown above the elution peaks. The STIM2 EF-SAM elution profile in the presence (*black line*) and absence (*red line*) of Ca^{2+} is shown (*lower panel*). The MALS-determined monomeric molecular weights for holo and apo STIM2 EF-SAM at 4°C are shown above the corresponding elution peaks. The 25°C elution profile of apo STIM2 EF-SAM is shown as a *broken blue line*. The *vertical black line* intersects the S200 void volume showing that only apo STIM1 and STIM2 EF-SAM access the oligomerized state [adapted from Stathopulos et al. (2006) and Zheng et al. (2008)]

(i.e., apo) of STIM1 and STIM2 EF-SAM are markedly less stable than the Ca^{2+}-loaded counterparts (i.e., holo), but also exhibit single, cooperative unfolding curves. The temperature midpoint of unfolding (i.e., T_m) in single transition unfolding curves is a good indicator of protein stability. Apo STIM1 EF-SAM shows a T_m of ~19°C compared to holo STIM1 EF-SAM with a T_m of ~45°C near physiological pH (Stathopulos et al. 2006). The holo and apo states of STIM2 EF-SAM are more stable than STIM1 with T_m values of ~50°C and 36°C, respectively (Fig. 2.2a) (Zheng et al. 2008).

In the case of STIM1 EF-SAM, the protein undergoes a conformational change from a highly α-helical state in the presence of Ca^{2+} to a less well-folded state in the absence of Ca^{2+} (Fig. 2.2b). STIM2 EF-SAM also loses α-helicity in response to Ca^{2+} depletion; however, the structural transition is less striking, with apo STIM2 EF-SAM retaining considerable α-helicity (Fig. 2.2b) (Zheng et al. 2008). The ability of STIM2 EF-SAM to preserve a high degree of α-helicity in the absence of Ca^{2+} is consistent with the lesser destabilization observed in temperature denaturation experiments.

Consistent with the presence of a single canonical binding loop, STIM1 EF-SAM becomes saturated with Ca^{2+} at a molar ratio of ~1. The equilibrium dissociation constant (i.e., K_D), indirectly calculated using Ca^{2+}-induced changes in tertiary structure via intrinsic aromatic fluorescence or changes in secondary structure via far-UV circular dichroism, as well as directly measured via $^{45}Ca^{2+}$ titration, is high (i.e., K_D between ~0.2 and 0.6 mM) compared to most vertebrate EF-hand proteins (Fig. 2.2c) (Stathopulos et al. 2006). The K_D of Ca^{2+} binding for STIM1 is temperature-dependent, with higher affinities (i.e., lower K_D) at lower temperatures. There is a high preference of the EF-hand for Ca^{2+} over Mg^{2+} since inclusion of Mg^{2+} in the experiments does not alter the binding curves. Grafting of the canonically defined EF-hand motif onto a stabilizing CD2 domain and solution nuclear magnetic resonance (NMR) titration with Ca^{2+} shows a K_D of ~0.5 mM for the STIM1 EF-hand motif in pseudo-isolation (Huang et al. 2009). Considering the high sequence similarity between the STIM1 and STIM2 canonical EF-hand binding loops, it is not surprising that STIM2 EF-SAM also exhibits a low affinity for Ca^{2+} in the same range as that determined for STIM1 (Fig. 2.2c). Nonetheless, it is important to note that similar probes of Ca^{2+} binding (i.e., fluorescence, CD, $^{45}Ca^{2+}$) applied to the STIM2 EF-SAM system produced considerably higher variability in the curves. This increased error and uncertainty is probably due to a somewhat lower Ca^{2+} affinity, albeit in the same sub-mM K_D range.

The inherent low affinity Ca^{2+} binding EF-hand motifs of STIMs are well suited to the relatively high Ca^{2+} levels in the ER/SR lumen. In resting, non-excitable cells, the ER luminal Ca^{2+} level is typically between 0.6 and 0.8 mM (Berridge et al. 2000, 2003). Following agonist-induced stimulation, the luminal Ca^{2+} level may decrease, at least locally, to ~0.2–0.4 mM; moreover, with a Ca^{2+} binding K_D of ~0.2–0.6 mM, STIM1 has evolved to proficiently equilibrate between the Ca^{2+}-loaded and Ca^{2+}-free states in response to these fluctuations in ER Ca^{2+}. The in vitro Ca^{2+} binding data for STIM2, at least qualitatively, suggest a decreased affinity compared to STIM1 (Fig. 2.2c). Consistent with this in vitro work, full-length

STIM2 forms puncta in response to smaller decreases in ER Ca^{2+} (i.e., at higher ER Ca^{2+} levels) compared to STIM1 (Bird et al. 2009; Brandman et al. 2007). This Ca^{2+} binding property of STIM2 facilitates a role for the ER-localized protein in regulating basal Ca^{2+} homeostasis since at resting ER Ca^{2+} concentrations, a significant fraction of STIM2 may be Ca^{2+}-depleted and coupled to Orai1. Contributory to the Ca^{2+} store-independent regulatory mode of STIM2 may be cytosolic STIM2 that is not anchored to the ER membrane (Graham et al. 2011). In the cytosol, the vast majority of STIM2 molecules would be maintained in a Ca^{2+}-depleted state due to a Ca^{2+} level that is several orders of magnitude lower than the ER lumen (Feske 2007).

Homotypic oligomerization of STIM proteins following ER luminal Ca^{2+} depletion is a vital initiation step in the activation of SOCE (Liou et al. 2007; Stathopulos et al. 2006). In excess Ca^{2+} concentrations, STIM1 EF-SAM strictly exists as a monomer at physiological-like pH and temperatures below the thermal unfolding transition (Fig. 2.2d). In concert with the marked destabilization and partial unfolding accompanying Ca^{2+} depletion, STIM1 EF-SAM undergoes a change in quaternary structure, forming dimers and oligomers at low (i.e., 4°C) and ambient (i.e., ~25°C) temperatures (Stathopulos et al. 2006). On the other hand, STIM2 EF-SAM exhibits a resistance to oligomerization at low temperature, despite undergoing Ca^{2+} depletion-induced destabilization, as observed for STIM1 (Fig. 2.2d) (Zheng et al. 2008). At ambient temperature, however, STIM2 EF-SAM oligomerizes, albeit with distinct kinetics compared to STIM1 EF-SAM (Stathopulos et al. 2009).

2.4 STIM1 and STIM2 EF-SAM Structure

NMR spectroscopy has been an invaluable tool for teasing out the precise atomic basis for the differences in the physicochemical properties between STIM1 and STIM2 EF-SAM. The Ca^{2+}-loaded form of STIM1 EF-SAM folds into a compact ten-helix structure (Fig. 2.3a, left panel) (Stathopulos et al. 2008). A second EF-hand, not identified in the primary sequence of STIM1, is adjacent to the canonical EF-hand, stabilizing the Ca^{2+} binding loop through hydrogen (H)-bonding. This H-bonding between the canonical and noncanonical EF-hand loops forms a short β-sheet. The three-dimensional (3D) structure also exposes a short α-helix which links the EF-hand pair to the five-helix bundle SAM domain. Overall, the majority of the compact structure exhibits an acidic surface charge at neutral pH, particularly concentrated over the EF-hand domain; however, a smaller patch of basic electrostatic potential exists over the SAM domain (Stathopulos et al. 2008).

The NMR structure reveals a basis for the compact nature of STIM1 EF-SAM. Internally within EF-SAM, the EF-hand pair forms a hydrophobic pocket in the Ca^{2+}-loaded state through the side chain orientation of at least 12 amino acids (i.e., Val68, Ile71, His72, Leu74, Met75, Val83, Leu92, Leu96, Lys104, Phe108, Ile115, Leu120) (Fig. 2.3b, left panel). This hydrophobic cleft serves as a dock for nonpolar side chains protruding on the distal end of the α10 helix on the SAM domain (i.e., Leu195 and Leu199) (Fig. 2.3a, left panel) (Stathopulos

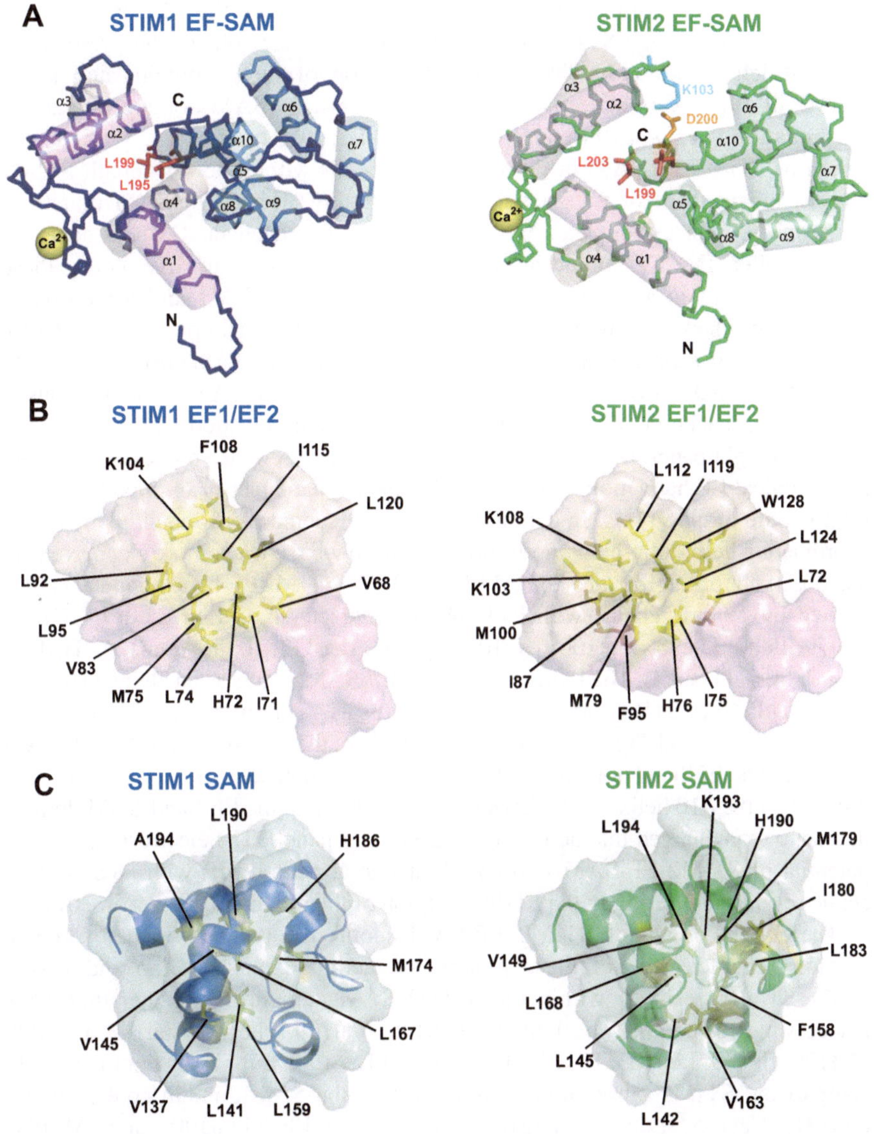

Fig. 2.3 *Atomic-resolution NMR structures of STIM1 and STIM2 EF-SAM.* (**a**) *Structural features of holo EF-SAM proteins.* The backbone atoms of Ca^{2+}-loaded STIM1 and STIM2 EF-SAM are traced in *blue* (*left structure*) and *green* (*right structure*), respectively. The canonical EF-hand helices are shown in *violet*, the noncanonical EF-hand helices are in *beige*, and the linker and SAM helices are shown as *green cylinders*. The Ca^{2+} ions coordinated in the canonical loop are shown as *yellow spheres* and the carboxy (C) and amino (N)-termini are indicated. The critical α10 SAM anchor side chains are presented as *red sticks*. Unique to the STIM2 EF-SAM structure is the complementary positioning of the basic Lys103 (*cyan sticks*) and acidic Asp200 (*orange sticks*) side chains. (**b**) *EF-hand cleft architecture of EF-SAM proteins*. A surface representation of the EF-hand domains is shown for STIM1 (*left*) and STIM2 (*right*) where coloring is consistent

et al. 2008). These intimate hydrophobic contacts between the EF-hand pair and the SAM domain lock STIM1 EF-SAM into a compact fold in the presence of Ca^{2+}, which auto-inhibits homotypic association of these domains and keeps STIM1 in a quiescent state. Disruption of the EF-hand-SAM domain interaction via mutation (i.e., Phe108Asp/Gly110Asp or Leu195Arg) facilitates oligomerization of STIM1 EF-SAM in vitro and in live cells within the full-length STIM1 context independent of Ca^{2+} levels. In vitro, these EF-hand or SAM mutations cause the protein to adopt an apo-like structure, based on far-UV CD data, but do not alter the Ca^{2+} binding properties of EF-SAM. In live cells, these mutations result in a constitutive puncta formation of STIM1 and activation of SOCs even when the ER luminal Ca^{2+} stores are full (Stathopulos et al. 2008).

As expected from the high sequence conservation between STIM1 and STIM2 EF-SAM, the 3D NMR structure of Ca^{2+}-loaded STIM2 EF-SAM is homologous to STIM1 [i.e., backbone Cα, NH, CO root mean square deviation (rmsd) of 2.7 Å] (Fig. 2.3a, right panel) (Zheng et al. 2011). STIM2 EF-SAM also possesses a second, noncanonical EF-hand motif, suggesting that the EF-hand pair is an important structural feature of all STIMs. In conjunction with the canonical EF-hand, the non-ion-coordinating helix-loop-helix motif forms a hydrophobic pocket, more extensive than that observed for Ca^{2+}-loaded STIM1 EF-SAM. The STIM2 EF-hand nonpolar cleft is created by 13 side chains with hydrophobic character (i.e., Leu72, Ile75, His76, Met79, Ile87, Phe95, Met100, Lys103, Lys108, Leu112, Ile119, Leu124, Trp128) (Fig. 2.3b, right panel). The Lys103 of STIM2 EF-SAM occurs as an aligned His (i.e., amino acid 99) in STIM1, and the Trp128 occurs as a conserved Trp124 in STIM1; moreover, His99 and Trp124 are directed away from the cleft in the STIM1 EF-SAM structure. The EF-hand domain Lys103 of STIM2 is oriented in close proximity with Asp200 on the α10 helix of the SAM domain, stabilizing the EF-hand-SAM domain interaction via charged interactions (Fig. 2.3a, right panel). The electrostatic surface potential of STIM2 EF-SAM is primarily acidic at neutral pH, with most surface acidic residues clustering in the EF-hand region of the protein (Zheng et al. 2011).

The SAM domain of STIM2 EF-SAM adopts the typical five-helix bundle topology characteristic of these protein interaction domains (Qiao and Bowie 2005). Twelve residues within the STIM2 SAM domain are greater than 95% inaccessible to solvent (i.e., Leu142, Leu145, Val149, Phe158, Val163, Leu168, Met179, Ile180, Leu183, His190, Lys193, and Leu194) (Fig. 2.3c, right panel). In comparison, STIM1 buries nine residues within the SAM domain core (Val137, Leu141, Val145, Leu159, Leu167, Met174, His186, Leu190, and Ala194) (Fig. 2.3c, left panel). The hydrophobic STIM2 Ile180 is not conserved in

Fig. 2.3 (continued) with panel (**a**). Residues (*sticks*) and associated surface forming the nonpolar clefts are shown in *yellow*. (**c**) *SAM domain hydrophobic core packing within EF-SAM proteins*. The STIM1 and STIM2 SAM backbones are shown as *blue* (*left*) and *green* (*right*) ribbon representations within the SAM domain surface, respectively. Side chains that are greater than 95% inaccessible to solvent are indicated with *yellow sticks*. STIM1 EF-SAM pdbID: 2K60.pdb; STIM2 EF-SAM pdbID: 2L5Y.pdb

STIM1. This bulky Ile side chain facilitates a rearrangement of residues resulting in the insertion of Phe158 and Lys193 into the STIM2 SAM core (Zheng et al. 2011). These conserved Phe (i.e., amino acid 154) and Lys (i.e., amino acid 189) residues are excluded from the core in STIM1. The STIM2 SAM domain protrusion residues essential for interaction with the EF-hand pocket include the conserved Leu199 and Leu203 of α10 (Fig. 2.3a, right panel). The enlarged EF-hand hydrophobic cleft, the orientation of basic Lys103 on the EF-hand domain in close proximity to acidic Asp200 on the SAM domain, and the enhanced hydrophobic core of the SAM domain all contribute to the augmented stability and attenuated oligomerization propensity observed for STIM2 EF-SAM compared to STIM1.

Interestingly, it has been recently demonstrated that mimicking glycosylation on the Asn131 and Asn171 sites of STIM1 EF-SAM causes structural changes which converge on the α8 core helix; moreover, these structural changes decrease the Ca^{2+} binding affinity and markedly destabilize EFSAM (Choi et al. 2016). While STIM1 EF-SAM contains the two consensus motifs (i.e., Asn-X-Ser/Thr) required for *N*-glycosylation at the Asn131 and Asn171 positions, STIM2 EF-SAM only conserves the Asn135 site (i.e., equivalent to STIM1 Asn131). The effects of STIM2 EF-SAM *N*-glycosylation at Asn135 are currently unknown.

2.5 Extraneous Luminal Regions Involved in STIM Ca^{2+} Sensing

Considerable primary sequence variability exists within the luminal-oriented regions of STIM1 and STIM2 outside the EF-SAM domains (Fig. 2.1b). These extraneous residues influence Ca^{2+} sensing and regulation of CRAC entry by STIMs. The extension of STIM1 EF-SAM to include all luminal residues outside the signal peptide (i.e., amino acids 23–213) enhances the stability in the presence and absence of Ca^{2+} (Stathopulos et al. 2009). A STIM2 construct engineered with similarly aligned boundaries is susceptible to C-terminal degradation; however, a somewhat shorter, degradation-resistant STIM2 protein (i.e., amino acids 15–205) also shows an enhanced stability compared to EF-SAM in vitro. Far-UV CD data of these extended EF-SAM constructs show less α-helicity per residue in the presence of Ca^{2+} compared to the minimal EF-SAMs, suggesting that the extraneous residues may have a more prominent effect on the unfolded states of EF-SAM (Stathopulos et al. 2009). In live cells, swapping STIM1 residues 1–65 (i.e., including the STIM1 signal peptide) with STIM2 residues 1–69 and vice versa exchanges the CRAC channel activation phenotype, where full-length STIM1 harboring the STIM2 N-terminal residues exhibits a delay in Orai1 activation and STIM2 fused to the STIM1 N-terminal residues demonstrates wild-type STIM1 CRAC channel activation kinetics (Zhou et al. 2009).

Two conserved cysteines (i.e., Cys49 and Cys56 in STIM1) are encoded in the N-terminal residues extraneous to EF-SAM. Oxidant-induced *S*-glutathionylation of Cys56 in STIM1 results in constitutive puncta formation and CRAC entry, suggesting that this Cys residue may afford STIM1 with an additional Ca^{2+}-independent, oxidant-dependent sensory function (Hawkins et al. 2010).

The open reading frame of STIM2 extends 87 residues upstream of the previously determined non-AUG translational start site (Fig. 2.1b) (Williams et al. 2001). Recent data suggests that these 87 upstream residues in addition to the predicted STIM2 ER signal peptide downstream of the non-AUG start site (i.e., residues 88–101; numbering 1–14 downstream of the non-AUG start site) are required for STIM2 to insert into the ER membrane (Graham et al. 2011). With this exceedingly long 101 residue signal peptide, a fraction of STIM2 remains cytosolic, activating PM Orai1 CRAC channels associated with basal Ca^{2+} homeostasis. Interestingly, the 101 amino acid signal peptide that is cleaved from the ER-inserted STIM2 pre-protein may have a function in Ca^{2+}- and SOCE-independent regulation of gene transcription (Graham et al. 2011).

2.6 STIM1 TM Structure

Recently, it has been shown that the STIM1 TM region is homodimerized in the resting state, and changes in the nature of the TM:TM interaction can modulate the position of the adjacent CC1 regions within STIM1 dimers (Ma et al. 2015). Specifically, solution NMR experiments of the STIM1 TM region (i.e., residues 209–237) reconstituted in bicelles show chemical shift perturbations for Gly225, Gly226, Trp228, Gln233, and Asn234 after incorporation of a constitutively active Cys227Trp mutation. Further, chemical cross-linking and fluorescence resonance energy transfer (FRET) experiments demonstrated that the N-terminal residues of the STIM1 TM region come into close proximity of one another in this mutation-mediated activated state while closely apposing the N-terminal region of the CC1 domains, consistent with EF-SAM-induced dimerization/oligomerization of the region (Ma et al. 2015).

2.7 STIM1 CC1 Structure

STIMs contain three conserved CC regions immediately after the TM domain (i.e., CC1, CC2, and CC3). By crystallography, the STIM1 Met244Leu/Leu321Met CC1 double mutant encompassing residues 237–340 adopts an elongated α-helical structure that extends from the N-terminus to the C-terminus and spans ~13 nm (Cui et al. 2013). This extended conformation would bridge the bulk of the necessary distance between the ER and PM for direct STIM1 coupling to Orai1 channels and STIM1 interaction with PM phosphoinositides (Calloway et al. 2011; Korzeniowski et al. 2009; Liou et al. 2007; Walsh et al. 2010), although this construct does not contain the necessary machinery for these heteromeric protein or lipid interactions. The CC1 molecules dimerize in the crystalline state in an antiparallel manner with the C-termini clustered close together; moreover, deletion of the C-terminal region of this construct (i.e., Δ311–340) results in persistent monomerization in solution. Within the crystal, hydrophobic interactions between Leu335, Leu328, Val324, and Ala317 of one subunit and Leu286, Ile290, Ala293, and Leu300 of the partner subunit; H-bonding between Leu303, Arg304, and Thr307 of one subunit and Glu310 and Gln314 of the partner subunit; and an

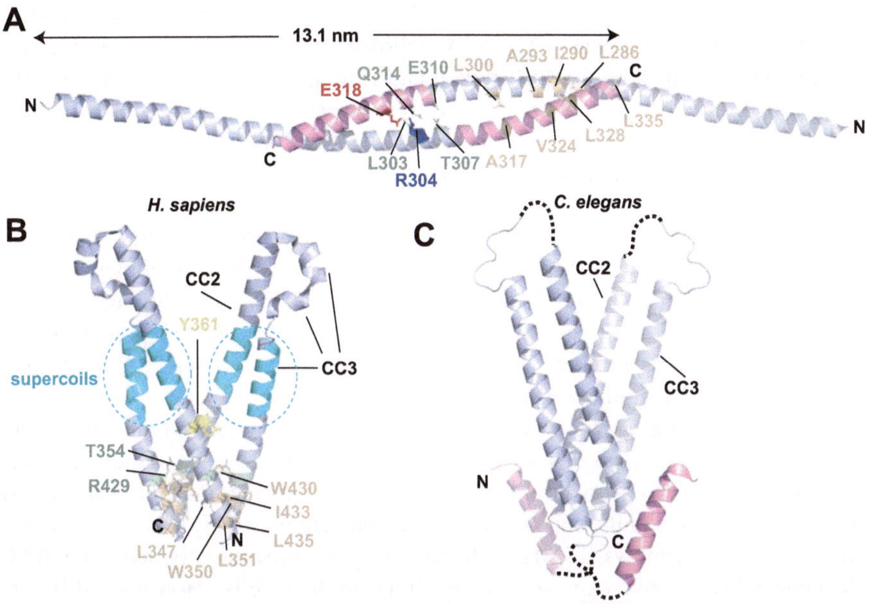

Fig. 2.4 *Atomic-resolution crystal structures of STIM CC domains.* (**a**) *Structural features of the extended human CC1 domain.* Residues participating in intermolecular hydrophobic interactions (*light orange*), H-bonding (*light green*), and ionic bonds (acidic, *red*; basic, *blue*) are presented as *sticks*. The portion of the CC1 helix previously demonstrated to play a role in modulating the activity of STIM1 is indicated (*light magenta*). (**b**) *Structural features of the human Leu374Met/Val419Ala/Cys436Thr SOAR/CAD/ccb9 domain.* The V-shaped dimer is shown with the regions between CC2 and CC3 that undergo supercoiling colored *cyan*. The intermolecular forces stabilizing the dimer are indicated with residues involved in hydrophobic interactions (*light orange*), H-bonding (*light green*), and aromatic side chain stacking (*yellow*) drawn as *sticks*. (**c**) *Structural features of the C. elegans CC1-CAD region.* The homologous CC1 region previously revealed to play a role in modulating the activity of human STIM1 is colored *light magenta* [i.e., similar to the region shown in **a**]. In **a**–**c**, the amino and carboxy termini are indicated by N and C, respectively

ionic bond involving Arg304 and Glu318 on opposite subunits are the main forces mediating dimerization through the C-terminal ends (Fig. 2.4a). The dimerization interface and proximity of the C-terminal ends (i.e., residues 310–340) observed in the CC1 double mutant crystal structure is inconsistent with the orientation of the same region in the *Caenorhabditis elegans* SOAR crystal structure (Yang et al. 2012) (see Sect. 2.8) where the structures overlap, implying that this helix within STIM1 is capable of undergoing extensive conformational dynamics.

2.8 STIM1 CC2–CC3 (SOAR/CAD/ccb9) Structure

The crystal structure of the human STIM1 SOAR/CAD/ccb9 region (i.e., residues 345–444) has been solved at ~1.9 Å (Yang et al. 2012) using a Leu374Met/Val419Ala/Cys436Thr triple mutant protein with inhibited propensity for aggregation and stabilized dimeric structure. This symmetric dimer buries ~1800 Å^2, with the two monomers arranged in a V shape (Fig. 2.4b). Each monomer in the human protein contains an extended CC2 helix, followed by two short and one extended α-helix, which makes up the CC3 region and is antiparallel to CC2. Although the V-shaped dimer architecture is structurally conserved, the three broken helices making up the CC3 region are distinct in the human homologue compared to the *C. elegans* structure which shows a continuous CC3 region (Fig. 2.4c). No intermolecular supercoiling exists between the CC regions; however, intramolecular supercoiling can be identified between the CC2 (i.e., residues Lys366 to Ala376) and CC3 (i.e., residues Ile409 to Ala419) regions of each monomer. Critical residues stabilizing the dimer interface include Leu347, Trp350, and Leu351 from one monomer hydrophobically interacting with Leu436, Ile433, and Trp430 of the partner subunit. Further, Thr354 forms an H-bond with Arg429 of the partner subunit, and Tyr361 homotypically stacks between subunits.

The structure of a *C. elegans* construct that is extended relative to the aforementioned human protein reveals an additional α-helix in the CC1 region, packing against CC2 and CC3 in an intramolecular and intermolecular manner, respectively (Fig. 2.4c). Deletion of the equivalent CC1 region in human STIM1 (i.e., Δ310–337 human residues) endows STIM1 with the ability to constitutively activate CRAC channels when co-overexpressed with Orai1 in live cells. However, different mutations in this CC1 helix, as opposed to full deletion, can constitutively activate as well as inhibit CRAC entry (Stathopulos et al. 2013; Yu et al. 2013). Thus, this region of CC1 plays an important modulatory role in the conformational change that occurs in the cytosolic domains, requisite for the recruitment and activation of Orai CRAC channels at ER-PM junctions.

2.9 *D. melanogaster* Orai Structure

A crystal structure of the *D. melanogaster* Orai homologue encompassing residues 132–341 (i.e., an N- and C-terminally truncated form of *D. melanogaster*) and carrying a Cys224Ser/Cys283Thr/Pro276Arg/Pro277Arg quadruple mutation to produce well-diffracting crystals (i.e., ~3.4 Å) has been elucidated (Hou et al. 2012). The TM regions of the *D. melanogaster* homologue are >70% identical with the human Orai1 form of the protein. Remarkably, this Orai crystal structure exhibits a sixfold central axis of symmetry along the pore and overall threefold symmetry due to differences in the TM4 orientation within the dimer building blocks of the hexamer (Fig. 2.5a). The hexameric quaternary state is in contrast to several studies suggesting that Orai1 assembles as a functional tetramer (Demuro et al. 2013; Maruyama et al. 2009; Mignen et al. 2008; Penna et al. 2008; Thompson and Shuttleworth 2013). TM1 helices line the pore in the oligomeric structure, and a circle composed of Glu178 corresponding to human Glu106 near the extracellular surface of the assembled channel is vital for Ca^{2+} binding and ion permeability (Fig. 2.5b). Central in the pore is a hydrophobic core made up of Leu167, Phe171, and Val174 which correspond to human Orai1 Leu95, Phe99, and Val102 (Fig. 2.5b). Finally, on the intracellular side, the pore surface of TM1 is lined by Arg155, Lys159, and Lys163, corresponding to Arg83, Lys87, and Arg91 in human

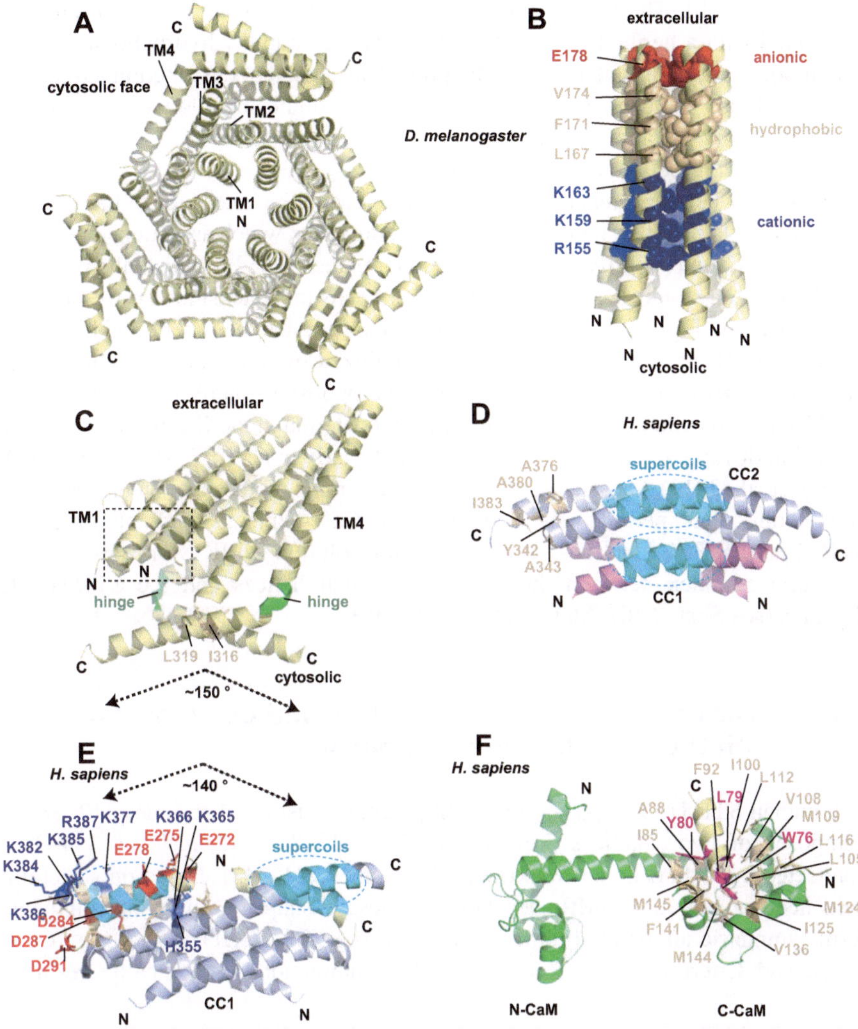

Fig. 2.5 *Atomic-resolution structures of Orai alone as well as in complex with the STIM1 CC region and CaM*. (**a**) *Hexamer architecture of truncated Cys224Ser/Cys283Thr/Pro276Arg/Pro277Arg D. melanogaster Orai revealed by the crystal structure*. The relative positions of TM1–TM4 within one monomer are indicated. The TM1 helices exhibit a sixfold pore symmetry, while the two possible C-terminal domain conformations result in an outer threefold symmetry. (**b**) *Anionic-hydrophobic-cationic pore architecture of D. melanogaster Orai*. Residues conferring anionic (*red*), hydrophobic (*light orange*), and cationic (*blue*) properties are indicated by *spheres*. (**c**) *Multiple C-terminal domain conformations and TM1 extensions in the dimer units of D. melanogaster Orai*. The hinge regions (*green*) in the TM4 extensions facilitating the antiparallel orientation of the C-terminal domains and the residues involved in stabilizing hydrophobic interactions of this region (*light orange sticks*) are shown. The interhelical angle of the C-terminal domain is highlighted by the *dashed arrows*. The TM1 N-terminal extension is indicated by a *dashed box*. (**d**) *Orai1-free solution conformation of the wild-type human STIM1 CC1–CC2 region*. The residues undergoing supercoiling in the CC1 and CC2 region are colored *cyan*. The CC1 region previously demonstrated to modulate the activity of STIM1 is colored *magenta* (i.e., analogous to Fig. 2.4a, c). Hydrophobic residues mediating intermolecular CC2:L1 interactions

Orai1 (Fig. 2.5b). The Lys163Trp mutation (i.e., Arg91Trp mutation in humans) associated with a heritable form of SCID introduces a bulky hydrophobic residue which points into the center axis of the pore. Thus, channel dysfunction associated with SCID may be due to an inability of the pore to dilate and allow the Ca^{2+} ion to move from the anionic entrance through the hydrophobic core and beyond the cationic base on the intracellular side of the pore.

Each TM1 helix extends in a linear manner beyond the inner face of the PM into the cytosol. On the other hand, the C-terminal domains, which are extensions of the TM4 helices, run primarily parallel to the inner plane of the PM due to the presence of a hinge (i.e., residues 305–308) region that bends the extensions in two possible conformations (Fig. 2.5c). The two conformations arrange the C-terminal helices in an antiparallel manner such that they are stabilized by inter-subunit hydrophobic interactions between Ile316 and Leu319 of the opposite subunit (i.e., Phe270 and Leu273 in human numbering). Considering that deletion of the C-terminal cytosolic extension disrupts interactions with STIM1 (McNally et al. 2013; Park et al. 2009; Zheng et al. 2013), the supercoiled helix pair may represent a vital structural architecture that permits interactions with Ca^{2+}-depleted and oligomerized STIM1 at ER-PM junctions, consistent with solution NMR work revealing a structural complex between the Orai1 C-terminal helices and the STIM1 CC domains (see Sect. 2.10) (Stathopulos et al. 2013).

2.10 STIM1 $CC1_{[TM\text{-}Distal]}$-CC2 Structure and Complexation with the Orai1 C-Terminal Domain

The human STIM1 $CC1_{[TM\text{-}distal]}$-CC2 fragment consisting of residues 312–387 is composed of two α-helices (i.e., α1, residues 313–340; α2, residues 344–382) connected by a short linker loop (i.e., L1, residues 341–343). Based on structural evidence from solution NMR spectroscopy, these U-shaped monomeric fragments form symmetric antiparallel dimers via α1:α1′ (i.e., CC1:CC1), α2:α2′ (i.e., CC2: CC2), and C-terminal α2:L1′ (i.e., CC2:L1) interactions. Specifically, residues

Fig. 2.5 (continued) that stabilize the Orai1-free conformation are shown as *light orange sticks*. (**e**) *Orai1 C-terminal domain-bound solution conformation of the wild-type human STIM1 CC1–CC2 region*. The hydrophobic (*light orange*) and basic residues (*blue*) making up the SOAP are shown as *sticks*. Hydrophobic residues making up the SOAP are not labeled for simplicity. The Orai1 C-terminal domain peptides (i.e., residues 272–292) are colored *yellow* and the intermolecular regions of supercoiling are *cyan*. The acidic residues on the Orai1 C-terminal domain peptides that interact with the basic rim are shown as *red sticks*. Only one of the two interacting Orai1 C-terminal peptides is annotated for simplicity. The interhelical angle of the human Orai1 C-terminal peptides is highlighted by the *dashed arrows*. (**f**) *Crystal structure of the human Orai1 N-terminal domain extension bound to CaM*. The dumbbell structure of CaM is shown in *green*. The residues forming the hydrophobic pocket (*light orange*) on the C-terminal domain of CaM (C-CaM) are indicated. The Orai1 N-terminal domain extension (i.e., residues 68–91) is colored *yellow*. The Orai1 N-terminal residues that anchor into the C-CaM cleft are illustrated in *magenta*. In **a–f**, the amino and carboxy termini are labeled N and C, respectively

320–331 of each monomer form the antiparallel CC1 supercoil, residues 355–369 form the antiparallel CC2 supercoil, and residues Ala376, Ala380, and Ile383 of C-terminal α2/CC2 interact with residues Tyr342′ and Ala343′ of L1′ (Fig. 2.5d).

Solution NMR spectroscopy also showed that this region of STIM1 can interact with an Orai1 C-terminal fragment consisting of residues 272–292. NMR titration revealed that residues Glu272, Leu273, Asn274, Ala277, Glu278, Ala280, Arg281, His288, and Arg289 of Orai1 $C_{272–292}$ undergo the most pronounced chemical shift perturbations upon binding to STIM1 $CC1_{[TM\text{-}distal]}$-CC2. Within dimeric STIM1 $CC1_{[TM\text{-}distal]}$-CC2, residues Pro344, Leu347, Leu351, His355, and Val359 of N-terminal α2 and residues Tyr362′, Lys366′, Ala369′, Leu373′, Ala376′, Ala380′, and Ile383′ of C-terminal α2′ all contribute to the binding of Orai1 $C_{272–292}$. Specifically, Orai1 Asn274, Ala277, Arg281, Gln285, and Arg289 residues interact with STIM1 N-terminal α2, whereas Orai1 Leu273, Leu276, Ala280, Gln283, and Leu286 residues interact with STIM1 C-terminal α2′. This STIM-Orai association pocket (SOAP) that has extensive hydrophobicity in the central cavity is also surrounded by a basic rim of residues for electrostatic interactions (Fig. 2.5e). His355, Lys365′, Lys366′, Lys377′, Lys382′, Lys384′, Lys385′, Lys386′, and Arg387′ of the α2 helix form the basic rim which complement the Glu272, Glu275, Glu278, Asp284, Asp287, and Asp291 acidic residues on the Orai1 $C_{272–292}$ surface.

The importance of each of the α1:α1′, α2:α2′, and α2:L1′ interfaces elucidated by the solution NMR structures in the activation of CRAC entry has been assessed. Interestingly, the Glu318Gln/Glu319Gln/Glu320Gln/Glu322Gln (4EQ) charge-neutralizing quadruple mutation stabilizes STIM1 α1:α1′ interface, enhances dimerization, increases its interaction with Orai1 $C_{272–292}$, and promotes spontaneous inward-rectifying currents; in contrast, the Val324Pro α1 helix-breaking mutation destabilizes the STIM1 α1:α1′ interface, attenuates dimerization, decreases its interaction with Orai1 $C_{272–292}$, and lowers the maximal inward currents. The STIM1 α2 Tyr361Lys/Tyr362Lys double mutant cannot activate CRAC channels due to destabilization of the $CC1_{[TM\text{-}distal]}$-CC2 dimer and abrogation of binding to Orai1 $C_{272–292}$. Therefore, the α1:α1′ interface plays a role in the efficiency and α2:α2′ interface is essential for CRAC channel activation. The Lys382Glu/Lys384Glu/Lys385Glu/Lys386Glu (4KE) quadruple mutation on the basic C-terminal α2 helix does not significantly affect the stability or dimerization of STIM1 $CC1_{[TM\text{-}distal]}$-CC2; however, this 4KE mutation prevents interactions with Orai1 $C_{272–292}$, indicating that the α2:L1′ interface only marginally affects STIM1 $CC1_{[TM\text{-}distal]}$-CC2 dimerization, but the C-terminal α2 residues play an important role in Orai1-mediated activation of CRAC channels.

Interestingly, there are striking similarities between the *D. melanogaster* Orai crystal structure and the solution structure for *H. sapiens* Orai1 $C_{272–292}$ in complex with STIM1 $CC1_{[TM\text{-}distal]}$-CC2. First, the *D. melanogaster* Orai C-terminal helices are in antiparallel conformation, as is observed with the human Orai1 $C_{272–292}$ peptides. Second, the interhelical angle between the *D. melanogaster* Orai cytosolic C-terminal helices is 152°, analogous to the obtuse angle of 136° revealed in the

H. sapiens complex structure (Fig. 2.5c, e). Finally, docking of three $CC1_{[TM\text{-}distal]}$-CC2:Orai1 $_{C272–292}$ dimers onto the *D. melanogaster* Orai hexamer by structural alignment of the homologous residues of the C-termini illustrates a steric compatibility at the assembled hexameric level (Stathopulos et al. 2013). Remarkably, the CC3 locations after the docking are adjacent to one another and poised for CC3: CC3′ interactions that enhance oligomerization, as previously observed for this region of STIM1 experimentally (Covington et al. 2010; Muik et al. 2009). Consistent with this mode of STIM1:Orai1 $C_{272–292}$ interaction, it has been shown that disruption of the bend architecture preceding the cytosolic C-termini facilitates the antiparallel Orai1 C-terminal domain arrangement that abolishes binding, and locking of the Orai1 C-terminal bend with the incorporation of a Cys-mediated disulfide still permits binding to STIM1 CC2–CC3 (Tirado-Lee et al. 2015). Additionally, C-terminal deletion mutants found to abrogate binding to STIM1, among several variable length deletions of the Orai1 C-terminal domain, are consistent with the interaction mode revealed by the solution NMR structure (Palty et al. 2015).

2.11 CaM Structure and Complexation with the Orai1 N-terminal Domain

CRAC channels undergo a Ca^{2+}-dependent inactivation (CDI) phenomenon via the actions of STIM1 and CaM (Derler et al. 2009; Mullins et al. 2009; Roos et al. 2005). Residues located C-terminal to the CC3 region of STIM1 (i.e., residues 470–491) have been linked to the fast CDI of CRAC channels (Mullins et al. 2009). Further, Ca^{2+}-loaded CaM binds to the polybasic tails of both STIM1 (i.e., residues 667–685) and STIM2 (i.e., residues 730–746) with μM affinity (Bauer et al. 2008). Therefore, CaM may downregulate SOCE activity by inhibiting and/or disrupting ER-PM relocalization of STIM molecules that occurs via interactions of the polybasic STIM C-termini with phosphoinositides (Calloway et al. 2011; Korzeniowski et al. 2009; Liou et al. 2007; Walsh et al. 2010; Yuan et al. 2009). Ca^{2+}-loaded CaM also binds to the Orai1 N-terminal domain (i.e., residues 68–91), and interaction of CaM with this Orai1 N-terminal region has been implicated in the CDI of CRAC channels (Mullins et al. 2009). A complex structure of a human Orai1 N-terminal fragment encompassing residues 69–91 interacting with CaM has been solved at 1.9 Å. This crystal structure shows Ca^{2+}-loaded CaM adopts the well-characterized dumbbell configuration where each lobe is separated by an extended α-helical linker; however, only the C-terminal lobe shows the presence of the Orai1-N peptide (Fig. 2.5f). The interaction observed in this crystal complex occurs primarily via Trp76, Leu79, and Tyr80 hydrophobic side chains packing into the CaM C-terminal domain cleft made up of Ile85, Ala88, Phe92, Ile100, Leu105, Val108, Met109, Leu112, Leu116, Met124, Ile125, Val136, Phe141, Met144, and Met145 (Fig. 2.5f).

Although the crystal structure does not reveal the mode of binding of the Orai1-N peptide to the CaM N-terminal domain, solution experiments indeed show that an interaction occurs with lower affinity on this CaM domain (Derler et al. 2013). Since the Orai1 N-terminal domain binds independently to both the CaM N- and C-terminal domains, it is possible that the CDI mechanism involves bridging two Orai1 N-termini within the Orai1 tetramer or hexamer complex (Liu et al. 2012).

2.12 Concluding Remarks

Four fundamental features contribute to the Ca^{2+}-sensing function of EF-SAM domains: (1) the binding affinity of the canonical EF-hand, (2) the nature of the EF-hand hydrophobic cleft mutually formed by the canonical and noncanonical EF-hand motifs, (3) the stability of the EF-hand-SAM domain interaction, and (4) the local stability of the SAM domain core. These four elements are not mutually exclusive, but are inter-dependent during Ca^{2+} sensing. Upon Ca^{2+} dissociation from the canonical EF-hand loop, the EF-hand-SAM domain interaction which auto-inhibits oligomerization of this domain is destabilized, resulting in a coupled oligomerization. This oligomerization instigates a TM region rearrangement, which closely apposes the N-terminal CC1 domains within STIM dimers and triggers a cytosolic CC reorganization which promotes interactions with the PM lipids at ER-PM junctions. Punctate STIM molecules directly couple to Orai1 C-and N-terminal domains in the recruitment of Orai1 and gating of assembled CRAC channels. Differences in the CAD/SOAR/ccb9 (i.e., CC2–CC3), CC1-CAD/SOAR/ccb9 (Yang et al. 2012), and CC1 (Cui et al. 2013) crystal structures and solution NMR data on CC1–CC2 fragments solved in the absence and presence of Orai (Stathopulos et al. 2013) are in line with the dynamic closed-to-open transition that the STIM1 cytosolic domains must undergo to adopt an Orai recruitment- and activation-competent conformation (Fahrner et al. 2014; Korzeniowski et al. 2011, 2016; Maus et al. 2015; Muik et al. 2011; Yu et al. 2013; Zhou et al. 2013). The antiparallel C-terminal domain extensions of TM4 (Hou et al. 2012) present Orai C-terminal binding sites to STIM molecules at ER-PM junctions during the activation process. Further, the Orai1 C-terminal antiparallel domain orientation of the *D. melanogaster* Orai structure is remarkably consistent with the geometry of assembly observed in the human STIM1 CC1–CC2:Orai1 $C_{272–292}$ complex structure (Stathopulos et al. 2013), providing a compelling coupling mechanism between the two principle players of SOCE.

Despite the significant progress in elucidating atomic details of CRAC channel regulation, several integral structural features remain unresolved. The precise structure of the Ca^{2+}-depleted EF-SAM domain is unknown; further the manner in which the wild-type human CAD/SOAR/ccb9 region interacts with human Orai1-constituted channels remains enigmatic. Additionally, the open conformation of the full-length native Orai1 channel has not been resolved; similarly, the closed native human Orai1 channel structure has not been elucidated. The structural basis for STIM homomeric intermolecular CC3 interactions involved in oligomerization and potential interactions with the Orai proteins during the activation process is also unknown. The basis for Orai1 N-terminal binding to STIM1, which has been implicated in channel recruitment and gating, has not been disseminated. Additionally, the precise stoichiometry of the functional STIM-Orai complex requires further clarification. Future investigations into these aforementioned features will reveal a more complete picture on the precise atomic-based mechanisms underlying CRAC channel function and open the door to effective new therapies and research tools.

Acknowledgments This work was supported by a Natural Sciences and Engineering Research Council of Canada Discovery Grant to P.B.S. and Canadian Institutes of Health Research Operating Grant to Q.F., P.B.S., and W.Y.L. We thank Mitsuhiko Ikura for his invaluable contributions in preparing the first edition version of this chapter and Steve Chung for his careful review of the manuscript.

References

Baba Y, Hayashi K, Fujii Y, Mizushima A, Watarai H, Wakamori M, Numaga T, Mori Y, Iino M, Hikida M, Kurosaki T (2006) Coupling of STIM1 to store-operated Ca2+ entry through its constitutive and inducible movement in the endoplasmic reticulum. Proc Natl Acad Sci U S A 103:16704–16709

Balasuriya D, Srivats S, Murrell-Lagnado RD, Edwardson JM (2014) Atomic force microscopy (AFM) imaging suggests that stromal interaction molecule 1 (STIM1) binds to Orai1 with sixfold symmetry. FEBS Lett 588:2874–2880

Bauer MC, O'Connell D, Cahill DJ, Linse S (2008) Calmodulin binding to the polybasic C-termini of STIM proteins involved in store-operated calcium entry. Biochemistry 47:6089–6091

Berna-Erro A, Braun A, Kraft R, Kleinschnitz C, Schuhmann MK, Stegner D, Wultsch T, Eilers J, Meuth SG, Stoll G, Nieswandt B (2009) STIM2 regulates capacitive Ca2+ entry in neurons and plays a key role in hypoxic neuronal cell death. Sci Signal 2:ra67

Berridge MJ (2009) Cell signaling biology. http://www.cellsignallingbiology.org

Berridge MJ, Lipp P, Bootman MD (2000) The versatility and universality of calcium signalling. Nat Rev Mol Cell Biol 1:11–21

Berridge MJ, Bootman MD, Roderick HL (2003) Calcium signalling: dynamics, homeostasis and remodelling. Nat Rev Mol Cell Biol 4:517–529

Bird GS, Hwang SY, Smyth JT, Fukushima M, Boyles RR, Putney JW Jr (2009) STIM1 is a calcium sensor specialized for digital signaling. Curr Biol 19:1724–1729

Bootman MD, Lipp P (2001) Calcium signalling and regulation of cell function. Encycl Life Sci 1–7. doi:10.1038/npg.els.0001265

Brandman O, Liou J, Park WS, Meyer T (2007) STIM2 is a feedback regulator that stabilizes basal cytosolic and endoplasmic reticulum Ca2+ levels. Cell 131:1327–1339

Cai X (2007) Molecular evolution and structural analysis of the Ca(2+) release-activated Ca(2+) channel subunit, Orai. J Mol Biol 368:1284–1291

Calloway N, Owens T, Corwith K, Rodgers W, Holowka D, Baird B (2011) Stimulated association of STIM1 and Orai1 is regulated by the balance of PtdIns(4,5)P(2) between distinct membrane pools. J Cell Sci 124:2602–2610

Choi YJ, Zhao Y, Bhattacharya M, Stathopulos PB (2016) Structural perturbations induced by Asn131 and Asn171 glycosylation converge within the EFSAM core and enhance stromal interaction molecule-1 mediated store operated calcium entry. Biochim Biophys Acta pii: S0167-4889(16)30301-9. doi: 10.1016/j.bbamcr.2016.11.015 [Epub ahead of print]

Collins SR, Meyer T (2011) Evolutionary origins of STIM1 and STIM2 within ancient Ca(2+) signaling systems. Trends Cell Biol 21:202–211

Covington ED, MM W, Lewis RS (2010) Essential role for the CRAC activation domain in store-dependent oligomerization of STIM1. Mol Biol Cell 21:1897–1907
Cui B, Yang X, Li S, Lin Z, Wang Z, Dong C, Shen Y (2013) The inhibitory helix controls the intramolecular conformational switching of the C-terminus of STIM1. PLoS One 8:e74735
Demuro A, Penna A, Safrina O, Yeromin AV, Amcheslavsky A, Cahalan MD, Parker I (2013) Subunit stoichiometry of human Orai1 and Orai3 channels in closed and open states. Proc Natl Acad Sci U S A 108:17832–17837
Derler I, Fahrner M, Muik M, Lackner B, Schindl R, Groschner K, Romanin C (2009) A CRAC modulatory domain (CMD) within STIM1 mediates fast Ca2+-dependent inactivation of ORAI1 channels. J Biol Chem 284:24933–24938
Derler I, Plenk P, Fahrner M, Muik M, Jardin I, Schindl R, Gruber HJ, Groschner K, Romanin C (2013) The extended transmembrane Orai1 N-terminal (ETON) region combines binding interface and gate for Orai1 activation by STIM1. J Biol Chem 288:29025–29034
Fahrner M, Muik M, Schindl R, Butorac C, Stathopulos P, Zheng L, Jardin I, Ikura M, Romanin C (2014) A coiled-coil clamp controls both conformation and clustering of stromal interaction molecule 1 (STIM1). J Biol Chem 289:33231–33244
Feske S (2007) Calcium signalling in lymphocyte activation and disease. Nat Rev Immunol 7:690–702
Feske S, Gwack Y, Prakriya M, Srikanth S, Puppel SH, Tanasa B, Hogan PG, Lewis RS, Daly M, Rao A (2006) A mutation in Orai1 causes immune deficiency by abrogating CRAC channel function. Nature 441:179–185
Frischauf I, Muik M, Derler I, Bergsmann J, Fahrner M, Schindl R, Groschner K, Romanin C (2009) Molecular determinants of the coupling between STIM1 and Orai channels: differential activation of Orai1-3 channels by a STIM1 coiled-coil mutant. J Biol Chem 284:21696–21706
Graham SJ, Dziadek MA, Johnstone LS (2011) A cytosolic STIM2 preprotein created by signal peptide inefficiency activates ORAI1 in a store-independent manner. J Biol Chem 286:16174–16185
Hawkins BJ, Irrinki KM, Mallilankaraman K, Lien YC, Wang Y, Bhanumathy CD, Subbiah R, Ritchie MF, Soboloff J, Baba Y, Kurosaki T, Joseph SK, Gill DL, Madesh M (2010) S-glutathionylation activates STIM1 and alters mitochondrial homeostasis. J Cell Biol 190:391–405
Hou X, Pedi L, Diver MM, Long SB (2012) Crystal structure of the calcium release-activated calcium channel Orai. Science 338:1308–1313
Huang GN, Zeng W, Kim JY, Yuan JP, Han L, Muallem S, Worley PF (2006) STIM1 carboxyl-terminus activates native SOC, I(crac) and TRPC1 channels. Nat Cell Biol 8:1003–1010
Huang Y, Zhou Y, Wong HC, Chen Y, Wang S, Castiblanco A, Liu A, Yang JJ (2009) A single EF-hand isolated from STIM1 forms dimer in the absence and presence of Ca2+. FEBS J 276:5589–5597
Korzeniowski MK, Popovic MA, Szentpetery Z, Varnai P, Stojilkovic SS, Balla T (2009) Dependence of STIM1/Orai1-mediated calcium entry on plasma membrane phosphoinositides. J Biol Chem 284:21027–21035
Korzeniowski MK, Manjarres IM, Varnai P, Balla T (2011) Activation of STIM1-Orai1 involves an intramolecular switching mechanism. Sci Signal 3:ra82
Korzeniowski MK, Baird B, Holowka D (2016) STIM1 activation is regulated by a 14 amino acid sequence adjacent to the CRAC activation domain. AIMS Biophys 3:99–118
Larkin MA, Blackshields G, Brown NP, Chenna R, McGettigan PA, McWilliam H, Valentin F, Wallace IM, Wilm A, Lopez R, Thompson JD, Gibson TJ, Higgins DG (2007) Clustal W and Clustal X version 2.0. Bioinformatics 23:2947–2948
Li Z, Lu J, Xu P, Xie X, Chen L, Xu T (2007) Mapping the interacting domains of STIM1 and Orai1 in Ca2+ release-activated Ca2+ channel activation. J Biol Chem 282:29448–29456
Liou J, Kim ML, Heo WD, Jones JT, Myers JW, Ferrell JE Jr, Meyer T (2005) STIM is a Ca2+ sensor essential for Ca2+-store-depletion-triggered Ca2+ influx. Curr Biol 15:1235–1241
Liou J, Fivaz M, Inoue T, Meyer T (2007) Live-cell imaging reveals sequential oligomerization and local plasma membrane targeting of stromal interaction molecule 1 after Ca2+ store depletion. Proc Natl Acad Sci U S A 104:9301–9306

Liu Y, Zheng X, Mueller GA, Sobhany M, DeRose EF, Zhang Y, London RE, Birnbaumer L (2012) Crystal structure of calmodulin binding domain of Orai1 in complex with Ca2+ calmodulin displays a unique binding mode. J Biol Chem 287:43030–43041

Luik RM, Wu MM, Buchanan J, Lewis RS (2006) The elementary unit of store-operated Ca2+ entry: local activation of CRAC channels by STIM1 at ER-plasma membrane junctions. J Cell Biol 174:815–825

Luik RM, Wang B, Prakriya M, Wu MM, Lewis RS (2008) Oligomerization of STIM1 couples ER calcium depletion to CRAC channel activation. Nature 454:538–542

Ma G, Wei M, He L, Liu C, Wu B, Zhang SL, Jing J, Liang X, Senes A, Tan P, Li S, Sun A, Bi Y, Zhong L, Si H, Shen Y, Li M, Lee MS, Zhou W, Wang J, Wang Y, Zhou Y (2015) Inside-out Ca(2+) signalling prompted by STIM1 conformational switch. Nat Commun 6:7826

Manji SS, Parker NJ, Williams RT, van Stekelenburg L, Pearson RB, Dziadek M, Smith PJ (2000) STIM1: a novel phosphoprotein located at the cell surface. Biochim Biophys Acta 1481:147–155

Maruyama Y, Ogura T, Mio K, Kato K, Kaneko T, Kiyonaka S, Mori Y, Sato C (2009) Tetrameric Orai1 is a teardrop-shaped molecule with a long, tapered cytoplasmic domain. J Biol Chem 284:13676–13685

Maus M, Jairaman A, Stathopulos PB, Muik M, Fahrner M, Weidinger C, Benson M, Fuchs S, Ehl S, Romanin C, Ikura M, Prakriya M, Feske S (2015) Missense mutation in immunodeficient patients shows the multifunctional roles of coiled-coil domain 3 (CC3) in STIM1 activation. Proc Natl Acad Sci U S A 112:6206–6211

McNally BA, Prakriya M (2012) Permeation, selectivity, and gating in store-operated CRAC channels. J Physiol 590:4179–4191

McNally BA, Yamashita M, Engh A, Prakriya M (2009) Structural determinants of ion permeation in CRAC channels. Proc Natl Acad Sci U S A 106:22516–22521

McNally BA, Somasundaram A, Yamashita M, Prakriya M (2012) Gated regulation of CRAC channel ion selectivity by STIM1. Nature 482:241–245

McNally BA, Somasundaram A, Jairaman A, Yamashita M, Prakriya M (2013) The C- and N-terminal STIM1 binding sites on Orai1 are required for both trapping and gating CRAC channels. J Physiol 591:2833–2850

Mercer JC, Dehaven WI, Smyth JT, Wedel B, Boyles RR, Bird GS, Putney JW Jr (2006) Large store-operated calcium-selective currents due to co-expression of Orai1 or Orai2 with the intracellular calcium sensor, Stim1. J Biol Chem 281:24979–24990

Mignen O, Thompson JL, Shuttleworth TJ (2008) Orai1 subunit stoichiometry of the mammalian CRAC channel pore. J Physiol 586:419–425

Muik M, Frischauf I, Derler I, Fahrner M, Bergsmann J, Eder P, Schindl R, Hesch C, Polzinger B, Fritsch R, Kahr H, Madl J, Gruber H, Groschner K, Romanin C (2008) Dynamic coupling of the putative coiled-coil domain of ORAI1 with STIM1 mediates ORAI1 channel activation. J Biol Chem 283:8014–8022

Muik M, Fahrner M, Derler I, Schindl R, Bergsmann J, Frischauf I, Groschner K, Romanin C (2009) A cytosolic homomerization and a modulatory domain within STIM1 C terminus determine coupling to ORAI1 channels. J Biol Chem 284:8421–8426

Muik M, Fahrner M, Schindl R, Stathopulos P, Frischauf I, Derler I, Plenk P, Lackner B, Groschner K, Ikura M, Romanin C (2011) STIM1 couples to ORAI1 via an intramolecular transition into an extended conformation. EMBO J 30:1678–1689

Muik M, Schindl R, Fahrner M, Romanin C (2012) Ca(2+) release-activated Ca(2+) (CRAC) current, structure, and function. Cell Mol Life Sci 69:4163–4176

Mullins FM, Park CY, Dolmetsch RE, Lewis RS (2009) STIM1 and calmodulin interact with Orai1 to induce Ca2+-dependent inactivation of CRAC channels. Proc Natl Acad Sci U S A 106:15495–15500

Oh-hora M, Rao A (2008) Calcium signaling in lymphocytes. Curr Opin Immunol 20:250–258

Palty R, Isacoff EY (2016) Cooperative binding of stromal interaction molecule 1 (STIM1) to the N and C termini of calcium release-activated calcium modulator 1 (Orai1). J Biol Chem 291:334–341

Palty R, Stanley C, Isacoff EY (2015) Critical role for Orai1 C-terminal domain and TM4 in CRAC channel gating. Cell Res 25:963–980

Park CY, Hoover PJ, Mullins FM, Bachhawat P, Covington ED, Raunser S, Walz T, Garcia KC, Dolmetsch RE, Lewis RS (2009) STIM1 clusters and activates CRAC channels via direct binding of a cytosolic domain to Orai1. Cell 136:876–890

Penna A, Demuro A, Yeromin AV, Zhang SL, Safrina O, Parker I, Cahalan MD (2008) The CRAC channel consists of a tetramer formed by Stim-induced dimerization of Orai dimers. Nature 456:116–120

Pozo-Guisado E, Campbell DG, Deak M, Alvarez-Barrientos A, Morrice NA, Alvarez IS, Alessi DR, Martin-Romero FJ (2010) Phosphorylation of STIM1 at ERK1/2 target sites modulates store-operated calcium entry. J Cell Sci 123:3084–3093

Prakriya M, Feske S, Gwack Y, Srikanth S, Rao A, Hogan PG (2006) Orai1 is an essential pore subunit of the CRAC channel. Nature 443:230–233

Putney JW Jr (1986) A model for receptor-regulated calcium entry. Cell Calcium 7:1–12

Qiao F, Bowie JU (2005) The many faces of SAM. Sci STKE 2005:re7

Roos J, DiGregorio PJ, Yeromin AV, Ohlsen K, Lioudyno M, Zhang S, Safrina O, Kozak JA, Wagner SL, Cahalan MD, Velicelebi G, Stauderman KA (2005) STIM1, an essential and conserved component of store-operated Ca2+ channel function. J Cell Biol 169:435–445

Smyth JT, Petranka JG, Boyles RR, DeHaven WI, Fukushima M, Johnson KL, Williams JG, Putney JW Jr (2009) Phosphorylation of STIM1 underlies suppression of store-operated calcium entry during mitosis. Nat Cell Biol 11:1465–1472

Soboloff J, Spassova MA, Tang XD, Hewavitharana T, Xu W, Gill DL (2006) Orai1 and STIM reconstitute store-operated calcium channel function. J Biol Chem 281:20661–20665

Spassova MA, Soboloff J, He LP, Xu W, Dziadek MA, Gill DL (2006) STIM1 has a plasma membrane role in the activation of store-operated Ca(2+) channels. Proc Natl Acad Sci U S A 103:4040–4045

Stathopulos PB, Li GY, Plevin MJ, Ames JB, Ikura M (2006) Stored Ca2+ depletion-induced oligomerization of stromal interaction molecule 1 (STIM1) via the EF-SAM region: an initiation mechanism for capacitive Ca2+ entry. J Biol Chem 281:35855–35862

Stathopulos PB, Zheng L, Li GY, Plevin MJ, Ikura M (2008) Structural and mechanistic insights into STIM1-mediated initiation of store-operated calcium entry. Cell 135:110–122

Stathopulos PB, Zheng L, Ikura M (2009) Stromal interaction molecule (STIM) 1 and STIM2 calcium sensing regions exhibit distinct unfolding and oligomerization kinetics. J Biol Chem 284:728–732

Stathopulos PB, Schindl R, Fahrner M, Zheng L, Gasmi-Seabrook GM, Muik M, Romanin C, Ikura M (2013) STIM1/Orai1 coiled-coil interplay in the regulation of store operated calcium entry. Nat Commun 4:2963

Thompson JL, Shuttleworth TJ (2013) How many Orai's does it take to make a CRAC channel? Sci Rep 3:1961

Tirado-Lee L, Yamashita M, Prakriya M (2015) Conformational changes in the Orai1 C-terminus evoked by STIM1 binding. PLoS One 10:e0128622

Varnai P, Toth B, Toth DJ, Hunyady L, Balla T (2007) Visualization and manipulation of plasma membrane-endoplasmic reticulum contact sites indicates the presence of additional molecular components within the STIM1-Orai1 complex. J Biol Chem 282:29678–29690

Vig M, Beck A, Billingsley JM, Lis A, Parvez S, Peinelt C, Koomoa DL, Soboloff J, Gill DL, Fleig A, Kinet JP, Penner R (2006a) CRACM1 multimers form the ion-selective pore of the CRAC channel. Curr Biol 16:2073–2079

Vig M, Peinelt C, Beck A, Koomoa DL, Rabah D, Koblan-Huberson M, Kraft S, Turner H, Fleig A, Penner R, Kinet JP (2006b) CRACM1 is a plasma membrane protein essential for store-operated Ca2+ entry. Science 312:1220–1223

Walsh CM, Chvanov M, Haynes LP, Petersen OH, Tepikin AV, Burgoyne RD (2010) Role of phosphoinositides in STIM1 dynamics and store-operated calcium entry. Biochem J 425:159–168

Williams RT, Manji SS, Parker NJ, Hancock MS, Van Stekelenburg L, Eid JP, Senior PV, Kazenwadel JS, Shandala T, Saint R, Smith PJ, Dziadek MA (2001) Identification and characterization of the STIM (stromal interaction molecule) gene family: coding for a novel class of transmembrane proteins. Biochem J 357:673–685

Williams RT, Senior PV, Van Stekelenburg L, Layton JE, Smith PJ, Dziadek MA (2002) Stromal interaction molecule 1 (STIM1), a transmembrane protein with growth suppressor activity, contains an extracellular SAM domain modified by N-linked glycosylation. Biochim Biophys Acta 1596:131–137

Wu MM, Buchanan J, Luik RM, Lewis RS (2006) Ca2+ store depletion causes STIM1 to accumulate in ER regions closely associated with the plasma membrane. J Cell Biol 174:803–813

Xu P, Lu J, Li Z, Yu X, Chen L, Xu T (2006) Aggregation of STIM1 underneath the plasma membrane induces clustering of Orai1. Biochem Biophys Res Commun 350:969–976

Yang X, Jin H, Cai X, Li S, Shen Y (2012) Structural and mechanistic insights into the activation of Stromal interaction molecule 1 (STIM1). Proc Natl Acad Sci U S A 109:5657–5662

Yeromin AV, Zhang SL, Jiang W, Yu Y, Safrina O, Cahalan MD (2006) Molecular identification of the CRAC channel by altered ion selectivity in a mutant of Orai. Nature 443:226–229

Yu F, Sun L, Hubrack S, Selvaraj S, Machaca K (2013) Intramolecular shielding maintains the ER Ca(2)(+) sensor STIM1 in an inactive conformation. J Cell Sci 126:2401–2410

Yuan JP, Zeng W, Dorwart MR, Choi YJ, Worley PF, Muallem S (2009) SOAR and the polybasic STIM1 domains gate and regulate Orai channels. Nat Cell Biol 11:337–343

Zhang SL, Yu Y, Roos J, Kozak JA, Deerinck TJ, Ellisman MH, Stauderman KA, Cahalan MD (2005) STIM1 is a Ca2+ sensor that activates CRAC channels and migrates from the Ca2+ store to the plasma membrane. Nature 437:902–905

Zhang SL, Yeromin AV, Zhang XH, Yu Y, Safrina O, Penna A, Roos J, Stauderman KA, Cahalan MD (2006) Genome-wide RNAi screen of Ca(2+) influx identifies genes that regulate Ca(2+) release-activated Ca(2+) channel activity. Proc Natl Acad Sci U S A 103:9357–9362

Zhang SL, Yeromin AV, Hu J, Amcheslavsky A, Zheng H, Cahalan MD (2011) Mutations in Orai1 transmembrane segment 1 cause STIM1-independent activation of Orai1 channels at glycine 98 and channel closure at arginine 91. Proc Natl Acad Sci U S A 108:17838–17843

Zheng L, Stathopulos PB, Li GY, Ikura M (2008) Biophysical characterization of the EF-hand and SAM domain containing Ca2+ sensory region of STIM1 and STIM2. Biochem Biophys Res Commun 369:240–246

Zheng L, Stathopulos PB, Schindl R, Li GY, Romanin C, Ikura M (2011) Auto-inhibitory role of the EF-SAM domain of STIM proteins in store-operated calcium entry. Proc Natl Acad Sci U S A 108:1337–1342

Zheng H, Zhou MH, Hu C, Kuo E, Peng X, Hu J, Kuo L, Zhang SL (2013) Differential roles of the C and N termini of Orai1 protein in interacting with stromal interaction molecule 1 (STIM1) for Ca2+ release-activated Ca2+ (CRAC) channel activation. J Biol Chem 288:11263–11272

Zhou Y, Mancarella S, Wang Y, Yue C, Ritchie M, Gill DL, Soboloff J (2009) The short N-terminal domains of STIM1 and STIM2 control the activation kinetics of Orai1 channels. J Biol Chem 284:19164–19168

Zhou Y, Meraner P, Kwon HT, Machnes D, Oh-hora M, Zimmer J, Huang Y, Stura A, Rao A, Hogan PG (2010a) STIM1 gates the store-operated calcium channel ORAI1 in vitro. Nat Struct Mol Biol 17:112–116

Zhou Y, Ramachandran S, Oh-Hora M, Rao A, Hogan PG (2010b) Pore architecture of the ORAI1 store-operated calcium channel. Proc Natl Acad Sci U S A 107:4896–4901

Zhou Y, Srinivasan P, Razavi S, Seymour S, Meraner P, Gudlur A, Stathopulos PB, Ikura M, Rao A, Hogan PG (2013) Initial activation of STIM1, the regulator of store-operated calcium entry. Nat Struct Mol Biol 20:973–981

The STIM-Orai Pathway: Orai, the Pore-Forming Subunit of the CRAC Channel

3

Aparna Gudlur and Patrick G. Hogan

Abstract

This chapter focuses on the Orai proteins, Orai1–Orai3, with special emphasis on Orai1, in humans and other mammals, and on the definitive evidence that Orai is the pore subunit of the CRAC channel. It begins by reviewing briefly the defining characteristics of the CRAC channel, then discusses the studies that implicated Orai as part of the store-operated Ca^{2+} entry pathway and as the CRAC channel pore subunit, and finally examines ongoing work that is providing insights into CRAC channel structure and gating.

Keywords

Orai1 • STIM1 • Calcium channel • Calcium conductance • CRAC channel • CRAC current • Gating

3.1 The Native CRAC Channel

The CRAC current was originally defined electrophysiologically in T cells and mast cells (Fig. 3.1) (Lewis and Cahalan 1989; Hoth and Penner 1992, 1993; McDonald et al. 1993; Zweifach and Lewis 1993, 1995). Its essential characteristics are that it is activated by a reduction in free Ca^{2+} concentration in endoplasmic reticulum (ER) stores and that under physiological conditions it exhibits a very high selectivity for Ca^{2+} over Na^+ and other ions. Its small unitary current, estimated at ~6 fA in normal physiological solution at −110 mV (Prakriya and Lewis 2006), speaks of an energetic barrier to ion passage through the pore. In part this barrier may represent a purely physical constraint, the narrow pore diameter (~0.39 nm)

A. Gudlur (✉) • P.G. Hogan (✉)
Division of Signalling and Gene Expression, La Jolla Institute for Allergy and Immunology, 9420 Athena Circle, La Jolla, CA, USA
e-mail: aparna@lji.org; phogan@liai.org

K. Groschner et al. (eds.), *Store-Operated Ca²⁺ Entry (SOCE) Pathways*, Advances in Experimental Medicine and Biology 993, DOI 10.1007/978-3-319-57732-6_3

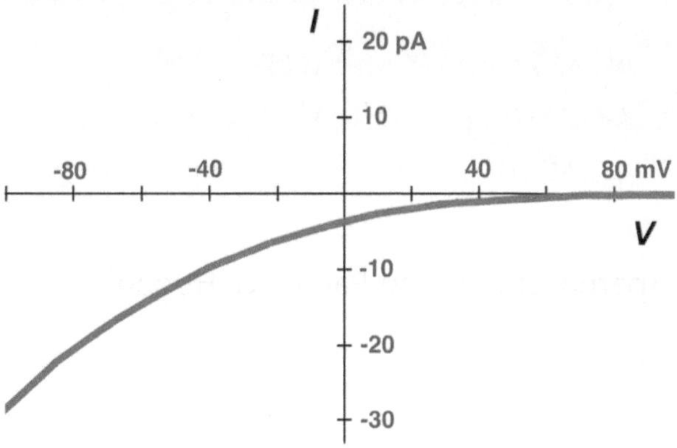

Fig. 3.1 Typical current–voltage (*I*–*V*) relation of the CRAC channel. The graph is an idealized rendering of the whole-cell current recorded from T cells or mast cells. In physiological solutions, the inward current at negative transmembrane potentials is carried by Ca^{2+}, and outward current at positive transmembrane potentials is negligible

inferred from permeation by a series of organic cations in the absence of Ca^{2+} and Mg^{2+} (Bakowski and Parekh 2002; Prakriya and Lewis 2006). Selective CRAC channel inhibitors have not been available until recently (Prakriya and Lewis 2015), but the classical pharmacological fingerprint of the channel includes blockade by low concentrations of lanthanides (Hoth and Penner 1993; Ross and Cahalan 1995; Aussel et al. 1996) and enhancement of current by low concentrations and block by higher concentrations of 2-aminoethoxydiphenyl borate (2-APB) (Prakriya and Lewis 2001).

CRAC current was observed electrophysiologically in cells other than T cells or mast cells (reviewed in Parekh and Putney 2005). These scattered observations foreshadowed the broad tissue distribution of Orai proteins (Gwack et al. 2007, 2008; Vig et al. 2008; McCarl et al. 2009) and the physiological role of Orai1 in tissues from the skin to secretory epithelia to muscle (Gwack et al. 2008; McCarl et al. 2009; Davis et al. 2015; Concepcion et al. 2016).

3.2 Identification of Orai

Orai was linked to store-operated Ca^{2+} entry by three RNAi screens in *Drosophila* S2 cells (Feske et al. 2006; Vig et al. 2006a; Zhang et al. 2006, reviewed in Hogan et al. 2010). One screen (Feske et al. 2006) scored nuclear localization of the transcription factor NFAT, visualized as a human NFAT-GFP fusion protein, to report on sustained Ca^{2+} influx in response to ER Ca^{2+} store depletion. RNAi treatment identified a handful of *Drosophila* genes whose depletion prevents nuclear import of NFAT-GFP, including a gene annotated at the time as *olf186-F*, now renamed *Drosophila Orai*. This finding meshed with the genetic mapping of severe combined immunodeficiency (SCID) trait in a human family to a region of human chromosome 12 containing *Orai1*, a human homolog of *Drosophila Orai* (Feske et al. 2006). Carriers of the SCID trait were found to be heterozygous for a point mutation that encoded an R91W replacement in the Orai1 protein, and the two affected SCID patients were homozygous. Reconstitution of T cells from a SCID patient with wildtype Orai1 restored CRAC current.

The two other RNAi screens (Vig et al. 2006a; Zhang et al. 2006), using cytoplasmic Ca^{2+} levels as readout, found a large number of *Drosophila* genes contributing to store-operated Ca^{2+} entry, including *Orai*. These laboratories verified that *Drosophila* Orai or its homolog Orai1 has a role in CRAC current by showing that there was a substantial increase in the store-operated current when Orai was coexpressed with its corresponding ER Ca^{2+} sensor protein, *Drosophila* STIM or human STIM1 (Zhang et al. 2006; Peinelt et al. 2006).

Orai1 deficit is not one of the common causes of inherited immunodeficiency in humans, but a few additional families have been identified with loss of CRAC current due to independent Orai1 mutations (see also Chap. 21, reviewed in Feske et al. 2010; Lacruz and Feske 2015). In addition, rare autosomal dominant mutations in Orai1 lead to increased or constitutive Ca^{2+} influx and cause skeletal myopathy (reviewed in Lacruz and Feske 2015). Studies with cells from *Orai1*$^{-/-}$ mice show that Orai1 is a main contributor to CRAC current in murine mast cells and mature effector T cells (Vig et al. 2008; Gwack et al. 2008). A residual CRAC-like current in naïve murine T cells has been ascribed to other Orai-family proteins (Vig et al. 2008; Gwack et al. 2008).

3.3 Recombinant Orai Currents

Orai is a plasma membrane protein of mass ~33 kDa with four transmembrane helices (Fig. 3.2). The multimeric Orai1 channel is distributed more or less uniformly in the plasma membrane of resting cells. Following store depletion, Orai1 redistributes to discrete sites on the cell surface, coincident with the STIM1 "puncta" that have been shown to mark sites of Ca^{2+} influx (Luik et al. 2006; Xu et al. 2006; Li et al. 2007; Muik et al. 2008; Navarro-Borelly et al. 2008; Calloway et al. 2009). The physical basis for this redistribution in a STIM-Orai protein-protein interaction is elaborated in Chap. 4. Overexpressed Orai collaborates with overexpressed STIM to produce large CRAC currents, exceeding native CRAC currents in some cases by two orders of magnitude (Zhang et al. 2006; Peinelt et al. 2006; Soboloff et al. 2006; Mercer et al. 2006). Thus, STIM and Orai are the only limiting components of the CRAC channel pathway in the mammalian and *Drosophila* cells tested.

Currents recorded from cells expressing recombinant human or *Drosophila* Orai have precisely the same characteristics as native CRAC currents (Fig. 3.3) (Feske et al. 2006; Prakriya et al. 2006; Zhang et al. 2006; Peinelt et al. 2006; Mercer et al. 2006; Li et al. 2007; Yamashita et al. 2007). The channels are activated by depletion of ER Ca^{2+} stores. They are highly selective for Ca^{2+}. The unitary current is miniscule, and the channel pore is narrow. The pharmacology is that of the CRAC current. Thus, expression of Orai1 gives functional CRAC channels.

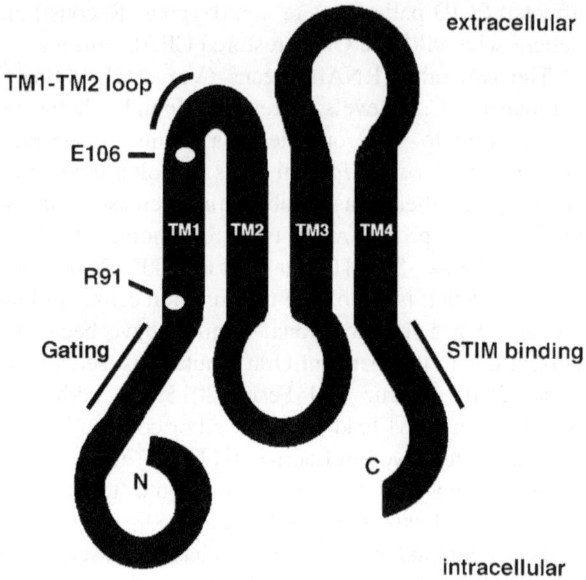

Fig. 3.2 The Orai1 monomer is the basic building block of the CRAC channel. The 301-residue Orai1 polypeptide has four transmembrane helices (TM1–TM4) and intracellular N and C termini. Features discussed in the text include STIM binding and gating segments in the cytoplasmic regions of Orai; E106 residues that constitute the principal Ca^{2+} binding site in the pore; the TM1–TM2 loop, whose acidic residues account for lanthanide binding and channel blockade; and R91, site of an R > W replacement that underlies an inherited immunodeficiency syndrome

3.4 Orai Is the Pore-Forming Subunit

Several point mutations in Orai1 are sensed by ions permeating the channel, implying that Orai contributes to the CRAC channel pore. The replacements E106A or E106Q, which eliminate the negative charge at E106 in transmembrane helix 1 (TM1), block Ca^{2+} current, even though the mutant Orai proteins are expressed at normal levels at the cell surface (Prakriya et al. 2006; Vig et al. 2006b; Spassova et al. 2008). More tellingly, the point mutation E106D in human Orai1 or the corresponding mutation, E180D, in *Drosophila* Orai alters ion selectivity (Yeromin et al. 2006; Prakriya et al. 2006; Vig et al. 2006b; Spassova et al. 2008) and reduces Ca^{2+} affinity for a site in the pore as measured by Ca^{2+} block of Na^+ currents (Prakriya et al. 2006; Yamashita et al. 2007). These electrophysiological data, together with biochemical and structural evidence discussed below, lead to the conclusion that E106 is part of a Ca^{2+} binding site or sites in the pore. Aside from the effect of mutations at E106, replacement of individual acidic residues in the TM1–TM2 loop alters lanthanide blockade of the channel (Yeromin et al. 2006; McNally et al. 2009), and the replacement E190Q in TM3 has allosteric effects on Ca^{2+} selectivity and pore diameter (Prakriya et al. 2006; Vig et al. 2006b; Yamashita et al. 2007; McNally et al. 2009; Zhou et al. 2010b). The latter findings are additional strong evidence that Orai1 is part of the channel, although the TM1–TM2 loop is not an essential Ca^{2+} binding site and E190 is not in the permeation pathway.

There is further evidence that only Orai is needed to assemble functional CRAC channels in the plasma membrane. In the STIM and Orai coexpression studies,

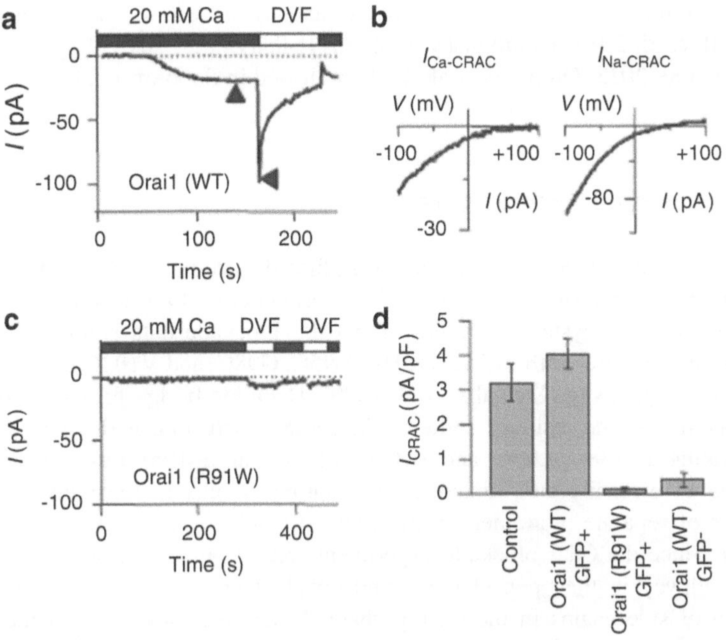

Fig. 3.3 Expression of Orai1 restores CRAC channel function in Orai1(R91W) SCID T cells. (**a**) Development of CRAC current upon store depletion in a SCID T cell expressing recombinant wildtype Orai1. Divalent-free (*DVF*) extracellular solution is used to examine the current carried by Na^+ in the absence of Ca^{2+} and Mg^{2+}. Na^+ does not carry appreciable current when divalent ions are present. (**b**) Typical inward rectifying *I–V* curves for CRAC current carried by Ca^{2+} or by Na^+ at the times indicated by *arrows* in **a**. (**c**) CRAC current is not observed upon store depletion in SCID T cells expressing recombinant Orai1(R91W). (**d**) Peak CRAC current densities in normal T cells, in SCID T cells expressing wildtype Orai1 or Orai1(R91W), and in cells from the cultures transduced with wildtype Orai1 that received little or no expression vector as indicated by the absence of the marker GFP (reproduced from Feske et al. 2006; Fig. 6)

labeled STIM stays in the ER (Mercer et al. 2006; Xu et al. 2006), so only Orai is overexpressed in the plasma membrane. Recombinant Orai1 in isolated yeast membrane vesicles is gated by a purified soluble STIM1 fragment (Zhou et al. 2010a). Because *S. cerevisiae* does not have a STIM-Orai Ca^{2+} signaling mechanism, nor indeed an ER-based Ca^{2+} signaling mechanism, the yeast expression host is unlikely to have contributed an essential channel subunit. Finally, purified recombinant Orai1 reconstituted into liposomes conducts Ca^{2+} when gated by soluble STIM1 (Gudlur et al. 2014). Recombinant Orai1 solubilized and purified from insect, HEK293, and yeast cells is a homomultimer (Park et al. 2009; Maruyama et al. 2009; Hogan 2012), indicating that the basic channel complex consists only of Orai. All of these studies involve overexpression of Orai, however, and they do not establish that overexpressed Orai is gated as efficiently as Orai in the native CRAC channel of T cells and mast cells. In mammalian cells, other associated proteins may modulate the efficiency of CRAC current activation (Srikanth et al. 2010; Krapivinsky et al. 2011; Palty et al. 2012; Jing et al. 2015; Sharma et al. 2013; Quintana et al. 2015; reviewed in Soboloff et al. 2012).

3.5 Channel Architecture

The Orai1 channel pore architecture was first deduced from Cd^{2+} blockade of current and from disulfide crosslinking experiments on Orai channels harboring single engineered cysteine residues. The Cd^{2+} blockade data showed that TM1 helices line the pore, with residues R91C, L95C, G98C, and V102C facing into the conductance pathway (McNally et al. 2009). Disulfide bridge formation between engineered cysteine residues came to the same conclusion, with sharp peaks of crosslinking at A88C, L95C, and V102C (Zhou et al. 2010b). Engineered E106C residues were readily crosslinked, demonstrating the physical proximity of E106 residues of separate monomers in the channel (McNally et al. 2009; Zhou et al. 2010b). Since the Cd^{2+} blockade experiments report on occlusion of the conductance pathway in the open channel, and covalent crosslinking on the preferred position of side chains in the resting channel, the close similarity of the results suggested that Orai gating movements are subtle. The deduced pore architecture has been confirmed by a 3.35 Å structure of an inactive *Drosophila* Orai channel (Hou et al. 2012) (Fig. 3.4). Importantly, the E106 residues form a negatively charged ring encircling the extracellular opening of the pore, and crystals soaked in the permeant ion Ba^{2+} or the pore blocker Gd^{3+} exhibit electron density for those ions at the E106 ring (Hou et al. 2012). (A second lanthanide-binding site that has been observed in electrophysiological experiments (Yeromin et al. 2006; McNally et al. 2009) is not observed in the crystal structure.) The biochemical and structural verification that E106 residues constitute a Ca^{2+} binding site or sites completes the argument that originated from electrophysiological studies.

The *Drosophila* channel structure also showed that Orai is a hexameric channel. This finding prompted considerable debate, as reviewed in detail by Amcheslavsky et al. (2015), since studies with Orai1 concatemers, as well as technically difficult single-molecule fluorescence experiments, had favored the idea that the channel was a tetramer. The conclusion from the crystal structure that Orai is a hexamer was reinforced by determination of the molecular mass of the purified channel complex using light scattering/UV absorbance/refractive index measurements and by protein cross-linking of *Drosophila* Orai expressed in mammalian HEK293 cell membranes that produced a clearly resolvable ladder of multimers up to the hexamer (Hou et al. 2012). Two careful new studies examining Orai1 concatemers explain the earlier concatemer results and support a hexameric structure (Yen et al. 2016; Cai et al. 2016).

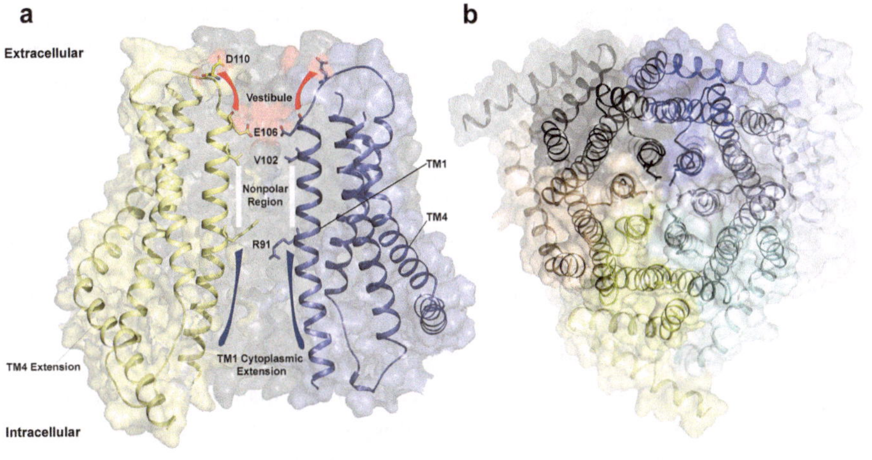

Fig. 3.4 Structural overview of the Orai channel. (**a**) A cutaway view of a human Orai1 model, generated using the *Drosophila* Orai structure PDB:4HKR, highlighting key features of the channel. Some of the subunits and surfaces have been culled for clarity. Surfaces corresponding to E106 and D110 are *shaded red*. The TM1–TM2 vestibule at the extracellular mouth of the channel is marked approximately by *red lines*, the narrow nonpolar region by *white lines*, and the TM1 cytoplasmic extension region by *blue lines*. Residues D110, E106, V102, and R91 are shown in *stick* representation. Note that parts of this apparently concrete model derive from in silico predictions, since the vestibule TM1–TM2 loops and the cytoplasmic TM2–TM3 loops were not resolved in the *Drosophila* Orai crystal structure, and the structure of the *Drosophila* TM3–TM4 loops is not informative for the human protein. (**b**) A snapshot of the channel as viewed from the extracellular side. Each subunit of the hexamer is represented in a *different color* to accent the hexameric organization of the channel. E106 residues are shown in *stick* representation

3.6 The Conductance Pathway

The conductance pathway from outside to inside comprises an outer vestibule, the Ca^{2+} binding site or sites in the vicinity of the E106 ring, a pore segment lined by nonpolar TM1 side chains, and a pore segment flanked by a more polar region of the TM1 helices and their cytoplasmic extensions (Fig. 3.4). Two of these regions have plausible connections to two defining properties of the CRAC channel—the E106 Ca^{2+} binding site(s) to the Ca^{2+} selectivity of the channel and the nonpolar pore segment to the very small single-channel current. However, important mechanistic details of Ca^{2+} ion conductance remain to be unraveled, as noted below.

The vestibule has not yet been defined structurally. The TM1–TM2 loops are not resolved in the *Drosophila* Orai structure—presumably because they are flexible and can adopt multiple conformations (McNally et al. 2009)—and the neighboring TM3–TM4 loops of *Drosophila* Orai, which are resolved, differ substantially in length and sequence from those of Orai1. What is known is that the vestibule immediately external to E106 is relatively wide, given that introduced cysteine residues 107C–110C are accessible to the reagent MTS-TEAE, which has an

8 Å-diameter headgroup (McNally et al. 2009). It has been established by comparison of wildtype and D110A channels that the negatively charged D110 side chains can facilitate Ca^{2+} delivery into the pore (Frischauf et al. 2015). In contrast, D > A replacements indicate that D112 and D114 individually do not contribute significantly to Ca^{2+} current (Frischauf et al. 2015). The latter finding is not surprising, since continuity of the protein backbone to TM2 implies that these residues are more distant from the pore opening than D110, in a region where their negative charge may be countered by the basic residues in the TM3–TM4 loop of human Orai1. The channel vestibule also appears to influence the configuration of the ion selectivity filter, directly or indirectly, since the more extensive vestibule substitutions D110A/D112A and D110A/D112A/D114A decrease Ca^{2+} selectivity and increase pore diameter (Yeromin et al. 2006; Vig et al. 2006b; Yamashita et al. 2007).

The Ca^{2+} binding site at the mouth of channel is a main determinant of Ca^{2+} selectivity. The blockade of Na^{+} currents through the wildtype CRAC channel by Ca^{2+} at low micromolar concentrations had been interpreted as a block by Ca^{2+} in transit through the pore (Lepple-Wienhues and Cahalan 1996; Bakowski and Parekh 2002; Prakriya and Lewis 2006) and can now be referred specifically to binding in the vicinity of the E106 ring. This has a functional parallel in the Ca^{2+} binding site that underlies discrimination between Ca^{2+} and Na^{+} in the L-type Ca^{2+} channel (Yang et al. 1993; Ellinor et al. 1995). There is another layer of complexity under this seemingly straightforward conclusion. Certain experimental observations, exemplified, for example, by the anomalous mole fraction behavior of Ca^{2+} and Ba^{2+} currents (Hoth 1995), are inconsistent with a model in which the CRAC channel binds only one Ca^{2+} at a time. Where could the additional Ca^{2+} ion (s) bind? One possibility is the E106 ring itself. Note that Ca^{2+} binding "site" of the L-type Ca^{2+} channel—sometimes more properly termed a Ca^{2+} binding "locus"—is understood to be capable of binding two Ca^{2+} ions simultaneously during Ca^{2+} influx (Yang et al. 1993; Ellinor et al. 1995). The E106 ring of the Orai channel, with two more acidic side chains than the acidic ring in the L-type channel, should arguably also be able to bind more than one Ca^{2+}. Another candidate is the Gd^{3+}/La^{3+} site (or, again, locus) in the channel vestibule, which has been rigorously documented in electrophysiological experiments (Yeromin et al. 2006; McNally et al. 2009). Gd^{3+} binding at this physiologically defined locus was not evident in the crystal structure, presumably either because the TM1–TM2 loop is disordered and Gd^{3+} binding has no one preferred configuration or because Gd^{3+} does not bind tightly to these sites in the closed channel. Defining how individual Ca^{2+} ions interact with available ligands as they traverse the channel will be challenging. The relevant configurations are necessarily fleeting and—given the low channel conductance—infrequent. Nonetheless, understanding the possible Ca^{2+} trajectories through the pore is inextricably linked to understanding the very high Ca^{2+} selectivity of the Orai channel.

The highly conserved nonpolar segment of TM1 spans three turns of the helix just internal to the E106 site. It is anchored by the region from residues 99–104, whose relative structural rigidity is evidenced by low rotational mobility in the intermonomer disulfide crosslinking assay and by low temperature factors in the corresponding region of the *Drosophila* Orai crystal structure (Zhou et al. 2010b; Hou et al. 2012). The nonpolar segment of the pore presents a barrier to ion flux that can be traced in silico in the free energy profile for Na^+ traversing the closed wildtype channel (Dong et al. 2013) and experimentally in the barrier that prevents constitutive Ca^{2+} flux in a channel truncated to remove other proposed barriers at R91 and in the TM1 cytoplasmic extensions (Gudlur et al. 2014). The several lines of evidence that this barrier is displaced during STIM-dependent channel gating are discussed below.

The role of the TM1 cytoplasmic extensions remains uncertain. Intermonomer crosslinking at residue A88C of the inactive channel (Zhou et al. 2010b), the partial Cd^{2+} blockade at residue R91C of the open channel (McNally et al. 2009), and the effect of diamide crosslinking on current through the R91C channel (Zhang et al. 2011) had established that separate TM1 helices can come into close apposition near the cytoplasmic boundary of the membrane. The *Drosophila* Orai crystal structure visualizes helices projecting roughly 2 nm beyond the membrane boundary, to a position corresponding to Q72 in human Orai1 (Hou et al. 2012). Any interpretation relating the TM1 extensions to channel conductance properties is tentative, because it is uncertain whether the configuration of the helices observed for channels in detergent-lipid micelles represents that in a native lipid environment and whether the position of the helices in the crystallized closed channel reflects their configuration in the open channel. In this connection, for instance, it is not obvious from the *Drosophila* Orai structural model how residues 74–83 would bind cholesterol and decrease channel activity (Derler et al. 2016). The TM1 extensions are stabilized in the crystallized form by an iron-containing anionic complex, and it is a further open question whether physiological anions give the same stabilization in cells. All these questions will be answered, of course, and the answers will lead to a clearer understanding of how the TM1 extensions contribute to channel function.

3.7 Channel Gating

Channel gating depends on a direct interaction of Orai1 channels with STIM (reviewed extensively in Gudlur et al. 2013). The initial interaction in cells that recruits Orai to junctions requires the C-terminal cytoplasmic regions of Orai (Li et al. 2007). The region of interaction in Orai1 has been mapped roughly to residues 267–283 by truncations and mutations (Muik et al. 2008; Navarro-Borelly et al. 2008; Yuan et al. 2009; Park et al. 2009; Frischauf et al. 2009; Lee et al. 2009) and by direct binding measurements with Orai1 peptide fragments (Muik et al. 2008; Yuan et al. 2009; Park et al. 2009; Zhou et al. 2010a). The solution NMR structure of a fragment of STIM1 complexed with Orai1(272–292) illustrates one way in which a STIM1 dimer could bind to a pair of adjacent Orai1 C-terminal helices (Stathopulos et al. 2013). Other experiments have pointed to an alternative binding model in which Orai C-terminal helices interact individually with STIM1 (Hou et al. 2012; Zhou et al. 2015, Tirado-Lee et al. 2015, Palty et al. 2015). It is, of course, conceivable that both modes of binding occur during the physiological interaction of STIM1 with the channel. The issue will be clarified by further experiments.

The N-terminal region of Orai1 spanning residues 73–91 is essential for STIM-dependent activation of the channel (Li et al. 2007; McNally et al. 2013; Derler et al. 2013; Zheng et al. 2013; Palty and Isacoff 2016). Some specific residues that contribute to gating have been mapped in mutational studies (Lis et al. 2010; Derler et al. 2013; Gudlur et al. 2014; Zhou et al. 2016). Most strikingly, the introduction of the three mutations ^{81}LSRAK^{85} > ^{81}AARAE^{85} or the single mutation L81A disrupts gating by STIM1, despite STIM-Orai binding through an intact Orai1 C-terminus (Gudlur et al. 2014; Zhou et al. 2016). The isolated peptides Orai1 (68–87) and Orai1(65–91) bind to STIM1 (Park et al. 2009; Zhou et al. 2010a), and deletions or mutations in this region decrease the interaction of Orai1 with full-length STIM1 or with a soluble STIM1 fragment in cells (McNally et al. 2013; Derler et al. 2013), suggesting that STIM binds directly to the N-terminal region to gate Orai. Other contrasting evidence favors the possibility that STIM binding to the Orai C-terminus is all that is required for gating (Zhou et al. 2015, 2016). In the latter case, the impairment of gating by mutations in the N-terminal segment might indicate that contacts between this segment and other parts of the Orai channel itself are necessary to stabilize the open conformation of the channel. A clearer view of the positioning of this region in the STIM-gated channel will be crucial in understanding physiological channel gating.

While the detailed STIM-Orai interactions and the detailed conformational rearrangements linking STIM binding to channel gating remain to be defined, there has been progress in delineating how gating affects the pore itself. Specifically, productive interaction with STIM1 leads to a rearrangement of the TM1 helices at E106 and at V102 in the nonpolar segment of the pore. The first direct evidence locating a gate in the ion-conducting path of Orai1 was the state-dependent accessibility of G98C to the covalent modifier MTSEA (McNally et al. 2012). More recently, a gating movement of the wildtype channel has been observed directly by monitoring luminescence of Tb^{3+} bound in the Orai1 Ca^{2+} binding site (Gudlur et al. 2014) (Fig. 3.5). The luminescence of bound Tb^{3+} increases when STIM1 binds under conditions that trigger ion flux, indicating that there is a structural rearrangement in the vicinity of E106. This in vitro conformational change is sensitive to the ^{81}LSRAK^{85} > ^{81}AARAE^{85} mutation and is blocked—as is STIM1-dependent current in cells—by the Orai N-terminal peptide Orai1(66–91). Additionally, an increase in intermonomer disulfide crosslinking triggered by soluble STIM1 in the V102C channel strengthens the argument that the hydrophobic region near V102 moves during gating (Gudlur et al. 2014). Completing the argument that the rearrangements detected are essential to gating, an intact L95–V102 segment largely blocks ion flux in a closed channel, even in the absence of R91 and the TM1 extensions (Gudlur et al. 2014), indicating that movement of the nonpolar segment is necessary for ion conductance.

It is informative to compare the physiologically gated wildtype channel with the constitutively conducting channels produced by certain V102X replacements. Replacement of V102 by the small and more polar residues C, S, T, G, or A results

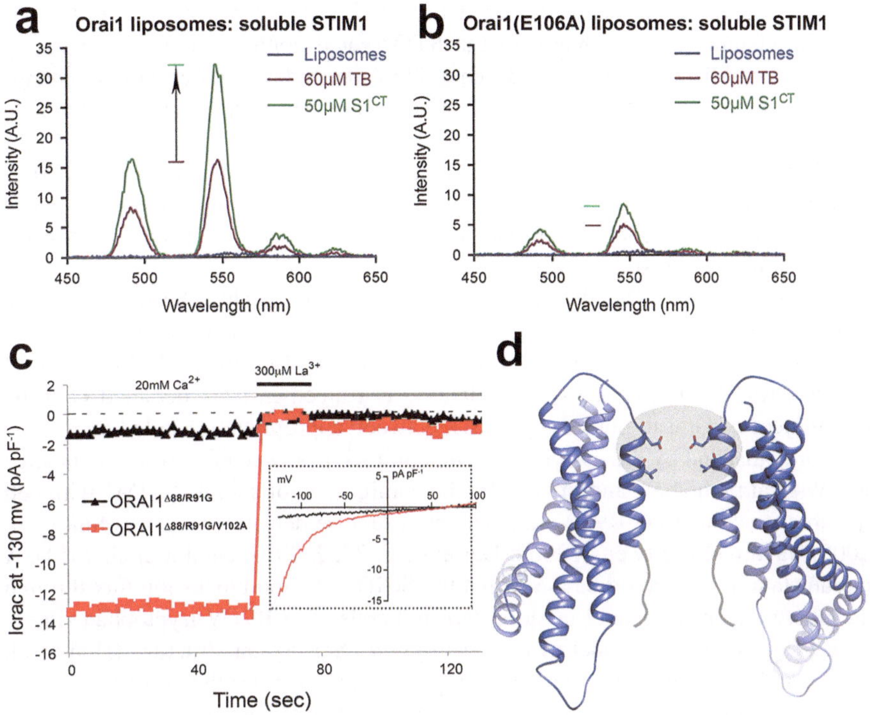

Fig. 3.5 A gating movement in the Orai channel. (**a**) Tb^{3+} binding to purified Orai1 channels reconstituted into liposomes is detected as a luminescence signal following Tb^{3+} addition (*red curve*). Soluble STIM1 ($S1^{CT}$) causes a further increase in luminescence of Tb^{3+} bound to Orai1 (*green curve*), indicating a structural rearrangement at the Tb^{3+} binding site. (**b**) The purified E106A channel in liposomes exhibits very little Tb^{3+} binding (*red curve*) and a minimal response following addition of soluble STIM1 (*green curve*), identifying the ring of E106 side chains as the main site of Tb^{3+} binding and STIM1-dependent rearrangement. Other evidence that STIM elicits movement of the TM1 segment G98–E106 is cited in the text. (**c**) Analysis of constitutive whole-cell currents in cells expressing truncated Orai channels with Orai residues 1–88 replaced by a short unstructured peptide and residue R91 replaced by glycine. Truncated V102A Orai1 channels (*red*) exhibit constitutive currents comparable in amplitude to those of full-length V102A channels expressed at the same level. In contrast, cells expressing corresponding truncated wildtype channels (*black*) have little or no constitutive current. Thus the nonpolar pore region of the Orai channel in itself forms a major barrier to ion permeation. *Inset*: *I–V* curves of constitutive currents in cells expressing the truncated wildtype and V102A Orai1 channels. (**d**) A schematic view of the pore of the truncated wildtype channel in panel **c**, showing the locus probed by Tb^{3+} in the experiment of panel **a** (*gray ellipse*) and the short unstructured regions replacing the TM1 helix extensions for the experiment of panel **c** (*gray*). Only two of the six channel subunits are depicted (**a**–**c** from Gudlur et al. 2014; Figs. 2b, 2d, and 6b–c)

in a constitutively conducting and less selective channel (McNally et al. 2012; Derler et al. 2013). In silico calculations for the *Drosophila* V174A channel—corresponding to a human V102A channel—suggest that it retains the closed pore conformation of the wildtype *Drosophila* channel but presents a markedly lower energetic barrier to ion conductance (Dong et al. 2013). Consistent with the idea that the channel is not already in the STIM-gated conformation, the relatively nonselective V102X channels undergo a STIM-dependent conformational change to more selective channels (McNally et al. 2012; Derler et al. 2013). The Tb^{3+} luminescence assay also detects a STIM-dependent conformational change in the V102A channel (Gudlur et al. 2014). This evidence indicating that V102X replacements produce a leaky channel without a gating conformational change further supports the conclusion that the nonpolar segment is a barrier to passage of ions in the closed state of the channel.

The SCID mutant R91W channel also offers some insight. The R91W protein inserts normally into the plasma membrane (Feske et al. 2006) and assembles into a multimeric Orai complex (Muik et al. 2008; Navarro-Borelly et al. 2008). STIM engages R91W Orai upon store depletion, as gauged in store-depleted cells by recruitment of Orai to puncta, by FRET between appropriately labeled STIM and Orai proteins, and by the detectable change in FRET between C-terminally labeled R91W Orai monomers and in vitro by the comigration of soluble STIM1 with cell membranes containing R91W channels on a sucrose density gradient (Muik et al. 2008; Navarro-Borelly et al. 2008; Derler et al. 2009, 2013; Gudlur et al. 2014). A circumstantial case could be made that the SCID mutation blocks ion flux through an otherwise open channel in store-depleted cells. The R91W tryptophan ring is seen to occlude the channel in the *Drosophila* Orai crystal structure (Hou et al. 2012), and experimentally the R91W replacement overrides the constitutive conductance of the V102C mutant (McNally et al. 2012). However, both observations refer to the closed conformation of the channel. Importantly, R91W channels do not exhibit STIM1-dependent gating as measured by Tb^{3+} luminescence or by V102C crosslinking (Gudlur et al. 2014). Therefore, the simplest explanation of the failure to conduct ions is that packing of the six tryptophan residues stabilizes the resting configuration of the Orai N-terminus and disallows a productive gating interaction of the N-terminal segment either with STIM1 or with other parts of the Orai channel.

Gating might additionally involve widening of the pore by an outward movement of residues 76–95 (Zhang et al. 2011; Hou et al. 2012; Derler et al. 2013; Rothberg et al. 2013). It has been proposed, specifically, that the several basic pore-facing residues in this region—R83, K87, and R91—constitute a barrier to ion flux that is repositioned upon gating. However, empirical support for this hypothesis remains equivocal. R91 has been the most prominent candidate to form a barrier, but the R91G, R91D, and R91E Orai1 channels are all closed at rest and open normally upon store depletion, seemingly ruling out an essential role for R91 in gating (Derler et al. 2009; Zhang et al. 2011). The absence of constitutive current through the Orai1(R83A/K87A) channel (Derler et al. 2013) further suggests that R83 and K87 are not required elements of the channel gate. The latter mutant raises some unanswered questions, though, since even the Orai1(R83A/K87A/V102A) channel does not conduct in resting cells, but it is opened by STIM1 to a Ca^{2+}-selective channel (Derler et al. 2013). Other substitutions at positions 83 and 87 might provide further insight. In any case, it is unlikely that the segment spanning residues 76–95 acts independently as a gate, given that the nonpolar segment from L95–V102 by itself, in a truncated channel lacking the basic residues and the TM1 helix extension, allows little or no constitutive ion flux (Gudlur et al. 2014) (Fig. 3.5).

Mechanistic studies of CRAC channel gating and kinetics have been hampered in the past by the inability to resolve either single-channel currents or gating currents. Recently developed alternative methods, such as detecting the gating conformational change with Tb^{3+} or other fluorescent probes (Gudlur et al. 2014) or recording optically from single Orai1 channels tagged with genetically encoded calcium indicators (Dynes et al. 2016), may offer a way around these limitations.

3.8 Orai2 and Orai3 Channels

Orai2 and Orai3 are reviewed in detail elsewhere (Hoth and Niemeyer 2013), and hence only a few essential facts with regard to pore formation and CRAC activity are outlined here. Both proteins exhibit high sequence similarity to Orai1, particularly in their transmembrane segments. Consistent with this similarity, overexpression of Orai2 or Orai3 together with STIM1 results in store-dependent Ca^{2+} influx and Ca^{2+}-selective currents (Gwack et al. 2007; Lis et al. 2007; DeHaven et al. 2007). Orai3 channels account for a major fraction of the CRAC current in the MCF-7 breast adenocarcinoma cell line (Motiani et al. 2010), and Orai1–Orai3 heteromultimers underlie an arachidonate/leukotriene C_4-regulated Ca^{2+}-selective current, or ARC current, elicited by receptor-phospholipase A_2 signaling (Mignen et al. 2008a, 2009; Gonzalez-Cobos et al. 2013; Zhang et al. 2014). Thus, the other Orai-family proteins are pore-forming channel subunits, and Orai3 has a documented physiological role as a CRAC channel subunit, but the spectrum of physiological Orai channels extends beyond the classical CRAC channel.

Orai3 channels, and to a lesser extent Orai1 channels, can be gated by 2-APB without the intervention of STIM (Peinelt et al. 2008; Zhang et al. 2008; DeHaven et al. 2008; Schindl et al. 2008). STIM and 2-APB trigger the same basic gating and permeation mechanism: The Orai3 correlate of the E106A replacement, E81A, depresses both store-operated current and 2-APB-dependent current (Zhang et al. 2008); and the Orai3 correlate of the R91W mutation, R66W, abolishes the inward current triggered by either stimulus (Schindl et al. 2008). However, in clear contrast to the STIM-gated Orai3 channel, which is selective for Ca^{2+}, the 2-APB-treated Orai3 channel undergoes pore dilation and loses selectivity for Ca^{2+} (Peinelt et al. 2008; Zhang et al. 2008; DeHaven et al. 2008; Schindl et al. 2008, Amcheslavsky et al. 2014). This difference may limit the usefulness of 2-APB as a tool for understanding physiological STIM1-dependent gating.

The close sequence similarity of Orai-family proteins has aided in constructing chimeric Orai proteins to dissect specific channel properties, an approach that has been especially fruitful in dissecting STIM-Orai coupling (Frischauf et al. 2009), as discussed in Chap. 4. The same approach has been applied to other regions of Orai proteins (Zhang et al. 2008; Frischauf et al. 2011).

3.9 Conclusions

Classical electrophysiological studies defined the CRAC current as a store-operated Ca^{2+} current responsible for physiological activation of T cells and mast cells. The CRAC channel is now known to be widely expressed. RNAi screens identified Orai as a protein essential for CRAC channel function, and a combination of studies using human SCID T cells, protein biochemistry, and electrophysiology established that Orai is the pore subunit of the channel. Ongoing work has illuminated the pore architecture of the channel and the basis for its signature electrophysiological features and has provided insights into channel gating. Despite these considerable advances, a thorough understanding of this specialized Ca^{2+} channel will require further high-resolution structural studies, biophysical probing of conducting Orai channels, and in silico simulations of channel gating and Ca^{2+} permeation.

Acknowledgments The authors' work on STIM-Orai signaling is funded by US National Institutes of Health grants AI084167 and GM110397.

References

Amcheslavsky A, Safrina O, Cahalan MD (2014) State-dependent block of Orai3 TM1 and TM3 cysteine mutants: insights into 2-APB activation. J Gen Physiol 143:621–631

Amcheslavsky A, Wood ML, Yeromin AV, Parker I, Freites JA, Tobias DJ, Cahalan MD (2015) Molecular biophysics of Orai store-operated Ca2+ channels. Biophys J 108:237–246

Aussel C, Marhaba R, Pelassy C, Breittmayer JP (1996) Submicromolar La3+ concentrations block the calcium release-activated channel, and impair CD69 and CD25 expression in CD3- or thapsigargin-activated jurkat cells. Biochem J 313:909–913

Bakowski D, Parekh AB (2002) Monovalent cation permeability and Ca2+ block of the store-operated Ca2+ current ICRAC in rat basophilic leukemia cells. Pflugers Arch 443:892–902

Cai X, Zhou Y, Nwokonko RM, Loktionova NA, Wang X, Xin P, Trebak M, Wang Y, Gill DL (2016) The Orai1 store-operated calcium channel functions as a hexamer. J Biol Chem 291:25764–25775

Calloway N, Vig M, Kinet JP, Holowka D, Baird B (2009) Molecular clustering of STIM1 with Orai1/CRACM1 at the plasma membrane depends dynamically on depletion of Ca2+ stores and on electrostatic interactions. Mol Biol Cell 20:389–399

Concepcion AR, Vaeth M, Wagner LE 2nd, Eckstein M, Hecht L, Yang J, Crottes D, Seidl M, Shin HP, Weidinger C, Cameron S, Turvey SE, Issekutz T, Meyts I, Lacruz RS, Cuk M, Yule DI, Feske S (2016) Store-operated Ca2+ entry regulates Ca2+-activated chloride channels and eccrine sweat gland function. J Clin Invest 126:4303–4318

Davis FM, Janoshazi A, Janardhan KS, Steinckwich N, D'Agostin DM, Petranka JG, Desai PN, Roberts-Thomson SJ, Bird GS, Tucker DK, Fenton SE, Feske S, Monteith GR, Putney JW Jr (2015) Essential role of Orai1 store-operated calcium channels in lactation. Proc Natl Acad Sci U S A 112:5827–5832

DeHaven WI, Smyth JT, Boyles RR, Putney JW Jr (2007) Calcium inhibition and calcium potentiation of Orai1, Orai2, and Orai3 calcium release-activated calcium channels. J Biol Chem 282:17548–17556
DeHaven WI, Smyth JT, Boyles RR, Bird GS, Putney JW Jr (2008) Complex actions of 2-aminoethyldiphenyl borate on store-operated calcium entry. J Biol Chem 283:19265–19273
Derler I, Fahrner M, Carugo O, Muik M, Bergsmann J, Schindl R, Frischauf I, Eshaghi S, Romanin C (2009) Increased hydrophobicity at the N terminus/membrane interface impairs gating of the severe combined immunodeficiency-related ORAI1 mutant. J Biol Chem 284:15903–15915
Derler I, Plenk P, Fahrner M, Muik M, Jardin I, Schindl R, Gruber HJ, Groschner K, Romanin C (2013) The extended transmembrane Orai1 N-terminal (ETON) region combines binding interface and gate for Orai1 activation by STIM1. J Biol Chem 288:29025–29034
Derler I, Jardin I, Stathopulos PB, Muik M, Fahrner M, Zayats V, Pandey SK, Poteser M, Lackner B, Absolonova M, Schindl R, Groschner K, Ettrich R, Ikura M, Romanin C (2016) Cholesterol modulates Orai1 channel function. Sci Signal 9:ra10
Dong H, Fiorin G, Carnevale V, Treptow W, Klein ML (2013) Pore waters regulate ion permeation in a calcium release-activated calcium channel. Proc Natl Acad Sci U S A 110:17332–17337
Dynes JL, Amcheslavsky A, Cahalan MD (2016) Genetically targeted single-channel optical recording reveals multiple Orai1 gating states and oscillations in calcium influx. Proc Natl Acad Sci U S A 113:440–445
Ellinor PT, Yang J, Sather WA, Zhang JF, Tsien RW (1995) Ca2+ channel selectivity at a single locus for high-affinity Ca2+ interactions. Neuron 15:1121–1132
Feske S, Gwack Y, Prakriya M, Srikanth S, Puppel SH, Tanasa B, Hogan PG, Lewis RS, Daly M, Rao A (2006) A mutation in Orai1 causes immune deficiency by abrogating CRAC channel function. Nature 441:179–185
Feske S, Picard C, Fischer A (2010) Immunodeficiency due to mutations in ORAI1 and STIM1. Clin Immunol 135:169–182
Frischauf I, Muik M, Derler I, Bergsmann J, Fahrner M, Schindl R, Groschner K, Romanin C (2009) Molecular determinants of the coupling between STIM1 and Orai channels: differential activation of Orai1-3 channels by a STIM1 coiled-coil mutant. J Biol Chem 284:21696–21706
Frischauf I, Schindl R, Bergsmann J, Derler I, Fahrner M, Muik M, Fritsch R, Lackner B, Groschner K, Romanin C (2011) Cooperativeness of Orai cytosolic domains tunes subtype-specific gating. J Biol Chem 286:8577–8584
Frischauf I, Zayats V, Deix M, Hochreiter A, Jardin I, Muik M, Lackner B, Svobodova B, Pammer T, Litvinukova M, Sridhar AA, Derler I, Bogeski I, Romanin C, Ettrich RH, Schindl R (2015) A calcium-accumulating region, CAR, in the channel Orai1 enhances Ca2+ permeation and SOCE-induced gene transcription. Sci Signal 8:ra131
Gonzalez-Cobos JC, Zhang X, Zhang W, Ruhle B, Motiani RK, Schindl R, Muik M, Spinelli AM, Bisaillon JM, Shinde AV, Fahrner M, Singer HA, Matrougui K, Barroso M, Romanin C, Trebak M (2013) Store-independent Orai1/3 channels activated by intracrine leukotriene C4: role in neointimal hyperplasia. Circ Res 112:1013–1025
Gudlur A, Zhou Y, Hogan PG (2013) STIM-ORAI interactions that control the CRAC channel. Curr Top Membr 71:33–58
Gudlur A, Quintana A, Zhou Y, Hirve N, Mahapatra S, Hogan PG (2014) STIM1 triggers a gating rearrangement at the extracellular mouth of the ORAI1 channel. Nat Commun 5:5164
Gwack Y, Srikanth S, Feske S, Cruz-Guilloty F, Oh-hora M, Neems DS, Hogan PG, Rao A (2007) Biochemical and functional characterization of Orai proteins. J Biol Chem 282:16232–16243
Gwack Y, Srikanth S, Oh-Hora M, Hogan PG, Lamperti ED, Yamashita M, Gelinas C, Neems DS, Sasaki Y, Feske S, Prakriya M, Rajewsky K, Rao A (2008) Hair loss and defective T- and B-cell function in mice lacking ORAI1. Mol Cell Biol 28:5209–5222
Hogan PG (2012) STIM1-ORAI1 store-operated calcium current. In: Egelman E (ed) Comprehensive biophysics, vol 6. Academic, Oxford, pp 223–233
Hogan PG, Lewis RS, Rao A (2010) Molecular basis of calcium signaling in lymphocytes: STIM and ORAI. Annu Rev Immunol 28:491–533

Hoth M (1995) Calcium and barium permeation through calcium release-activated calcium (CRAC) channels. Pflugers Arch 430:315–322
Hoth M, Niemeyer BA (2013) The neglected CRAC proteins: Orai2, Orai3, and STIM2. Curr Top Membr 71:237–271
Hoth M, Penner R (1992) Depletion of intracellular calcium stores activates a calcium current in mast cells. Nature 355:353–356
Hoth M, Penner R (1993) Calcium release-activated calcium current in rat mast cells. J Physiol 465:359–386
Hou X, Pedi L, Diver MM, Long SB (2012) Crystal structure of the calcium release-activated calcium channel Orai. Science 338:1308–1313
Jing J, He L, Sun A, Quintana A, Ding Y, Ma G, Tan P, Liang X, Zheng X, Chen L, Shi X, Zhang SL, Zhong L, Huang Y, Dong MQ, Walker CL, Hogan PG, Wang Y, Zhou Y (2015) Proteomic mapping of ER-PM junctions identifies STIMATE as a regulator of Ca^{2+} influx. Nat Cell Biol 17:1339–1347
Krapivinsky G, Krapivinsky L, Stotz SC, Manasian Y, Clapham DE (2011) POST, partner of stromal interaction molecule 1 (STIM1), targets STIM1 to multiple transporters. Proc Natl Acad Sci U S A 108:19234–19239
Lacruz RS, Feske S (2015) Diseases caused by mutations in ORAI1 and STIM1. Ann N Y Acad Sci 1356:45–79
Lee KP, Yuan JP, Zeng W, So I, Worley PF, Muallem S (2009) Molecular determinants of fast Ca2+-dependent inactivation and gating of the orai channels. Proc Natl Acad Sci U S A 106:14687–14692
Lepple-Wienhues A, Cahalan MD (1996) Conductance and permeation of monovalent cations through depletion-activated Ca2+ channels (ICRAC) in Jurkat T cells. Biophys J 71:787–794
Lewis RS, Cahalan MD (1989) Mitogen-induced oscillations of cytosolic Ca2+ and transmembrane Ca2+ current in human leukemic T cells. Cell Regul 1:99–112
Li Z, Lu J, Xu P, Xie X, Chen L, Xu T (2007) Mapping the interacting domains of STIM1 and Orai1 in Ca2+ release-activated Ca2+ channel activation. J Biol Chem 282:29448–29456
Lis A, Peinelt C, Beck A, Parvez S, Monteilh-Zoller M, Fleig A, Penner R (2007) CRACM1, CRACM2, and CRACM3 are store-operated Ca2+ channels with distinct functional properties. Curr Biol 17:794–800
Lis A, Zierler S, Peinelt C, Fleig A, Penner R (2010) A single lysine in the N-terminal region of store-operated channels is critical for STIM1-mediated gating. J Gen Physiol 136:673–686
Luik RM, Wu MM, Buchanan J, Lewis RS (2006) The elementary unit of store-operated Ca2+ entry: local activation of CRAC channels by STIM1 at ER-plasma membrane junctions. J Cell Biol 174:815–825
Maruyama Y, Ogura T, Mio K, Kato K, Kaneko T, Kiyonaka S, Mori Y, Sato C (2009) Tetrameric Orai1 is a teardrop-shaped molecule with a long, tapered cytoplasmic domain. J Biol Chem 284:13676–13685
McCarl CA, Picard C, Khalil S, Kawasaki T, Röther J, Papolos A, Kutok J, Hivroz C, Ledeist F, Plogmann K, Ehl S, Notheis G, Albert MH, Belohradsky BH, Kirschner J, Rao A, Fischer A, Feske S (2009) ORAI1 deficiency and lack of store-operated Ca2+ entry cause immunodeficiency, myopathy, and ectodermal dysplasia. J Allergy Clin Immunol 124:1311–1318
McDonald TV, Premack BA, Gardner P (1993) Flash photolysis of caged inositol 1,4,5-trisphosphate activates plasma membrane calcium current in human T cells. J Biol Chem 268:3889–3896
McNally BA, Yamashita M, Engh A, Prakriya M (2009) Structural determinants of ion permeation in CRAC channels. Proc Natl Acad Sci U S A 106:22516–22521
McNally BA, Somasundaram A, Yamashita M, Prakriya M (2012) Gated regulation of CRAC channel ion selectivity by STIM1. Nature 482:241–245
McNally BA, Somasundaram A, Jairaman A, Yamashita M, Prakriya M (2013) The C- and N-terminal STIM1 binding sites on Orai1 are required for both trapping and gating CRAC channels. J Physiol 591:2833–2850

Mercer JC, Dehaven WI, Smyth JT, Wedel B, Boyles RR, Bird GS, Putney JW Jr (2006) Large store-operated calcium selective currents due to co-expression of Orai1 or Orai2 with the intracellular calcium sensor, Stim1. J Biol Chem 281:24979–24990
Mignen O, Thompson JL, Shuttleworth TJ (2008a) Both Orai1 and Orai3 are essential components of the arachidonate-regulated Ca2+-selective (ARC) channels. J Physiol 586:185–195
Mignen O, Thompson JL, Shuttleworth TJ (2009) The molecular architecture of the arachidonate-regulated Ca2+-selective ARC channel is a pentameric assembly of Orai1 and Orai3 subunits. J Physiol 587:4181–4197
Motiani RK, Abdullaev IF, Trebak M (2010) A novel native store-operated calcium channel encoded by Orai3: selective requirement of Orai3 versus Orai1 in estrogen receptor-positive versus estrogen receptor-negative breast cancer cells. J Biol Chem 285:19173–19183
Muik M, Frischauf I, Derler I, Fahrner M, Bergsmann J, Eder P, Schindl R, Hesch C, Polzinger B, Fritsch R, Kahr H, Madl J, Gruber H, Groschner K, Romanin C (2008) Dynamic coupling of the putative coiled-coil domain of ORAI1 with STIM1 mediates ORAI1 channel activation. J Biol Chem 283:8014–8022
Navarro-Borelly L, Somasundaram A, Yamashita M, Ren D, Miller RJ, Prakriya M (2008) STIM1-Orai1 interactions and Orai1 conformational changes revealed by live-cell FRET microscopy. J Physiol 586:5383–5401
Palty R, Isacoff EY (2016) Cooperative binding of stromal interaction molecule 1 (STIM1) to the N and C termini of calcium release-activated calcium modulator 1 (Orai1). J Biol Chem 291:334–341
Palty R, Raveh A, Kamisky I, Meller R, Reuveny E (2012) SARAF inactivates the store operated calcium entry machinery to prevent excess calcium refilling. Cell 149:425–438
Palty R, Stanley C, Isacoff EY (2015) Critical role for Orai1 C-terminal domain and TM4 in CRAC channel gating. Cell Res 25:963–980
Parekh AB, Putney JW Jr (2005) Store-operated calcium channels. Physiol Rev 85:757–810
Park CY, Hoover PJ, Mullins FM, Bachhawat P, Covington ED, Raunser S, Walz T, Garcia KC, Dolmetsch RE, Lewis RS (2009) STIM1 clusters and activates CRAC channels via direct binding of a cytosolic domain to Orai1. Cell 136:876–890
Peinelt C, Vig M, Koomoa DL, Beck A, Nadler MJ, Koblan-Huberson M, Lis A, Fleig A, Penner R, Kinet JP (2006) Amplification of CRAC current by STIM1 and CRACM1 (Orai1). Nat Cell Biol 8:771–773
Peinelt C, Lis A, Beck A, Fleig A, Penner R (2008) 2-Aminoethoxydiphenyl borate directly facilitates and indirectly inhibits STIM1-dependent gating of CRAC channels. J Physiol 586:3061–3073
Prakriya M, Lewis RS (2001) Potentiation and inhibition of Ca2+ release-activated Ca2+ channels by 2-aminoethyldiphenyl borate (2-APB) occurs independently of IP(3) receptors. J Physiol 536:3–19
Prakriya M, Lewis RS (2006) Regulation of CRAC channel activity by recruitment of silent channels to a high open-probability gating mode. J Gen Physiol 128:373–386
Prakriya M, Lewis RS (2015) Store-operated calcium channels. Physiol Rev 95:1383–1436
Prakriya M, Feske S, Gwack Y, Srikanth S, Rao A, Hogan PG (2006) Orai1 is an essential pore subunit of the CRAC channel. Nature 443:230–233
Quintana A, Rajanikanth V, Farber-Katz S, Gudlur A, Zhang C, Jing J, Zhou Y, Rao A, Hogan PG (2015) TMEM110 regulates the maintenance and remodeling of mammalian ER-plasma membrane junctions competent for STIM-ORAI signaling. Proc Natl Acad Sci U S A 112: E7083–E7092
Ross PE, Cahalan MD (1995) Ca2+ influx pathways mediated by swelling or stores depletion in mouse thymocytes. J Gen Physiol 106:415–444
Rothberg BS, Wang Y, Gill DL (2013) Orai channel pore properties and gating by STIM: implications from the Orai crystal structure. Sci Signal 6:pe9
Schindl R, Bergsmann J, Frischauf I, Derler I, Fahrner M, Muik M, Fritsch R, Groschner K, Romanin C (2008) 2-Aminoethoxydiphenyl borate alters selectivity of Orai3 channels by increasing their pore size. J Biol Chem 283:20261–20267

Sharma S, Quintana A, Findlay GM, Mettlen M, Baust B, Jain M, Nilsson R, Rao A, Hogan PG (2013) An siRNA screen for NFAT activation identifies septins as coordinators of store-operated Ca2+ entry. Nature 499:238–242

Soboloff J, Spassova MA, Tang XD, Hewavitharana T, Xu W, Gill DL (2006) Orai1 and STIM reconstitute store-operated calcium channel function. J Biol Chem 281:20661–20665

Soboloff J, Rothberg BS, Madesh M, Gill DL (2012) STIM proteins: dynamic calcium signal transducers. Nat Rev Mol Cell Biol 213:549–565

Spassova MA, Hewavitharana T, Fandino RA, Kaya A, Tanaka J, Gill DL (2008) Voltage gating at the selectivity filter of the Ca2+ release-activated Ca2+ channel induced by mutation of the Orai1 protein. J Biol Chem 283:14938–14945

Srikanth S, Jung HJ, Kim KD, Souda P, Whitelegge J, Gwack Y (2010) A novel EF-hand protein, CRACR2A, is a cytosolic Ca2+ sensor that stabilizes CRAC channels in T cells. Nat Cell Biol 12:436–446

Stathopulos PB, Schindl R, Fahrner M, Zheng L, Gasmi-Seabrook GM, Muik M, Romanin C, Ikura M (2013) STIM1/Orai1 coiled-coil interplay in the regulation of store-operated calcium entry. Nat Commun 4:2963

Tirado-Lee L, Yamashita M, Prakriya M (2015) Conformational changes in the Orai1 C-terminus evoked by STIM1 binding. PLoS One 10:e0128622

Vig M, Peinelt C, Beck A, Koomoa DL, Rabah D, Koblan-Huberson M, Kraft S, Turner H, Fleig A, Penner R, Kinet JP (2006a) CRACM1 is a plasma membrane protein essential for store-operated Ca2+ entry. Science 312:1220–1223

Vig M, Beck A, Billingsley JM, Lis A, Parvez S, Peinelt C, Koomoa DL, Soboloff J, Gill DL, Fleig A, Kinet JP, Penner R (2006b) CRACM1 multimers form the ion-selective pore of the CRAC channel. Curr Biol 16:2073–2079

Vig M, DeHaven WI, Bird GS, Billingsley JM, Wang H, Rao PE, Hutchings AB, Jouvin MH, Putney JW Jr, Kinet JP (2008) Defective mast cell effector functions in mice lacking the CRACM1 pore subunit of store-operated calcium release-activated calcium channels. Nat Immunol 9:89–96

Xu P, Lu J, Li Z, Yu X, Chen L, Xu T (2006) Aggregation of STIM1 underneath the plasma membrane induces clustering of Orai1. Biochem Biophys Res Commun 350:969–976

Yamashita M, Navarro-Borelly L, McNally BA, Prakriya M (2007) Orai1 mutations alter ion permeation and Ca2+-dependent fast inactivation of CRAC channels: evidence for coupling of permeation and gating. J Gen Physiol 130:525–540

Yang J, Ellinor PT, Sather WA, Zhang JF, Tsien RW (1993) Molecular determinants of Ca2+ selectivity and ion permeation in L-type Ca2+ channels. Nature 366:158–161

Yen M, Lokteva LA, Lewis RS (2016) Functional analysis of Orai1 concatemers supports a hexameric stoichiometry for the CRAC channel. Biophys J 111:1897–1907

Yeromin AV, Zhang SL, Jiang W, Yu Y, Safrina O, Cahalan MD (2006) Molecular identification of the CRAC channel by altered ion selectivity in a mutant of Orai. Nature 443:226–229

Yuan JP, Zeng W, Dorwart MR, Choi YJ, Worley PF, Muallem S (2009) SOAR and the polybasic STIM1 domains gate and regulate Orai channels. Nat Cell Biol 11:337–343

Zhang SL, Yeromin AV, Zhang XH, Yu Y, Safrina O, Penna A, Roos J, Stauderman KA, Cahalan MD (2006) Genome-wide RNAi screen of Ca2+ influx identifies genes that regulate Ca2+ release-activated Ca2+ channel activity. Proc Natl Acad Sci U S A 103:9357–9362

Zhang SL, Kozak JA, Jiang W, Yeromin AV, Chen J, Yu Y, Penna A, Shen W, Chi V, Cahalan MD (2008) Store-dependent and -independent modes regulating Ca2+ release-activated Ca2+ channel activity of human Orai1 and Orai3. J Biol Chem 283:17662–17671

Zhang SL, Yeromin AV, Hu J, Amcheslavsky A, Zheng H, Cahalan MD (2011) Mutations in Orai1 transmembrane segment 1 cause STIM1-independent activation of Orai1 channels at glycine 98 and channel closure at arginine 91. Proc Natl Acad Sci U S A 108:17838–17843

Zhang X, Zhang W, Gonzalez-Cobos JC, Jardin I, Romanin C, Matrougui K, Trebak M (2014) Complex role of STIM1 in the activation of store-independent Orai1/3 channels. J Gen Physiol 143:345–359

Zheng H, Zhou MH, Hu C, Kuo E, Peng X, Hu J, Kuo L, Zhang SL (2013) Differential roles of the C and N termini of Orai1 protein in interacting with stromal interaction molecule 1 (STIM1) for Ca2+ release-activated Ca2+ (CRAC) channel activation. J Biol Chem 288:11263–11272
Zhou Y, Meraner P, Kwon HT, Machnes D, Oh-hora M, Zimmer J, Huang Y, Stura A, Rao A, Hogan PG (2010a) STIM1 gates the store-operated calcium channel ORAI1 in vitro. Nat Struct Mol Biol 17:112–116
Zhou Y, Ramachandran S, Oh-Hora M, Rao A, Hogan PG (2010b) Pore architecture of the ORAI1 store-operated calcium channel. Proc Natl Acad Sci U S A 107:4896–4901
Zhou Y, Wang X, Wang X, Loktionova NA, Cai X, Nwokonko RM, Vrana E, Wang Y, Rothberg BS, Gill DL (2015) STIM1 dimers undergo unimolecular coupling to activate Orai1 channels. Nat Commun 6:8395
Zhou Y, Cai X, Loktionova NA, Wang X, Nwokonko RM, Wang X, Wang Y, Rothberg BS, Trebak M, Gill DL (2016) The STIM1-binding site nexus remotely controls Orai1 channel gating. Nat Commun 7:13725
Zweifach A, Lewis RS (1993) Mitogen-regulated Ca2+ current of T lymphocytes is activated by depletion of intracellular Ca2+ stores. Proc Natl Acad Sci U S A 90:6295–6299
Zweifach A, Lewis RS (1995) Rapid inactivation of depletion-activated calcium current (ICRAC) due to local calcium feedback. J Gen Physiol 105:209–226

The STIM-Orai Pathway: The Interactions Between STIM and Orai

4

Marc Fahrner, Rainer Schindl, Martin Muik, Isabella Derler, and Christoph Romanin

Abstract

A primary Ca^{2+} entry pathway in non-excitable cells is established by the Ca^{2+} release-activated Ca^{2+} channels. Their two limiting molecular components include the Ca^{2+}-sensor protein STIM1 located in the endoplasmic reticulum and the Orai channel in the plasma membrane. STIM1 senses the luminal Ca^{2+} content, and store depletion induces its oligomerization into puncta-like structures, thereby triggering coupling to as well as activation of Orai channels. A C-terminal STIM1 domain is assumed to couple to both C- and N-terminal, cytosolic strands of Orai, accomplishing gating of the channel. Here we highlight the inter- and intramolecular steps of the STIM1-Orai signaling cascade together with critical sites of the pore structure that accomplishes Ca^{2+} permeation.

Keywords

STIM-Orai coupling • STIM oligomerization • Orai nexus • Orai activation • Orai pore

4.1 Introduction

The Ca^{2+} release-activated Ca^{2+} (CRAC) channel is the best characterized store-operated channel (Parekh and Putney 2005). Function-based genetic screen by systematic RNA interference (RNAi) has revealed solid evidence that the STIM1 (stromal interaction molecule) and the Orai (also termed CRACM) family represent the key molecular components of CRAC channels (Feske et al. 2006; Liou et al. 2005; Roos et al. 2005; Vig et al. 2006b; Zhang et al. 2006). STIM1 has been

M. Fahrner (✉) • R. Schindl • M. Muik • I. Derler • C. Romanin (✉)
Institute of Biophysics, Johannes Kepler University Linz, 4020 Linz, Austria
e-mail: marc.fahrner@jku.at; christoph.romanin@jku.at

K. Groschner et al. (eds.), *Store-Operated Ca^{2+} Entry (SOCE) Pathways*, Advances in Experimental Medicine and Biology 993, DOI 10.1007/978-3-319-57732-6_4

identified as endoplasmic reticulum (ER)-located Ca^{2+} sensor (Liou et al. 2005; Roos et al. 2005) which senses ER Ca^{2+} content by its luminal EF-hand in the N-terminus. Further it contains a single transmembrane domain followed by a long C-terminal strand for Orai1 binding. Each of the three Orai 1–3 channels forms a hexameric channel complex with a highly Ca^{2+}-selective pore located in the plasma membrane (Cai et al. 2016; Hou et al. 2012; Yen et al. 2016), whereas each Orai subunit contains four transmembrane domains with cytosolic N- and C-termini (Prakriya et al. 2006; Schindl et al. 2008; Vig et al. 2006b; Yeromin et al. 2006). The Orai isoforms are known to homo- and heteromerize (Gwack et al. 2007; Lis et al. 2007; Schindl et al. 2009; Zhang et al. 2008) among which Orai1-Orai3 heteromers have been found to promote cancer cell proliferation (Dubois et al. 2014) and migration of vascular smooth muscle cells (Zhang et al. 2015). The respective Orai channels are distinct by their inactivation profiles and 2-APB (2-aminoethyldiphenyl borate) sensitivity (DeHaven et al. 2007; Lis et al. 2007).

STIM1 is uniformly distributed within the ER membrane at resting state. Upon store depletion, it forms oligomers and translocates into punctate clusters close to the plasma membrane (Baba et al. 2006; Liou et al. 2005, 2007; Luik et al. 2006; Mercer et al. 2006; Soboloff et al. 2006; Wu et al. 2006; Xu et al. 2006; Zhang et al. 2005), thereby activating Orai/CRAC channels.

4.2 STIM1

The single-pass transmembrane (TM) protein STIM1, which is located in the ER membrane, displays constitutive movement along microtubuli (MT) in resting cells. ER calcium store depletion dramatically changes the intracellular STIM1 localization by inducing the formation of STIM1 clusters in the cell periphery in close apposition to the PM (Baba et al. 2006; Smyth et al. 2008; Soboloff et al. 2012). Early studies analyzing this calcium-dependent effect revealed that the constitutive dynamic movement along microtubules is dependent on the cytosolic coiled-coil (CC) domains and the S-/P-rich domain in the C-terminal part of STIM1, whereas the N-terminal portion of STIM1, which is located in the ER lumen, functions as calcium sensor triggering STIM1 oligomerization in response to $[Ca^{2+}]_{ER}$ decrease (Baba et al. 2006; Stathopulos et al. 2006, 2008). Examination of microtubule-associated tracking molecules revealed that EB1 (a MT plus-end-tracking molecule) is involved in connecting STIM1 to MT. Both EB1 and STIM1 play a role for TAC (tip attachment complex)-mediated ER tubule extension in combination with a growing microtubule (Grigoriev et al. 2008; Honnappa et al. 2009).

Initially, STIM1 oligomerization starts upon dissociation of Ca^{2+} from the ER luminal part of STIM1, a signal that is transduced through the transmembrane segment reaching the cytosolic CC domains resulting in higher-order STIM1 oligomerization (Luik et al. 2008). In aggregate, early studies on the activation mechanism of STIM1 have all in common that a decrease of $[Ca^{2+}]_{ER}$ is the initial trigger for STIM1 oligomerization of its luminal part followed by oligomerization of the cytosolic STIM1 portions (Baba et al. 2006, 2008; Liou et al. 2005, 2007;

Luik et al. 2008; Malli et al. 2008; Muik et al. 2008; Roos et al. 2005; Wu et al. 2006). In detail, the luminal part of STIM1 (corresponding to STIM1 N-terminus) contains a canonical and a hidden EF-hand followed by a sterile-α motif (SAM) (Fig. 4.1) (Stathopulos and Ikura 2010; Stathopulos et al. 2006, 2008). The STIM1 EF-hand domain shows the classic helix-loop-helix motif containing negatively charged D and E residues able to bind Ca^{2+} within a K_D range of 200–600 μM in the presence of high $[Ca^{2+}]_{ER}$. Decreasing $[Ca^{2+}]_{ER}$ consequently results in EF-hand–Ca^{2+} dissociation structurally destabilizing the entire EF-SAM entity (Stathopulos et al. 2006). Under high $[Ca^{2+}]_{ER}$ conditions, the STIM1 EF-SAM domain contains high α-helicity in contrast to STIM1 EF-SAM in low $[Ca^{2+}]_{ER}$. Stoichiometric analyses revealed a prevalent monomeric form of STIM1 EF-SAM in the presence of high $[Ca^{2+}]_{ER}$ and a predominant dimeric form after Ca^{2+} dissociation (Stathopulos et al. 2006, 2008). In line, Covington et al. (2010) have shown that STIM1 C-terminal deletion mutants containing the ER luminal part and the transmembrane domain resulted in increased FRET upon ER Ca^{2+} store depletion. Analysis of the EF-SAM domain of the homologous protein STIM2 revealed differences in conformational stability with respect to STIM1 in line with the fact that STIM2 responds faster to changes of $[Ca^{2+}]_{ER}$ (Brandman et al. 2007; Zheng et al. 2008, 2011).

The STIM1 N-terminus is the domain initially responding to store depletion; however, the signal has to be transduced to the cytosolic STIM1 domains for the subsequent structural rearrangements, cluster formation, and interaction with Orai. STIM1 contains one α-helical transmembrane helix spanning the ER membrane. Within this domain, Ma et al. (2015) have identified a crucial STIM1 TM mutant (STIM1 C227W) which elicits constitutive CRAC channel activation, comprising the advantage of an activated STIM1 protein without store depletion or manipulation of STIM1 ER luminal or cytosolic domains. Under full store conditions, the mutant C227W revealed proximity of the N-terminal parts of the TM domains of STIM1 similar to oligomerized EF-SAM domains upon store depletion. Using FRET, a decreased STIM1 CC1-SOAR (STIM-Orai-activating region) interaction was observed in case of C227W, which indicates the formation of the activated STIM1 form (Ma et al. 2015). Finally, Ma et al. (2015) present an unexpected STIM1 TM structure revealing a dimeric form which, however, is not perfectly parallel, but exhibits a crossing point at a specific position within the TM domain, comprising a crossing angle big enough allowing the separation of the luminal portions at high ER Ca^{2+} condition. Depending on store conditions, the angle of the TM crossing point changes, allowing store depletion-induced oligomerization of ER luminal as well as cytosolic STIM1 domains (Ma et al. 2015; Zhou et al. 2013). Furthermore, three glycines (G223, G225, and G226) are present in the STIM1 TM domain comprising high flexibility needed for conformational changes transducing the activation signal from the ER luminal part to the cytosolic STIM1 domains (Dong et al. 2012; Ma et al. 2015). Ultimately following the STIM1 TM domain, the cytosolic part contains three coiled-coil regions (CC1, CC2, and CC3), a CRAC modulatory domain (CMD), and S-/P- and a K-rich region (Fig. 4.1a) (Fahrner et al. 2009; Soboloff et al. 2012). The STIM1 C-terminus (cytosolic STIM1 part

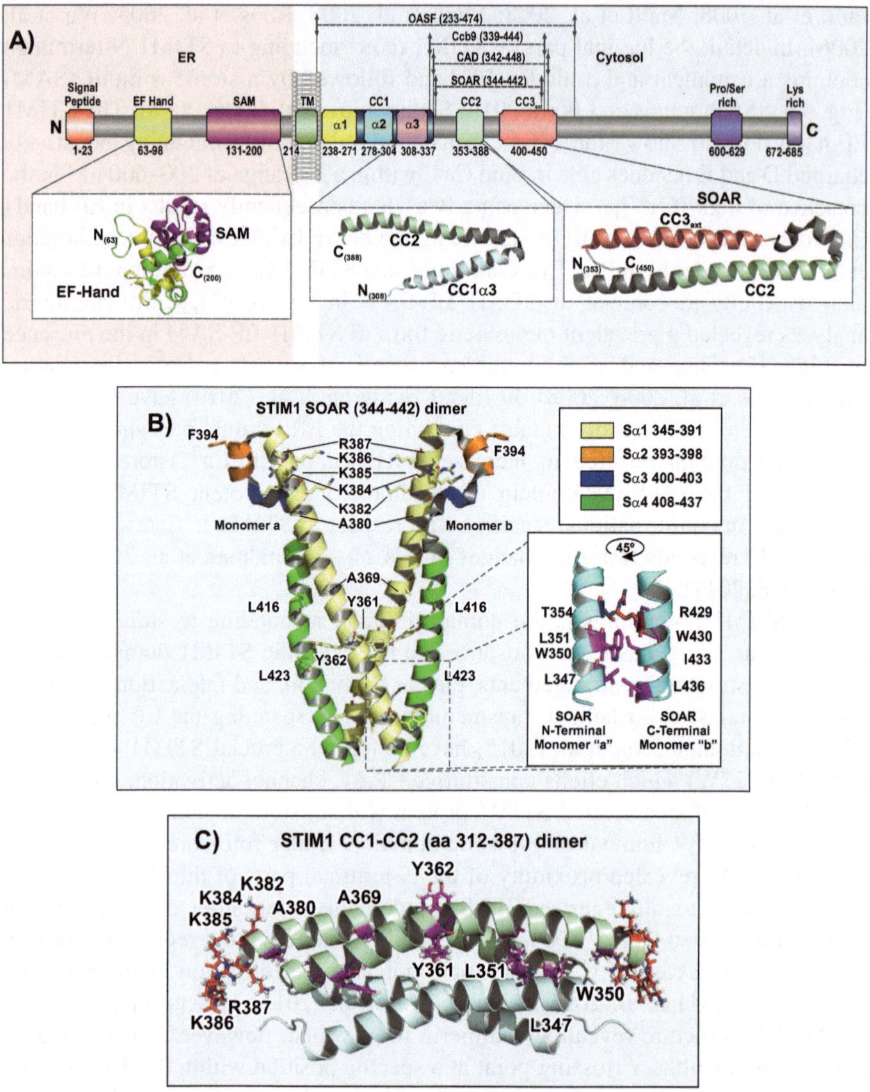

Fig. 4.1 (A) STIM1 scheme depicting essential structural/functional regions. NMR and crystal structures of the EF-SAM domain, the STIM1 CC1α3-CC2 fragment (312–387), as well as the STIM1 SOAR (344–444) fragment are highlighted in *insets*. (B) Dimeric STIM1 SOAR (344–444) crystallographic resolution structure. The *V-shaped* form includes the coiled-coil 2 (CC2) and CC3 domains. The *color code* specifies Sα1, Sα2, Sα3, and Sα4. Interface between SOAR N-terminal monomer "a" and C-terminal monomer "b" is highlighted. (C) NMR structure of a STIM1 CC1α3-CC2 dimer (312–387)

aa233–685) co-expressed with Orai1 has proven to be sufficient to activate Orai1 independent of ER Ca^{2+} store depletion (Huang et al. 2006; Muik et al. 2008). Subsequently, several overlapping STIM1 C-terminal fragments able to activate

Orai1 have been identified: SOAR (aa344–442), CAD (aa342–448), OASF (aa233–450), and Ccb9 (aa339–444) (Kawasaki et al. 2009; Muik et al. 2009; Park et al. 2009; Yuan et al. 2009) (Fig. 4.1a). All these Orai-activating STIM1 fragments suggest STIM1 CC2 (aa363–389), CC3 (aa399–423), and the STIM1 homomerization domain SHD (~aa421–450) as essential to fulfill the functions of oligomerization and coupling to and activation of CRAC channels (Muik et al. 2009).

Finally, structural insights into C-terminal portions of STIM1 have been provided by two independent studies. Yang et al. (2012) have presented the X-ray crystallographic structure of hSOAR (345–444 $_{L374M,\ V419A,\ C437T}$) (Fig. 4.1b), while Stathopulos et al. (2013) reported an NMR structure of a SOAR overlapping fragment (aa312–387). The X-ray structure of hSOAR represents a dimer with several intra- as well as intermolecular interactions; however, it is still not fully clear which activation state is represented by this structure. The monomeric SOAR resembles the capital letter "R" formed by antiparallel CC2 and CC3 with two inter-adjacent short α-helices (Fig. 4.1b). The nomenclatures of these regions are Sα1 (345–391), Sα2 (393–398), Sα3 (400–403), and Sα4 (409–437). Two SOAR molecules form the structurally resolved V-shaped dimer, which comprises intermolecular interactions between the CC2-associated residues T354, L351, W350, and L347 from one monomer and CC3-associated residues R429, W430, I433, and L436 of the second monomer (Fig. 4.1b). Both CC2 domains in the dimeric structure cross at amino acid Y361. Furthermore, the basic amino acid cluster (K382, K384, K385, K386, R387), which has been functionally analyzed by Korzeniowski et al. (2010) (see below), is located at the tips of the V-shaped SOAR dimer (Yang et al. 2012). The NMR structure of STIM1 aa312–387, comprising CC1α3 and CC2, bends between these two helical regions by almost 180° and forms dimers assembled in an antiparallel manner (Fig. 4.1c) (Stathopulos et al. 2013). This STIM1 NMR structure proposes dimeric interactions via regions of CC1α3 (E320–A331) as well as CC2 (H355–A369). Consistent to the hSOAR structure, also the STIM1 CC1α3-CC2 structure displays a crossing point at Y361/Y362 (Fig. 4.1c). Nonetheless, the X-ray compared to the NMR structure shows substantial differences.

Unfortunately, the STIM1 CC1 (aa238–343) domain is not part of the crystallized hSOAR structure and only partially visible in the NMR structure. CC1 ultimately links the STIM1 TM domain with CC2 and comprises three α-helical segments (CC1α1 aa238–271, CC1α2 aa278–304, and CC1α3 aa308–337) (Fahrner et al. 2014; Soboloff et al. 2012). STIM1 CC1's role is not unique to bridge the distance at ER-PM junctions to allow STIM1-Orai coupling, but has been shown to play an essential role in keeping STIM1 in a compact and inactive conformation at high $[Ca^{2+}]_{ER}$ conditions (Fahrner et al. 2014; Ma et al. 2015; Muik et al. 2011; Zhou et al. 2013). As described above, lowering $[Ca^{2+}]_{ER}$ results in luminal oligomerization and rearrangement of the TM domains, which consequently induces a conformational change in the cytosolic domains of STIM1. The most prominent structural change of the STIM1 cytosolic portions is the switch from the inactive tight state to the activated extended state, where CC1 releases the clamp with CC3, therefore exposing SOAR (Fig. 4.2a–c) (Fahrner et al. 2014; Ma et al. 2015; Zhou et al. 2013). An early study by Covington et al. (2010) using FRET

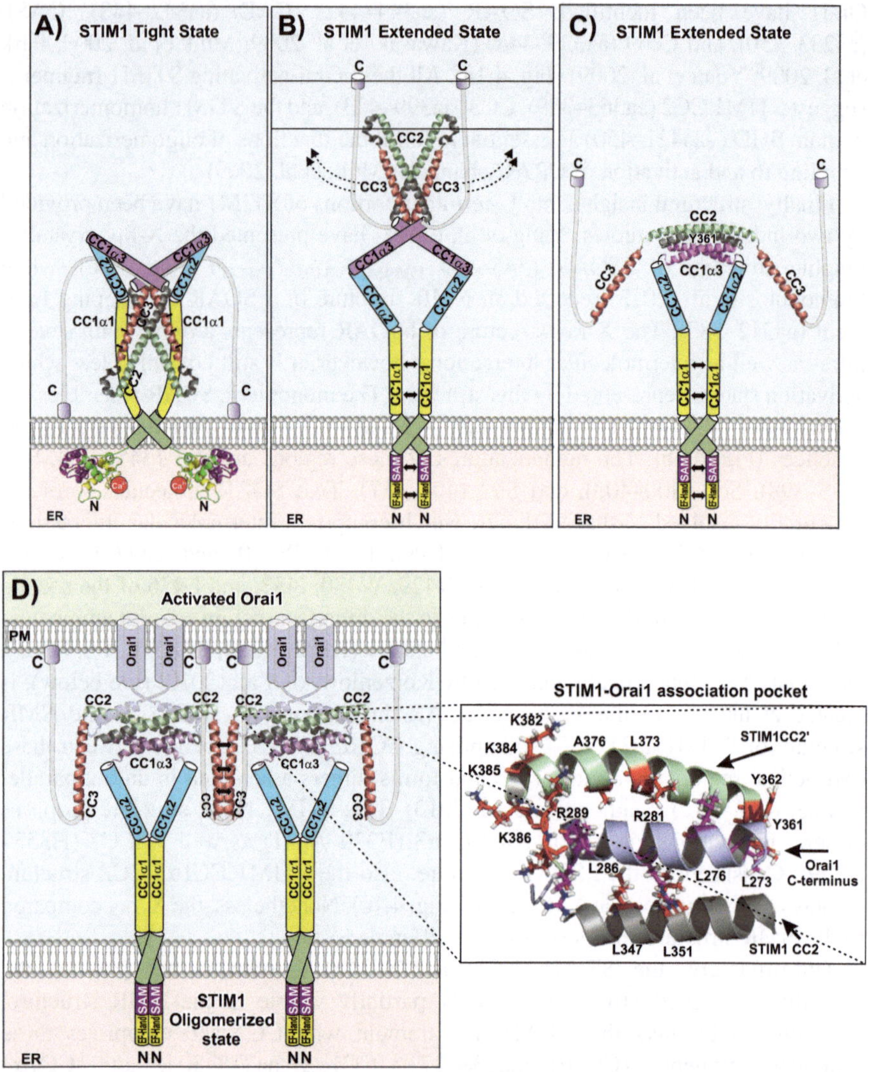

Fig. 4.2 Hypothetical model showing STIM1 conformational changes in the process of STIM1 activation and coupling to Orai upon store depletion. (**A**) STIM1 quiescent state is stabilized by STIM1 CC1/CC3 interaction (full ER Ca^{2+} stores). Store depletion results in intraluminal STIM1 EF-SAM domains losing bound Ca^{2+} that triggers a conformational rearrangement in the luminal part of STIM1. This in turn changes the angle of the TM domains relaying the signal to the STIM1 cytosolic parts. (**B**) CC1α1 homomerization is accompanied with the release of the CC1/CC3 clamp leading to an extended conformation with SOAR flipping to the top of the protein. (**C**) CC1α3 and CC2 in the dimeric STIM1 assembly change angle and release CC3 to form SOAP (STIM-Orai association pocket) in combination with Orai1 C-terminus. (**D**) STIM1 CC3 including SHD (STIM1 homomerization domain) connects STIM1 dimers forming higher-order oligomers. *Inset* highlights intermolecular interactions between STIM1 and Orai1 C-terminus

has shown that CC1 and CC3/SHD are both involved in homomerization. Subsequently, it has been demonstrated by other groups that STIM1 CC1 additionally comprises an essential regulatory role in the activation cascade of STIM1 upon store depletion. Several studies and hypotheses have been attributed to the topic how CC1 regulates the switch between the quiescent and the activated state of STIM1. An early hypothetical model presented by Korzeniowski et al. (2010) suggests an autoinhibitory clamp based on electrostatic attraction between the polybasic cluster in CC2 (aa382–387) and the acidic portion within CC1α3 (aa308–337). The latter domain is also called "inhibitory helix" and is part of the *C. elegans* extended SOAR crystal structure, which, however, shows that the polybasic cluster is not in the correct position relative to the CC1α3 to form the proposed electrostatic interaction (Korzeniowski et al. 2010; Yang et al. 2012). The name "inhibitory helix" originates from Yang et al. (2012) who proposed that residues within CC1α3 interact with amino acids of CC2 as well as CC3 in a dimeric structure, hypothetically representing the STIM1 quiescent state (Yang et al. 2012). This intramolecular shielding by CC1α3 in the STIM1 quiescent state was further examined by the use of point mutations or deletions within this region, which suggested that contribution of electrostatic or coiled-coil interactions is rather unlikely (Yu et al. 2013). However, Yu et al. (2013) have reported that multiple substitutions affecting the amphipathic character of CC1α3 have an impact on STIM1 activation and concluded that CC1α3's amphipathic nature is responsible for its regulatory role (Yu et al. 2013). OASF, the Orai-activating small fragment, comprises the cytosolic STIM1 portion 233–474, which is composed of CC1 + SOAR with C-terminal extension to aa474. The double-labeled (N-terminal YFP and C-terminal CFP) form of OASF has been engineered to monitor intramolecular conformational changes by FRET (Muik et al. 2011). Wild-type YFP-OASF-CFP reveals robust FRET, suggesting a compact structure with close proximity of the fluorophores. Specific point mutations within OASF induce a conformational change which results in different intramolecular FRET signals. Most prominently are point mutations within CC1 or CC3, which lead to an extended OASF structure (Muik et al. 2011). In detail, substitutions within CC1α1 (L251S) and CC3 (L416S L423S) reveal strongly reduced intramolecular FRET values. The same mutations in full-length STIM1 induce constitutive STIM1 activity and CRAC channel activation independent of $[Ca^{2+}]_{ER}$, indicating that the extended OASF structure equals the activated state of STIM1 (Fahrner et al. 2014; Muik et al. 2011). In summary, these data suggest that STIM1 CC1 interacts with CC3, forming an intramolecular clamp, which fixes the quiescent state. Store depletion or specific CC1 or CC3 point mutations release the clamp, resulting in an extended, active STIM1 structure. In line, using Tb^{3+}-acceptor energy transfer measurements, Zhou et al. (2013) have presented that the whole STIM1 C-terminus (233–685) represents a tight conformation in the quiescent, inactive STIM1 state. Extension of STIM1 C-terminus was monitored by artificial cross-linking of the CC1 domains or by introducing the "activating" mutation L251S (Zhou et al. 2013). Both studies finally reveal the essential role of CC1 in regulating the transition of STIM1 from the quiescent, tight state to the active, extended state, which involves CC1 homomerization as well as CC1–CC3 clamp release. To further elucidate the structure-function relationship of the STIM1 quiescent and active states, a FRET-based method called “FIRE” (FRET Interaction in a Restricted Environment) was engineered. Using FIRE, parts of CC1 have been explored in detail for their role in controlling the STIM1 activation mechanism (Fahrner et al. 2014). Indeed, a direct interaction involving CC1α1–CC3 proved to be essential for the tight, inactive STIM1 state. The substitution L251S, already known from previous studies to activate STIM1, led to a disrupted CC1α1–CC3 interaction, as expected. The “inhibitory helix” CC1α3, which has been previously hypothesized to be the key structure in controlling and regulating STIM1 (Yang et al. 2012), revealed to be less dominant in comparison to CC1α1, as analyzed with full-length STIM1 deletion mutants and STIM1 fragments using the FIRE system (Fahrner et al. 2014). Furthermore, this study proofs a destabilizing role of CC1α2, as the STIM1 Δα2 mutant-activated Orai1 with a significant delay in time, suggesting that the tight state (CC1α1-CC3 clamp) is more stabilized in the absence of CC1α2 (Fahrner et al. 2014). The Stormorken syndrome-associated mutant STIM1 R304W, which carries the mutation at the end of CC1α2, is characterized by constitutive STIM1 cluster formation and CRAC channel activation (Misceo et al. 2014; Morin et al. 2014; Nesin et al. 2014). Hypothetically, the CC1α2 inherent function, destabilizing the tight state upon store depletion, is enhanced by the R304W mutation resulting in a permanently oligomerized and activated STIM1.

In summary, integrating all structural and functional data, a hypothetical model of the dynamic STIM1 activation process, which is triggered by low $[Ca^{2+}]_{ER}$, is shown in Fig. 4.2. A resting cell with full ER Ca^{2+} stores comprises inactive STIM1 proteins with tightly packed C-termini. In the present model, inactive STIM1 is most likely dimeric. The tight state is stabilized by the intramolecular CC1–CC3 clamp formation (Fahrner et al. 2014; Muik et al. 2011). Ca^{2+} flux from the ER lumen into the cytosol represents the key step inducing the luminal STIM1 conformational change (Stathopulos and Ikura 2010; Stathopulos et al. 2006), a signal which is transmitted through the crossing transmembrane domains to the cytosolic STIM1 portions (Ma et al. 2015). Subsequently, the CC1–CC3 clamp is released, involving enhanced CC1 homomerization and SOAR exposure (Fahrner et al. 2014; Ma et al. 2015; Muik et al. 2011; Stathopulos et al. 2013; Zhou et al. 2013). Via CC3 the STIM1 dimers finally form higher-order oligomers (Fahrner et al. 2014; Muik et al. 2009). How in detail the Stormorken syndrome-associated STIM1 R304W mutation modifies the STIM1 activation cascade is still not understood; however, a process inducing enforced CC1 homomerization is very likely to be involved (Fahrner et al. unpublished data).

4.3 STIM1-Orai1 Coupling

Store depletion-induced conformational rearrangement of STIM1 to a more extended or open state is currently thought to facilitate spanning the distance between the plasma and ER membranes culminating in the formation of ER-PM junctions (Muik et al. 2011). There, STIM1 recruits and aggregates to prior evenly distributed Orai channels, which lead to the formation of clusters known as "hot spots" of local Ca^{2+} entry (Luik et al. 2006). FRET microscopy has enabled to visualize a close proximity between STIM1 and Orai1 in store-depleted cells and therefore suggests a rather direct communication of these two proteins (Barr et al. 2008; Muik et al. 2008; Navarro-Borelly et al. 2008) enabling Ca^{2+} entry across Orai channels into the cell. The fact that the two proteins are sufficient to reconstitute CRAC currents (Gudlur et al. 2014) and, furthermore, that activation of Orai channels occurs even with purified C-terminal STIM1 fragments under isolated conditions attributes a more modulatory role to additional components that have so far been reported to be involved in the CRAC channel signaling complex (Zhou et al. 2010a).

Upon store depletion, STIM1 faces the cytosolic side of the Orai1 channel which likely provides its N- and C-terminus as well as the second loop, which connects TM2 with TM3, as putative STIM1-coupling domains (Fig. 4.3a). The CAD fragment has been shown to directly bind to both flanking strands of Orai1, while an interaction with a fragment of the second loop has not been detected (Park et al. 2009; Yuan et al. 2009). A series of studies attribute an indispensable role to the Orai1 C-terminus for proper coupling to STIM1. All three Orai homologues show certain probabilities for the formation of C-terminally embedded coiled-coil domains, whereas that of Orai1 is much lower than the estimated probabilities for the other two isoforms. Orai1 contains the amino acids L273 as well as L276 as hydrophobic "a" and "d" core positions crucial for the formation of coiled-coil regions. Their substitution to more hydrophilic residues (S/D) and the truncation of the Orai1 C-terminus interfere with co-clustering and coupling to STIM1 (Li et al. 2007; Muik et al. 2008). In contrast to Orai1, where single-point mutations are sufficient to destabilize the interaction with STIM1, the situation is apparently different for Orai2 and Orai3. Here, mutations of at least two hydrophobic core positions are required to fully interrupt communication with STIM1 (Frischauf et al. 2009). This might be attributed to the more canonical nature of the embedded coiled-coil domains. Since the STIM1/Orai interaction data (Li et al. 2007; Muik et al. 2008) point to the involvement of a heteromeric coiled-coil assembly, the search for putative counterparts within STIM1 was quite obvious. Accordingly, the systematic performance of hydrophobic point mutations with a special focus on the second coiled-coil domain revealed a crucial role for L373 as well as A376. The hydrophilic substitution of either one (L373S, A376K) or both of these amino acids (L373S, A376S) exhibits reduced or fully abolished STIM1-Orai coupling similar as observed with the mutations in the Orai C-terminus (Covington et al. 2010; Frischauf et al. 2009). In addition, there are essential STIM1 residue mutations within CC2 (L347A, Q348A) or CC3 (C437G), which retain co-clustering with Orai1, but result in a complete loss of channel activation (Park et al. 2009; Yuan

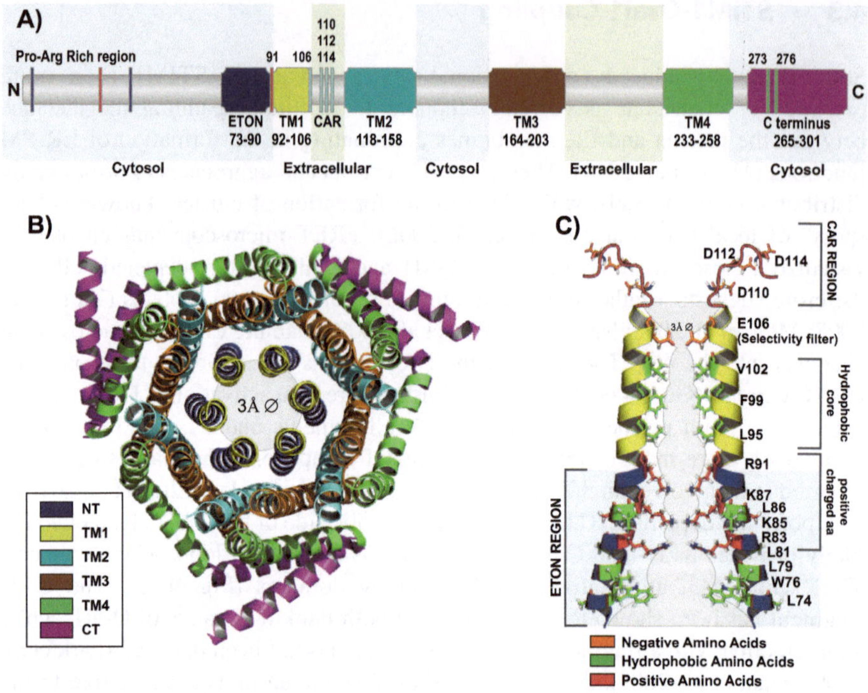

Fig. 4.3 Orai1: (**A**) schematic representing human Orai1 specifying essential structural/functional regions and residues required for Orai1 function. (**B**) Cartoon based on the X-ray crystallographic resolution of a *Drosophila* dOrai channel depicting the top view of Orai subunits revealing a hexameric assembly. The six transmembrane domains (TM)1 form the inner ring surrounding the ion-conducting pore, while the concentric ring formed by domains TM2 and TM3 of the six subunits separates the TM1 ring from the lipid environment of the PM. The *color code* specifies the different helical regions of the Orai channel. (**C**) Cartoon showing two TM1 strands depicting the human Orai1 pore with the extracellular CAR at the C-terminal end of TM1 and the cytosolic N-terminal helical extensions including the conserved ETON region. The selectivity filter, the hydrophobic core, as well as charged residues within the CAR and the ETON region are highlighted (amino acid numbering refers to human Orai1)

et al. 2009). Besides the relevance of hydrophobic residues for proper formation of a coiled-coil region and for the interplay of coiled-coil domains, the contribution of charged amino acids to stabilization of coiled-coil regions and their interactions in STIM1/Orai proteins should not be neglected. With respect to Orai1, a series of negatively charged residues within a certain area of the C-terminus (272–291) have been shown to fulfill such a role. Also, STIM1 C-terminus includes a basic cluster (aa382–387) which when substituted by neutral residues disrupts coupling to and activation of Orai1 (Calloway et al. 2009; Korzeniowski et al. 2010).

Altogether these results strongly suggest a model of direct STIM1/Orai coupling via coiled-coil domains. Conclusive evidence was provided by the NMR structure of a STIM1 fragment comprising the latter part of CC1 as well as the whole CC2 domain (aa312–387) together with an Orai1 C-terminal fragment (aa272–292) (Fig. 4.2d) (Stathopulos et al. 2013). These data have revealed for the first time the C-terminal interaction of STIM1 with Orai1 (STIM1-Orai1-Association-Pocket, SOAP) on a structural basis. The structure uncovers both the core hydrophobic interactions and the involvement of charged residues. With respect to Orai1, both of the leucines, which were discussed earlier (L273, L276), are involved in hydrophobic interactions with STIM1. Additionally, negatively charged aspartates (D284, D287, D291) are able to form electrostatic interactions with complementary basic STIM1 regions (K382, K384, K385, K386). A search for hydrophobic key residues within STIM1 has highlighted multiple positions (L347, L351, Y362, L373, A376) that are in close proximity to Orai1 C-terminus and most likely involved in coupling. Hence, the NMR structure of SOAP has confirmed the so far established functional hypothesis and complements it with structural insights.

The dOrai X-ray structure exhibits an antiparallel crossing of Orai C-termini within each Orai dimer that is supported by hydrophobic interactions of L273 and L276 (Hou et al. 2012). In accordance with their critical role in the Orai1 activation, mutation of those residues to less hydrophobic ones abolishes Orai1 activation (Palty et al. 2015). Further cysteine cross-linking of those positions (L273C, L276C) stunts STIM1 binding that is rapidly reversed upon disulfide bond breakage (Tirado-Lee et al. 2015).

TM4 and C-terminus of Orai1 are connected via a flexible hinge region that is crucial for correct orientation of the C-termini enabling STIM1 coupling (Palty et al. 2015; Tirado-Lee et al. 2015; Zhou et al. 2016). STIM1 binding to Orai1 hinge mutants is significantly reduced compared to wild-type, but only fully abolished upon additional Orai1 C-terminus mutation (Palty et al. 2015; Zhou et al. 2016), suggesting that the hinge brings the C-termini in the optimal position required for accurate STIM1 binding. Intriguingly, in contrast to several hinge mutants exhibiting loss of function (Palty et al. 2015; Tirado-Lee et al. 2015; Zhou et al. 2016), Zhou et al. (2016) elucidated a few constitutively active hinge mutants, exhibiting widely comparable biophysical characteristics like wild-type Orai1. Constitutive, highly Ca^{2+}-selective currents together with reduced STIM1 binding led Zhou and colleagues to conclude that these hinge mutations induce conformational changes within Orai1 similar to those occurring upon STIM1 binding, thus accomplishing signal propagation to open the channel.

Regarding STIM1-induced alterations of the orientation of Orai C-termini, the NMR structure of SOAP reveals an angle of 136° of the crossing Orai1 C-termini, while the dOrai crystal structure exhibits an angle of 152° (Fig. 4.3b). Potentially, the angle of 136° mirrors C-terminal dimers in the Orai active state, while the C-terminal crossing angle of 152° represents the closed state. Alternatively, another model (Hou et al. 2012) suggests the TM4-C-terminus connection more straightened upon Orai activation.

Taking together the structural and functional data, STIM1 binding is predominantly achieved by coupling to the Orai1 C-terminus. However, the presence of the N-terminus is absolutely necessary for channel gating, since diverse N-terminal truncation and point mutants exhibit loss of Orai function (Derler et al. 2013; Muik et al. 2008; Zhou et al. 2016). Although with a weaker affinity, the STIM1-Orai1 N-terminus interaction is also detectable in biochemical assays and involves a conserved sequence (aa73–90) adjacent to the first transmembrane segment, the so-called extended transmembrane Orai1 N-terminal (ETON) region (Fig. 4.3c) (Derler et al. 2013; Park et al. 2009). It incorporates both hydrophobic (L74, W76, L81) and several charged residues (R83, K85, K87) that are supposed to be relevant for STIM1 binding as well as Orai1 gating. In line, truncations of the N-terminus result either in reduced store-operated activation (Δ1–74/75) or in a complete loss of function if a larger portion of the ETON region (Δ1–76) is deleted. It is worth to note that the function of the three Orai homologues is differently affected upon N-terminal deletions till equivalent positions in the ETON region. Despite specific analogue Orai N-truncation mutants contain a fully conserved portion of the ETON region, Orai3 retains activity upon deletion of larger N-terminal fragments than Orai1. In terms of the latter, additional yet unknown structures might be responsible for this effect.

In contrast to the well-established structural insights on the binding sites between STIM1 and Orai1 C-terminus coupling to each other, the STIM1-Orai1 N-terminus coupling sites are not well characterized. A recently published study assumed a region located between the second and third coiled-coil domain (aa393–403) (Fig. 4.1a, b) to function as a coupling partner for Orai N-terminus. Specifically, they suppose the residue F394, residing within this sequence, to bind to a hydrophobic counterpart of the ETON region as it abolishes the activation of Orai1 if mutated to a more hydrophilic amino acid (Wang et al. 2014). Although this provides a first indication on where to focus, further studies are definitely required to finally identify and characterize the STIM1 binding site to the N-terminus of Orai1.

With respect to stoichiometric requirements, the Orai channel complex has been thought for a long time to form a tetramer (Demuro et al. 2011; Maruyama et al. 2009; Mignen et al. 2008; Penna et al. 2008). However, the crystal structure of *Drosophila melanogaster* dOrai (Fig. 4.3) (Hou et al. 2012) in line with recent concatemeric studies has suggested a hexameric complex (Cai et al. 2016). While the hexameric Orai channel is widely accepted, the exact stoichiometry of the functional STIM1-Orai complex has remained controversial. Electrophysiological studies revealed that Orai activation depends on the number of coupled STIM1 proteins (Hoover and Lewis 2011; Li et al. 2011; Scrimgeour et al. 2009). Especially, the degree of inactivation of CRAC currents depends on the amount of STIM1 molecules, in that more pronounced inactivation requires higher STIM1 compared to Orai1 levels (Scrimgeour et al. 2014). Independent approaches investigated the stoichiometric requirements not only via a variation of the ratio of expressed STIM1 versus Orai1 but took also advantage of Orai fusion proteins directly connected to STIM1 fragments (Hoover and Lewis 2011). Studies with

Orai proteins linked to tandem dimers of CAD fragments indicate that eight STIM1 molecules are required for maximal activation and inactivation (Li et al. 2011) suggesting that CRAC activation occurs rather by a gradual process involving a complex interplay of cytosolic domains, but not in an "all or none" manner. In line, Ca^{2+} selectivity is enhanced with an increasing amount of coupled STIM1 fragments, as shown by Orai1 and Orai1 V102A proteins attached to single and tandem CAD/SOAR domains (McNally et al. 2012). From the perspective of the crystal structure, it is conceivable that six, but not eight, STIM1 molecules bind to a hexamer. The recent NMR-SOAP structure supports a 1:1 ratio of STIM1 to Orai1 in an active channel complex (Stathopulos et al. 2013), suggesting six STIM1 proteins attach probably to six Orai subunits or three STIM1 dimers couple to the three Orai dimers within a hexamer. In order to fulfill a 2:1 STIM1-to-Orai1 ratio, 12 STIM1 proteins would be required for Orai activation. In contrast to the so far suggested bimolecular models for STIM1-Orai coupling, a recent report proposed a unimolecular coupling model sufficient to activate the Orai channel. These conclusions are based on the fact that SOAR tandems containing F394H, critical for Orai activation, only within one monomer lead to Orai activation to comparable levels like a wild-type SOAR tandem. However, SOAR tandems containing in both monomers the F394H substitution impair Orai activation. Moreover, the interaction of a SOAR tandem containing one F394H mutation with an expressed PM-fixed Orai1-C-terminus construct occurs to a half-maximal extent with a SOAR wild-type tandem. Consequently, a unimolecular interaction between one Orai channel subunit and one STIM1 of a STIM1 dimer has been proposed. This unimolecular coupling led further to the assumption that the second STIM1 within a STIM1 dimer couples to an Orai monomer of an adjacent Orai hexameric assembly, thus enabling clustering of Orai channels. This novel mechanism, however, leaves unresolved how a single STIM1 protein induces alterations in the Orai C-terminus enabling CRAC activation. With respect to the clustering of Orai channels, it is also conceivable that it occurs via a bimolecular STIM1-Orai1 binding, as STIM1 proteins are able to form clusters via the $CC3_{ext}$ region in the active state (Fahrner et al. 2014). To proof the bi- or unimolecular coupling model, additional experimental verification is necessary.

In aggregate, coupling of STIM1 occurs initially and predominantly to the C-terminus of Orai1. This coupling most likely involves a conformational rearrangement of subdomains within Orai1 that is accompanied by a bridging to the Orai N-terminus resulting in full channel activation. However, at this juncture, further functional and structural evidence is required to fully understand the underlying mechanism of the STIM1-Orai1 coupling and oligomeric cluster formation.

4.4 The Ca^{2+}-Selective Orai1 Pore

In the absence of STIM1 binding and ER store depletion, all three homologue Orai channels remain in a closed conformation upon heterologous expression in cell lines (Lis et al. 2007; Schindl et al. 2009). Orai1, Orai2, and Orai3 can form homo- or heteromeric subunit assemblies. Upon store depletion, the functional STIM1/Orai channel complex mediates Ca^{2+}-selective currents with a remarkably low single-channel conductance (Mercer et al. 2006; Peinelt et al. 2006). The Ca^{2+} ions permeate through a central pore formed by a ring of TM1 helices (Hou et al. 2012; McNally et al. 2009; Zhou et al. 2010a, b). The crystal structure of dOrai1 revealed considerable insights into the channel architecture (Fig. 4.3b). The Orai channel consists of a symmetric hexameric structure, formed by three rings of transmembrane (TM) helices (Fig. 4.3b). The central pore is formed by TM1, TM2, and TM3 helices surrounding the pore, and TM4 helices form the outer ring (Hou et al. 2012). TM4 helices extend into the cytosol and form pair-wise coiled-coil helices oriented in an antiparallel manner enabling STIM1 binding (Stathopulos et al. 2013). The external loop between the TM1 and TM2 includes negatively charged amino acids that form an initial Ca^{2+} accumulation region (CAR) close to the external side of the Orai1 pore (Frischauf et al. 2015) (Fig. 4.3c). CAR can bind up to three Ca^{2+} ions at the same time, one to two nanometers away from the selectivity filter. Hence, CAR locally increases external Ca^{2+} levels (Frischauf et al. 2015) and drives Ca^{2+} ions to the Orai1 pore (Jha and Muallem 2016). Recent structural insights revealed that other Ca^{2+}-selective channels also have a similar negatively charged segment just a few nanometers apart from the selectivity filter. A potential extracellular Ca^{2+} accumulation region is predicted in the voltage-gated Cav1.1 channel (Wu et al. 2015) and the TRPV6 channel (Saotome et al. 2016). In addition, a luminal Ca^{2+} binding site in the ryanodine receptor has been suggested (Zalk et al. 2015). Mutations of CAR residues in Orai1 lowered Ca^{2+} permeation and interfered with the channel function together with gene regulation by NFAT, the nuclear factor of activated T cells (Frischauf et al. 2015). CAR is also a binding site for lanthanides that block Ca^{2+} channels efficiently (Yeromin et al. 2006). Additionally, CAR is very sensitive to alterations of pH, and a lowered negative charge of CAR decreases the Orai1 currents (Beck et al. 2014).

While CAR is located in a flexible loop, the selectivity filter is placed in the helical TM1 pore (Fig. 4.3c). Six glutamates, one E106 per TM1 helix, form a highly selective Ca^{2+} selectivity filter (Hou et al. 2012). A conservative Orai1-E106D mutation yielded nonselective currents, while other engineered mutations fully blocked Orai1 currents (McNally et al. 2009; Prakriya et al. 2006; Vig et al. 2006a). More central in the pore, a hydrophobic segment of three helical turns hinders Ca^{2+} permeation in the closed channel conformation (Fig. 4.3c). Specifically, a mutation of valine 102 one helix turn next to the selectivity filter yielded constitutive Ca^{2+} influx but less selective currents (McNally et al. 2012). Remarkable, co-expression of STIM1 and the Orai1-V102A mutation shifted the Ca^{2+} selectivity back to that of wt-Orai1. Introduction of small amino acids, like glycine or cysteine residues at position V102, also turned the channel in a constitutively active conformation. The pore-facing positioning of V102 side chain seems to depend on STIM1, as access by Cd^{+} to an inner pore residue is only possible upon STIM1 binding (McNally et al. 2012). These experiments determine that V102 acts as a gating site in the Orai1 channel and might be switched by STIM1 out of the pore center. Additionally, Orai1 pore dynamics by STIM1 could be directly visualized by Tb^{3+} luminescence that was mainly bound to the selectivity filter but also to CAR (Gudlur et al. 2014). The Orai1 Tb^{3+} fluorescence was largely increased upon interaction with STIM1, suggesting a conformational change between the selectivity filter and a tyrosine Y115 next to CAR that mediates the energy transfer for Tb^{3+} ions. A similar STIM1-dependent Tb^{3+} increase was also observed for an Orai1-V102C (Gudlur et al. 2014), suggesting that the main dynamics were rather induced by STIM1 but not by the Orai1-V102C mutation itself. Intriguingly (Yamashita et al. 2017), recent evidence based on Cd^{+} and Ag^{+} experiments revealed that V102C remains pore-facing in the absence and presence of STIM1 corresponding to the resting and active conformation, respectively, which suggests another gating mechanism than the movement of V102 out of the pore upon STIM1 binding. Prakriya's lab only lately suggested that Orai gating involves a rotation of TM1 moving F99 away from the pore, while G98 captures a more pore-facing orientation. Molecular dynamic simulations gave a first energetic explanation for the V102- and F99-dependent gating. In these simulations, a valine to alanine mutation or a phenylalanine to tyrosine increases the amount of water molecules in the whole hydrophobic pore region (Dong et al. 2013; Yeung et al. 2016). Water molecules arranged around a Ca^{2+} ion might create a water shell that would decrease the net charge of Ca^{2+}. As a consequence, the energetic barrier formed by the whole hydrophobic segment including the valine and the phenylalanine is largely reduced. The remaining barrier occurs due to a positively charged cluster formed of three side chains per channel subunit that are facing into the pore center (Dong et al. 2013; Hou et al. 2012) (Fig. 4.3c). It is of note that in the crystallized Orai structure, a negative plug was bound to this positively charged cluster (Hou et al. 2012). Binding and unbinding of a small negatively charged molecule could act as a gating plug. The predominant role in this cluster is mediated by R91 side chains. Initial evidence that R91 forms a gate was obtained with another pore mutation that kept the channel active, an Orai1-G98P or G98D mutation (Zhang et al. 2011). These glycine mutations extended the cytosolic pore segment, as R91 substitutions to cysteines did not longer cross-link with each other, either due to a larger distance or a rotational movement of engineered cysteine residues (Zhang et al. 2011). However, point mutations that substituted R91 with small or negatively charged side chains did not turn the Orai1 channel in a constitutively active conformation (Derler et al. 2009). Moreover, substitution of the whole positively charged pore segment with an unrelated amino acid stretch resulted in only small constitutively active Orai1 currents (Gudlur et al. 2014). The Orai channel was still functional, as an additional V102A mutation largely increased constitutive Orai1 channel activity. A further critical role of R91 was observed by engineering large hydrophobic substitutions of R91 that blocked store-operated Orai1 channel activation (Derler et al. 2009). An Orai1-R91W mutation in patients causes severe combined immune deficiency (Feske et al. 2006). Crystal structure of Orai1 with the R91W substitution determined that the pore is fully occluded. These experiments fit together best when two gating sites are assumed, V102 and R91. Besides remarkable insight into gating of this unique Ca^{2+} channel, most experiments suffer from the fact that pore residues are mutated, with large impacts on energy barriers along the whole pore. Herein, channel simulations determined that Orai1-V102A channels even impacted the energetic barrier of the R91 gate (Dong et al. 2013). Consequently, Orai1 mutations outside the pore that keep the Orai1 channel in an active conformation will further help to understand this unique gating mechanism.

Indeed, pathophysiological Orai1 mutations within the TM helices surrounding the pore have been observed to turn the channel in an active conformation. An Orai1-L138F mutation in the TM2 helix caused tubular aggregate myopathy disease (Endo et al. 2015). Further, an Orai1-P245L also resulted in myopathy disease as well as congenital miosis (Nesin et al. 2014). Expression of Orai1-P245L alone resulted in constitutively active currents that were less Ca^{2+}selective (Palty et al. 2015). Similar, as reported for Orai1-V102A, co-expression of STIM1 and Orai1-P245L recovered the high Ca^{2+} selectivity but also fast Ca^{2+}-dependent inactivation (Palty et al. 2015). While the reason for constitutive Orai1-L138F channel activity is unknown, the proline 245 in Orai1 seems to play a major function in breaking the TM4 helix, as every tested amino acid substitution turned the Orai1 channel in a constitutively active form (Palty et al. 2015). Recently, specific Orai1 hinge mutants containing substitutions in the flexible region connecting TM4 and the C-terminus revealed constitutive currents with widely, but not fully comparable, biophysical characteristics like those of CRAC currents (Zhou et al. 2016). Further the constitutive hinge mutants displayed reduced coupling to STIM1, which finally led to the hypothesis that sole STIM1 coupling to Orai C-terminus leads to structural changes allowing pore opening (Zhou et al. 2016). Signal propagation initiated via STIM1-Orai-C-terminus coupling assumedly occurs via allosteric interactions of the closely positioned TM3 and TM2 regions (Zhou et al. 2017) that are finally transduced to TM1 probably there inducing a rotation of TM1 allowing pore opening (Yeung et al. 2016). However, mechanistic insight on how the mutations regulate the Orai1 gates is currently undetermined.

4.5 Perspectives

Within the past decade, considerable advances have been made to elucidate the key molecular players of the CRAC channel family as well as the 3D atomic resolutions of their structures. Nevertheless, further analysis is still required to answer several still unresolved questions.

With respect to STIM1, 3D atomic structures are so far only available for STIM1 C-terminal fragments, i.e., the CAD/SOAR region and a portion (CC2) of it, which, however, exhibit substantial differences. For detailed clarification of intra- and intermolecular interactions which convert STIM1 from its tight, inactive to its extended, active conformation, especially the structural resolution of larger portions of STIM1 C-terminal fragments, at least including the regulatory CC1 region, would be required. The extended or the tight conformation might be locked by specific point mutations like L251S or R426L potentially facilitating the resolution of the active or inactive state of STIM1, respectively.

Regarding Orai channels, the dOrai X-ray structure exhibits the closed state, while for the open state, a structural resolution is so far not available. Store-operated coupling of STIM1 to cytosolic strands of the Orai channel initiates its gating to an open conformation. However, while binding of STIM1 to the Orai1 C-terminus as resolved in the NMR-SOAP structure is widely accepted as a critical step for Orai activation, STIM1 coupling to the N-terminus for opening the pore is currently controversial and needs to be further clarified experimentally. Alternatively, the Orai1 N-terminus together with the pore may couple to other structures within the Orai1 channel, thus being allosterically affected upon STIM1 binding to Orai C-terminus. Furthermore, critical regions within STIM1 that couple to other cytosolic portions than the C-terminus of Orai1 are not yet resolved. The F394 residue within STIM1 critical for activation of Orai1 channels has been originally supposed to form hydrophobic contacts with specific residues in the N-terminus of Orai1 (Wang et al. 2014), while an alternative model proposes a role of STIM1 F394 in the coupling to residues within or close to the hinge of Orai1 (Zhou et al. 2017).

Besides the assertive step of STIM1 coupling initiating Orai1 activation, STIM1-induced conformational rearrangements leading to an open pore are only partially elucidated. Constitutively active mutants (e.g., Orai1 P245L, Orai1 ANSGA) inducing an open state of Orai1 may help to structurally resolve open Orai channel conformations, and they would provide hints on how the conformational change from the closed to the open state may take place. Nonetheless, as to how these mutationally induced open conformations of Orai1 will mimic that triggered by interaction with STIM1 remains to be seen. STIM1 coupling to either the C-terminus or both the N- and C-terminal strands likely induces signal propagation to the pore via alterations in the orientation and/or interaction of the four TM helices. Here, it needs to be clarified if the direction of signal propagation for opening the pore occurs either from the C-terminal to the N-terminal sides, only the N-terminal side, or from both sides. Ultimately, a structural resolution of a STIM1/Orai1 complex is required to finally understand the molecular mechanisms underlying Orai gating and linking conformational changes in the C-terminus induced via STIM1 binding to a rotation of the TM1 helix (Yeung et al. 2016) for opening of the pore.

The increasingly emerging optogenetic engineering approaches on the generation and test of photoactivatable CRAC channels will provide, via a sophisticated control by light, more detailed insights into the regulatory mechanisms of the STIM-Orai downstream signaling pathways (He et al. 2015; Ishii et al. 2015).

While STIM1 and Orai1 have been shown to be sufficient to fully reconstitute CRAC channel, a series of proteins and lipids have been proposed to modulate the interaction and activation of the STIM1-Orai complex. Here, STIMATE (Jing et al. 2015), septin (Sharma et al. 2013), SARAF (Palty et al. 2012), CRACR2A (Srikanth et al. 2010), and cholesterol (Derler et al. 2016; Pacheco et al. 2016) enhance the complexity of the native CRAC channel system.

Acknowledgments This work was supported by the Austrian Science Fund projects P28123 (to MF), P26067 (to RS), P28498 (to MM), P25210 as well as P27641 (to ID), and P27263 (to CR).

References

Baba Y, Hayashit K, Fujii Y, Mizushima A, Watarai H, Wakamori M, Numaga T, Mori Y, Iino M, Hikida M, Kurosaki T (2006) Coupling of STIM1 to store-operated Ca2+ entry through its constitutive and inducible movement in the endoplasmic reticulum. Proc Natl Acad Sci U S A 103:16704–16709

Baba Y, Nishida K, Fujii Y, Hirano T, Hikida M, Kurosaki T (2008) Essential function for the calcium sensor STIM1 in mast cell activation and anaphylactic responses. Nat Immunol 9:81–88

Barr VA, Bernot KM, Srikanth S, Gwack Y, Balagopalan L, Regan CK, Helman DJ, Sommers CL, Oh-Hora M, Rao A, Samelson LE (2008) Dynamic movement of the calcium sensor STIM1 and the calcium channel Orai1 in activated T-cells: puncta and distal caps. Mol Biol Cell 19:2802–2817

Beck A, Fleig A, Penner R, Peinelt C (2014) Regulation of endogenous and heterologous Ca(2)(+) release-activated Ca(2)(+) currents by pH. Cell Calcium 56:235–243

Brandman O, Liou J, Park WS, Meyer T (2007) STIM2 is a feedback regulator that stabilizes basal cytosolic and endoplasmic reticulum Ca(2+) levels. Cell 131:1327–1339

Cai X, Zhou Y, Nwokonko RM, Loktionova NA, Wang X, Xin P, Trebak M, Wang Y, Gill DL (2016) The Orai1 store-operated calcium channel functions as a hexamer. J Biol Chem 291:25764–25775

Calloway N, Vig M, Kinet JP, Holowka D, Baird B (2009) Molecular clustering of STIM1 with Orai1/CRACM1 at the plasma membrane depends dynamically on depletion of Ca2+ stores and on electrostatic interactions. Mol Biol Cell 20:389–399

Covington ED, Wu MM, Lewis RS (2010) Essential role for the CRAC activation domain in store-dependent oligomerization of STIM1. Mol Biol Cell 21:1897–1907

De Haven WI, Smyth JT, Boyles RR, Putney JW Jr (2007) Calcium inhibition and calcium potentiation of Orai1, Orai2, and Orai3 calcium release-activated calcium channels. J Biol Chem 282:17548–17556

Demuro A, Penna A, Safrina O, Yeromin AV, Amcheslavsky A, Cahalan MD, Parker I (2011) Subunit stoichiometry of human Orai1 and Orai3 channels in closed and open states. Proc Natl Acad Sci U S A 108:17832–17837

Derler I, Fahrner M, Carugo O, Muik M, Bergsmann J, Schindl R, Frischauf I, Eshaghi S, Romanin C (2009) Increased hydrophobicity at the N terminus/membrane interface impairs gating of the severe combined immunodeficiency-related ORAI1 mutant. J Biol Chem 284:15903–15915

Derler I, Plenk P, Fahrner M, Muik M, Jardin I, Schindl R, Gruber HJ, Groschner K, Romanin C (2013) The extended transmembrane Orai1 N-terminal (ETON) region combines binding interface and gate for Orai1 activation by STIM1. J Biol Chem 288:29025–29034

Derler I, Jardin I, Stathopulos PB, Muik M, Fahrner M, Zayats V, Pandey SK, Poteser M, Lackner B, Absolonova M, Schindl R, Groschner K, Ettrich R, Ikura M, Romanin C (2016) Cholesterol modulates Orai1 channel function. Sci Signal 9:ra10

Dong H, Sharma M, Zhou HX, Cross TA (2012) Glycines: role in alpha-helical membrane protein structures and a potential indicator of native conformation. Biochemistry 51:4779–4789

Dong H, Fiorin G, Carnevale V, Treptow W, Klein ML (2013) Pore waters regulate ion permeation in a calcium release-activated calcium channel. Proc Natl Acad Sci U S A 110:17332–17337
Dubois C, Vanden Abeele F, Lehen'kyi V, Gkika D, Guarmit B, Lepage G, Slomianny C, Borowiec AS, Bidaux G, Benahmed M, Shuba Y, Prevarskaya N (2014) Remodeling of channel-forming ORAI proteins determines an oncogenic switch in prostate cancer. Cancer Cell 26:19–32
Endo Y, Noguchi S, Hara Y, Hayashi YK, Motomura K, Miyatake S, Murakami N, Tanaka S, Yamashita S, Kizu R, Bamba M, Goto Y, Matsumoto N, Nonaka I, Nishino I (2015) Dominant mutations in ORAI1 cause tubular aggregate myopathy with hypocalcemia via constitutive activation of store-operated Ca(2)(+) channels. Hum Mol Genet 24:637–648
Fahrner M, Muik M, Derler I, Schindl R, Fritsch R, Frischauf I, Romanin C (2009) Mechanistic view on domains mediating STIM1-Orai coupling. Immunol Rev 231:99–112
Fahrner M, Muik M, Schindl R, Butorac C, Stathopulos P, Zheng L, Jardin I, Ikura M, Romanin C (2014) A coiled-coil clamp controls both conformation and clustering of stromal interaction molecule 1 (STIM1). J Biol Chem 289:33231–33244
Feske S, Gwack Y, Prakriya M, Srikanth S, Puppel SH, Tanasa B, Hogan PG, Lewis RS, Daly M, Rao A (2006) A mutation in Orai1 causes immune deficiency by abrogating CRAC channel function. Nature 441:179–185
Frischauf I, Muik M, Derler I, Bergsmann J, Fahrner M, Schindl R, Groschner K, Romanin C (2009) Molecular determinants of the coupling between STIM1 and Orai channels: differential activation of Orai1-3 channels by a STIM1 coiled-coil mutant. J Biol Chem 284:21696–21706
Frischauf I, Zayats V, Deix M, Hochreiter A, Jardin I, Muik M, Lackner B, Svobodova B, Pammer T, Litvinukova M, Sridhar AA, Derler I, Bogeski I, Romanin C, Ettrich RH, Schindl R (2015) A calcium-accumulating region, CAR, in the channel Orai1 enhances Ca(2+) permeation and SOCE-induced gene transcription. Sci Signal 8:ra131
Grigoriev I, Gouveia SM, van der Vaart B, Demmers J, Smyth JT, Honnappa S, Splinter D, Steinmetz MO, Putney JW Jr, Hoogenraad CC, Akhmanova A (2008) STIM1 is a MT-plus-end-tracking protein involved in remodeling of the ER. Curr Biol 18:177–182
Gudlur A, Quintana A, Zhou Y, Hirve N, Mahapatra S, Hogan PG (2014) STIM1 triggers a gating rearrangement at the extracellular mouth of the ORAI1 channel. Nat Commun 5:5164
Gwack Y, Srikanth S, Feske S, Cruz-Guilloty F, Oh-hora M, Neems DS, Hogan PG, Rao A (2007) Biochemical and functional characterization of Orai proteins. J Biol Chem 282:16232–16243
He L, Zhang Y, Ma G, Tan P, Li Z, Zang S, Wu X, Jing J, Fang S, Zhou L, Wang Y, Huang Y, Hogan PG, Han G, Zhou Y (2015) Near-infrared photoactivatable control of Ca(2+) signaling and optogenetic immunomodulation. Elife 4. doi:10.7554/eLife.10024
Honnappa S, Gouveia SM, Weisbrich A, Damberger FF, Bhavesh NS, Jawhari H, Grigoriev I, van Rijssel FJ, Buey RM, Lawera A, Jelesarov I, Winkler FK, Wuthrich K, Akhmanova A, Steinmetz MO (2009) An EB1-binding motif acts as a microtubule tip localization signal. Cell 138:366–376
Hoover PJ, Lewis RS (2011) Stoichiometric requirements for trapping and gating of Ca2+ release-activated Ca2+ (CRAC) channels by stromal interaction molecule 1 (STIM1). Proc Natl Acad Sci U S A 108:13299–13304
Hou X, Pedi L, Diver MM, Long SB (2012) Crystal structure of the calcium release-activated calcium channel Orai. Science 338:1308–1313
Huang GN, Zeng W, Kim JY, Yuan JP, Han L, Muallem S, Worley PF (2006) STIM1 carboxyl-terminus activates native SOC, I(crac) and TRPC1 channels. Nat Cell Biol 8:1003–1010
Ishii T, Sato K, Kakumoto T, Miura S, Touhara K, Takeuchi S, Nakata T (2015) Light generation of intracellular Ca(2+) signals by a genetically encoded protein BACCS. Nat Commun 6:8021
Jha A, Muallem S (2016) The CAR that drives Ca2+ to Orai1. Sci Signal 9:fs5
Jing J, He L, Sun A, Quintana A, Ding Y, Ma G, Tan P, Liang X, Zheng X, Chen L, Shi X, Zhang SL, Zhong L, Huang Y, Dong MQ, Walker CL, Hogan PG, Wang Y, Zhou Y (2015) Proteomic mapping of ER-PM junctions identifies STIMATE as a regulator of Ca(2)(+) influx. Nat Cell Biol 17:1339–1347

Kawasaki T, Lange I, Feske S (2009) A minimal regulatory domain in the C terminus of STIM1 binds to and activates ORAI1 CRAC channels. Biochem Biophys Res Commun 385:49–54

Korzeniowski MK, Manjarres IM, Varnai P, Balla T (2010) Activation of STIM1-Orai1 involves an intramolecular switching mechanism. Sci Signal 3:ra82

Li Z, Lu J, Xu P, Xie X, Chen L, Xu T (2007) Mapping the interacting domains of STIM1 and Orai1 in Ca2+ release-activated Ca2+ channel activation. J Biol Chem 282:29448–29456

Li Z, Liu L, Deng Y, Ji W, Du W, Xu P, Chen L, Xu T (2011) Graded activation of CRAC channel by binding of different numbers of STIM1 to Orai1 subunits. Cell Res 21:305–315

Liou J, Kim ML, Heo WD, Jones JT, Myers JW, Ferrell JE Jr, Meyer T (2005) STIM is a Ca2+ sensor essential for Ca2+-store-depletion-triggered Ca2+ influx. Curr Biol 15:1235–1241

Liou J, Fivaz M, Inoue T, Meyer T (2007) Live-cell imaging reveals sequential oligomerization and local plasma membrane targeting of stromal interaction molecule 1 after Ca2+ store depletion. Proc Natl Acad Sci U S A 104:9301–9306

Lis A, Peinelt C, Beck A, Parvez S, Monteilh-Zoller M, Fleig A, Penner R (2007) CRACM1, CRACM2, and CRACM3 are store-operated Ca2+ channels with distinct functional properties. Curr Biol 17:794–800

Luik RM, Wu MM, Buchanan J, Lewis RS (2006) The elementary unit of store-operated Ca2+ entry: local activation of CRAC channels by STIM1 at ER-plasma membrane junctions. J Cell Biol 174:815–825

Luik RM, Wang B, Prakriya M, Wu MM, Lewis RS (2008) Oligomerization of STIM1 couples ER calcium depletion to CRAC channel activation. Nature 454:538–542

Ma G, Wei M, He L, Liu C, Wu B, Zhang SL, Jing J, Liang X, Senes A, Tan P, Li S, Sun A, Bi Y, Zhong L, Si H, Shen Y, Li M, Lee MS, Zhou W, Wang J, Wang Y, Zhou Y (2015) Inside-out Ca signalling prompted by STIM1 conformational switch. Nat Commun 6:7826

Malli R, Naghdi S, Romanin C, Graier WF (2008) Cytosolic Ca2+ prevents the subplasmalemmal clustering of STIM1: an intrinsic mechanism to avoid Ca2+ overload. J Cell Sci 121:3133–3139

Maruyama Y, Ogura T, Mio K, Kato K, Kaneko T, Kiyonaka S, Mori Y, Sato C (2009) Tetrameric Orai1 is a teardrop-shaped molecule with a long, tapered cytoplasmic domain. J Biol Chem 284:13676–13685

McNally BA, Yamashita M, Engh A, Prakriya M (2009) Structural determinants of ion permeation in CRAC channels. Proc Natl Acad Sci U S A 106:22516–22521

McNally BA, Somasundaram A, Yamashita M, Prakriya M (2012) Gated regulation of CRAC channel ion selectivity by STIM1. Nature 482:241–245

Mercer JC, Dehaven WI, Smyth JT, Wedel B, Boyles RR, Bird GS, Putney JW Jr (2006) Large store-operated calcium selective currents due to co-expression of Orai1 or Orai2 with the intracellular calcium sensor, Stim1. J Biol Chem 281:24979–14990

Mignen O, Thompson JL, Shuttleworth TJ (2008) Orai1 subunit stoichiometry of the mammalian CRAC channel pore. J Physiol 586:419–425

Misceo D, Holmgren A, Louch WE, Holme PA, Mizobuchi M, Morales RJ, De Paula AM, Stray-Pedersen A, Lyle R, Dalhus B, Christensen G, Stormorken H, Tjonnfjord GE, Frengen E (2014) A dominant STIM1 mutation causes Stormorken syndrome. Hum Mutat 35:556–564

Morin G, Bruechle NO, Singh AR, Knopp C, Jedraszak G, Elbracht M, Bremond-Gignac D, Hartmann K, Sevestre H, Deutz P, Herent D, Nurnberg P, Romeo B, Konrad K, Mathieu-Dramard M, Oldenburg J, Bourges-Petit E, Shen Y, Zerres K, Ouadid-Ahidouch H, Rochette J (2014) Gain-of-Function mutation in STIM1 (p.R304W) is associated with Stormorken syndrome. Hum Mutat 35:1221–1232

Muik M, Frischauf I, Derler I, Fahrner M, Bergsmann J, Eder P, Schindl R, Hesch C, Polzinger B, Fritsch R, Kahr H, Madl J, Gruber H, Groschner K, Romanin C (2008) Dynamic coupling of the putative coiled-coil domain of ORAI1 with STIM1 mediates ORAI1 channel activation. J Biol Chem 283:8014–8022

Muik M, Fahrner M, Derler I, Schindl R, Bergsmann J, Frischauf I, Groschner K, Romanin C (2009) A cytosolic homomerization and a modulatory domain within STIM1 C terminus determine coupling to ORAI1 channels. J Biol Chem 284:8421–8426

Muik M, Fahrner M, Schindl R, Stathopulos P, Frischauf I, Derler I, Plenk P, Lackner B, Groschner K, Ikura M, Romanin C (2011) STIM1 couples to ORAI1 via an intramolecular transition into an extended conformation. EMBO J 30:1678–1689

Navarro-Borelly L, Somasundaram A, Yamashita M, Ren D, Miller RJ, Prakriya M (2008) STIM1-Orai1 interactions and Orai1 conformational changes revealed by live-cell FRET microscopy. J Physiol 586:5383–5401

Nesin V, Wiley G, Kousi M, Ong EC, Lehmann T, Nicholl DJ, Suri M, Shahrizaila N, Katsanis N, Gaffney PM, Wierenga KJ, Tsiokas L (2014) Activating mutations in STIM1 and ORAI1 cause overlapping syndromes of tubular myopathy and congenital miosis. Proc Natl Acad Sci U S A 111:4197–4202

Pacheco J, Dominguez L, Bohorquez-Hernandez A, Asanov A, Vaca L (2016) A cholesterol-binding domain in STIM1 modulates STIM1-Orai1 physical and functional interactions. Sci Rep 6:29634

Palty R, Raveh A, Kaminsky I, Meller R, Reuveny E (2012) SARAF inactivates the store operated calcium entry machinery to prevent excess calcium refilling. Cell 149:425–438

Palty R, Stanley C, Isacoff EY (2015) Critical role for Orai1 C-terminal domain and TM4 in CRAC channel gating. Cell Res 25:963–980

Parekh AB, Putney JW Jr (2005) Store-operated calcium channels. Physiol Rev 85:757–810

Park CY, Hoover PJ, Mullins FM, Bachhawat P, Covington ED, Raunser S, Walz T, Garcia KC, Dolmetsch RE, Lewis RS (2009) STIM1 clusters and activates CRAC channels via direct binding of a cytosolic domain to Orai1. Cell 136:876–890

Peinelt C, Vig M, Koomoa DL, Beck A, Nadler MJ, Koblan-Huberson M, Lis A, Fleig A, Penner R, Kinet JP (2006) Amplification of CRAC current by STIM1 and CRACM1 (Orai1). Nat Cell Biol 8:771–773

Penna A, Demuro A, Yeromin AV, Zhang SL, Safrina O, Parker I, Cahalan MD (2008) The CRAC channel consists of a tetramer formed by Stim-induced dimerization of Orai dimers. Nature 456:116–120

Prakriya M, Feske S, Gwack Y, Srikanth S, Rao A, Hogan PG (2006) Orai1 is an essential pore subunit of the CRAC channel. Nature 443:230–233

Roos J, DiGregorio PJ, Yeromin AV, Ohlsen K, Lioudyno M, Zhang S, Safrina O, Kozak JA, Wagner SL, Cahalan MD, Velicelebi G, Stauderman KA (2005) STIM1, an essential and conserved component of store-operated Ca2+ channel function. J Cell Biol 169:435–445

Saotome K, Singh AK, Yelshanskaya MV, Sobolevsky AI (2016) Crystal structure of the epithelial calcium channel TRPV6. Nature 534:506–511

Schindl R, Bergsmann J, Frischauf I, Derler I, Fahrner M, Muik M, Fritsch R, Groschner K, Romanin C (2008) 2-aminoethoxydiphenyl borate alters selectivity of Orai3 channels by increasing their pore size. J Biol Chem 283:20261–20267

Schindl R, Frischauf I, Bergsmann J, Muik M, Derler I, Lackner B, Groschner K, Romanin C (2009) Plasticity in Ca2+ selectivity of Orai1/Orai3 heteromeric channel. Proc Natl Acad Sci U S A 106:19623–19628

Scrimgeour N, Litjens T, Ma L, Barritt GJ, Rychkov GY (2009) Properties of Orai1 mediated store-operated current depend on the expression levels of STIM1 and Orai1 proteins. J Physiol 587:2903–2918

Scrimgeour NR, Wilson DP, Barritt GJ, Rychkov GY (2014) Structural and stoichiometric determinants of Ca release-activated Ca (CRAC) channel Ca-dependent inactivation. Biochim Biophys Acta 1838:1281–1287

Sharma S, Quintana A, Findlay GM, Mettlen M, Baust B, Jain M, Nilsson R, Rao A, Hogan PG (2013) An siRNA screen for NFAT activation identifies septins as coordinators of store-operated Ca2+ entry. Nature 499:238–242

Smyth JT, Dehaven WI, Bird GS, Putney JW Jr (2008) Ca2+-store-dependent and -independent reversal of Stim1 localization and function. J Cell Sci 121:762–772

Soboloff J, Spassova MA, Hewavitharana T, He LP, Xu W, Johnstone LS, Dziadek MA, Gill DL (2006) STIM2 is an inhibitor of STIM1-mediated store-operated Ca(2+) entry. Curr Biol 16:1465–1470

Soboloff J, Rothberg BS, Madesh M, Gill DL (2012) STIM proteins: dynamic calcium signal transducers. Nat Rev Mol Cell Biol 13:549–565

Srikanth S, Jung HJ, Kim KD, Souda P, Whitelegge J, Gwack Y (2010) A novel EF-hand protein, CRACR2A, is a cytosolic Ca2+ sensor that stabilizes CRAC channels in T cells. Nat Cell Biol 12:436–446

Stathopulos PB, Ikura M (2010) Partial unfolding and oligomerization of stromal interaction molecules as an initiation mechanism of store operated calcium entry. Biochem Cell Biol 88:175–183

Stathopulos PB, Li GY, Plevin MJ, Ames JB, Ikura M (2006) Stored Ca2+ depletion-induced oligomerization of STIM1 via the EF-SAM region: an initiation mechanism for capacitive Ca2 + entry. J Biol Chem 281:35855–35862

Stathopulos PB, Zheng L, Li GY, Plevin MJ, Ikura M (2008) Structural and mechanistic insights into STIM1-mediated initiation of store-operated calcium entry. Cell 135:110–122

Stathopulos PB, Schindl R, Fahrner M, Zheng L, Gasmi-Seabrook GM, Muik M, Romanin C, Ikura M (2013) STIM1/Orai1 coiled-coil interplay in the regulation of store-operated calcium entry. Nat Commun 4:2963

Tirado-Lee L, Yamashita M, Prakriya M (2015) Conformational changes in the Orai1 C-terminus evoked by STIM1 binding. PLoS One 10:e0128622

Vig M, Beck A, Billingsley JM, Lis A, Parvez S, Peinelt C, Koomoa DL, Soboloff J, Gill DL, Fleig A, Kinet JP, Penner R (2006a) CRACM1 multimers form the ion-selective pore of the CRAC channel. Curr Biol 16:2073–2079

Vig M, Peinelt C, Beck A, Koomoa DL, Rabah D, Koblan-Huberson M, Kraft S, Turner H, Fleig A, Penner R, Kinet JP (2006b) CRACM1 is a plasma membrane protein essential for store-operated Ca2+ entry. Science 312:1220–1223

Wang X, Wang Y, Zhou Y, Hendron E, Mancarella S, Andrake MD, Rothberg BS, Soboloff J, Gill DL (2014) Distinct Orai-coupling domains in STIM1 and STIM2 define the Orai-activating site. Nat Commun 5:3183

Wu MM, Buchanan J, Luik RM, Lewis RS (2006) Ca2+ store depletion causes STIM1 to accumulate in ER regions closely associated with the plasma membrane. J Cell Biol 174:803–813

Wu J, Yan Z, Li Z, Yan C, Lu S, Dong M, Yan N (2015) Structure of the voltage-gated calcium channel Cav1.1 complex. Science 350:aad2395

Xu P, Lu J, Li Z, Yu X, Chen L, Xu T (2006) Aggregation of STIM1 underneath the plasma membrane induces clustering of Orai1. Biochem Biophys Res Commun 350:969–976

Yamashita M, Yeung PS, Ing CI, McNally BA, Pomes R, Prakrya M (2017) STIM1 activates CRAC channels through rotation of the pore helix to open a hydrophobic gate. Nat Commun. doi:10.1038/ncomms14512

Yang X, Jin H, Cai X, Li S, Shen Y (2012) Structural and mechanistic insights into the activation of Stromal interaction molecule 1 (STIM1). Proc Natl Acad Sci U S A 109:5657–5662

Yen M, Lokteva LA, Lewis RS (2016) Functional analysis of Orai1 concatemers supports a hexameric stoichiometry for the CRAC channel. Biophys J 111:1897–1907

Yeromin AV, Zhang SL, Jiang W, Yu Y, Safrina O, Cahalan MD (2006) Molecular identification of the CRAC channel by altered ion selectivity in a mutant of Orai. Nature 443:226–229

Yeung PS, Yamashita M, Prakriya M (2016) Pore opening mechanism of CRAC channels. Cell Calcium. doi:10.1016/j.ceca.2016.12.006

Yu F, Sun L, Hubrack S, Selvaraj S, Machaca K (2013) Intramolecular shielding maintains the ER Ca(2)(+) sensor STIM1 in an inactive conformation. J Cell Sci 126:2401–2410

Yuan JP, Zeng W, Dorwart MR, Choi YJ, Worley PF, Muallem S (2009) SOAR and the polybasic STIM1 domains gate and regulate Orai channels. Nat Cell Biol 11:337–343

Zalk R, Clarke OB, des Georges A, Grassucci RA, Reiken S, Mancia F, Hendrickson WA, Frank J, Marks AR (2015) Structure of a mammalian ryanodine receptor. Nature 517:44–49

Zhang SL, Yu Y, Roos J, Kozak JA, Deerinck TJ, Ellisman MH, Stauderman KA, Cahalan MD (2005) STIM1 is a Ca2+ sensor that activates CRAC channels and migrates from the Ca2+ store to the plasma membrane. Nature 437:902–905

Zhang SL, Yeromin AV, Zhang XH, Yu Y, Safrina O, Penna A, Roos J, Stauderman KA, Cahalan MD (2006) Genome-wide RNAi screen of Ca(2+) influx identifies genes that regulate Ca(2+) release-activated Ca(2+) channel activity. Proc Natl Acad Sci U S A 103:9357–9362

Zhang SL, Kozak JA, Jiang W, Yeromin AV, Chen J, Yu Y, Penna A, Shen W, Chi V, Cahalan MD (2008) Store-dependent and -independent modes regulating Ca2+ release-activated Ca2+ channel activity of human Orai1 and Orai3. J Biol Chem 283:17662–17671

Zhang SL, Yeromin AV, Hu J, Amcheslavsky A, Zheng H, Cahalan MD (2011) Mutations in Orai1 transmembrane segment 1 cause STIM1-independent activation of Orai1 channels at glycine 98 and channel closure at arginine 91. Proc Natl Acad Sci U S A 108:17838–17843

Zhang W, Zhang X, Gonzalez-Cobos JC, Stolwijk JA, Matrougui K, Trebak M (2015) Leukotriene-C4 synthase, a critical enzyme in the activation of store-independent Orai1/Orai3 channels, is required for neointimal hyperplasia. J Biol Chem 290:5015–5027

Zheng L, Stathopulos PB, Li GY, Ikura M (2008) Biophysical characterization of the EF-hand and SAM domain containing Ca2+ sensory region of STIM1 and STIM2. Biochem Biophys Res Commun 369:240–246

Zheng L, Stathopulos PB, Schindl R, Li GY, Romanin C, Ikura M (2011) Auto-inhibitory role of the EF-SAM domain of STIM proteins in store-operated calcium entry. Proc Natl Acad Sci U S A 108:1337–1342

Zhou Y, Meraner P, Kwon HT, Machnes D, Oh-hora M, Zimmer J, Huang Y, Stura A, Rao A, Hogan PG (2010a) STIM1 gates the store-operated calcium channel ORAI1 in vitro. Nat Struct Mol Biol 17:112–116

Zhou Y, Ramachandran S, Oh-Hora M, Rao A, Hogan PG (2010b) Pore architecture of the ORAI1 store-operated calcium channel. Proc Natl Acad Sci U S A 107:4896–4901

Zhou Y, Srinivasan P, Razavi S, Seymour S, Meraner P, Gudlur A, Stathopulos PB, Ikura M, Rao A, Hogan PG (2013) Initial activation of STIM1, the regulator of store-operated calcium entry. Nat Struct Mol Biol 20:973–981

Zhou Y, Cai X, Loktionova NA, Wang X, Nwokonko RM, Wang X, Wang Y, Rothberg BS, Trebak M, Gill DL (2016) The STIM1-binding site nexus remotely controls Orai1 channel gating. Nat Commun 7:13725

Zhou Y, Cai X, Nwokonko RM, Loktionova NA, Wang Y, Gill DL (2017) The STIM-Orai coupling interface and gating of the Orai1 channel. Cell Calcium. doi:10.1016/j.ceca.2017.01.001

The STIM-Orai Pathway: Conformational Coupling Between STIM and Orai in the Activation of Store-Operated Ca^{2+} Entry

5

Robert M. Nwokonko, Xiangyu Cai, Natalia A. Loktionova, Youjun Wang, Yandong Zhou, and Donald L. Gill

Abstract

Store-operated Ca^{2+} entry fulfills a crucial role in controlling Ca^{2+} signals in almost all cells. The Ca^{2+}-sensing stromal interaction molecule (STIM) proteins in the endoplasmic reticulum (ER) undergo complex conformational changes in response to depleted ER luminal Ca^{2+}, allowing them to unfold and become trapped in ER-plasma membrane (PM) junctions. Dimers of STIM proteins trap and gate the plasma membrane Orai Ca^{2+} channels within these junctions to generate discrete zones of high Ca^{2+} and regulate sensitive Ca^{2+}-dependent intracellular signaling pathways. The STIM-Orai activating region (SOAR) of STIM1 becomes exposed upon store depletion and promotes trapping of Orai1 at the PM. Residue Phe-394 within SOAR forms an integral part of the high-affinity Orai1-interacting site. Our results demonstrate that only a single active site within the dimeric SOAR domain of STIM1 is required for the activation of Orai1 channel activity. This unimolecular model is strongly supported by evidence of variable STIM1:Orai1 stoichiometry reported in many studies. We hypothesize that unimolecular coupling promotes cross-linking of channels, localizing Ca^{2+} signals, and regulating channel activity. We have also identified a key "nexus" region in Orai1 near the C-terminal STIM1-binding site that can be mutated to constitutively activate Ca^{2+} entry, mimicking STIM1 activated channels. This suggests that STIM1 mediates gating of Orai1 in an allosteric manner via interaction with the Orai1 C-terminus alone. This model suggests the dual role of STIM1 in regulating both localization and gating of Orai1 channels

R.M. Nwokonko • X. Cai • N.A. Loktionova • Y. Zhou (✉) • D.L. Gill (✉)
Department of Cellular and Molecular Physiology, The Pennsylvania State University College of Medicine, Hershey, PA 17033, USA
e-mail: zhouyd@psu.edu; dongill@psu.edu

Y. Wang
Beijing Key Laboratory of Gene Resources and Molecular Development, College of Life Sciences, Beijing Normal University, Beijing 100875, People's Republic of China

K. Groschner et al. (eds.), *Store-Operated Ca^{2+} Entry (SOCE) Pathways*, Advances in Experimental Medicine and Biology 993, DOI 10.1007/978-3-319-57732-6_5

and has important implications for the regulation of SOCE-mediated downstream signaling and the kinetics of channel activation.

Keywords
Calcium • Channels • Orai1 • STIM1 • Channel gating • Calcium sensor

5.1 Introduction

Store-operated Ca^{2+} entry regulates a diverse set of signaling pathways and is essential in virtually all eukaryotes (Prakriya and Lewis 2015; Amcheslavsky et al. 2015; Putney et al. 2016). Transient depletion of ER Ca^{2+} stores is sensed by the Ca^{2+}-binding EF hands of the single-pass transmembrane protein STIM1 (Soboloff et al. 2012). Upon sensing these changes, STIM1 undergoes an intricate unfolding process that allows the lysine-rich C-terminus to elongate and interact with PM phosphatidylinositol 4,5-bisphosphate residues (Prakriya and Lewis 2015; Amcheslavsky et al. 2015; Soboloff et al. 2012; Park et al. 2009; Korzeniowski et al. 2009; Shim et al. 2015). Interactions with these phospholipids lead to STIM1 trapping within ER-PM junctions and promote subsequent tethering and activation of Orai1 channels that passively migrate through the PM (Park et al. 2009). This dynamic coupling process is critical to the formation of discrete Ca^{2+} microdomains within ER-PM junctions (Prakriya and Lewis 2015; Amcheslavsky et al. 2015; Soboloff et al. 2012). Ca^{2+} signals generated through active Orai1 channels control a diverse array of cellular processes including cell growth, cell motility, transcription, secretion, and the maintenance of intracellular Ca^{2+} homeostasis. Ca^{2+} actively pumped back into the ER via the sarco-/endoplasmic reticulum ATPase (SERCA) pump allows stores to be replenished. In recent years, much progress has been made toward deciphering the key details behind the STIM1-Orai1 interaction, specifically regarding Orai1 activation. What is unclear, however, is precisely how STIM1 coupling to Orai1 leads to channel gating and subsequent Ca^{2+} entry (Amcheslavsky et al. 2015; Shim et al. 2015; Rothberg et al. 2013). We will use this chapter to describe some recent advances in our laboratory toward better understanding STIM1-Orai1 coupling and channel gating.

5.2 STIM Proteins and the STIM-Orai Activating Region

Two major components of the store-operated Ca^{2+} entry pathway are the membrane-spanning ER STIM proteins (STIM1 and STIM2) and the family of Orai Ca^{2+} entry channels (Soboloff et al. 2012). The STIM proteins naturally exist as dimers at resting ER Ca^{2+} concentrations. In their inactive forms, the C-terminal cytosolic portions of STIM proteins are folded in upon themselves, occluding the Orai1-binding domain and preventing trapping within ER-PM junctions (Fig. 5.1) (Muik et al. 2009; Covington et al. 2010; Zhou et al. 2013). Decreases in ER luminal Ca^{2+} concentration are recognized by the two N-terminal Ca^{2+}-binding

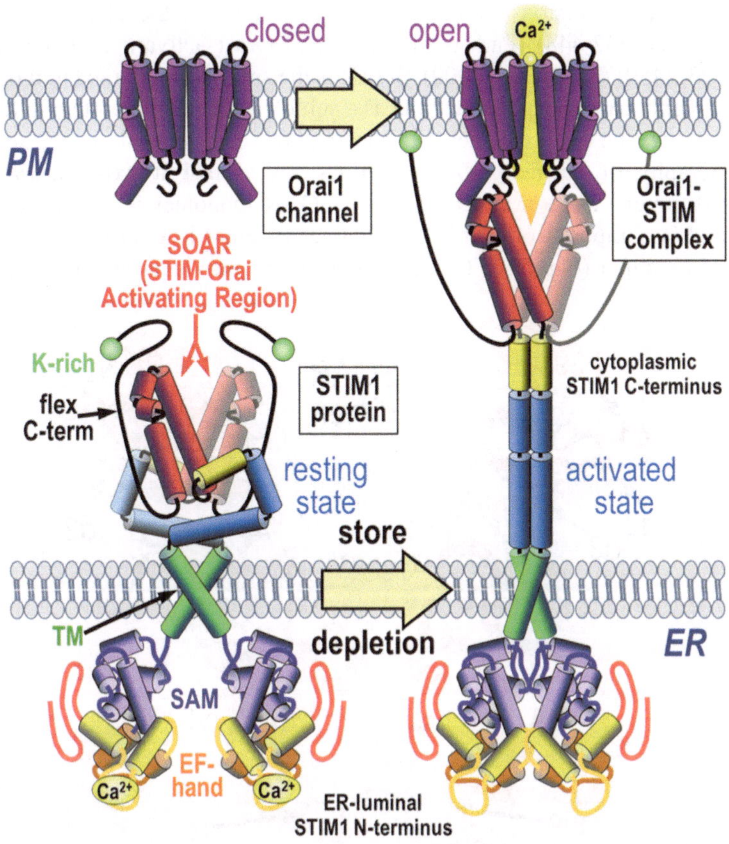

Fig. 5.1 The STIM1-Orai1 interaction mechanism. The inactive form (*left*) of the STIM1 dimer remains folded in on itself at resting Ca^{2+} concentrations in the endoplasmic reticulum (ER). High concentrations of ER luminal Ca^{2+} maintain inactivation of the STIM proteins through binding to the EF-hand domain of STIM. The closed form of Orai1 (*left*) can freely diffuse throughout the plasma membrane (PM). Upon ER Ca^{2+} store depletion, STIM proteins unfold and extend to associate with the PM and bind Orai1 directly (*right*) to cause Ca^{2+} entry into ER-PM junctions. The cytosolic-facing segments of STIM1 include the lysine-rich (K-rich) C-terminal domain that binds to phospholipids in the PM and the flexible C-terminal domains (flex C-term) that aid in shielding the STIM-Orai activating region (SOAR) during store replete conditions. The luminal/intramembranous segments include the transmembrane region (TM) that transmits conformational changes induced by the sterile-α-motif (SAM) and Ca^{2+}-binding EF-hand domain

EF-hand domains that reside in the ER lumen. Transient ER luminal Ca^{2+} depletion results in a major STIM protein conformational change whereby the luminal N-termini of the dimeric proteins become more closely associated. This conformational change induces the C-terminal cytosolic portion of the dimeric STIM protein to unfold and extend into the cytosol and be able to associate with the PM in ER-PM junctions (Ma et al. 2015). The C-termini of the dimeric STIM proteins are highly conserved between the two STIM isoforms particularly in a region termed the STIM-Orai activating region (SOAR; 344-442 in STIM1) (Yuan et al. 2009).

Slightly extended versions of this region are known as the channel-activating domain (CAD; 342-448) (Park et al. 2009) and Orai1-activating small fragment (OASF; 233-450) (Muik et al. 2009). SOAR is the smallest functional unit of STIM and remarkably can be expressed alone as a soluble cytoplasmic protein that is able to fully activate the Orai1 channel (Yuan et al. 2009). Structures derived from X-ray crystallography of purified SOAR peptides reveal that it exists as a highly structured dimer (Yang et al. 2012). Each SOAR monomer is mainly helical in nature and is composed of four core α-helices or eight in total between the two subunits in a dimer (Fig. 5.2). This core helical structure is purportedly conserved

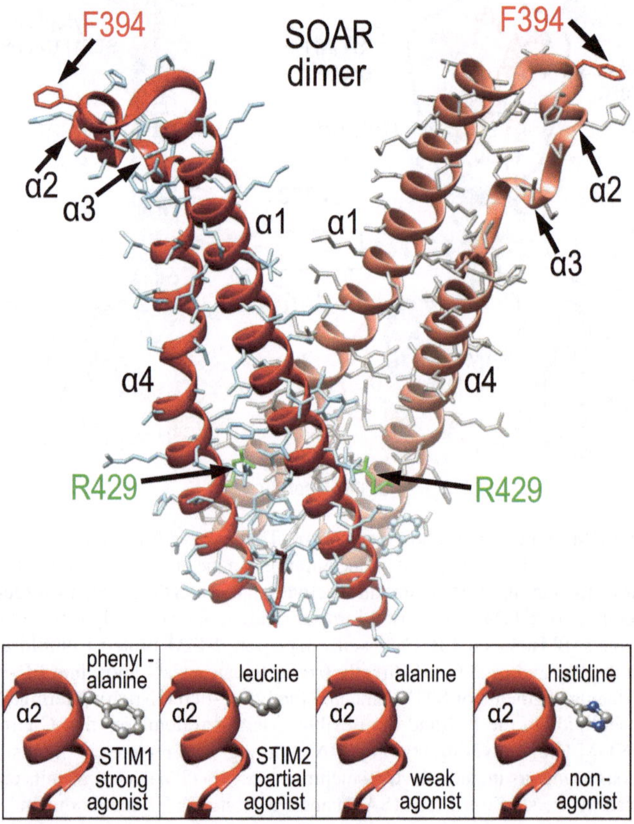

Fig. 5.2 The crystal structure of the dimeric SOAR domain. The dimeric SOAR domain comprises four α-helical regions (α1, α2, α3, α4). The purported strong Orai1 activation site resides between helixes α1 and α3 and includes the Phe-394 residue. The α4 helical region is critical for maintaining the structure of SOAR dimers, as mutations of the α4-residing R429 residue prevent SOCE and initiate aberrant STIM1 unfolding, accounting for the loss of function in STIM1 observed in human patients with the R429C STIM1 mutation. Our studies have substituted residue Phe-394 to measure its importance to the STIM1-Orai1 interaction (*lower panel*). Changing the residue to leucine (the equivalent residue in STIM2), alanine, or histidine results in abrogation of the interaction with Orai1 and prevention of SOCE. The histidine mutation results in a STIM1 protein completely devoid of Orai1 channel gating activity

within the C-terminus of full-length extended STIM1, and empirical evidence suggests that it is critical for dimer formation and maintaining the dimeric interaction between monomers of STIM1 (Muik et al. 2009; Covington et al. 2010; Zhou et al. 2013; Yang et al. 2012). The exposed SOAR domain in the activated STIM1 protein is free to bind Orai1 at the PM. Binding to Orai1 channels causes transient trapping and activation of the channels within ER-PM junctions (Zhou et al. 2013; Yang et al. 2012; Ma et al. 2015). Ca^{2+} brought in through Orai1 trapped in ER-PM junctions potentiates calcineurin-induced NFAT dephosphorylation, which regulates transcription of a large number of genes in many different cell types (Kar and Parekh 2015; Zhou et al. 2015a).

5.3 Structure and Function of the Orai Channel

The structure of the *Drosophila* Orai channel (dOrai) is also now known in some detail (Fig. 5.3). X-ray crystallographic structures of the purified PM localized dOrai channel reveal it to be a homo-hexamer (Hou et al. 2012), with each monomeric Orai1 subunit being composed of four transmembrane (TM)-spanning domains (TM1–TM4) (Hou et al. 2012). The central pore of the hexamer is lined by the N-terminal TM1 (Hou et al. 2012). TM2 and TM3 are tightly packed around the pore, while the C-terminal TM4 is less tightly packed and in fact extends partially into the cytosol and is the strong binding site for the STIM proteins (Park et al. 2009; Hou et al. 2012). Unfortunately there are not as yet any solved structures of STIM bound to Orai1.

Considerable uncertainty has surrounded the question of whether the Orai channel functions as a hexamer. Earlier evidence from a number of labs had suggested the channel might be tetrameric (see Cai et al. 2016). We undertook an in-depth study of this question using a series of concatenated Orai1 constructs. We were surprised to observe that dimers, trimers, tetramers, pentamers, and hexamers of Orai1 all behaved similarly to expression of Orai1 monomers (Cai et al. 2016). Thus all concatemers were expressed almost exclusively at the PM, and all gave rise to store-operated Ca^{2+} entry and Ca^{2+} release-activated Ca^{2+} (CRAC) currents indistinguishable from those observed after Orai1 monomer expression. However, using Orai1 concatemeric constructs in which we systematically replaced wild-type Orai1 subunits with subunits defective in the crucial Glu-106 selectivity filter residue, we observed that the hexamer is a special case. Thus, function of the concatemeric hexamer is sensitive to replacement of each of the six residues with the pore-dead E106A mutation indicating that all the residues are required for channel function. Interestingly, the function of the other shorter concatemers (pentamers, tetramers, trimers, and dimers) resulted from the first two N-terminal subunits being unexpectedly fed into a hexameric structure composed of a "trimer of dimers." Our results strongly suggested that the functional Orai1 channel is indeed a hexamer (Cai et al. 2016) agreeing with other studies published at the same time (Yen et al. 2016)

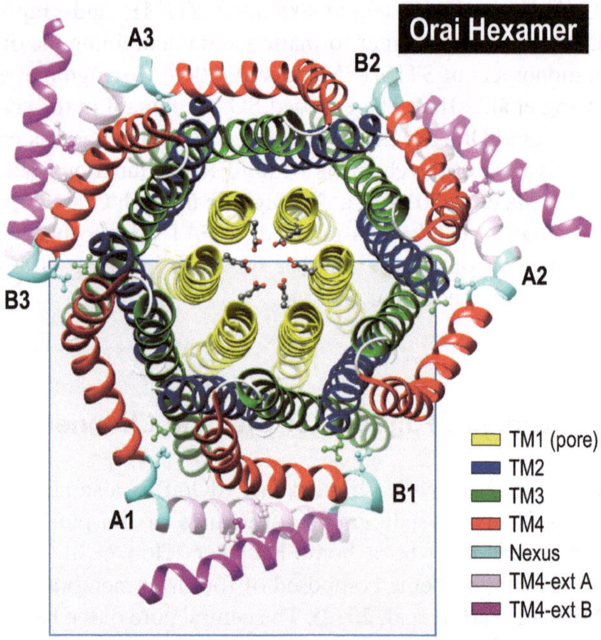

Fig. 5.3 Structure of the *Drosophila* Orai (dOrai) channel. The dOrai channel has close homology with the mammalian Orai1 channel and likely very similar structure. The crystal structure does not include the N-terminal cytosolic sequence (equivalent to the first 60 residues in human Orai1) which is largely redundant to channel function and activation by STIM1. The Orai channel largely has sixfold symmetry, with six subunits each with identical sequences. Each subunit has four transmembrane helices (TM1 through TM4). The TM1 helix forms the channel pore (*yellow*), and the outer channel selectivity filter glutamate is shown (E178 in dOrai1, or E106 in human Orai1). The other TM helices are packed around the central pore helix, with TM4 (*red*) toward the outside. The C-terminal helical extension of TM4 (TM4-ext) extends at the periphery of the channel and is joined to the TM4 through a "nexus" region (*light blue*). This nexus includes a hinge that is configured in two different ways: three TM4 helices are in a "straight" hinge configuration (A1, A2, and A3), and the three intervening subunits are in a bent configuration (B1, B2, and B3) as shown in the diagram. Thus, the Orai hexamer includes three dimers, each dimer comprising one "A" and one "B" form of the channel. As are result, the two adjacent TM4-ext A and B helices in each dimer are closely aligned in an antiparallel manner and linked through hydrophobic reactions between the L273 and L276 residues. These residues and the outer Orai TM4-ext configuration which constitutes the STIM-binding site are described further in Fig. 5.5

5.4 Coupling Models for the STIM-Orai Interaction

The precise features of the STIM-Orai coupling interaction are still unclear and currently a major focus of many labs (Prakriya and Lewis 2015; Amcheslavsky et al. 2015; Shim et al. 2015; Rothberg et al. 2013; Derler et al. 2016). Nevertheless, a number of studies using a combination of approaches including analysis of fluorescently labeled Orai and STIM fragments, basic electrophysiology and biochemistry, have led to some significant advances in understanding the stoichiometry of STIM1-Orai1 coupling. Interestingly, these experiments have led to the conclusion that the stoichiometry of interaction can vary, with maximum activation occurring at a STIM1:Orai1 ratio of 2:1 (Scrimgeour et al. 2009; Hoover and Lewis 2011; Li et al. 2011). However, recent structural studies utilizing partial SOAR and C-terminal STIM1-binding Orai1 TM4 fragments suggested that binding occurs in a bimolecular fashion, whereby a STIM1 dimer will bind to two adjacent Orai1 subunits within the same Orai1 hexamer (Stathopulos et al. 2013; Fahrner et al. 2014; Maus et al. 2015). This model would indicate a stoichiometry of 1:1, yet it cannot explain the variable stoichiometry seen in previous experiments. Follow-up studies in our laboratory are described in detail below and lend support to the unimolecular model, indicating a 2:1 STIM1:Orai1 stoichiometry.

To better understand the STIM1-Orai1 coupling interaction, we undertook a comparative analysis of the SOAR regions of STIM1 and STIM2 because of the natural differences in affinity for Orai1 between STIM1 and STIM2, with STIM1 being the higher-affinity isoform. We observed almost complete homology between the two isoforms except at position Phe-394 in STIM1 (Fig. 5.2) and the equivalent position Leu-485 in STIM2. Experiments in our lab indicate that this residue plays an important role in the reduced affinity that STIM2 has for Orai1 (Wang et al. 2014). Mutation of Phe-394 in STIM1 to alanine resulted in a strong reduction of channel activity and strong decreases in the interaction. Surprisingly, the relatively conservative mutation to histidine (Fig. 5.2, lower) almost completely abolished STIM1 binding to Orai1 and completely abolished Ca^{2+} entry (Wang et al. 2014), yet had no effect on the resting state of STIM1 or its ability to be activated by transient ER Ca^{2+} depletion.

5.5 SOAR-Derived Probes to Understand STIM-Orai Coupling

Our lab has developed novel SOAR probes that included the mutation at residue Phe-394. Using these probes, we identified this residue as being a vital component of the Orai1-binding site (Zhou et al. 2015b). By genetically encoding a flexible linker between each SOAR subunit, we were able to make efficiently expressed SOAR dimers and precisely control the stoichiometry of mutant to wild-type SOAR within a construct (Fig. 5.4a). We constructed a series of concatenated dimers containing either one or both of the Phe-394 residues mutated to histidine in each SOAR peptide and used these to test the foundations of the bimolecular interaction model. This model would predict that heterodimers containing one subunit with the F394H mutation would be nonfunctional; however to our surprise we observed similar activity between wild-type homodimers and mutant heterodimers (Zhou et al. 2015b). This result suggests that Orai1 channel activity is dependent on only one functionally active site within a SOAR dimer, which is not predicted by the bimolecular model.

While SOAR is always a dimer in both STIM1 and concatenated peptides, only one functional subunit appears required for interaction and full activation of Orai1 channels. Nevertheless, dimerization of STIM1 is key to its function in activating

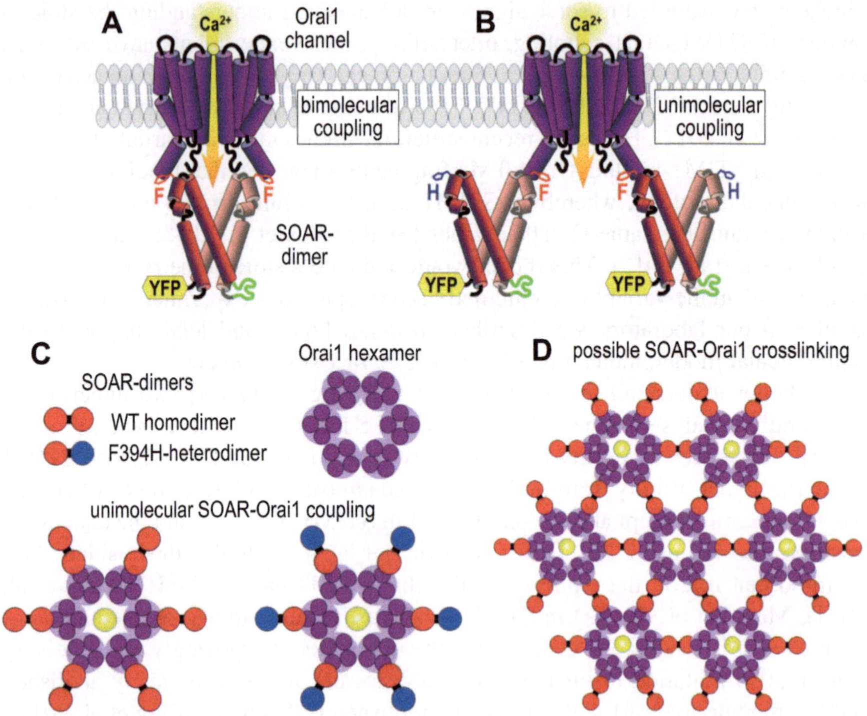

Fig. 5.4 An alternative unimolecular STIM1-Orai1 coupling model. Previous studies have suggested a bimolecular STIM1-Orai1 model for coupling (**a**) model, in which the two active sites of the SOAR dimer unit must both interact with two adjacent Orai1 subunits of the hexamer in order to open the channel. Only two Orai1 subunits from the Orai1 channel hexamer are shown. The SOAR dimer is depicted as the concatemer with two Phe-394 wild-type residues with YFP attached. In our unimolecular STIM1-Orai1 model for coupling (**b**), just one of the SOAR subunits in the dimer is necessary to bind to a single subunit of the Orai1 hexamer to gate the channel. The diagram shows a concatenated SOAR dimer with one wild type and one mutated F394H SOAR subunit. (**c**) Depiction of the theoretical interaction of SOAR dimers with the hexameric Orai1 channel. A unimolecular model for coupling envisages that a wild-type SOAR homodimer and the heterodimer of SOAR with one SOAR-F394H mutant unit (*blue*) would both be able to interact with and completely activate Orai1. (**d**) Possible clustering of hexameric Orai1 channels mediated by cross-linking by wild-type SOAR dimers

Orai1. Mutations that alter dimerization have a major impact on the ability of STIM1 to couple with and activate Orai1 (Yang et al. 2012). An interesting mutation in the SOAR segment (Fig. 5.2), R429C, reveals that this residue is crucial for maintaining the structural integrity of SOAR even though mutating it does not cause dimer dissociation. Patients expressing STIM1 containing the naturally occurring but rare R429C mutation experience a complete loss in STIM1 function (Maus et al. 2015). This residue was initially identified as being structurally important in the crystal structure of SOAR, as it resides within the α4-helix within an area involved in maintaining intermolecular protein interactions (Fig. 5.2). The occurrence of the mutant in patients highlighted its importance in promoting the stability of the SOAR dimer region, which has been shown to be critical for SOAR binding to Orai1 channels (Yang et al. 2012; Stathopulos et al. 2013; Maus et al. 2015). Subsequent experiments from constructs expressing STIM1 (R429C) demonstrated that mutant had a dual effect on blocking STIM1-Orai1 coupling and also caused constitutive localization of STIM1 into ER-PM junctions (Maus et al. 2015). This suggests that the Arg-429 residue is critical to maintaining the resting folded state of STIM1 and prevents STIM1 from unfolding without the appropriate signal from depleted ER Ca^{2+} stores.

The results from our studies using the F394H mutation in SOAR indicate that its mode of function is entirely different from R429C (Zhou et al. 2015b). While the R429C mutant is constitutively aggregated into ER-PM junctions, STIM1 F394H protein maintains its folded state until stores are depleted. Only after store depletion does the F394H mutant move into junctions where it is functionally deficient in coupling to Orai1 channels which is necessary for channel gating. Based on the crystal structure, the Phe-394 residue is located toward the apex of each SOAR monomer (Fig. 5.2), whereas the Arg-429 residue is located deep within the structure and appears to be solely involved in intramolecular interaction between peptide helices (Fig. 5.2).

Experiments utilizing novel SOAR probes suggest a coupling mechanism entirely different than that previously proposed. The previous bimolecular model dictates that SOAR dimers bind across two adjacent Orai1 subunits within the same hexamer to activate the channel (Fig. 5.4a). This was supported by NMR studies that used shorter fragments of the STIM-Orai1-binding domain and a portion of the C-terminus STIM1 binding site of Orai1 (Stathopulos et al. 2013; Fahrner et al. 2014). These studies using short SOAR-derived peptides, however, selected regions of SOAR that were unable to activate Orai1 when co-expressed and did not include the high-affinity region surrounding the Phe-394 residue that was shown to be critical for STIM1-Orai1 coupling in our experiments (Stathopulos et al. 2013; Fahrner et al. 2014). Results from our lab that utilized SOAR concatemeric dimers containing F394H mutations to one or both subunits suggest that a "unimolecular" interaction between SOAR/STIM and Orai1 is necessary and sufficient for channel coupling and gating (Fig. 5.4b). These results were the first demonstration that a single binding site within a SOAR dimer is sufficient for binding to adjacent Orai1 hexamers and not within the same hexamer as predicted by the "bimolecular model."

To examine the actual binding in more detail, and to clarify the unimolecular interaction model, we probed the ability of each SOAR dimer to independently interact with the C-terminal STIM1-binding site on TM4 of Orai1 channels. The cytosolic C-terminal peptide of Orai1 has long been established as the strong binding site for STIM1/SOAR (Park et al. 2009; McNally et al. 2013). We designed a novel construct encompassing 35 amino acids from the C-terminal helix (Orai1CT; residues 267-301) labeled with CFP. Within the N-terminus of CFP, we encoded a PM-directed single transmembrane-spanning helical peptide (PMP) and expressed these constructs in HEK cells along with our assortment of YFP-labeled SOAR concatemer probes (constructs comprising either two wild-type binding domains, one mutant F394H in either domain, or a double mutant

F394H) (Fig. 5.4b, c) (Zhou et al. 2015b). We used quantitative instrument-independent Förster resonance energy transfer efficiency (E-FRET) to analyze the proximity between PMP-CFP-Orai1CT and the different YFP-SOAR concatemeric constructs. To our surprise, we found that homodimers of the wild-type SOAR-binding domain bound roughly two PMP-CFP-Orai1CT molecules, whereas the heterodimers were only able to bind to one. These results definitively show that SOAR dimers can bind across at least two separate STIM-binding sites located on the C-termini of Orai1 subunits.

Our results from the quantitative E-FRET experiments led us to question the need for a second site for STIM1 to bind on Orai1 channels and strongly corroborated earlier functional studies undertaken in our lab. Interestingly, if a SOAR dimer within full-length STIM1 or soluble SOAR concatemers can bind across two Orai1 C-termini, then they may play a role in bridging separate Orai1 channels within ER-PM junctions. This bridge between Orai1 channels may also be a way of cooperatively regulating Orai1 channel activity, both through activation and inactivation, although there is no data to support this hypothesis yet. We hypothesize, however, that the SOAR domain within STIM1 can cross-link adjacent Orai1 channels to promote channel clustering (Fig. 5.4d). Along with our E-FRET data supporting the notion that SOAR dimers can bind across two Orai1 C-termini, other labs have previously shown that the longer CAD fragment of STIM1 could cluster Orai1 channels as indicated by electron microscopy (Park et al. 2009). Another study utilized freeze-fracture electron microscopy to quantitatively measure the lateral distances between Orai1 channels and STIM1 after store depletion. The authors found that the mean space between Orai1 channels in ER-PM junctions was very similar to the mean distance (13–18 nm) between STIM1 in the ER, implying that there may be a mechanism to promote organization of Orai1 and STIM in these junctions (Perni et al. 2015). Based on our results, we would conclude that it is the act of STIM1 binding to Orai1 that can promote organization into clusters within the ER-PM junction (Kar and Parekh 2015; Zhou et al. 2015a).

As mentioned earlier, previous studies have noted the variable stoichiometry between STIM1 and Orai1 (Scrimgeour et al. 2009; Hoover and Lewis 2011; Li et al. 2011). The bimolecular coupling model has difficulty explaining variable stoichiometry between STIM1 and Orai1 because it mandates that the ratio of STIM1:Orai1 be 1:1, or six molecules of STIM1 per hexameric Orai1 channel. On the other hand, the unimolecular coupling can easily explain these observations. Using a hypothetical lattice between Orai1 and STIM1 dimers (Fig. 5.4d), we can easily predict that the stoichiometry varies as the size of a channel cluster gets bigger. Based on the size of the cluster, we would expect to see a variable STIM1-Orai1 stoichiometry ranging from 1:1 to 2:1. The stoichiometry would approach 1:1, the larger the cluster grows. The lattice model has interesting implications on the regulation of activation and deactivation kinetics of channels and also on regulating the spatial disruption of channels to restrict Ca^{2+} entry signals to the ER-PM junctions. This may be a mechanism to discriminate complex signaling pathways as Ca^{2+} is a highly potent and cross-reactive second messenger.

5.6 Transducing STIM Activation into Orai Channel Gating and Ca^{2+} Signals

Besides studying the interactions between STIM1 and Orai1, our lab has also focused on identifying how Orai1 channels transmit C-terminal binding of STIM1 into an open channel. STIM1 specifically binds to the C-terminal helical cytosolic extension of Orai1 TM4 (TM4-ext) (Zhou et al. 2016). Using a mutational screen, we identified a key region linking TM4 to TM3 and termed it the "nexus." This nexus region is only five-amino acids long (residues 261-265 in hOrai1; LVSHK) and can transmit the allosteric binding signal from STIM1 to the channel core to initiate gating (Fig. 5.5). We found that mutating this nexus region from LVSHK to ANSGA caused the channel to be constitutively active and display all of the properties of a channel opened naturally by STIM1. We believe that this mutation exactly replicates the conformational change of TM4-ext normally caused

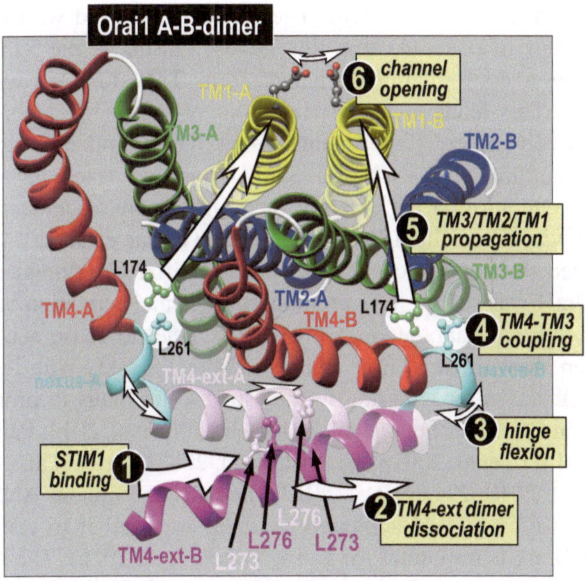

Fig. 5.5 Structure of *Drosophila* Orai1 dimers. Hydrophobic cross interactions between Leu-273 and Leu-276 between antiparallel TM4 extensions (TM4-ext) of adjacent Orai1 dimers help hold the molecule together. The crystal reveals that each monomer is structurally homologous in each transmembrane (TM) region except for the TM4-ext between adjacent Orai1 monomers within a dimer. In the diagram above, TM4-A (*light pink*) is straight, and TM4-B (*fuchsia*) is bent. The TM4-ext of each Orai1 monomer is linked to the rest of the molecule at the hinge/nexus region (*cyan*). Our results from experiments utilizing the Orai1 nexus mutant (Orai1-ANSGA) suggest that STIM1 binding (1) causes dissociation of the TM4 extensions (2) and flexion on the hinge/nexus (3). This provides the force to displace the hinge plate residue (Leu-261) and subsequent displacement of the closely apposed Leu-174 residue on TM3 (4). Finally, displacement of Leu-174 would lead to conformational repositioning of core helixes TM3, TM2, and TM1 (5) and dilation of E106 selectivity filter in the Orai1 pore (6) to allow for selective Ca^{2+} entry

by coupling with STIM1. The nexus is a discrete sequence and is composed of two fundamental regions. The first is the "hinge" domain (Ser-263, His-264, Lys-265) (Hou et al. 2012; Palty et al. 2015; Tirado-Lee et al. 2015) and, the second, the "hinge plate" (Leu-261, Val-262). Specifically, the residues of the hinge plate appear to be the hydrophobic attachment to TM3 and are the key transducers of the STIM1-binding signal. Previous groups have studied the hinge region by substituting with proline or cross-linking with cysteine through a redox reaction. These two alterations will lock the hinge and prevent the binding of STIM1 and Ca^{2+} entry through Orai1 (Palty et al. 2015; Tirado-Lee et al. 2015). However, no group has studied the "hinge plate" region, which based on the crystal structure of Orai1 appears to be the main contact point between TM4 and TM3, specifically at residues Leu-261 from TM4 and Leu-174 from TM3. These two leucines form hydrophobic contacts that are critical for normal channel function. Substitution of either Leu-174 or Leu-261 with charged residues results in a block in STIM1-induced activation of Orai1 (Zhou et al. 2016). Mutating these residues to cysteines and using a redox reagent to cross-link them enhanced channel activation (Zhou et al. 2016). These results, coupled with the reports by other labs, lead us to conclude that this nexus transduces the STIM1 coupling signal to trigger a conformational change from TM4 to TM3 and eventually TM2 and TM1. This leads to a rearrangement of the channel pore that allows Ca^{2+} to selectively enter. Experiments using Tb^{3+} luminescence in purified Orai1 protein confirm that cytosolic STIM1 binding leads to an immediate conformational alteration at the external face of the pore, specifically movement near the selectivity filter Glu-106 and Val-102 (Gudlur et al. 2014).

We next used this model to ask whether the initial C-terminal interaction of STIM1 with Orai1 was sufficient to promote full channel activation or does STIM1 require interaction with another region of the channel to cause activation. It has previously been shown that STIM1 can also interact weakly with the N-terminus (TM1) of Orai1 only after tethering to the Orai1 C-terminus to promote channel gating (Park et al. 2009; Rothberg et al. 2013; McNally et al. 2013; Palty et al. 2015; Gudlur et al. 2014; Li et al. 2007; Takahashi et al. 2007; Derler et al. 2013; Lis et al. 2010; Zhou et al. 2010; Zheng et al. 2013). Because the mutant ANSGA channel mimics the wild-type STIM1-activated channel, we used it to test whether the N-terminus of Orai1 is important for gating (Zhou et al. 2016). The specific claim has been that residues within TM1 weakly interact with STIM1 to "pull open" the channel pore to allow Ca^{2+} entry (Gudlur et al. 2014). Specifically, Orai1 residues Leu-81, Ser-82, and Lys-85 have been implicated in this STIM1 interaction and are located within TM1 of Orai1. We introduced each of these mutations individually into the STIM1-idependent, constitutively opened Orai1-ANSGA channels and also developed a construct containing all three mutations in the ANSGA Orai1 channel. We found that channel inhibition was similar in the ANSGA background to what was seen in wild-type STIM1-activated channels, suggesting that STIM1 interaction with these three TM1 residues is not necessary for channel gating but, rather, may be necessary for the integrity of the channel pore (Zhou et al. 2016). It is more likely that mutation to these three residues disrupts channel function due to disruption of the pore architecture. These results support the notion that STIM1 interactions with the strong C-terminus binding site are necessary and sufficient for channel gating and Ca^{2+} entry. However we do not rule out the possibility that STIM1 interactions with the N-terminus are important for other mechanisms. Orai1-ANSGA channels do not experience Ca^{2+}-dependent inhibition, which has been linked to the N-terminal interaction by STIM1 (Zhou et al. 2016; Mullins et al. 2016; Mullins and Lewis 2016). Thus, it is likely that the N-terminal is important for mediating other aspects of channel activity but not the initial gating activity.

The discovery of the Orai1-ANSGA mutation has been extremely useful since it is a powerful tool for studying the role of STIM1 in promoting the Orai1 open-channel conformation. It is also useful for determining the potential role that STIM1 plays in regulating other important aspects of channel activity such as Ca^{2+}-dependent inhibition and the effect that localizing channels has on downstream signaling. This mutation is a superior alternative to other constitutively active Orai1 mutants as it exhibits channel characteristics almost identical to the wild-type STIM1-activated channel. Our lab was able to use the ANSGA construct as a tool to work out that STIM1 gating of Orai1 occurs primarily through its interactions with the C-terminal extension adjacent to the Orai1 nexus (Figs. 5.3 and 5.5). Based on the Orai1 crystal structure, we know that adjacent Orai1 monomers interact in a nonsymmetrical antiparallel manner through their TM4-extension sequences. Specifically, the two adjacent TM4 extensions differ in their resting state conformations (Fig. 5.3). For one monomer, the hinge region is almost entirely straight (TM4-A), whereas in the other it is sharply bent (TM4-B) (Fig. 5.3). These extensions appear to be held together through hydrophobic interactions between Leu-273 and Leu-276, which are required for STIM1 binding (Fig. 5.5) (Tirado-Lee et al. 2015). We propose that STIM1 binding to the Orai1 C-terminus acts by prying the TM4-extensions apart, perhaps through interactions with the Phe-394 residue in the SOAR active site that we described earlier. Displacement of these two M4-extensions may provide the force to flex the SHK hinges, which initiates a force on the LV hinge plate and displacement of the TM4 Leu-261 and TM3 Leu-174 residues. An analogy of this would be a classic lever system where STIM1 acts as the load, the C-terminus of Orai1 acts as the load arm, and the LVSHK hinge/hinge-plate region acts as the fulcrum. The concerted effort by these "lever" components generates a force that is transmitted through the LVSHK fulcrum through the tightly packed TM3/TM2/TM1 helices and allows for pore dilation and Ca^{2+} entry. Specifically, we believe that this force eventually leads to the reconfiguration of the extracellular-facing selectivity filter (Glu-106) to open the channel and occurs entirely through an allosteric mechanism that does not require any direct interaction by STIM1 to the N-terminus of Orai1. Further studies will determine the authenticity of this model.

5.7 Conclusions

In conclusion, we propose a model for STIM1-induced Orai1 channel gating that exclusively involves the C-terminus of Orai1 (Zhou et al. 2016). This model is different from previously suggested models involving both the C- and N-termini of Orai1 (Park et al. 2009; Rothberg et al. 2013; McNally et al. 2013; Palty et al. 2015; Gudlur et al. 2014; Li et al. 2007; Takahashi et al. 2007; Derler et al. 2013; Lis et al. 2010; Zhou et al. 2010; Zheng et al. 2013). We would suggest that development of the novel Orai1-ANSGA mutant provides us with a powerful tool to accurately study the role of STIM1 in Orai1 channel gating in ways that were previously impossible. We recognize the importance that the purported N-terminal STIM1-binding site (Leu-81, Ser-82, and Lys-85) may play on other channel properties, such as general maintenance of pore structure and/or Ca^{2+}-dependent inhibition. However, results from our lab show that STIM1 interactions with this N-terminal region are not necessary for channel gating. Our model provides a holistic view of the STIM1-Orai1 interaction mechanism (Zhou et al. 2015b, 2016). In summary, we propose that Orai1 channel activity is regulated by the unimolecular interaction between the monomeric active SOAR subunit in STIM1 dimers with the antiparallel cytoplasmic C-terminal TM4-extensions of Orai1 channels. Clearly further study is required to better understand how allosteric interactions between STIM1 and the Orai1 C-terminus lead to conformational changes at the Orai1 pore and transition into an active, Ca^{2+}-selective Orai1 channel.

References

Amcheslavsky A, Wood ML, Yeromin AV, Parker I, Freites JA, Tobias DJ, Cahalan MD (2015) Molecular biophysics of Orai store-operated Ca2+ channels. Biophys J 108:237–246

Cai X, Zhou Y, Nwokonko RM, Loktionova NA, Wang X, Xin P, Trebak M, Wang Y, Gill DL (2016) The Orai1 store-operated calcium channel functions as a hexamer. J Biol Chem 291:25764–25775

Covington ED, MM W, Lewis RS (2010) Essential role for the CRAC activation domain in store-dependent oligomerization of STIM1. Mol Biol Cell 21:1897–1907

Derler I, Plenk P, Fahrner M, Muik M, Jardin I, Schindl R, Gruber HJ, Groschner K, Romanin C (2013) The extended transmembrane Orai1 N-terminal (ETON) region combines binding interface and gate for Orai1 activation by STIM1. J Biol Chem 288:29025–29034

Derler I, Jardin I, Romanin C (2016) Molecular mechanisms of STIM/Orai communication. Am J Physiol Cell Physiol 310:C643–C662

Fahrner M, Muik M, Schindl R, Butorac C, Stathopulos P, Zheng L, Jardin I, Ikura M, Romanin C (2014) A coiled-coil clamp controls both conformation and clustering of stromal interaction molecule 1 (STIM1). J Biol Chem 289:33231–33244

Gudlur A, Quintana A, Zhou Y, Hirve N, Mahapatra S, Hogan PG (2014) STIM1 triggers a gating rearrangement at the extracellular mouth of the ORAI1 channel. Nat Commun 5:5164

Hoover PJ, Lewis RS (2011) Stoichiometric requirements for trapping and gating of Ca2+ release-activated Ca2+ (CRAC) channels by stromal interaction molecule 1 (STIM1). Proc Natl Acad Sci U S A 108:13299–13304

Hou X, Pedi L, Diver MM, Long SB (2012) Crystal structure of the calcium release-activated calcium channel Orai. Science 338:1308–1313

Kar P, Parekh AB (2015) Distinct spatial Ca2+ signatures selectively activate different NFAT transcription factor isoforms. Mol Cell 58:232–243

Korzeniowski MK, Popovic MA, Szentpetery Z, Varnai P, Stojilkovic SS, Balla T (2009) Dependence of STIM1/Orai1-mediated calcium entry on plasma membrane phosphoinositides. J Biol Chem 284:21027–21035

Li Z, Lu J, Xu P, Xie X, Chen L, Xu T (2007) Mapping the interacting domains of STIM1 and Orai1 in CRAC channel activation. J Biol Chem 282:29448–20456

Li Z, Liu L, Deng Y, Ji W, Du W, Xu P, Chen L, Xu T (2011) Graded activation of CRAC channel by binding of different numbers of STIM1 to Orai1 subunits. Cell Res 21:305–315

Lis A, Zierler S, Peinelt C, Fleig A, Penner R (2010) A single lysine in the N-terminal region of store-operated channels is critical for STIM1-mediated gating. J Gen Physiol 136:673–686

Ma G, Wei M, He L, Liu C, Wu B, Zhang SL, Jing J, Liang X, Senes A, Tan P, Li S, Sun A, Bi Y, Zhong L, Si H, Shen Y, Li M, Lee MS, Zhou W, Wang J, Wang Y, Zhou Y (2015) Inside-out Ca(2+) signalling prompted by STIM1 conformational switch. Nat Commun 6:7826

Maus M, Jairaman A, Stathopulos PB, Muik M, Fahrner M, Weidinger C, Benson M, Fuchs S, Ehl S, Romanin C, Ikura M, Prakriya M, Feske S (2015) Missense mutation in immunodeficient patients shows the multifunctional roles of coiled-coil domain 3 (CC3) in STIM1 activation. Proc Natl Acad Sci U S A 112:6206–6211

McNally BA, Somasundaram A, Jairaman A, Yamashita M, Prakriya M (2013) The C- and N-terminal STIM1 binding sites on Orai1 are required for both trapping and gating CRAC channels. J Physiol 591:2833–2850

Muik M, Fahrner M, Derler I, Schindl R, Bergsmann J, Frischauf I, Groschner K, Romanin C (2009) A cytosolic homomerization and a modulatory domain within STIM1 C-terminus determine coupling to ORAI1 channels. J Biol Chem 284:8421–8426

Mullins FM, Lewis RS (2016) The inactivation domain of STIM1 is functionally coupled with the Orai1 pore to enable Ca2+-dependent inactivation. J Gen Physiol 147:153–164

Mullins FM, Yen M, Lewis RS (2016) Orai1 pore residues control CRAC channel inactivation independently of calmodulin. J Gen Physiol 147:137–152

Palty R, Stanley C, Isacoff EY (2015) Critical role for Orai1 C-terminal domain and TM4 in CRAC channel gating. Cell Res 25:963–980

Park CY, Hoover PJ, Mullins FM, Bachhawat P, Covington ED, Raunser S, Walz T, Garcia KC, Dolmetsch RE, Lewis RS (2009) STIM1 clusters and activates CRAC channels via direct binding of a cytosolic domain to Orai1. Cell 136:876–890

Perni S, Dynes JL, Yeromin AV, Cahalan MD, Franzini-Armstrong C (2015) Nanoscale patterning of STIM1 and Orai1 during store-operated Ca2+ entry. Proc Natl Acad Sci U S A 112:E5533–E5542

Prakriya M, Lewis RS (2015) Store-operated calcium channels. Physiol Rev 95:1383–1436

Putney JW Jr., Steinckwich-Besancon N, Numaga-Tomita T, Davis FM, Desai PN, D'Agostin DM, Wu S, Bird GS (2016) The functions of store-operated calcium channels. Biochim Biophys Acta. doi: 10.1016/j.bbamcr.2016.11.028

Rothberg BS, Wang Y, Gill DL (2013) Orai channel pore properties and gating by STIM: implications from the Orai crystal structure. Sci Signal 6:pe9

Scrimgeour N, Litjens T, Ma L, Barritt GJ, Rychkov GY (2009) Properties of Orai1 mediated store-operated current depend on the expression levels of STIM1 and Orai1 proteins. J Physiol 587:2903–2918

Shim AH, Tirado-Lee L, Prakriya M (2015) Structural and functional mechanisms of CRAC channel regulation. J Mol Biol 427:77–93

Soboloff J, Rothberg BS, Madesh M, Gill DL (2012) STIM proteins: dynamic calcium signal transducers. Nat Rev Mol Cell Biol 13:549–565

Stathopulos PB, Schindl R, Fahrner M, Zheng L, Gasmi-Seabrook GM, Muik M, Romanin C, Ikura M (2013) STIM1/Orai1 coiled-coil interplay in the regulation of store-operated calcium entry. Nat Commun 4:2963

Takahashi Y, Murakami M, Watanabe H, Hasegawa H, Ohba T, Munehisa Y, Nobori K, Ono K, Iijima T, Ito H (2007) Essential role of the N-terminus of murine Orai1 in store-operated Ca2+ entry. Biochem Biophys Res Commun 356:45–52

Tirado-Lee L, Yamashita M, Prakriya M (2015) Conformational changes in the Orai1 C-terminus evoked by STIM1 binding. PLoS One 10:e0128622

Wang X, Wang Y, Zhou Y, Hendron E, Mancarella S, Andrake MD, Rothberg BS, Soboloff J, Gill DL (2014) Distinct Orai-coupling domains in STIM1 and STIM2 define the Orai-activating site. Nat Commun 5:3183

Yang X, Jin H, Cai X, Li S, Shen Y (2012) Structural and mechanistic insights into the activation of Stromal interaction molecule 1 (STIM1). Proc Natl Acad Sci U S A 109:5657–5662

Yen M, Lokteva LA, Lewis RS (2016) Functional analysis of Orai1 concatemers supports a hexameric stoichiometry for the CRAC channel. Biophys J 111:1897–1907

Yuan JP, Zeng W, Dorwart MR, Choi YJ, Worley PF, Muallem S (2009) SOAR and the polybasic STIM1 domains gate and regulate Orai channels. Nat Cell Biol 11:337–343

Zheng H, Zhou MH, Hu C, Kuo E, Peng X, Hu J, Kuo L, Zhang SL (2013) Differential roles of the C and N termini of Orai1 protein in interacting with stromal interaction molecule 1 (STIM1) for Ca2+ release-activated Ca2+ (CRAC) channel activation. J Biol Chem 288:11263–11272

Zhou Y, Meraner P, Kwon HT, Machnes D, Oh-hora M, Zimmer J, Huang Y, Stura A, Rao A, Hogan PG (2010) STIM1 gates the store-operated calcium channel ORAI1 in vitro. Nat Struct Mol Biol 17:112–116

Zhou Y, Srinivasan P, Razavi S, Seymour S, Meraner P, Gudlur A, Stathopulos PB, Ikura M, Rao A, Hogan PG (2013) Initial activation of STIM1, the regulator of store-operated calcium entry. Nat Struct Mol Biol 20:973–981

Zhou Y, Trebak M, Gill DL (2015a) Calcium signals tune the fidelity of transcriptional responses. Mol Cell 58:197–199

Zhou Y, Wang X, Wang X, Loktionova NA, Cai X, Nwokonko RM, Vrana E, Wang Y, Rothberg BS, Gill DL (2015b) STIM1 dimers undergo unimolecular coupling to activate Orai1 channels. Nat Commun 6:8395

Zhou Y, Cai X, Loktionova NA, Wang X, Nwokonko RM, Wang X, Wang Y, Rothberg BS, Trebak M, Gill DL (2016) The STIM1-binding site nexus remotely controls Orai1 channel gating. Nat Commun 7:13725

The STIM-Orai Pathway: Regulation of STIM and Orai by Thiol Modifications

6

Barbara A. Niemeyer

Abstract

Cysteines are among the least abundant amino acids found in proteins. Due to their unique nucleophilic thiol group, they are able to undergo a broad range of chemical modifications besides their known role in disulfide formation, such as S-sulfenylation (-SOH), S-sulfinylation (-SO(2)H), S-sufonylation (-SO(3)H), S-glutathionylation (-SSG), and S-sulfhydration (-SSH), among others. These posttranslational modifications can be irreversible and act as transitional modifiers or as reversible on-off switches for the function of proteins. Disturbances of the redox homeostasis, for example, in situations of increased oxidative stress, can contribute to a range of diseases. Because Ca^{2+} signaling mediated by store-operated calcium entry (SOCE) is involved in a plethora of cellular responses, the cross-talk between reactive oxygen species (ROS) and Ca^{2+} is critical for homeostatic control. Identification of calcium regulatory protein targets of thiol redox modifications is needed to understand their role in biology and disease.

Keywords

CRAC • SOCE • ROS • Oxidation • Oxidomimetic

6.1 Introduction

The initiation of the acquired immune response requires antigen-presenting cells (APCs) such as dendritic cells to migrate to primary lymphoid organs where they present the antigenic peptide to naïve T cells. Efficient T cell activation is achieved

B.A. Niemeyer (✉)
Molecular Biophysics, Center for Integrative Physiology and Molecular Medicine, Saarland University, Homburg, Germany
e-mail: barbara.niemeyer@uks.eu

K. Groschner et al. (eds.), *Store-Operated Ca^{2+} Entry (SOCE) Pathways*, Advances in Experimental Medicine and Biology 993, DOI 10.1007/978-3-319-57732-6_6

by formation of an immunological synapse (IS) between T cells and APCs. Activation of the T cell receptor (TCR) complex initiates a signaling cascade that results in a rapid increase in intracellular calcium ($[Ca^{2+}]_i$) which is necessary for T cell activation, differentiation, and proliferation (Feske 2007; Lewis 2001). In T cells, this store-operated Ca^{2+} entry (SOCE) is almost exclusively governed by plasma membrane localized Ca^{2+} release-activated Ca^{2+} (CRAC) channels (Parekh and Putney 2005). CRAC channels are activated by receptor-initiated inositol-1,4,5 trisphosphate (IP_3)-triggered Ca^{2+} release from the endoplasmic reticulum (ER) (Hoth and Penner 1992). The concomitant decrease in ER luminal Ca^{2+} triggers accumulation of the ER Ca^{2+} sensor protein STIM1 and STIM2 into puncta close to the plasma membrane (Liou et al. 2005; Zhang et al. 2005). Depending on the strength and duration of receptor activation, this Ca^{2+} efflux results in a graded depletion of the luminal (400–500 μM) $[Ca^{2+}]$ concentration within the ER. Hereby, a small decrease will be sensed mainly by ER-resident STIM2 molecules, and a decrease below ~100 μM ER $[Ca^{2+}]$ will be sensed by STIM1 molecules (Brandman et al. 2007). Clustered STIM proteins directly couple and activate Ca^{2+}-influx through CRAC channels encoded by the Orai genes (Feske et al. 2006; Vig et al. 2006; Zhang et al. 2006). In T lymphocytes CRAC channels are mainly formed by Orai1 (Feske et al. 2006) although heteromeric channels can assemble with Orai2 or Orai3 (Schindl et al. 2009). Binding of STIM1 channel-activating domain (CAD) domains traps and activates Orai channels leading to store-operated Ca^{2+} entry (SOCE) (reviewed in Prakriya and Lewis 2015). Besides the major role of SOCE in immune cells, STIM and Orai proteins are ubiquitously expressed and contribute to Ca^{2+} homeostasis in most cell types and tissues.

Humans with specific mutations in Orai1 or STIM1 have drastically impaired SOCE and consequently defective T cell proliferation and cytokine production. These patients develop severe immunodeficiency, myopathy, ectodermal dysplasia, and autoimmune and lymphoproliferative diseases (Feske 2009). In mice, deletion of Orai1 and STIM1 results in prenatal lethality. However, gene-targeted T cell specific deletion of Orai1 or Stim1 leads, as in humans, to impaired SOCE in $CD4^+$ and $CD8^+$ T cells, B cells, and mast cells. Although some differences exist in the degree of T cell malfunction, these mice suffer from very similar immunodeficiencies as humans (Feske 2009). Interestingly, a number of pathological syndromes have recently been described that are also due to gain of function mutations in STIM1/Orai1 with patients presenting bleeding disorders and tubular aggregate myopathy (TAM), a skeletal muscle disorder causing muscular pain, weakness, and cramping with exercise (reviewed in Lacruz and Feske 2015).

Activated T cells proliferate and differentiate into effector T cells which subsequently migrate toward inflamed tissues and either kill ($CD8^+$) target cells or help ($CD4^+$) locally infiltrated macrophages and B cells to eliminate the invading pathogens. In these environments and particularly in tumor environments, T cells are exposed to highly oxidizing agents (reactive oxygen species, ROS) for extended time periods and serve as an excellent system (Segal 1996) to study the influence of oxidation on cellular function (Droge 2002b; Reth 2002). In addition, B-cells, monocytes, and macrophages themselves produce ROS, and patients with deficient

NADPH oxidase 2, one of the major ROS-producing enzyme complexes, suffer from chronic granulomatous disease (CGD) (Babior and Curnutte 1987) and may have a higher susceptibility to autoimmune diseases, such as idiopathic thrombocytopenic purpura, rheumatoid arthritis, and lupus (see below). The molecular mechanisms causing this increased susceptibility are not well understood, but increased antibody production may be one reason for less efficient selection against self-directed antibodies. In humans, ROS have been connected with many pathological conditions, such as cancer, cardiovascular diseases, arteriosclerosis, hypertension, and Parkinson's and Alzheimer's disease, and are considered to be a major factor of aging (Droge 2002a; Kirkwood and Kowald 2012). In addition, it has been suggested that oxidative stress also has negative or dual roles in the pathogenesis of several autoimmune diseases such as rheumatoid arthritis (RA), multiple sclerosis (MS), and diabetes mainly due to their pro-inflammatory nature (Bashan et al. 2009; Gilgun-Sherki et al. 2004; Griffiths 2008). RA and MS are complex chronic autoimmune diseases dependent on both genetic and environmental factors and are characterized by an autoimmune response against the joints or myelin sheaths. The etiology of the diseases is still to a large extent unknown although genetic and environmental causes have been suggested. Also, the complexity and heterogeneity of the diseases hinder the clarification of the mechanisms behind the development of RA and MS (Coenen and Gregersen 2009; Griffiths 2008; Sospedra and Martin 2005). Several reports suggested a protective role for ROS against RA. These findings are based on the fact that ncf1 (phox 47, subunit of NOX2) KO mice were more susceptible toward RA than the wild type, resembling similar conditions seen in humans where defects in ROS-producing machinery (NOX2) lead to autoimmune chronic granulomatous disease (CDG) (Curnutte and Babior 1987; Hultqvist et al. 2004; Segal 1996). Furthermore, it has been shown that ROS are important determinants for many physiological processes and that they act as intracellular second messengers regulating various signaling pathways (Droge 2002b; Reth 2002; Rhee 2006).

6.2 Sources of ROS

Environmental sources of ROS include factors such as heat, radiation, and chemical substances, which generate ROS, for example, through the radiolysis of H_2O. The major biological sources of ROS are NADPH oxidases (NOX) producing ROS into luminal (phagosomes) or into the extracellular space, while mitochondria generate high amounts of oxidants in the mitochondrial intermembrane space (Bedard and Krause 2007; Starkov 2008). The NOX family of enzymes consists of seven members (NOX 1–5 and two dual oxidases, DUOX 1 and 2). Many cells express some or more subtypes of NOX, but the highest relative expression can be detected in phagocytes (NOX2). When a phagocyte is activated by bacterial peptides or a range of immunological stimuli, NOX enzymes assemble at the membrane of phagosomes or at the plasma membrane and generate $O_2^{\bullet-}$ by transferring electrons from the cytosolic NADPH to the free oxygen. Subsequently, the highly unstable

$O_2^{\bullet-}$ is rapidly transformed into H_2O_2 (Bedard and Krause 2007). While the ROS production by NOX is considered to be "physiological," mitochondrial ROS are generally considered as byproducts and often are indicators of some pathological conditions. ROS in mitochondria are generated when electrons escape the mitochondrial electron transfer chain and react with the free oxygen again leading to $O_2^{\bullet-}$ production (Brookes et al. 2004; Starkov 2008). In addition, ROS are also produced by 5-lipooxygenases (LOX), cyclooxygenases (COX), and xanthine oxidases. ROS include more than 20 substances, of which superoxide ($O_2^{\bullet-}$) and hydrogen peroxide (H_2O_2) are probably the biologically most important ones (Droge 2002b). Superoxide lifetime is extremely short, and it is rapidly converted to the more stable H_2O_2 by superoxide dismutases (SOD).

H_2O_2 is increasingly recognized as a second messenger, regulating gene expression through alterations of transcription factors, likely affecting epigenetic patterns, and fine-tuning signal transduction, migration, and tissue function (Clempus and Griendling 2006; Davalli et al. 2016; Morgan and Liu 2011; Rhee 2006). One of the major obstacles in measuring ROS is that they are also quickly eliminated either by intracellular glutathione reductases, catalases, or superoxide dismutase (Couto et al. 2016) or in the extracellular space through membrane-associated catalases and superoxide dismutases (e.g., SOD3 (Kwon et al. 2012)). Although a number of genetically modified ROS sensors have been developed, accurate quantitative and time-resolved measurements remain a challenge (Gibhardt et al. 2016; Oparka et al. 2016). Here, a very recent method applying rare-earth nanoparticles is able to provide quantitative, localized H_2O_2 detection with micromolar sensitivity and much improved high time resolution (Abdesselem et al. 2017). Extracellular and intracellular ROS can pass membranes through passive diffusion or through aquaporins (Appenzeller-Herzog et al. 2016; Watanabe et al. 2016) and are usually quickly degraded in the cytosol through the action of enzymes such as catalase, superoxide dismutase, or glutathione reductase system. While low concentrations (likely in the nanomolar to low micromolar range) of ROS trigger or influence local signaling cascades and are indeed physiological, higher concentrations can also cause damage to nucleic acids, proteins, or lipids (see reviews Bogeski et al. 2011; Schieber and Chandel 2014).

6.3 Cysteine Biochemistry

Cysteines are one of the least abundant amino acid found in proteins (Marino and Gladyshev 2010). Their functional significance in proteins and the soft acid/base thiol chemistry that facilitates cysteine modification have recently been summarized in a perspective by Wible and Sutter (2017). Reactive cysteine residues are major targets of ROS-induced modification in proteins, and modified cysteines are an emerging posttranslational modification capable of altering function, conformation, or localization of proteins. Depending on the local environment (pK_a), the thiolate group (S-) reacts with H_2O_2 with rates ranging from 10 to 10^5 M^{-1} s^{-1}, while the thiol groups (SH) do not react physiologically with H_2O_2 unless the reaction is catalyzed (Forman et al. 2004). The thiolate undergoes reversible (to sulfenic) or irreversible (to sulfinic and sulfonic acid) covalent modifications upon oxidation with protein S-sulfenylation being a fleeting molecular switch (Beedle et al. 2016), which depending on the substrate can undergo a wide array of chemical modifications such as glutathionylation, nitrosylation, or covalent modifications by electrophilic agents. Reversible oxidation of cysteines is a process relatively well understood and utilized, for example, by thioredoxin proteins within the ER to facilitate formation of disulfide bonds but is increasingly recognized as a potent means to alter protein function as an on-off switch (Conway and Lee 2015; Leichert and Dick 2015). Additionally, mild oxidation can induce reversible cysteine disulfide bond formation and thus prevent further irreversible cysteine modifications (Droge 2002b) (Fig. 6.1).

Oxidation and consequent structural changes such as intermolecular cross-linking can modify the function of the target proteins (Bogeski et al. 2011; Kourie 1998). Very high concentrations of ROS can cause a wider range of modifications, including the oxidation of arginine and lysine residues to aldehydes or of methionine to the corresponding sulfoxide or sulfone. Oxidation can also lead to intra- and intermolecular cross-linking and trapping of proteins in multimolecular complexes. Prolonged exposure to high oxidant concentrations lead to modifications which are essentially irreversible, abolish protein function, and eventually lead to degradation. On the contrary, physiologically relevant concentrations of ROS can react with specific regulatory amino acids within the polypeptide chain. This selective action depends on the accessibility within the tertiary structure, on the local concentration and nature of the reacting redox species, and on the local redox potential.

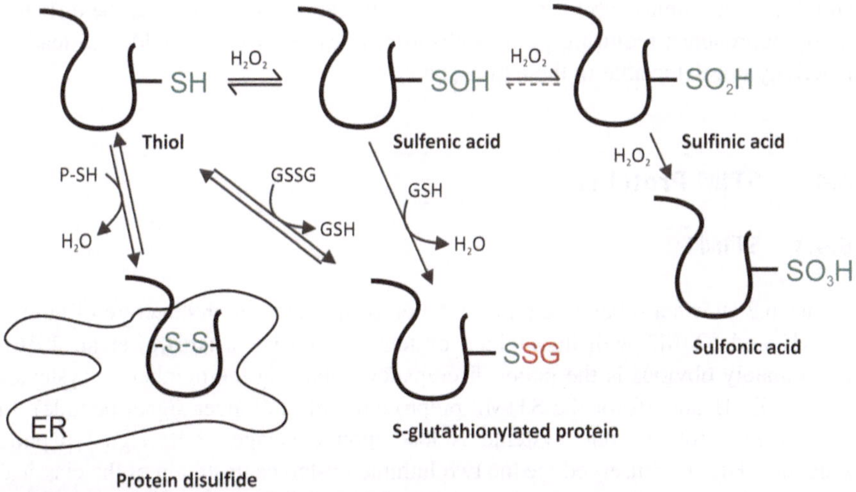

Fig. 6.1 Examples of reversible or irreversible thiol modifications. Catalyzing enzymes are thioredoxins (Trx), glutaredoxins (Grx); GSH glutathione. Not shown: S-nitrosylation or metal bound thiols. Disulfide bridge formation might also happen in the cytosol when catalyzed

To clearly delineate the physiological effects of redox-mediated cysteine oxidation on SOCE, also recently reviewed by Bhardwaj et al. (2016), the ideal system would not contain any endogenous STIM or Orai proteins to then be replaced by SOCE components lacking reactive cysteines without alteration of the structural function of these residues. Many studies that investigated the effects of exogenously added ROS, indirect ROS production after hypoxia, or other redox active substances on SOCE (reviewed in Bhardwaj et al. 2016; Bogeski et al. 2011; Nunes and Demaurex 2014) suffer from the complexity of ROS effects on many different target proteins that will directly or indirectly affect SOCE (i.e. activation of IP_3 receptors, inhibition of SERCA), but this does not necessarily mean that STIM or Orai proteins act as primary sensors (drivers) for an oxidative stress response. A recent report which confirmed the SOCE enhancing effects of hypoxia mediated by ROS (Mungai et al. 2011) showed that these effects were due to upregulation of STIM1, Orai1, and TRPC1 protein expression (Chen et al. 2017). As an unbiased starting point of analyses, the first consideration for SOCE proteins as ROS sensors concerns the number of cysteines of the given protein. Secondly, and more difficult to determine is the potential reactivity of a given cysteine, which is mainly dependent of the local pK_a (see above). Several biochemical techniques use electrophilic addition of compounds such as MTSEA- or DTNB-biotin to reactive cysteines with subsequent specific protein detection by Western Blot. Increasingly, thiol-specific proteomic approaches using, for example, clickable electrophile probes in combination with a quantitative chemoproteomic platform enable global, in situ, site-specific profiling of redox-sensitive thiols in complex proteomes, described, for example, in Chouchani et al. (2011); Yang et al. (2016) and offer unique possibilities to determine exact cysteine modifications. In the case of intracellular or intraluminal and hidden cysteines (Marino and Gladyshev 2010) and due to the highly labile and dynamic modifications, detection can be difficult. Many approaches requiring permeabilization or lysis of cells could also lead to reactivity not detectable in the native state.

6.4 STIM Proteins

6.4.1 STIM1

Figure 6.2 shows a schematic representation of the domain architecture of human STIM1 and STIM2 with highlighted cysteines (see also Bhardwaj et al. 2016). Immediately obvious is the major discrepancy regarding the number of cysteines (5 for STIM1 and 16 for the STIM2 preprotein with its longer signal peptide). In both STIM proteins, one cysteine is lost upon cleavage of its signal peptide sequence. Highly conserved are the two luminal cysteines upstream of the classical calcium binding EF hand (Cys-49 and Cys-56 of hSTIM1) where D76 is the first acidic side chain involved in Ca^{2+}-binding (Stathopulos et al. 2008). No overall

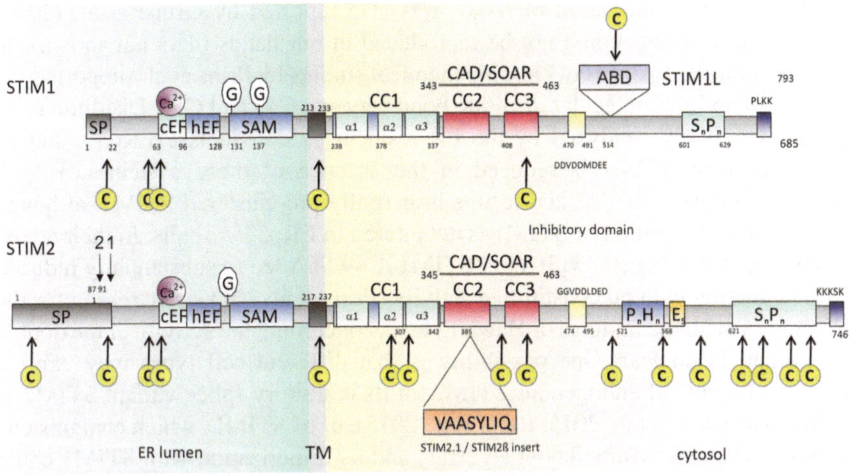

Fig. 6.2 Simplified domain structure of hSTIM1 and hSTIM2. *SP* signal peptide. *Arrow 1* indicates cleavage by signal peptide peptidase, and *arrow 2* indicates an additional cleavage which results in the release of a cytosolic signal peptide fragment (Graham et al. 2011). *cEF* classical EF hand, *hEF* hidden EF hand, *SAM* sterile a motif, G denotes consensus sites for N-glycosylation, *TM* transmembrane spanning region, *CC* coiled-coil domain, *CAD* channel-activating domain, *ABD* spliced-in actin binding domain of STIM1L (Darbellay et al. 2011); inhibitory exon 9 spliced in STIM2.1 or STIM2β (Miederer et al. 2015; Rana et al. 2015). Cysteines marked in *yellow* (modified from Hoth and Niemeyer 2013)

consensus as to the reactivity of these cysteines has been reached, and mutational analysis of these cysteines may affect the folding of the calcium binding EF hand, with altered or absent Ca^{2+} chelation leading to STIM1 clustering and activation of Orai1. Hawkins et al. expressed the EF-SAM domain containing these cysteines in bacteria and show that these cysteines are not susceptible to oxidation-/reduction-induced shifts in MW; however, this is not surprising as they would be unlikely to form disulfide bridges upon bacterial expression. Prolonged, treatment (24 h with BSO or 30 min with H_2O_2) of COS-7 cells induced significant S-glutathionylation of STIM1 and reduced its Ca^{2+} binding affinity leading to clustering of full length STIM1 and activation of SOCE (Hawkins et al. 2010). Of interest, but not investigated in the context of modification of STIMs, is the localization of NOX4 to the ER membrane of non-phagocytic cells (Chen et al. 2008). S-glutathionylation of STIM1 was abolished by mutation of C56 and should have prevented glutathionylation-induced SOCE activation; however Cys-49/Cys-56 double mutants lead to constitutive Ca^{2+} entry in STIM1$^{-/-}$ DT40 cells and in HEK cells upon coexpression with Orai1, possibly due to allosteric effects on the Ca^{2+} binding affinity of the EF hand (Hawkins et al. 2010). H_2O_2 triggered,

STIM1-dependent activation of I_{CRAC} was also reported by Grupe et al. (2010) although these results could not be reproduced in our hands (data not shown). In contrast to the results of Hawkins, biochemical studies by Prins et al. supported the presence of an intramolecular disulfide bond between C49 and C56. Disulfide bond formation is likely catalyzed by the ER resident oxidoreductase ERp57, and its interaction with STIM1 is reduced in the absence of these cysteines. ERp57 deficiency increases SOCE and results in partially pre-clustered STIM1, although disulfide bond formation of STIM1 is not altered in ERp57$^{-/-}$ cells. In their study, rescue of STIM1$^{-/-}$ cells with YFP–STIM1–C49/56A led to substantially reduced STIM1 function with no constitutive activity (Prins et al. 2011). The reason for the discrepancy with the findings of Hawkins et al. concerning the activity of the double Cys mutant is unclear. One possibility is that different cell types may express different amounts of endogenous STIM2, of its inhibitory splice variant STIM2.1/STIM2β (Miederer et al. 2015; Rana et al. 2015), or of STIML, which contains one additional cysteine (Darbellay et al. 2011) and by dimerization with STIM1 could affect the outcome of ROS induced stress. Besides the abovementioned analysis of Cys-49 and Cys-56 of STIM1, no further analysis of STIM1 cysteine reactivity has been performed. Cys-227, the ER-transmembrane domain conserved between STIM1 and STIM2, has been mutated to show that its helical region is critical in transmitting the conformational state of the luminal EF-hands upon store depletion to the cytosolic region (Ma et al. 2015), but whether this cysteine is or can be oxidized is unclear. Cys-437 at the end of the channel-activating domain (CAD/SOAR) has been mutated (Park et al. 2009), and its effects on the rate of SOCE are summarized in detail by Bhardwaj et al. (2016); however it has not been investigated under ROS challenge or mutated to simulate an oxidomimetic amino acid (see Fig. 6.3). As mentioned below, the finding that ROS preincubation of HEK cells expressing STIM1 and a ROS-insensitive Orai1 show no overt alteration in I_{CRAC} (Bogeski et al. 2010) argues against a modulatory role of altered STIM1 cysteines but does not exclude an activating effect. Effects of hypoxia, which may be concomitant or are followed by increased ROS production after reperfusion, on STIM1 have also been extensively reviewed by Bhardwaj et al. (2016) and Bogeski et al. (2011, 2012). Because many treatments lead to ER stress and calcium efflux from the ER, which will then lead to activation of STIM1 and SOCE, STIM1 is likely a "passenger" of ROS-altered intracellular milieu changes.

6.4.2 STIM2

Given the vastly increased number of cysteines found in STIM2 (15 vs. 4 for STIM1 in the mature protein), see Fig. 6.2, it appears likely that some of these are reactive and may thus drive changes in STIM2 function upon oxidation. Because they are localized within a well-buffered cytosol, oxidation possible only occurs in cellular microdomains (i.e., near the PM upon LOX and NOX activation or in regions of close proximity of ER to mitochondria). A strong hint that this might be the case is due to the fact that STIM2$^{-/-}$ neurons are protected from

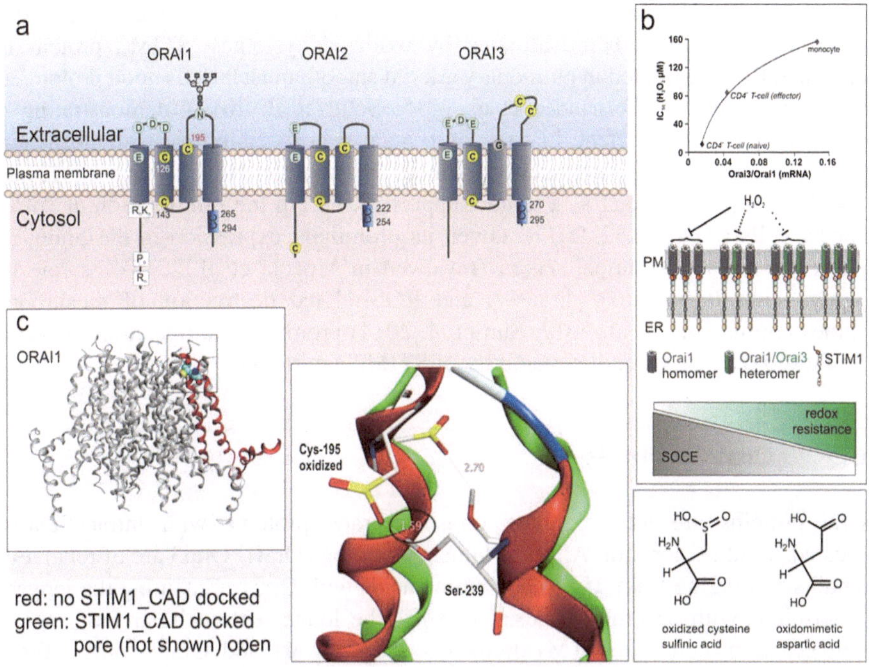

Fig. 6.3 (**a**) Schematic representation of Orai homologs with cysteines highlighted in *yellow* (**b**) IC50 of ROS induced inhibition of SOCE in different human primary cells with a model of the underlying relative Orai1/Orai3 ratios shown below. (**c**) Representation of morphed hOrai1 onto the crystal structure of *Drosophila* Orai1, in *red* TM3 and TM4. Expanded view shows the location of C195 and S239 in the closed (*red*) and open (*green*) MD simulation. Oxidizing C195 to sulfinic acid reduces the distance to S239 to 1.59 Å, enabling hydrogen bonding (part **b** reproduced from Saul et al. 2016, part **c** reproduced from Alansary et al. 2016)

hypoxic cell death (Berna-Erro et al. 2009), although a clear link to oxidized STIM2 cysteines is not given. In addition to STIM2, STIM2.1 (also called STIM2β), an inhibitory splice variant, which shows insertion of 8 amino acid (VAASYLIQ) into the channel-activating domain (CAD/SOAR) has been identified (Miederer et al. 2015; Rana et al. 2015). Both groups show that insertion reduces interaction with Orai1 and that STIM2.1 can act as a dominant negative regulator of both STIM1 and STIM2 function. Miederer et al. also demonstrated a Ca^{2+} dependency of the inhibitory function; however, whether oxidative stress is able to affect inhibition or splicing is unknown. STIM2 contains several potential calmodulin (CaM) binding sites, a high affinity site partially overlapping with the CAD domain which shows splice-specific affinities of 9.1 ± 2.7 nM for STIM2.1 compared with 40.7 ± 10.5 nM for STIM2.2 (Miederer et al. 2015) and could be the reason for the slow and biphasic activation kinetics of STIM2 (Parvez et al. 2008). A second, but low affinity (~1 μM) binding site can be found at the lysine-rich C-terminal domain of STIM2, with binding of CaM inhibiting its interaction with plasma membrane phospholipids (Bauer et al. 2008; Bhardwaj et al. 2013), possibly facilitating the detachment of the C-terminus with elevated cytosolic calcium, the latter site potentially being affected by oxidized cysteines. STIM2 protein is prominently upregulated in pulmonary arterial smooth muscle cells upon prolonged hypoxic conditions (Fernandez et al. 2015; Song et al. 2011), demonstrating a possible "passenger" effect. In contrast to many reports claiming pro-tumorigenic upregulation of SOCE components in cancer (reviewed in Hoth 2016), STIM2 has also been reported to act as a tumor suppressor, i.e., in the proliferation of colon cancer cells (Aytes et al. 2012). Given its prominent expression in the brain, in particular the hippocampal region (reviewed in Moccia et al. 2015), its role in aging, neurodegenerative diseases, and ROS-related destruction of mushroom spines (Berna-Erro et al. 2017; Sun et al. 2014) promises exciting future results. So far, no data exists on the reactivity of STIM2's cysteines.

6.5 Orai Channels

Orai proteins are tetra-spanning transmembrane proteins with intracellularly located N- and C-termini. All three human homologs Orai1–Orai3 are of relatively small size ranging from 254 to 301 amino acids with Orai1 containing the longest N-terminus with 85 amino acids. Orai3 has the longer extracellular loop region between transmembrane (TM) domain 3 and 4. Orai1 and Orai2 contain three highly conserved cysteines, with the following nomenclature referring to human Orai1 (BC015369): Cys-126 in the second transmembrane domain, Cys-143 at the border of TM2 and the intracellular loop region, and Cys-195 located at the extracellular facing exit of the third transmembrane domain, in close proximity of Lys-198, which may affect its reactivity. Orai2 also contains an N-terminal cytosolic Cys-16. Orai3 lacks the Cys-195 homolog but contains two extracellular cysteines within the extracellular loop II (Cys-226 and Cys-232) (Fig. 6.3a), which may be involved in Orai3-mediated detection of the gaseous signaling molecule hydrogen sulfide (H2S) via sulfhydration of these cysteines (Velmurugan et al. 2015).

6.5.1 Orai1 and Orai3

Biochemical experiments using thiol-reactive DTNB-biotin to pull down overexpressed Orai1 suggested that Cys-195 is the major reactive cysteine of Orai1 and may represent the target for increased environmental oxidative stress inhibition of Orai1-mediated SOCE seen in $CD4^+$ T cells, Jurkat cells, and HEK cells overexpressing STIM1 and Orai1 (Bogeski et al. 2010). Electrophilic addition to Orai1's C195 is also the main reason for the inhibitory effect of curcumin and caffeic acid phenethyl ester (CAPE) on I_{CRAC} (Shin et al. 2012). Orai3, one of the two Orai1 paralogs in humans, is redox insensitive and lacks the cysteine at the homologous position to Cys-195. Mutation Cys-195 to a nonreactive serine greatly reduces Orai1's redox sensitivity and mutation of all three cysteines to serines abolished redox sensitivity. Vice versa, mutation of Gly-170 to Cys at the homologous position in Orai3 resulted in gain of function of redox sensitivity and also increased current density by the nonoxidized channels. The finding that a triple cysteine mutant of Orai1 lacks all redox sensitivity of STIM1-mediated peak I_{CRAC} current density suggests that possible concomitant oxidation of STIM1's cysteines does not inhibit SOCE; however, ROS may still slightly alter the kinetics of current development. STIM1-Orai1-mediated currents are not inhibited by ROS once the channel complex is formed; thus a scenario where intracellular ROS leads to STIM activation and subsequent Orai activation is possible. Differentiation of human naïve CD4 T helper cells into effector cells (Th_{eff}) upon TCR stimulation shows an upregulation of the ROS-resistant paralogue Orai3 and of intracellular antioxidant enzymes with cytokine production and proliferation of effector cells becoming more resistant to inhibition by H_2O_2. The oxidant-mediated inhibition of SOCE in Th_{eff} cells displays an increased IC_{50} when compared to naïve cells (Bogeski et al. 2010). Differential ROS resistance of SOCE due to altered Orai3 expression has also been confirmed for primary prostate epithelial cells versus cells derived from prostate cancers (Holzmann et al. 2015) and also for ROS-producing monocytes (Saul et al. 2016), which show high endogenous Orai3 expression and little susceptibility toward ROS-mediated inhibition of SOCE (Fig. 6.3b). In addition, upon bacterial challenge, the Orai3/Orai1 ratio increases even further and allows for tuning a feedback adaptation optimizing Ca^{2+}-dependent ROS production (Saul et al. 2016). These findings show that Orai1 mediates the H_2O_2 inhibition of I_{CRAC} and that upregulation of Orai3 can tune and reduce or even abolish this I_{CRAC} inhibition by increasing homomeric/heteromeric complexes containing Orai3. In a study using heteromeric concatenated channel subunits (i.e., one Orai3 and three Orai1), a single subunit of Orai3 was able to confer redox resistance to Orai1-Orai3 heterotetrameric channels (Alansary et al. 2015). However, structural and physiological analyses show that Orai channels function as hexamers (Hou et al. 2012; Yen et al. 2016) and a recent study by the Gill lab suggested that only the first two subunits of concatenated tetramers participate in the formation of a hexameric pore (Yen et al. 2016), possibly necessitating a reevaluation of the exact Orai1:Orai3 ratio requirements required for redox resistance. While Orai1 is a complex glycosylated membrane protein whose function can be tuned in a cell type-specific manner through changes in its glycosylation pattern (Dorr et al. 2016), neither Orai2 nor Orai3 contain N-glycosylation sites and Orai3 but not Orai2 recruitment to STIM1 clusters at ER-PM junctions in HEK293 cells and to an artificial immunological synapse in Jurkat T cells is facilitated by Orai1 (Alansary et al. 2015). This dependency is obliterated when the C-terminal region of Orai3 is mutated to include the C-terminal "EFA" motif of Orai1, lowering the strength of Orai3's C-terminal coiled-coil prediction but increasing STIM1-Orai3 interaction (FRET), suggesting either a stabilizing effect of these residues for the interaction of docked STIM1 or being a required "structural element" to differentially expose upstream or downstream residues for a stable STIM1-Orai interaction (Alansary et al. 2015). Altogether these findings further support the notion that Orai3 functions together with Orai1 in a heteromeric complex to bring about resistance of SOCE toward oxidative stress and give rise to a model where increasing Orai3 expression decreases the overall amount of SOCE but increases its resistance toward environmental oxidative stress (Fig. 6.3b). In this context it is quite interesting that a number of studies on tumor cells, including breast, prostate, and lung cancer, have reported the involvement of Orai3 in the complex machinery of carcinogenesis, in which Orai3 also mediates non-store-operated calcium entry (Benzerdjeb et al. 2016; Faouzi et al. 2013; Motiani et al. 2010; Zhang et al. 2014). ROS elevated in the tumor microenvironment are associated with tumor-induced immunosuppression (Zhang et al. 2014).

Mechanistically puzzling remained the finding that neither SOCE nor I_{CRAC} was inhibited by H_2O_2 after formation of functional STIM1-Orai1 complexes, suggesting that either H_2O_2 retains the channels in a closed configuration or prevents STIM1 from gating the channels. C195 is located distant to the ion conduction pore encoded by residues of TM1 with E106 constituting the selectivity filter at the outer vestibule of the pore (reviewed in Prakriya and Lewis 2015). To better understand the mechanism of how ROS inhibits generation of SOCE, Alansary et al. combined a number of measurements including analysis of cluster formation, diffusional parameters, and Forster resonance energy transfer (FRET) analysis with MD simulations of hSTIM1 CAD domain docked- or undocked Orai1 and mutational analysis of potentially interacting residues (Alansary et al. 2016). While docking of STIM1 to Orai1 was actually enhanced after ROS preincubation, in contrast to a decreased FRET after hypoxia and cytosolic acidosis seen by Mancarella et al. (2011), Orai1 subunit interaction was decreased. MD simulations of Orai1 with our without docked STIM1 CAD with structural information of hOrai1 morphed onto the published crystal structure of *Drosophila* Orai1 (Hou et al. 2012), suggested that oxidized Cys-195 may interact with a highly conserved Ser-239 from the opposing TM4 region but only in the closed state. Replacing the thiol group in the model by a sulfinyl or sulfonyl group, expected to be formed upon irreversible oxidation of C195, reduces the distance to S239 from 2.70 to 1.59 Å in the closed state, whereas this distance still remained too large in the open state. The shorter distance between C195 and S239 together with the lower pK_a value of the sulfinic acid, which will lead to a deprotonated state under physiological conditions, enables formation of a stable hydrogen bond between the sulfinyl group and S239 in the closed state (Alansary et al. 2016).

Figure 6.3c shows a model of the residues surrounding Cys-195. Indeed, also an oxidomimetic mutation Cys-195/Asp-195 substituting Cys with a negatively charged amino acid was able to interact in the model and phenocopied the ROS-inhibited channel (Alansary et al. 2016). Mutation of Ser-239 to alanine rescued the oxidomimetic current inhibition and prevented ROS-mediated inhibition of I_{CRAC}, confirming that docking of STIM1 at the "nexus" region of Orai1 (see also Zhou et al. 2016, 2017) initiates small rotational shifts in the transmembrane helices that are transmitted toward TM1 and ultimately widen the pore region around the selectivity filter. Oxidation of Cys-158 thus locks the channels in a closed configuration and prevents the stepwise transmission of TM helical movements toward the pore (see also Derler et al. 2012; Palty et al. 2015; Yeung et al. 2016; Zhou et al. 2017). This also implies that small alterations in the TM4 helix, i.e., by heteromerization or possibly association with other proteins may suffice to prevent oxidized Cys-195 to interact and may also explain differential ROS sensitivity of SOCE in different cell types. Whether an oxidomimetic Cys to Asp mutation will be useful in the analysis of other potentially reactive cysteines requires further investigation.

6.5.2 Orai2

The physiological role of Orai2 is much less studied than Orai1 (Hoth and Niemeyer 2013), but Orai2 currents alone show differential Ca^{2+}-dependent inactivation and give rise to smaller currents when compared with Orai1 (DeHaven et al. 2007; Frischauf et al. 2011; Gross et al. 2007; Inayama et al. 2015; Lee et al. 2009; Lis et al. 2007). In Jurkat T cells, Orai2 reduces Orai1 function (Alansary et al. 2015), but in some cells SOCE is likely mediated predominantly by Orai2, as it is prominently expressed in B cells, chondrocytes, and in the brain among others (Diez-Bello et al. 2016; Fernandez et al. 2015; Kito et al. 2015). In Ca^{2+} imaging experiments, Orai2 also displays strong inhibition by preincubation with H_2O_2 (Bogeski et al. 2010); however, the redox sensitivity or Orai1/Orai2 heteromeric channels or of Orai2/Orai3 heteromeric channels has not been studied. In many cases high *Orai2* expression correlates with high *STIM2* expression, and more work is required to figure out the redox regulation of these alternate SOCE protein configurations.

6.6 Summary and Outlook

Understanding the complex interplay between Ca^{2+} and H_2O_2 second messenger systems is an emerging field of research. Tools to accurately measure changes in cellular ROS concentration concomitantly to changes in Ca^{2+} concentration remain a major challenge. Redox proteomics is rapidly developing and will facilitate the identification of candidate ROS regulated proteins involved in Ca^{2+} signaling. How these systems are balanced in health and disease and if STIM2 will potentially emerge as a "driver" for ROS-mediated alterations remains to be seen. Sensing extracellular ROS by Orai1 leading to locked channels and upregulation of the ROS insensitive Orai3 as a physiological adaptation for cells invading ROS rich environments is one of many possible adaptive mechanisms. It will also be interesting to see if mutational analyses of other reactive cysteines confirm an oxidomimetic Cys/Asp mutation as a tool to study cysteine function.

Acknowledgments The author thanks M. Hoth for critical reading of the manuscript and Drs. Alansary and Bogeski for scientific input. This work was supported by grants from the Deutsche Forschungsgemeinschaft (DFG) SFB1027 C4, C7, SFB894 A2 and FOR2289 to B.A.N.

References

Abdesselem M, Ramodiharilafy R, Devys L, Gacoin T, Alexandrou A, Bouzigues CI (2017) Fast quantitative ROS detection based on dual-color single rare-earth nanoparticle imaging reveals signaling pathway kinetics in living cells. Nanoscale 9:656–665

Alansary D, Bogeski I, Niemeyer BA (2015) Facilitation of Orai3 targeting and store-operated function by Orai1. Biochim Biophys Acta 1853:1541–1550

Alansary D, Schmidt B, Dorr K, Bogeski I, Rieger H, Kless A, Niemeyer BA (2016) Thiol dependent intramolecular locking of Orai1 channels. Sci Rep 6:33347

Appenzeller-Herzog C, Banhegyi G, Bogeski I, Davies KJ, Delaunay-Moisan A, Forman HJ, Gorlach A, Kietzmann T, Laurindo F, Margittai E, Meyer AJ, Riemer J, Rutzler M, Simmen T, Sitia R, Toledano MB, Touw IP (2016) Transit of H2O2 across the endoplasmic reticulum membrane is not sluggish. Free Radic Biol Med 94:157–160

Aytes A, Mollevi DG, Martinez-Iniesta M, Nadal M, Vidal A, Morales A, Salazar R, Capella G, Villanueva A (2012) Stromal interaction molecule 2 (STIM2) is frequently overexpressed in colorectal tumors and confers a tumor cell growth suppressor phenotype. Mol Carcinog 51:746–753

Babior BM, Curnutte JT (1987) Chronic granulomatous disease—pieces of a cellular and molecular puzzle. Blood Rev 1:215–218

Bashan N, Kovsan J, Kachko I, Ovadia H, Rudich A (2009) Positive and negative regulation of insulin signaling by reactive oxygen and nitrogen species. Physiol Rev 89:27–71

Bauer MC, O'Connell D, Cahill DJ, Linse S (2008) Calmodulin binding to the polybasic C-termini of STIM proteins involved in store-operated calcium entry. Biochemistry 47:6089–6091

Bedard K, Krause KH (2007) The NOX family of ROS-generating NADPH oxidases: physiology and pathophysiology. Physiol Rev 87:245–313

Beedle AE, Lynham S, Garcia-Manyes S (2016) Protein S-sulfenylation is a fleeting molecular switch that regulates non-enzymatic oxidative folding. Nat Commun 7:12490

Benzerdjeb N, Sevestre H, Ahidouch A, Ouadid-Ahidouch H (2016) Orai3 is a predictive marker of metastasis and survival in resectable lung adenocarcinoma. Oncotarget 7:81588–81597

Berna-Erro A, Braun A, Kraft R, Kleinschnitz C, Schuhmann MK, Stegner D, Wultsch T, Eilers J, Meuth SG, Stoll G, Nieswandt B (2009) STIM2 regulates capacitive Ca2+ entry in neurons and plays a key role in hypoxic neuronal cell death. Sci Signal 2:ra67

Berna-Erro A, Jardin I, Salido GM, Rosado JA (2017) Role of STIM2 in cell function and physiopathology. J Physiol, Jan 14. doi:10.1113/JP273889

Bhardwaj R, Muller HM, Nickel W, Seedorf M (2013) Oligomerization and Ca2+/calmodulin control binding of the ER Ca2+-sensors STIM1 and STIM2 to plasma membrane lipids. Biosci Rep 33(5):e00077. doi:10.1042/BSR20130089

Bhardwaj R, Hediger MA, Demaurex N (2016) Redox modulation of STIM-ORAI signaling. Cell Calcium 60:142–152

Bogeski I, Kummerow C, Al-Ansary D, Schwarz EC, Koehler R, Kozai D, Takahashi N, Peinelt C, Griesemer D, Bozem M, Mori Y, Hoth M, Niemeyer BA (2010) Differential redox regulation of ORAI ion channels: a mechanism to tune cellular calcium signaling. Sci Signal 3:ra24

Bogeski I, Kappl R, Kummerow C, Gulaboski R, Hoth M, Niemeyer BA (2011) Redox regulation of calcium ion channels: chemical and physiological aspects. Cell Calcium 50:407–423

Bogeski I, Kilch T, Niemeyer BA (2012) ROS and SOCE: recent advances and controversies in the regulation of STIM and Orai. J Physiol 590:4193–4200

Brandman O, Liou J, Park WS, Meyer T (2007) STIM2 is a feedback regulator that stabilizes basal cytosolic and endoplasmic reticulum Ca2+ levels. Cell 131:1327–1339

Brookes PS, Yoon Y, Robotham JL, Anders MW, Sheu SS (2004) Calcium, ATP, and ROS: a mitochondrial love-hate triangle. Am J Physiol Cell Physiol 287:C817–C833

Chen K, Kirber MT, Xiao H, Yang Y, Keaney JF Jr (2008) Regulation of ROS signal transduction by NADPH oxidase 4 localization. J Cell Biol 181:1129–1139

Chen T, Xu X, Zhao Z, Zhao F, Gao Y, Yan X, Wan Y (2017) Hydrogen peroxide is a critical regulator of the hypoxia-induced alterations of store-operated Ca2+ entry into rat pulmonary arterial smooth muscle cells. Am J Physiol Lung Cell Mol Physiol 312(4):L477–L487

Chouchani ET, James AM, Fearnley IM, Lilley KS, Murphy MP (2011) Proteomic approaches to the characterization of protein thiol modification. Curr Opin Chem Biol 15:120–128

Clempus RE, Griendling KK (2006) Reactive oxygen species signaling in vascular smooth muscle cells. Cardiovasc Res 71:216–225

Coenen MJ, Gregersen PK (2009) Rheumatoid arthritis: a view of the current genetic landscape. Genes Immun 10:101–111

Conway ME, Lee C (2015) The redox switch that regulates molecular chaperones. Biomol Concepts 6:269–284

Couto N, Wood J, Barber J (2016) The role of glutathione reductase and related enzymes on cellular redox homeostasis network. Free Radic Biol Med 95:27–42

Curnutte JT, Babior BM (1987) Chronic granulomatous disease. Adv Hum Genet 16:229–297

Darbellay B, Arnaudeau S, Bader CR, Konig S, Bernheim L (2011) STIM1L is a new actin-binding splice variant involved in fast repetitive Ca2+ release. J Cell Biol 194:335–346

Davalli P, Mitic T, Caporali A, Lauriola A, D'Arca D (2016) ROS, cell senescence, and novel molecular mechanisms in aging and age-related diseases. Oxidative Med Cell Longev 2016:3565127

DeHaven WI, Smyth JT, Boyles RR, Putney JW Jr (2007) Calcium inhibition and calcium potentiation of Orai1, Orai2, and Orai3 calcium release-activated calcium channels. J Biol Chem 282:17548–17556

Derler I, Madl J, Schutz G, Romanin C (2012) Structure, regulation and biophysics of I(CRAC), STIM/Orai1. Adv Exp Med Biol 740:383–410

Diez-Bello R, Jardin I, Salido GM, Rosado JA (2016) Orai1 and Orai2 mediate store-operated calcium entry that regulates HL60 cell migration and FAK phosphorylation. Biochim Biophys Acta 16:30300–7. doi:10.1016/j.bbamcr.2016.11.014

Dorr K, Kilch T, Kappel S, Alansary D, Schwar G, Niemeyer BA, Peinelt C (2016) Cell type-specific glycosylation of Orai1 modulates store-operated Ca2+ entry. Sci Signal 9:ra25

Droge W (2002a) Aging-related changes in the thiol/disulfide redox state: implications for the use of thiol antioxidants. Exp Gerontol 37:1333–1345

Droge W (2002b) Free radicals in the physiological control of cell function. Physiol Rev 82:47–95

Faouzi M, Kischel P, Hague F, Ahidouch A, Benzerdjeb N, Sevestre H, Penner R, Ouadid-Ahidouch H (2013) ORAI3 silencing alters cell proliferation and cell cycle progression via c-myc pathway in breast cancer cells. Biochim Biophys Acta 1833:752–760

Fernandez RA, Wan J, Song S, Smith KA, Gu Y, Tauseef M, Tang H, Makino A, Mehta D, Yuan JX (2015) Upregulated expression of STIM2, TRPC6, and Orai2 contributes to the transition of pulmonary arterial smooth muscle cells from a contractile to proliferative phenotype. Am J Physiol Cell Physiol 308:C581–C593

Feske S (2007) Calcium signalling in lymphocyte activation and disease. Nat Rev Immunol 7:690–702

Feske S (2009) ORAI1 and STIM1 deficiency in human and mice: roles of store-operated Ca2+ entry in the immune system and beyond. Immunol Rev 231:189–209

Feske S, Gwack Y, Prakriya M, Srikanth S, Puppel SH, Tanasa B, Hogan PG, Lewis RS, Daly M, Rao A (2006) A mutation in Orai1 causes immune deficiency by abrogating CRAC channel function. Nature 441:179–185

Forman HJ, Fukuto JM, Torres M (2004) Redox signaling: thiol chemistry defines which reactive oxygen and nitrogen species can act as second messengers. Am J Physiol Cell Physiol 287: C246–C256

Frischauf I, Schindl R, Bergsmann J, Derler I, Fahrner M, Muik M, Fritsch R, Lackner B, Groschner K, Romanin C (2011) Cooperativeness of Orai cytosolic domains tunes subtype-specific gating. J Biol Chem 286:8577–8584

Gibhardt CS, Zimmermann KM, Zhang X, Belousov VV, Bogeski I (2016) Imaging calcium and redox signals using genetically encoded fluorescent indicators. Cell Calcium 60:55–64

Gilgun-Sherki Y, Melamed E, Offen D (2004) The role of oxidative stress in the pathogenesis of multiple sclerosis: the need for effective antioxidant therapy. J Neurol 251:261–268
Graham SJ, Dziadek MA, Johnstone LS (2011) A cytosolic STIM2 preprotein created by signal peptide inefficiency activates ORAI1 in a store-independent manner. J Biol Chem 286:16174–16185
Griffiths HR (2008) Is the generation of neo-antigenic determinants by free radicals central to the development of autoimmune rheumatoid disease? Autoimmun Rev 7:544–549
Gross SA, Wissenbach U, Philipp SE, Freichel M, Cavalie A, Flockerzi V (2007) Murine ORAI2 splice variants form functional Ca2+ release-activated Ca2+ (CRAC) channels. J Biol Chem 282:19375–19384
Grupe M, Myers G, Penner R, Fleig A (2010) Activation of store-operated I(CRAC) by hydrogen peroxide. Cell Calcium 48:1–9
Hawkins BJ, Irrinki KM, Mallilankaraman K, Lien YC, Wang Y, Bhanumathy CD, Subbiah R, Ritchie MF, Soboloff J, Baba Y, Kurosaki T, Joseph SK, Gill DL, Madesh M (2010) S-glutathionylation activates STIM1 and alters mitochondrial homeostasis. J Cell Biol 190:391–405
Holzmann C, Kilch T, Kappel S, Dorr K, Jung V, Stockle M, Bogeski I, Peinelt C (2015) Differential redox regulation of Ca(2)(+) signaling and viability in normal and malignant prostate cells. Biophys J 109:1410–1419
Hoth M (2016) CRAC channels, calcium, and cancer in light of the driver and passenger concept. Biochim Biophys Acta 1863:1408–1417
Hoth M, Niemeyer BA (2013) The neglected CRAC proteins: Orai2, Orai3, and STIM2. Curr Top Membr 71:237–271
Hoth M, Penner R (1992) Depletion of intracellular calcium stores activates a calcium current in mast cells. Nature 355:353–356
Hou X, Pedi L, Diver MM, Long SB (2012) Crystal structure of the calcium release-activated calcium channel Orai. Science 338:1308–1313
Hultqvist M, Olofsson P, Holmberg J, Backstrom BT, Tordsson J, Holmdahl R (2004) Enhanced autoimmunity, arthritis, and encephalomyelitis in mice with a reduced oxidative burst due to a mutation in the Ncf1 gene. Proc Natl Acad Sci U S A 101:12646–12651
Inayama M, Suzuki Y, Yamada S, Kurita T, Yamamura H, Ohya S, Giles WR, Imaizumi Y (2015) Orai1-Orai2 complex is involved in store-operated calcium entry in chondrocyte cell lines. Cell Calcium 57:337–347
Kirkwood TB, Kowald A (2012) The free-radical theory of ageing—older, wiser and still alive: modelling positional effects of the primary targets of ROS reveals new support. Bioessays 34:692–700
Kito H, Yamamura H, Suzuki Y, Ohya S, Asai K, Imaizumi Y (2015) Regulation of store-operated Ca2+ entry activity by cell cycle dependent up-regulation of Orai2 in brain capillary endothelial cells. Biochem Biophys Res Commun 459:457–462
Kourie JI (1998) Effects of ATP-sensitive potassium channel regulators on chloride channels in the sarcoplasmic reticulum vesicles from rabbit skeletal muscle. J Membr Biol 164:47–58
Kwon MJ, Kim B, Lee YS, Kim TY (2012) Role of superoxide dismutase 3 in skin inflammation. J Dermatol Sci 67:81–87
Lacruz RS, Feske S (2015) Diseases caused by mutations in ORAI1 and STIM1. Ann N Y Acad Sci 1356(1):45–79
Lee KP, Yuan JP, Zeng W, So I, Worley PF, Muallem S (2009) Molecular determinants of fast Ca2+-dependent inactivation and gating of the Orai channels. Proc Natl Acad Sci U S A 106:14687–14692
Leichert LI, Dick TP (2015) Incidence and physiological relevance of protein thiol switches. Biol Chem 396:389–399
Lewis RS (2001) Calcium signaling mechanisms in T lymphocytes. Annu Rev Immunol 19:497–521
Liou J, Kim ML, Heo WD, Jones JT, Myers JW, Ferrell JE Jr, Meyer T (2005) STIM is a Ca2+ sensor essential for Ca2+-store-depletion-triggered Ca2+ influx. Curr Biol 15:1235–1241

Lis A, Peinelt C, Beck A, Parvez S, Monteilh-Zoller M, Fleig A, Penner R (2007) CRACM1, CRACM2, and CRACM3 are store-operated Ca2+ channels with distinct functional properties. Curr Biol 17:794–800
Ma G, Wei M, He L, Liu C, Wu B, Zhang SL, Jing J, Liang X, Senes A, Tan P, Li S, Sun A, Bi Y, Zhong L, Si H, Shen Y, Li M, Lee MS, Zhou W, Wang J, Wang Y, Zhou Y (2015) Inside-out Ca(2+) signalling prompted by STIM1 conformational switch. Nat Commun 6:7826
Mancarella S, Wang Y, Deng X, Landesberg G, Scalia R, Panettieri RA, Mallilankaraman K, Tang XD, Madesh M, Gill DL (2011) Hypoxia-induced acidosis uncouples the STIM-Orai calcium signaling complex. J Biol Chem 286:44788–44798
Marino SM, Gladyshev VN (2010) Cysteine function governs its conservation and degeneration and restricts its utilization on protein surfaces. J Mol Biol 404:902–916
Miederer AM, Alansary D, Schwar G, Lee PH, Jung M, Helms V, Niemeyer BA (2015) A STIM2 splice variant negatively regulates store-operated calcium entry. Nat Commun 6:6899
Moccia F, Zuccolo E, Soda T, Tanzi F, Guerra G, Mapelli L, Lodola F, D'Angelo E (2015) Stim and Orai proteins in neuronal Ca(2+) signaling and excitability. Front Cell Neurosci 9:153
Morgan MJ, Liu ZG (2011) Crosstalk of reactive oxygen species and NF-kappaB signaling. Cell Res 21:103–115
Motiani RK, Abdullaev IF, Trebak M (2010) A novel native store-operated calcium channel encoded by Orai3: selective requirement of Orai3 versus Orai1 in estrogen receptor-positive versus estrogen receptor-negative breast cancer cells. J Biol Chem 285:19173–19183
Mungai PT, Waypa GB, Jairaman A, Prakriya M, Dokic D, Ball MK, Schumacker PT (2011) Hypoxia triggers AMPK activation through reactive oxygen species-mediated activation of calcium release-activated calcium channels. Mol Cell Biol 31:3531–3545
Nunes P, Demaurex N (2014) Redox regulation of store-operated Ca2+ entry. Antioxid Redox Signal 21:915–932
Oparka M, Walczak J, Malinska D, van Oppen LM, Szczepanowska J, Koopman WJ, Wieckowski MR (2016) Quantifying ROS levels using CM-H2DCFDA and HyPer. Methods 109:3–11
Palty R, Stanley C, Isacoff EY (2015) Critical role for Orai1 C-terminal domain and TM4 in CRAC channel gating. Cell Res 25:963–980
Parekh AB, Putney JW Jr (2005) Store-operated calcium channels. Physiol Rev 85:757–810
Park CY, Hoover PJ, Mullins FM, Bachhawat P, Covington ED, Raunser S, Walz T, Garcia KC, Dolmetsch RE, Lewis RS (2009) STIM1 clusters and activates CRAC channels via direct binding of a cytosolic domain to Orai1. Cell 136:876–890
Parvez S, Beck A, Peinelt C, Soboloff J, Lis A, Monteilh-Zoller M, Gill DL, Fleig A, Penner R (2008) STIM2 protein mediates distinct store-dependent and store-independent modes of CRAC channel activation. FASEB J 22:752–761
Prakriya M, Lewis RS (2015) Store-operated calcium channels. Physiol Rev 95:1383–1436
Prins D, Groenendyk J, Touret N, Michalak M (2011) Modulation of STIM1 and capacitative Ca2 + entry by the endoplasmic reticulum luminal oxidoreductase ERp57. EMBO Rep 12:1182–1188
Rana A, Yen M, Sadaghiani AM, Malmersjo S, Park CY, Dolmetsch RE, Lewis RS (2015) Alternative splicing converts STIM2 from an activator to an inhibitor of store-operated calcium channels. J Cell Biol 209:653–669
Reth M (2002) Hydrogen peroxide as second messenger in lymphocyte activation. Nat Immunol 3:1129–1134
Rhee SG (2006) Cell signaling. H2O2, a necessary evil for cell signaling. Science 312:1882–1883
Saul S, Gibhardt CS, Schmidt B, Lis A, Pasieka B, Conrad D, Jung P, Gaupp R, Wonnenberg B, Diler E, Stanisz H, Vogt T, Schwarz EC, Bischoff M, Herrmann M, Tschernig T, Kappl R, Rieger H, Niemeyer BA, Bogeski I (2016) A calcium-redox feedback loop controls human monocyte immune responses: the role of ORAI Ca2+ channels. Sci Signal 9:ra26
Schieber M, Chandel NS (2014) ROS function in redox signaling and oxidative stress. Curr Biol 24:R453–R462
Schindl R, Frischauf I, Bergsmann J, Muik M, Derler I, Lackner B, Groschner K, Romanin C (2009) Plasticity in Ca2+ selectivity of Orai1/Orai3 heteromeric channel. Proc Natl Acad Sci U S A 106:19623–19628

Segal AW (1996) The NADPH oxidase and chronic granulomatous disease. Mol Med Today 2:129–135
Shin DH, Nam JH, Lee ES, Zhang Y, Kim SJ (2012) Inhibition of Ca(2+) release-activated Ca(2+) channel (CRAC) by curcumin and caffeic acid phenethyl ester (CAPE) via electrophilic addition to a cysteine residue of Orai1. Biochem Biophys Res Commun 428:56–61
Song MY, Makino A, Yuan JX (2011) STIM2 contributes to enhanced store-operated Ca2+ entry in pulmonary artery smooth muscle cells from patients with idiopathic pulmonary arterial hypertension. Pulm Circ 1:84–94
Sospedra M, Martin R (2005) Immunology of multiple sclerosis. Annu Rev Immunol 23:683–747
Starkov AA (2008) The role of mitochondria in reactive oxygen species metabolism and signaling. Ann N Y Acad Sci 1147:37–52
Stathopulos PB, Zheng L, Li GY, Plevin MJ, Ikura M (2008) Structural and mechanistic insights into STIM1-mediated initiation of store-operated calcium entry. Cell 135:110–122
Sun S, Zhang H, Liu J, Popugaeva E, Xu NJ, Feske S, White CL 3rd, Bezprozvanny I (2014) Reduced synaptic STIM2 expression and impaired store-operated calcium entry cause destabilization of mature spines in mutant presenilin mice. Neuron 82:79–93
Velmurugan GV, Huang H, Sun H, Candela J, Jaiswal MK, Beaman KD, Yamashita M, Prakriya M, White C (2015) Depletion of H2S during obesity enhances store-operated Ca2+ entry in adipose tissue macrophages to increase cytokine production. Sci Signal 8:ra128
Vig M, Peinelt C, Beck A, Koomoa DL, Rabah D, Koblan-Huberson M, Kraft S, Turner H, Fleig A, Penner R, Kinet JP (2006) CRACM1 is a plasma membrane protein essential for store-operated Ca2+ entry. Science 312:1220–1223
Watanabe S, Moniaga CS, Nielsen S, Hara-Chikuma M (2016) Aquaporin-9 facilitates membrane transport of hydrogen peroxide in mammalian cells. Biochem Biophys Res Commun 471:191–197
Wible RS, Sutter TR (2017) Soft Cysteine Signaling Network: the functional significance of cysteine in protein function and the soft acid/base thiol chemistry that facilitates cysteine modification. Chem Res Toxicol 30(3):729–762
Yang J, Carroll KS, Liebler DC (2016) The expanding landscape of the thiol redox proteome. Mol Cell Proteomics 15:1–11
Yen M, Lokteva LA, Lewis RS (2016) Functional analysis of Orai1 concatemers supports a hexameric stoichiometry for the CRAC channel. Biophys J 111:1897–1907
Yeung PS, Yamashita M, Prakriya M (2016) Pore opening mechanism of CRAC channels. Cell Calcium 16:30212–3. doi:10.1016/j.ceca.2016.12.006
Zhang SL, Yu Y, Roos J, Kozak JA, Deerinck TJ, Ellisman MH, Stauderman KA, Cahalan MD (2005) STIM1 is a Ca2+ sensor that activates CRAC channels and migrates from the Ca2+ store to the plasma membrane. Nature 437:902–905
Zhang SL, Yeromin AV, Zhang XH, Yu Y, Safrina O, Penna A, Roos J, Stauderman KA, Cahalan MD (2006) Genome-wide RNAi screen of Ca(2+) influx identifies genes that regulate Ca(2+) release-activated Ca(2+) channel activity. Proc Natl Acad Sci U S A 103:9357–9362
Zhang X, Zhang W, Gonzalez-Cobos JC, Jardin I, Romanin C, Matrougui K, Trebak M (2014) Complex role of STIM1 in the activation of store-independent Orai1/3 channels. J Gen Physiol 143:345–359
Zhou Y, Cai X, Loktionova NA, Wang X, Nwokonko RM, Wang Y, Rothberg BS, Trebak M, Gill DL (2016) The STIM1-binding site nexus remotely controls Orai1 channel gating. Nat Commun 7:13725
Zhou Y, Cai X, Nwokonko RM, Loktionova NA, Wang Y, Gill DL (2017) The STIM-Orai coupling interface and gating of the Orai1 channel. Cell Calcium 16:30224–X. doi:10.1016/j.ceca.2017.01.001

The STIM-Orai Pathway: Light-Operated Ca^{2+} Entry Through Engineered CRAC Channels

7

Guolin Ma, Shufan Wen, Yun Huang, and Yubin Zhou

Abstract

Ca^{2+} signals regulate a plethora of cellular functions that include muscle contraction, heart beating, hormone secretion, lymphocyte activation, gene expression, and metabolism. To study the impact of Ca^{2+} signals on biological processes, pharmacological tools and caged compounds have been commonly applied to induce fluctuations of intracellular Ca^{2+} concentrations. These conventional approaches, nonetheless, lack rapid reversibility and high spatiotemporal resolution. To overcome these disadvantages, we and others have devised a series of photoactivatable genetically encoded Ca^{2+} actuators (GECAs) by installing light sensitivities into a bona fide highly selective Ca^{2+} channel, the Ca^{2+} release-activated Ca^{2+} (CRAC) channel. Store-operated CRAC channel serves as a major route for Ca^{2+} entry in many cell types. These GECAs enable remote and precise manipulation of Ca^{2+} signaling in both excitable and non-excitable cells. When combined with nanotechnology, it becomes feasible to wirelessly photo-modulate Ca^{2+}-dependent activities in vivo. In this chapter,

G. Ma (✉) • S. Wen
Center for Translational Cancer Research, Institute of Biosciences and Technology, Texas A&M University, Houston, TX 77030, USA
e-mail: gma@ibt.tamhsc.edu

Y. Huang
Center for Epigenetics and Disease Prevention, Institute of Biosciences and Technology, Texas A&M University, Houston, TX 77030, USA

Y. Zhou (✉)
Center for Translational Cancer Research, Institute of Biosciences and Technology, Texas A&M University, Houston, TX 77030, USA

Department of Medical Physiology, College of Medicine, Texas A&M University, Temple, TX, USA
e-mail: yzhou@ibt.tamhsc.edu

K. Groschner et al. (eds.), *Store-Operated Ca^{2+} Entry (SOCE) Pathways*, Advances in Experimental Medicine and Biology 993, DOI 10.1007/978-3-319-57732-6_7

we briefly review most recent advances in engineering CRAC channels to achieve optical control over Ca^{2+} signaling, outline their design principles and kinetic features, and present exemplary applications of GECAs engineered from CRAC channels.

Keywords
Optogenetics • Calcium release-activated calcium (CRAC) channel • Immune response • STIM1 • LOV2 • Cryptochrome

7.1 Background

Optogenetics is a powerful technique that combines genetics and optics to remotely control biological processes in living cells, tissues, and organism (Fenno et al. 2011). Over the past decade, the use of microbial opsin-based optogenetics has given rise to important advances in neuroscience by allowing neuroscientists to interrogate neural circuits with unprecedented precision (Deisseroth 2011, 2015; Fenno et al. 2011). In parallel, non-opsin-based optogenetic tools have been gaining rapid momentum in applications beyond neuroscience (Kianianmomeni 2015; Tan et al. 2016; Tischer and Weiner 2014; Zhang and Cui 2015). Very recently, a set of tools tailored for optical control of Ca^{2+} signaling (designated as "genetically encoded Ca^{2+} actuators" or GECAs) have been created by endowing a highly Ca^{2+}-selective channel, the Ca^{2+} release-activated Ca^{2+} (CRAC) channel, with light sensitivity. These innovative tools are built upon photosensory domains derived from *Arabidopsis thaliana* cryptochrome 2 (AtCRY2) (Kyung et al. 2015) or the light-oxygen-voltage-sensing (LOV) domain of *Avena sativa* phototropin 1 (He et al. 2015; Ishii et al. 2015; Tan et al. 2016). GECAs engineered from CRAC channels, when used in combination with genetically encoded Ca^{2+} indicators (GECIs), offer enormous untapped potential for simultaneously perturbing and recording Ca^{2+} signals at real time in living cells, tissues, and even organisms.

Store-operated Ca^{2+} release-activated Ca^{2+} (CRAC) channels serve as a major route for Ca^{2+} entry in many cell types (Cahalan et al. 2007; Hogan et al. 2010; Parekh and Putney 2005; Prakriya and Lewis 2015; Putney 1986; Soboloff et al. 2012; Vig and Kinet 2007). CRAC channels comprise two major protein families, the stromal interaction molecules (STIM1 and STIM2) as ER luminal Ca^{2+} sensors (Liou et al. 2005; Roos et al. 2005; Zhang et al. 2005) and the pore-forming subunit ORAI (ORAI1, ORAI2, and ORAI3) situated on the plasma membrane (PM) (Feske et al. 2006; Vig et al. 2006). The activation of CRAC channel is initiated by the depletion of the ER Ca^{2+} store that successively induces the oligomerization of the STIM1 luminal domain (Stathopulos et al. 2008), rearrangement of the single-pass transmembrane domain (Ma et al. 2015; Zhou et al. 2013), and a conformational switch within the STIM1 cytoplasmic domain (STIM1ct; Fig. 7.1a) (Fahrner et al. 2014; Ma et al. 2015; Muik et al. 2011; Zhou et al. 2013),

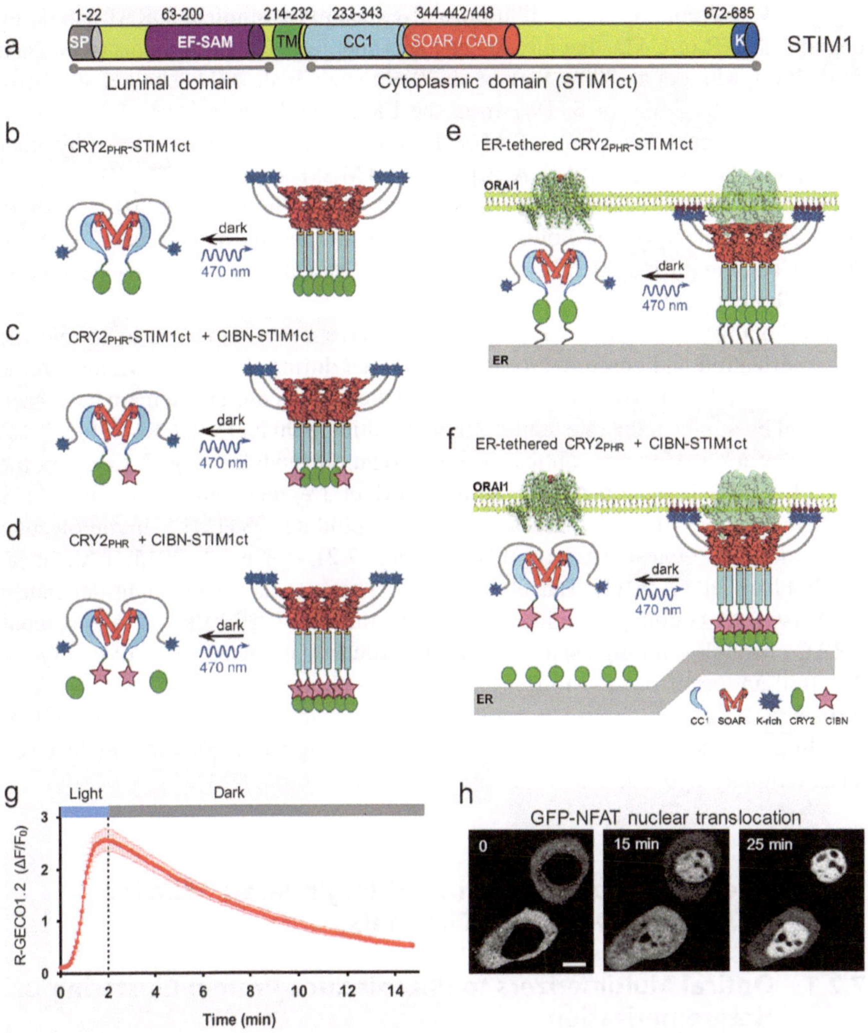

Fig. 7.1 Photoinducible clustering of STIM1ct to elicit Ca^{2+} influx through ORAI channels. (**a**) Domain architecture of human STIM1. *SP* signal peptide, *EF* the helix-loop-helix EF-hand motif, *SAM* sterile alpha motif, *TM* transmembrane region, *CC1* predicted coiled-coil region 1, *SOAR/CAD* STIM1-ORAI-activating region/CRAC activation domain, *K* polybasic C-tail. (**b–f**) Cartoon illustration of CRY2-/CIBN-STIM1ct chimeras. Photoactivatable Ca^{2+} influx can be achieved by using the following strategies: (**b**) $CRY2_{PHR}$-STIM1ct homo-oligomerization, (**c**) heteromerization between $CRY2_{PHR}$-STIM1ct and CIBN-STIM1ct, (**d**) heteromerization between cytosolic $CRY2_{PHR}$ and CIBN-STIM1ct, (**e**) homo-association of ER-anchored $CRY2_{PHR}$-STIM1ct, and (**f**) heteromerization between ER-tethered $CRY2_{PHR}$ and cytosolic CIBN-STIM1ct. (**g**) A typical light-activated Ca^{2+} response curve recorded in HeLa cells co-expressing $CRY2_{PHR}$-$STIM1_{233–685}$ and the red-emitting GECI, R-GECO1.2. Transfected cells were illuminated with blue light at 470 nm with a power density of 40 $\mu W/mm^2$ for 2 min and then returned to the dark. Upon exposure to blue light, the cytosolic Ca^{2+} rapidly increased with a halftime of 38.4 ± 2.1 s. The Ca^{2+} signals decayed with a halftime of 320 ± 37.2 s after switching off the blue light. (**h**)

ultimately overcoming autoinhibition and exposing a minimal ORAI-activating domain, SOAR or CAD, that directly engages and gates ORAI channels (Kawasaki et al. 2009; Muik et al. 2009; Park et al. 2009; Yuan et al. 2009; Zhou et al. 2010). The efficient targeting of STIM1 from the ER network toward the plasma membrane, as well as the dynamic ORAI-STIM coupling, is further facilitated by other protein regulators (e.g., CRAC2A, septin, and TMEM110/STIMATE) (Hooper and Soboloff 2015; Jing et al. 2015; Quintana et al. 2015; Sharma et al. 2013; Srikanth et al. 2010) and bioactive lipids embedded in the plasma membrane (Cao et al. 2015; Chang and Liou 2016; Derler et al. 2016; Maleth et al. 2014; Pacheco et al. 2016; Zhou et al. 2013).

Two optogenetic engineering approaches have been employed to mimic the oligomerization and conformational switch steps during STIM1 activation. In the first strategy, the STIM1 luminal domain and the transmembrane domain have been replaced by a light-inducible homo-oligomerization domain from CRY2 (Fig. 7.1b) (Kyung et al. 2015) or by optical heteromerization modules (Fig. 7.1c–e). In the second approach, an ORAI-activating STIM1ct fragment has been fused to a photoswitchable LOV2 domain to recapitulate STIM1 intramolecular autoinhibition (Opto-CRAC or BACCS; Fig. 7.2) (He et al. 2015; Ishii et al. 2015; Pham et al. 2011; Tan et al. 2016). These tools enable light-dependent STIM1-ORAI coupling to elicit Ca^{2+} influx through Ca^{2+}-selective endogenous CRAC channels with high spatiotemporal resolution in a wide range of cell types. The trail-and-error process to test and optimize photoactivatable CRAC channels further yields insights into the regulatory mechanism of functional STIM-ORAI coupling. In this chapter, we briefly overview the design principles, kinetic features, and exemplary applications of GECAs engineered from STIM1.

7.2 Photoinducible Clustering of Engineered STIM1ct to Activate ORAI Ca^{2+} Channels

7.2.1 Optical Multimerizers to Photoinduce Protein Clustering or Heteromerization

Optical homo-/heteromerization systems (termed as "optical multimerizers") are based on photosensory domains that respond to light with the excitation wavelengths ranging from 350 nm (UV) to 750 nm (far-red). These modules include ultraviolet-B receptors (UVR8), cryptochromes (CRY2), LOV domains, and phytochromes (PhyB) (Kianianmomeni 2015; Pudasaini et al. 2015; Tan et al. 2016; Tischer and Weiner 2014; Zhang and Cui 2015; Zoltowski et al. 2009). An optogenetic heteromerization system typically contains two components:

Fig. 7.1 (continued) Confocal images showing light-inducible nuclear translocation of GFP-NFAT1 in HeLa cells expressing CRY2-STIM1$_{233–685}$. Transfected cells were continuously illuminated by a 470-nm blue light LED (power density: 40 μW/mm^2). *Scale bar* 10 μm

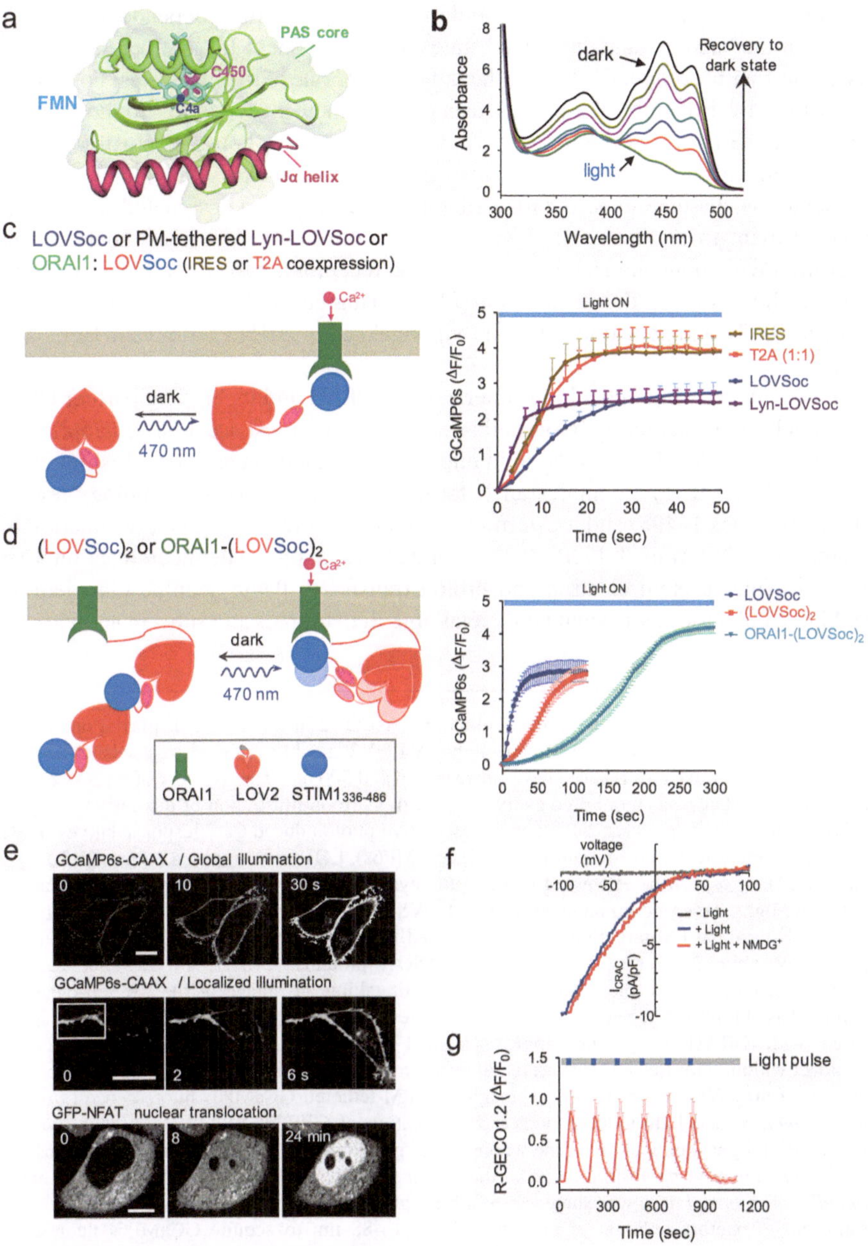

Fig. 7.2 LOV2-STIM1ct chimeras designed to photoactivate Ca^{2+} entry with high spatiotemporal precision in mammalian cells. (**a**) The three-dimensional structure of the dark-state LOV2 domain from *Avena sativa* phototropin 1 (PDB entry: 2V0W), with the PAS core highlighted in *green* and the C-terminal helix in *red*. The FMN chromophore is shown in *cyan* as *sticks* (with the C4a position shown as a *blue dot*), whereas the conserved cysteine residue is shown in *magenta*. (**b**) UV-visible absorbance profiles of purified recombinant AsLOV2 before and after photoactivation.

Component I as a photosensory module that undergoes light-dependent switches between the "active" and "inactive" states and Component II which preferentially associates with Component I when the latter adopts one of the two states. Following the withdrawal of light, the photosensory Component I reverts to ground state to terminate the light-dependent reactions within seconds to hours. By fusing signaling proteins of interest to Components I and II, one can easily manipulate protein associations, dissociations, and reversible translocations to a designated subcellular compartment to regulate cell signaling and control cellular events, such as cell motility (Weitzman and Hahn 2014), tyrosine receptor kinase signaling (Kim et al. 2014), phosphoinositide metabolism (Idevall-Hagren et al. 2012; Kakumoto and Nakata 2013), and gene transcription (Nihongaki et al. 2015; Polstein and Gersbach 2015).

Of all the existing optical multimerizers, AtCRY2 and its binding partner CIB1 (cryptochrome-interacting basic-helix-loop-helix) are most successfully applied to control intermolecular association in mammals (Kennedy et al. 2010; Taslimi et al. 2014, 2016). AtCRY2 contains an N-terminal DNA photolyase homology region (PHR; residues 1–498) and a C-terminal domain required for signal transduction in plants. The N-terminal PHR domain undergoes a photochemical reaction that involves intra-protein electron and proton transfer to flavin adenine dinucleotide (FAD), a cofactor that is abundantly available in mammals and stays in an oxidized

---

Fig. 7.2 (continued) Upon blue light illumination, AsLOV2 in its lit-state displayed one major peak at 390 nm. In the absence of blue light, AsLOV2 quickly returned to its dark state, as characterized by three absorbance peaks between 400 and 500 nm. The recovery of AsLOV2 from the lit to dark states was monitored every 25 s. (**c**, **d**) Cartoon illustration of the design of Opto-CRAC constructs (*left panels*) and their representative photoinduced Ca^{2+} response curves (*right panels*; reported by the green-emitting GECI, GCaMP6s). LOVSoc composed of $AsLOV2_{404–546}$ and $STIM1_{336–486}$ can be expressed alone in the cytosol to activate endogenous ORAI channel with blue light. To speed up photoactivation, LOVSoc can be tethered to the plasma membrane with an N-terminal PM-targeting sequence derived from Lyn kinase (Lyn-LOVSoc). To enable photoactivatable Ca^{2+} entry in cells lacking ORAI proteins, ORAI1 and LOVSoc can be co-expressed by using a bicistronic IRES vector or taking advantage of the 2A self-cleaving peptide-based multiple gene expression system. Alternatively, a dimeric form of LOVSoc can be fused with ORAI1 to produce more sustained Ca^{2+} influx but with delayed onset of photoactivation. Transfected cells were subjected to light stimulation at 470 nm with a power density of 40 μW/mm^2. (**e**) Confocal imaging of PM-tethered GCaMP6s fluorescence (*top* and *middle panels*) and light-induced nuclear translocation of GFP-NFAT1 in HeLa cells (*bottom panel*). Blue light illumination on the whole imaging field induced a global increase of cytosolic Ca^{2+} (*top*). Focused photostimulation in the *boxed area* induced local activation of Ca^{2+} entry (*middle*). The *boxed area* was subjected to a brief photoexcitation with a 488-nm laser for 10 s, followed by photostimulation of the whole field at 488 nm to acquire GCaMP6s fluorescent signals. *Scale bar*, 10 μm. (**f**) Blue light-generated current-voltage relations of Opto-CRAC currents in HEK293-ORAI1 cells expressing LOVSoc. mCherry-positive cells were subjected to whole-cell patch clamp by a ramp protocol ranging from −100 to 100 mV. For the *red curve*, extracellular Na^+ was substituted by a non-permeant ion $NMDG^+$ to assess the ion selectivity. (**g**) A representative example of light-inducible Ca^{2+} oscillation pattern generated by HeLa cells expressing Opto-CRAC when exposed to repeated light-dark cycles (30 s ON and 120 s OFF; 470 nm with a power density of 40 μW/mm^2)

state in the dark (Giovani et al. 2003). Upon blue light illumination, photoreduction of FAD results in the formation of a neutral radical redox state (Banerjee et al. 2007; Kondoh et al. 2011) and further enables light-dependent association between CRY2 and CIB1, with the minimal interacting domains mapped to the N-terminal domains of both CRY2 (aa 1–498) and CIB1 (Kennedy et al. 2010; Taslimi et al. 2016). The lifetime of this intermediate redox state correlates with that of the biologically active "lit" state for CRY2 (Procopio et al. 2016) and can be tuned through introduction of mutations into key positions surrounding the FAD-binding pocket (Taslimi et al. 2016). In addition to light-inducible CRY2-CIB1 heterodimerization, CRY2 by itself undergoes homo-oligomerization when exposed to blue light (Lee et al. 2014; Park et al. 2016), which can be further enhanced by introducing a single mutation into CRY2 (E490G; termed "CRY2olig") (Taslimi et al. 2014). Collectively, AtCRY2 can be used as a dual-purpose tool to induce both protein clustering and protein-protein heterodimerization to initiate or terminate cell signaling. Here we use CRY2- or CIB1-STIM1ct chimeras as an example to demonstrate the design and application of CRY2-based optogenetic tools to control Ca^{2+} entry into mammalian cells.

7.2.2 Design and Biophysical Characteristics of CRY2-/CIBN-STIM1ct Chimeras

Oligomerization of the STIM1 luminal domain following ER Ca^{2+} depletion is regarded as a decisive initiation step during the activation of CRAC channel (Stathopulos and Ikura 2016; Stathopulos et al. 2008). Upon store depletion, Ca^{2+} dissociates from the STIM1 luminal EF-hand Ca^{2+}-binding motif and results in the destabilization and subsequent multimerization of EF-SAM, which in turn rearranges the STIM1 transmembrane domains to enable an inside-out signal transduction toward the cytoplasmic side. The activation process of STIM1 has been recapitulated by two chemogenetic approaches that obviate the depletion of ER Ca^{2+} store. First, replacement of the STIM1 luminal domain with a rapamycin-inducible dimerization module composed of FRB and FKBP12 can initiate STIM1 activation upon addition of rapamycin in cellulo (Luik et al. 2008). Second, close apposition of STIM1ct through chemical crosslining at its N-terminus (residue 233), the position where the transmembrane domain ends and the cytoplasmic domain of STIM1 merges from the ER membrane, can switch STIM1ct from a less active folded-back conformation to an activated and more extended configuration in vitro (Feske and Prakriya 2013; Ma et al. 2015; Zhou et al. 2013). These studies have clearly demonstrated that bringing the N-terminus of STIM1ct molecules into close proximity, much like Ca^{2+} depletion-induced EF-SAM oligomerization in the ER lumen, can trigger conformational changes within STIM1ct to ultimately activate ORAI Ca^{2+} channel on the plasma membrane.

In addition to chemogenetic manipulations, this process can be recapitulated by conferring light sensitivity to STIM1ct using the photosensitive PHR domain of CRY2 (Kyung et al. 2015) or the CRY2-CIB1 optical multimerizer (Fig. 7.1b–f).

A set of CRY2- or CIB1-STIM1ct chimeras were constructed by fusing STIM1ct (233–685 or 238–685) with $CRY2_{PHR}$ or the N-terminal domain of CIB1 (CIBN; aa 1–171). With these tools, we can achieve photoinducible Ca^{2+} influx through endogenous ORAI channels by using different $CRY2_{PHR}$/CIBN combinations: (1) expression of $CRY2_{PHR}$-STIM1ct alone (Fig. 7.1b), (2) co-expression of $CRY2_{PHR}$-STIM1ct with CIBN-STIM1ct (Fig. 7.1c), and (3) co-expression of $CRY2_{PHR}$ with CIBN-STIM1ct (Fig. 7.1d). The cytosolic $CRY2_{PHR}$ acts a genetically encoded photocrosslinker to force the close apposition of the N-terminus of CIBN-STIM1ct and subsequently switch STIM1ct into an activated configuration to engage and gate ORAI Ca^{2+} channels. To fully mimic localized Ca^{2+} influx at ER-PM junctions, we have also tested the idea of either expressing ER-tethered $CRY2_{PHR}$-STIM1ct (Fig. 7.1e) or co-expressing CIBN-STIM1ct with an ER-tethered CRY2 that faces toward the cytosolic side of the ER membrane (Fig. 7.1f). Following blue light illumination, ER-tethered CRY2 forms oligomers to co-cluster and activate STIM1ct proteins, thereby trapping and gating PM-resident ORAI channels to enable the generation of Ca^{2+} microdomains (Ma G and Zhou Y, unpublished data). All these strategies could efficiently induce STIM1ct clustering to evoke Ca^{2+} influx through endogenous ORAI channels in mammalian cells, with an activation halftime (the time to reach half-maximal Ca^{2+}-responsive fluorescence intensity, $t_{1/2}$) of ~0.5 min and a deactivation halftime of ~5–6 min upon the withdrawal of blue light (Fig. 7.1g and Table 7.1). The generated Ca^{2+} signals (up to the submicromolar range) are robust enough to drive the efficient translocation of a downstream Ca^{2+}-responsive transcriptional factor, the nuclear factor of activated T cells (NFAT), into the nuclei (Fig. 7.1h).

The magnitude of light-generated Ca^{2+} signals can be further tuned by using a variety of STIM1ct fragments (see Table 7.1). Some of the CRY2-STIM1ct chimeras (238–685 or 238–463) displayed light-dependent translocation to the plasma membrane to induce Ca^{2+} influx (Kyung et al. 2015), while others (238–416 or 342–448) formed abundant clusters driven by CRY2 homo-oligomerization upon light illumination. Constructs with deletion of the polybasic domain (238–670) or partial deletion of the SOAR domain (238–416) failed to evoke Ca^{2+} influx, thereby implying the importance of these structural elements in mediating STIM-ORAI functional coupling. Moreover, the expression levels of CRY2-STIM1ct chimeras influenced the Ca^{2+} influx signals because the homo-oligomerization kinetics of CRY2 is partially dependent on the intracellular CRY2 concentrations. In parallel, the deactivation kinetics can be altered by introducing photocycle-sensitive mutations into CRY2. For example, the L348F mutant has a prolonged halftime of approximately 24 min, whereas the W349R mutant shortens the deactivation halftime to ~2.5 min (Kyung et al. 2015; Taslimi et al. 2014). Together, GECAs based on CRY2-STIM1ct chimeras are most useful in applications requiring relatively prolonged Ca^{2+} signals that fluctuate at the time scale of minutes to hours. To generate GECAs with faster kinetics, we resorted to the LOV2 photoswitch as described below.

Table 7.1 Summary of photoactivatable CRAC channels designed for optogenetic control of Ca^{2+} signaling

Constructs	Photosensitive module(s)	Linker	STIM fragment	ON ($t_{1/2}$, s)	OFF ($t_{1/2}$, s)	Ca^{2+} reporter	Fold change ($\Delta F/F_0$)	Cell types
CRY2-STIM1ct-based GECAs[a]								
	$CRY2_{1–498}$	24 aa	238–685	64.5 ± 4.8	274 ± 23.7	R-GECO1	11[b]	HeLa
	$CRY2_{1–498}$	24 aa	238–670				No response	HeLa
	$CRY2_{1–498}$	24 aa	238–463	43.2 ± 3.8	383 ± 52.3	R-GECO 1	~10[b]	HeLa
	$CRY2_{1–498}$	24 aa	238–416				No response	HeLa
	$CRY2_{1–498}$	24 aa	342–685	48.2 ± 5.4	344 ± 34.6	R-GECO 1	~10[b]	HeLa
	$CRY2_{1–498}$	26 aa	233–685	38.4 ± 2.1	320 ± 37.2	R-GECO1.2	2.9	HeLa
	ER-tethered $CRY2_{1–498}$	26 aa	233–685	42.9 ± 3.9	101 ± 17.2	R-GECO1.2	3.5	HeLa
CRY2-/CIBN-STIM1ct-based GECAs[c]								
$CRY2_{1–498}$-STIM1ct + CIBN-STIM1ct		26 aa	233–685	48.6 ± 3.8	305 ± 31.8	R-GECO1.2	~4	HeLa
$CRY2_{1–498}$ + CIBN-STIM1ct		26 aa	233–685	55.8 ± 2.3	293 ± 26.3	R-GECO1.2	~3	HeLa
$CRY2_{1–498,\ L348F}$ + CIBN-STIM1ct		26 aa	233–685	~60	>1400	R-GECO1.2	~3	HeLa
$CRY2_{1–498,\ W349R}$ + CIBN-STIM1ct		26 aa	233–685	~60	<180	R-GECO1.2	~3	HeLa
$CRY2_{1–498,\ E490G}$ + CIBN-STIM1ct		26 aa	233–685	13.3 ± 2.3	511.1 ± 66.8	R-GECO1.2	~3	HeLa
LOV2-STIM1ct-based GECAs[d]								
Opto-CRAC	$LOV2_{404–546}$	KL	336–486	23.4 ± 4.2	24.9 ± 4.8	R-GECO 1.2	3.0	HeLa
				36.4 ± 2.4	30.0 ± 1.1	Fura-2 AM	1.2	HeLa
				31.1 ± 4.9	34.7 ± 4.9	R-CaMP2	0.2	HeLa
				10.2 ± 3.0		GCaMP6s	4.1	HeLa
	Lyn-$LOV2_{404–546}$	KL	336–486	2.9 ± 1.1		GCaMP6s	2.5	HeLa
	2*$LOV2_{404–546}$	KL	336–486	46.2 ± 8.5		GCaMP6s	2.8	HeLa

(continued)

Table 7.1 (continued)

Constructs	Photosensitive module(s)	Linker	STIM fragment	ON ($t_{1/2}$, s)	OFF ($t_{1/2}$, s)	Ca^{2+} reporter	Fold change ($\Delta F/F_0$)	Cell types
	$LOV2_{404-546}$		336–486			GCaMP6s	1.4	HeLa
	$LOV2_{404-546}$	KLAAA	336–486			GCaMP6s	1.0	HeLa
	$LOV2_{404-546}$	KL	336–450			GCaMP6s	1.9	HeLa
	$LOV2_{404-546}$	KL	336–460			GCaMP6s	2.0	HeLa
	$LOV2_{404-546}$	KL	336–473			GCaMP6s	1.0	HeLa
	$LOV2_{404-546}$	KL	342–486			GCaMP6s	1.8	HeLa
	$LOV2_{404-546}$	KL	344–486			GCaMP6s	3.5	HeLa
	$LOV2_{404-546}$	KL	344–442			GCaMP6s	No response	HeLa
	$LOV2_{404-546, C450A}$	KL	336–486			GCaMP6s	0.1	HeLa
	$LOV2_{404-546, G528A}$	KL	336–486			GCaMP6s	2.4	HeLa
	$LOV2_{404-546, I532E}$	KL	336–486			GCaMP6s	0.5	HeLa
	$LOV2_{404-546, N538E}$	KL	336–486			GCaMP6s	0.3	HeLa
	$LOV2_{404-546, I539E}$	KL	336–486			GCaMP6s	No response	HeLa
LOVS1K	$LOV2_{404-546}$	KL	233–450			GCaMP6s	1.8	HeLa
hBACCS1	$LOV2_{404-538}$		347–448			Fluo-8 AM	~1.8	HEK293T

hBACCS2	2 × $LOV2_{404–538}$		347–448			Fluo-8 AM	~3.0	HEK293T
dmBACCS2	2 × $LOV2_{404–538}$		413–514			Fluo-8 AM	No response	HEK293T
dmBACCS2	2 × $LOV2_{404–538}$		413–514	<30		Fluo-4 AM	~1.6	S2
dmBACCS2 + dmOrai	2 × $LOV2_{404–538}$		413–514	<30		Fluo-8 AM	~5.4	HEK293T
dmBACCS2 + dmOrai	2 × $LOV_{404–538}$		413–514	<30	~50–60	Rhod-3 AM		HEK293T
dmBACCS2NS + dmOrai	2 × $LOV2_{404–538, N425S}$		413–514	<30	<30	Rhod-3 AM		HEK293T
dmBACCS2VL + dmOrai	2 × $LOV2_{404–538, V416L}$		413–514	<30	>480	Rhod-3 AM		HEK293T

GECAs, genetically encoded Ca^{2+} actuators; h, *Homo sapiens*; dm, *Drosophila melanogaster*

[a]Based on reference Kyung et al. (2015) and unpublished data by Ma G and Zhou Y; Tested in HeLa, NIH3T3, COS7, hESC, HEK293, HUVEC, astrocyte, zebra fish embryo, and mouse hippocampus

[b]R-GECO1 fluorescence normalized to the intensity at $t = 0$ (F_0) and values calculated as F_{max}/F_o. Other values were presented as $(F_{max} - F_0)/F_0$ or $\Delta F/F_0$

[c]Unpublished data by Ma G and Zhou Y

[d]According to references He et al. (2015); Ishii et al. (2015); Pham et al. (2011); Tan et al. (2016). Tested in HEK293, COS7, MEF, HIT-T15, S2 cells, hippocampal neurons, macrophage, dendritic cells, primary and leukemic T cells, and cancer cell lines derived from the ovary, prostate, breast, skin, and brain, as well as in olfactory epithelium and lymphoid organs

7.3 Photoswitchable Exposure of SOAR/CAD to Evoke Ca^{2+} Influx Through ORAI Channels

7.3.1 Optogenetic Tools for Intramolecular Control of Protein Function

LOV domains were initially characterized as photoreactive modules involved in the regulation of plant phototropism (Huala et al. 1997). The LOV2 domain from *Avena sativa* phototropin 1 (aa 404–546; MW = 16.5 kDa) is most widely used in optogenetic applications (Zimmerman et al. 2016). AsLOV2 contains a PAS (Per-ARNT-Sim) core that binds a photoreactive cofactor flavin mononucleotide (FMN), followed by a C-terminal 24-residue long Jα helix (Fig. 7.2a). In the dark, the Jα helix tightly docks to the PAS core with the bound FMN in its oxidized ground state (Harper et al. 2003). After blue light illumination, a covalent bond forms between the C4a position of FMN and a conserved cysteine (C450) of LOV2 and subsequently results in a conformational change within PAS to cause the dissociation and unfolding of the Jα helix within a few seconds (Pudasaini et al. 2015; Tischer and Weiner 2014; Zoltowski et al. 2009). Upon the withdrawal of light, the FMN-cysteine photoadduct undergoes hydrolysis and, therefore, allows the Jα helix to refold and reassociate with the PAS core. This reversible process can be conveniently monitored by UV-visible spectroscopy as shown in Fig. 7.2b. The photoreactive property can be exploited to regulate the activity of an effector domain located downstream of the Jα helix of AsLOV2 through steric hindrance and/or conformational restriction (Pathak et al. 2013; Tan et al. 2016; Tischer and Weiner 2014; Weitzman and Hahn 2014; Wu et al. 2009; Zhang and Cui 2015), as best exemplified by the design of a photoactivatable GTPase PA-Rac1 (Wu et al. 2009). In PA-Rac1, the LOV2 domain sterically blocks the Rac1-effector interaction interface in the dark and results in a tenfold reduction in the affinity for effector binding (Wu et al. 2009). Following blue light irradiation, the unwinding and unfolding of the Jα helix results in the removal of the steric hindrance imposed upon Rac1 to restore the function of Rac1. This strategy can likewise be extended to mimic the intramolecular autoinhibition within STIM1ct, which is mediated through coiled-coil interplays between CC1 and SOAR/CAD.

7.3.2 Design and Biophysical Characteristics of GECAs Based on the LOV2 Photoswitch

Through a series of truncation and deletion studies on STIM1ct, minimal ORAI-activating fragments such as SOAR (STIM1-ORAI-activating region; aa 344–442) or CAD (CRAC activation domain; 342–448) have been identified by multiple groups (Kawasaki et al. 2009; Muik et al. 2009; Park et al. 2009; Yuan et al. 2009; Zhou et al. 2010). When expressed alone in mammalian cells, these truncated STIM1ct fragments physically associated with ORAI1 and elicited Ca^{2+} influx without store depletion. By contrast, the expression of STIM1ct only led to a marginal constitutive Ca^{2+} influx, implying the existence of an intramolecular autoinhibitory mechanism to sequester SOAR/CAD and thus keep STIM1 quiescent at rest. Indeed, adjacent regions upstream (CC1, 233–343) or downstream (448–490) of SOAR/CAD turned out to be involved in maintaining the intramolecular trapping of STIM1 (Prakriya and Lewis 2015; Soboloff et al. 2012). Particularly, the coiled-coil interactions between CC1 and SOAR play a crucial role in keeping STIM1ct in a folded-back and less active conformation at rest (Fahrner et al. 2014; Ma et al. 2015). These findings promoted us to test the idea of substituting the STIM1-CC1 with AsLOV2 to control the exposure of SOAR/CAD by harnessing the power of light (He et al. 2015). After testing >100 chimeric constructs (designated as "Opto-CRAC") by varying the length of STIM1ct fragments, introducing mutations into AsLVO2, and optimizing the linker between the two domains, we identified the best construct "LOVSoc" ($STIM1_{336–486}$ fused to the C-terminus of $AsLVO2_{404–546}$) that exhibited the highest dynamic range in photoactivating Ca^{2+} influx. This Opto-CRAC construct underwent rapid translocation between the cytosol and the PM in response to blue light illumination ($t_{1/2,\ on} = 6.8 \pm 2.3$ s; $t_{1/2,\ off} = 28.7 \pm 6.5$ s), when expressed in HEK293-ORAI1 stable cells. This process could be reversed multiple times without significant loss in the magnitude of response. The degree of Ca^{2+} influx could be tuned by changing the power densities of blue light. To confer more flexibility to the Opto-CRAC system, we further explored the use of co-expression, membrane tethering, and concatemerization strategies to generate more variants with varying kinetic properties (Fig. 7.2c, d). Their photoactivation halftimes range from 3 to 50 s, and the deactivation halftimes largely fall in the range of 25–35 s (summarized in Table 7.1). Light-generated Ca^{2+} signals could efficiently drive the nuclear translocation of NFAT within 30 min (Fig. 7.2e). Furthermore, Opto-CRAC can be expressed in mammalian cells to generate highly Ca^{2+}-selective currents, and the substitution of extracellular Na^+ by a non-permeant ion $NMDG^+$ did not alter the amplitude or overall shape of the current (Fig. 7.2f). With Opto-CRAC, different oscillatory patterns of Ca^{2+} signals could be produced by altering the blue light pulses (Fig. 7.2g). Moreover, localized light stimulation can be applied to achieve the local activation of Ca^{2+} influx at a defined spatial resolution (Fig. 7.2e), thereby providing a new approach to dissect how the spatial profiles of Ca^{2+}signals impact on biological processes.

Following a very similar strategy, Ishii et al. independently developed a "blue light-activated Ca^{2+} channel switch" (BACCS) by fusing $AsLOV2_{404–538}$ with a different STIM1ct fragment spanning residues 347–448 (Ishii et al. 2015). A dimeric form of BACCS (termed "BACCS2") was further created to enable more efficient photoswitch and deliver more robust Ca^{2+} signals. Interestingly, the concatemeric form of LOV2-STIM1ct chimeras (BACCS2 or the dimeric LOVSoc developed by us; Fig. 7.2d) substantially delayed the onset of photo-responsive Ca^{2+} influx. To overcome this weakness, a *Drosophila* version of BACCS2 (dmBACCS2) was generated. dmBACCS2 failed to activate human ORAI1, but it can be used in combination with *Drosophila* Orai to photoinduce rapid Ca^{2+} entry in mammalian cells (Table 7.1). Both the BACCS2 and Opto-CRAC systems exhibit higher sensitivity and lower basal activity when compared to LOVS1K (Pham et al. 2011), an earlier version of LOV2-STIM1ct chimera composed of $AsLOV2_{404-546}$ and $STIM1_{233-450}$. We speculate that the inclusion of the CC1 region in LOVS1K might antagonize the steric hindrance imposed by LOV2 to weaken the photocaging capacity upon SOAR/CAD, thereby causing a higher degree of background activation in the dark.

7.4 Applications of Photoactivatable CRAC Channels

Ca^{2+} is regarded as a versatile second messenger that regulates a myriad of biological processes, ranging from short-term muscle contraction or neurotransmitter release (within seconds) to long-term processes such as gene expression and metabolism (lasting for hours or even days) (Berridge et al. 1998, 2000). The location, amplitude, and frequency of Ca^{2+} signals in mammalian cells are exquisitely programmed to meet the diverse requirements of Ca^{2+}-modulated events. Compared to pharmacological tools, the aforementioned GECAs offer two obvious advantages when delivering Ca^{2+} signals in the cellular context. First, the high temporal resolution of light-generated Ca^{2+} signals enables the dissection of kinetic requirements of Ca^{2+} signals during mechanistic studies of Ca^{2+}-dependent processes. Second, GECAs can be used to conveniently program the spatial profiles of Ca^{2+} signals in a reversible manner. Global or local Ca^{2+} signals can be remotely generated to modulate Ca^{2+}-dependent activities at subcellular precision by applying a focused beam of light on the whole-cell or at user-defined areas. The development of photoactivatable CRAC channels that are capable of delivering Ca^{2+} signals with user-defined spatial and temporal properties opens exciting new opportunities to photo-manipulate Ca^{2+}-dependent cellular events both in vitro and in vivo.

7.4.1 Compatibility of Photoactivatable CRAC Channels with Both Excitable and Non-excitable Cell Types

Given the wide distribution of endogenous ORAI channels in both excitable and non-excitable tissues in mammals, CRY2-STIM1ct or LOV2-STIM1ct chimeras can be used as a single-component system to photoinduce Ca^{2+} influx in many cell types, including but not limited to cells of the immune system, neurons, embryonic stem cells, endothelial cells, and cancer cell lines derived from various tissues (He et al. 2015; Ishii et al. 2015; Kyung et al. 2015). They can be further used to photoinduce Ca^{2+} entry in model organisms including *Drosophila melanogaster* (fruit fly), *Danio rerio* (zebra fish), and *Caenorhabditis elegans* (worm). The amplitudes of light-generated Ca^{2+} signals in those cells may differ because of varying expression levels of endogenous ORAI. The relatively small sizes of LOV2-STIM1ct constructs (<3 kb) make them compatible with almost all existing viral packaging systems. Lentiviruses expressing Opto-CRAC or OptoSTIM1 have been produced to transduce cells of the immune system (He et al. 2015) and the hippocampus of mice (Kyung et al. 2015). Adenoviruses encoding BACCS2 have been used to infect mouse olfactory epithelium to photo-manipulate the activities of olfactory sensory neurons (Ishii et al. 2015). Furthermore, retroviruses encoding Opto-CRAC have been successfully transduced into mouse T cells, macrophages, and bone marrow-derived dendritic cells (DCs) to induce Ca^{2+} influx (He et al. 2015). For cells lacking sufficient amounts of endogenous ORAI channels, we recommend the co-expression of ORAI1 with STIM1-derived GECAs. Nonetheless, overexpression of ORAI1 tends to cause basal activation of OptoSTIM1- or LOV2-based GECAs prior to photoactivation. A potential solution to solve this problem is to tightly sequester LOV2-STIM1ct chimeras to subcellular compartments (e.g., mitochondria) by using the LOVTRAP system (Wang et al. 2016), thus preventing its association with ORAI1 in the dark. With blue light illumination, LOV2-STIM1ct will be released from the anchor compartment to restore its ORAI-activating function.

7.4.2 Light-Inducible Control of Ca^{2+}-Dependent Hallmark Cellular Responses

Photoactivatable CRAC channels can be used to trigger hallmark cellular responses with a simple flash of light. Ca^{2+} is well known to regulate the activities of Ca^{2+}-/calmodulin-dependent phosphatase or kinases to control the phosphorylation status and subcellular localization of Ca^{2+}-responsive transcriptional factors (Feske 2007; Rao 2009). Photoinduced Ca^{2+} influx has been successfully used to drive the nuclear translocation of NFAT (Figs. 7.1h and 7.2e) and CREB in HeLa or HEK293 cells expressing OptoSTIM1 (Kyung et al. 2015) or Opto-CRAC (He et al. 2015). Opto-CRAC has been further used to stimulate NFAT-dependent luciferase or insulin expression in HeLa and HEK293T cells, as well as cytokine productions (e.g., IL-2 and IFN-γ) in primary mouse T lymphocytes (He et al. 2015). In addition to driving effector T cell activation, Opto-CRAC has been expressed in macrophages to enable light-dependent boost in the production of IL-1β and processed caspase-1, thus demonstrating the feasibility of photo-tunable amplification of inflammasome activation (He et al. 2015). Moreover, OptoSTIM1 has been used to generate Ca^{2+} signals to induce reversible F-actin disassembly at the cell cortex, thereby causing light-dependent remodeling of actin cytoskeleton structure and cell morphological changes (Kyung et al. 2015).

7.4.3 Photoactivatable CRAC Channels for Remote Immunomodulation and Neuromodulation In Vivo

The application of optogenetic tools in vivo is hampered by their inability to stimulate deep within tissues without the use of invasive indwelling fiber optic probes. To overcome this bottleneck, we resorted to lanthanide-doped upconversion nanoparticles (UCNPs; Fig. 7.3a), which can be functionalized by streptavidin to enable targeted delivery to the surface of cells expressing an engineered ORAI1 protein that contains a streptavidin-binding tag in its second extracellular loop (Fig. 7.3b). Owing to interatomic energy transfer between metal ions, cell-surface-anchored UCNPs can act as nanotransducers to efficiently convert near-infrared (NIR) light into blue light, thus serving as a mobile light source to activate CRY2- or LOV2-based optogenetic constructs (Chen et al. 2014; He et al. 2015; Shen et al. 2013; Tan et al. 2016; Zhang et al. 2016). We have successfully demonstrated the use of Opto-CRAC, in conjugation with UCNPs, to achieve in vivo NIR photoactivatable Ca^{2+} influx (Fig. 7.3c, d) and Ca^{2+}-/NFAT-dependent luciferase expression (Fig. 7.3e) without causing significant heat generation (He et al. 2015). The NIR light-stimulable Opto-CRAC system has been more rigorously tested in a disease-relevant context. He et al. explored the use of the Opto-CRAC system with dendritic cell (DC)-mediated immunotherapy in the B16-OVA mouse model of melanoma. They found that NIR light illumination resulted in twofold to eightfold increase in the surface expression of MHC-I/MHC-II, CD86, and CCR7, which are characteristic of matured DCs that are capable of homing to the tumor draining lymph nodes (dLNs). DCs residing in dLNs are able to interact with T cells and sensitize T cells toward tumor antigens to boost antitumor immune response (Fig. 7.3f). To further validate the immunomodulatory function of this optogenetic system in vivo, He et al. injected UCNPs-Stv/OVA loaded Opto-CRAC DCs to the B16-OVA murine model of melanoma. Pulsed NIR light stimulation significantly suppressed the tumor growth with diminished tumor volume (Fig. 7.3f) and reduced numbers of tumor foci in the lungs. Collectively, the NIR-stimulable Opto-CRAC system acts as a genetically encoded "photoactivatable adjuvant" to promote DC maturation and boost antigen-specific immune response both in vitro and in vivo (He et al. 2015). This exciting progress attests to the high feasibility of combining optogenetics with immunoengineering to develop light-controllable immunomodulatory therapeutics (Tan et al. 2016).

In addition to optogenetic immunomodulation, poof-of-concept experiments have been carried out by Kyung et al. to demonstrate its potential in remote neuromodulation. An OptoSTIM1-expressing lentivirus under the control of a Ca^{2+}-/calmodulin-dependent protein kinase II promoter was targeted to the CA1 region of the hippocampus of mice. The infected mice were implanted with an optical fiber through their skulls to effectively deliver blue light. Light-induced hippocampal Ca^{2+} influx mediated by OptoSTIM1 was shown to affect contextual, but not auditory-cued, fear memory formation in mice, as reflected by the increase in time dedicated to freezing behavior in photostimulated animals (Kyung et al. 2015). This in vivo finding points to the possibility of applying OptoSTIM1 in vivo to modulate the neural circuits and enhance the learning capacity of mammals (Kyung et al. 2015; LeDoux 2000). Moreover, a transgenic mouse strain with a subset of olfactory sensory neurons expressing the ORAI1::hBACCS2 has been recently generated, thus paving the way for future in vivo studies on the mouse olfactory epithelium (Ishii et al. 2015).

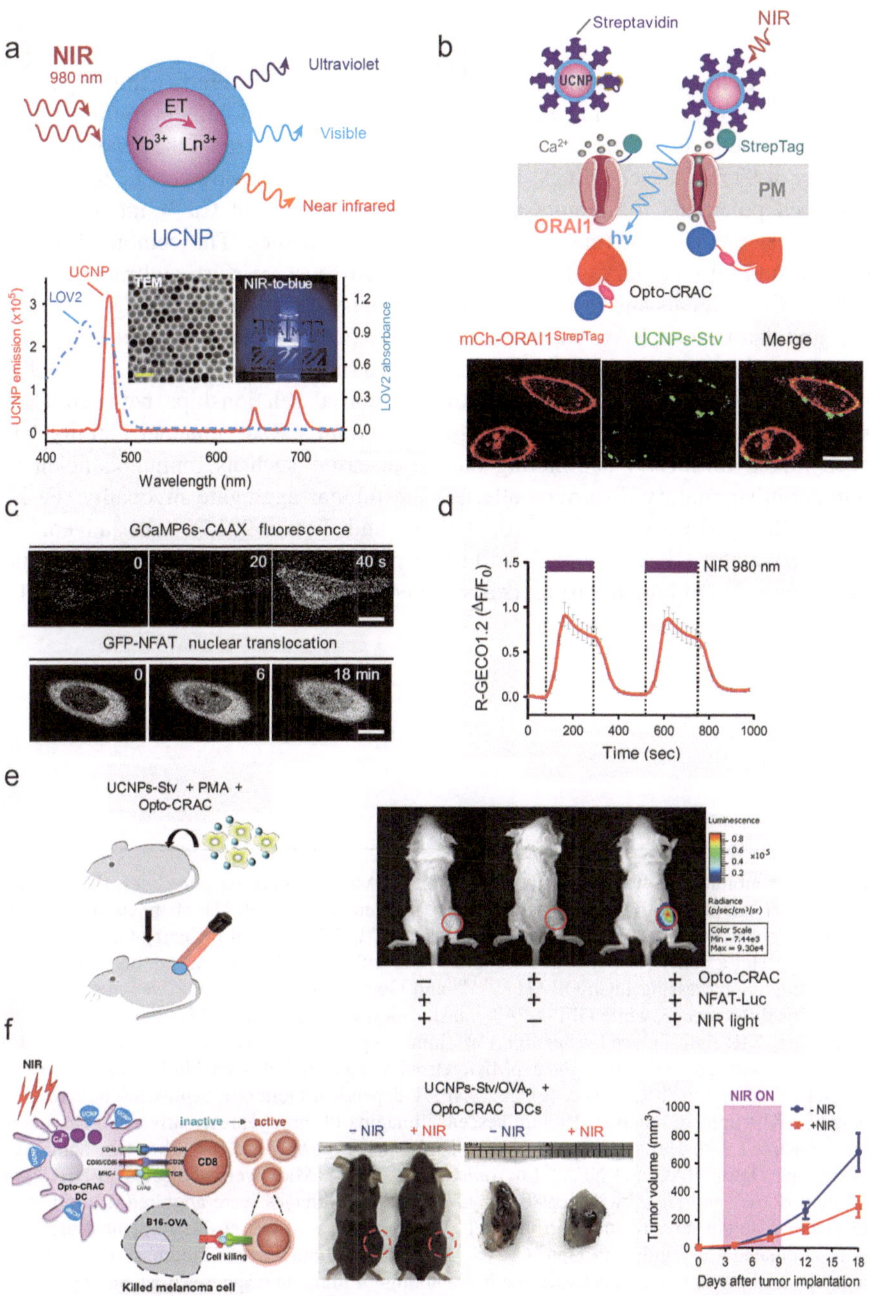

Fig. 7.3 NIR light control of Opto-CRAC by UCNPs to drive gene expression and aid dendritic cell-based immunotherapy against melanoma. (**a**) Schematic illustration of the core/shell structure and interatomic energy transfer (ET) between lanthanide ions in upconversion nanoparticles (UCNPs). The *lower panel* showed the emission spectrum of UCNPs (*solid red line*) superimposed

7.5 Summary

Optogenetics has enabled scientists to harness the power of light to achieve gain or loss of function of defined events in a specific population of cells in living tissues. The optogenetic applications described above have demonstrated the general applicability of STIM1-based GECAs in mammalian cells. These optogenetic tools offer untapped potentials for remote and noninvasive control of Ca^{2+} signaling with subcellular precision at time scales of seconds to minutes. The combined use of GECAs and GECIs makes it possible to simultaneously perturb and record intracellular Ca^{2+} signals and provides novel approaches to aid the screening of chemical modulators for CRAC channels in the near future. These technical leaps will greatly facilitate the study of Ca^{2+} signaling in a more quantitative and predictable manner, thereby enabling the dissection of causal relationships between Ca^{2+} signals and cellular or animal behaviors. Aberrant CRAC channel activity has been linked to several debilitating human diseases, such as immunodeficiency, immunoinflammatory disorders, allergy, and tubular aggregate myopathy (Feske et al. 2015; Hogan et al. 2010; Lacruz and Feske 2015). The advent of photoactivatable CRAC channels will likely make it possible to remotely intervene the progression of human disorders associated with dysfunctional CRAC channels.

Fig. 7.3 (continued) by the absorbance spectrum of AsLOV2 (*dashed blue line*). (**b**) Specific targeting of streptavidin (Stv)-conjugated UCNPs to engineered ORAI1 channels on the cell surface. Confocal images showed the accumulation of UCNPs-Stv on mCherry-ORAI1^{StrepTag}-expressing HeLa cells. (**c**) NIR light-triggered Ca^{2+} influx and NFAT nuclear translocation in HeLa cells co-expressing mCh-ORAI1^{StrepTag} and Opto-CRAC. Ca^{2+} influx was monitored by GCaMP6s fluorescence, while GFP-NFAT translocation was reported by GFP signals. *Scale bar*, 10 μm. (**d**) NIR light-induced reversible Ca^{2+} influx reported by R-GECO1.2. HeLa cells were transfected with an IRES bicistronic pMIG retroviral construct that enabled co-expression of ORAI1^{StrepTag} and mCh-LOVSoc. (**e**) Ca^{2+}-/NFAT-dependent luciferase expression in vivo triggered by NIR light stimulation. Bioluminescence imaging of three representative BALB/c mice, one implanted with HeLa cells expressing NFAT-Luc only (*left*) and the other two with cells expressing Opto-CRAC and NFAT-Luc (*middle* and *right*). Mice were subjected to NIR light irradiation (*left* and *right*) with a 980 nm CW laser. The images were acquired 20 min after receiving a single dose of luciferin (100 μL, 15 mg/mL, *s.c.*). *Red circle*, implanted area. (**f**) NIR-stimulated Ca^{2+} influx in Opto-CRAC DCs prompts immature DC maturation and OVA antigen cross-presentation to activate and boost antitumor immune responses mediated by CD8 T cells (cytotoxic T lymphocytes, CTLs), thereby sensitizing tumor cells to antigen-specific, CTL-mediated killing in the B16-OVA melanoma model. OVA peptide (OVAp, $_{257}$SIINFEKL$_{264}$) is used here as a surrogate tumor antigen. Tumor-inoculated sites (*red circles*) were isolated from tumor-bearing mice shielded or exposed to NIR light. The tumor sizes were measured at indicated time points shown in the growth curve after tumor implantation

Acknowledgments This work was supported by grants from the Welch Foundation (BE-1913 to Y.Z.), the American Cancer Society (RSG-16-215-01-TBE to Y.Z.), and the National Institutes of Health (R01GM112003 to Y.Z.).

References

Banerjee R, Schleicher E, Meier S, Viana RM, Pokorny R, Ahmad M, Bittl R, Batschauer A (2007) The signaling state of Arabidopsis cryptochrome 2 contains flavin semiquinone. J Biol Chem 282:14916–14922

Berridge MJ, Bootman MD, Lipp P (1998) Calcium—a life and death signal. Nature 395:645–648

Berridge MJ, Lipp P, Bootman MD (2000) The versatility and universality of calcium signalling. Nat Rev Mol Cell Biol 1:11–21

Cahalan MD, Zhang SL, Yeromin AV, Ohlsen K, Roos J, Stauderman KA (2007) Molecular basis of the CRAC channel. Cell Calcium 42:133–144

Cao X, Choi S, Maleth JJ, Park S, Ahuja M, Muallem S (2015) The ER/PM microdomain, PI(4,5)P(2) and the regulation of STIM1-Orai1 channel function. Cell Calcium 58:342–348

Chang CL, Liou J (2016) Homeostatic regulation of the PI(4,5)P2-Ca(2+) signaling system at ER-PM junctions. Biochim Biophys Acta 1861:862–873

Chen G, Qiu H, Prasad PN, Chen X (2014) Upconversion nanoparticles: design, nanochemistry, and applications in theranostics. Chem Rev 114:5161–5214

Deisseroth K (2011) Optogenetics. Nat Methods 8:26–29

Deisseroth K (2015) Optogenetics: 10 years of microbial opsins in neuroscience. Nat Neurosci 18:1213–1225

Derler I, Jardin I, Stathopulos PB, Muik M, Fahrner M, Zayats V, Pandey SK, Poteser M, Lackner B, Absolonova M, Schindl R, Groschner K, Ettrich R, Ikura M, Romanin C (2016) Cholesterol modulates orai1 channel function. Sci Signal 9(412):ra10

Fahrner M, Muik M, Schindl R, Butorac C, Stathopulos P, Zheng L, Jardin I, Ikura M, Romanin C (2014) A coiled-coil clamp controls both conformation and clustering of stromal interaction molecule 1 (STIM1). J Biol Chem 289:33231–33244

Fenno L, Yizhar O, Deisseroth K (2011) The development and application of optogenetics. Annu Rev Neurosci 34:389–412

Feske S (2007) Calcium signalling in lymphocyte activation and disease. Nat Rev Immunol 7:690–702

Feske S, Prakriya M (2013) Conformational dynamics of STIM1 activation. Nat Struct Mol Biol 20:918–919

Feske S, Gwack Y, Prakriya M, Srikanth S, Puppel SH, Tanasa B, Hogan PG, Lewis RS, Daly M, Rao A (2006) A mutation in Orai1 causes immune deficiency by abrogating CRAC channel function. Nature 441:179–185

Feske S, Wulff H, Skolnik EY (2015) Ion channels in innate and adaptive immunity. Annu Rev Immunol 33:291–353

Giovani B, Byrdin M, Ahmad M, Brettel K (2003) Light-induced electron transfer in a cryptochrome blue-light photoreceptor. Nat Struct Biol 10:489–490

Harper SM, Neil LC, Gardner KH (2003) Structural basis of a phototropin light switch. Science 301:1541–1544

He L, Zhang Y, Ma G, Tan P, Li Z, Zang S, Wu X, Jing J, Fang S, Zhou L, Wang Y, Huang Y, Hogan PG, Han G, Zhou Y (2015) Near-infrared photoactivatable control of Ca(2+) signaling and optogenetic immunomodulation. eLife 4:e10024

Hogan PG, Lewis RS, Rao A (2010) Molecular basis of calcium signaling in lymphocytes: STIM and ORAI. Annu Rev Immunol 28:491–533

Hooper R, Soboloff J (2015) STIMATE reveals a STIM1 transitional state. Nat Cell Biol 17:1232–1234

Huala E, Oeller PW, Liscum E, Han IS, Larsen E, Briggs WR (1997) Arabidopsis NPH1: a protein kinase with a putative redox-sensing domain. Science 278:2120–2123
Idevall-Hagren O, Dickson EJ, Hille B, Toomre DK, De Camilli P (2012) Optogenetic control of phosphoinositide metabolism. Proc Natl Acad Sci U S A 109:E2316–E2323
Ishii T, Sato K, Kakumoto T, Miura S, Touhara K, Takeuchi S, Nakata T (2015) Light generation of intracellular Ca(2+) signals by a genetically encoded protein BACCS. Nat Commun 6:8021
Jing J, He L, Sun A, Quintana A, Ding Y, Ma G, Tan P, Liang X, Zheng X, Chen L, Shi X, Zhang SL, Zhong L, Huang Y, Dong MQ, Walker CL, Hogan PG, Wang Y, Zhou Y (2015) Proteomic mapping of ER-PM junctions identifies STIMATE as a regulator of Ca(2)(+) influx. Nat Cell Biol 17:1339–1347
Kakumoto T, Nakata T (2013) Optogenetic control of PIP3: PIP3 is sufficient to induce the actin-based active part of growth cones and is regulated via endocytosis. PLoS One 8:e70861
Kawasaki T, Lange I, Feske S (2009) A minimal regulatory domain in the C terminus of STIM1 binds to and activates ORAI1 CRAC channels. Biochem Biophys Res Commun 385:49–54
Kennedy MJ, Hughes RM, Peteya LA, Schwartz JW, Ehlers MD, Tucker CL (2010) Rapid blue-light-mediated induction of protein interactions in living cells. Nat Methods 7:973–975
Kianianmomeni A (2015) UVB-based optogenetic tools. Trends Biotechnol 33:59–61
Kim N, Kim JM, Lee M, Kim CY, Chang KY, Heo WD (2014) Spatiotemporal control of fibroblast growth factor receptor signals by blue light. Chem Biol 21:903–912
Kondoh M, Shiraishi C, Muller P, Ahmad M, Hitomi K, Getzoff ED, Terazima M (2011) Light-induced conformational changes in full-length Arabidopsis thaliana cryptochrome. J Mol Biol 413:128–137
Kyung T, Lee S, Kim JE, Cho T, Park H, Jeong YM, Kim D, Shin A, Kim S, Baek J, Kim J, Kim NY, Woo D, Chae S, Kim CH, Shin HS, Han YM, Kim D, Heo WD (2015) Optogenetic control of endogenous Ca(2+) channels in vivo. Nat Biotechnol 33:1092–1096
Lacruz RS, Feske S (2015) Diseases caused by mutations in ORAI1 and STIM1. Ann N Y Acad Sci 1356:45–79
LeDoux JE (2000) Emotion circuits in the brain. Annu Rev Neurosci 23:155–184
Lee S, Park H, Kyung T, Kim NY, Kim S, Kim J, Heo WD (2014) Reversible protein inactivation by optogenetic trapping in cells. Nat Methods 11:633–636
Liou J, Kim ML, Heo WD, Jones JT, Myers JW, Ferrell JE Jr, Meyer T (2005) STIM is a Ca2+ sensor essential for Ca2+-store-depletion-triggered Ca2+ influx. Curr Biol 15:1235–1241
Luik RM, Wang B, Prakriya M, Wu MM, Lewis RS (2008) Oligomerization of STIM1 couples ER calcium depletion to CRAC channel activation. Nature 454:538–542
Ma G, Wei M, He L, Liu C, Wu B, Zhang SL, Jing J, Liang X, Senes A, Tan P, Li S, Sun A, Bi Y, Zhong L, Si H, Shen Y, Li M, Lee MS, Zhou W, Wang J, Wang Y, Zhou Y (2015) Inside-out Ca(2+) signalling prompted by STIM1 conformational switch. Nat Commun 6:7826
Maleth J, Choi S, Muallem S, Ahuja M (2014) Translocation between PI(4,5)P2-poor and PI(4,5)P2-rich microdomains during store depletion determines STIM1 conformation and Orai1 gating. Nat Commun 5:5843
Muik M, Fahrner M, Derler I, Schindl R, Bergsmann J, Frischauf I, Groschner K, Romanin C (2009) A cytosolic homomerization and a modulatory domain within STIM1 C terminus determine coupling to ORAI1 channels. J Biol Chem 284:8421–8426
Muik M, Fahrner M, Schindl R, Stathopulos P, Frischauf I, Derler I, Plenk P, Lackner B, Groschner K, Ikura M, Romanin C (2011) STIM1 couples to ORAI1 via an intramolecular transition into an extended conformation. EMBO J 30:1678–1689
Nihongaki Y, Yamamoto S, Kawano F, Suzuki H, Sato M (2015) CRISPR-Cas9-based photoactivatable transcription system. Chem Biol 22:169–174
Pacheco J, Dominguez L, Bohorquez-Hernandez A, Asanov A, Vaca L (2016) A cholesterol-binding domain in STIM1 modulates STIM1-Orai1 physical and functional interactions. Sci Rep 6:29634
Parekh AB, Putney JW Jr (2005) Store-operated calcium channels. Physiol Rev 85:757–810

Park CY, Hoover PJ, Mullins FM, Bachhawat P, Covington ED, Raunser S, Walz T, Garcia KC, Dolmetsch RE, Lewis RS (2009) STIM1 clusters and activates CRAC channels via direct binding of a cytosolic domain to Orai1. Cell 136:876–890

Park H, Lee S, Heo WD (2016) Protein inactivation by optogenetic trapping in living cells. Methods Mol Biol 1408:363–376

Pathak GP, Vrana JD, Tucker CL (2013) Optogenetic control of cell function using engineered photoreceptors. Biol Cell 105:59–72

Pham E, Mills E, Truong K (2011) A synthetic photoactivated protein to generate local or global Ca(2+) signals. Chem Biol 18:880–890

Polstein LR, Gersbach CA (2015) A light-inducible CRISPR-Cas9 system for control of endogenous gene activation. Nat Chem Biol 11:198–200

Prakriya M, Lewis RS (2015) Store-operated calcium channels. Physiol Rev 95:1383–1436

Procopio M, Link J, Engle D, Witczak J, Ritz T, Ahmad M (2016) Kinetic modeling of the Arabidopsis cryptochrome photocycle: FADH(o) accumulation correlates with biological activity. Front Plant Sci 7:888

Pudasaini A, El-Arab KK, Zoltowski BD (2015) LOV-based optogenetic devices: light-driven modules to impart photoregulated control of cellular signaling. Front Mol Biosci 2:18

Putney JW Jr (1986) A model for receptor-regulated calcium entry. Cell Calcium 7:1–12

Quintana A, Rajanikanth V, Farber-Katz S, Gudlur A, Zhang C, Jing J, Zhou Y, Rao A, Hogan PG (2015) TMEM110 regulates the maintenance and remodeling of mammalian ER–plasma membrane junctions competent for STIM–ORAI signaling. Proc Natl Acad Sci U S A 112: E7083–E7092

Rao A (2009) Signaling to gene expression: calcium, calcineurin and NFAT. Nat Immunol 10:3–5

Roos J, DiGregorio PJ, Yeromin AV, Ohlsen K, Lioudyno M, Zhang S, Safrina O, Kozak JA, Wagner SL, Cahalan MD, Velicelebi G, Stauderman KA (2005) STIM1, an essential and conserved component of store-operated Ca2+ channel function. J Cell Biol 169:435–445

Sharma S, Quintana A, Findlay GM, Mettlen M, Baust B, Jain M, Nilsson R, Rao A, Hogan PG (2013) An siRNA screen for NFAT activation identifies septins as coordinators of store-operated Ca2+ entry. Nature 499:238–242

Shen J, Zhao L, Han G (2013) Lanthanide-doped upconverting luminescent nanoparticle platforms for optical imaging-guided drug delivery and therapy. Adv Drug Deliv Rev 65:744–755

Soboloff J, Rothberg BS, Madesh M, Gill DL (2012) STIM proteins: dynamic calcium signal transducers. Nat Rev Mol Cell Biol 13:549–565

Srikanth S, Jung HJ, Kim KD, Souda P, Whitelegge J, Gwack Y (2010) A novel EF-hand protein, CRACR2A, is a cytosolic Ca2+ sensor that stabilizes CRAC channels in T cells. Nat Cell Biol 12:436–446

Stathopulos PB, Ikura M (2016) Store operated calcium entry: from concept to structural mechanisms. Cell Calcium. PMID: 27914753, doi: 10.1016/j.ceca.2016.11.005

Stathopulos PB, Zheng L, Li GY, Plevin MJ, Ikura M (2008) Structural and mechanistic insights into STIM1-mediated initiation of store-operated calcium entry. Cell 135:110–122

Tan P, He L, Han G, Zhou Y (2016) Optogenetic immunomodulation: shedding light on antitumor immunity. Trends Biotechnol 35(3):215–226

Taslimi A, Vrana JD, Chen D, Borinskaya S, Mayer BJ, Kennedy MJ, Tucker CL (2014) An optimized optogenetic clustering tool for probing protein interaction and function. Nat Commun 5:4925

Taslimi A, Zoltowski B, Miranda JG, Pathak GP, Hughes RM, Tucker CL (2016) Optimized second-generation CRY2-CIB dimerizers and photoactivatable Cre recombinase. Nat Chem Biol 12:425–430

Tischer D, Weiner OD (2014) Illuminating cell signalling with optogenetic tools. Nat Rev Mol Cell Biol 15:551–558

Vig M, Kinet JP (2007) The long and arduous road to CRAC. Cell Calcium 42:157–162

Vig M, Peinelt C, Beck A, Koomoa DL, Rabah D, Koblan-Huberson M, Kraft S, Turner H, Fleig A, Penner R, Kinet JP (2006) CRACM1 is a plasma membrane protein essential for store-operated Ca2+ entry. Science 312:1220–1223

Wang H, Vilela M, Winkler A, Tarnawski M, Schlichting I, Yumerefendi H, Kuhlman B, Liu R, Danuser G, Hahn KM (2016) LOVTRAP: an optogenetic system for photoinduced protein dissociation. Nat Methods 13:755–758

Weitzman M, Hahn KM (2014) Optogenetic approaches to cell migration and beyond. Curr Opin Cell Biol 30:112–120

Wu YI, Frey D, Lungu OI, Jaehrig A, Schlichting I, Kuhlman B, Hahn KM (2009) A genetically encoded photoactivatable Rac controls the motility of living cells. Nature 461:104–108

Yuan JP, Zeng W, Dorwart MR, Choi YJ, Worley PF, Muallem S (2009) SOAR and the polybasic STIM1 domains gate and regulate Orai channels. Nat Cell Biol 11:337–343

Zhang K, Cui B (2015) Optogenetic control of intracellular signaling pathways. Trends Biotechnol 33:92–100

Zhang SL, Yu Y, Roos J, Kozak JA, Deerinck TJ, Ellisman MH, Stauderman KA, Cahalan MD (2005) STIM1 is a Ca2+ sensor that activates CRAC channels and migrates from the Ca2+ store to the plasma membrane. Nature 437:902–905

Zhang Y, Huang L, Li Z, Ma G, Zhou Y, Han G (2016) Illuminating cell signaling with near-infrared light-responsive nanomaterials. ACS Nano 10:3881–3885

Zhou Y, Meraner P, Kwon HT, Machnes D, Oh-hora M, Zimmer J, Huang Y, Stura A, Rao A, Hogan PG (2010) STIM1 gates the store-operated calcium channel ORAI1 in vitro. Nat Struct Mol Biol 17:112–116

Zhou Y, Srinivasan P, Razavi S, Seymour S, Meraner P, Gudlur A, Stathopulos PB, Ikura M, Rao A, Hogan PG (2013) Initial activation of STIM1, the regulator of store-operated calcium entry. Nat Struct Mol Biol 20:973–981

Zimmerman SP, Kuhlman B, Yumerefendi H (2016) Engineering and application of LOV2-based photoswitches. Methods Enzymol 580:169–190

Zoltowski BD, Vaccaro B, Crane BR (2009) Mechanism-based tuning of a LOV domain photoreceptor. Nat Chem Biol 5:827–834

STIM-TRP Pathways and Microdomain Organization: Ca^{2+} Influx Channels – The Orai-STIM1-TRPC Complexes

8

Dora Bodnar, Woo Young Chung, Dongki Yang, Jeong Hee Hong, Archana Jha, and Shmuel Muallem

Abstract

Ca^{2+} influx by plasma membrane Ca^{2+} channels is the crucial component of the receptor-evoked Ca^{2+} signal. The two main Ca^{2+} influx channels of non-excitable cells are the Orai and TRPC families of Ca^{2+} channels. These channels are activated in response to cell stimulation and Ca^{2+} release from the endoplasmic reticulum (ER). The protein that conveys the Ca^{2+} content of the ER to the plasma membrane is the ER Ca^{2+} sensor STIM1. STIM1 activates the Orai channels and is obligatory for channel opening. TRPC channels can function in two modes, as STIM1-dependent and STIM1-independent. When activated by STIM1, both channel types function at the ER/PM (plasma membrane) junctions. This chapter describes the properties and regulation of the channels by STIM1, with emphasis how and when TRPC channels function as STIM1-dependent and STIM1-independent modes and their unique Ca^{2+}-dependent physiological functions that are not shared with the Orai channels.

Keywords

TRPC channels • Orai channels • STIM1 • Complexes • Regulation • ER-PM junctions

D. Bodnar • W.Y. Chung • A. Jha • S. Muallem (✉)
Epithelial Signaling and Transport Section, National Institute of Dental and Craniofacial Research – NIDCR, National Institute of Health, Building 10, Room 1N-112, Bethesda, MD 20892, USA
e-mail: Shmuel.muallem@nih.gov

D. Yang • J.H. Hong
Department of Physiology, Collage of Medicine, Gachon University, 155 Gaetbeol-ro, Yeonsu-gu, Incheon City 21999, Korea

K. Groschner et al. (eds.), *Store-Operated Ca^{2+} Entry (SOCE) Pathways*, Advances in Experimental Medicine and Biology 993, DOI 10.1007/978-3-319-57732-6_8

8.1 Introduction

Ca^{2+} influx channels are crucial component of the receptor-evoked Ca^{2+} signal. Receptor stimulation activates several types of Ca^{2+} channels depending on the tissue. However, common to all cells in most eukaryotes are the channel that are activated in response to Ca^{2+} release from the endoplasmic reticulum (ER) and thus called store-operated Ca^{2+} channels (SOCs) (Parekh and Putney 2005). Stimulation of surface receptors that triggers the hydrolysis of $PI(4,5)P_2$ to generate IP_3 causes activation of the ER resident IP_3 receptors (IP_3Rs) Ca^{2+} channels that release the Ca^{2+} stored in the ER. Ca^{2+} release from the ER provides a signal to activate the Ca^{2+} influx channels at the plasma membrane. Activation of the ER and plasma membrane Ca^{2+} pumps by the increase in free cytoplasmic Ca^{2+} (Ca^{2+}) completes the cycle and sets the resting $[Ca^{2+}]_i$ during cell stimulation (Cao et al. 2015).

Ca^{2+} influx by the SOC channels has multiple roles in Ca^{2+} signaling and cell function. During the physiological response of Ca^{2+} oscillations, the SOC-mediated Ca^{2+} influx replenishes the stores between Ca^{2+} spikes, as evident from decline of oscillations frequency until their termination in the absence of SOC-mediated Ca^{2+} influx. SOC-mediated Ca^{2+} influx reloads the ER and all other mobilized Ca^{2+} stores at the end of the stimulation period. In fact, the SOCs remain active long after removal of the stimulus, including by addition of antagonist, until all stores are fully loaded with Ca^{2+} (Parekh and Putney 2005; Muallem et al. 1988; Pandol et al. 1987). Ca^{2+} signals usually initiate at discreet cellular domain and then propagate in the form of Ca^{2+} waves (Kasai and Augustine 1990; Thorn et al. 1993). SOC-mediated Ca^{2+} influx accelerates the rate of Ca^{2+} wave propagation (Lee et al. 1997). Because of their crucial roles in maintaining the receptor-generated Ca^{2+} signals, Ca^{2+} influx by SOCs is essential for all Ca^{2+}-regulated cell functions. Excess $[Ca^{2+}]_i$ due to uncontrolled Ca^{2+} signaling is highly toxic. Invariably, Ca^{2+} toxicity is due to excessive activity of the SOCs, either due to excessive Ca^{2+} store depletion of other means of channels opening (Lee et al. 2010a; Petersen et al. 2006).

The concept of store-operated channels was conceived (Putney 1986) and then shown to operate when SERCA pump inhibitors became available to deplete the stores independent of cell stimulation (Takemura et al. 1989). A distinctive highly Ca^{2+} selective inward-rectifying current associated with activation of a channel in response to store depletion that was latter named the CRAC (Ca^{2+} release-activated Ca^{2+}) current was then found (Hoth and Penner 1992). At the same time, several studies reported activation of nonselective cation current in response to store depletion that mediated part and some time the bulk of the receptor-mediated Ca^{2+} influx (Lee et al. 2010a). However, the molecular identity of the SOCs remained a mystery until the identification of first TRPC channels (Zhu et al. 1995, 1996; Wes et al. 1995), then STIM1 (Roos et al. 2005; Liou et al. 2005), and finally Orai1 (Zhang et al. 2006; Prakriya et al. 2006; Vig et al. 2006). The contribution and role of the TRPC channels to receptor-stimulated Ca^{2+} influx has been extensively studied (Pandol et al. 1987; Lee et al. 1997). The TRPC channels are associated with the nonselective cation current, while the Orai channels (Zhang et al. 2006;

Vig et al. 2006; Feske et al. 2006) mediate the CRAC current. STIM1 functions as the ER Ca^{2+} sensor and is required for the function of both channel types (Roos et al. 2005; Liou et al. 2005). Here, we discuss the gating of TRPC and Orai channels by STIM1 and the relationship between the TRPC and Orai channels in the endoplasmic reticulum/plasma membrane junctions (ER/PM junctions).

8.2 STIM1

Since STIM1 gates both TRPC and Orai channels, we discuss first the STIM1 domains and its mode of interaction with the TRPC and Orai channels. The SOC concept required a mechanism for communication between the Ca^{2+} load of the ER and channels at the plasma membrane. This was provided with the discovery that the ER Ca^{2+} content sensor is the multidomain protein STIM1 (Roos et al. 2005; Liou et al. 2005). Since the structure of full-length STIM1 is not available yet, most of what we know about STIM1 domains is based on mutation and functional analysis. The functional STIM1 domains and the helices of SOAR (STIM1 Orai1-activating region) are illustrated in Fig. 8.1a. The STIM1 EF hand and SAM (sterile α motif) domains reside in the ER lumen. The EF hand binds Ca^{2+}

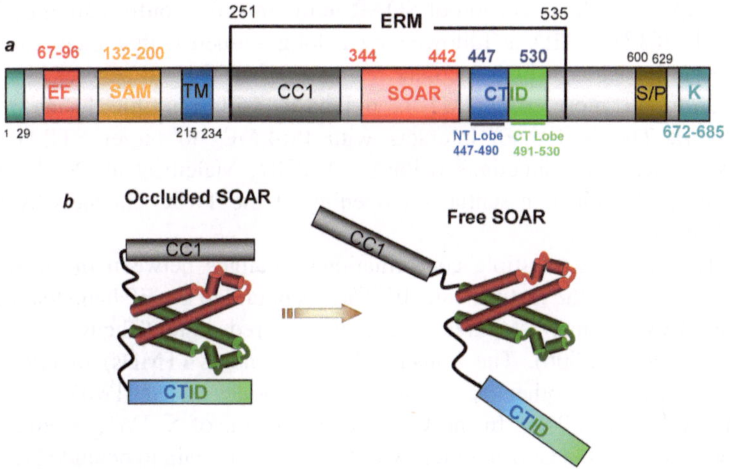

Fig. 8.1 The STIM1 domains are shown in (**a**) and include EF; Ca^{2+}-binding EF hand; SAM, sterile α motif; TM, transmembrane; ERM, ezrin/radixin/moesin domain; CC, coiled-coil; SOAR, STIM1 Orai1-activating region; CTID, C-terminus inhibitory domain; S/P, serine- and/or proline-rich segment; K, lysine-rich domain. Also shown is SOAR occluded by CC1 and CTID in resting state and its release by Ca^{2+} dissociation from the EF hand (**b**). (**b**) The SOAR domain has an R shape and exists as a dimer (the *red* and *green* monomers). In the resting state (*left image*), each SOAR monomer is occluded by the first coiled-coil domain (CC1) and the CTID domain that are shown only for the *green* monomer. In the resting occluded state, SOAR cannot interact with Orai or TRPC channels. Cell stimulation and Ca^{2+} release from the ER results in a conformational change in STIM1 and release of SOAR (*right* extended state), which can now access the channels and activate them

at normal store filling to keep STIM1 in non-clustered and away from the ER/PM junctions. The SAM domain participates in and facilitates STIM1 clustering (Stathopulos et al. 2008; Zheng et al. 2008). Cysteins in STIM1 luminal domain function in ER redox sensing through S-glutathionylation of cysteine 56 (Hawkins et al. 2010) and regulation of STIM1 function by ROS (Bhardwaj et al. 2016). The SAM domain is followed by a short transmembrane domain (TMD) that reorganizes in response to Ca^{2+} store depletion to promote the STIM1 active conformation (Ma et al. 2015). The cytoplasmic domain of STIM1 mediates the opening of the TRPC and Orai channels. In fact, the first channel shown to be activated by STIM1, and more specifically by STIM1 cytoplasmic domain, is TRPC1 (Huang et al. 2006). The cytoplasmic domain starts with a coiled-coil domain (CC1) composed of three distinctive helices and includes a short helix at its C-terminus that functions as an inhibitory helix (IH) (Yu et al. 2013; Yang et al. 2012; Derler et al. 2016). CC1 is followed by the SOAR domain (Yuan et al. 2009) (similar domain was identified as CAD (Park et al. 2009) or CCb9 (Feske et al. 2010)) that is the only STIM1 domain needed to fully activate Orai1 (Yuan et al. 2009; Park et al. 2009; Feske et al. 2010) and mediates the binding of STIM1 to of TRPC channels (Lee et al. 2014). The dimeric structure of the SOAR domain (Yang et al. 2012) is shown in Fig. 8.1b. SOAR is followed by a highly conserved domain immediately C-terminus to SOAR named CTID (for C-terminus inhibitory domain) that participates in the occlusion of SOAR in the inactive conformation of STIM1 (Jha et al. 2013). CTID is followed by a long sequence that includes several phosphorylation sites, which mediate response of STIM1 to several kinases. In the C terminal, most of STIM1 is polybasic with multiple lysine residues (K-domain). The K-domain interacts with $PI(4,5)P_2$ to target STIM1 to the ER/plasma membrane junctions (Liou et al. 2007; Maleth et al. 2014). Significantly, the K-domain is essential for opening of the TRPC channels by STIM1 (Zeng et al. 2008).

STIM1 undergoes multiple conformational changes between the resting and stimulated states. In the resting state, ER Ca^{2+} is bound to the EF hand that together with the SAM domain stabilizes the non-clustered state (Zheng et al. 2011; Stathopulos et al. 2006). The transmembrane domains (TMDs) of two STIM1 appear to cross each other by interaction of residues in the TMDs C terminal portions (Ma et al. 2015). In the C-terminus domain of STIM1, several studies provided evidence that CC1 interacts with the SOAR domain to occlude SOAR and prevent it from accessing the SOC channels (Yu et al. 2013; Korzeniowski et al. 2010; Muik et al. 2011). FRET measurement of a construct including SOAR and CC1 provided direct evidence for interaction of CC1 and SOAR likely by hydrophobic interaction (Muik et al. 2011), which may be mediated in part by glutamates in CC1 (Yu et al. 2013). Analysis of regions C terminal to SOAR identified the CTID domain as a domain required to keep SOAR in the occluded state (Jha et al. 2013).

Activation of STIM1 in response to Ca^{2+} store depletion involves major conformational rearrangements of STIM1 and likely of SOAR as well. These are discussed in details in a recent review by Derler et al. (2016) and will not be discussed here. In brief, store depletion results in clustering of STIM1 and Orai1 into puncta at the ER/PM junctions. The puncta formation and stabilization require intact ER/PM junctions that are formed by tether proteins like the extended synaptotagmines (E-Syts) (Maleth et al. 2014; Giordano et al. 2013; Chang et al. 2013), cytoskeletal and compartment separators like the septins and caveolin (Maleth et al. 2014; Sharma et al. 2013), and the STIM1 clustering regulator STIMATE (Jing et al. 2015; Quintana et al. 2015). The conformational change in STIM1 starts with unfolding of the EF hand-SAM domain upon Ca^{2+} release from the EF hand (Zheng et al. 2011), which is transmitted to the STIM1 cytoplasmic domain by the TMD to unfold the compact CC1-SOAR-CTID to release and unfold SOAR (Derler et al. 2016). Interestingly, STIM1 clusters first at a $PI(4,5)P_2$-poor region of the plasma membrane before it translocates to a $PI(4,5)P_2$-rich domain or before $PI(4,5)P_2$-rich domain forms around STIM1-Orai1 and perhaps STIM1-TRPC channel complexes (Maleth et al. 2014).

In addition to STIMATE, several other regulators of STIM1 action have been identified. That include STIM2 (Bhardwaj et al. 2016), the microtubule-plus-end-tracking protein end binding (EB1) (Grigoriev et al. 2008), and the SOCE-associated regulatory factor (SARAF) (Palty et al. 2012). The exact role of STIM2 in Ca^{2+} signaling is not fully understood. STIM2 exists in two splice variants, one that inhibits and one that facilitates Ca^{2+} influx by Orai1 (Miederer et al. 2015; Rana et al. 2015). The STIM2 variant that enhances Ca^{2+} influx does so by facilitating STIM1 clustering and interaction with Orai1 in response to minimal Ca^{2+} store depletion and thus causes STIM1 to respond to physiological stimulus intensity (Ong et al. 2015). As such, STIM2 appears to have critical role in the physiological and not the pathological Ca^{2+} influx. STIM1 interaction with microtubules is prominent at the resting state and is mediated by EB1 to form comet-like accumulation at the microtubule-ER contact sites (Grigoriev et al. 2008). The interaction is regulated by phosphorylation of C-terminus STIM1 residues and need to be dissociated for activation of STIM1 (Smyth et al. 2012; Pozo-Guisado et al. 2013). Excess Ca^{2+} influx is highly toxic to the cells, and both the Orai and TRPC channels are inhibited by high $[Ca^{2+}]_i$ that occurs when the channels are overactivated. Although not well understood yet, the mechanism of inactivation by high $[Ca^{2+}]_i$ is mediated by SARAF (Palty et al. 2012). SARAF has a single transmembrane domain, ER lumen resident N-terminus, and cytoplasmic C-terminus. SARAF interacts with STIM1 but not with Orai1 in response to ER Ca^{2+} release to facilitate inactivation of Orai1 by Ca^{2+} (Maleth et al. 2014; Palty et al. 2012). SARAF interacts specifically with SOAR, and the interaction is regulated by CTID to mediate the Ca^{2+}-dependent inactivation of current by Orai1 (Jha et al. 2013). Particularly, SARAF does not interact with STIM1 in the resting state. Rather, after store depletion, STIM1 interacts with Orai1 to form the puncta at a $PI(4,5)P_2$-poor domain, and only after formation of the STIM1-Orai1 complex and formation of $PI(4,5)P_2$-rich domain at the ER/PM junctions is SARAF recruited by STIM1 to mediate the slow Ca^{2+}-dependent inactivation of Orai1 (Maleth et al. 2014). This likely ensures maximal Ca^{2+} influx upon cell activation, and only after sufficient Ca^{2+} influx takes place does SARAF interacts with STIM1 to reduce the influx to protect against cell toxicity by excess Ca^{2+} influx.

8.3 Gating of Orai1 by STIM1

This topic is covered by other chapters in this book (Part I) and is discussed here only briefly when the information is relevant to understanding regulation of TRPC channels by STIM1. Further details can be found in recent comprehensive reviews (Bhardwaj et al. 2016; Derler et al. 2016; Criddle 2016; Lacruz and Feske 2015; Hogan 2015). STIM1 activation of Orai1 mediates the Ca^{2+} release-activated Ca^{2+} (CRAC) current (Zhang et al. 2006). The Orai channels are hexamers of four transmembrane span protein subunits with cytoplasmic N- and C-termini (Hou et al. 2012). The channel is arranged as trimers of dimers with the cytoplasmic extensions of the fourth transmembrane domain of two neighboring subunits fold into coiled-coil (CC) domains. Functional (Muik et al. 2008; Frischauf et al. 2009) and the structural (Hou et al. 2012) information indicate that CC domain at the C-terminus of the Orai channels interacts with the SOAR domain of STIM1 to open the channels (Lee et al. 2009). A second interaction of SOAR with the N-terminus of Orai1 (Park et al. 2009) functions to modulate channel opening (McNally et al. 2013). Interaction of SOAR with both the Orai1 N- and C-termini is required for determining Orai1 channel selectivity by STIM1 (McNally et al. 2012, 2013).

8.4 TRPC Channels and STIM1

Although the first channel activity shown to be regulated by STIM1 is TRPC1, how STIM1 regulates TRPC channels is poorly understood and sparsely studied. Several properties of TRPC channel function and their regulation by STIM1 are generally accepted. TRPC channels are activated by receptor stimulation and mediate significant part of the receptor-stimulated Ca^{2+} influx. TRPC3, TRPC6, and TRPC7 are activated by the lipid diacylglycerol, while TRPC1, TRPC4, and TRPC5 are not, although it is not known whether heteromultimers like those formed between TRPC1-TRPC3 and TRPC4-TRPC6 are activated by diacylglycerol. This is important since multimerization plays a key role in gating of the channels by STIM1 (see Sect. 8.6). TRPC channels are ubiquitous and have multiple roles in virtually every cell type (Parekh and Putney 2005; Pedersen et al. 2005; Freichel et al. 2005). Interaction between TRPC channels and STIM1 is a consistent finding, both with expressed and native proteins. Moreover, consistent knockdown or deletion of TRPC channels reduces Ca^{2+} influx and knockdown or knockout of Orai1, or STIM1 eliminates the store-dependent and receptor-activated Ca^{2+} influx (Lee et al. 2010a). For example, as shown in Fig. 8.2, in salivary gland cells, TRPC1 and TRPC3 form a complex with STIM1 in response to cell stimulation (Fig. 8.2a) and passive store depletion (Lee et al. 2014; Hong et al. 2011; Kim et al. 2009a). Deletion of TRPC1 and TRPC3 similarly reduces receptor-stimulated (Fig. 8.2b)

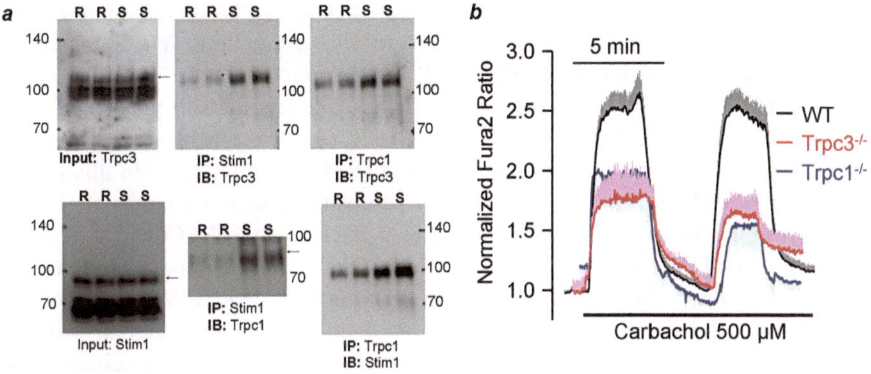

Fig. 8.2 Interaction of native STIM1 and TRPC1. Panel (**a**) shows the Co-IP of native salivary glands STIM1, TRPC1, and TRPC3 and its enhancement by cell stimulation (S). Panel (**b**) shows that knockdown of TRPC1 and TRPC3 similarly reduces Ca^{2+} influx. The results are reproduced from Lee et al. (2014)

and passive store depletion (SOCs)-mediated Ca^{2+} influx (Lee et al. 2014; Hong et al. 2011; Kim et al. 2009a). Additional examples of the effect of TRPC channel knockout/knockdown on Ca^{2+} influx have been reported (Lee et al. 2010a; Ng et al. 2012; Antigny et al. 2013; Zhang et al. 2010; Rao et al. 2012; Sundivakkam et al. 2012). Since deletion of Orai1 or STIM1 completely inhibits SOCs and at the same time deletion of TRPC channels partially inhibits the same, SOCs strongly suggest that TRPC channels form complexes with STIM1 and Orai1 and that Ca^{2+} influx by TRPC channels requires Orai1 function and/or Ca^{2+} influx by Orai1 requires TRPC channel function. The relationship between the channels has been documented in numerous studies (for few examples, see Lee et al. 2014; Hong et al. 2011; Kim et al. 2009b; Ong and Ambudkar 2015; Ong et al. 2007; Sabourin et al. 2015; Almirza et al. 2012; Cioffi et al. 2012; Ng et al. 2012; Lu et al. 2010; Molnar et al. 2016; Asanov et al. 2015) and is discussed in more detail in Sect. 8.6.

8.5 STIM1-Dependent and STIM1-Independent Function of TRPC Channels

A complication in understanding gating of TRPC channels by STIM1 is that when activated by receptor stimulation, they can function in STIM1-independent and STIM1-dependent modes, depending on the complement of TRPC channels present in the cells. This leads to underappreciation of the importance of gating of TRPC channels by STIM1 in understanding the roles of STIM1 in Ca^{2+} signaling and cell function, in spite of the constant observations that knockout/knockdown of TRPC channels reduces SOC-mediated Ca^{2+} influx. The evidence for interaction between STIM1 and TRPC channels mostly rely on co-immunoprecipitation (Co-IP) and co-localization experiments. When expressed alone, TRPC1, TRPC2, TRPC4, and

TRPC5 Co-IP with STIM1, while TRPC3, TRPC6, and TRPC7 do not (Huang et al. 2006). However, when they are present in the same cells TRPC channels heteromultimarize, including TRPC1-TRPC3 (Yuan et al. 2007; Liu et al. 2005) and TRPC4-TRPC6 (Yuan et al. 2007). TRPC channels that do not Co-IP with STIM1 when expressed alone (TRPC3 and TRPC6) do so when in complex with TRPC channels that do interact with STIM1 (TRPC1 and TRPC4). In this mode, TRPC3 and TRPC6 function as STIM1-dependent channels (Yuan et al. 2007). Notably, in the absence of TRPC1 and TRPC4, TRPC3 and TRPC6 are active and function in a STIM1-independent mode. Hence, the cellular composition of TRPC channels and their ratio determine whether they function in a STIM1-dependent or STIM1-independent modes. A common problem in the field is that without controlling expression of the channels to obtain the ratio required to observe the STIM1-dependent mode leads to the erroneous conclusion that the channels are not gated by STIM1, although sometimes the same studies show that knockout/knock-down of the channels reduces SOC.

Additional consideration when studying gating of TRPC channels by STIM1 must take into account the dependence of the SOC activity of TRPC channels on the Orai1-STIM1 complex. Orai and TRPC channels can function independent of each other to mediate the Orai1 distinctive CRAC current (Zeng et al. 2008) and the TRPC channels nonselective, Ca^{2+}-permeable current (Pedersen et al. 2005; Freichel et al. 2005; Nilius et al. 2007). Nevertheless, the native SOC is dependent on both TRPC and Orai1 channels that affect the activity of each other. As indicated above, Orai1 and STIM1 form complexes with TRPC channels and enhance TRPC channels store dependence (Liao et al. 2007, 2008, 2009). Most notably, functional Orai1 and TRPC channels are required to restore the physiological store-mediated Ca^{2+} influx, with channel-dead Orai1 or TRPC channel mutants unable to do so (Kim et al. 2009b). This important observation was confirmed recently in insulin-secreting cells and its physiological function demonstrated (Sabourin et al. 2015). Two mechanisms can account for the channel interdependence. Insertion of the channels in the plasma membrane from intracellular pool and formation of the Orai1-STIM1-TRPC complexes in at the ER/PM junctions (Kim et al. 2009b; Ong et al. 2007; Cheng et al. 2011; de Souza et al. 2015). Importantly, modulation of TRPC channels by the Orais and reconstitution of SOCs were observed only when the channels are expressed at close to physiological levels (Kim et al. 2009b; Liao et al. 2007, 2008, 2009). This further stresses the importance of the Orai1/TRPC channel ratio in observing the STIM1-dependent mode of function of the TRPC channels.

It is clear that particular caution is needed in selecting the system and conditions to study regulation of TRPC channel current and Ca^{2+} influx by STIM1. Nevertheless, to understand better the molecular and physiological function of STIM1 and the TRPC channels, we must understand better the molecular mechanism by which STIM1 gates the TRPC channels.

8.6 Gating TRPC Channels by STIM1

Gating TRPC channels by STIM1 relies on co-expression of the proteins and demonstration of increased TRPC channel current by STIM1. More direct and strong evidence for gating of TRPC channels by STIM1 was obtained using expressed channel and a recombinant purified SOAR domain (Asanov et al. 2015). SOAR interacts with TRPC channels in a manner very similar to STIM1, showing Co-IP with TRPC1, TRPC4, and TRPC5, but not with TRPC3 and TRPC6 when expressed alone (Lee et al. 2014). The use of excised patches from cells expressing the TRPC channels and recombinant purified SOAR was used to show that SOAR activates TRPC1, TRPC4, and TRPC5, but not TRPC3 and TRPC6 that are activated by diacylglycerol, when expressed individually. Moreover, single-molecule photobleaching analysis suggested that two molecules of SOAR activate a TRPC channel tetramer (Asanov et al. 2015). Further strong evidence for direct gating of TRPC channels by STIM1 is the finding that STIM1 gates TRPC channels through specific residues in the two proteins (Zeng et al. 2008). Early observations showed that deletion of the PI(4,5)P_2-binding STIM1 K-domain prevents activation of TRPC1 by STIM1 (Huang et al. 2006). Search for a domain in TRPC channels that may interact with or influenced by the STIM1 K-domain identified two conserved negatively charged residues (DD/DE/EE) at the C-terminus of all TRPC channels (Park et al. 2009). Mutation of the negatively charged residues (D^{639} and D^{640} in TRPC1) suggested that they communicate with the last two positively charged lysines (K^{684} and K^{685}) of STIM1. Moreover, notably, switching the charges in STIM1 and the TRPC channels showed that gating of TRPC channels by STIM1 occurs whether the positive or negative charges are on STIM1 or TRPC channels, as long as the negative charges on one protein are matched with positive charges on the other protein (Zeng et al. 2008). Gating by STIM1 K^{684} and K^{685} and the negative charges in TRPC channels was interpreted as gating by electrostatic interaction between STIM1 and the TRPC channels rather than affecting the TRPC channels pore (Zeng et al. 2008). However, how such interaction results in channel opening is not clear. An attractive possibility is that STIM1 interact with the TRPC channels to recruit them to a PI(4,5)P_2-rich domain, similar to the interaction with Orai1, and recruitment to a PI(4,5)P_2-rich domain is essential for keeping the TRPC channels in an active state. Indeed, several studies reported regulation of TRPC channels by PI(4,5)P_2 (Itsuki et al. 2014; Shi et al. 2014; Kim et al. 2013; Imai et al. 2012).

Gating by STIM1 K^{684} and K^{685} and the negative charges in TRPC channels C-terminus was shown to operate with TRPC3, TRPC4, TRPC5, and TRPC6 as well (Lee et al. 2010b). In the case of TRPC3 and TRPC6, their function as STIM1-gated channels required interaction with TRPC1 and TRPC4, respectively (Yuan et al. 2007). Since STIM1 K^{684} and K^{685} communicate with C-terminus-negative charges of TRPC channel, the question then is how the TRPC1 and TRPC4 confer the STIM1-dependent mode on TRPC3 and TRPC6. This was examined in detail in Lee et al. (2014), and some of the results are shown in Fig. 8.3. The use of CFP-TRPC3-YFP (TRPC3 tagged at the N- and C-termini with CFP and YFP,

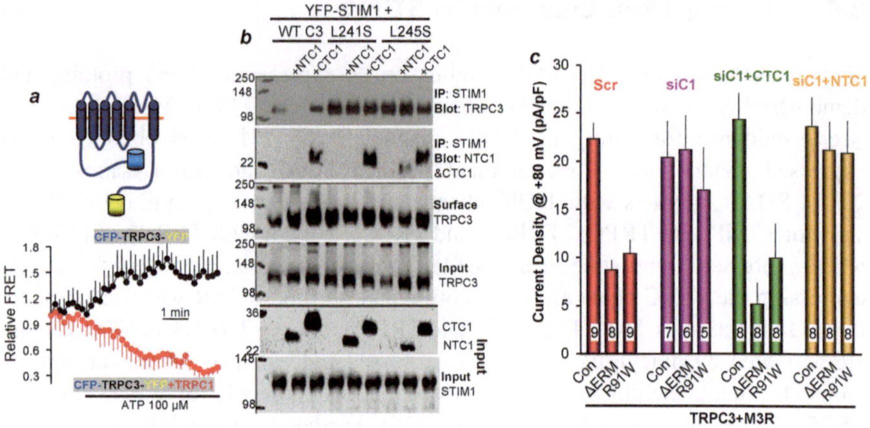

Fig. 8.3 TRPC1 C terminal coiled-coil domain dissociates C and N terminal coiled. Coil domains TRPC3 to confer STIM1 access to TRPC3. (**a**) CFP-TRPC3-YFP channel tagged at both the N-terminus with CFP and the C-terminus with YFP was transfected alone or together with AH-tagged TRPC1 in HeLa cells. FRET between CFP and YFP was measured before and after stimulation with 100 μM ATP. Note dissociation of N and C TRPC3 coiled-coil domains interaction with TRPC1. (**b**) Interaction of TRPC1 C but not N-terminus TRPC1 coiled-coil domain with TRPC3 and TRPC3 mutants with enhanced STIM1 binding. (**c**) TRPC1 CT-CCD but not NT-CCD confers STIM1-dependence mode to TRPC2 function (reproduced from Lee et al. 2014)

respectively) when expressed alone and measurement of FRET showed that the N- and C-termini are in close proximity/contact and the interaction is enhanced by cell stimulation. Importantly, when CFP-TRPC3-YFP is expressed with TRPC1, now cell stimulation dissociates the interaction between the TRPC3 N- and C-termini (Fig. 8.3a). Co-IP assay with isolated TRPC1 N- and C-termini CC domains showed that only TRPC1 C-terminus CC domain interacts with TRPC3 (Fig. 8.3b). The functional consequence of these interactions is shown in Fig. 8.3c. Measurement of receptor-stimulated TRPC3 current shows that it can be strongly inhibited by knockdown of STIM1 (Lee et al. 2014) and by scavenging STIM1 with ΔERM-STIM1 or by the Orai1(R91W) mutant that has no current but does bind and cluster with STIM1 (red columns in Fig. 8.3c). The STIM1 dependence of TRPC3 is eliminated by knockdown of TRPC1 (pink columns) and restored by expression of the C-terminus (green columns) but not the N-terminus (orange columns) TRPC1 CC domain.

The model in Fig. 8.4 summarizes the currently available results on the gating of TRPC channel isoforms by STIM1 using TRPC1 and TRPC3 as examples. TRPC channels have minimal spontaneous activity, and in the resting state their N and C terminal CC domains interact to keep the channels inactive. Cell stimulation causes dissociation in the interaction between the N- and C-termini CC domains of TRPC1 (and TRPC4 and TRPC5) to allow the SOAR domain of STIM1 to access and bind to the C terminus CC domain and STIM1 to escort TRPC1 to a $PI(4,5)P_2$-rich domain at the ER/PM junctions. Binding of SOAR to the C-terminus CC domain

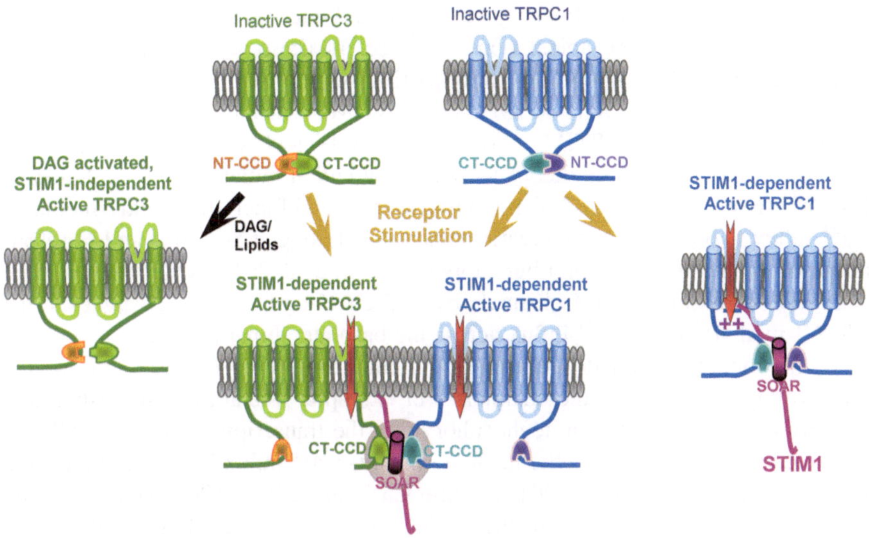

Fig. 8.4 A model depicting potential mechanisms for gating of TRPC1 and the TRPC1-TRPC3 heterodimer by STIM1

stabilizes the channel open conformation. When TRPC3 is present in cells that do not express high level of TRPC1, cell stimulation results in a conformational change that is not sufficient to dissociate between the CC domains but can be activated by DAG. Exogenous DAG stabilizes similar conformation. When TRPC3 is present in a cell together with TRPC1, cell stimulation results in interaction between TRPC3 and TRPC1 to allow the free C-terminus CC domain of TRPC1 to interact with the C-terminus CC domain of TRPC3 and expose it to the SOAR domain of STIM1. Now STIM1 escorts the TRPC1-TRPC3 complex to the PI(4,5) P_2-rich domain at the ER/PM junctions and determines the open state of TRPC3. This is the STIM1-dependent code of TRPC3. This model needs to be examined further by molecular and structural assays.

8.7 Specific Physiological Roles of TRPC Channels

Both the Orai and TRPC channels are activated by receptor stimulation and are activated by STIM1 to mediate Ca^{2+} influx. Nevertheless, it appears that Ca^{2+} influx by the two channels mediates specific cell functions. Several physiological functions appear to be mediated specifically by TRPC channels, and in some cases when tested, it appears that Ca^{2+} influx by Orai1 is less efficient or cannot stimulate the same functions. Below we describe some physiological functions that are strongly associated with Ca^{2+} influx by TRPC and not Orai1 channels. One way to achieve specificity is through differential localization. There is little information on the expression pattern of native STIM1, Orai1, and TRPC channels. In the polarized pancreatic and salivary gland acinar and duct cells, Orai1 shows preferential localization at the apical pool, in close proximity with IP_3Rs, while the localization of TRPC1 and STIM1 only partially overlaps with that of IP_3Rs. This suggests formation of Orai1-STIM1, Orai1-STIM1-TRPC1, and STIM1-TRPC1 complexes. The TRPC1-STIM1 and Orai1-STIM1-TRPC channel complexes are low in resting cells but markedly increased by cell stimulation and depletion of ER Ca^{2+} (Hong et al. 2011). Polarized localization and function of Orai1-mediated Ca^{2+} influx is observed in pancreatic acinar cells (Hong et al. 2011) and in serous airway Calu-3 cells grown on filter support (Balghi et al. 2011). On the other hand, activation of basolateral K^+ channels by TRPC1 (Ong et al. 2012) suggests that TRPC1 mediates most Ca^{2+} influx across the basolateral membrane.

Formation of STIM1-dependent Ca^{2+} influx complexes can account for the non-overlapping effect of the channels. For example, it is amply established that activation of calcineurin that dephosphorylates the transcription factor NFAT and promotes its translocation to the nucleus requires Ca^{2+} influx by Orai1 (Shaw and Feske 2012; Darbellay et al. 2010), while Ca^{2+} influx by TRPC channels has no effect on NFAT translocation to the nucleus (Cheng et al. 2011; Darbellay et al. 2010). Conversely, Ca^{2+} influx by TRPC1 activates the transcription factor NFκB, whereas Ca^{2+} influx by Orai1 is less effective (Cheng et al. 2011; Darbellay et al. 2010). While Ca^{2+} influx by TRPC1 activates the Ca^{2+}-activated K^+ channel (Cheng et al. 2011), Ca^{2+} influx by Orai1 does not activate these channels. Ca^{2+} influx by Orai1 activates the Ca^{2+}-activated Cl^- channel ANO1 (Forrest et al. 2010). In secretory cells, ANO1 is in the luminal membrane (Yang et al. 2008; Lee et al. 2012).

Ca^{2+} influx by Orai1 activates the transcription factor MEF2 and myogenin to promote myoblast differentiation (Hong et al. 2011; Balghi et al. 2011), while Ca^{2+} influx by TRPC1 and TRPC4 mediates MEF2-dependent myoblast fusion that could not be mediated by Ca^{2+} influx through Orai1 (Antigny et al. 2013). Ca^{2+} influx by TRPC1 and TRPC5 mediates restoration of neuronal progenitor cells proliferation (Yao et al. 2012; Shin et al. 2010). Loss of dopaminergic neurons in a Parkinson's disease model is associated with decreased TRPC1 expression and its interaction with STIM1 that is required for activating the AKT pathway to mediate the neuroprotection (Selvaraj et al. 2012). Vascular permeability is controlled by endothelial cells lining the vessels. The integrity of the endothelial barrier and endothelial cells function is strictly dependent of Ca^{2+} influx by STIM1-activated TRPC1 and TRPC4 (Di and Malik 2010). Knockdown of Trpc4 or STIM1 inhibits endothelial cell Ca^{2+} influx and disrupts the endothelial barrier, while knockdown of Orai1 or Orai3 in these cells had minimal effect and overexpression of Orai1 did not restore barrier function in the $Trpc4^{-/-}$ mice (Sundivakkam et al. 2012). Interaction between TRPC3 and STIM1 mediates mGluR-dependent synaptic transmission in cerebellar Purkinje cells and is crucial for motor coordination (Hartmann et al. 2008, 2014). Finally, several diseases not discussed here have been associated with mutations in TRPC channels (for review, see Nilius and Owsianik 2011).

The discussion above stresses the specific functions of TRPC channels, their regulation by STIM1, and the urgent need to better understand gating of TRPC channels by STIM1. In particular, since TRPC channels should be an important therapeutical target. While mutations in Orai1 and STIM1 have severe physiological phenotypes, including immune, cardiac, and neuronal deficiencies (Lacruz and Feske 2015), knockout of any of the TRPC channels has minor phenotypes (Liao et al. 2014). Development of specific TRPC channel inhibitors and activators should be of great value for better understanding their acute physiological roles and therapeutically.

8.8 Conclusion

This chapter emphasizes the regulation of TRPC channels by STIM1 and their specialized function in cell physiology. The consistent findings of reduced SOCs activity in cells with deleted TRPC channels and the unique dependence of TRPC channels activity on the STIM1 polylysine domain are probably the strongest evidence for regulation of TRPC channels by STIM1 and their contribution to SOCs. The need for precise ratio of TRPC channel isoforms complements and STIM1 to reconstitute regulation of TRPC channels by STIM1 in expression systems is highlighted that also explains the confusion in understanding the role of STIM1-TRPC channel function in cell physiology. With better understanding of cell signaling at the ER/PM junctions and the prominent role of STIM1 in directing Orai1 and TRPC channels to the junctions, future studies will likely further understanding of TRPC channel regulation by STIM1 by examining behavior of the TRPC channels when at or outside the ER/PM junctions. Localization in and outside the junctions may determine functional modes of TRPC channels. Further understanding of TRPC channel function is critical for further clarification of their physiological roles. TRPC channels should be a preferred therapeutic target because of their role in critical physiological functions, while deletion of any or combination of TRPC channels in mice had minor apparent deleterious physiological effects while changing the intended examined physiological function.

References

Almirza WH, Peters PH, van Zoelen EJ, Theuvenet AP (2012) Role of Trpc channels, Stim1 and Orai1 in PGF(2alpha)-induced calcium signaling in NRK fibroblasts. Cell Calcium 51 (1):12–21

Antigny F, Koenig S, Bernheim L, Frieden M (2013) During post-natal human myogenesis, normal myotube size requires TRPC1 and TRPC4 mediated Ca2+ entry. J Cell Sci 126 (Pt 11):2525–2533

Asanov A, Sampieri A, Moreno C, Pacheco J, Salgado A, Sherry R, Vaca L (2015) Combined single channel and single molecule detection identifies subunit composition of STIM1-activated transient receptor potential canonical (TRPC) channels. Cell Calcium 57(1):1–13

Balghi H, Robert R, Rappaz B, Zhang X, Wohlhuter-Haddad A, Evagelidis A, Luo Y, Goepp J, Ferraro P, Romeo P, Trebak M, Wiseman PW, Thomas DY, Hanrahan JW (2011) Enhanced Ca2+ entry due to Orai1 plasma membrane insertion increases IL-8 secretion by cystic fibrosis airways. FASEB J 25(12):4274–4291

Bhardwaj R, Hediger MA, Demaurex N (2016) Redox modulation of STIM-ORAI signaling. Cell Calcium 60(2):142–152

Cao X, Choi S, Maleth JJ, Park S, Ahuja M, Muallem S (2015) The ER/PM microdomain, PI(4,5)P(2) and the regulation of STIM1-Orai1 channel function. Cell Calcium 58(4):342–348

Chang CL, Hsieh TS, Yang TT, Rothberg KG, Azizoglu DB, Volk E, Liao JC, Liou J (2013) Feedback regulation of receptor-induced Ca2+ signaling mediated by E-Syt1 and Nir2 at endoplasmic reticulum-plasma membrane junctions. Cell Rep 5(3):813–825

Cheng KT, Liu X, Ong HL, Swaim W, Ambudkar IS (2011) Local Ca2+ entry via Orai1 regulates plasma membrane recruitment of TRPC1 and controls cytosolic Ca2+ signals required for specific cell functions. PLoS Biol 9(3):e1001025

Cioffi DL, Wu S, Chen H, Alexeyev M, St Croix CM, Pitt BR, Uhlig S, Stevens T (2012) Orai1 determines calcium selectivity of an endogenous TRPC heterotetramer channel. Circ Res 110(11):1435–1444

Criddle DN (2016) Reactive oxygen species, Ca2+ stores and acute pancreatitis; a step closer to therapy? Cell Calcium 60(3):180–189

Darbellay B, Arnaudeau S, Ceroni D, Bader CR, Konig S, Bernheim L (2010) Human muscle economy myoblast differentiation and excitation-contraction coupling use the same molecular partners, STIM1 and STIM2. J Biol Chem 285(29):22437–22447

Derler I, Jardin I, Romanin C (2016) Molecular mechanisms of STIM/Orai communication. Am J Physiol Cell Physiol 310(8):C643–C662

de Souza LB, Ong HL, Liu X, Ambudkar IS (2015) Fast endocytic recycling determines TRPC1-STIM1 clustering in ER-PM junctions and plasma membrane function of the channel. Biochim Biophys Acta 1853(10 Pt A):2709–2721

Di A, Malik AB (2010) TRP channels and the control of vascular function. Curr Opin Pharmacol 10(2):127–132

Feske S, Gwack Y, Prakriya M, Srikanth S, Puppel SH, Tanasa B, Hogan PG, Lewis RS, Daly M, Rao A (2006) A mutation in Orai1 causes immune deficiency by abrogating CRAC channel function. Nature 441(7090):179–185

Feske S, Picard C, Fischer A (2010) Immunodeficiency due to mutations in ORAI1 and STIM1. Clin Immunol 135(2):169–182

Forrest AS, Angermann JE, Raghunathan R, Lachendro C, Greenwood IA, Leblanc N (2010) Intricate interaction between store-operated calcium entry and calcium-activated chloride channels in pulmonary artery smooth muscle cells. Adv Exp Med Biol 661:31–55

Freichel M, Vennekens R, Olausson J, Stolz S, Philipp SE, Weissgerber P, Flockerzi V (2005) Functional role of TRPC proteins in native systems: implications from knockout and knock-down studies. J Physiol 567(Pt 1):59–66

Frischauf I, Muik M, Derler I, Bergsmann J, Fahrner M, Schindl R, Groschner K, Romanin C (2009) Molecular determinants of the coupling between STIM1 and Orai channels: differential activation of Orai1-3 channels by a STIM1 coiled-coil mutant. J Biol Chem 284(32):21696–21706

Giordano F, Saheki Y, Idevall-Hagren O, Colombo SF, Pirruccello M, Milosevic I, Gracheva EO, Bagriantsev SN, Borgese N, De Camilli P (2013) PI(4,5)P(2)-dependent and Ca(2+)-regulated ER-PM interactions mediated by the extended synaptotagmins. Cell 153(7):1494–1509

Grigoriev I, Gouveia SM, van der Vaart B, Demmers J, Smyth JT, Honnappa S, Splinter D, Steinmetz MO, Putney JW Jr, Hoogenraad CC, Akhmanova A (2008) STIM1 is a MT-plus-end-tracking protein involved in remodeling of the ER. Curr Biol 18(3):177–182

Hartmann J, Dragicevic E, Adelsberger H, Henning HA, Sumser M, Abramowitz J, Blum R, Dietrich A, Freichel M, Flockerzi V, Birnbaumer L, Konnerth A (2008) TRPC3 channels are required for synaptic transmission and motor coordination. Neuron 59(3):392–398

Hartmann J, Karl RM, Alexander RP, Adelsberger H, Brill MS, Rühlmann C, Ansel A, Sakimura K, Baba Y, Kurosaki T, Misgeld T, Konnerth A (2014) STIM1 controls neuronal Ca(2)(+) signaling, mGluR1-dependent synaptic transmission, and cerebellar motor behavior. Neuron 82(3):635–644

Hawkins BJ, Irrinki KM, Mallilankaraman K, Lien YC, Wang Y, Bhanumathy CD, Subbiah R, Ritchie MF, Soboloff J, Baba Y, Kurosaki T, Joseph SK, Gill DL, Madesh M (2010) S-glutathionylation activates STIM1 and alters mitochondrial homeostasis. J Cell Biol 190 (3):391–405

Hogan PG (2015) The STIM1-ORAI1 microdomain. Cell Calcium 58(4):357–367

Hong JH, Li Q, Kim MS, Shin DM, Feske S, Birnbaumer L, Cheng KT, Ambudkar IS, Muallem S (2011) Polarized but differential localization and recruitment of STIM1, Orai1 and TRPC channels in secretory cells. Traffic 12(2):232–245

Hoth M, Penner R (1992) Depletion of intracellular calcium stores activates a calcium current in mast cells. Nature 355(6358):353–356

Hou X, Pedi L, Diver MM, Long SB (2012) Crystal structure of the calcium release-activated calcium channel Orai. Science 338(6112):1308–1313

Huang GN, Zeng W, Kim JY, Yuan JP, Han L, Muallem S, Worley PF (2006) STIM1 carboxyl-terminus activates native SOC, I(crac) and TRPC1 channels. Nat Cell Biol 8(9):1003–1010

Imai Y, Itsuki K, Okamura Y, Inoue R, Mori MX (2012) A self-limiting regulation of vasoconstrictor-activated TRPC3/C6/C7 channels coupled to PI(4,5)P(2)-diacylglycerol signalling. J Physiol 590(5):1101–1119

Itsuki K, Imai Y, Hase H, Okamura Y, Inoue R, Mori MX (2014) PLC-mediated PI(4,5)P2 hydrolysis regulates activation and inactivation of TRPC6/7 channels. J Gen Physiol 143 (2):183–201

Jha A, Ahuja M, Maléth J, Moreno CM, Yuan JP, Kim MS, Muallem S (2013) The STIM1 CTID domain determines access of SARAF to SOAR to regulate Orai1 channel function. J Cell Biol 202(1):71–79

Jing J, He L, Sun A, Quintana A, Ding Y, Ma G, Tan P, Liang X, Zheng X, Chen L, Shi X, Zhang SL, Zhong L, Huang Y, Dong MQ, Walker CL, Hogan PG, Wang Y, Zhou Y (2015) Proteomic mapping of ER-PM junctions identifies STIMATE as a regulator of Ca(2)(+) influx. Nat Cell Biol 17(10):1339–1347

Kasai H, Augustine GJ (1990) Cytosolic Ca2+ gradients triggering unidirectional fluid secretion from exocrine pancreas. Nature 348(6303):735–738

Kim MS, Hong JH, Li Q, Shin DM, Abramowitz J, Birnbaumer L, Muallem S (2009a) Deletion of TRPC3 in mice reduces store-operated Ca2+ influx and the severity of acute pancreatitis. Gastroenterology 137(4):1509–1517

Kim MS, Zeng W, Yuan JP, Shin DM, Worley PF, Muallem S (2009b) Native store-operated Ca2+ influx requires the channel function of Orai1 and TRPC1. J Biol Chem 284(15):9733–9741

Kim H, Jeon JP, Hong C, Kim J, Myeong J, Jeon JH, So I (2013) An essential role of PI(4,5)P (2) for maintaining the activity of the transient receptor potential canonical (TRPC)4beta. Pflugers Arch 465(7):1011–1021

Korzeniowski MK, Manjarres IM, Varnai P, Balla T (2010) Activation of STIM1-Orai1 involves an intramolecular switching mechanism. Sci Signal 3(148):ra82

Lacruz RS, Feske S (2015) Diseases caused by mutations in ORAI1 and STIM1. Ann N Y Acad Sci 1356:45–79

Lee MG, Xu X, Zeng W, Diaz J, Wojcikiewicz RJ, Kuo TH, Wuytack F, Racymaekers L, Muallem S (1997) Polarized expression of Ca2+ channels in pancreatic and salivary gland cells. Correlation with initiation and propagation of [Ca2+]i waves. J Biol Chem 272 (25):15765–15770

Lee KP, Yuan JP, Zeng W, So I, Worley PF, Muallem S (2009) Molecular determinants of fast Ca2 +-dependent inactivation and gating of the Orai channels. Proc Natl Acad Sci U S A 106 (34):14687–14692

Lee KP, Yuan JP, Hong JH, So I, Worley PF, Muallem S (2010a) An endoplasmic reticulum/plasma membrane junction: STIM1/Orai1/TRPCs. FEBS Lett 584(10):2022–2027

Lee KP, Yuan JP, So I, Worley PF, Muallem S (2010b) STIM1-dependent and STIM1-independent function of transient receptor potential canonical (TRPC) channels tunes their store-operated mode. J Biol Chem 285(49):38666–38673

Lee MG, Ohana E, Park HW, Yang D, Muallem S (2012) Molecular mechanism of pancreatic and salivary gland fluid and HCO3 secretion. Physiol Rev 92(1):39–74

Lee KP, Choi S, Hong JH, Ahuja M, Graham S, Ma R, So I, Shin DM, Muallem S, Yuan JP (2014) Molecular determinants mediating gating of Transient Receptor Potential Canonical (TRPC) channels by stromal interaction molecule 1 (STIM1). J Biol Chem 289(10):6372–6382

Liao Y, Erxleben C, Yildirim E, Abramowitz J, Armstrong DL, Birnbaumer L (2007) Orai proteins interact with TRPC channels and confer responsiveness to store depletion. Proc Natl Acad Sci U S A 104(11):4682–4687

Liao Y, Erxleben C, Abramowitz J, Flockerzi V, Zhu MX, Armstrong DL, Birnbaumer L (2008) Functional interactions among Orai1, TRPCs, and STIM1 suggest a STIM-regulated heteromeric Orai/TRPC model for SOCE/Icrac channels. Proc Natl Acad Sci U S A 105 (8):2895–2900

Liao Y, Plummer NW, George MD, Abramowitz J, Zhu MX, Birnbaumer L (2009) A role for Orai in TRPC-mediated Ca2+ entry suggests that a TRPC:Orai complex may mediate store and receptor operated Ca2+ entry. Proc Natl Acad Sci U S A 106(9):3202–3206

Liao Y, Abramowitz J, Birnbaumer L (2014) The TRPC family of TRP channels: roles inferred (mostly) from knockout mice and relationship to ORAI proteins. Handb Exp Pharmacol 223:1055–1075

Liou J, Kim ML, Heo WD, Jones JT, Myers JW, Ferrell JE Jr, Meyer T (2005) STIM is a Ca2+ sensor essential for Ca2+-store-depletion-triggered Ca2+ influx. Curr Biol 15(13):1235–1241

Liou J, Fivaz M, Inoue T, Meyer T (2007) Live-cell imaging reveals sequential oligomerization and local plasma membrane targeting of stromal interaction molecule 1 after Ca2+ store depletion. Proc Natl Acad Sci U S A 104(22):9301–9306

Liu X, Bandyopadhyay BC, Singh BB, Groschner K, Ambudkar IS (2005) Molecular analysis of a store-operated and 2-acetyl-sn-glycerol-sensitive non-selective cation channel. Heteromeric assembly of TRPC1-TRPC3. J Biol Chem 280(22):21600–21606

Lu M, Bränström R, Berglund E, Höög A, Björklund P, Westin G, Larsson C, Farnebo LO, Forsberg L (2010) Expression and association of TRPC subtypes with Orai1 and STIM1 in human parathyroid. J Mol Endocrinol 44(5):285–294

Ma G, Wei M, He L, Liu C, Wu B, Zhang SL, Jing J, Liang X, Senes A, Tan P, Li S, Sun A, Bi Y, Zhong L, Si H, Shen Y, Li M, Lee MS, Zhou W, Wang J, Wang Y, Zhou Y (2015) Inside-out Ca(2+) signalling prompted by STIM1 conformational switch. Nat Commun 6:7826

Maleth J, Choi S, Muallem S, Ahuja M (2014) Translocation between PI(4,5)P2-poor and PI(4,5) P2-rich microdomains during store depletion determines STIM1 conformation and Orai1 gating. Nat Commun 5:5843

McNally BA, Somasundaram A, Yamashita M, Prakriya M (2012) Gated regulation of CRAC channel ion selectivity by STIM1. Nature 482(7384):241–245

McNally BA, Somasundaram A, Jairaman A, Yamashita M, Prakriya M (2013) The C- and N-terminal STIM1 binding sites on Orai1 are required for both trapping and gating CRAC channels. J Physiol 591(Pt 11):2833–2850

Miederer AM, Alansary D, Schwär G, Lee PH, Jung M, Helms V, Niemeyer BA (2015) A STIM2 splice variant negatively regulates store-operated calcium entry. Nat Commun 6:6899

Molnar T, Yarishkin O, Iuso A, Barabas P, Jones B, Marc RE, Phuong TT, Krizaj D (2016) Store-operated calcium entry in Muller glia is controlled by synergistic activation of TRPC and Orai channels. J Neurosci 36(11):3184–3198

Muallem S, Schoeffield MS, Fimmel CJ, Pandol SJ (1988) Agonist-sensitive calcium pool in the pancreatic acinar cell. II. Characterization of reloading. Am J Physiol 255(2 Pt 1):G229–G235

Muik M, Frischauf I, Derler I, Fahrner M, Bergsmann J, Eder P, Schindl R, Hesch C, Polzinger B, Fritsch R, Kahr H, Madl J, Gruber H, Groschner K, Romanin C (2008) Dynamic coupling of the putative coiled-coil domain of ORAI1 with STIM1 mediates ORAI1 channel activation. J Biol Chem 283(12):8014–8022

Muik M, Fahrner M, Schindl R, Stathopulos P, Frischauf I, Derler I, Plenk P, Lackner B, Groschner K, Ikura M, Romanin C (2011) STIM1 couples to ORAI1 via an intramolecular transition into an extended conformation. EMBO J 30(9):1678–1689

Ng LC, O'Neill KG, French D, Airey JA, Singer CA, Tian H, Shen XM, Hume JR (2012) TRPC1 and Orai1 interact with STIM1 and mediate capacitative Ca(2+) entry caused by acute hypoxia in mouse pulmonary arterial smooth muscle cells. Am J Physiol Cell Physiol 303(11):C1156–C1172

Nilius B, Owsianik G (2011) The transient receptor potential family of ion channels. Genome Biol 12(3):218

Nilius B, Owsianik G, Voets T, Peters JA (2007) Transient receptor potential cation channels in disease. Physiol Rev 87(1):165–217

Ong HL, Ambudkar IS (2015) Molecular determinants of TRPC1 regulation within ER-PM junctions. Cell Calcium 58(4):376–386

Ong HL, Cheng KT, Liu X, Bandyopadhyay BC, Paria BC, Soboloff J, Pani B, Gwack Y, Srikanth S, Singh BB, Gill DL, Ambudkar IS (2007) Dynamic assembly of TRPC1/STIM1/Orai1 ternary complex is involved in store operated calcium influx: evidence for similarities in SOC and CRAC channel components. J Biol Chem 282(12):9105–9116

Ong HL, Jang SI, Ambudkar IS (2012) Distinct contributions of Orai1 and TRPC1 to agonist-induced [Ca(2+)](i) signals determine specificity of Ca(2+)-dependent gene expression. PLoS One 7(10):e47146

Ong HL, de Souza LB, Zheng C, Cheng KT, Liu X, Goldsmith CM, Feske S, Ambudkar IS (2015) STIM2 enhances receptor-stimulated Ca(2)(+) signaling by promoting recruitment of STIM1 to the endoplasmic reticulum-plasma membrane junctions. Sci Signal 8(359):ra3

Palty R, Raveh A, Kaminsky I, Meller R, Reuveny E (2012) SARAF inactivates the store operated calcium entry machinery to prevent excess calcium refilling. Cell 149(2):425–438

Pandol SJ, Schoeffield MS, Fimmel CJ, Muallem S (1987) The agonist-sensitive calcium pool in the pancreatic acinar cell. Activation of plasma membrane Ca2+ influx mechanism. J Biol Chem 262(35):16963–16968

Parekh AB, Putney JW Jr (2005) Store-operated calcium channels. Physiol Rev 85(2):757–810

Park CY, Hoover PJ, Mullins FM, Bachhawat P, Covington ED, Raunser S, Walz T, Garcia KC, Dolmetsch RE, Lewis RS (2009) STIM1 clusters and activates CRAC channels via direct binding of a cytosolic domain to Orai1. Cell 136(5):876–890

Pedersen SF, Owsianik G, Nilius B (2005) TRP channels: an overview. Cell Calcium 38 (3–4):233–252

Petersen OH, Sutton R, Criddle DN (2006) Failure of calcium microdomain generation and pathological consequences. Cell Calcium 40(5–6):593–600

Pozo-Guisado E, Casas-Rua V, Tomas-Martin P, Lopez-Guerrero AM, Alvarez-Barrientos A, Martin-Romero FJ (2013) Phosphorylation of STIM1 at ERK1/2 target sites regulates interaction with the microtubule plus-end binding protein EB1. J Cell Sci 126(Pt 14):3170–3180

Prakriya M, Feske S, Gwack Y, Srikanth S, Rao A, Hogan PG (2006) Orai1 is an essential pore subunit of the CRAC channel. Nature 443(7108):230–233

Putney JW Jr (1986) A model for receptor-regulated calcium entry. Cell Calcium 7(1):1–12

Quintana A, Rajanikanth V, Farber-Katz S, Gudlur A, Zhang C, Jing J, Zhou Y, Rao A, Hogan PG (2015) TMEM110 regulates the maintenance and remodeling of mammalian ER-plasma membrane junctions competent for STIM-ORAI signaling. Proc Natl Acad Sci U S A 112 (51):E7083–E7092

Rana A, Yen M, Sadaghiani AM, Malmersjö S, Park CY, Dolmetsch RE, Lewis RS (2015) Alternative splicing converts STIM2 from an activator to an inhibitor of store-operated calcium channels. J Cell Biol 209(5):653–670

Rao JN, Rathor N, Zhuang R, Zou T, Liu L, Xiao L, Turner DJ, Wang JY (2012) Polyamines regulate intestinal epithelial restitution through TRPC1-mediated Ca(2)+ signaling by differentially modulating STIM1 and STIM2. Am J Physiol Cell Physiol 303(3):C308–C317

Roos J, DiGregorio PJ, Yeromin AV, Ohlsen K, Lioudyno M, Zhang S, Safrina O, Kozak JA, Wagner SL, Cahalan MD, Veliçelebi G, Stauderman KA (2005) STIM1, an essential and conserved component of store-operated Ca2+ channel function. J Cell Biol 169(3):435–445

Sabourin J, Le Gal L, Saurwein L, Haefliger JA, Raddatz E, Allagnat F (2015) Store-operated Ca2 + entry mediated by Orai1 and TRPC1 participates to insulin secretion in rat beta-cells. J Biol Chem 290(51):30530–30539

Selvaraj S, Sun Y, Watt JA, Wang S, Lei S, Birnbaumer L, Singh BB (2012) Neurotoxin-induced ER stress in mouse dopaminergic neurons involves downregulation of TRPC1 and inhibition of AKT/mTOR signaling. J Clin Invest 122(4):1354–1367

Sharma S, Quintana A, Findlay GM, Mettlen M, Baust B, Jain M, Nilsson R, Rao A, Hogan PG (2013) An siRNA screen for NFAT activation identifies septins as coordinators of store-operated Ca2+ entry. Nature 499(7457):238–242

Shaw PJ, Feske S (2012) Physiological and pathophysiological functions of SOCE in the immune system. Front Biosci (Elite Ed) 4:2253–2268

Shi J, Birnbaumer L, Large WA, Albert AP (2014) Myristoylated alanine-rich C kinase substrate coordinates native TRPC1 channel activation by phosphatidylinositol 4,5-bisphosphate and protein kinase C in vascular smooth muscle. FASEB J 28(1):244–255

Shin HY, Hong YH, Jang SS, Chae HG, Paek SL, Moon HE, Kim DG, Kim J, Paek SH, Kim SJ (2010) A role of canonical transient receptor potential 5 channel in neuronal differentiation from A2B5 neural progenitor cells. PLoS One 5(5):e10359

Smyth JT, Beg AM, Wu S, Putney JW Jr, Rusan NM (2012) Phosphoregulation of STIM1 leads to exclusion of the endoplasmic reticulum from the mitotic spindle. Curr Biol 22(16):1487–1493

Stathopulos PB, Li GY, Plevin MJ, Ames JB, Ikura M (2006) Stored Ca2+ depletion-induced oligomerization of stromal interaction molecule 1 (STIM1) via the EF-SAM region: an initiation mechanism for capacitive Ca2+ entry. J Biol Chem 281(47):35855–35862

Stathopulos PB, Zheng L, Li GY, Plevin MJ, Ikura M (2008) Structural and mechanistic insights into STIM1-mediated initiation of store-operated calcium entry. Cell 135(1):110–122

Sundivakkam PC, Freichel M, Singh V, Yuan JP, Vogel SM, Flockerzi V, Malik AB, Tiruppathi C (2012) The Ca(2+) sensor stromal interaction molecule 1 (STIM1) is necessary and sufficient for the store-operated Ca(2+) entry function of transient receptor potential canonical (TRPC) 1 and 4 channels in endothelial cells. Mol Pharmacol 81(4):510–526

Takemura H, Hughes AR, Thastrup O, Putney JW Jr (1989) Activation of calcium entry by the tumor promoter thapsigargin in parotid acinar cells. Evidence that an intracellular calcium pool and not an inositol phosphate regulates calcium fluxes at the plasma membrane. J Biol Chem 264(21):12266–12271

Thorn P, Lawrie AM, Smith PM, Gallacher DV, Petersen OH (1993) Local and global cytosolic Ca2+ oscillations in exocrine cells evoked by agonists and inositol trisphosphate. Cell 74 (4):661–668

Vig M, Peinelt C, Beck A, Koomoa DL, Rabah D, Koblan-Huberson M, Kraft S, Turner H, Fleig A, Penner R, Kinet JP (2006) CRACM1 is a plasma membrane protein essential for store-operated Ca2+ entry. Science 312(5777):1220–1223

Wes PD, Chevesich J, Jeromin A, Rosenberg C, Stetten G, Montell C (1995) TRPC1, a human homolog of a Drosophila store-operated channel. Proc Natl Acad Sci U S A 92(21):9652–9656

Yang YD, Cho H, Koo JY, Tak MH, Cho Y, Shim WS, Park SP, Lee J, Lee B, Kim BM, Raouf R, Shin YK, Oh U (2008) TMEM16A confers receptor-activated calcium-dependent chloride conductance. Nature 455(7217):1210–1215

Yang X, Jin H, Cai X, Li S, Shen Y (2012) Structural and mechanistic insights into the activation of Stromal interaction molecule 1 (STIM1). Proc Natl Acad Sci U S A 109(15):5657–5662

Yao H, Duan M, Yang L, Buch S (2012) Platelet-derived growth factor-BB restores human immunodeficiency virus Tat-cocaine-mediated impairment of neurogenesis: role of TRPC1 channels. J Neurosci 32(29):9835–9847

Yu F, Sun L, Hubrack S, Selvaraj S, Machaca K (2013) Intramolecular shielding maintains STIM1 in an inactive conformation. J Cell Sci 126(Pt 11):2401–2410

Yuan JP, Zeng W, Huang GN, Worley PF, Muallem S (2007) STIM1 heteromultimerizes TRPC channels to determine their function as store-operated channels. Nat Cell Biol 9(6):636–645

Yuan JP, Zeng W, Dorwart MR, Choi YJ, Worley PF, Muallem S (2009) SOAR and the polybasic STIM1 domains gate and regulate Orai channels. Nat Cell Biol 11(3):337–343

Zeng W, Yuan JP, Kim MS, Choi YJ, Huang GN, Worley PF, Muallem S (2008) STIM1 gates TRPC channels, but not Orai1, by electrostatic interaction. Mol Cell 32(3):439–448

Zhang SL, Yeromin AV, Zhang XH, Yu Y, Safrina O, Penna A, Roos J, Stauderman KA, Cahalan MD (2006) Genome-wide RNAi screen of Ca(2+) influx identifies genes that regulate Ca(2+) release-activated Ca(2+) channel activity. Proc Natl Acad Sci U S A 103(24):9357–9362

Zhang ZY, Pan LJ, Zhang ZM (2010) Functional interactions among STIM1, Orai1 and TRPC1 on the activation of SOCs in HL-7702 cells. Amino Acids 39(1):195–204

Zheng L, Stathopulos PB, Li GY, Ikura M (2008) Biophysical characterization of the EF-hand and SAM domain containing Ca2+ sensory region of STIM1 and STIM2. Biochem Biophys Res Commun 369(1):240–246

Zheng L, Stathopulos PB, Schindl R, Li GY, Romanin C, Ikura M (2011) Auto-inhibitory role of the EF-SAM domain of STIM proteins in store-operated calcium entry. Proc Natl Acad Sci U S A 108(4):1337–1342

Zhu X, Chu PB, Peyton M, Birnbaumer L (1995) Molecular cloning of a widely expressed human homologue for the Drosophila trp gene. FEBS Lett 373(3):193–198

Zhu X, Jiang M, Peyton M, Boulay G, Hurst R, Stefani E, Birnbaumer L (1996) trp, a novel mammalian gene family essential for agonist-activated capacitative Ca2+ entry. Cell 85 (5):661–671

STIM-TRP Pathways and Microdomain Organization: Contribution of TRPC1 in Store-Operated Ca^{2+} Entry – Impact on Ca^{2+} Signaling and Cell Function

9

Hwei Ling Ong and Indu S. Ambudkar

Abstract

Store-operated calcium entry (SOCE) is a ubiquitous Ca^{2+} entry pathway that is activated in response to depletion of ER-Ca^{2+} stores and critically controls the regulation of physiological functions in a wide variety of cell types. The transient receptor potential canonical (TRPC) channels (TRPCs 1–7), which are activated by stimuli leading to PIP_2 hydrolysis, were first identified as molecular components of SOCE channels. While TRPC1 was associated with SOCE and regulation of function in several cell types, none of the TRPC members displayed I_{CRAC}, the store-operated current identified in lymphocytes and mast cells. Intensive search finally led to the identification of Orai1 and STIM1 as the primary components of the CRAC channel. Orai1 was established as the pore-forming channel protein and STIM1 as the ER-Ca^{2+} sensor protein involved in activation of Orai1. STIM1 also activates TRPC1 via a distinct domain in its C-terminus. However, TRPC1 function depends on Orai1-mediated Ca^{2+} entry, which triggers recruitment of TRPC1 into the plasma membrane where it is activated by STIM1. TRPC1 and Orai1 form distinct store-operated Ca^{2+} channels that regulate specific cellular functions. It is now clearly established that regulation of TRPC1 trafficking can change plasma membrane levels of the channel, the phenotype of the store-operated Ca^{2+} current, as well as pattern of SOCE-mediated $[Ca^{2+}]_i$ signals. Thus, TRPC1 is activated downstream of Orai1 and modifies the initial $[Ca^{2+}]_i$ signal generated by Orai1. This review will highlight current concepts of the activation and regulation of TRPC1 channels and its impact on cell function.

H.L. Ong (✉) • I.S. Ambudkar (✉)
Secretory Physiology Section, Molecular Physiology and Therapeutics Branch, National Institute of Dental and Craniofacial Research – NIDCR, National Institutes of Health – NIH, Building 10, Room 1N-113, Bethesda, MD 20892, USA
e-mail: ongh@mail.nih.gov; indu.ambudkar@nih.gov

K. Groschner et al. (eds.), *Store-Operated Ca^{2+} Entry (SOCE) Pathways*, Advances in Experimental Medicine and Biology 993, DOI 10.1007/978-3-319-57732-6_9

Keywords
TRPC • STIM1 • Orai1 • SOCE • ER-PM junctions • Lipid rafts • Caveolin

9.1 Introduction

Depletion of the endoplasmic reticulum (ER)-Ca^{2+} stores leads to activation of Ca^{2+} entry via plasma membrane Ca^{2+} channels. This process, termed store-operated calcium entry (SOCE), was first described almost 30 years ago. Physiologically SOCE is activated in response to stimulation of cells with agonists that lead to hydrolysis of phosphatidylinositol 4,5-bisphosphate (PIP_2), generation of inositol 1,4,5-triphosphate (IP_3), and IP_3-mediated Ca^{2+} release from the ER. The resulting decrease in ER-[Ca^{2+}] is the primary signal for activation of SOCE. This was further established by the seminal finding that SOCE can also be activated by passive depletion of ER-Ca^{2+} stores that occurs when Ca^{2+} uptake into the ER is blocked by treating cells with inhibitors of the ER-Ca^{2+} pump. Thus, receptor-coupled mechanisms upstream of ER-Ca^{2+} store depletion are not essential for SOCE. The first store-operated calcium channel to be identified was the calcium release-activated channel (CRAC) which mediates the Ca^{2+}-release-activated Ca^{2+} current (I_{CRAC}). This current is highly Ca^{2+} selective and exhibits a characteristic inward rectification with a reversal potential (E_{rev}) > +40 mV (Cahalan and Lewis 1990; Hoth and Penner 1992, 1993; Lewis and Cahalan 1989; Matthews et al. 1989; Parekh et al. 1997; Penner et al. 1988). Later studies revealed cell-specific differences in calcium currents activated by ER-Ca^{2+} store depletion (Cheng et al. 2013; Liu et al. 2004; Parekh and Putney 2005) which gave rise to the suggestion that different channel components or regulatory proteins might also be involved in SOCE. Elucidating the molecular identity of the CRAC and other SOCE channels as well as the regulatory protein(s) involved in relaying the status of ER-[Ca^{2+}] to the plasma membrane proved to be a challenge to investigators in the field for almost two decades.

The finding that *Drosophila* TRP channel, a Ca^{2+}-permeable channel, is activated downstream of receptor-coupled PIP_2 hydrolysis led to the suggestion that TRP channels are candidate SOCE channels and fueled extensive search for mammalian homologues of the channel. The transient receptor potential canonical 1 was the first mammalian member of the TRPC channel family to be cloned (Wes et al. 1995; Zhu et al. 1995). Subsequently, seven members (TRPCs 1–7) were identified, which exhibit diverse channel properties and physiological functions in a wide range of cells. Based on biochemical and functional similarities, the TRPC channels are further subdivided into two groups: TRPC1/TRPC4/TRPC5 and TRPC3/TRPC6/TRPC7 (Dietrich and Gudermann 2014; Freichel et al. 2014; Lichtenegger and Groschner 2014; Miller 2014; Nesin and Tsiokas 2014; Zhang and Trebak 2014; Zholos 2014). TRPC1 was shown to mediate SOCE, contribute to the calcium-permeable currents generated by store depletion, and regulate physiological function in many cell types (Ong et al. 2014). However, the electrophysiological characteristics of TRPC1-mediated, relatively Ca^{2+}-selective currents did not resemble those reported for the CRAC channel. Hence, the TRPC1-mediated current was termed store-operated calcium current (I_{SOC}) to differentiate it from I_{CRAC} (Liu et al. 2003). Further, it was also established that TRPC1 was not a component of CRAC channels. Continued search to identify CRAC channel components resulted, almost two decades later, in the discovery of a four-transmembrane protein known as Orai1 which was confirmed as the primary pore-forming component of CRAC channel. Orai1 was independently discovered by two groups using two very different approaches (Vig et al. 2006, Zhang et al. 2006). Importantly, siRNA screening led to identification of the ER-resident Ca^{2+}-sensing protein, STIM1, as the primary activator of Orai1. STIM1 is an ER protein with an EF hand domain that is located in the lumen of the ER. In response to store depletion, the Ca^{2+} is released from the EF hand domain which triggers extensive intramolecular and intermolecular rearrangement of STIM1 causing it to oligomerize and translocate to the cell periphery. Here, STIM1 aggregates within ER-plasma membrane (ER-PM) junctions, where the ER membrane is juxtaposed with the plasma membrane. Orai1 is recruited by STIM1 into these junctions where the channel is gated by STIM1. Several studies have reported that Orai1 and STIM1 are sufficient to generate I_{CRAC} (Feske et al. 2006; Gwack et al. 2007; Hogan et al. 2010; Liou et al. 2005; Prakriya et al. 2006; Roos et al. 2005; Vig et al. 2006; Zhang et al. 2005). The STIM-Orai1-activating region (SOAR) in the C-terminal CC2-CC3 region binds to and gates Orai1. An increasing amount of data have now established the functionally relevant domains in STIM1 as well as identified accessory proteins that are recruited to the domain to regulate the assembly of the channels or architecture of the domain.

Critical studies demonstrate that STIM1 also activates TRPC1 via a domain which is distinct from that involved in gating of Orai1. The C-terminal end of STIM1 (^{684}KK^{689}) interacts electrostatically with ^{639}DD^{640} in the C-terminus of TRPC1 resulting in gating of the channel. Nonetheless, activation of TRPC1 requires a crucial functional interaction with Orai1 (Cheng et al. 2008; Huang et al. 2006; Kim et al. 2009). Herein, we will discuss current concepts regarding the molecular components and mechanisms involved in regulating TRPC1 function, its contribution to SOCE, as well as the impact of this on cytosolic Ca^{2+} signaling and regulation of cell function.

9.2 Functional Properties of TRPC1

Given the relatively widespread expression of TRPC1 in many tissues, the channel has been shown to contribute to SOCE and mediate diverse physiological functions in a wide range of cells. TRPC1 is suggested to assemble as homomeric or heteromeric channels, the latter involving TRPCs, TRPV, and Orai1 (recently reviewed in Ong et al. 2014). These studies mainly utilize heterologous expression of the channels, and thus endogenous status of TRPC1 channels has not yet been established. However, the electrophysiological characteristics of TRPC1-

associated currents measured in different cell types appear to depend on the respective TRPC channel composition. For example, store depletion in the salivary gland cell line, HSG cells, triggers a TRPC1-mediated, relatively Ca^{2+}-selective current with a slight inward rectification and a $E_{rev} \approx +15$ mV with little contribution from other TRPC channels (Liu et al. 2003, 2004). In contrast, TRPC1 forms a heteromeric complex with TRPC3 in human parotid gland ductal cells, generating a nonselective, linear current (28 pS) with E_{rev} of +0 mV (Liu et al. 2004, 2005). Store-dependent TRPC1-associated currents have also been measured in endothelial cells and smooth muscle cells (Ahmmed et al. 2004; Byron 1996; Paria et al. 2004; Seth et al. 2009). The function of TRPC1 has been assessed in cell lines and primary cells by using knockdown or overexpression approaches or using knockout mouse model. Loss of TRPC1 significantly decreases SOCE in miscellaneous cell types, including HSG cells, smooth muscle cells, endothelial cells, and platelets (Brownlow et al. 2004; Dietrich et al. 2006; Liu et al. 2000, 2003; Mehta et al. 2003; Rosado et al. 2002; Tiruppathi et al. 2006), and in preparations of salivary gland acinar cells as well as aortic endothelial cells from TRPC1 knockout ($TRPC1^{-/-}$) mice (Hong et al. 2011; Liu et al. 2007; Ma et al. 2011a).

TRPC1 is associated with the regulation of physiological functions in many tissues. It promotes migration of intestinal epithelial cells (Bomben et al. 2011; Rao et al. 2006) and human malignant gliomas (Cuddapah et al. 2013), proliferation of neural stem and hippocampal neural progenitor cells (Fiorio Pla et al. 2005; Li et al. 2012; McGurk et al. 2011) and contributes to synaptic plasticity of neuromuscular junctions (McGurk et al. 2011). TRPC1 also plays a crucial role in maintenance of the permeability of endothelial cell barrier, wound healing within the intestinal epithelial layer following tissue injury, protection against cytotoxicity, contraction of glomerular mesangial cells, and regulation of osteoclast formation and function (Bollimuntha et al. 2005b; Du et al. 2007; Ong et al. 2013; Paria et al. 2004; Sours et al. 2006). As shown in studies with $TRPC1^{-/-}$ mice, TRPC1 is a vital SOCE component in several tissues. While $TRPC1^{-/-}$ mice display normal viability, development, and behavior, preparations of acinar cells from salivary glands and pancreas and aortic endothelial cells exhibit dramatically attenuated SOCE. There is a corresponding reduction in Ca^{2+}-dependent processes such as the activation of Ca^{2+}-dependent potassium (K_{Ca}) and Ca^{2+}-activated Cl^- channels. Importantly, stimulated fluid and protein secretion from the salivary glands and pancreas, as well as vasorelaxation of the aorta, is adversely impacted (Dietrich et al. 2014; Hong et al. 2011; Liu et al. 2007; Ma et al. 2011a; Pani et al. 2013; Sun et al. 2015). It is worth noting that the levels of Orai1 expression in salivary gland and pancreatic acinar cells are similar in control and $TRPC1^{-/-}$ mice. Hence, loss of TRPC1-mediated SOCE is the primary cause for the decrease in fluid secretion from the salivary glands and pancreas. More importantly, Orai1 does not compensate for the loss of TRPC1 function in these cells, clearly demarcating physiological functions attributed to Ca^{2+} entry mediated by either channel.

In endothelial cells, TRPC1 forms a heteromeric channel with TRPV4. Store depletion promoted trafficking of the TRPC1-TRPV4 complex, resulting in enhanced channel insertion into the plasma membrane and increased the magnitude of SOCE (Ma et al. 2011a, b). The increased Ca^{2+} entry via TRPC1-TRPV4 then triggered the production of nitric oxide and subsequent activation of guanylate cyclase, resulting in elevated intracellular cGMP levels. Channel complex activity was negatively regulated by cGMP, which induced the translocation of protein kinase G1α to the plasma membrane, where it associated with and phosphorylated TRPC1 (Zhang et al. 2016).

TRPC1-mediated SOCE has been associated with the regulation of vascular tone. An early study with vascular smooth muscle cells (VSMCs) showed the association and colocalization of TRPC1 with the large conductance, Ca^{2+}-sensitive K^+ channel (BK_{Ca}) in these cells. Following store depletion, Ca^{2+} entry via TRPC1 activates BK_{Ca} to promote membrane hyperpolarization and prevent excessive agonist-induced contraction of the cells (Kwan et al. 2009). Interestingly, a recent study (Avila-Medina et al. 2016) showed that TRPC1 associates with both Orai1 and the voltage-dependent $Ca_v1.2$ L-type Ca^{2+} channel in resting VSMCs, with an increase in the interaction following store depletion. Ca^{2+} entry via both Orai1 and TRPC1 caused depolarization of the plasma membrane, which then activated $Ca_v1.2$ and induced vasoconstriction (Avila-Medina et al. 2016). While both studies utilized primary cultures of dispersed VSMC from the thoracic aorta, the former study harvested the aorta from rats, whereas the latter used mice. Species difference may underlie the different roles attributed to TRPC1 in VSMCs, although contributions of TRPC1-interacting proteins that may affect channel function cannot be completely discounted. Intriguingly, STIM1 has been shown to reciprocally regulate Orai1 and $Ca_v1.2$, promoting the function of the former but inhibiting the latter (Wang et al. 2010). How TRPC1 interactions with these diverse ion channels are determined is not yet clear. One possible mechanism would be via compartmentalization of the proteins into a signaling complex in discrete cellular locations.

TRPC1 interacts with numerous signaling, scaffolding, and trafficking proteins which ensures proper assembly, targeting, and regulation of the channel (recently reviewed in Ong et al. 2014). Thus, modulating the expression or regulation of these proteins will significantly impact TRPC1 function and SOCE. For example, the caveolae-residing protein, caveolin-1 (Cav-1), functions as a plasma membrane scaffold for TRPC1, and knockdown of Cav-1 results in mislocalization of TRPC1, preventing STIM1 from binding to and gating the channel (Pani et al. 2009). Further, Cav-$1^{-/-}$ mice exhibit reduced fluid secretion from salivary glands and disruption of TRPC1 localization in acinar cell preparations (Pani et al. 2013) and endothelial cells (Lin et al. 2007; Murata et al. 2007a). Homer1 is a scaffolding protein that mediates TRPC1 interaction with the IP_3R. Homer$1^{-/-}$ mice have aberrant TRPC1-mediated SOCE that was shown to underlie the development of skeletal myopathy in these mice (Stiber et al. 2008). Studies with a Parkinson's disease mouse model showed that TRPC1 is crucial for maintaining calcium homeostasis controls' autophagy and promotes the survival of neurons while limiting neuronal degeneration (Selvaraj et al. 2009, 2012). While lack of TRPC1 has been implicated in aberrant vasorelaxation (Kochukov et al. 2013), physical movement disorder (He et al. 2016), muscle fatigue and slower regeneration after muscle injury (Zanou et al. 2010, 2012), myopathies in patients with Duchenne muscular dystrophy, and *mdx* mice lacking dystrophin (Sabourin et al. 2009a, b; Stiber et al. 2008; Vandebrouck et al. 2007; Williams and Allen 2007), it remains to be seen whether such effects are a consequence of aberrant TRPC1-mediated SOCE.

9.3 TRPC1 Interactions with Orai1 and STIM1

The ER-resident STIM proteins, STIM1 and STIM2, are required for SOCE as loss of either protein results in a significant decrease of function. Both proteins have a Ca^{2+}-sensing domain in their N-termini, an EF hand domain that is bound to Ca^{2+} when ER-[Ca^{2+}] is relatively high, i.e., under resting/unstimulated conditions, and respond to decrease in ER-[Ca^{2+}] by undergoing molecular rearrangements, aggregation, and translocation to the ER-PM junctions at the cell periphery (Hogan et al. 2010; Liou et al. 2005; Roos et al. 2005). The role of STIM2 in TRPC1-mediated SOCE has not yet been studied in detail. However, loss of STIM1 eliminates endogenous TRPC1-mediated SOCE and Ca^{2+} current, whereas exogenous co-expression of STIM1 with TRPC1 increases SOCE (Cheng et al. 2008, 2011; Huang et al. 2006). TRPC1 strongly interacts and colocalizes with STIM1 following store depletion, as shown by co-immunoprecipitation, FRET, and TIRFM experiments. Conversely, refilling of the ER-Ca^{2+} stores causes dissociation of STIM1 from TRPC1 and inactivation of TRPC1 (Huang et al. 2006; Lopez et al. 2006; Ong et al. 2007; Pani et al. 2008; Yuan et al. 2007; Zeng et al. 2008). The gating of TRPC1 by STIM1 involves electrostatic interactions between the negatively charged aspartate residues in TRPC1 (^{639}DD^{640}) with the positively charged lysines in the STIM1 polybasic domain (^{684}KK^{685}). Charge swap (STIM1-^{684}EE^{685}) or deletion of polybasic domain (STIM1ΔK; deletion of amino acids (aa) 672–685) in STIM1 eliminates its ability to activate TRPC1. However, these STIM1 mutants retain their ability to gate Orai1. Conversely, the charge-swap mutation in TRPC1 (TRPC1-^{639}KK^{640}) abolishes gating by STIM1. Swapping the charged residues that are involved in STIM1 with those in TRPC1 triggers recovery in gating, indicating that these residues control gating of the channel (Yuan et al. 2009).

A considerable number of studies describe the changes in STIM1 configuration and intramolecular rearrangements that are required for optimal STIM1-Orai1 interactions and activation of Orai1 (recently reviewed in Shim et al. 2015). Such detailed studies for STIM1-TRPC1 interactions have not yet been conducted. The EF hand mutation of STIM1 (STIM1D76A), which promotes an extended conformation of the protein, spontaneously clusters with and activates TRPC1 independent of store depletion, similar to its spontaneous interaction with and activation of Orai1 (Huang et al. 2006; Pani et al. 2008). This suggests that the open conformation of STIM1 promotes gating of TRPC1 (Zeng et al. 2008). However, STIM1ΔK does not activate TRPC1, but it can still bind to the channel (Zeng et al. 2008). The binding domain in STIM1 is suggested to involve the ezrin/radixin/moesin (ERM) domain (aa 251–535) (Huang et al. 2006), but the exact residues within ERM that are involved have not yet been identified. More studies will be required to elucidate what conformational changes of STIM1 are required for activation of TRPC1. With regard to TRPC1 domains involved in interaction with STIM1, coiled-coil (CC) regions located in the N- and C-terminal domains of the channel have been implicated. Exogenously expressed SOAR has been reported to strongly interact with these TRPC1 domains (Lee et al. 2014; Yuan et al. 2007). Another study showed that the STIM1-SOAR domain and calmodulin (CaM) reciprocally regulate TRPC1 channel activity. Each TRPC1 tetramer is bound to two STIM1 via the SOAR domains, and this is suggested to enable interaction with STIM1-KK domain and activation. In contrast, binding of four CaM leads to inactivation of the channel. Interestingly, STIM1-SOAR domain remains bound to TRPC1 during the initial stages of channel inactivation by CaM with subsequent dissociation of SOAR from TRPC1 further increasing CaM-dependent TRPC1 inactivation (Asanov et al. 2015). If SOAR binds to both Orai1 and TRPC1, then the question arises as to how such interactions occur under stimulatory conditions and whether there is competition between the channels for STIM1. Cav-1 is another component that contributes to the mechanisms involved in TRPC1 gating. Cav-1 binds to TRPC1, retaining the channel in an inactive state. Interaction of STIM1 with TRPC1 displaces Cav-1 resulting in channel activation. Conversely, Cav-1 reassociates with TRPC1 as the channel is inactivated following refill of ER-Ca^{2+} stores (Pani et al. 2009). The exact TRPC1 domains involved in these interactions are not yet known.

In addition to the STIM1 requirement for gating, TRPC1 channel function displays an absolute requirement for Orai1. This was first reported by Ambudkar and colleagues in HSG cells (Cheng et al. 2008, 2011, Ong et al. 2007) and subsequently by other groups in various cell types (Desai et al. 2015, Jardin et al. 2008a, Kim et al. 2009). The initial observation showed that knockdown of TRPC1 reduced SOCE by about 60%, while loss of Orai1 or STIM1 induced complete elimination of SOCE. Expression of WT-Orai1, but not a pore-deficient mutant of Orai1 (E106Q), supported TRPC1 activation following ER-Ca^{2+} store depletion. Further, overexpression of STIM1 and TRPC1 could not induce recovery of SOCE in the absence of Orai1. These data conclusively established that Orai1 was required for TRPC1 function. Several models were hypothesized to explain this crucial functional interaction: (1) TRPC1 and Orai1 form a heteromeric channel; (2) Ca^{2+} entering cells via Orai1 directly activates TRPC1; and (3) other mechanism downstream from Orai1-mediated Ca^{2+} entry regulates TRPC1. This rather contentious puzzle was resolved by studies showing that store depletion triggered the assembly of TRPC1 in a dynamic complex with Orai1 and STIM1 that is localized in ER-PM junctions. In addition to HSG cells (Ong et al. 2007), this has now been shown in human parathyroid cells (Lu et al. 2010), human liver cells (Zhang et al. 2010), mouse pulmonary arterial smooth muscle cells (Ng et al. 2009), rat kidney fibroblasts (Almirza et al. 2012), and mouse acinar cells from the pancreas and salivary glands (Hong et al. 2011; Pani et al. 2013). Further, assembly of the TRPC1/Orai1/STIM1 complex was dependent on the clustering of STIM1 in ER-PM junctions and was eliminated by knockdown of STIM1. Final resolution of the role of Orai1 in TRPC1 function came from the observation that Orai1-mediated Ca^{2+} entry was involved in recruitment of TRPC1 into the plasma membrane where it is activated by STIM1. Nonfunctional Orai1 mutants, either Orai1E106Q or Orai1R91W, did not support the regulated surface expression of TRPC1 or activation (Cheng et al. 2008; Kim et al. 2009; Ong et al. 2007). Importantly, TRPC1 is recruited to the plasma membrane in cells expressing STIM1-EE mutant, but is not activated. These studies also demonstrated that under conditions where TRPC1 function is suppressed, cells expressing both TRPC1 and Orai1 show a phenotypic change, displaying I_{CRAC} instead of I_{SOC} in response to stimulation. These findings suggest that STIM1-TRPC1 forms a distinct channel which assembles in close proximity to STIM1-Orai1 channels. Importantly, Ca^{2+} entry via the two channels differentially regulates cellular function with specific downstream targets (Cheng et al. 2011). It is worth noting that the true TRPC1-mediated current has not yet been measured in any cell type as most reported currents associated with TRPC1, e.g., I_{SOC}, include those generated by both TRPC1+STIM1 and Orai1+STIM1 channels (Cheng et al. 2011) (Fig. 9.1).

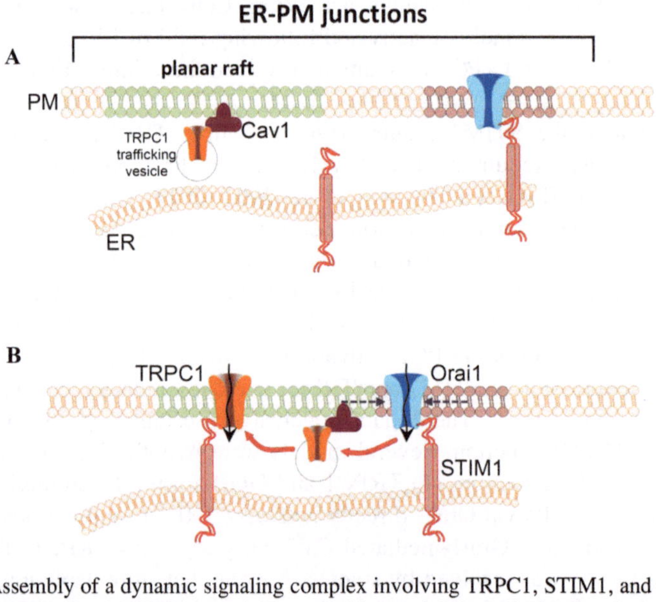

Fig. 9.1 Assembly of a dynamic signaling complex involving TRPC1, STIM1, and Orai1 in the plasma membrane. (**a**) A TRPC1-containing trafficking vesicle is docked at the plasma membrane (PM) via TRPC1–Cav-1 interactions within the Cav-1-containing planar raft. Following store depletion, Orai1 associates with and is activated by STIM1 possibly in a separate nanodomain of lipid rafts (non-caveolae). (**b**) Following Orai1-mediated Ca^{2+} entry, TRPC1 is inserted into the PM and subsequently gated by STIM1. We hypothesize that this involves a lateral movement of the Cav-1-containing planar raft and the non-caveolae nanodomain, resulting in a fusion of both domains (Ong and Ambudkar 2015)

9.4 STIM2 Functions in SOCE

STIM2 shares considerable homology with STIM1, where the EF hand, CC, SOAR, and polybasic domains are fairly well conserved. However, several significant and critical amino acid differences between the two proteins account for their distinct properties and functions. A 3 aa difference within the EF hand domain of STIM2 is sufficient to confer a lower sensitivity for Ca^{2+} when compared to STIM1 (Stathopulos et al. 2009; Zheng et al. 2008, 2011). This enables STIM2 to sense and respond to smaller changes in the ER-[Ca^{2+}], even those within the range of ambient [Ca^{2+}] in the ER. In contrast, STIM1-EF hand has a higher Ca^{2+} affinity and requires substantial decrease in ER-[Ca^{2+}] to trigger a response (Collins and Meyer 2011). Thus, compared to STIM1, STIM2 can respond to relatively low stimulus intensities, which induce less decrease in ER-[Ca^{2+}] (Ong et al. 2015). STIM2 has also been proposed to maintain resting [Ca^{2+}]$_i$ in cells (Brandman et al. 2007). A recent study proposed a novel role for STIM2 in the regulation of SOCE. The findings showed that STIM2 promotes clustering of STIM1 in ER-PM junctions which is physiologically relevant when cells are stimulated with low [agonist], causing small decrease in ER-[Ca^{2+}] that is not sensed by STIM1 (Ong et al. 2015). The data demonstrated that STIM2 forms punctae in response to stimulation with low [carbachol (CCh)] while STIM1 responds only at higher [CCh]. However, when both STIMs are co-expressed in cells, STIM1 and STIM2 co-clustered even at low [CCh]. In contrast, expression of Orai1 did not rescue the STIM1 clustering at low [CCh]. Additionally, Orai1 was co-localized with STIM2 and STIM1 within the same ER-PM junctions. The polybasic domain of STIM2 was vital for localizing STIM2 as well as co-clustering with STIM1 and Orai1 in these junctions. Importantly, loss of STIM2 decreased the sensitivity of SOCE activation by agonist, with a marked increase in the apparent affinity for the agonist. However, knockdown of STIM1 abolished SOCE at both low and relatively high [CCh]. In aggregate, this study suggested that STIM2 serves to tune the agonist sensitivity of STIM1-Orai1 interactions and the associated Ca^{2+} signals (Ong et al. 2015), while STIM1 is the primary activator of Orai1. Thus, STIM2 plays a vital role in modulation of agonist-induced Ca^{2+} signals. At the single-cell level, low [CCh] primarily induces oscillatory increases in [Ca^{2+}]$_i$ calcium signals, whereas a more sustained increase is observed at higher [CCh]. Knockdown of endogenous STIM2 shifted the sustained [Ca^{2+}]$_i$ induced by high [CCh] to an oscillatory pattern, more like that seen with lower [CCh]. Thus, STIM2 increases the sensitivity of SOCE response to store depletion.

There have been very few studies assessing the role of STIM2 in TRPC1 channel function. Targeted knockout of STIM2 in mouse salivary glands decreased agonist-induced fluid secretion, which is strongly dependent on TRPC1 channel function. This decrease was observed only with low pilocarpine stimulation, highlighting the importance of STIM2 in modulating responses to smaller stimulus intensities. However, it is not known whether STIM2 directly interacts with TRPC1 or mediates its effects via STIM1-Orai1 clustering in the salivary glands (Ong et al. 2015). As mentioned earlier, TRPC1-mediated SOCE is crucial in promoting migration of intestinal epithelial cells after tissue injury to promote wound healing (Bomben et al. 2011, Rao et al. 2006). The process underlying rapid restitution of the epithelial layer is a complex process regulated by modulating cellular levels of polyamines (Rao et al. 2012). Increase in intracellular [polyamine] stimulated STIM1 but inhibited STIM2 expression, promoting the formation of TRPC1 +STIM1 complexes that resulted in greater SOCE and epithelial restitution. Conversely, decrease in [polyamine] increased STIM2 but decreased STIM1 levels, leading to fewer TRPC1+STIM1 complexes, attenuation of SOCE, and abrogation of wound healing. Co-immunoprecipitation studies showed that TRPC1 interacted with STIM1 but not STIM2 in intestinal epithelial cells. Interestingly, exogenously expressed STIM2 inhibited translocation of STIM1 to the plasma membrane and thus decreased the number of STIM1 molecules available to activate TRPC1 (Rao et al. 2012). In contrast, STIM2 has been shown to directly regulate two different TRPC1-associated store-dependent currents in HEK293 cells, the TRPC1-mediated I_{max} current and the I_{min} current. While activation of the TRPC1-mediated I_{max} requires both STIM2 and STIM1, I_{min} current was suggested to be solely activated by STIM2 (Shalygin et al. 2015). Further studies are required to elucidate the interactions between TRPC1, STIM1, and STIM2 and how such interactions affect the function and regulation of TRPC1.

9.5 Modulation of TRPC1 Expression and Activity at the Plasma Membrane

TRPC1 function in the plasma membrane is determined by proper targeting of the channel to specific regions in the cell periphery, near ER-PM junctions where it can be regulated by STIM1 and Orai1. Mechanisms regulating insertion and retention of TRPC1 in the plasma membrane, as well as those involved in internalization of the channel for degradation or recycling, all contribute to the level of functional TRPC1 in the plasma membrane. There are few studies examining these critical mechanisms that impact TRPC1 function.

9.5.1 Intracellular Recycling of TRPC1

The exact biosynthetic pathway of TRPC1 has not yet been identified. As for other plasma membrane proteins, TRPC1 has been suggested to be trafficked from the ER through the Golgi apparatus and then to the plasma membrane via transport vesicles. Studies have shown that levels of this channel in the plasma membrane are low in unstimulated cells and increase upon activation of Orai1 after store depletion. However, TRPC1-containing vesicles are retained in the sub-plasma membrane region in close proximity to Orai1 and STIM1 so that they can detect the local Ca^{2+} signal generated by Orai1. Its transport to this sub-plasma membrane region, as well as its internalization, is determined by exocytic and endocytic vesicular trafficking pathways, respectively. Typical pathways involved in endocytosis are mediated by clathrin, ADP-ribosylation factor 6 (Arf6), and caveolin. Once endocytosed, proteins are sorted into the early endosomal compartment from where they are either delivered to lysosomes for degradation or recycled back to the plasma membrane through recycling vesicles (Doherty and McMahon 2009; Maxfield and McGraw 2004). The fast recycling pathway mediates trafficking of the channels directly from the early endosomes to the plasma membrane. In addition, a slow recycling pathway is also present in cells that routes the proteins from early endosomes to a recycling compartment and then to the plasma membrane. The third pathway, the retrograde pathway, first transports proteins to the trans-Golgi network (TGN) from where they reach the plasma membrane via the secretory pathway (Doherty and McMahon 2009; Grant and Donaldson 2009; Taguchi 2013). In all three pathways, channel insertion into the plasma membrane requires an exocytic step that regulates vesicle fusion with the plasma membrane. Additionally, vesicular transport of the channels is tightly controlled by vesicle-associated docking proteins and sensors that regulate precise delivery of the channel-containing vesicles to the correct location and the rate of exocytosis (de Souza and Ambudkar 2014). For example, internalization into the early endosome is regulated by Rab5 and its effectors, early endosome antigen 1 and Rabenosyn-5. From this compartment, proteins are recycled to the plasma membrane via Rab4- or Rab35-mediated fast recycling pathway or via Rab11-, Rab15-, Rab22-, or Rab25-dependent slow recycling pathway (Grant and Donaldson 2009; Stenmark 2009; Zerial and McBride 2001). Routing from early endosome to the late endosome is regulated by Rab7 and Rab9 from where it reaches Golgi network via a Rab9-mediated mechanism (de Souza and Ambudkar 2014; Grant and Donaldson 2009; Stenmark 2009).

A few studies have examined intracellular trafficking of TRPC channels. A recent study identified the vesicular compartments and Rab proteins involved in regulating TRPC1 trafficking (de Souza et al. 2015). This study demonstrated that TRPC1 is recycled via a fast recycling pathway which involves Rab5 and Rab4. Importantly, this recycling achieves clustering of TRPC1 within ER-PM junctions where STIM1 clusters in response to ER-Ca^{2+} store depletion. The authors reported that expression of Rab5 increased the retention of TRPC1 in early endosomes, leading to reduced surface expression and decreased TRPC1-mediated SOCE. Importantly, co-expression of Rab4 with Rab5, but not STIM1 or Rab11, rescued routing of TRPC1 to the plasma membrane. Notably, while STIM1 recruited TRPC1 into the plasma membrane after cell stimulation, it is not involved in intracellular trafficking of the channel (de Souza et al. 2015). Moreover, internalization of TRPC1 occurs via an endocytic pathway mediated by ARF6 that is independent of clathrin and Cav-1. Overexpression of ARF6 suppressed TRPC1-mediated SOCE by attenuating the store-dependent increase in TRPC1 expression at the plasma membrane. It is worth mentioning that in unstimulated cells, ARF6 did not affect resting levels of TRPC1 in the plasma membrane, whereas Rab5 decreased those levels. More importantly, the effects of Rab5 and Arf6 were highly specific for TRPC1, leaving Orai1 and STIM1 unaffected. Generation of the TRPC1-mediated I_{SOC} was suppressed in HSG cells expressing either Rab5 or

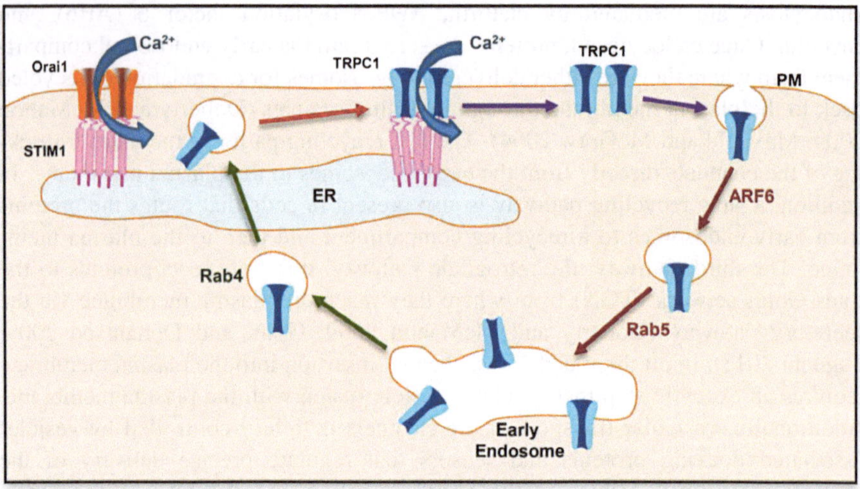

Fig. 9.2 Model depicting the role of a rapid recycling pathway in the regulation of TRPC1 trafficking and function. TRPC1 is endocytosed from the plasma membrane by Arf6, is sorted to Rab5-early endosomes, and then is recycled back to the plasma membrane by Rab4-dependent fast recycling endosomes. This recycling carries TRPC1 to the cellular region near the plasma membrane where STIM1 clusters in response to ER-Ca^{2+} depletion. Within these ER-PM junctions, STIM1 interacts with and activates Orai1. Ca^{2+} entry via Orai1 triggers recruitment of TRPC1 from Rab4 vesicles into the plasma membrane, where the channel interacts with STIM1 and is activated. Thus, endocytic recycling via Rab4 determines clustering of TRPC1with STIM1, as well as plasma membrane insertion and function of the channel (modified from de Souza et al. 2015)

ARF6, resulting instead in the generation of the Orai1-mediated I_{CRAC} (de Souza et al. 2015). Thus, the fast endocytic recycling pathway is a critical determinant of the level and function of TRPC1 in the plasma membrane (Fig. 9.2).

9.5.2 Role of Caveolin and Other Scaffolding Proteins

Many studies have reported the interactions of miscellaneous scaffolding proteins with TRPC1 (recently reviewed in Ong et al. 2014). Among these TRPC1-interacting proteins is Cav-1, a cholesterol-binding protein that is enriched in caveolar lipid rafts and plays a vital role in proper localization of TRPC1 in the plasma membrane. TRPC1 has been shown to co-immunoprecipitate with Cav-1 in HSG cells (Brazer et al. 2003; Lockwich et al. 2000; Pani and Singh 2009), mouse submandibular gland preparations (Pani et al. 2013), and pulmonary artery endothelial cells (Bergdahl et al. 2003; Kwiatek et al. 2006). Under resting/unstimulated conditions, TRPC1 is targeted to the plasma membrane via constitutive trafficking mechanism(s) and interacts with Cav-1 in the plasma membrane to be retained intracellularly close to the plasma membrane. The Cav-1-bound TRPC1 is inactive. Following depletion of the ER-Ca^{2+} stores, STIM1 aggregates and translocates to the ER-PM junctions where it recruits and gates Orai1. The STIM1-Orai1 channel complex is assembled in the region very close to where TRPC1 is anchored. Indeed, recent studies have shown that Cav-1 might also affect Orai1 function (Yeh and Parekh 2015; Yu et al. 2010).

The pivotal trigger for initiating TRPC1 function is Orai1-mediated Ca^{2+} entry which triggers insertion of TRPC1 into the plasma membrane following which the channel is activated by STIM1. As noted above, STIM1-TRPC1 association accompanies TRPC1-Cav-1 dissociation. When the ER-Ca^{2+} stores are refilled, STIM1 dissociates from TRPC1, likely moving back into the ER. Under these conditions, TRPC1 reassociates with Cav-1 (Cheng et al. 2011; Pani et al. 2009). TRPC1 has two binding sites for Cav-1 in its N- and C-termini. There are two putative Cav-1-binding sites on the C-terminus (626–635aa and 781–789aa), and while their function has not yet been established, they do appear to have some effects on TRPC1 activity (Brazer et al. 2003; Kwiatek et al. 2006; Lockwich et al. 2000; Pani et al. 2009; Pani and Singh 2009). The N-terminal Cav-1-binding site (322–349 aa) is suggested to regulate the scaffolding and localization of TRPC1 at the plasma membrane (Brazer et al. 2003; Murata et al. 2007b). Knockdown of endogenous Cav-1 or mutations that impair Cav-1 binding to TRPC1 causes mislocalization of TRPC1 resulting in decreased surface expression of TRPC1 and abrogation of SOCE and I_{SOC} (Brazer et al. 2003; Kwiatek et al. 2006; Lockwich et al. 2000; Pani et al. 2009, 2013; Rathor et al. 2014; Sundivakkam et al. 2009; Weihuang et al. 2015). Thus, caveolin is not involved in internalization of TRPC1, but rather serves as a plasma membrane scaffold routing TRPC1 to specific lipid domains where it can interact with STIM1 (see Sect. 9.5.2). How scaffolding of TRPC1 by Cav-1 fits into the Rab4-Rab5-Arf6 pathway that determines channel recycling remains to be clearly defined. Since Cav-1 is localized in the plasma membrane, Cav-1 may scaffold TRPC1 after the channel has been trafficked to the plasma membrane via the Rab4-Rab5 pathway and prior to channel insertion. This would provide a readily available pool of TRPC1 that can be recruited soon after activation of SOCE.

Homer1 regulates TRPC1 activity by promoting TRPC1-IP_3R association and forming a dynamic complex with both proteins. TRPC1 has a putative binding site for Homer1 in its C-terminus, located at aa 645–648. It has been proposed that in some cell types, Homer1b/Homer1c mediates TRPC1-IP_3R interaction under resting conditions, which maintains the channel in an inactive state. Mutations that abolish TRPC1-Homer1 binding render the channel spontaneously active by uncoupling TRPC1 from the IP_3R. Interestingly, following channel activation, Homer1b/Homer1bc dissociates from IP_3R but remains bound to TRPC1. It has been suggested that this enables Homer1b/c to rapidly reassemble the TRPC1-Homer1-IP3R complex when the ER-Ca^{2+} stores are refilled and TRPC1 channel is inactivated (Worley et al. 2007; Yuan et al. 2003, 2012). Notably, remodeling of TRPC1-Homer-IP3R interactions is reminiscent of those involving Cav-1-TRPC1-STIM1. Furthermore, the binding sites for Cav-1, STIM1, and Homer1 in the TRPC1 C-terminus are separated only by a few amino acids. TRPC1 interaction with either protein needs to be highly synchronized as each plays a distinct role in regulating channel function. It is also possible that binding of TRPC1 to one of the proteins might affect the binding to another, just because of the proximity of these binding domains. Further studies are required to elucidate the steps by which such synchronous interactions regulate the TRPC1 channel function and what remodeling takes place within the TRPC1 C-terminus.

9.5.3 TRPC1 Insertion into the Membrane

There is little information on the exact mechanisms that are involved in the final step of TRPC1 insertion into the membrane. Other than the important finding that it is a Ca^{2+}-dependent process, triggered by local $[Ca^{2+}]_i$ elevations mediated by Orai1, there is no information on the identity of the Ca^{2+} sensors on TRPC1-containing vesicles or other proteins involved in driving the fusion, such as v-SNAREs and t-SNAREs (Cheng et al. 2011). It is important to add that this final critical step is essential for generation of functional TRPC1 channels in the plasma membrane. Based on currently available knowledge, we can surmise that vesicle-associated docking and Ca^{2+}-sensor proteins will be involved in determining where and how fast the channel-containing vesicles fuse with the plasma membrane. Further, in the case of other exocytotic vesicles, when the stimulus is removed, the vesicles are internalized. For example, Ca^{2+} regulates a kiss-and-run mechanism whereby the vesicles are fused to the plasma membrane in the presence of Ca^{2+} but internalized when Ca^{2+} is removed (Alabi and Tsien 2013; Grant and Donaldson 2009; Leitz and Kavalali 2011; Richards 2010). The study reported by Cheng et al. (2011) argues against this mechanism for TRPC1 trafficking as removal of external Ca^{2+} after channel insertion did not alter plasma membrane levels of the channel (Cheng et al. 2011). The fate of the channel after ER-Ca^{2+} stores are refilled and STIM1 puncta dissociate is another important aspect that needs to be further examined.

Some details of the final step in TRPC1 trafficking come from studies which show that synaptosome-associated protein (SNAP-25) which regulates membrane fusion within intracellular compartments or between vesicles, and plasma membrane is involved in trafficking of the channel. Treatment of human platelets with botulinum toxin (BTx) significantly decreased SOCE by causing the cleavage and inactivation of SNAP-25. TRPC1 interaction with the IP_3R was clearly reduced following BTx treatment, uncoupling channel activation from any changes in the ER-$[Ca^{2+}]$. While the surface levels of TRPC1 in resting cells were reported to be unaffected by BTx, it remains to be shown whether store-dependent recruitment of TRPC1 to the cell surface was affected (Redondo et al. 2004). In endothelial cells, stimulation with thrombin induced assembly of TRPC1 with RhoA and IP_3R in a complex and subsequent translocation of the complex to the plasma membrane. Inhibition of RhoA reduced expression of TRPC1 in the plasma membrane and also adversely affected TRPC1-IP_3R association, resulting in attenuation of SOCE. RhoA is a monomeric GTPase protein that regulates the actin cytoskeleton which suggests that spatial rearrangement of the actin filaments could be involved in controlling TRPC1-IP_3R association. This was demonstrated by treating cells with latrunculin which prevented actin polymerization, inhibited TRPC1-IP_3R association, and attenuated SOCE (Mehta et al. 2003). The neuronal-specific cytoskeletal protein, β-tubulin, has been reported to play a similar role as RhoA in retinal epithelial cells. Disruption of tubulin by colchicine reduced both the surface expression of TRPC1 and corresponding SOCE (Bollimuntha et al. 2005a). Despite lack of conclusive data regarding the molecular components involved in the final vesicle fusion step, it is clear that the spatial and temporal aspects of intracellular trafficking of TRPC1, as well as its insertion into the plasma membrane, are absolutely critical in regulating channel function. The exact domain where the functional TRPC1 channel is assembled controls generation of specific Ca^{2+} signals that govern several distinct downstream cellular functions. Any adverse impact on any step in the trafficking or insertion could affect TRPC1 localization, activation, and regulation of cell function. Notably, aberrant TRPC channel function and trafficking have been associated with disease and dysfunction.

In addition to scaffolding and trafficking proteins, TRPC1 activity can also be modulated by its interactions with signaling proteins. As described in the previous section, STIM1 is the primary gating moiety of TRPC1 in many different cell types. Nonetheless, a few studies have proposed other proteins or lipids as a gating ligand for TRPC1 in VSMCs. In resting/unstimulated cells, the TRPC1 channel was kept inactive by interacting with the PIP_2-binding MARCKS protein that acts as a localized PIP_2 buffer. Following store depletion, TRPC1 was proposed to first form a complex with STIM1. The Gαq proteins and PLCδ were then recruited to and associated with the TRPC1-STIM1 complex, resulting in the activation of PKC. The authors suggested that TRPC1 is activated by PKC-dependent phosphorylation to induce channel dissociation and release of PIP_2 from MARCKS (Shi et al. 2014, 2016, 2017). Whether PIP_2 also gates TRPC1 in non-excitable cells such as HEK293 and HSG cells remains to be seen. As further described below, remodeling of PIP_2 at the plasma membrane may also aid in the compartmentalization of TRPC1 into discrete cellular regions. Hence, PIP_2 may have dual roles in modulating TRPC1 channel activity (Ma et al. 2011a, b; Ong et al. 2014; Zhang et al. 2016).

9.6 Compartmentalization of TRPC1 and SOCE

Compartmentalization of a channel enhances protein-protein and protein-lipid interactions that are involved in precise regulation of its function. In addition, such compartmentalization can also determine the amplitude, rate, and duration of the Ca^{2+} signal generated. In addition, effector proteins scaffolded near the region of Ca^{2+} influx can rapidly detect and respond to local Ca^{2+} signals before they dissipate. Numerous studies have focused on the ER-PM junctions where SOCE occurs, identifying the molecular components that are enriched within these junctions and the structural components required for channel assembly and stability within the domain. The ER-PM junctions refer to distinct regions of the plasma membrane where the ER membrane is in close apposition (<30 nm) (Luik et al. 2006; Wu et al. 2006). While it is well known that TRPC1 and Orai1 aggregate within these junctions where they can interact with STIM1, how Orai1 is targeted to these junctions is not known. Orai1 is proposed to be recruited to the STIM1 punctae by a diffusion trapping method (Wu et al. 2014). As noted above, TRPC1 is delivered to these ER-PM junctions via a fast Rab4-dependent recycling pathway. Furthermore, plasma membrane lipid domains are also involved in partitioning of TRPC1 into specific plasma membrane locations. Lipid Raft Domains (LRDs), enriched in cholesterol and sphingolipids (Ambudkar et al. 2010; Dart 2010; Lingwood and Simons 2010; Ong and Ambudkar 2012; Pani and Singh 2009), have been proposed to provide a stable platform for the assembly of calcium signaling complexes and ensure that the channels are localized in close proximity to the proteins that regulate their function. In addition, dynamic, smaller rafts can coalesce into larger, more stable domains by interacting with various proteins and lipids in the plasma membrane and the cytoskeletal network underlying the plasma membrane. Such mechanisms could bring proteins that are segregated from each other into the same plasma membrane domain (Ong and Ambudkar 2015). This could also bring proteins together into a larger complex. The structural integrity of LRD appears to be vital for SOCE as disruption using methyl-β-cyclodextrin (mβCD) attenuates SOCE in HSG cells (Lockwich et al. 2000; Pani et al. 2008), glioma cells (Bomben et al. 2011), platelets (Brownlow et al. 2004; Jardin et al. 2008b), and neutrophils (Kannan et al. 2007) due to impaired association of TRPC1 with Orai1, STIM1, and IP_3R2 (Brownlow et al. 2004; Brownlow and Sage 2005; Galan et al. 2010; Pani et al. 2008).

Partitioning of TRPC1 into LRDs following store depletion was first reported by Indu Ambudkar and colleagues (Lockwich et al. 2000). Importantly, subsequent studies showed that STIM1 also partitions into the same domains after store depletion. In addition to HSG cells (Lockwich et al. 2000; Pani et al. 2008), partitioning of TRPC1 into LRDs has also been reported in polymorphonuclear neutrophils (Kannan et al. 2007), C2C12 skeletal myoblasts (Formigli et al. 2009), and THP-1 monocytic cells (Berthier et al. 2004). Recruitment of TRPC1 into LRDs is mediated by Cav-1, whereas STIM1 directly interacts with the plasma membrane phospholipids via its C-terminal polybasic domain. Cholesterol depletion by mβCD treatment decreased the partitioning of both TRPC1 and STIM1 into raft fractions following store depletion, significantly reducing TRPC1+STIM1-mediated SOCE (Alicia et al. 2008; Pani et al. 2008). Both STIM1 and STIM2 have C-terminal polybasic domains that anchor the proteins to the plasma membrane. Abrogation of STIM1 puncta formation has been observed in HSG cells treated with mβCD, but not by knocking down Cav-1 (Pani et al. 2008, 2009). Recruitment of SARAF to the ER-PM junctions, where it interacts with STIM1 to mediate the inactivation of Orai1, has been shown to require Cav-1. Moreover, the Orai1-STIM1 complex is recruited into a microdomain that is PIP_2 rich and contains Cav-1 (Jha et al. 2013; Maleth et al. 2014). Hence, it is proposed that TRPC1 interacts with STIM1 and Orai1 in a Cav-1-containing LRD within the ER-PM junctions.

Physiological functions attributed to TRPC1 were similarly impaired following cholesterol depletion. Treatment with mβCD attenuated TRPC1-mediated SOCE and currents in glioma cells (Bomben et al. 2011) and C2C12 skeletal myoblasts (Formigli et al. 2009), leading to reduced growth factor-induced chemotaxis and cell differentiation, respectively. It is worth mentioning that in cancer cells, Orai1 is recruited into the lipid rafts by the SK3 channel, resulting in the development of constitutive Ca^{2+} entry that promotes cell migration and development of bone metastases (Chantome et al. 2013). However, the effect of mβCD treatment on Orai1 function appears to be more variable. One study suggested that cholesterol depletion affects the activation of Orai1-SOCE with HEK293 cells, although no effect was observed if cells were treated after SOCE activation (Galan et al. 2010). Another report showed that cholesterol depletion in RBL and HEK293 cells enhanced SOCE and I_{CRAC}. Notably, Orai1 was suggested to bind cholesterol directly via a cholesterol recognition motif located within the ETON (extended transmembrane Orai1 N-terminal) region. Point mutation within the cholesterol recognition motif mimicked the effects observed with mβCD treatment (Derler et al. 2016). Additionally, Orai1 also binds to Cav-1 and is internalized during meiosis via the Cav-1-/dynamin-dependent endocytic pathway (Yu et al. 2010).

Recently, there has been a considerable advance in our understanding of the formation and stabilization of ER-PM junctions involved in SOCE. Several proteins have been identified that determine either the physical structure of the domain or promote STIM1 clustering within the domains. Extended synaptotagmins (E-Syts 1–3) have been proposed to function as tethers holding the ER membrane and plasma membrane together within the junctions (Chang et al. 2013; Chang and Liou 2016; Min et al. 2007). Other proteins such as junctophilins (Takeshima et al. 2015) and junctate (Srikanth et al. 2012; Treves et al. 2004) have been proposed to function as ER-PM junctional tethers. Additionally, septins are localized within these junctions which appear to bind PIP_2 concentrating them in plasma membrane microdomains (Chao et al. 2014; Sharma et al. 2013). STIMATE (encoded by the *TMEM110* gene) regulates activation and translocation of STIM1 as well as anchoring in ER-PM junctions (Jing et al. 2015; Quintana et al. 2015). Presently there are no data to demonstrate that STIM1-TRPC1 assembly is directly controlled by any of the proteins described above. Nevertheless, given how vital Orai1 and STIM1 are to TRPC1 function, proteins that help promote and stabilize STIM1-Orai1 interactions and/or ER-PM junctions would be expected to also influence TRPC1 function. Additionally, since TRPC1 is localized within intracellular vesicles which need to be positioned close to the STIM1-Orai1 complex in the sub-plasma membrane regions, the architecture of the ER-PM junctional domain could have a huge impact on the activation of TRPC1. Both spatial and temporal issues would be critical for the scaffolding of TRPC1 vesicles in this region, as well as sensing the initial Orai1-generated Ca^{2+} signals. It is reasonable to expect that the very narrow junctional space between the ER and PM might impede trafficking of TRPC1 vesicles into the ER-PM junctions, and thus remodeling of the domains might be involved in regulating TRPC1 trafficking and exocytosis. Further studies are required to resolve these very important processes and the status of TRPC1 within ER-PM junctions.

There is increasing evidence that ER-PM junctions are dynamic, being formed, and remodeled during assembly and regulation of SOCE channels. It has been shown that the ER moves to the PM during the formation of STIM1 puncta to form nascent junctions (Hartzell et al. 2016; Poteser et al. 2016). Further, proteins such as E-Syt control the size of the ER-PM junctions, with some E-Syt members being regulated by Ca^{2+} entry (Chang et al. 2013; Idevall-Hagren et al. 2015; Liou and Chang 2015; Prinz 2014). The proximity of the ER to the PM is critical for the interaction of STIM1 in the ER with the plasma membrane lipids, and so the distance between the two membranes has to be exquisitely controlled. The spanning of STIM1 within this domain is also critical for its interaction with Orai1. Plasma membrane PIP_2 domains and the actin cytoskeleton have a crucial role in this process. Current models have proposed a key role for the cortical actin cytoskeleton that is closely apposed to the plasma membrane. This actin layer acts as impedance for interactions between the PM and ER, although there could be spaces within this actin network where the ER could move toward the plasma membrane. It can be hypothesized that the cytoskeleton would likely need to be depolymerized locally to allow ER-PM interactions and then be remodeled to stabilize the junctions. Interestingly, actin, tubulin, and myosin have all been suggested to control the formation and stability of the LRDs. Reorganization of LRDs may also be mediated by actin polymerization. Actin can be tethered to the plasma membrane PIP_2 via actin-binding proteins that have PIP_2-binding domains. Thus, changes in the actin structure can be expected to cause changes in the PIP_2 domains as well (concentrate or make more diffuse) (Chichili and Rodgers 2009; Liu and Fletcher 2006; Niggli 2001; Yin and Janmey 2003). As proposed in a recent review (Ong and Ambudkar 2015), caveolar and non-caveolar lipid rafts may form nanodomains that interact with the underlying actin cytoskeletal network. These interactions determine the movements of raft nanodomains within the plasma membrane. Signaling events triggered after cell stimulation can induce reorganization of the actin filaments near the plasma membrane, which subsequently can induce remodeling of plasma membrane PIP_2 and drive fusion of smaller nanodomains into larger macrodomains. Alternatively, rearrangement of plasma membrane phospholipids can also trigger actin remodeling (Ong and Ambudkar 2015). Septin-induced remodeling of the plasma membrane PIP_2 which regulates SOCE (Sharma et al. 2013) is likely to be accompanied by changes in the underlying actin layer. It is worth noting that there are no data to unequivocally establish which occurs first, phospholipid or cytoskeleton remodeling.

As mentioned earlier, there is relatively little information regarding the status or remodeling of the ER-PM junction during assembly of STIM1-TRPC1 channels (Ong et al. 2014). Advanced imaging techniques using super-resolution microscopy could be useful to elucidate the formation of a TRPC1-containing nanodomain and how it may interact with other proteins or lipids in the ER-PM junctions. Also of importance will be knowledge of how the ER-PM domain architecture affects TRPC1 trafficking to these domains, its insertion into the plasma membrane, as well as its interaction with and activation by STIM1. Whether the domain remodels during inactivation of the channel is also a very interesting and as yet unknown aspect of TRPC1 function in cells.

9.7 Distinct Ca^{2+} Signals and Functional Specificity of TRPC1

Physiologically relevant Ca^{2+} signals display specific temporal and spatial patterns, in addition to the amplitude. For example, agonist-stimulated increase in $[Ca^{2+}]_i$ displays an oscillatory pattern, depending on the cell type and stimulus intensity. In some cells, the oscillations are driven by Ca^{2+} entry, while in others, they are controlled by IP_3-mediated Ca^{2+}, release from the ER controls the oscillations. Importantly, while global $[Ca^{2+}]_i$ increases in cells are easily detected, changes in local $[Ca^{2+}]_i$ occurring close to the channel may be the relevant factor in regulating cell function. Such local Ca^{2+} signals are detected by Ca^{2+} sensor proteins localized within this microdomain and near to the channel. Localization of the Ca^{2+} channels very often will determine the exact site where Ca^{2+} signals are generated. For example, in acinar cells from exocrine gland tissues, Orai1 is localized near the apical region of the cell, while TRPC1 is more concentrated along the basolateral regions. In these cells, initial Ca^{2+} release occurs near the apical pole and $[Ca^{2+}]_i$ increase spreads to the basal region of the cell. Sustained $[Ca^{2+}]_i$ elevation depends on activation of Orai1 and TRPC1. However, Ca^{2+} entry mediated by the two channels is utilized by the cell for regulation of distinct cellular functions. A major point that needs to be addressed is how does the cell sense and differentiate the $[Ca^{2+}]_i$ increase originating from each channel. It is well established that Orai1-mediated SOCE activates NFAT in an "all or none" manner. Local $[Ca^{2+}]$ increase near the Orai1 channel activates calcineurin within the ER-PM junctions, which then triggers dephosphorylation and nuclear translocation of NFAT (Parekh 2011). While TRPC1 function modifies the $[Ca^{2+}]_i$ increase triggered by Orai1, it does not affect NFAT activation. In contrast, Ca^{2+} entry via TRPC1 contributes to the activation of NFκB (Cheng et al. 2011; Ong et al. 2012) and K_{Ca} channel in salivary glands and Ca^{2+}-dependent Cl^- channel in pancreatic ducts (Hong et al. 2011). Thus, unlike NFAT, TRPC1 is the main source of Ca^{2+} for these cell functions and cannot be compensated by Orai1-mediated Ca^{2+} entry in the absence of TRPC1.

Noteworthy also is the fact that Orai1 and TRPC1 channels are not evenly distributed in the salivary gland acinar cells (*c.f.* uniform distribution in the plasma membrane of cell lines). TRPC1 is primarily localized in the lateral membrane with some in the basal membrane of the acinar cell, whereas Orai1 is mainly concentrated in the lateral membrane toward the apical pole. Following ER-Ca^{2+} depletion, STIM1 translocates to the lateral membranes where it colocalizes with both channels (Hong et al. 2011). Since the initial increase in $[Ca^{2+}]_i$ in acinar cells occurs at the apical pole, it can be suggested that Orai1 channels that are localized closer to this site would be activated first. Regulation of secretory processes in both the salivary gland (fluid) and pancreas (protein) requires Ca^{2+}-dependent activation of various ion channel and transporters as well as granule fusion (pancreas) in different regions of the cell. Therefore, the spread of Ca^{2+} from the apical to the basal pole of the cell is critical (Cheng et al. 2013). The currently proposed model suggests that Orai1 provides an initial trigger pool of Ca^{2+} that induces recruitment of TRPC1 into the plasma membrane. Activation of TRPC1 by STIM1 provides Ca^{2+} influx which drives the spread of $[Ca^{2+}]_i$ to the basal pole and results in a global elevation of $[Ca^{2+}]_i$ that sustains ion channel activity and secretion. It will be important to determine how TRPC1 channels that are localized at sites away from Orai1 are inserted into the plasma membrane. It also remains to be established whether intracellular Ca^{2+} release or even Ca^{2+} entry via TRPC1 itself provides Ca^{2+} to recruit neighboring TRPC1 channels. Alternately, it is possible that Orai1 channels are located all along the lateral membrane, but at concentrations that are below detection by immunofluorescence. Irrespective of the underlying mechanism that is involved in TRPC1 recruitment, it is clear that the spatiotemporal aspects of $[Ca^{2+}]_i$ signaling required to drive exocrine secretion are determined by the respective localization and assembly of TRPC1, Orai1, and STIM1. Very little is known about the localization of endogenous Orai1 or TRPC1 and the types of Ca^{2+} signals that are generated by the channels in various cell types. Indeed, it will be very important to examine these in greater detail to understand the relevance of their respective Ca^{2+} signal characteristics in the context of the cell function that each channel regulates (Fig. 9.3).

9.8 Conclusions and Future Directions

SOCE involves multiple coordinate processes that allow cells to display dynamic regulation of Ca^{2+} signaling under physiological conditions. Substantial data reported over the past 10 years have provided novel information regarding the multiplicity of channels, accessory proteins, and protein-protein interactions that underlie the Ca^{2+} signals generated in response to activation of SOCE. Additional spatial constraints are provided by compartmentalization of the channels as well as their regulatory proteins. While Orai1 and STIM1 generate the basic Ca^{2+} signal and represent the primary SOCE channels, other channels like TRPC1 are also recruited to ER-PM junctions and are regulated via interaction with STIM1. TRPC1 activation modifies the initial Ca^{2+} signal leading to the regulation of additional cell functions that cannot be accomplished by Orai1 alone. It is worth noting that splice variants have been identified for STIM2, STIM1, and Orai1 (recently reviewed in Niemeyer 2016). While the exact physiological functions of these have not yet been established, these variants, when present in cells, could modify SOCE and Ca^{2+} signaling processes. Thus, the pattern of $[Ca^{2+}]_i$ signals including frequency, amplitude, as well as spatial patterning is key to the regulation of diverse cell functions. Activation of a channel such as TRPC1 increases the dynamic range of SOCE. Fine regulation of TRPC1 trafficking at several levels and plasma membrane expression of the channel controls the modulation of $[Ca^{2+}]_i$. Further tuning of SOCE is provided by STIM2 which increases the sensitivity and detection of changes in ER-$[Ca^{2+}]$. The exact sequence of molecular events involved in the

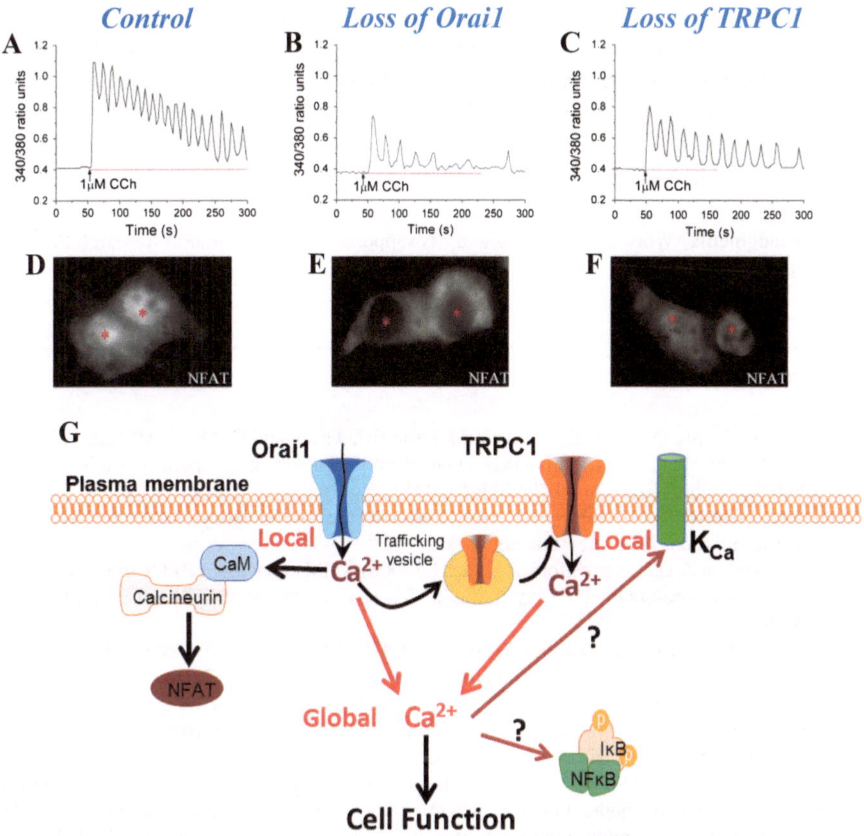

Fig. 9.3 Distinct Ca^{2+} signals and currents, as well as downstream cellular events associated with Orai1- and TRPC1-mediated Ca^{2+} entry. (**a–c**) Distinct Ca^{2+} signals induced by 1 μM CCh. The pattern shown in control cells reflects the activation of both Orai1 and TRPC1 channels. No response was seen following knockdown of Orai1. In the case of cells lacking TRPC1, baseline oscillations were seen, and these are attributed to the still functional Orai1 channel. (**d–f**) Activation of NFAT translocation from the cytoplasm into the nucleus (denoted by *asterisk*) following store depletion. Nuclear translocation of a GFP-tagged NFAT could be clearly seen in control cells but not in cells lacking Orai1, pointing to Ca^{2+} entry via Orai1 as the primary determinant of NFAT activation. This was further supported by the observation of nuclear translocation of NFAT in cells lacking TRPC1, where the Orai1 channel could still be activated following store depletion (modified from Cheng et al. 2013). (**g**) Physiological cell functions that are dependent on SOCE. Stimulation with agonists generates $[Ca^{2+}]_i$ changes that occur locally (i.e., close to the channel pore) and globally (i.e., throughout the cell cytosol). Local SOCE mediated by Orai1 has been shown to activate calcineurin, which subsequently induces NFAT translocation into the nucleus to drive gene expression. Local Orai1-SOCE also promotes insertion of TRPC1 into the plasma membrane. Ca^{2+} entry via both Orai1 and TRPC1 contributes to increase in global $[Ca^{2+}]_i$, which has been shown to activate NFκB and induce cell functions such as fluid secretion from salivary gland acinar cells. While the Ca^{2+}-activated ion channels in the plasma membrane are also activated by global $[Ca^{2+}]_i$, it is not clear whether the activating Ca^{2+} comes from those situated in the vicinity of neighboring Orai1 and TRPC1 channels and/or from the deeper regions of the cell cytosol (modified from Ong et al. 2016)

regulation and function of TRPC channels in response to ER-Ca^{2+} depletion remains to be established. Special emphasis needs to be placed on the elucidation and possible molecular rearrangements of STIM1 that are required for activation of TRPC1 and the possible role of STIM2 in this process. Finally, the critical trafficking of TRPC1 within ER-PM junctions as well as the determinants of vesicle fusion needs to be further clarified.

Acknowledgments Work in ISA's laboratory is supported by the Intramural Research Program of the NIH, NIDCR.

References

Ahmmed GU, Mehta D, Vogel S, Holinstat M, Paria BC, Tiruppathi C, Malik AB (2004) Protein kinase Ca phosphorylates the TRPC1 channel and regulates store-operated Ca2+ entry in endothelial cells. J Biol Chem 279:20941–20949

Alabi AA, Tsien RW (2013) Perspectives on kiss-and-run: role in exocytosis, endocytosis, and neurotransmission. Annu Rev Physiol 75:393–422

Alicia S, Angelica Z, Carlos S, Alfonso S, Vaca L (2008) STIM1 converts TRPC1 from a receptor-operated to a store-operated channel: moving TRPC1 in and out of lipid rafts. Cell Calcium 44:479–491

Almirza WH, Peters PH, van Zoelen EJ, Theuvenet AP (2012) Role of Trpc channels, Stim1 and Orai1 in PGF2a-induced calcium signaling in NRK fibroblasts. Cell Calcium 51:12–21

Ambudkar IS, Ong HL, Singh BB (2010) Molecular and functional determinants of Ca2+ signaling microdomains. In: Sitaramayya A (ed) Signal transduction: pathways, mechanisms and diseases. Springer, Heidelberg, pp 237–253

Asanov A, Sampieri A, Moreno C, Pacheco J, Salgado A, Sherry R, Vaca L (2015) Combined single channel and single molecule detection identifies subunit composition of STIM1-activated transient receptor potential canonical (TRPC) channels. Cell Calcium 57:1–13

Avila-Medina J, Calderon-Sanchez E, Gonzalez-Rodriguez P, Monje-Quiroga F, Rosado JA, Castellano A, Ordonez A, Smani T (2016) Orai1 and TRPC1 proteins co-localize with CaV1.2 channels to form a signal complex in vascular smooth muscle cells. J Biol Chem 291:21148–21159

Bergdahl A, Gomez MF, Dreja K, SZ X, Adner M, Beech DJ, Broman J, Hellstrand P, Sward K (2003) Cholesterol depletion impairs vascular reactivity to endothelin-1 by reducing store-operated Ca2+ entry dependent on TRPC1. Circ Res 93:839–847

Berthier A, Lemaire-Ewing S, Prunet C, Monier S, Athias A, Bessede G, Pais de Barros JP, Laubriet A, Gambert P, Lizard G, Neel D (2004) Involvement of a calcium-dependent dephosphorylation of BAD associated with the localization of Trpc-1 within lipid rafts in 7-ketocholesterol-induced THP-1 cell apoptosis. Cell Death Differ 11:897–905

Bollimuntha S, Cornatzer E, Singh BB (2005a) Plasma membrane localization and function of TRPC1 is dependent on its interaction with b-tubulin in retinal epithelium cells. Vis Neurosci 22:163–170

Bollimuntha S, Singh BB, Shavali S, Sharma SK, Ebadi M (2005b) TRPC1-mediated inhibition of 1-methyl-4-phenylpyridinium ion neurotoxicity in human SH-SY5Y neuroblastoma cells. J Biol Chem 280:2132–2140

Bomben VC, Turner KL, Barclay TT, Sontheimer H (2011) Transient receptor potential canonical channels are essential for chemotactic migration of human malignant gliomas. J Cell Physiol 226:1879–1888

Brandman O, Liou J, Park WS, Meyer T (2007) STIM2 is a feedback regulator that stabilizes basal cytosolic and endoplasmic reticulum Ca2+ levels. Cell 131:1327–1339

Brazer SC, Singh BB, Liu X, Swaim W, Ambudkar IS (2003) Caveolin-1 contributes to assembly of store-operated Ca2+ influx channels by regulating plasma membrane localization of TRPC1. J Biol Chem 278:27208–27215

Brownlow SL, Sage SO (2005) Transient receptor potential protein subunit assembly and membrane distribution in human platelets. Thromb Haemost 94:839–845

Brownlow SL, Harper AG, Harper MT, Sage SO (2004) A role for hTRPC1 and lipid raft domains in store-mediated calcium entry in human platelets. Cell Calcium 35:107–113

Byron KL (1996) Vasopressin stimulates Ca2+ spiking activity in A7r5 vascular smooth muscle cells via activation of phospholipase A2. Circ Res 78:813–820

Cahalan MD, Lewis RS (1990) Functional roles of ion channels in lymphocytes. Semin Immunol 2:107–117

Chang CL, Liou J (2016) Homeostatic regulation of the PI(4,5)P2-Ca2+ signaling system at ER–PM junctions. Biochim Biophys Acta 1861:862–873

Chang CL, Hsieh TS, Yang TT, Rothberg KG, Azizoglu DB, Volk E, Liao JC, Liou J (2013) Feedback regulation of receptor-induced Ca2+ signaling mediated by E-Syt1 and Nir2 at endoplasmic reticulum-plasma membrane junctions. Cell Rep 5:813–825

Chantome A, Potier-Cartereau M, Clarysse L, Fromont G, Marionneau-Lambot S, Gueguinou M, Pages JC, Collin C, Oullier T, Girault A, Arbion F, Haelters JP, Jaffres PA, Pinault M, Besson P, Joulin V, Bougnoux P, Vandier C (2013) Pivotal role of the lipid Raft SK3-Orai1 complex in human cancer cell migration and bone metastases. Cancer Res 73:4852–4861

Chao JT, Wong AK, Tavassoli S, Young BP, Chruscicki A, Fang NN, Howe LJ, Mayor T, Foster LJ, Loewen CJ (2014) Polarization of the endoplasmic reticulum by ER-septin tethering. Cell 158:620–632

Cheng KT, Liu X, Ong HL, Ambudkar IS (2008) Functional requirement for Orai1 in store-operated TRPC1-STIM1 channels. J Biol Chem 283:12935–12940

Cheng KT, Liu X, Ong HL, Swaim W, Ambudkar IS (2011) Local Ca2+ entry via Orai1 regulates plasma membrane recruitment of TRPC1 and controls cytosolic Ca2+ signals required for specific cell functions. PLoS Biol 9:e1001025

Cheng KT, Ong HL, Liu X, Ambudkar IS (2013) Contribution and regulation of TRPC channels in store-operated Ca2+ entry. In: Prakriya M (ed) Store-operated calcium channels. Elsevier, Amsterdam, pp 149–179

Chichili GR, Rodgers W (2009) Cytoskeleton-membrane interactions in membrane raft structure. Cell Mol Life Sci 66:2319–2328

Collins SR, Meyer T (2011) Evolutionary origins of STIM1 and STIM2 within ancient Ca2+ signaling systems. Trends Cell Biol 21:202–211

Cuddapah VA, Turner KL, Sontheimer H (2013) Calcium entry via TRPC1 channels activates chloride currents in human glioma cells. Cell Calcium 53:187–194

Dart C (2010) Lipid microdomains and the regulation of ion channel function. J Physiol 588:3169–3178

Derler I, Jardin I, Stathopulos PB, Muik M, Fahrner M, Zayats V, Pandey SK, Poteser M, Lackner B, Absolonova M, Schindl R, Groschner K, Ettrich R, Ikura M, Romanin C (2016) Cholesterol modulates Orai1 channel function. Sci Signal 9:ra10

Desai PN, Zhang X, Wu S, Janoshazi A, Bolimuntha S, Putney JW Jr, Trebak M (2015) Multiple types of calcium channels arising from alternative translation initiation of the Orai1 message. Sci Signal 8:ra74

de Souza LB, Ambudkar IS (2014) Trafficking mechanisms and regulation of TRPC channels. Cell Calcium 56:43–50

de Souza LB, Ong HL, Liu X, Ambudkar IS (2015) Fast endocytic recycling determines TRPC1-STIM1 clustering in ER-PM junctions and plasma membrane function of the channel. Biochim Biophys Acta 1853:2709–2721

Dietrich A, Gudermann T (2014) TRPC6: physiological function and pathophysiological relevance. Handb Exp Pharmacol 222:157–188

Dietrich A, Chubanov V, Kalwa H, Rost BR, Gudermann T (2006) Cation channels of the transient receptor potential superfamily: their role in physiological and pathophysiological processes of smooth muscle cells. Pharmacol Ther 112:744–760

Dietrich A, Fahlbusch M, Gudermann T (2014) Classical transient receptor potential 1 (TRPC1): channel or channel regulator? Cell 3:939–962

Doherty GJ, McMahon HT (2009) Mechanisms of endocytosis. Annu Rev Biochem 78:857–902

Du J, Sours-Brothers S, Coleman R, Ding M, Graham S, Kong DH, Ma R (2007) Canonical transient receptor potential 1 channel is involved in contractile function of glomerular mesangial cells. J Am Soc Nephrol 18:1437–1445

Feske S, Gwack Y, Prakriya M, Srikanth S, Puppel SH, Tanasa B, Hogan PG, Lewis RS, Daly M, Rao A (2006) A mutation in Orai1 causes immune deficiency by abrogating CRAC channel function. Nature 441:179–185

Fiorio Pla A, Maric D, Brazer SC, Giacobini P, Liu X, Chang YH, Ambudkar IS, Barker JL (2005) Canonical transient receptor potential 1 plays a role in basic fibroblast growth factor (bFGF)/FGF receptor-1-induced Ca2+ entry and embryonic rat neural stem cell proliferation. J Neurosci 25:2687–2701

Formigli L, Sassoli C, Squecco R, Bini F, Martinesi M, Chellini F, Luciani G, Sbrana F, Zecchi-Orlandini S, Francini F, Meacci E (2009) Regulation of transient receptor potential canonical channel 1 (TRPC1) by sphingosine 1-phosphate in C2C12 myoblasts and its relevance for a role of mechanotransduction in skeletal muscle differentiation. J Cell Sci 122:1322–1333

Freichel M, Tsvilovskyy V, Camacho-Londono JE (2014) TRPC4- and TRPC4-containing channels. Handb Exp Pharmacol 222:85–128

Galan C, Woodard GE, Dionisio N, Salido GM, Rosado JA (2010) Lipid rafts modulate the activation but not the maintenance of store-operated Ca2+ entry. Biochim Biophys Acta 1803:1083–1093

Grant BD, Donaldson JG (2009) Pathways and mechanisms of endocytic recycling. Nat Rev Mol Cell Biol 10:597–608

Gwack Y, Srikanth S, Feske S, Cruz-Guilloty F, Oh-hora M, Neems DS, Hogan PG, Rao A (2007) Biochemical and functional characterization of Orai proteins. J Biol Chem 282:16232–16243

Hartzell CA, Jankowska KI, Burkhardt JK, Lewis RS (2016) Calcium influx through CRAC channels controls actin organization and dynamics at the immune synapse. eLife 5. doi:10.7554/eLife.14850

He K, Qi F, Guo C, Zhan S, Xu H, Liu J, Yang X (2016) Movement deficits and neuronal loss in basal ganglia in TRPC1 deficient mice. Oncotarget 7:69337–69346

Hogan PG, Lewis RS, Rao A (2010) Molecular basis of calcium signaling in lymphocytes: STIM and ORAI. Annu Rev Immunol 28:491–533

Hong JH, Li Q, Kim MS, Shin DM, Feske S, Birnbaumer L, Cheng KT, Ambudkar IS, Muallem S (2011) Polarized but differential localization and recruitment of STIM1, Orai1 and TRPC channels in secretory cells. Traffic 12:232–245

Hoth M, Penner R (1992) Depletion of intracellular calcium stores activates a calcium current in mast cells. Nature 355:353–356

Hoth M, Penner R (1993) Calcium release-activated calcium current in rat mast cells. J Physiol 465:359–386

Huang GN, Zeng W, Kim JY, Yuan JP, Han L, Muallem S, Worley PF (2006) STIM1 carboxyl-terminus activates native SOC, Icrac and TRPC1 channels. Nat Cell Biol 8:1003–1010

Idevall-Hagren O, Lu A, Xie B, De Camilli P (2015) Triggered Ca2+ influx is required for extended synaptotagmin 1-induced ER-plasma membrane tethering. EMBO J 34:2291–2305

Jardin I, Lopez JJ, Salido GM, Rosado JA (2008a) Orai1 mediates the interaction between STIM1 and hTRPC1 and regulates the mode of activation of hTRPC1-forming Ca2+ channels. J Biol Chem 283:25296–25304

Jardin I, Salido GM, Rosado JA (2008b) Role of lipid rafts in the interaction between hTRPC1, Orai1 and STIM1. Channels 2:401–403

Jha A, Ahuja M, Maleth J, Moreno CM, Yuan JP, Kim MS, Muallem S (2013) The STIM1 CTID domain determines access of SARAF to SOAR to regulate Orai1 channel function. J Cell Biol 202:71–79

Jing J, He L, Sun A, Quintana A, Ding Y, Ma G, Tan P, Liang X, Zheng X, Chen L, Shi X, Zhang SL, Zhong L, Huang Y, Dong MQ, Walker CL, Hogan PG, Wang Y, Zhou Y (2015) Proteomic mapping of ER-PM junctions identifies STIMATE as a regulator of Ca influx. Nat Cell Biol 17:1339–1347

Kannan KB, Barlos D, Hauser CJ (2007) Free cholesterol alters lipid raft structure and function regulating neutrophil Ca2+ entry and respiratory burst: correlations with calcium channel raft trafficking. J Immunol 178:5253–5261

Kim MS, Zeng W, Yuan JP, Shin DM, Worley PF, Muallem S (2009) Native store-operated Ca2+ influx requires the channel function of Orai1 and TRPC1. J Biol Chem 284:9733–9741

Kochukov MY, Balasubramanian A, Noel RC, Marrelli SP (2013) Role of TRPC1 and TRPC3 channels in contraction and relaxation of mouse thoracic aorta. J Vasc Res 50:11–20

Kwan HY, Shen B, Ma X, Kwok YC, Huang Y, Man YB, Yu S, Yao X (2009) TRPC1 associates with BKCa channel to form a signal complex in vascular smooth muscle cells. Circ Res 104:670–678

Kwiatek AM, Minshall RD, Cool DR, Skidgel RA, Malik AB, Tiruppathi C (2006) Caveolin-1 regulates store-operated Ca2+ influx by binding of its scaffolding domain to transient receptor potential channel-1 in endothelial cells. Mol Pharmacol 70:1174–1183

Lee KP, Choi S, Hong JH, Ahuja M, Graham S, Ma R, So I, Shin DM, Muallem S, Yuan JP (2014) Molecular determinants mediating gating of transient receptor potential canonical (TRPC) channels by stromal interaction molecule 1 (STIM1). J Biol Chem 289:6372–6382

Leitz J, Kavalali ET (2011) Ca2+ influx slows single synaptic vesicle endocytosis. J Neurosci 31:16318–16326

Lewis RS, Cahalan MD (1989) Mitogen-induced oscillations of cytosolic Ca2+ and transmembrane Ca2+ current in human leukemic T cells. Cell Regul 1:99–112

Li M, Chen C, Zhou Z, Xu S, Yu Z (2012) A TRPC1-mediated increase in store-operated Ca2+ entry is required for the proliferation of adult hippocampal neural progenitor cells. Cell Calcium 51:486–496

Lichtenegger M, Groschner K (2014) TRPC3: a multifunctional signaling molecule. Handb Exp Pharmacol 222:67–84

Lin MI, Yu J, Murata T, Sessa WC (2007) Caveolin-1-deficient mice have increased tumor microvascular permeability, angiogenesis, and growth. Cancer Res 67:2849–2856

Lingwood D, Simons K (2010) Lipid rafts as a membrane-organizing principle. Science 327:46–50

Liou J, Chang CL (2015) Unveiling physiological functions of extended synaptotagmins. Cell Cycle 14:799–800

Liou J, Kim ML, Heo WD, Jones JT, Myers JW, Ferrell JE Jr, Meyer T (2005) STIM is a Ca2+ sensor essential for Ca2+-store-depletion-triggered Ca2+ influx. Curr Biol 15:1235–1241

Liu AP, Fletcher DA (2006) Actin polymerization serves as a membrane domain switch in model lipid bilayers. Biophys J 91:4064–4070

Liu X, Wang W, Singh BB, Lockwich T, Jadlowiec J, O'Connell B, Wellner R, Zhu MX, Ambudkar IS (2000) Trp1, a candidate protein for the store-operated Ca2+ influx mechanism in salivary gland cells. J Biol Chem 275:3403–3411

Liu X, Singh BB, Ambudkar IS (2003) TRPC1 is required for functional store-operated Ca2+ channels. Role of acidic amino acid residues in the S5–S6 region. J Biol Chem 278:11337–11343

Liu X, Groschner K, Ambudkar IS (2004) Distinct Ca2+-permeable cation currents are activated by internal Ca2+-store depletion in RBL-2H3 cells and human salivary gland cells, HSG and HSY. J Membr Biol 200:93–104

Liu X, Bandyopadhyay BC, Singh BB, Groschner K, Ambudkar IS (2005) Molecular analysis of a store-operated and 2-acetyl-sn-glycerol-sensitive non-selective cation channel. Heteromeric assembly of TRPC1-TRPC3. J Biol Chem 280:21600–21606

Liu X, Cheng KT, Bandyopadhyay BC, Pani B, Dietrich A, Paria BC, Swaim WD, Beech D, Yildrim E, Singh BB, Birnbaumer L, Ambudkar IS (2007) Attenuation of store-operated Ca2+ current impairs salivary gland fluid secretion in TRPC1−/− mice. Proc Natl Acad Sci U S A 104:17542–17547

Lockwich TP, Liu X, Singh BB, Jadlowiec J, Weiland S, Ambudkar IS (2000) Assembly of Trp1 in a signaling complex associated with caveolin-scaffolding lipid raft domains. J Biol Chem 275:11934–11942

Lopez JJ, Salido GM, Pariente JA, Rosado JA (2006) Interaction of STIM1 with endogenously expressed human canonical TRP1 upon depletion of intracellular Ca2+ stores. J Biol Chem 281:28254–28264

Lu M, Branstrom R, Berglund E, Hoog A, Bjorklund P, Westin G, Larsson C, Farnebo LO, Forsberg L (2010) Expression and association of TRPC subtypes with Orai1 and STIM1 in human parathyroid. J Mol Endocrinol 44:285–294

Luik RM, MM W, Buchanan J, Lewis RS (2006) The elementary unit of store-operated Ca2+ entry: local activation of CRAC channels by STIM1 at ER-plasma membrane junctions. J Cell Biol 174:815–825

Ma X, Cheng KT, Wong CO, O'Neil RG, Birnbaumer L, Ambudkar IS, Yao X (2011a) Heteromeric TRPV4-C1 channels contribute to store-operated Ca2+ entry in vascular endothelial cells. Cell Calcium 50:502–509

Ma X, Nilius B, Wong JW, Huang Y, Yao X (2011b) Electrophysiological properties of heteromeric TRPV4-C1 channels. Biochim Biophys Acta 1808:2789–2797

Maleth J, Choi S, Muallem S, Ahuja M (2014) Translocation between PI(4,5)P2-poor and PI(4,5)P2-rich microdomains during store depletion determines STIM1 conformation and Orai1 gating. Nat Commun 5:5843

Matthews G, Neher E, Penner R (1989) Second messenger-activated calcium influx in rat peritoneal mast cells. J Physiol 418:105–130

Maxfield FR, McGraw TE (2004) Endocytic recycling. Nat Rev Mol Cell Biol 5:121–132

McGurk JS, Shim S, Kim JY, Wen Z, Song H, Ming GL (2011) Postsynaptic TRPC1 function contributes to BDNF-induced synaptic potentiation at the developing neuromuscular junction. J Neurosci 31:14754–14762

Mehta D, Ahmmed GU, Paria BC, Holinstat M, Voyno-Yasenetskaya T, Tiruppathi C, Minshall RD, Malik AB (2003) RhoA interaction with inositol 1,4,5-triphosphate receptor and transient receptor potential channel-1 regulates Ca2+ entry. J Biol Chem 278:33492–33500

Miller BA (2014) Trpc2. Handb Exp Pharmacol 222:53–65

Min SW, Chang WP, Sudhof TC (2007) E-Syts, a family of membranous Ca2+-sensor proteins with multiple C2 domains. Proc Natl Acad Sci U S A 104:3823–3828

Murata T, Lin MI, Huang Y, Yu J, Bauer PM, Giordano FJ, Sessa WC (2007a) Reexpression of caveolin-1 in endothelium rescues the vascular, cardiac, and pulmonary defects in global caveolin-1 knockout mice. J Exp Med 204:2373–2382

Murata T, Lin MI, Stan RV, Bauer PM, Yu J, Sessa WC (2007b) Genetic evidence supporting caveolae microdomain regulation of calcium entry in endothelial cells. J Biol Chem 282:16631–16643

Nesin V, Tsiokas L (2014) Trpc1. Handb Exp Pharmacol 222:15–51

Ng LC, McCormack MD, Airey JA, Singer CA, Keller PS, Shen XM, Hume JR (2009) TRPC1 and STIM1 mediate capacitative Ca2+ entry in mouse pulmonary arterial smooth muscle cells. J Physiol 587:2429–2442

Niemeyer BA (2016) Changing calcium: CRAC channel (STIM and Orai) expression, splicing, and posttranslational modifiers. Am J Physiol Cell Physiol 310:C701–C709

Niggli V (2001) Structural properties of lipid-binding sites in cytoskeletal proteins. Trends Biochem Sci 26:604–611

Ong HL, Ambudkar IS (2012) Role of lipid rafts in the regulation of store-operated Ca2+ channels. In: Levitan I, Barrantes F (eds) Cholesterol regulation of ion channels and receptors. Wiley, Hoboken, NJ, pp 69–90

Ong HL, Ambudkar IS (2015) Molecular determinants of TRPC1 regulation within ER-PM junctions. Cell Calcium 58:376–386

Ong HL, Cheng KT, Liu X, Bandyopadhyay BC, Paria BC, Soboloff J, Pani B, Gwack Y, Srikanth S, Singh BB, Gill DL, Ambudkar IS (2007) Dynamic assembly of TRPC1-STIM1-Orai1 ternary complex is involved in store-operated calcium influx. Evidence for similarities in store-operated and calcium release-activated calcium channel components. J Biol Chem 282:9105–9116

Ong HL, Jang SI, Ambudkar IS (2012) Distinct contributions of Orai1 and TRPC1 to agonist-induced [Ca2+]i signals determine specificity of Ca2+-dependent gene expression. PLoS One 7:e47146

Ong EC, Nesin V, Long CL, Bai CX, Guz JL, Ivanov IP, Abramowitz J, Birnbaumer L, Humphrey MB, Tsiokas L (2013) A TRPC1-dependent pathway regulates osteoclast formation and function. J Biol Chem 288:22219–22231

Ong HL, de Souza LB, Cheng KT, Ambudkar IS (2014) Physiological functions and regulation of TRPC channels. Handb Exp Pharmacol 223:1005–1034

Ong HL, de Souza LB, Zheng C, Cheng KT, Liu X, Goldsmith CM, Feske S, Ambudkar IS (2015) STIM2 enhances receptor-stimulated Ca2+ signaling by promoting recruitment of STIM1 to the endoplasmic reticulum-plasma membrane junctions. Sci Signal 8:ra3

Ong HL, de Souza LB, Ambudkar IS (2016) Role of TRPC channels in store-operated calcium entry. Adv Exp Med Biol 898:87–109

Pani B, Singh BB (2009) Lipid rafts/caveolae as microdomains of calcium signaling. Cell Calcium 45:625–633

Pani B, Ong HL, Liu X, Rauser K, Ambudkar IS, Singh BB (2008) Lipid rafts determine clustering of STIM1 in endoplasmic reticulum-plasma membrane junctions and regulation of store-operated Ca2+ entry (SOCE). J Biol Chem 283:17333–17340

Pani B, Ong HL, Brazer SC, Liu X, Rauser K, Singh BB, Ambudkar IS (2009) Activation of TRPC1 by STIM1 in ER-PM microdomains involves release of the channel from its scaffold caveolin-1. Proc Natl Acad Sci U S A 106:20087–20092

Pani B, Liu X, Bollimuntha S, Cheng KT, Niesman IR, Zheng C, Achen VR, Patel HH, Ambudkar IS, Singh BB (2013) Impairment of TRPC1-STIM1 channel assembly and AQP5 translocation compromise agonist-stimulated fluid secretion in mice lacking caveolin1. J Cell Sci 126:667–675

Parekh AB (2011) Decoding cytosolic Ca2+ oscillations. Trends Biochem Sci 36:78–87

Parekh AB, Putney JW Jr (2005) Store-operated calcium channels. Physiol Rev 85:757–810

Parekh AB, Fleig A, Penner R (1997) The store-operated calcium current I_{CRAC}: nonlinear activation by InsP3 and dissociation from calcium release. Cell 89:973–980

Paria BC, Vogel SM, Ahmmed GU, Alamgir S, Shroff J, Malik AB, Tiruppathi C (2004) Tumor necrosis factor-a-induced TRPC1 expression amplifies store-operated Ca2+ influx and endothelial permeability. Am J Physiol Lung Cell Mol Physiol 287:L1303–L1313

Penner R, Matthews G, Neher E (1988) Regulation of calcium influx by second messengers in rat mast cells. Nature 334:499–504

Poteser M, Leitinger G, Pritz E, Platzer D, Frischauf I, Romanin C, Groschner K (2016) Live-cell imaging of ER-PM contact architecture by a novel TIRFM approach reveals extension of junctions in response to store-operated Ca2+-entry. Sci Rep 6:35656

Prakriya M, Feske S, Gwack Y, Srikanth S, Rao A, Hogan PG (2006) Orai1 is an essential pore subunit of the CRAC channel. Nature 443:230–233

Prinz WA (2014) Bridging the gap: membrane contact sites in signaling, metabolism, and organelle dynamics. J Cell Biol 205:759–769

Quintana A, Rajanikanth V, Farber-Katz S, Gudlur A, Zhang C, Jing J, Zhou Y, Rao A, Hogan PG (2015) TMEM110 regulates the maintenance and remodeling of mammalian ER-plasma membrane junctions competent for STIM-ORAI signaling. Proc Natl Acad Sci U S A 112: E7083–E7092

Rao JN, Platoshyn O, Golovina VA, Liu L, Zou T, Marasa BS, Turner DJ, Yuan JX, Wang JY (2006) TRPC1 functions as a store-operated Ca2+ channel in intestinal epithelial cells and regulates early mucosal restitution after wounding. Am J Physiol Gastrointest Liver Physiol 290:G782–G792

Rao JN, Rathor N, Zhuang R, Zou T, Liu L, Xiao L, Turner DJ, Wang JY (2012) Polyamines regulate intestinal epithelial restitution through TRPC1-mediated Ca2+ signaling by differentially modulating STIM1 and STIM2. Am J Physiol Cell Physiol 303:C308–C317

Rathor N, Chung HK, Wang SR, Wang JY, Turner DJ, Rao JN (2014) Caveolin-1 enhances rapid mucosal restitution by activating TRPC1-mediated Ca2+ signaling. Physiol Rep 2:e1219

Redondo PC, Harper AG, Salido GM, Pariente JA, Sage SO, Rosado JA (2004) A role for SNAP-25 but not VAMPs in store-mediated Ca2+ entry in human platelets. J Physiol 558:99–109

Richards DA (2010) Regulation of exocytic mode in hippocampal neurons by intra-bouton calcium concentration. J Physiol 588:4927–4936

Roos J, DiGregorio PJ, Yeromin AV, Ohlsen K, Lioudyno M, Zhang S, Safrina O, Kozak JA, Wagner SL, Cahalan MD, Velicelebi G, Stauderman KA (2005) STIM1, an essential and conserved component of store-operated Ca2+ channel function. J Cell Biol 169:435–445

Rosado JA, Brownlow SL, Sage SO (2002) Endogenously expressed Trp1 is involved in store-mediated Ca2+ entry by conformational coupling in human platelets. J Biol Chem 277:42157–42163

Sabourin J, Cognard C, Constantin B (2009a) Regulation by scaffolding proteins of canonical transient receptor potential channels in striated muscle. J Muscle Res Cell Motil 30:289–297

Sabourin J, Lamiche C, Vandebrouck A, Magaud C, Rivet J, Cognard C, Bourmeyster N, Constantin B (2009b) Regulation of TRPC1 and TRPC4 cation channels requires an a1-syntrophin-dependent complex in skeletal mouse myotubes. J Biol Chem 284:36248–36261

Selvaraj S, Watt JA, Singh BB (2009) TRPC1 inhibits apoptotic cell degeneration induced by dopaminergic neurotoxin MPTP/MPP+. Cell Calcium 46:209–218

Selvaraj S, Sun Y, Watt JA, Wang S, Lei S, Birnbaumer L, Singh BB (2012) Neurotoxin-induced ER stress in mouse dopaminergic neurons involves downregulation of TRPC1 and inhibition of AKT/mTOR signaling. J Clin Invest 122:1354–1367

Seth M, Zhang ZS, Mao L, Graham V, Burch J, Stiber J, Tsiokas L, Winn M, Abramowitz J, Rockman HA, Birnbaumer L, Rosenberg P (2009) TRPC1 channels are critical for hypertrophic signaling in the heart. Circ Res 105:1023–1030

Shalygin A, Skopin A, Kalinina V, Zimina O, Glushankova L, Mozhayeva GN, Kaznacheyeva E (2015) STIM1 and STIM2 proteins differently regulate endogenous store-operated channels in HEK293 cells. J Biol Chem 290:4717–4727

Sharma S, Quintana A, Findlay GM, Mettlen M, Baust B, Jain M, Nilsson R, Rao A, Hogan PG (2013) An siRNA screen for NFAT activation identifies septins as coordinators of store-operated Ca2+ entry. Nature 499:238–242

Shi J, Birnbaumer L, Large WA, Albert AP (2014) Myristoylated alanine-rich C kinase substrate coordinates native TRPC1 channel activation by phosphatidylinositol 4,5-bisphosphate and protein kinase C in vascular smooth muscle. FASEB J 28:244–255

Shi J, Miralles F, Birnbaumer L, Large WA, Albert AP (2016) Store depletion induces Galphaq-mediated PLCb1 activity to stimulate TRPC1 channels in vascular smooth muscle cells. FASEB J 30:702–715

Shi J, Miralles F, Birnbaumer L, Large WA, Albert AP (2017) Store-operated STIM1 translocation and interaction with TRPC1 at the plasma membrane stimulates PLC activity to induce channel gating in vascular smooth muscle cells. J Physiol 595(4):1039–1058

Shim AH, Tirado-Lee L, Prakriya M (2015) Structural and functional mechanisms of CRAC channel regulation. J Mol Biol 427:77–93

Sours S, Du J, Chu S, Ding M, Zhou XJ, Ma R (2006) Expression of canonical transient receptor potential (TRPC) proteins in human glomerular mesangial cells. Am J Physiol Renal Physiol 290:F1507–F1515

Srikanth S, Jew M, Kim KD, Yee MK, Abramson J, Gwack Y (2012) Junctate is a Ca2+-sensing structural component of Orai1 and stromal interaction molecule 1 (STIM1). Proc Natl Acad Sci U S A 109:8682–8687

Stathopulos PB, Zheng L, Ikura M (2009) Stromal interaction molecule (STIM) 1 and STIM2 calcium sensing regions exhibit distinct unfolding and oligomerization kinetics. J Biol Chem 284:728–732

Stenmark H (2009) Rab GTPases as coordinators of vesicle traffic. Nat Rev Mol Cell Biol 10:513–525

Stiber JA, Zhang ZS, Burch J, JP E, Zhang S, Truskey GA, Seth M, Yamaguchi N, Meissner G, Shah R, Worley PF, Williams RS, Rosenberg PB (2008) Mice lacking Homer 1 exhibit a skeletal myopathy characterized by abnormal transient receptor potential channel activity. Mol Cell Biol 28:2637–2647

Sun Y, Birnbaumer L, Singh BB (2015) TRPC1 regulates calcium-activated chloride channels in salivary gland cells. J Cell Physiol 230:2848–2856

Sundivakkam PC, Kwiatek AM, Sharma TT, Minshall RD, Malik AB, Tiruppathi C (2009) Caveolin-1 scaffold domain interacts with TRPC1 and IP3R3 to regulate Ca2+ store release-induced Ca2+ entry in endothelial cells. Am J Physiol Cell Physiol 296:C403–C413

Taguchi T (2013) Emerging roles of recycling endosomes. J Biochem 153:505–510

Takeshima H, Hoshijima M, Song LS (2015) Ca2+ microdomains organized by junctophilins. Cell Calcium 58:349–356

Tiruppathi C, Ahmmed GU, Vogel SM, Malik AB (2006) Ca2+ signaling, TRP channels, and endothelial permeability. Microcirculation 13:693–708

Treves S, Franzini-Armstrong C, Moccagatta L, Arnoult C, Grasso C, Schrum A, Ducreux S, Zhu MX, Mikoshiba K, Girard T, Smida-Rezgui S, Ronjat M, Zorzato F (2004) Junctate is a key element in calcium entry induced by activation of InsP3 receptors and/or calcium store depletion. J Cell Biol 166:537–548

Vandebrouck A, Sabourin J, Rivet J, Balghi H, Sebille S, Kitzis A, Raymond G, Cognard C, Bourmeyster N, Constantin B (2007) Regulation of capacitative calcium entries by a1-syntrophin: association of TRPC1 with dystrophin complex and the PDZ domain of a1-syntrophin. FASEB J 20:136–138

Vig M, Beck A, Billingsley JM, Lis A, Parvez S, Peinelt C, Koomoa DL, Soboloff J, Gill DL, Fleig A, Kinet JP, Penner R (2006) CRACM1 multimers form the ion-selective pore of the CRAC channel. Curr Biol 16:2073–2079

Wang Y, Deng X, Mancarella S, Hendron E, Eguchi S, Soboloff J, Tang XD, Gill DL (2010) The calcium store sensor, STIM1, reciprocally controls Orai and CaV1.2 channels. Science 330:105–109

Weihuang Y, Chang SJ, Harn HI, Huang HT, Lin HH, Shen MR, Tang MJ, Chiu WT (2015) Mechanosensitive store-operated calcium entry regulates the formation of cell polarity. J Cell Physiol 230:2086–2097

Wes PD, Chevesich J, Jeromin A, Rosenberg C, Stetten S, Montell C (1995) TRPC1, a human homolog of a Drosophila store-operated channel. Proc Natl Acad Sci USA 92:9652–9656

Williams IA, Allen DG (2007) Intracellular calcium handling in ventricular myocytes from mdx mice. Am J Physiol Heart Circ Physiol 292:H846–H855

Worley PF, Zeng W, Huang G, Kim JY, Shin DM, Kim MS, Yuan JP, Kiselyov K, Muallem S (2007) Homer proteins in Ca2+ signaling by excitable and non-excitable cells. Cell Calcium 42:363–371

Wu MM, Buchanan J, Luik RM, Lewis RS (2006) Ca2+ store depletion causes STIM1 to accumulate in ER regions closely associated with the plasma membrane. J Cell Biol 174:803–813

Wu MM, Covington ED, Lewis RS (2014) Single-molecule analysis of diffusion and trapping of STIM1 and Orai1 at endoplasmic reticulum-plasma membrane junctions. Mol Biol Cell 25:3672–3685

Yeh YC, Parekh AB (2015) Distinct structural domains of caveolin-1 independently regulate CRAC channels and Ca2+ microdomain-dependent gene expression. Mol Cell Biol 35:1341–1349

Yin HL, Janmey PA (2003) Phosphoinositide regulation of the actin cytoskeleton. Annu Rev Physiol 65:761–789

Yu F, Sun L, Machaca K (2010) Constitutive recycling of the store-operated Ca2+channel Orai1 and its internalization during meiosis. J Cell Biol 191:523–535

Yuan JP, Kiselyov K, Shin DM, Chen J, Shcheynikov N, Kang SH, Dehoff MH, Schwarz MK, Seeburg PH, Muallem S, Worley PF (2003) Homer binds TRPC family channels and is required for gating of TRPC1 by IP3 receptors. Cell 114:777–789

Yuan JP, Zeng W, Huang GN, Worley PF, Muallem S (2007) STIM1 heteromultimerizes TRPC channels to determine their function as store-operated channels. Nat Cell Biol 9:636–645

Yuan JP, Zeng W, Dorwart MR, Choi YJ, Worley PF, Muallem S (2009) SOAR and the polybasic STIM1 domains gate and regulate Orai channels. Nat Cell Biol 11:337–343

Yuan JP, Lee KP, Hong JH, Muallem S (2012) The closing and opening of TRPC channels by Homer1 and STIM1. Acta Physiol 204:238–247

Zanou N, Shapovalov G, Louis M, Tajeddine N, Gallo C, Van Schoor M, Anguish I, Cao ML, Schakman O, Dietrich A, Lebacq J, Ruegg U, Roulet E, Birnbaumer L, Gailly P (2010) Role of TRPC1 channel in skeletal muscle function. Am J Physiol Cell Physiol 298:C149–C162

Zanou N, Schakman O, Louis P, Ruegg UT, Dietrich A, Birnbaumer L, Gailly P (2012) Trpc1 ion channel modulates phosphatidylinositol 3-kinase/Akt pathway during myoblast differentiation and muscle regeneration. J Biol Chem 287:14524–14534

Zeng W, Yuan JP, Kim MS, Choi YJ, Huang GN, Worley PF, Muallem S (2008) STIM1 gates TRPC channels, but not Orai1, by electrostatic interaction. Mol Cell 32:439–448

Zerial M, McBride H (2001) Rab proteins as membrane organizers. Nat Rev Mol Cell Biol 2:107–117

Zhang X, Trebak M (2014) Transient receptor potential canonical 7: a diacylglycerol-activated non-selective cation channel. Handb Exp Pharmacol 222:189–204

Zhang SL, Yu Y, Roos J, Kozak JA, Deerinck TJ, Ellisman MH, Stauderman KA, Cahalan MD (2005) STIM1 is a Ca2+ sensor that activates CRAC channels and migrates from the Ca2+ store to the plasma membrane. Nature 437:902–905

Zhang SL, Yeromin AV, Zhang XH, Yu Y, Safrina O, Penna A, Roos J, Stauderman KA, Cahalan MD (2006) Genome-wide RNAi screen of Ca2+ influx identifies genes that regulate Ca2+ release-activated Ca2+ channel activity. Proc Natl Acad Sci U S A 103:9357–9362

Zhang ZY, Pan LJ, Zhang ZM (2010) Functional interactions among STIM1, Orai1 and TRPC1 on the activation of SOCs in HL-7702 cells. Amino Acids 39:195–204

Zhang P, Mao AQ, Sun CY, Zhang XD, Pan QX, Yang DT, Jin J, Tang CL, Yang ZY, Yao XQ, Lu XJ, Ma X (2016) Translocation of PKG1a acts on TRPV4-C1 heteromeric channels to inhibit endothelial Ca2+ entry. Acta Pharmacol Sin 37:1199–1207

Zheng L, Stathopulos PB, Li GY, Ikura M (2008) Biophysical characterization of the EF-hand and SAM domain containing Ca2+ sensory region of STIM1 and STIM2. Biochem Biophys Res Commun 369:240–246

Zheng L, Stathopulos PB, Schindl R, Li GY, Romanin C, Ikura M (2011) Auto-inhibitory role of the EF-SAM domain of STIM proteins in store-operated calcium entry. Proc Natl Acad Sci U S A 108:1337–1342

Zholos AV (2014) Trpc5. Handb Exp Pharmacol 222:129–156

Zhu X, Chu PB, Peyton M, Birnbaumer L (1995) Molecular cloning of a widely expressed human homologue for the Drosophila trp gene. FEBS Lett 373:193–198

STIM-TRP Pathways and Microdomain Organization: Auxiliary Proteins of the STIM/Orai Complex

10

Jonathan Pacheco and Luis Vaca

Abstract

The basic paradigm of a mechanism for calcium influx triggered after a reduction on calcium store content implies a sensor of calcium concentration on the endoplasmic reticulum (the stores) and a calcium channel immersed on the plasma membrane. These two basic components are STIM and Orai, the most fundamental and minimal molecular constituents of the store-operated calcium entry mechanism. However, even when minimal components can be reduced to these two proteins, the intricate process involved in approximating two cellular membranes (endoplasmic reticulum, ER and plasma membrane, PM) require the participation of several other components, many of which remain unidentified to this date. Here we review several of the proteins identified as constituents of the so-called store-operated calcium influx complex (SOCIC) and discuss their role in modulating this complex phenomenon.

Keywords

STIM • Orai • SOCE • STIMATE • SARAF • CRACR2A • ER-PM junctions • Junctate • Interaction

J. Pacheco • L. Vaca (✉)
Instituto de Fisiología Celular, Universidad Nacional Autónoma de México, Ciudad Universitaria, México, DF 04510, Mexico
e-mail: epacheco@email.ifc.unam.mx; lvaca@ifc.unam.mx

K. Groschner et al. (eds.), *Store-Operated Ca^{2+} Entry (SOCE) Pathways*, Advances in Experimental Medicine and Biology 993, DOI 10.1007/978-3-319-57732-6_10

10.1 Introduction

The idea of the store-operated calcium entry (SOCE) would have its origins from the early beginnings after the discovery of IP3 and its well-known effect promoting transient calcium (Ca^{2+}) release from the endoplasmic reticulum (ER). The Ca^{2+} release is followed by a sustained Ca^{2+} influx from the extracellular space (Berridge and Irvine 1984; Berridge 1993). Intense research was conducted in order to elucidate the relationship between Ca^{2+} release from ER and Ca^{2+} entry through the plasma membrane (PM). Novel concepts emerged, and new models were proposed (Prakriya and Lewis 2015). However, the molecular identity of the components responsible for sensing luminal Ca^{2+} and allowing Ca^{2+} influx remained undefined for almost two decades. In 2005, using large-scale RNAi-based screening, two groups identified the previously described protein STIM as a key element sensing Ca^{2+} depletion from the stores (Roos et al. 2005; Liou et al. 2005). One year later, Orai or CRACM1 was identified as the pore-forming unit responsible for SOCE in patients with a form of hereditary severe combined immune deficiency (SCID) and through wide-genome RNAi screenings (Feske et al. 2006; Vig et al. 2006).

In mammals, the Orai channel family has three homologs (Orai1–3), all of them highly selective for Ca^{2+} but with distinct kinetics (Lis et al. 2007; Lee et al. 2009). On the other hand, STIM has two homologs (STIM1–2) (Liou et al. 2005). STIM1 and STIM2 differ in their sensitivity to luminal Ca^{2+} (Zheng et al. 2008). STIM2 is a weaker activator of Orai1 when compared to STIM1 (Bird et al. 2009).

STIM1 and Orai1 are enough to reconstitute the current responsible of SOCE, also named I_{CRAC} (Ca^{2+} release-activated Ca^{2+} current). However, shortly after the discovery of STIM and Orai, a plethora of new partners emerged as regulators of STIM1-Orai1 function. The idea of SOCE implies a close apposition of membranes from the endoplasmic reticulum (ER) and the plasma membrane (PM), presumably requiring a variety of bridge molecules to maintain integrity of the so-called ER-PM junctions. In that way, SOCE elicits Ca^{2+} elevations in the vicinity of where the microdomain is recruited, triggering signaling pathways.

In this chapter we discuss the intricate relationships between STIM1 and Orai1 with their associated members, including structural proteins involved in the dynamic regulation of ER-PM junctions. Here we provide a picture of the new and not-so-new regulators of STIM1 and Orai1 during basal and activated states. Finally we mention the proteins associated to SOCE microdomain that are regulated by Ca^{2+}-ions flowing through Orai1.

10.2 Regulators of STIM1/Orai1 Proteins During Store-Filled Conditions

The cytosolic region of STIM1 covers most of 60% of the total protein sequence; here are located domains in charge of activating Orai1 and other domains important to recruit and oligomerize STIM1 molecules during its activation (Soboloff et al.

2012). In consequence there is higher abundance of regulators interacting with this region (Fig. 10.1), such is the case of SARAF (SOCE-associated regulatory factor), which constitutively interacts with STIM1. SARAF is an ER single membrane-spanning protein that controls Ca^{2+} homeostasis, including Ca^{2+} concentration on cytosol and into the ER. During store full conditions, overexpression of SARAF decreases cytosolic and ER Ca^{2+} levels, whereas silencing SARAF increases Ca^{2+} on the cytoplasm and the ER (Palty et al. 2012). The evidence suggests that SARAF prevents spontaneous SOCE activation by inhibiting STIM2 under basal conditions (Palty et al. 2012). SARAF is binding to a short conserved motif containing 4 lysines (384–389) within SOAR (STIM Orai-activating region) domain (Jha et al. 2013), which is the minimal region of STIM1 required to activate Orai1 channels (Yuan et al. 2009). Notwithstanding, the mechanism of SARAF regulation under store full conditions is not well understood.

STIM1 distribution at the ER is highly dynamic, presenting "comet-like" movements when observed by microscopy methods fused to fluorescent proteins (Grigoriev et al. 2008). The responsible of the dynamic movement of STIM1 is the end-binding 1 (EB1) protein. EB1 is a plus-end-tracking protein (+TIP) that regulates the assembling and disassembling of microtubules (Vitre et al. 2008). Thus, EB1 is the link between ER and cytoskeleton. The region of STIM1 that binds to EB1 encompasses residues 642–645 (Thr-Arg-Ile-Pro, or TRIP); this region is a conserved microtubule-interacting domain regulated by phosphorylations (Honnappa et al. 2009). In addition, the phosphorylation of STIM1 at the TRIP motif is regulated by Ca^{2+} store depletion (Pozo-Guisado et al. 2013). A synchronic interplay between Ca^{2+} stores depletion, independently from Ca^{2+} influx from extracellular space, promotes activation of ERK1/2 that in turn phosphorylates STIM1 to disrupt interaction with EB1 (Pozo-Guisado et al. 2013; Sampieri et al. 2009). STIM1 has also been reported to be phosphorylated by CDK1, an important kinase involved during progression of mitosis (Smyth et al. 2009). Previously, there was a strong evidence showing the drastic suppression of SOCE during mitosis and in particular in the course of metaphase state (Preston et al. 1991). The phosphorylation of STIM1 at important serines precludes the cluster formation of STIM1 after store depletion (Smyth et al. 2009). Nevertheless, in *Xenopus laevis*, oocyte phosphorylation of STIM1 has not been detected, arguing to an alternative mechanism to inhibit SOCE during meiosis (Yu et al. 2009).

In contrast to the cytosolic part, the luminal region of STIM1 interacts with ERp57 (Prins et al. 2011), an ER oxidoreductase member of the protein disulfide isomerase family (Turano et al. 2011). ERp57 is involved in the assembly of the major histocompatibility complex (MHC) class I (Lindquist et al. 1998) and in the correct folding of newly synthetized glycoproteins (Molinari and Helenius 1999; Oliver et al. 1997). ERp57 modulates the folding of glycoproteins by means of disulfide bonds between newly synthetized proteins and complexes with calreticulin and calnexin (Oliver et al. 1999). In addition, ERp57-deficient cells show a marked increase in SOCE. ERp57 interacts with STIM1 through cysteines at the amino terminal of STIM1 (C49 and C56). Mutations of both cysteines decrease ERp57-STIM1 interaction and results in the inhibition of Ca^{2+} entry (Prins et al. 2011). Controversially, other reports have shown the opposite effect when C56 is

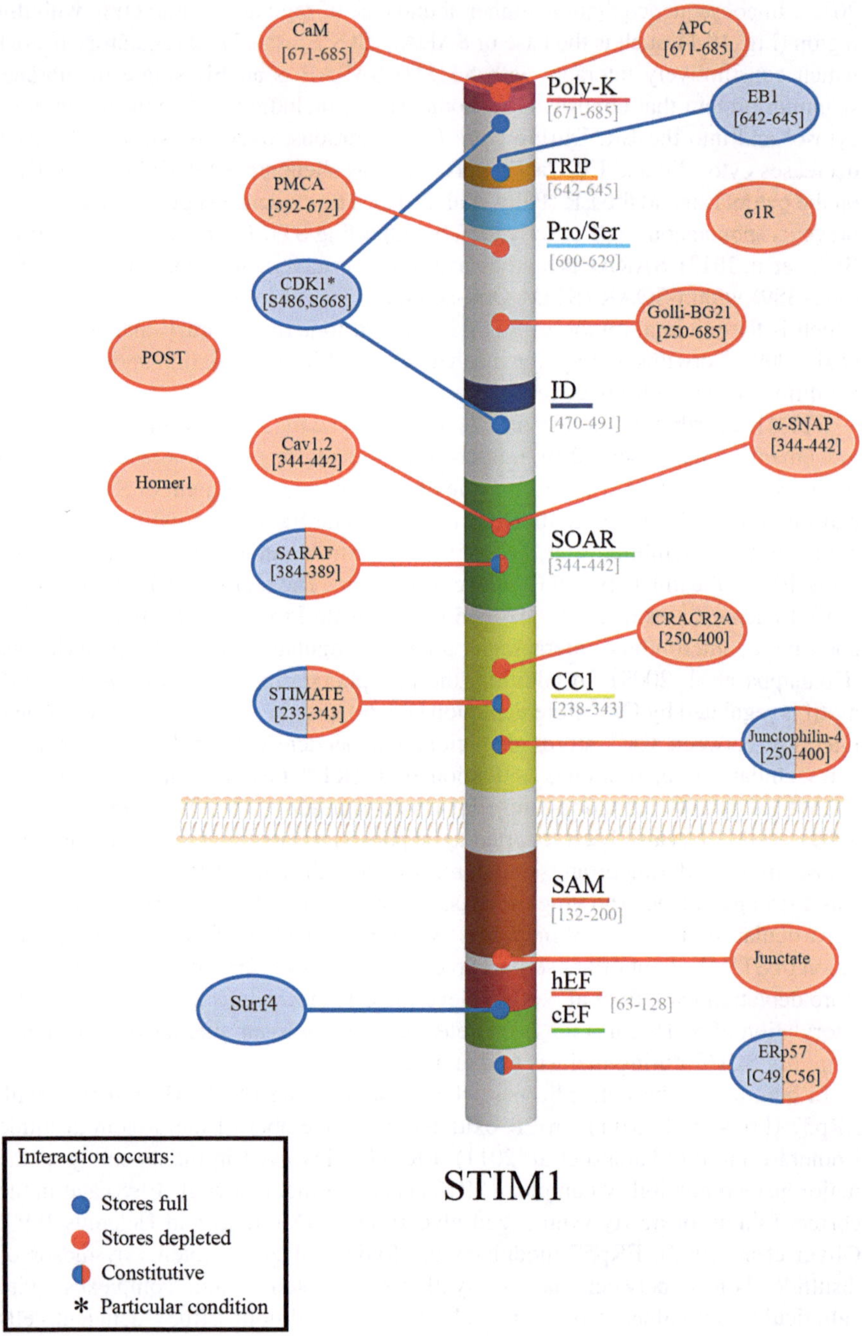

Fig. 10.1 STIM1 associated partners. Topology of STIM1 and the main components that interact with it. *Underlined words* remark the principal domains on STIM1. *Blue circles* show proteins associated to STIM1 under store full conditions, *red circles* show proteins associated to STIM1 under store depletion, and *blue-red circles* show proteins with constitutive interactions. *Asterisks* represent interacting partners under specific condition or under debate (refer to main text)

mutated (Hawkins et al. 2010). It shows that C56 is S-glutathionylated under oxidant stress, which results in a constitutive Ca^{2+} entry independent of intracellular Ca^{2+} stores (Hawkins et al. 2010). This discrepancy could be attributed to different endogenous expressions of antioxidant proteins inherent to different cellular systems (Bogeski et al. 2012). Another luminal protein that interacts with STIM1 is the surfeit locus protein 4 (Surf4) (Fujii et al. 2012), the mammalian orthologue of the yeast cargo receptor Erv29p (Mitrovic et al. 2008). Surf4 deletion produces an increased Ca^{2+} entry as a consequence of a dramatic increment of STIM1clusters. Surf4 interacts with STIM1 by its luminal region. Mutations in STIM1 EF-hand decrease interactions with Surf4, suggesting a differential association of Surf4 to STIM1 active and inactive forms or that the altered relocation of STIM1 at ER-PM junctions impairs the interaction with Surf4 (Fujii et al. 2012).

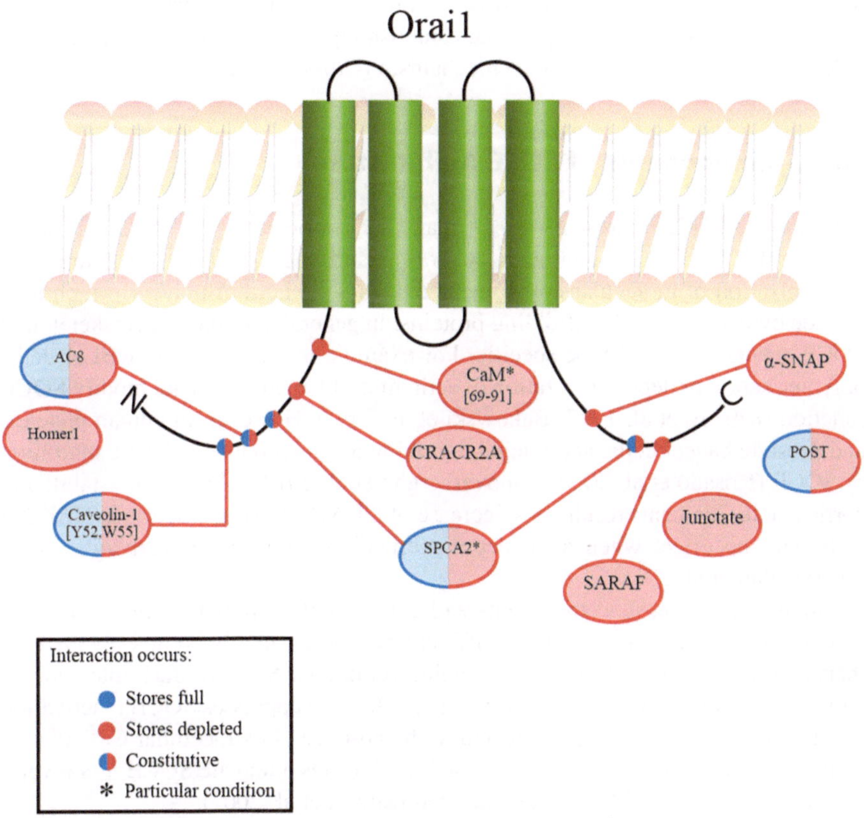

Fig. 10.2 Orai1 associated partners. Topology of the Orai1 channel. *Blue circles* show proteins associated to Orai1 under store full conditions, *red circles* show proteins associated to Orai1 under store depleted and *blue-red circles* proteins with a constitutive interaction. *Asterisks* represent interacting partners under specific condition or under debate (refer to main text). The exact location where junctate interacts with Orai1 remains undefined (Srikanth et al. 2012)

On the other hand, in breast cancer cells, Orai1 is constitutively activated by SPCA2 (secretory pathway Ca^{2+}-ATPase) in a store-independent manner (Fig. 10.2). SPCA2 is a Golgi Ca^{2+} pump expressed primarily in the gastrointestinal tract, rectum, and mammary gland (Vanoevelen et al. 2005). However, SPCA2 promotes activation of Orai1 independently of Golgi stores, STIM1 and SPCA2 ATPase activity. Interaction and activation of Orai1 is achieved through both amino and carboxyl terminus of SPCA2 that in turn interact with N- and C-terminal of Orai1. The binding is sensitive to the mutation L273S (Feng et al. 2010), which previously has been reported to disrupt the coiled-coil domain of the C-terminal from Orai1 (Muik et al. 2008). In addition, SPCA2 presents a putative PDZ-binding domain. Mutation of this region precludes Orai1 activation, suggesting a mechanism dependent on scaffold proteins. Very interestingly, SPCA2 is overexpressed in patients with breast cancer, increasing cell proliferation and tumorigenesis by a Ca^{2+}-dependent mechanism (Feng et al. 2010). But Ca^{2+} influx mediated by SPCA2-Orai1 interaction seems to be a normal physiological process to support milk secretion in mammary epithelia, establishing a new role of Orai1 as a store-independent calcium entry (SICE) mechanism (Cross et al. 2013).

10.3 Components of SOCE Architecture

Structurally, SOCE represents a Ca^{2+} pathway restricted at the close apposition between ER and PM membranes (Carrasco and Meyer 2011). Such dramatic reorganization of molecular components entails a plethora of players, which may include cytoskeletal and scaffolding proteins. In general, the role of cytoskeleton in SOCE seems to be cell type-specific. For example in NIH 3T3 and RBL-1 cells, polymerization or depolymerization of actin microfilaments does not modify SOCE function (Ribeiro et al. 1997; Bakowski et al. 2001). However, in human platelets and prostate cancer cells, actin depolymerizing agents produce a drastic inhibition of SOCE (Rosado et al. 2000; Vanoverberghe et al. 2012). Meanwhile, stabilizing cortical actin filaments results in a decrease of STIM1/Orai1 interaction in HEK293 cells with no effect when actin depolymerization occurs (Vanoverberghe et al. 2012; Galan et al. 2011).

On the other side, short treatments with microtubule-depolymerizing agents like colchicine or nocodazole induce different responses in Ca^{2+} entry evoked by thapsigargin. In HEK293 cells, colchicine reinforces STIM1/Orai1 interactions. Otherwise, microtubular stabilization with paclitaxel suppresses SOCE if activation of STIM1/Orai1 complex is conducted in the absence of extracellular Ca^{2+} (Galan et al. 2011). But, under extracellular Ca^{2+}, both drugs attenuate SOCE in a reversible manner when STIM1 is overexpressed (Smyth et al. 2007).

The molecular mechanism of cytoskeleton regulation over STIM1/Orai1 is poorly understood. Notwithstanding, EB1 protein is referred as the link between STIM1 and microtubules (Grigoriev et al. 2008). In addition, EB1 presents several +TIP partners that interact by a short polypeptide motif, Ser-x-Ile-Pro (SxIP), including STIM1 (Honnappa et al. 2009; Galjart 2010). Another important partner of EB1 is the protein adenomatous polyposis coli (APC), a tumor suppressor protein first identified from patients with a condition of polyps in the intestine (Groden et al. 1991). Patients with mutations in APC show a high prevalence of colon cancer (Nathke 2004). Our group showed that STIM1 interacts with APC after STIM1-EB1 dissociation. Furthermore, silencing APC with RNAi produces a diminished Ca^{2+} entry evoked by thapsigargin (TG). In addition, reduced expression of APC seems to disrupt tubular structures of STIM1 during store depletion. The binding region of STIM1 to interact with APC is located at carboxyl terminus (Asanov et al. 2013). The same region in STIM1 has been reported to bind phosphoinositides from the PM (Liou et al. 2007; Walsh et al. 2010a).

Currently, intense research is conducted to elucidate the molecular mechanism underlying the relocation of STIM1 to ER-PM junctions. Most attempts have resulted in the identification of proteins important to stabilize the connection between ER and the PM. Such is the case of extended synaptotagmin 1 (E-SYT1), which has shown to tether ER membrane at sites of the PM enriched in PIP2 (Maleth et al. 2014). Furthermore, an interesting feedback promoted by Ca^{2+} signals triggers translocation of E-SYT1 to ER-PM junctions, which promotes the replenishment of PIP2 in a process dependent of Nir2, a phosphatidylinositol-transfer protein (Chang et al. 2013). Other structural components of ER-PM junctions are septin proteins, which play a key role targeting STIM1 at ER-PM junctions (Sharma et al. 2013). Coordinated interplay between components of ER-PM junctions are platforms to gather molecules involved in Orai1 regulation (Maleth et al. 2014). Other proteins reported to recruit STIM1 to ER-PM junctions include junctate, which enhances junctional regions for clustering of STIM1 and Orai1. Most interestingly, STIM1 lacking the polybasic region forms puncta when junctate is overexpressed, suggesting a recruitment of STIM1 independent of phosphoinositides (Srikanth et al. 2012). Moreover, junctophilin-4, a junctional protein expressed in excitable and immune cells, interacts with STIM1 and junctate. Silencing junctophilin-4 alters ER-Ca^{2+} homeostasis and impairs clustering of STIM1 (Woo et al. 2016).

Recently, a novel integral approach to dissect proteome of intact ER-PM junctions came to the identification of the new adaptor family member STIMATE (STIM-activating enhancer, encoded by TMEM110 gene) (Jing et al. 2015). STIMATE is an ER-resident protein with multiple spanning transmembrane regions and a polybasic domain at carboxyl terminus. Knockout of STIMATE results in the inhibition of STIM1 puncta formation. At the same time STIMATE has been shown to be an important regulator in the maintenance of cortical ER (Quintana et al. 2015; Jing et al. 2015). Furthermore, STIMATE promotes conformational changes of STIM1 important to fully activate Orai1 channels (Jing et al. 2015), and this finding will be tackled on the next section. Proteome analysis that led to the identification of STIMATE also gave a complex picture of proteins expressed at ER-PM junctions, resulting in the identification of more than 70 different proteins (Jing et al. 2015).

ER-PM junctions are the platforms that recruit SOCE components; moreover caveolae also function as a center for trafficking and organization of Ca^{2+} signals

(Laude and Prior 2004; Pani and Singh 2009). Caveolae are "flask-shaped" structures of the PM, typically measuring from 25 to 150 nm (Pike 2004). Caveolae are enriched in Caveolin1 (Cav-1) protein, an integral membrane protein essential for caveolar biogenesis (Cheng and Nichols 2016). Cav-1 together with dynamin has shown to control endocytosis of Orai1 during meiosis. In fact Orai1 presents a well-recognized caveolin interaction domain at its amino terminus (Yu et al. 2010). There is enough evidence showing that STIM1-Orai1 complex is established and stabilized on caveolae through Cav-1 (Pani and Singh 2009; Yeh and Parekh 2015). In airway smooth muscle, overexpression of Cav-1 increases Orai1 expression accompanied by an increase of Ca^{2+} entry (Sathish et al. 2012). Meanwhile silencing Cav-1 reduces STIM1 puncta (Pani et al. 2009).

Cytosolic proteins important in the allocation of STIM1-Orai1 complex to ER-PM junctions include Homer1, a family member of adaptor proteins highly expressed in the brain and upregulated by synaptic activity (Yuan et al. 2012). Homer1 interacts with Orai1 and STIM1 in a Ca^{2+}-dependent manner. However, interaction of STIM1 with Homer1 is well defined, but Orai1-Homer1 interaction seems to be mediated by STIM1. Competitive addition of the peptide PPKKFR, which precludes Homer1 function, produces a suppression of SOCE in human platelets (Jardin et al. 2012). At the same time, Homer1 regulates functional interactions between STIM1 and Cav1.2, a voltage-gated calcium channel (VGCC) sensitive to nifedipine (Dionisio et al. 2015). Another important cytosolic protein that stabilizes SOCE junctional structures is α-SNAP (α-soluble NSF attachment protein). α-SNAP together with NSF (*N*-ethylmaleimide-sensitive factor) mediates disassembling and recycling the SNARE complex, a basic mechanism for intracellular vesicle fusion events (Winter et al. 2009; Chang et al. 2012). Silencing α-SNAP results in diminished Ca^{2+} entry independently of NSF. α-SNAP bolsters STIM1-Orai1 interaction at ER-PM junctions by direct binding with the SOAR domain in STIM1 and with the C-terminal of Orai1 (Miao et al. 2013) (Figs. 10.1 and 10.2).

Given the above picture, it can be suggested a mechanism for junctional location of STIM1 that is dependent of anionic lipids, in particular PIP2. Adaptor proteins may be anchored to the PM via lipid targeting domains. This pathway includes E-SYT1 that interacts with the PM through its conserved C2 domain and cytosolic Ca^{2+} (Giordano et al. 2013). Septins and STIMATE interact with phosphoinositides via polybasic domain present also in STIM1 (Zhang et al. 1999; Bertin et al. 2010; Jing et al. 2015). On the other hand, of particular importance in excitable cells, junctophilin-4 assembled with junctate at ER-PM junctions through repeated MORN motifs (membrane occupation and recognition nexus) recognize and bind phospholipids (Takeshima et al. 2015). Finally, translocation of STIM1 is phosphoinositide-independent when Orai1 is overexpressed, suggesting a stronger interaction between Orai1 and STIM1 (Walsh et al. 2010a).

Table 10.1 Components of the store-operated calcium influx complex (SOCIC)

Accessory protein	Interaction with Orai1 or STIM1	Role in SOCE	Identification method	Cellular localization	Cell type reported	Reference
α-SNAP	Orai1 and STIM1	Promotes clustering of STIM1/Orai1	Pulldown and co-immunoprecipitation	Cytoplasm at ER-MP junctions	Kc cells, HEK293 and Jurkat T-cells	Miao et al. (2013)
Adenylyl cyclase 8 (AC8)	Orai1	AC8 is regulated by SOCE	FRET, peptide array, pulldown	PM (lipid rafts)	HEK293	Willoughby et al. (2012)
Adenomatous polyposis coli (APC)	STIM1	Help to recruit STIM1 to the plasma membrane	Colocalization, co-immunoprecipitation	Cytoplasm	HEK293	Asanov et al. (2013)
Caveolin-1 (Cav1)	Orai1	Silencing Cav1 decreases SOCE; otherwise overexpression of Cav1 increases SOCE	Co-immunoprecipitation	PM and endocytic pathway	Xenopus laevis oocytes, RBL-1 AND HEK293	Yu et al. (2010); Yeh and Parekh (2015)
Calmodulin (CaM)	Orai1 and STIM1	Promotes calcium-dependent inactivation of Orai	Pull-down and co-immunoprecipitation for Orai1 Isothermal titration calorimetry for STIM1	Cytoplasm	HEK293	Liu et al. (2012); Mullins et al. (2009); Frischauf et al. (2008)
CDK1	STIM1	Inhibits clustering of STIM1	Western blot and functional assays with mutagenesis	Cytoplasm	HEK293, HELA	Smyth et al. (2009)
CRACR2A	Orai1 and STIM1	Promotes clustering of STIM1/Orai1	Affinity protein purification, pulldown, and co-immunoprecipitation	Cytoplasm	HeLa, HEK293, and T-cells	Srikanth et al. (2010)

(continued)

Table 10.1 (continued)

Accessory protein	Interaction with Orai1 or STIM1	Role in SOCE	Identification method	Cellular localization	Cell type reported	Reference
EB1	STIM1	Regulates dynamic movement of STIM1	Pull-down, FRET, and co-immunoprecipitation	Microtubules	HEK293, HeLa, and MRC5-SV	Grigoriev et al. (2008); Pozo-Guisado et al. (2013); Sampieri et al. (2009)
ERp57	STIM1	Regulates STIM1	SPR and FRET	Lumen of ER	MEF cells	Prins et al. (2011)
E-Syt1	Not reported	Tethered protein recruits SARAF to promote SCDI	Not reported	ER-PM junctions	HEK293	Maleth et al. (2014)
Golli-BG21	STIM1	Overexpression of golli reduces SOCE	Pull-down and BiFC	Lipid rafts associated	HeLa	Walsh et al. (2010b)
Junctate	Orai1 and STIM1	Promotes recruitment of STIM1 independently of store depletion	Affinity protein purification, pulldown, and co-immunoprecipitation	ER-PM junctions	HeLa and Jurkat T-cells	Srikanth et al. (2012)
Junctophilin-4	STIM1	Promotes recruitment of STIM1 in complex with junctate	Pulldown and co-immunoprecipitation	ER-PM junctions	T-cells	Woo et al. (2016)
Homer1	Orai1 and STIM1	Allows full STIM1-Orai1 interaction	Co-immunoprecipitation	Cytoplasm	Human platelets	Jardin et al. (2012)
POST	Orai1 and STIM1	None. Induces decrease in PMCA activity	Affinity protein purification and co-immunoprecipitation	ER-PM junctions	HEK293	Krapivinsky et al. (2011)
Polycystin-1 cleavage product (P100)	STIM1	Inhibits SOCE by suppression of STIM1 translocation	Co-immunoprecipitation	Membrane of ER	Xenopus laevis oocytes and CHO cells	Woodward et al. (2010)

PMCA	STIM1	None. STIM1 reduces PMCA activity	Co-immunoprecipitation	PM at immunological synapse	Jurkat T-cells	Ritchie et al. (2012)
SARAF	STIM1 and Orai1	Regulates SOCE by means of deoligomerization of STIM1	Co-immunoprecipitation, FRET, and proximity ligation assay	Membrane of ER	HEK293, MEG-01, and HeLa	Palty et al. (2012); Albarran et al. (2016)
Septins	Not reported	Microdomain organization for STIM1-Orai-1	Not reported	PM at enriched PIP2 domains	HeLa and Jurkat T-cells	Sharma et al. (2013)
Sigma 1 receptor	STIM1	Suppression of SOCE	Co-immunoprecipitation and AFM	ER-PM junctions	HEK293 and CHO cells	Srivats et al. (2016)
SPCA2	Orai1	Store-independent activation of Orai1	Co-immunoprecipitation and pulldown	Membrane of Golgi apparatus	Human mammary tumor cells and HEK293	Feng et al. (2010)
STIMATE	STIM1	Promotes conformational changes of STIM1	BiFC, FRET, co-immunoprecipitation, pulldown, and SPR	Membrane of ER	HEK293, HeLa, and COS-7	Jing et al. (2015)
Stanniocalcin 2 (STC2)	STIM1	Negative modulator of SOCE	Co-immunoprecipitation	Membrane of ER	MEF and COS cells	Zeiger et al. (2011)
Surf4	STIM1	Regulates STIM1 oligomerization	Co-immunoprecipitation	Lumen of ER	DT40B-cells	Fujii et al. (2012)
VGCC ($Ca_v1.2$)	STIM1	None, STIM1 suppresses $Ca_v1.2$ activity	Co-immunoprecipitation	MP	(A7r5) Smooth muscle cells and HEK293	Wang et al. (2010)

10.4 Regulators of STIM1/Orai1 Proteins During Depleted Ca^{2+} Conditions

The activated STIM1-Orai1 complex gathers a plethora of components that we have called store-operated calcium influx complex (SOCIC) (Vaca 2010) (Table 10.1). Here, important regulators control Orai1 activation through Ca^{2+}-dependent mechanisms. These modulatory effects avoid excessive and lethal Ca^{2+} influx to protect the cell. In fact, calcium itself promotes inactivation of Orai1 channels by two modes: one is developed several minutes after Orai1 activation and is called slow Ca^{+2}-dependent inactivation (SCDI) (Parekh 1998). The other mode occurs within a range of milliseconds after Orai1 is gated and is called fast Ca^{+2}-dependent inactivation (FCDI) (Hoth and Penner 1993; Fierro and Parekh 1999).

The most prominent component to mediate fast Ca^{2+}-dependent inactivation of Orai1 is calmodulin (CaM) (Mullins et al. 2009). In silico analysis (Frischauf et al. 2008), structural (Liu et al. 2012), biochemical, and physicochemical data (Mullins et al. 2009; Liu et al. 2012) support the evidence of a CaM-binding domain in Orai1 N-terminus, specifically in the region covered from amino acids 68 to 91 (Mullins et al. 2009; Liu et al. 2012). CaM-Orai1 functional interaction requires Ca^{2+} to occur (Mullins et al. 2009; Liu et al. 2012). Notwithstanding, recently Lewis' group began to reconsider the CaM-Orai1 model. Based on the crystal structure of dOrai (from *Drosophila*) (Hou et al. 2012), Lewis' group showed a steric clash of CaM with the hexameric Orai pore. By means of dominant-negative CaM mutants and through elegant studies with mutants on Orai1 N-terminal, they showed a FCDI mechanism independent of CaM. Discrepancies with previous works could be attributed to the ratio of STIM1-Orai1 expression (Mullins et al. 2016). In contrast to Orai1, STIM1 and STIM2 bind to CaM through their polybasic tail (Bauer et al. 2008; Miederer et al. 2015). However, physiological or molecular consequences of CaM-STIM1 interaction remain largely unexplored.

On the other hand, SARAF mediates a SOCE reduction through induction of SCDI (slow calcium-dependent inactivation) of Orai1 channels (Palty et al. 2012; Jha et al. 2013). The intricate relationship of SARAF with SOCE components is elusive after store depletion, where SARAF translocates at ER-PM junctions and promotes STIM1 deoligomerization (Palty et al. 2012). SARAF presents a dynamic interaction with both, STIM1 and Orai1, on a synchronous fashion. First, SARAF-Orai1 interaction is enhanced after 30 s of TG application; on the other side, STIM1-SARAF interaction diminishes. Thereafter association of SARAF-Orai1 decreases and STIM1 strengthens its interaction with SARAF. The C-terminal of Orai1 is the region that associates to SARAF (Albarran et al. 2016). In counterpart, the mapped region to SARAF interaction with STIM1 involves its cytosolic region. Meanwhile SARAF associates to STIM1 by direct binding to SOAR (at lysine-rich region) and a downstream region from the SOAR domain (which covers from 490 to 521 residues in STIM1) (Jha et al. 2013). However, functional interactions between SARAF and STIM1 require the polybasic domain of STIM1 (Maleth et al. 2014), inasmuch as the polybasic domain of STIM1 is involved in interactions with phosphoinositides (Korzeniowski et al. 2009; Walsh et al. 2010a).

Thus, STIM1, Orai1, and SARAF interaction is established at ER-PM junctions enriched in PIP2. Here, ancillary proteins (E-Syt1 and septins) form a macromolecular complex playing a critical role in SCDI of Orai1 (Maleth et al. 2014).

Another component involved in Orai1 regulation is CRACR2A. CRACR2A stabilize STIM1 and Orai1 complex at ER-PM junctions. CRACR2A is present in the cytosol and displays two putative EF-hand domains. Interaction between CRACR2A and the STIM1-Orai1 complex is favored in the absence of Ca^{2+}. Silencing CRACR2A produces a decrease of Ca^{2+} entry, whereas silencing of the homologous CRACR2B, which presents a 36% identity to CRACR2A, produces a mild SOCE inhibition. Otherwise, mutation of EF-hand domains from CRACR2A promotes clustering of STIM1 independently of Ca^{2+} store content (Srikanth et al. 2010). In a similar way, mutation on the EF-hand domain of the protein junctate also increases the probability of STIM1 clustering into the PM-proximal area independently of store depletion. Junctate interacts with both STIM1 and Orai1. Interaction between junctate and STIM1 takes place between their luminal regions. Meanwhile interaction with Orai1 occurs via its cytoplasmic region. Junctate is an ER single pass membrane protein with the EF-hand domain located at its luminal portion (Srikanth et al. 2012). These mechanisms denote the fine regulation of the STIM1-Orai1 complex by Ca^{2+}, not only by Ca^{2+} in the vicinity of Orai1 intracellular pore but also by Ca^{2+} elevations arising from other regions of the cell and even by the extracellular Ca^{2+} (Zweifach and Lewis 1996; Frischauf et al. 2015).

In the previous section, we mentioned that STIMATE is important for STIM1 puncta formation (Jing et al. 2015). But clustering of STIM1 is a previous step needed for Orai1 activation. Before channel activation, STIM1 must release Ca^{2+} from its intraluminal EF-hand domain. The unbinding of Ca^{2+} triggers complex conformational changes on the cytosolic portion STIM1. At rest, STIM1 avoids spurious activation of Orai1, by preventing SOAR exposure through direct self-association with the first coiled-coil region (CC1, residues 233–343) (Muik et al. 2011; Yu et al. 2013). Molecular switching is performed with assistance of STIMATE, which promotes the conformational change of STIM1 by direct association with the CC1 region of STIM1 (Fig. 10.1). Thus, the role of STIMATE is to modulate the interaction between two domains in STIM1: the auto-inhibitory domain (located at CC1) and SOAR (Jing et al. 2015).

The cytosolic region of STIM1 is a target of intense regulation (Fig. 10.1). An example includes the isoform BG21 of golli (gene in the oligodendrocyte lineage), an alternative spliced form of myelin basic protein (MBP) expressed ubiquitously and involved in myelination of nerves. BG21 disrupts also T-cell activation when overexpressed (Feng et al. 2004). Golli-BG21 is present a PM distribution thanks to a myristoylation on its N-terminus (Feng et al. 2006). Golli-BG21 interacts with the cytosolic region of STIM1 after Ca^{2+} store depletion, and its overexpression produces inhibition of SOCE, which is reversible upon overexpression of STIM1 (Walsh et al. 2010b). These data suggest that golli-BG21 competes for STIM1 C-terminus at the region in close proximity to the PM, most likely on lipid rafts (Levental et al. 2010).

Polycystin proteins (PCs) have important functions in calcium homeostasis, and they are members of the TRP superfamily of channels (Nilius and Owsianik 2011). Dysregulation or mutation on them can lead to cyst formation in the mammalian nephron, resulting in the autosomal dominant polycystic kidney disease (ADPKD) (Mangolini et al. 2016). Polycystin-1 (PC1) suffers several cleavages to trigger important signaling pathways. One PC1 cleavage product (P100) interacts with STIM1 to suppress oligomerization of STIM1 and in consequence disruption of Orai1 activation (Woodward et al. 2010).

Stanniocalcins are a family of proteins involved in Ca^{2+} and phosphate regulation in fish gill. In mammals, the role of stanniocalcins remains poorly understood. Some evidence remarks its role in the development of several cancers, but Stanniocalcin 2 (STC2) knockout animals have no phenotype (Zeiger et al. 2011; Volland et al. 2009; Na et al. 2015; Wu et al. 2015). STC2 has been reported to interact and diminish STIM1 function in a clustering-independent manner. Knockout cells for STC2 present an increase in SOCE. In addition, STC2 also promotes protection from oxidative stress through STIM1 inhibition avoiding Ca^{+2} overloading (Zeiger et al. 2011). The sigma 1 receptors (σ1R), also involved in certain types of cancer and with a wide range of ligands including alcohol, cocaine, methamphetamine, and antidepressive drugs, have been recently reported to regulate SOCE (Maurice and Su 2009). The σ1R are expressed in ER membranes, nuclear envelope, and the PM. These receptors transfer Ca^{2+} from the ER to mitochondria. The σ1R inhibits SOCE by interacting with STIM1. Such interaction depends on the activation of the receptor with its ligand. In counterpart, antagonists of σ1R prevent its function as negative modulator of SOCE. The σ1R interacts with the STIM1-Orai1 complex after store depletion, but the formation of this macromolecular complex requires STIM1 (Srivats et al. 2016).

10.5 Proteins Regulated by SOCE

The most well-known protein to be regulated by calcium from SOCE is the nuclear factor of activated T-cells (NFAT), an important transcription factor needed to induce cytokine production, T-cell differentiation, and T-cell tolerance among other functions (Macian 2005; Srikanth and Gwack 2013). The NFAT family has five members (from 1 to 5), all of them regulated by Ca^{2+}, with the exception of NFAT5, which is activated by osmotic stress (Macian 2005). The first step in NFAT activation requires calmodulin-binding Ca^{2+} coming from Orai1. This event promotes the formation of the complex calmodulin-calcineurin. The last step includes a phosphatase, which in turn dephosphorylates the cytoplasmic NFAT. The NFAT dephosphorylation triggers its translocation to the nucleus where it induces transcription of its target genes (Srikanth and Gwack 2013). There is strong evidence showing that the majority of the Ca^{2+} required for NFAT activation arises from SOCE (Kar et al. 2011). For example NFAT1 is activated through Ca^{2+} flowing through Orai1 channels, whereas NFAT4 requires a Ca^{2+} influx and a nuclear Ca^{2+} rise component (Kar and Parekh 2015). The robustness, frequency, and speed of Ca^{2+} signaling coming through Orai1 are essential to trigger specific NFAT activation (Dolmetsch et al. 1998; Kar and Parekh 2015). In that regard, Orai1 channels must remain active for this process. In order to do that, T-cells coordinate multiple players to maintain low subplasmalemmal Ca^{2+} concentration restricted at the immunological synapse (IS) as to prevent Ca^{2+}-dependent inactivation of Orai1. Mitochondria re-localize at the vicinity of the IS to buffer Ca^{2+} (Quintana et al. 2011). Meanwhile STIM1 interacts with the plasma membrane-Ca^{2+} ATPase (PMCA) and reduces the rate of PMCA Ca^{2+} extrusion increasing global Ca^{2+} in the cytosol (Ritchie et al. 2012; Quintana et al. 2011). This effect is promoted by the protein POST (partner of STIM1) that possesses 10 transmembrane regions and is present at the ER and PM. Although POST does not affect Orai1 or STIM1 function, it induces a reduction in the activity of the PMCA. POST interacts with STIM1 and Orai1 after store depletion, but in addition also presents interactions with PMCA, SERCA2, the Na/K-ATPase, as well as nuclear transporters (importins and exportins) (Krapivinsky et al. 2011). The relationship between nuclear transporters and SOCE components is not well understood. However, it has been proposed a role of POST as a scaffolding protein to facilitate interactions between STIM1 and SERCA or Na/K-ATPase after store depletion (Krapivinsky et al. 2011).

The relevance of Ca^{2+} microdomains arising from SOCE has been recognized by activation of the adenylyl cyclase type 8 (AC8) (Willoughby et al. 2012). It is a well-known protein in charge of cyclic adenosine monophosphate (cAMP) production, a very important second messenger (Willoughby and Cooper 2007). Adenylyl cyclases present 10 isoforms (9 associated to the PM and 1 soluble). From these, AC1 and AC8 are synergistically activated by Ca^{2+} and via G-protein coupled receptors. On other hand, AC5 and AC6 are inhibited by Ca^{2+} (Cooper 2015). AC8 shows a high selectivity of Ca^{2+} from SOCE over other sources of Ca^{2+}. AC8 discern from different sources of Ca^{2+} through a constitutive interaction with the amino terminal region of Orai1 (Willoughby et al. 2012) (Fig. 10.2). These studies bring an interconnection between the two major second messenger signaling pathways, Ca^{2+} and cAMP.

SOCE has also direct effects on microtubules. For example in mast cells, Ca^{2+} influx coming from SOCE produces microtubule protrusions, a key event during degranulation. Knockout cells for STIM1 completely suppress microtubule protrusions and display modified microtubule dynamics (Hajkova et al. 2011).

Finally, the ways in which SOCE components regulate other proteins are not limited through Orai1 Ca^{2+} pathways. An example is the regulation of voltage-gated calcium channels (Cav1.2) by STIM1. In addition to its role activating Orai channels, STIM1 also mediates suppression of Cav1.2 channel activity in a coordinated way that occurs after ER Ca^{2+} depletion. The accumulation of STIM1 at ER-PM junctions promotes the exposure of SOAR, which in turn activates Orai1 but also induces inhibition of Cav1.2 via direct interactions (Wang et al. 2010) (Fig. 10.1).

10.6 Summary

Since the discovery of STIM and Orai proteins as key components of the store-operated calcium entry, intense research have been conducted, and today the view of this calcium influx mechanism is better understood. Our compression has not been limited to understanding STIM-Orai functional interactions. In addition, a new series of regulators, scaffolding proteins, and even proteins regulated by STIM1-Orai1 complex have been identified. The challenge remains not only at identifying the detailed molecular mechanism of each individual protein but also to put together the big picture of the intricate relationships between all the protein members of the store-operated Ca^{2+} influx complex (SOCIC). Finally, understanding the mechanisms participating in the selective up- or downregulation of any of the multiple SOCIC components to modify responses in particular physiological or pathological events becomes extremely relevant.

References

Albarran L, Lopez JJ, Amor NB, Martin-Cano FE, Berna-Erro A, Smani T, Salido GM, Rosado JA (2016) Dynamic interaction of SARAF with STIM1 and Orai1 to modulate store-operated calcium entry. Sci Rep 6:24452

Asanov A, Sherry R, Sampieri A, Vaca L (2013) A relay mechanism between EB1 and APC facilitate STIM1 puncta assembly at endoplasmic reticulum-plasma membrane junctions. Cell Calcium 54(3):246–256

Bakowski D, Glitsch MD, Parekh AB (2001) An examination of the secretion-like coupling model for the activation of the Ca2+ release-activated Ca2+ current I(CRAC) in RBL-1 cells. J Physiol 532(Pt 1):55–71

Bauer MC, O'Connell D, Cahill DJ, Linse S (2008) Calmodulin binding to the polybasic C-termini of STIM proteins involved in store-operated calcium entry. Biochemistry 47(23):6089–6091

Berridge MJ (1993) Inositol trisphosphate and calcium signalling. Nature 361(6410):315–325

Berridge MJ, Irvine RF (1984) Inositol trisphosphate, a novel second messenger in cellular signal transduction. Nature 312(5992):315–321

Bertin A, McMurray MA, Thai L, Garcia G 3rd, Votin V, Grob P, Allyn T, Thorner J, Nogales E (2010) Phosphatidylinositol-4,5-bisphosphate promotes budding yeast septin filament assembly and organization. J Mol Biol 404(4):711–731

Bird GS, Hwang SY, Smyth JT, Fukushima M, Boyles RR, Putney JW Jr (2009) STIM1 is a calcium sensor specialized for digital signaling. Curr Biol 19(20):1724–1729

Bogeski I, Kilch T, Niemeyer BA (2012) ROS and SOCE: recent advances and controversies in the regulation of STIM and Orai. J Physiol 590(17):4193–4200

Carrasco S, Meyer T (2011) STIM proteins and the endoplasmic reticulum-plasma membrane junctions. Annu Rev Biochem 80:973–1000

Chang LF, Chen S, Liu CC, Pan X, Jiang J, Bai XC, Xie X, Wang HW, Sui SF (2012) Structural characterization of full-length NSF and 20S particles. Nat Struct Mol Biol 19(3):268–275

Chang CL, Hsieh TS, Yang TT, Rothberg KG, Azizoglu DB, Volk E, Liao JC, Liou J (2013) Feedback regulation of receptor-induced Ca2+ signaling mediated by E-Syt1 and Nir2 at endoplasmic reticulum-plasma membrane junctions. Cell Rep 5(3):813–825

Cheng JP, Nichols BJ (2016) Caveolae: one function or many? Trends Cell Biol 26(3):177–189

Cooper DM (2015) Store-operated Ca(2)(+)-entry and adenylyl cyclase. Cell Calcium 58 (4):368–375

Cross BM, Hack A, Reinhardt TA, Rao R (2013) SPCA2 regulates Orai1 trafficking and store independent Ca2+ entry in a model of lactation. PLoS One 8(6):e67348

Dionisio N, Smani T, Woodard GE, Castellano A, Salido GM, Rosado JA (2015) Homer proteins mediate the interaction between STIM1 and Cav1.2 channels. Biochim Biophys Acta 1853 (5):1145–1153

Dolmetsch RE, Xu K, Lewis RS (1998) Calcium oscillations increase the efficiency and specificity of gene expression. Nature 392(6679):933–936

Feng JM, Fernandes AO, Campagnoni CW, Hu YH, Campagnoni AT (2004) The golli-myelin basic protein negatively regulates signal transduction in T lymphocytes. J Neuroimmunol 152 (1–2):57–66

Feng JM, YK H, Xie LH, Colwell CS, Shao XM, Sun XP, Chen B, Tang H, Campagnoni AT (2006) Golli protein negatively regulates store depletion-induced calcium influx in T cells. Immunity 24(6):717–727

Feng M, Grice DM, Faddy HM, Nguyen N, Leitch S, Wang Y, Muend S, Kenny PA, Sukumar S, Roberts-Thomson SJ, Monteith GR, Rao R (2010) Store-independent activation of Orai1 by SPCA2 in mammary tumors. Cell 143(1):84–98

Feske S, Gwack Y, Prakriya M, Srikanth S, Puppel SH, Tanasa B, Hogan PG, Lewis RS, Daly M, Rao A (2006) A mutation in Orai1 causes immune deficiency by abrogating CRAC channel function. Nature 441(7090):179–185

Fierro L, Parekh AB (1999) Fast calcium-dependent inactivation of calcium release-activated calcium current (CRAC) in RBL-1 cells. J Membr Biol 168(1):9–17

Frischauf I, Schindl R, Derler I, Bergsmann J, Fahrner M, Romanin C (2008) The STIM/Orai coupling machinery. Channels 2(4):261–268

Frischauf I, Zayats V, Deix M, Hochreiter A, Jardin I, Muik M, Lackner B, Svobodova B, Pammer T, Litvinukova M, Sridhar AA, Derler I, Bogeski I, Romanin C, Ettrich RH, Schindl R (2015) A calcium-accumulating region, CAR, in the channel Orai1 enhances Ca2+ permeation and SOCE-induced gene transcription. Sci Signal 8(408):ra131

Fujii Y, Shiota M, Ohkawa Y, Baba A, Wanibuchi H, Kinashi T, Kurosaki T, Baba Y (2012) Surf4 modulates STIM1-dependent calcium entry. Biochem Biophys Res Commun 422(4):615–620

Galan C, Dionisio N, Smani T, Salido GM, Rosado JA (2011) The cytoskeleton plays a modulatory role in the association between STIM1 and the Ca2+ channel subunits Orai1 and TRPC1. Biochem Pharmacol 82(4):400–410

Galjart N (2010) Plus-end-tracking proteins and their interactions at microtubule ends. Curr Biol 20(12):R528–R537

Giordano F, Saheki Y, Idevall-Hagren O, Colombo SF, Pirruccello M, Milosevic I, Gracheva EO, Bagriantsev SN, Borgese N, De Camilli P (2013) PI(4,5)P(2)-dependent and Ca(2+)-regulated ER-PM interactions mediated by the extended synaptotagmins. Cell 153(7):1494–1509

Grigoriev I, Gouveia SM, van der Vaart B, Demmers J, Smyth JT, Honnappa S, Splinter D, Steinmetz MO, Putney JW Jr, Hoogenraad CC, Akhmanova A (2008) STIM1 is a MT-plus-end-tracking protein involved in remodeling of the ER. Curr Biol 18(3):177–182

Groden J, Thliveris A, Samowitz W, Carlson M, Gelbert L, Albertsen H, Joslyn G, Stevens J, Spirio L, Robertson M et al (1991) Identification and characterization of the familial adenomatous polyposis coli gene. Cell 66(3):589–600

Hajkova Z, Bugajev V, Draberova E, Vinopal S, Draberova L, Janacek J, Draber P, Draber P (2011) STIM1-directed reorganization of microtubules in activated mast cells. J Immunol 186 (2):913–923

Hawkins BJ, Irrinki KM, Mallilankaraman K, Lien YC, Wang Y, Bhanumathy CD, Subbiah R, Ritchie MF, Soboloff J, Baba Y, Kurosaki T, Joseph SK, Gill DL, Madesh M (2010) S-glutathionylation activates STIM1 and alters mitochondrial homeostasis. J Cell Biol 190 (3):391–405

Honnappa S, Gouveia SM, Weisbrich A, Damberger FF, Bhavesh NS, Jawhari H, Grigoriev I, van Rijssel FJ, Buey RM, Lawera A, Jelesarov I, Winkler FK, Wuthrich K, Akhmanova A,

Steinmetz MO (2009) An EB1-binding motif acts as a microtubule tip localization signal. Cell 138(2):366–376

Hoth M, Penner R (1993) Calcium release-activated calcium current in rat mast cells. J Physiol 465:359–386

Hou X, Pedi L, Diver MM, Long SB (2012) Crystal structure of the calcium release-activated calcium channel Orai. Science 338(6112):1308–1313

Jardin I, Albarran L, Bermejo N, Salido GM, Rosado JA (2012) Homers regulate calcium entry and aggregation in human platelets: a role for Homers in the association between STIM1 and Orai1. Biochem J 445(1):29–38

Jha A, Ahuja M, Maleth J, Moreno CM, Yuan JP, Kim MS, Muallem S (2013) The STIM1 CTID domain determines access of SARAF to SOAR to regulate Orai1 channel function. J Cell Biol 202(1):71–79

Jing J, He L, Sun A, Quintana A, Ding Y, Ma G, Tan P, Liang X, Zheng X, Chen L, Shi X, Zhang SL, Zhong L, Huang Y, Dong MQ, Walker CL, Hogan PG, Wang Y, Zhou Y (2015) Proteomic mapping of ER-PM junctions identifies STIMATE as a regulator of Ca influx. Nat Cell Biol 17 (10):1339–1347

Kar P, Parekh AB (2015) Distinct spatial Ca2+ signatures selectively activate different NFAT transcription factor isoforms. Mol Cell 58(2):232–243

Kar P, Nelson C, Parekh AB (2011) Selective activation of the transcription factor NFAT1 by calcium microdomains near Ca2+ release-activated Ca2+ (CRAC) channels. J Biol Chem 286 (17):14795–14803

Korzeniowski MK, Popovic MA, Szentpetery Z, Varnai P, Stojilkovic SS, Balla T (2009) Dependence of STIM1/Orai1-mediated calcium entry on plasma membrane phosphoinositides. J Biol Chem 284(31):21027–21035

Krapivinsky G, Krapivinsky L, Stotz SC, Manasian Y, Clapham DE (2011) POST, partner of stromal interaction molecule 1 (STIM1), targets STIM1 to multiple transporters. Proc Natl Acad Sci U S A 108(48):19234–19239

Laude AJ, Prior IA (2004) Plasma membrane microdomains: organization, function and trafficking. Mol Membr Biol 21(3):193–205

Lee KP, Yuan JP, Zeng W, So I, Worley PF, Muallem S (2009) Molecular determinants of fast Ca2+-dependent inactivation and gating of the Orai channels. Proc Natl Acad Sci U S A 106 (34):14687–14692

Levental I, Grzybek M, Simons K (2010) Greasing their way: lipid modifications determine protein association with membrane rafts. Biochemistry 49(30):6305–6316

Lindquist JA, Jensen ON, Mann M, Hammerling GJ (1998) ER-60, a chaperone with thiol-dependent reductase activity involved in MHC class I assembly. EMBO J 17(8):2186–2195

Liou J, Kim ML, Heo WD, Jones JT, Myers JW, Ferrell JE, Meyer T (2005) STIM is a Ca2+ sensor essential for Ca2+-store-depletion-triggered Ca2+ influx. Curr Biol 15(13):1235–1241

Liou J, Fivaz M, Inoue T, Meyer T (2007) Live-cell imaging reveals sequential oligomerization and local plasma membrane targeting of stromal interaction molecule 1 after Ca2+ store depletion. Proc Natl Acad Sci U S A 104(22):9301–9306

Lis A, Peinelt C, Beck A, Parvez S, Monteilh-Zoller M, Fleig A, Penner R (2007) CRACM1, CRACM2, and CRACM3 are store-operated Ca2+ channels with distinct functional properties. Curr Biol 17(9):794–800

Liu Y, Zheng X, Mueller GA, Sobhany M, DeRose EF, Zhang Y, London RE, Birnbaumer L (2012) Crystal structure of calmodulin binding domain of orai1 in complex with Ca2+ calmodulin displays a unique binding mode. J Biol Chem 287(51):43030–43041

Macian F (2005) NFAT proteins: key regulators of T-cell development and function. Nat Rev Immunol 5(6):472–484

Maleth J, Choi S, Muallem S, Ahuja M (2014) Translocation between PI(4,5)P2-poor and PI(4,5)P2-rich microdomains during store depletion determines STIM1 conformation and Orai1 gating. Nat Commun 5:5843

Mangolini A, de Stephanis L, Aguiari G (2016) Role of calcium in polycystic kidney disease: from signaling to pathology. World J Nephrol 5(1):76–83

Maurice T, Su TP (2009) The pharmacology of sigma-1 receptors. Pharmacol Ther 124 (2):195–206

Miao Y, Miner C, Zhang L, Hanson PI, Dani A, Vig M (2013) An essential and NSF independent role for alpha-SNAP in store-operated calcium entry. elife 2:e00802

Miederer AM, Alansary D, Schwar G, Lee PH, Jung M, Helms V, Niemeyer BA (2015) A STIM2 splice variant negatively regulates store-operated calcium entry. Nat Commun 6:6899

Mitrovic S, Ben-Tekaya H, Koegler E, Gruenberg J, Hauri HP (2008) The cargo receptors Surf4, endoplasmic reticulum-Golgi intermediate compartment (ERGIC)-53, and p25 are required to maintain the architecture of ERGIC and Golgi. Mol Biol Cell 19(5):1976–1990

Molinari M, Helenius A (1999) Glycoproteins form mixed disulphides with oxidoreductases during folding in living cells. Nature 402(6757):90–93

Muik M, Frischauf I, Derler I, Fahrner M, Bergsmann J, Eder P, Schindl R, Hesch C, Polzinger B, Fritsch R, Kahr H, Madl J, Gruber H, Groschner K, Romanin C (2008) Dynamic coupling of the putative coiled-coil domain of ORAI1 with STIM1 mediates ORAI1 channel activation. J Biol Chem 283(12):8014–8022

Muik M, Fahrner M, Schindl R, Stathopulos P, Frischauf I, Derler I, Plenk P, Lackner B, Groschner K, Ikura M, Romanin C (2011) STIM1 couples to ORAI1 via an intramolecular transition into an extended conformation. EMBO J 30(9):1678–1689

Mullins FM, Park CY, Dolmetsch RE, Lewis RS (2009) STIM1 and calmodulin interact with Orai1 to induce Ca2+-dependent inactivation of CRAC channels. Proc Natl Acad Sci U S A 106(36):15495–15500

Mullins FM, Yen M, Lewis RS (2016) Orai1 pore residues control CRAC channel inactivation independently of calmodulin. J Gen Physiol 147(2):137–152

Na SS, Aldonza MB, Sung HJ, Kim YI, Son YS, Cho S, Cho JY (2015) Stanniocalcin-2 (STC2): a potential lung cancer biomarker promotes lung cancer metastasis and progression. Biochim Biophys Acta 1854(6):668–676

Nathke IS (2004) The adenomatous polyposis coli protein: the Achilles heel of the gut epithelium. Annu Rev Cell Dev Biol 20:337–366

Nilius B, Owsianik G (2011) The transient receptor potential family of ion channels. Genome Biol 12(3):218

Oliver JD, van der Wal FJ, Bulleid NJ, High S (1997) Interaction of the thiol-dependent reductase ERp57 with nascent glycoproteins. Science 275(5296):86–88

Oliver JD, Roderick HL, Llewellyn DH, High S (1999) ERp57 functions as a subunit of specific complexes formed with the ER lectins calreticulin and calnexin. Mol Biol Cell 10 (8):2573–2582

Palty R, Raveh A, Kaminsky I, Meller R, Reuveny E (2012) SARAF inactivates the store operated calcium entry machinery to prevent excess calcium refilling. Cell 149(2):425–438

Pani B, Singh BB (2009) Lipid rafts/caveolae as microdomains of calcium signaling. Cell Calcium 45(6):625–633

Pani B, Ong HL, Brazer SC, Liu X, Rauser K, Singh BB, Ambudkar IS (2009) Activation of TRPC1 by STIM1 in ER-PM microdomains involves release of the channel from its scaffold caveolin-1. Proc Natl Acad Sci U S A 106(47):20087–20092

Parekh AB (1998) Slow feedback inhibition of calcium release-activated calcium current by calcium entry. J Biol Chem 273(24):14925–14932

Pike LJ (2004) Lipid rafts: heterogeneity on the high seas. Biochem J 378(Pt 2):281–292

Pozo-Guisado E, Casas-Rua V, Tomas-Martin P, Lopez-Guerrero AM, Alvarez-Barrientos A, Martin-Romero FJ (2013) Phosphorylation of STIM1 at ERK1/2 target sites regulates interaction with the microtubule plus-end binding protein EB1. J Cell Sci 126(Pt 14):3170–3180

Prakriya M, Lewis RS (2015) Store-operated calcium channels. Physiol Rev 95(4):1383–1436

Preston SF, Sha'afi RI, Berlin RD (1991) Regulation of Ca2+ influx during mitosis: Ca2+ influx and depletion of intracellular Ca2+ stores are coupled in interphase but not mitosis. Cell Regul 2(11):915–925

Prins D, Groenendyk J, Touret N, Michalak M (2011) Modulation of STIM1 and capacitative Ca2 + entry by the endoplasmic reticulum luminal oxidoreductase ERp57. EMBO Rep 12 (11):1182–1188

Quintana A, Pasche M, Junker C, Al-Ansary D, Rieger H, Kummerow C, Nunez L, Villalobos C, Meraner P, Becherer U, Rettig J, Niemeyer BA, Hoth M (2011) Calcium microdomains at the immunological synapse: how ORAI channels, mitochondria and calcium pumps generate local calcium signals for efficient T-cell activation. EMBO J 30(19):3895–3912

Quintana A, Rajanikanth V, Farber-Katz S, Gudlur A, Zhang C, Jing J, Zhou Y, Rao A, Hogan PG (2015) TMEM110 regulates the maintenance and remodeling of mammalian ER-plasma membrane junctions competent for STIM-ORAI signaling. Proc Natl Acad Sci U S A 112 (51):E7083–E7092

Ribeiro CM, Reece J, Putney JW Jr (1997) Role of the cytoskeleton in calcium signaling in NIH 3T3 cells. An intact cytoskeleton is required for agonist-induced [Ca2+]i signaling, but not for capacitative calcium entry. J Biol Chem 272(42):26555–26561

Ritchie MF, Samakai E, Soboloff J (2012) STIM1 is required for attenuation of PMCA-mediated Ca2+ clearance during T-cell activation. EMBO J 31(5):1123–1133

Roos J, DiGregorio PJ, Yeromin AV, Ohlsen K, Lioudyno M, Zhang S, Safrina O, Kozak JA, Wagner SL, Cahalan MD, Velicelebi G, Stauderman KA (2005) STIM1, an essential and conserved component of store-operated Ca2+ channel function. J Cell Biol 169(3):435–445

Rosado JA, Jenner S, Sage SO (2000) A role for the actin cytoskeleton in the initiation and maintenance of store-mediated calcium entry in human platelets. Evidence for conformational coupling. J Biol Chem 275(11):7527–7533

Sampieri A, Zepeda A, Asanov A, Vaca L (2009) Visualizing the store-operated channel complex assembly in real time: identification of SERCA2 as a new member. Cell Calcium 45 (5):439–446

Sathish V, Abcejo AJ, Thompson MA, Sieck GC, Prakash YS, Pabelick CM (2012) Caveolin-1 regulation of store-operated Ca(2+) influx in human airway smooth muscle. Eur Respir J 40 (2):470–478

Sharma S, Quintana A, Findlay GM, Mettlen M, Baust B, Jain M, Nilsson R, Rao A, Hogan PG (2013) An siRNA screen for NFAT activation identifies septins as coordinators of store-operated Ca2+ entry. Nature 499(7457):238–242

Smyth JT, DeHaven WI, Bird GS, Putney JW Jr (2007) Role of the microtubule cytoskeleton in the function of the store-operated Ca2+ channel activator STIM1. J Cell Sci 120(Pt 21):3762–3771

Smyth JT, Petranka JG, Boyles RR, DeHaven WI, Fukushima M, Johnson KL, Williams JG, Putney JW Jr (2009) Phosphorylation of STIM1 underlies suppression of store-operated calcium entry during mitosis. Nat Cell Biol 11(12):1465–1472

Soboloff J, Rothberg BS, Madesh M, Gill DL (2012) STIM proteins: dynamic calcium signal transducers. Nat Rev Mol Cell Biol 13(9):549–565

Srikanth S, Gwack Y (2013) Orai1-NFAT signalling pathway triggered by T cell receptor stimulation. Mol Cell 35(3):182–194

Srikanth S, Jung HJ, Kim KD, Souda P, Whitelegge J, Gwack Y (2010) A novel EF-hand protein, CRACR2A, is a cytosolic Ca2+ sensor that stabilizes CRAC channels in T cells. Nat Cell Biol 12(5):436–446

Srikanth S, Jew M, Kim KD, Yee MK, Abramson J, Gwack Y (2012) Junctate is a Ca2+-sensing structural component of Orai1 and stromal interaction molecule 1 (STIM1). Proc Natl Acad Sci U S A 109(22):8682–8687

Srivats S, Balasuriya D, Pasche M, Vistal G, Edwardson JM, Taylor CW, Murrell-Lagnado RD (2016) Sigma1 receptors inhibit store-operated Ca2+ entry by attenuating coupling of STIM1 to Orai1. J Cell Biol 213(1):65–79

Takeshima H, Hoshijima M, Song LS (2015) Ca(2)(+) microdomains organized by junctophilins. Cell Calcium 58(4):349–356

Turano C, Gaucci E, Grillo C, Chichiarelli S (2011) ERp57/GRP58: a protein with multiple functions. Cell Mol Biol Lett 16(4):539–563

Vaca L (2010) SOCIC: the store-operated calcium influx complex. Cell Calcium 47(3):199–209

Vanoevelen J, Dode L, Van Baelen K, Fairclough RJ, Missiaen L, Raeymaekers L, Wuytack F (2005) The secretory pathway Ca2+/Mn2+-ATPase 2 is a Golgi-localized pump with high affinity for Ca2+ ions. J Biol Chem 280(24):22800–22808

Vanoverberghe K, Lehen'kyi V, Thebault S, Raphael M, Vanden Abeele F, Slomianny C, Mariot P, Prevarskaya N (2012) Cytoskeleton reorganization as an alternative mechanism of store-operated calcium entry control in neuroendocrine-differentiated cells. PLoS One 7(9): e45615

Vig M, Peinelt C, Beck A, Koomoa DL, Rabah D, Koblan-Huberson M, Kraft S, Turner H, Fleig A, Penner R, Kinet JP (2006) CRACM1 is a plasma membrane protein essential for store-operated Ca2+ entry. Science 312(5777):1220–1223

Vitre B, Coquelle FM, Heichette C, Garnier C, Chretien D, Arnal I (2008) EB1 regulates microtubule dynamics and tubulin sheet closure in vitro. Nat Cell Biol 10(4):415–421

Volland S, Kugler W, Schweigerer L, Wilting J, Becker J (2009) Stanniocalcin 2 promotes invasion and is associated with metastatic stages in neuroblastoma. Int J Cancer 125 (9):2049–2057

Walsh CM, Chvanov M, Haynes LP, Petersen OH, Tepikin AV, Burgoyne RD (2010a) Role of phosphoinositides in STIM1 dynamics and store-operated calcium entry. Biochem J 425 (1):159–168

Walsh CM, Doherty MK, Tepikin AV, Burgoyne RD (2010b) Evidence for an interaction between Golli and STIM1 in store-operated calcium entry. Biochem J 430(3):453–460

Wang Y, Deng X, Mancarella S, Hendron E, Eguchi S, Soboloff J, Tang XD, Gill DL (2010) The calcium store sensor, STIM1, reciprocally controls Orai and CaV1.2 channels. Science 330 (6000):105–109

Willoughby D, Cooper DM (2007) Organization and Ca2+ regulation of adenylyl cyclases in cAMP microdomains. Physiol Rev 87(3):965–1010

Willoughby D, Everett KL, Halls ML, Pacheco J, Skroblin P, Vaca L, Klussmann E, Cooper DM (2012) Direct binding between Orai1 and AC8 mediates dynamic interplay between Ca2+ and cAMP signaling. Sci Signal 5(219):ra29

Winter U, Chen X, Fasshauer D (2009) A conserved membrane attachment site in alpha-SNAP facilitates N-ethylmaleimide-sensitive factor (NSF)-driven SNARE complex disassembly. J Biol Chem 284(46):31817–31826

Woo JS, Srikanth S, Nishi M, Ping P, Takeshima H, Gwack Y (2016) Junctophilin-4, a component of the endoplasmic reticulum-plasma membrane junctions, regulates Ca2+ dynamics in T cells. Proc Natl Acad Sci U S A 113(10):2762–2767

Woodward OM, Li Y, Yu S, Greenwell P, Wodarczyk C, Boletta A, Guggino WB, Qian F (2010) Identification of a polycystin-1 cleavage product, P100, that regulates store operated Ca entry through interactions with STIM1. PLoS One 5(8):e12305

Wu J, Lai M, Shao C, Wang J, Wei JJ (2015) STC2 overexpression mediated by HMGA2 is a biomarker for aggressiveness of high-grade serous ovarian cancer. Oncol Rep 34 (3):1494–1502

Yeh YC, Parekh AB (2015) Distinct structural domains of caveolin-1 independently regulate Ca2+ release-activated Ca2+ channels and Ca2+ microdomain-dependent gene expression. Mol Cell Biol 35(8):1341–1349

Yu F, Sun L, Machaca K (2009) Orai1 internalization and STIM1 clustering inhibition modulate SOCE inactivation during meiosis. Proc Natl Acad Sci U S A 106(41):17401–17406

Yu F, Sun L, Machaca K (2010) Constitutive recycling of the store-operated Ca2+ channel Orai1 and its internalization during meiosis. J Cell Biol 191(3):523–535

Yu F, Sun L, Hubrack S, Selvaraj S, Machaca K (2013) Intramolecular shielding maintains the ER Ca(2)(+) sensor STIM1 in an inactive conformation. J Cell Sci 126(Pt 11):2401–2410
Yuan JP, Zeng W, Dorwart MR, Choi YJ, Worley PF, Muallem S (2009) SOAR and the polybasic STIM1 domains gate and regulate Orai channels. Nat Cell Biol 11(3):337–343
Yuan JP, Lee KP, Hong JH, Muallem S (2012) The closing and opening of TRPC channels by Homer1 and STIM1. Acta Physiol (Oxf) 204(2):238–247
Zeiger W, Ito D, Swetlik C, Oh-hora M, Villereal ML, Thinakaran G (2011) Stanniocalcin 2 is a negative modulator of store-operated calcium entry. Mol Cell Biol 31(18):3710–3722
Zhang J, Kong C, Xie H, McPherson PS, Grinstein S, Trimble WS (1999) Phosphatidylinositol polyphosphate binding to the mammalian septin H5 is modulated by GTP. Curr Biol 9 (24):1458–1467
Zheng L, Stathopulos PB, Li GY, Ikura M (2008) Biophysical characterization of the EF-hand and SAM domain containing Ca2+ sensory region of STIM1 and STIM2. Biochem Biophys Res Commun 369(1):240–246
Zweifach A, Lewis RS (1996) Calcium-dependent potentiation of store-operated calcium channels in T lymphocytes. J Gen Physiol 107(5):597–610

Part II

SOCE: Crosstalk Between Organelle and Cellular Compartments

Introduction 11

Alexei V. Tepikin

Abstract

In the title of this part of the book, the tail is wagging not just in a single dog but multiple dogs; in other words, a single process SOCE (tail) somehow involves a cross talk of (wagging) large and powerful organelle and cellular compartments (dogs). So how is this possible? Is this really necessary? Is the title actually appropriate?

SOCE is a rather special process, it allows efficient signaling based on a ubiquitous second messenger (Ca^{2+}) in multiple cell and tissue types, it has specific signaling modality (i.e., some downstream reactions depend specifically on SOCE and not just on global Ca^{2+} increase), it is vital for the normal functioning of multiple types of cells and tissues, and when misregulated it induces important pathological processes. The reader hopefully agree that such an important "tail" is more appropriate for a kangaroo than for a Chihuahua and that it has awesome wagging capacity.

Keywords

Store-operated Ca^{2+} entry • ER-PM junctions • Membrane contact sites • Organelle contact sites • Ca^{2+} signaling • Ca^{2+} influx

11.1 SOCE, Organelles, and This Part of the Book

Another justification for the title of this part is that SOCE cannot occur without direct contact and interaction of two cellular organelles—the endoplasmic reticulum (ER) and the plasma membrane (PM). The proteins responsible for activating

A.V. Tepikin (✉)
The Physiological Laboratory, Department of Cellular and Molecular Physiology, Institute of Translational Medicine, University of Liverpool, Crown Street, Liverpool L69 3BX, UK
e-mail: A.Tepikin@liverpool.ac.uk

K. Groschner et al. (eds.), *Store-Operated Ca^{2+} Entry (SOCE) Pathways*, Advances in Experimental Medicine and Biology 993, DOI 10.1007/978-3-319-57732-6_11

SOCE are localized in the ER membrane, while channel-forming proteins (e.g., Orai1) are located in the plasma membrane (reviewed in Carrasco and Meyer 2011; Hogan et al. 2010; Putney 2007). The direct interaction between the two relatively small proteins, which are essential for SOCE, is only possible in the regions of close contacts between the ER and PM termed ER-PM junctions (reviewed in Carrasco and Meyer 2011; Hogan et al. 2010). The organelle interaction is therefore prerequisite for SOCE, and ER-PM junctions are the platforms for this form of Ca^{2+} entry. A number of processes related to formation, stabilization, and dissolution of the ER-PM junctions have been identified during the recent few years (reviewed in Carrasco and Meyer 2011; Okeke et al. 2016; Prinz 2014). The term "cross talk" (between the ER and PM) is very appropriate to summarize these dynamic interactions of the cellular organelles. During the recent decade, a number of proteins involved in such cross talk have been discovered (reviewed in Carrasco and Meyer 2011; Okeke et al. 2016; Prinz 2014; Reinisch and De Camilli 2016). This part of the book will discuss ER-PM junctions, their role in SOCE, and their composition as well as the cross talk of these cellular organelles leading to the formation of the junctions (see Chaps. 12, 14, and 15).

Long before the molecular players responsible for SOCE have been identified, the importance of the third organelle—the mitochondrion—for SOCE has been established. In a number of cell types, mitochondria have a major modulatory effect on SOCE (e.g., Frieden et al. 2005; Glitsch et al. 2002; Hoth et al. 1997; Malli et al. 2003). The relationship between SOCE and mitochondria is a component of a more general stimulus-metabolism coupling process. Inhibitors of mitochondrial functions, such as inhibitors of the mitochondrial electron transport chain, suppress SOCE (e.g., Barrow et al. 2008). Recent studies specified the role of the mitochondrial calcium uniporter (MCU) in SOCE regulation (e.g., Samanta et al. 2014). The progress in this research subfield will be undoubtedly accelerated by the recent identification of the molecular mechanisms responsible for mitochondrial Ca^{2+} entry (Baughman et al. 2011; De Stefani et al. 2011; reviewed in De Stefani et al. 2015; Kamer and Mootha, 2015] and the identification of the molecular mechanism of SOCE (Feske et al. 2006; Liou et al. 2005; Luik et al. 2006; Prakriya et al. 2006; Roos et al. 2005; Vig et al. 2006; reviewed in Hogan et al. 2010; Putney 2007). It is also likely that mitochondria influence SOCE by locally changing the levels of reactive oxygen species (reviewed in Bogeski et al. 2012) and possibly by changing the levels of ATP in the proximity to the SOCE channels (Montalvo et al. 2006). The relationships between SOCE and mitochondria form an important research avenue in both bioenergetics and Ca^{2+} signaling; it is covered in this part of the book by laboratories that have made important contributions to this research area (Chaps. 14 and 16).

The role of TRPC channels in SOCE is a hotly debated topic and a rapidly expanding research subfield (reviewed in Ambudkar et al. 2016; Lee et al. 2010). Properties of TRPC channels and their roles in linking Ca^{2+} signaling in intracellular organelles to Ca^{2+} entry via the plasma membrane are discussed in Chap. 13.

A new type of organellar junctions and a new type of SOCE were recently identified by Demaurex's laboratory (Nunes et al. 2012). In this case, the junctions are formed between the ER strands and phagosomal membrane. In these junctions, phagosomes serve as sources of Ca^{2+} for periphagosomal Ca^{2+} microdomains. This novel type of organellar junctions and the corresponding form of Ca^{2+} signaling are discussed in Chap. 15. Contact sites between the ER and endosomes/lysosomes are also discussed in this chapter. The discussion includes the role of these structures in lipid and Ca^{2+} signaling.

This part of the book also contains the three important case studies dealing with the specifics of SOCE and organelle arrangements/cross talk in smooth muscle cells (Chap. 17), endothelial cells (Chap. 18), and cardiac myocytes (Chap. 19). These examples are important illustrations of how these relationships between the SOCE and cellular organelles are formed in primary cells and of the consequences of SOCE/organelle cross talk for cell physiology and pathophysiology. Chapter 17 also contains elegant quantitative modeling of Ca^{2+} signals and processes occurring in organellar junctions.

The studies of SOCE had a major effect on the research field investigating organellar interactions and particularly junctions between cellular organelles. The interface of the two research fields is the focus of this part of the book.

References

Ambudkar IS, de Souza LB, Ong HL (2016) TRPC1, Orai1, and STIM1 in SOCE: friends in tight spaces. Cell Calcium. doi:10.1016/j.ceca.2016.12.009

Barrow SL, Voronina SG, da Silva XG, Chvanov MA, Longbottom RE, Gerasimenko OV, Petersen OH, Rutter GA, Tepikin AV (2008) ATP depletion inhibits Ca2+ release, influx and extrusion in pancreatic acinar cells but not pathological Ca2+ responses induced by bile. Pflugers Arch 455:1025–1039

Baughman JM, Perocchi F, Girgis HS, Plovanich M, Belcher-Timme CA, Sancak Y, Bao XR, Strittmatter L, Goldberger O, Bogorad RL, Koteliansky V, Mootha VK (2011) Integrative genomics identifies MCU as an essential component of the mitochondrial calcium uniporter. Nature 476:341–345

Bogeski I, Kilch T, Niemeyer BA (2012) ROS and SOCE: recent advances and controversies in the regulation of STIM and Orai. J Physiol 590:4193–4200

Carrasco S, Meyer T (2011) STIM proteins and the endoplasmic reticulum-plasma membrane junctions. Annu Rev Biochem 80:973–1000

De Stefani D, Raffaello A, Teardo E, Szabo I, Rizzuto R (2011) A forty-kilodalton protein of the inner membrane is the mitochondrial calcium uniporter. Nature 476:336–340

De Stefani D, Patron M, Rizzuto R (2015) Structure and function of the mitochondrial calcium uniporter complex. Biochim Biophys Acta 1853:2006–2011

Feske S, Gwack Y, Prakriya M, Srikanth S, Puppel SH, Tanasa B, Hogan PG, Lewis RS, Daly M, Rao A (2006) A mutation in Orai1 causes immune deficiency by abrogating CRAC channel function. Nature 441:179–185

Frieden M, Arnaudeau S, Castelbou C, Demaurex N (2005) Subplasmalemmal mitochondria modulate the activity of plasma membrane Ca2+-ATPases. J Biol Chem 280:43198–43208

Glitsch MD, Bakowski D, Parekh AB (2002) Store-operated Ca2+ entry depends on mitochondrial Ca2+ uptake. EMBO J 21:6744–6754

Hogan PG, Lewis RS, Rao A (2010) Molecular basis of calcium signaling in lymphocytes: STIM and ORAI. Annu Rev Immunol 28:491–533

Hoth M, Fanger CM, Lewis RS (1997) Mitochondrial regulation of store-operated calcium signaling in T lymphocytes. J Cell Biol 137:633–648

Kamer KJ, Mootha VK (2015) The molecular era of the mitochondrial calcium uniporter. Nat Rev Mol Cell Biol 16:545–553

Lee KP, Yuan JP, So I, Worley PF, Muallem S (2010) STIM1-dependent and STIM1-independent function of transient receptor potential canonical (TRPC) channels tunes their store-operated mode. J Biol Chem 285:38666–38673

Liou J, Kim ML, Heo WD, Jones JT, Myers JW, Ferrell JE Jr, Meyer T (2005) STIM is a Ca2+ sensor essential for Ca2+-store-depletion-triggered Ca2+ influx. Curr Biol 15:1235–1241

Luik RM, Wu MM, Buchanan J, Lewis RS (2006) The elementary unit of store-operated Ca2+ entry: local activation of CRAC channels by STIM1 at ER-plasma membrane junctions. J Cell Biol 174:815–825

Malli R, Frieden M, Osibow K, Zoratti C, Mayer M, Demaurex N, Graier WF (2003) Sustained Ca2+ transfer across mitochondria is Essential for mitochondrial Ca2+ buffering, sore-operated Ca2+ entry, and Ca2+ store refilling. J Biol Chem 278:44769–44779

Montalvo GB, Artalejo AR, Gilabert JA (2006) ATP from subplasmalemmal mitochondria controls Ca2+-dependent inactivation of CRAC channels. J Biol Chem 281:35616–35623

Nunes P, Cornut D, Bochet V, Hasler U, Oh-Hora M, Waldburger JM, Demaurex N (2012) STIM1 juxtaposes ER to phagosomes, generating Ca(2)(+) hotspots that boost phagocytosis. Curr Biol 22:1990–1997

Okeke E, Dingsdale H, Parker T, Voronina S, Tepikin AV (2016) Endoplasmic reticulum-plasma membrane junctions: structure, function and dynamics. J Physiol (Lond) 594:2837–2847

Prakriya M, Feske S, Gwack Y, Srikanth S, Rao A, Hogan PG (2006) Orai1 is an essential pore subunit of the CRAC channel. Nature 443:230–233

Prinz WA (2014) Bridging the gap: membrane contact sites in signaling, metabolism, and organelle dynamics. J Cell Biol 205:759–769

Putney JW Jr (2007) Recent breakthroughs in the molecular mechanism of capacitative calcium entry (with thoughts on how we got here). Cell Calcium 42:103–110

Reinisch KM, De Camilli P (2016) SMP-domain proteins at membrane contact sites: structure and function. Biochim Biophys Acta 1861:924–927

Roos J, DiGregorio PJ, Yeromin AV, Ohlsen K, Lioudyno M, Zhang S, Safrina O, Kozak JA, Wagner SL, Cahalan MD, Veliçelebi G, Stauderman KA (2005) STIM1, an essential and conserved component of store-operated Ca2+ channel function. J Cell Biol 169:435–445

Samanta K, Douglas S, Parekh AB (2014) Mitochondrial calcium uniporter MCU supports cytoplasmic Ca2+ oscillations, store-operated Ca2+ entry and Ca2+-dependent gene expression in response to receptor stimulation. PLoS One 9:e101188

Vig M, Peinelt C, Beck A, Koomoa DL, Rabah D, Koblan-Huberson M, Kraft S, Turner H, Fleig A, Penner R, Kinet JP (2006) CRACM1 is a plasma membrane protein essential for store-operated Ca2+ entry. Science 312:1220–1223

New Aspects of the Contribution of ER to SOCE Regulation: The Role of the ER and ER-Plasma Membrane Junctions in the Regulation of SOCE

12

Hayley Dingsdale, Emmanuel Okeke, Lee Haynes, Gyorgy Lur, and Alexei V. Tepikin

Abstract

The junctions between the endoplasmic reticulum and the plasma membrane are essential platforms for the activation of store-operated Ca^{2+} influx. These junctions have specific dimensions and are nonuniformly distributed in polarized cells. The mechanisms involved in the formation of the junctions are currently undergoing vigorous investigation, and significant progress was attained in this research area during the last 10 years. Some cell types display stationary junctions, while in other cells, new junctions can form rapidly following cytosolic Ca^{2+} signals and/or the reduction of the Ca^{2+} concentration in the lumen of the endoplasmic reticulum; furthermore, in moving cells, junctions can undergo saltatory formation, long distance sliding, and dissolution. The proteins involved in the activation of the Ca^{2+} influx could be also involved in the formation of the junctions. The architecture, dynamics, and localization of the junctions are important for the regulation of Ca^{2+} signaling cascades and their downstream events.

Keywords

Store-operated Ca^{2+} entry • ER-PM junctions • Ca^{2+} influx • Ca^{2+} signaling • STIM • Orai

---

H. Dingsdale (✉)
School of Biosciences, Cardiff University, Cardiff, UK
e-mail: dingsdaleh@cardiff.ac.uk

E. Okeke • L. Haynes • A.V. Tepikin (✉)
The Physiological Laboratory, Department of Cellular and Molecular Physiology, Institute of Translational Medicine, The University of Liverpool, Crown Street, Liverpool L69 3BX, UK
e-mail: a.tepikin@liverpool.ac.uk

G. Lur
Department of Neurobiology, Yale School of Medicine, New Haven, CT, USA

K. Groschner et al. (eds.), *Store-Operated Ca^{2+} Entry (SOCE) Pathways*, Advances in Experimental Medicine and Biology 993, DOI 10.1007/978-3-319-57732-6_12

12.1 Introduction

The endoplasmic reticulum (ER) is an intracellular organelle present in all eukaryotic cells, which is critical for both the survival and correct functioning of all cell types. Its functions include the synthesis, modification, and transport of proteins, and, centrally to this chapter, it also acts as the main intracellular calcium (Ca^{2+}) store (Berridge et al. 2000, 2003; Berridge 2009).

These diverse functions all require an ER Ca^{2+} concentration ($[Ca^{2+}]_{ER}$) of approximately 100–800 μM at rest; a sustained depletion of ER Ca^{2+} stores is detrimental to cells and can lead to cell stress, inhibition of protein synthesis, and apoptosis (Burdakov et al. 2005; Groenendyk and Michalak 2005). Cells therefore employ a number of mechanisms to prevent prolonged store depletion, using Ca^{2+}-binding proteins in the ER to buffer Ca^{2+} levels (Michalak et al. 2002) and sarcoplasmic/endoplasmic reticulum Ca^{2+}-ATPases (SERCAs) to actively reload the store (Carafoli 2002). Crucially, a reduction of the $[Ca^{2+}]_{ER}$ will also activate store-operated Ca^{2+} entry (SOCE) (Putney 2007), which involves direct interaction between ER-localized stromal interaction molecule (STIM) proteins and channel-forming Orai proteins located in the plasma membrane (PM) (Liou et al. 2005; Roos et al. 2005; Feske et al. 2006; Luik et al. 2006; Prakriya et al. 2006; Vig et al. 2006; Park et al. 2009; Yuan et al. 2009). The interaction between the two types of proteins occurs at junctions between the ER and the plasma membrane (ER-PM junctions). Recently, it has become clear that in addition to a subservient role in ER Ca^{2+} reloading, SOCE has its own signaling purposes, and ER-PM junctions accumulate proteins that are essential for such SOCE-specific signaling (Kar et al. 2011, 2014; Kar and Parekh 2015). This chapter will describe the basic structural and functional characteristics of the ER and will look in depth at ER-PM junctions and the role of the ER structure in the regulation of SOCE.

12.2 The Endoplasmic Reticulum

12.2.1 ER Structure

The ER is an intracellular organelle with a bilayer membrane and a multitude of different functions. Some of these functions (e.g., protein synthesis) have specific structural requirements; cells cater for this by utilizing three distinct types of ER—rough ER, smooth ER, and the nuclear envelope (Voeltz et al. 2002). Additionally, some ER classifiers mention transitional ER, a specialized ER domain involved in communication with the Golgi apparatus (Jamieson and Palade 1967; Voeltz et al. 2002). The surface of rough ER is characteristically studded with ribosomes, nucleoprotein complexes responsible for protein synthesis. These large structures (approximately 26 nm in diameter (Lur et al. 2009)) give the ER a "rough" appearance when seen on moderate magnification electron micrographs and thus give the rough ER its name (Palade 1955). Ribosomes are absent from smooth ER, which is primarily involved in the synthesis of lipids and sterols. A specialized form of smooth ER—the sarcoplasmic reticulum—plays a key role in Ca^{2+} signaling and excitation-contraction coupling in myocytes (Rossi and Dirksen 2006). The nuclear envelope is a double bilayer structure (containing large protein complexes—nuclear pores) that encapsulates the nucleus in eukaryotic cells; although it is continuous with the rough ER, it does not usually stretch to the cell periphery (Hetzer et al. 2005) and will not be covered further in this chapter. Although both rough ER and smooth ER can be found in every cell, the abundance of these two forms is tailored to the requirements of individual cells. Accordingly, in cells with a high protein production requirement—such as those secreting digestive enzymes—the majority of ER is ribosome decorated (Bolender 1974); in others smooth ER is predominant (Voeltz et al. 2002).

Regardless of the ratio of rough to smooth ER, in most cells this organelle forms a wide-reaching network, spreading throughout the cell and all the way to the cell periphery. This distribution allows the ER to play a key role in coordinating intracellular machinery by allowing it to communicate with other cellular organelles (Levine and Loewen 2006; English et al. 2009; Prinz 2014). Communication between the ER and other organelles allows inter-organelle transfer of lipids (Holthuis and Levine 2005; Chang et al. 2013; Chang and Liou 2015; Chung et al. 2015; Moser von Filseck et al. 2015), phosphatidic acid (Kim et al. 2015), and sterols (Schulz et al. 2009; Gatta et al. 2015). Communication is achieved both through vesicular transport (e.g., ER-Golgi communication (Lee et al. 2004)) and through physical junctions between ER and other organelles including ER and mitochondria (Copeland and Dalton 1959; Rizzuto et al. 1998; Csordas et al. 2006, 2010), ER and PM (Reger 1961; Rosenbluth 1962; Brandt et al. 1965; Franzini-Armstron 1974; Gardiner and Grey 1983; Wu et al. 2006; Lur et al. 2009; Fernandez-Busnadiego et al. 2015; recently reviewed in Prinz 2014; Okeke et al. 2016a), ER and endosomes (Eden et al. 2010; Henne et al. 2015), ER and lysosomes (Kilpatrick et al. 2013; Fameli et al. 2014), and ER and phagosomes (Nunes et al. 2012). Interorganellar communication involving ER includes spatially restricted signaling at all these contact sites.

12.2.2 The ER and Ca^{2+} Signaling

The pervasive distribution of the ER network throughout individual cells is of particular importance in cellular Ca^{2+} signaling, as it enables the ER to act as the intracellular hub for Ca^{2+} signaling events. Agonists binding PM receptors can stimulate the production of second messengers such as inositol 1,4,5-trisphosphate (IP_3; Berridge 1993) and cyclic ADP-ribose (Galione et al. 1993; Lee 2011). These bind receptors located on the ER (IP_3 receptors or ryanodine receptors) and stimulate the release of Ca^{2+}. In excitable cells, some agonists also act by opening ligand- or voltage-gated PM Ca^{2+} channels, which often leads to further release of ER Ca^{2+} through Ca^{2+}-induced Ca^{2+} release (Berridge et al. 2003). The subsequent increase in the cytosolic Ca^{2+} concentration ($[Ca^{2+}]_c$) can trigger a vast array of downstream cellular responses, e.g., secretion (Petersen and Tepikin 2008), cell contraction (Bers 2002), and gene expression (Dolmetsch et al. 1998; Parekh and Muallem 2011), determined by the particular cell type and spatiotemporal aspects of the Ca^{2+} signal. Many physiological responses require prolonged periods of Ca^{2+} signaling, yet sustained elevations in the $[Ca^{2+}]_c$ can lead to cell death via apoptosis or necrosis (Orrenius et al. 2003). This conundrum is solved by the use of oscillatory Ca^{2+} responses—rhythmic increases and decreases in the $[Ca^{2+}]_c$ that can be maintained over long time periods with no harmful side effects to the cell (Parekh 2011). In many cell types, the ER plays a critical role in both forming and sustaining oscillatory responses, as the transitory nature of each rise and fall in the $[Ca^{2+}]_c$ is largely dependent on the uptake and release of Ca^{2+} from the ER.

Not only is the presence of Ca^{2+} clearly fundamental to the ER's role in Ca^{2+} signaling, it is also essential for many of its other functions; for example, folding of newly synthesized proteins is carried out by Ca^{2+}-dependent chaperones (Michalak et al. 2002). Sustained depletion of the intracellular store leads to ER stress—a phenomenon caused partly by the buildup of incorrectly folded proteins—and can lead to stimulation of apoptosis (Rutkowski and Kaufman 2004; Groenendyk and Michalak 2005; Tabas and Ron 2011). As such, strict maintenance of the ER Ca^{2+} pool is a high priority for cells. SERCA pumps are crucial for the maintenance of the ER Ca^{2+} store. These ER-located ATPases retrieve Ca^{2+} from the cytosol and pump it back into the ER, counteracting Ca^{2+} lost during both cell signaling events and via the passive Ca^{2+} leak pathway. These pumps also help shape intracellular signals by ensuring the rapid return of $[Ca^{2+}]_c$ to resting levels (Wuytack et al. 2002). Ca^{2+} transport pathways at the PM—e.g., PM Ca^{2+}-ATPases (PMCA) and Na^+/Ca^{2+} exchangers—export Ca^{2+} from the cytosol into the extracellular environment (Blaustein and Lederer 1999; Di Leva et al. 2008), the rate of which increases considerably during periods of Ca^{2+} signaling (e.g., Tepikin et al. 1992; Camello et al. 1996). This can result in an overall loss of intracellular Ca^{2+} and hence must be counteracted if depletion of intracellular stores is to be avoided. SOCE is stimulated by a decrease in $[Ca^{2+}]_{ER}$ and acts to refill these stores. This makes it vital for the maintenance of sustained Ca^{2+} oscillations which would otherwise gradually diminish the ER Ca^{2+} stores due to extrusion of Ca^{2+} to the external environment. Indeed, numerous studies have shown that physiological cytosolic Ca^{2+} oscillations cannot be maintained over long time periods in the absence of extracellular Ca^{2+}, with strong evidence existing to suggest that SOCE plays a key role in propping up ER Ca^{2+} stores to sustain these repetitive Ca^{2+} transients (e.g., Tepikin et al. 1992; Bird and Putney 2005; reviewed in Putney and Bird 2008).

Experiments involving local reloading, fluorescence recovery after photobleaching, and uncaging of photolabile compounds (including caged Ca^{2+}) in the ER lumen revealed a considerable continuity throughout the ER and also the ability of Ca^{2+} to travel substantial distances via the ER lumen. This allows the transfer of Ca^{2+} signals by "tunneling" through the ER, so that local Ca^{2+} depletion in one region of the cell can be communicated to Ca^{2+} reloading mechanisms located in another part of the cell and SOCE elements under the PM (Mogami et al. 1997; Park et al. 2000). This also indicates that Ca^{2+} entering the cell via

SOCE channels can be taken locally into the ER lumen and then transferred—by tunneling—to refill Ca^{2+} stores near the release sites.

Replenishing the ER Ca^{2+} store is not the only function of SOCE, however, as it also plays a more direct role in intracellular signaling. SOCE-induced subplasmalemmal Ca^{2+} microdomains have been shown to activate Ca^{2+}-dependent adenylyl cyclase 8, a signaling phenomenon with a multitude of downstream effects (Willoughby et al. 2010, 2012). Recent papers by P. Kar and colleagues from A. Parekh's laboratory have also demonstrated the ability of SOCE-induced microdomains to specifically control gene transcription via the transcription factor NFAT1 (nuclear factor of activated T-cells 1) (Kar et al. 2011, 2014; Kar and Parekh 2015). Furthermore, an elegant recent study by K. Lefkimmiatis and colleagues from A. Hofer's laboratory characterized a cAMP signaling pathway activated by depletion of the ER Ca^{2+} store alone (Lefkimmiatis et al. 2009). In a later study from the same group, adenylyl cyclase 3 was identified as the protein responsible for this signaling modality (Maiellaro et al. 2012).

The ER, and particularly subplasmalemmal regions of the ER, therefore plays a crucial role in ER Ca^{2+} reloading and in the activation and/or maintenance of SOCE-dependent signaling events, as well as in signaling events induced by ER depletion.

12.2.3 ER-Plasma Membrane Junctions

SOCE occurs at specialized regions where the plasma membrane is closely aligned with the ER, i.e., at ER-PM junctions (see Fig. 12.1). Electron micrographic images of closely apposed ER and PM can be seen as early as 1957 (Porter and Palade 1957). In the following decade, ER-PM junctions were identified in a number of different cell types (Engstrom 1958; Fawcett and Revel 1961; Reger 1961; Rosenbluth 1962; Brandt et al. 1965; Franzini-Armstron 1974). In 1962, Rosenbluth described many key features of ER-PM junctions, though the functions of these structures were at this stage mostly unknown. He described a characteristic absence of ribosomes between the ER and PM, a high frequency of mitochondria in the vicinity of junctions, and a distance of 4–10 nm between the two membranes (Rosenbluth 1962).

ER-PM junctions do not only serve as sites for SOCE but also as sites for other types of Ca^{2+} responses. To date, the most well-characterized ER-PM junctions are found in striated muscle. The close juxtaposition between the sarcoplasmic reticulum and PM invaginations (T tubules) in these cells is a key requirement for excitation-contraction coupling (Eisenberg and Eisenberg 1982; Fill and Copello 2002; Rossi and Dirksen 2006; reviewed by Takeshima et al. 2015). Recently, ER-PM junctions were also characterized in other primary mammalian cell types including secretory epithelial cells (e.g., Lur et al. 2009; see Fig. 12.1) and primary neurons (e.g., Fernandez-Busnadiego et al. 2015).

ER-PM junctions are not restricted to mammalian cells. Considerable progress in characterizing the mechanisms of the formation of ER-PM junctions was attained

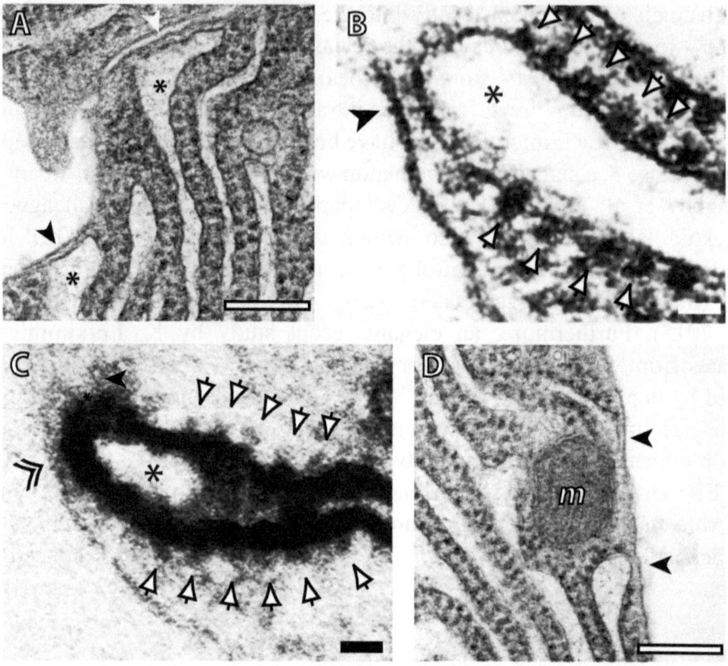

Fig. 12.1 ER-PM junctions—the platforms for store-operated Ca^{2+} entry. (**a**) Electron micrograph shows strands of ER approaching the plasma membrane and forming ER-PM junctions (indicated by *arrowheads*). The lumens of the ER strands participating in the junction are labeled with *asterisks*. *Scale bar* corresponds to 200 nm. (**b**) High magnification electron micrograph of an ER-PM junction. A ribosome (*black* and *white arrows*) decorated strand of ER (lumen labeled with an *asterisk*) approaches the plasma membrane to form the junction (*black arrowhead*). The lack of ribosomes in the junction allows a spectacularly close distance (12–13 nm) between the organelles. The electron density in the junctional area indicates high protein content between the ER and plasma membrane. *Scale bar* corresponds to 25 nm. (**c**) Diaminobenzidine histochemistry reveals the presence of horseradish peroxidase-tagged STIM1 in the ER-PM junction. Electron dense black staining highlights the membrane of the ER strand (*asterisk*) approaching the plasma membrane to form a junction (*double arrowhead*). The visible shadows of ribosomes indicate that the junction is formed by rough ER. *Scale bar* is 25 nm. (**d**) Electron micrograph shows a subplasmalemmal mitochondrion (*m*) in the close vicinity of two ER-PM junctions (*arrowheads*). *Scale bar* indicates 200 nm (parts of the figure were adapted from Lur et al. 2009)

in experiments conducted on yeast cells where six proteins responsible for tethering of the two organelles (Scs2, Scs22, Ist2, and tricalbins 1, 2, and 3) were identified (Loewen et al. 2007; Manford et al. 2012). This success in identifying the tethering proteins in yeasts facilitated identification of the junction-forming proteins in mammalian cells. Two groups recently characterized ER-PM junctions containing extended synaptotagmins (E-Syt(s), mammalian analogues of tricalbins) (Chang et al. 2013; Giordano et al. 2013; Fernandez-Busnadiego et al. 2015) and highlighted a putative role of these proteins in the formation of ER-PM junctions.

12.3 SOCE and ER-Plasma Membrane Junctions

12.3.1 Conformational Coupling and Its Participants

ER-PM junctions have more recently received attention as platforms for SOCE. It was in 1986 that Putney first proposed the concept of store-operated Ca^{2+} entry (originally termed capacitative Ca^{2+} entry), the entry of Ca^{2+} across the PM in response to depletion of intracellular stores (Putney 1986). A few years later, a conformational coupling hypothesis for SOCE was formulated. This hypothesis suggested direct coupling between an ER Ca^{2+} sensor (initially considered to be IP_3 receptors) and Ca^{2+} influx channels in the PM (Irvine 1990; Berridge 1995). IP_3 receptors were later found not to be essential for SOCE (Broad et al. 2001; Prakriya and Lewis 2001; Taylor et al. 2009), but the conformational coupling mechanism was proven to be crucial for SOCE activation (Putney 2007; Park et al. 2009). At the beginning of the twenty-first century, two proteins involved in SOCE—STIM1 and Orai1—were identified (Liou et al. 2005; Roos et al. 2005; Feske et al. 2006; Luik et al. 2006; Vig et al. 2006). The formation of functional Ca^{2+}-conducting SOCE channels was found to require a direct interaction between these two proteins (Mercer et al. 2006; Luik et al. 2008; Park et al. 2009; Yuan et al. 2009). Considerable progress was recently achieved in identifying specific protein domains involved in the STIM1-Orai1 interaction (Park et al. 2009; Yuan et al. 2009) and in characterizing auxiliary proteins involved in SOCE modulation (Palty et al. 2012; Sharma et al. 2013; Maleth et al. 2014). Other members of STIM and Orai families (correspondingly STIM2 and Orai2 and 3) have been shown to participate in the SOCE process (recently reviewed in Hoth and Niemeyer 2013).

STIM is the Ca^{2+} sensor located in the ER membrane (Liou et al. 2005; Roos et al. 2005; Zhang et al. 2005; Collins and Meyer 2011), while Orai (located in the PM) is the pore-forming component of the channel (Feske et al. 2006; Gwack et al. 2007). Close contacts (i.e., junctions) between the ER and the PM are therefore essential for the interaction of STIM and Orai and the activation of SOCE. Specialized ER strands act as scaffolds for SOCE, forming junctions with the PM and providing a location where all the required components can come together and thus enable Ca^{2+} influx (Luik et al. 2006; Lewis 2007; Lur et al. 2009; Orci et al. 2009; Lee et al. 2010; Carrasco and Meyer 2011).

12.3.2 The Formation of ER-Plasma Membrane Junctions

An uncertainty that still lingers on is whether these ER-PM junctions are formed as a result of Ca^{2+} signaling (i.e., induced by cytosolic Ca^{2+} responses or intracellular Ca^{2+} store depletion) or are preexisting—recruiting STIM and Orai only as and when the ER is emptied. In primary exocrine cells, depletion of ER Ca^{2+} stores did not increase the number of the junctions (as visualized using electron microscopy) suggesting that these structures are preformed (Lur et al. 2009) and stationary. A similar conclusion can be drawn from experiments by J. Smyth and colleagues from

J. Putney's laboratory. In this study on living cells examined using total internal reflection fluorescence microscopy, repeated store depletions (interrupted by periods of store refilling) resulted in the formation of STIM1 puncta in the same (or very similar) locations, suggesting that STIM1 translocates in and out of stable junctions (Smyth et al. 2008). However, in other cell types, a large number of junctions spectacularly developed as a result of Ca^{2+} store depletion in addition to the preformed junctions (Wu et al. 2006). Furthermore cytosolic Ca^{2+} rise has been also shown to induce junction formation (Chang et al. 2013; Giordano et al. 2013; Chang and Liou 2015; Idevall-Hagren et al. 2015). Most probably all mechanisms of junction formation exist, and their balance depends on the type and the status of the cell.

It is possible that STIM is not only involved in ER Ca^{2+} sensing but also plays a role in the construction of the ER-PM junctions. The presence of STIM in junctions was revealed using both high-resolution video imaging microscopy (Wu et al. 2006; Chang et al. 2013; Okeke et al. 2016b) and electron microscopy (Wu et al. 2006; Lur et al. 2009; Orci et al. 2009 and Fig. 12.1c). Overexpression of STIM1 has been shown on numerous occasions to increase both the number (Lur et al. 2009) and lateral length of the junctions (Orci et al. 2009), and similarities between STIM and Ist2—a protein involved in the formation of cortical ER in yeast cells—also suggest an additional structural role for STIM in junctions (Lavieu et al. 2010). STIM is therefore likely to participate in the biogenesis of ER-PM junctions, though it should be noted that overexpression of STIM also grossly affects ER morphology (Varnai et al. 2007). STIM1 has been shown to be targeted to the PM via an interaction between its cytosol-facing C-terminus polybasic sequence motif and polyphosphoinositides in the PM. Liou et al. (2007) suggested that the oligomerization of STIM1, which occurs as a result of ER store depletion, results in an increased potency of its interaction with the PM (because multiple polybasic sequences available in the STIM1 oligomers should bind the PM more efficiently) and translocation of STIM1 to subplasmalemmal junctions. The notion that STIM1 can form puncta via its interaction with polyphosphoinositides was confirmed by Walsh et al. (2010) who showed that the depletion of polyphosphoinositides drastically reduces the number of STIM1 puncta formed as a result of ER Ca^{2+} depletion. STIM-dependent trafficking could bring an ER strand into close proximity of the PM; the interaction between STIM1 (located in the ER membrane) and polyphosphoinositides (in the PM) could then allow the initial tethering of the two membranes. The contact sites could be later reinforced by protein-protein interactions, e.g., interactions between STIM and Orai proteins. Finally, more permanent fixtures involving linkers similar to those reported for ER junctions with mitochondria (Csordas et al. 2006) could develop and further strengthen the junctions. This STIM-centric and largely hypothetical sequence of events is clearly in need of further experimental evidence.

E-Syt1 is a Ca^{2+}-binding protein, and it is therefore not surprising that an increase of cytosolic Ca^{2+} leads to the formation of E-Syt1-dependent junctions (Chang et al. 2013; Giordano et al. 2013; Chang and Liou 2015; Fernandez-Busnadiego et al. 2015; Idevall-Hagren et al. 2015). Importantly, some E-Syt1-

decorated junctions persist in unstimulated cells, suggesting that a stabilization process occurs that converts nascent Ca^{2+}-induced, E-Syt1-dependent junctions into more permanent structures. The mechanism stabilizing E-Syt1-decorated junctions is not currently known and is an exciting area for further investigations.

12.3.3 The Dimensions of ER-Plasma Membrane Junctions

To allow direct interactions between STIM and Orai, the ER and PM must be positioned within the junctions at a very specific distance. This distance cannot be too small (otherwise the components of SOCE machinery could not maneuver into the space between the membranes) or too large (otherwise the two proteins would not be able to interact). Using rapamycin-inducible ER-PM linkers, Varnai et al. (2007) have demonstrated that a space of 6 nm between the membranes in the junction prevents the formation of STIM1-Orai complexes, while 11–14 nm is sufficient for the interaction of the proteins. Wu et al. (2006) estimated the distance between the membranes in the junction to be 17 nm. Both values are in good agreement to the distance reported by Lur et al. (2009) (12–13 nm). A recent study of E-Syt-mediated junctions reported greater distances between the ER and PM membranes of approximately 19 nm and 22 nm in cells transfected with E-Syt3 and E-Syt1, respectively (Fernandez-Busnadiego et al. 2015).

The distances between the ER and PM membranes are too small to allow the inclusion of ribosomes, which are roughly 26 nm wide (Lur et al. 2009); furthermore, the intermembrane space is filled with material described as "intermediate densities" and "filamentous structures" (Fernandez-Busnadiego et al. 2015) which should also prevent entry of ribosomes into the junctions. In some cell types, the requirement for the exclusion of ribosomes is easily fulfilled, as the ribosome-free smooth ER is used to create junctions (Gardiner and Grey 1983; Orci et al. 2009). A problem arises in cells that use ribosome-studded rough ER to form ER-PM junctions, such as pancreatic acinar cells (Lur et al. 2009) and frog ganglion cells (Watanabe and Burnstock 1976). Especially in pancreatic acinar cells, STIM puncta form basolaterally (Lur et al. 2009), where mainly rough ER is present (Bolender 1974; Lur et al. 2009). The potential problem of ribosomes preventing the formation of SOCE-competent junctions is elegantly solved by acinar cells, in which ribosomes are stripped off the rough ER as it approaches the PM (Fig. 12.1b), while those on the cytosolic side (in the immediate vicinity to the junctions) are retained (Lur et al. 2009). The mechanism by which this exclusion occurs is as yet undetermined, though ribosomes are not the only ER residents excluded from the junctions. GPR78 (78 kDa glucose-regulated protein; also known as binding immunoglobulin protein (BiP)) is a Ca^{2+}-sensitive chaperone that has also been shown to be excluded from ER-PM junctions (Orci et al. 2009). Furthermore, experiments using expression of an artificial, ER-targeted construct—a GFP-KDEL protein—demonstrated removal of this fluorescent protein from the ER strands upon approach to ER-PM junctions (Orci et al. 2009). Taken together, these results indicate that both surface structures (ribosomes) and luminal residents (GPR78

(BiP) and GFP-KDEL) of the ER can be excluded from the ER strands in the vicinity of junctions (Lur et al. 2009; Orci et al. 2009). Physiologically, aside from ribosomes and the BiP chaperone, it is not yet known if any other proteins are specifically excluded from these junctional sites. The reported lateral (parallel to the plasma membrane) sizes of ER-PM junctions vary significantly, probably because the junctions are usually visualized using exogenously expressed proteins that modify their properties. However, a fluorescently labeled protein probe termed MAPPER ("membrane-attached peripheral ER"), recently developed by the J. Liou laboratory, was shown to accumulate in ER-PM junctions but did not modify the density of the junctions (Chang et al. 2013). The lateral size of the junctions revealed by MAPPER was approximately 200–300 nm (Chang et al. 2013). The lateral size of the junction is important since it is likely to determine the influence of the local signaling network (e.g., the maximal number of Orai channels and consequently the maximal Ca^{2+} flux attainable in the junction; similar considerations will apply to other signaling modalities). It is difficult to exclude that MAPPERS will have no effect on the structure or dynamics of the ER-PM junctions in other cell types (or other experimental conditions), and therefore, careful controls, similar to those described by Chang et al. (2013), are necessary to verify the suitability of the selected junction-revealing probe for the planned experiments.

12.3.4 ER Dynamics: STIM and the Cytoskeleton

Perhaps the most important role the ER plays in the regulation of SOCE is its function in the localization of STIM (Fig. 12.1c). STIM is an ER-resident protein, and as such its distribution is highly dependent on ER structure. In resting conditions, STIM is distributed throughout the ER; once the ER is depleted of Ca^{2+}, this protein translocates to the junctions between the ER and the PM, forming characteristic puncta in fluorescent images (Liou et al. 2005; Xu et al. 2006; Smyth et al. 2008; Cahalan 2009). The positioning of STIM can change both due to its translocation within the ER and also due to movement of the ER itself. The ER is a dynamic organelle, moving—for the most part—along microtubules, and in this way, it can modify the distribution of its resident proteins (Baba et al. 2006; Borgese et al. 2006; Friedman et al. 2010). Interestingly, STIM1 is also reciprocally involved in regulating the movement of ER strands. Grigoriev et al. (2008) have demonstrated that STIM1 binds EB1 proteins located at the polymerizing end of microtubules and can potentiate ER transport. The physical movement of ER can lead to an accumulation of ER strands under the PM and the formation of new ER-PM junctions (Wu et al. 2006). The ability of an ER strand to form a junction probably partly depends on the local cytoskeletal architecture and particularly on the availability of convenient fenestrations in the subplasmalemmal actin network. In this respect, it is interesting to note that earlier studies of SOCE have demonstrated that manipulation of the actin network (particularly stabilization with jasplakinolide) reduces SOCE (Patterson et al. 1999; Rosado et al. 2004); it is conceivable that this reduction was due to a decrease in the size or number of junctions.

Not all ER-PM junctions are labile and dynamic. Stable junctions have been characterized in pancreatic acinar cells (Lur et al. 2009). Classical ER-PM junctions first characterized in muscle cells (Porter and Palade 1957) also belong to this "stable" category. This stability could be advantageous because it could allow the cell to attract protein complexes and cellular organelles to the vicinity of the ER-PM junctions and produce sophisticated multimodal signaling responses. There are however situations when junctions simply cannot become stable—an obvious example are migrating cells. In migrating cells, junctions continuously undergo formation and breakdown (Dingsdale et al. 2013; see Fig. 12.2). The ER-PM junctions and associated Ca^{2+} signaling complexes are found close to the components of the migratory apparatus and are important for migration (Tsai and Meyer 2012; Dingsdale et al. 2013; Tsai et al. 2014; Okeke et al. 2016b).

The ability of STIM to translocate to junctions is also affected during cell division; during which the ER undergoes substantial reorganization and is highly dynamic. STIM1 translocation to ER-PM junctions is inhibited during both meiosis (Yu et al. 2010) and mitosis (Smyth et al. 2009), the latter of which is due to phosphorylation of STIM preventing its movement to the PM. The phosphorylation of STIM1 is also important for shaping the structure of the ER during mitosis and specifically for the exclusion of the ER from the mitotic spindle (Smyth et al. 2012). The effect of STIM proteins on the structure of the ER and particularly on ER-PM junctions in other dynamic cellular responses including migration, invasion, and contraction is largely unknown and deserves further investigation.

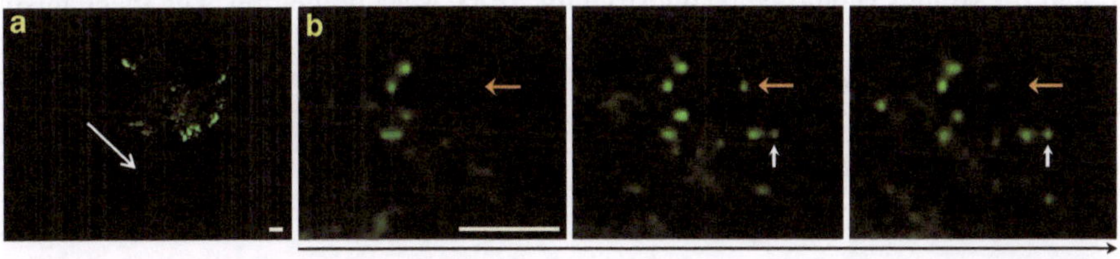

Fig. 12.2 The dynamics of ER-PM junctions in migrating cells. (**a**) Confocal microscopy revealed localization of YFP-labeled STIM1 puncta (*green color*) in migrating pancreatic cancer cells (PANC-1 cells). The direction of migration is indicated by *white arrow*. A high density of STIM1 puncta (i.e., STIM1-competent ER-PM junctions) was found close to the leading edge of the cell. *Scale bar* represents 10 μm. (**b**) Formation and dissolution of STIM1-labeled ER-PM junctions at the leading edge of a migrating PANC-1 cell. Images show the fragment of the leading edge of a migrating PANC1 cell (different cell from part **a**). The direction of migration is indicated by *black arrow* (under the images). *Scale bar* represents 10 μm. *Orange arrow* shows the junction that appears near the leading edge of the cell and then disappears before the last (*right*) image. *White arrow* indicates a STIM1-competent junction that appears and then stabilizes (increases in brightness) close to the leading edge of the cell. The components of the figure were adapted (with modifications) from Dingsdale et al. (2013) and Okeke et al. (2016b)

12.4 ER-Plasma Membrane Junctions and the Distribution of Auxiliary Ca^{2+}-Regulating Structures

12.4.1 ER-Plasma Membrane Junctions and SERCA Pumps

Although STIM and Orai are the key requirements for SOCE, other proteins are also needed for optimal Ca^{2+} influx, and the junctional part of the ER is vital to ensure their correct localization. SERCA pumps play a significant role in SOCE allowing Ca^{2+} entering from the extracellular environment to refill the ER. The functional advantages of concentrating SERCA pumps close to the Ca^{2+} influx sites in ER-PM junctions are clear; by providing increased local Ca^{2+} buffering and accelerated local Ca^{2+} transport, SERCA pumps can prevent Ca^{2+} inhibition of SOCE and facilitate Ca^{2+} influx. Increased SERCA density in the vicinity of the Ca^{2+} influx channels can also limit Ca^{2+} signals to microdomains and facilitate Ca^{2+} reloading of the ER. Hence, the correct positioning of the SERCA pumps in respect to SOCE junctions is probably very important. Indeed, there is evidence to suggest that these pumps are actually located at or near the junctional complex (Jousset et al. 2007; Manjarres et al. 2010, 2011). Manjarres and colleagues presented evidence of "SERCA forming an outer ring around Orai-STIM aggregates" (Manjarres et al. 2010; Alonso et al. 2012). The basis for this elegant model was the close proximity of SERCA with both STIM and Orai (revealed by optical microscopy), but the absence of FRET between SERCA and either STIM or Orai. It is not clear if SERCA actually enters the junctions (i.e., positioned in the cleft between the two membranes) or instead concentrates in regions of the ER strands just outside the junctions. In both instances, specialized domains of junctional ER strands should be responsible for SERCA accumulation. The mechanism for positioning the Ca^{2+} pumps at or near the junctional ER is currently unknown. It is interesting to note that another Ca^{2+} transporting pump PMCA is inhibited by STIM1 preventing futile Ca^{2+} cycles and facilitating Ca^{2+} reloading (Krapivinsky et al. 2011; Ritchie et al. 2012).

12.4.2 ER-Plasma Membrane Junctions and Mitochondria

Mitochondria are also frequently found in close proximity to ER-PM junctions (Fig. 12.1d), and the role of mitochondria in SOCE is well documented (Hoth et al. 1997; Parekh 2003, 2008; Barrow et al. 2008; Walsh et al. 2009; Singaravelu et al. 2011). Close localization of mitochondria next to ER-PM junctions may also be a mechanism by which ATP (Montalvo et al. 2006) or other substances released by mitochondria (e.g., reactive oxygen species (Bogeski et al. 2010)) could influence SOCE (reviewed by Walsh et al. 2009; Nunes and Demaurex 2014). The proximity between SOCE channels and mitochondria is likely to be crucial for the efficiency of Ca^{2+} transfer into subplasmalemmal mitochondria, particularly considering Ca^{2+}-dependent gating of the mitochondrial Ca^{2+} influx machinery (recently reviewed by De Stefani et al. 2015; Kamer and Mootha 2015). The complex responsible for Ca^{2+}

entry into mitochondria is composed of the mitochondrial Ca^{2+} uniporter (MCU), mitochondrial Ca^{2+} uptake proteins 1 and 2 (MICU1 and MICU2), and other components (Perocchi et al. 2010; Baughman et al. 2011; De Stefani et al. 2011; reviewed in Kamer and Mootha 2015). MICU1 and MICU2 are Ca^{2+}-binding proteins that serve as gatekeepers preventing Ca^{2+} entry at low-resting $[Ca^{2+}]_c$ and allowing rapid Ca^{2+} entry at higher $[Ca^{2+}]_c$ (Csordas et al. 2013; Kamer and Mootha 2014; Patron et al. 2014; reviewed in Kamer and Mootha 2015). This nonlinearity of Ca^{2+} influx suggests that distances between SOCE channels and mitochondria should be important for mitochondrial Ca^{2+} entry and for the ability of mitochondria to influence the local Ca^{2+} environment in the vicinity of the channels. The ER strands may play a key role in the positioning of mitochondria near influx sites by way of mitochondrial-ER tethers (Csordas et al. 2010). A recent study revealed a key role of mitochondria in STIM1 trafficking to ER-PM junctions, as depolarization of mitochondria prevented STIM1 translocation. The mitochondrial protein mitofusin 2 was identified as an important contributor to STIM1 trafficking (Singaravelu et al. 2011); the subplasmalemmal mitochondria should be particularly important in this process. The effect of mitochondrial Ca^{2+} uptake systems on STIM1 oligomerization and SOCE development was recently reported by Deak et al. (2014), confirming the significance of mitochondria to this form of Ca^{2+} entry. The importance of the proximity between mitochondria and ER-PM junctions for SOCE has been challenged (Naghdi et al. 2010) indicating that the effect of mitochondria positioning could be cell-type specific.

12.5 Conclusion

The importance of the ER-PM interface for SOCE, the spectacular STIM puncta formation, and the co-clustering of STIM1 and Orai1 has put the ER-PM junctions in the limelight of the Ca^{2+} signaling field. New research tools developed to study the STIM-containing junctions (e.g., recombinant STIM proteins labeled with fluorescent proteins, STIM combined with horseradish peroxidase, inducible ER-PM linkers, and MAPPERS) have allowed considerable progress in characterizing the dynamics, distribution, and structure of SOCE-competent ER-PM junctions (Wu et al. 2006, 2014; Varnai et al. 2007; Chang et al. 2013). However, numerous questions related to the dynamics, functions, and properties of these junctions remain to be resolved. We hope that this chapter has highlighted some of these questions and will stimulate further research into the mechanism of formation, protein composition, and characterization of ER-PM junctions, as well as their physiological and pathophysiological roles within the cell.

References

Alonso MT, Manjarres IM, Garcia-Sancho J (2012) Privileged coupling between Ca2+ entry through plasma membrane store-operated Ca2+ channels and the endoplasmic reticulum Ca2+ pump. Mol Cell Endocrinol 353:37–44

Baba Y, Hayashi K, Fujii Y, Mizushima A, Watarai H, Wakamori M, Numaga T, Mori Y, Iino M, Hikida M, Kurosaki T (2006) Coupling of STIM1 to store-operated Ca2+ entry through its constitutive and inducible movement in the endoplasmic reticulum. Proc Natl Acad Sci U S A 103:16704–16709

Barrow SL, Voronina SG, da Silva Xavier G, Chvanov MA, Longbottom RE, Gerasimenko OV, Petersen OH, Rutter GA, Tepikin AV (2008) ATP depletion inhibits Ca2+ release, influx and extrusion in pancreatic acinar cells but not pathological Ca2+ responses induced by bile. Pflugers Arch 455:1025–1039

Baughman JM, Perocchi F, Girgis HS, Plovanich M, Belcher-Timme CA, Sancak Y, Bao XR, Strittmatter L, Goldberger O, Bogorad RL, Koteliansky V, Mootha VK (2011) Integrative genomics identifies MCU as an essential component of the mitochondrial calcium uniporter. Nature 476:341–345

Berridge MJ (1993) Inositol trisphosphate and calcium signalling. Nature 361:315–325

Berridge MJ (1995) Capacitative calcium entry. Biochem J 312(Pt 1):1–11

Berridge MJ (2009) Inositol trisphosphate and calcium signalling mechanisms. Biochim Biophys Acta 1793:933–940

Berridge MJ, Lipp P, Bootman MD (2000) The versatility and universality of calcium signalling. Nat Rev Mol Cell Biol 1:11–21

Berridge MJ, Bootman MD, Roderick HL (2003) Calcium signalling: dynamics, homeostasis and remodelling. Nat Rev Mol Cell Biol 4:517–529

Bers DM (2002) Cardiac excitation-contraction coupling. Nature 415:198–205

Bird GS, Putney JW Jr (2005) Capacitative calcium entry supports calcium oscillations in human embryonic kidney cells. J Physiol 562:697–706

Blaustein MP, Lederer WJ (1999) Sodium/calcium exchange: its physiological implications. Physiol Rev 79:763–854

Bogeski I, Kummerow C, Al-Ansary D, Schwarz EC, Koehler R, Kozai D, Takahashi N, Peinelt C, Griesemer D, Bozem M, Mori Y, Hoth M, Niemeyer BA (2010) Differential redox regulation of ORAI ion channels: a mechanism to tune cellular calcium signaling. Sci Signal 3:24

Bolender RP (1974) Stereological analysis of the guinea pig pancreas. I. Analytical model and quantitative description of nonstimulated pancreatic exocrine cells. J Cell Biol 61:269–287

Borgese N, Francolini M, Snapp E (2006) Endoplasmic reticulum architecture: structures in flux. Curr Opin Cell Biol 18:358–364

Brandt PW, Reuben JP, Girardier L, Grundfest H (1965) Correlated morphological and physiological studies on isolated single muscle fibers. I. Fine structure of the crayfish muscle fiber. J Cell Biol 25(Suppl):233–260

Broad LM, Braun FJ, Lievremont JP, Bird GS, Kurosaki T, Putney JW Jr (2001) Role of the phospholipase C-inositol 1,4,5-trisphosphate pathway in calcium release-activated calcium current and capacitative calcium entry. J Biol Chem 276:15945–15952

Burdakov D, Petersen OH, Verkhratsky A (2005) Intraluminal calcium as a primary regulator of endoplasmic reticulum function. Cell Calcium 38:303–310

Cahalan MD (2009) STIMulating store-operated Ca2+ entry. Nat Cell Biol 11:669–677

Camello P, Gardner J, Petersen OH, Tepikin AV (1996) Calcium dependence of calcium extrusion and calcium uptake in mouse pancreatic acinar cells. J Physiol 490(Pt 3):585–593

Carafoli E (2002) Calcium signaling: a tale for all seasons. Proc Natl Acad Sci U S A 99:1115–1122

Carrasco S, Meyer T (2011) STIM proteins and the endoplasmic reticulum-plasma membrane junctions. Annu Rev Biochem 80:973–1000

Chang CL, Liou J (2015) Phosphatidylinositol 4,5-bisphosphate homeostasis regulated by Nir2 and Nir3 proteins at endoplasmic reticulum-plasma membrane junctions. J Biol Chem 290:14289–14301

Chang CL, Hsieh TS, Yang TT, Rothberg KG, Azizoglu DB, Volk E, Liao JC, Liou J (2013) Feedback regulation of receptor-induced Ca2+ signaling mediated by E-Syt1 and Nir2 at endoplasmic reticulum-plasma membrane junctions. Cell Rep 5:813–825

Chung J, Torta F, Masai K, Lucast L, Czapla H, Tanner LB, Narayanaswamy P, Wenk MR, Nakatsu F, De Camilli P (2015) INTRACELLULAR TRANSPORT. PI4P/phosphatidylserine countertransport at ORP5- and ORP8-mediated ER-plasma membrane contacts. Science 349:428–432

Collins SR, Meyer T (2011) Evolutionary origins of STIM1 and STIM2 within ancient Ca2+ signaling systems. Trends Cell Biol 21:202–211

Copeland DE, Dalton AJ (1959) An association between mitochondria and the endoplasmic reticulum in cells of the pseudobranch gland of a teleost. J Biophys Biochem Cytol 5:393–396

Csordas G, Renken C, Varnai P, Walter L, Weaver D, Buttle KF, Balla T, Mannella CA, Hajnoczky G (2006) Structural and functional features and significance of the physical linkage between ER and mitochondria. J Cell Biol 174:915–921

Csordas G, Varnai P, Golenar T, Roy S, Purkins G, Schneider TG, Balla T, Hajnoczky G (2010) Imaging interorganelle contacts and local calcium dynamics at the ER-mitochondrial interface. Mol Cell 39:121–132

Csordas G, Golenar T, Seifert EL, Kamer KJ, Sancak Y, Perocchi F, Moffat C, Weaver D, Perez SD, Bogorad R, Koteliansky V, Adijanto J, Mootha VK, Hajnoczky GR (2013) MICU1 controls both the threshold and cooperative activation of the mitochondrial Ca2+ uniporter. Cell Metab 17:976–987

Deak AT, Blass S, Khan MJ, Groschner LN, Waldeck-Weiermair M, Hallstrom S, Graier WF, Malli R (2014) IP3-mediated STIM1 oligomerization requires intact mitochondrial Ca2+ uptake. J Cell Sci 127:2944–2955

De Stefani D, Raffaello A, Teardo E, Szabo I, Rizzuto R (2011) A forty-kilodalton protein of the inner membrane is the mitochondrial calcium uniporter. Nature 476:336–340

De Stefani D, Patron M, Rizzuto R (2015) Structure and function of the mitochondrial calcium uniporter complex. Biochim Biophys Acta 1853:2006–2011

Di Leva F, Domi T, Fedrizzi L, Lim D, Carafoli E (2008) The plasma membrane Ca2+ ATPase of animal cells: structure, function and regulation. Arch Biochem Biophys 476:65–74

Dingsdale H, Okeke E, Awais M, Haynes L, Criddle DN, Sutton R, Tepikin AV (2013) Saltatory formation, sliding and dissolution of ER-PM junctions in migrating cancer cells. Biochem J 451:25–32

Dolmetsch RE, Xu KL, Lewis RS (1998) Calcium oscillations increase the efficiency and specificity of gene expression. Nature 392:933–936

Eden ER, White IJ, Tsapara A, Futter CE (2010) Membrane contacts between endosomes and ER provide sites for PTP1B-epidermal growth factor receptor interaction. Nat Cell Biol 12:267–272

Eisenberg BR, Eisenberg RS (1982) The T-SR junction in contracting single skeletal muscle fibers. J Gen Physiol 79:1–19

English AR, Zurek N, Voeltz GK (2009) Peripheral ER structure and function. Curr Opin Cell Biol 21:596–602

Engstrom H (1958) On the double innervation of the sensory epithelia of the inner ear. Acta Otolaryngol 49:109–118

Fameli N, Ogunbayo OA, van Breemen C, Evans AM (2014) Cytoplasmic nanojunctions between lysosomes and sarcoplasmic reticulum are required for specific calcium signaling. F1000Res 3:93

Fawcett DW, Revel JP (1961) The sarcoplasmic reticulum of a fast-acting fish muscle. J Biophys Biochem Cytol 10(4, Suppl):89–109

Fernandez-Busnadiego R, Saheki Y, De Camilli P (2015) Three-dimensional architecture of extended synaptotagmin-mediated endoplasmic reticulum-plasma membrane contact sites. Proc Natl Acad Sci U S A 112:E2004–E2013

Feske S, Gwack Y, Prakriya M, Srikanth S, Puppel S, Tanasa B, Hogan P, Lewis RS, Daly M, Rao A (2006) A mutation in Orai1 causes immune deficiency by abrogating CRAC channel function. Nature 441:179–185

Fill M, Copello JA (2002) Ryanodine receptor calcium release channels. Physiol Rev 82:893–922

Franzini-Armstron C (1974) Freeze fracture of skeletal muscle from the Tarantula spider. Structural differentiations of sarcoplasmic reticulum and transverse tubular system membranes. J Cell Biol 61:501–513

Friedman JR, Webster BM, Mastronarde DN, Verhey KJ, Voeltz GK (2010) ER sliding dynamics and ER-mitochondrial contacts occur on acetylated microtubules. J Cell Biol 190:363–375

Galione A, White A, Willmott N, Turner M, Potter BV, Watson SP (1993) cGMP mobilizes intracellular Ca2+ in sea urchin eggs by stimulating cyclic ADP-ribose synthesis. Nature 365:456–459

Gardiner DM, Grey RD (1983) Membrane junctions in Xenopus eggs: their distribution suggests a role in calcium regulation. J Cell Biol 96:1159–1163

Gatta AT, Wong LH, Sere YY, Calderon-Norena DM, Cockcroft S, Menon AK, Levine TP (2015) A new family of StART domain proteins at membrane contact sites has a role in ER-PM sterol transport. eLife 4:e07253

Giordano F, Saheki Y, Idevall-Hagren O, Colombo SF, Pirruccello M, Milosevic I, Gracheva EO, Bagriantsev SN, Borgese N, De Camilli P (2013) PI(4,5)P(2)-dependent and Ca(2+)-regulated ER-PM interactions mediated by the extended synaptotagmins. Cell 153:1494–1509

Grigoriev I, Gouveia SM, van der Vaart B, Demmers J, Smyth JT, Honnappa S, Splinter D, Steinmetz MO, Putney JW Jr, Hoogenraad CC, Akhmanova A (2008) STIM1 is a MT-plus-end-tracking protein involved in remodeling of the ER. Curr Biol 18:177–182

Groenendyk J, Michalak M (2005) Endoplasmic reticulum quality control and apoptosis. Acta Biochim Pol 52:381–395

Gwack Y, Srikanth S, Feske S, Cruz-Guilloty F, Oh-hora M, Neems D, Hogan P, Rao A (2007) Biochemical and functional characterization of Orai proteins. J Biol Chem 282:16232–16243

Henne WM, Zhu L, Balogi Z, Stefan C, Pleiss JA, Emr SD (2015) Mdm1/Snx13 is a novel ER-endolysosomal interorganelle tethering protein. J Cell Biol 210:541–551

Hetzer MW, Walther TC, Mattaj IW (2005) Pushing the envelope: structure, function, and dynamics of the nuclear periphery. Annu Rev Cell Dev Biol 21:347–380

Holthuis JC, Levine TP (2005) Lipid traffic: floppy drives and a superhighway. Nat Rev Mol Cell Biol 6:209–220

Hoth M, Niemeyer BA (2013) The neglected CRAC proteins: Orai2, Orai3, and STIM2. Curr Top Membr 71:237–271

Hoth M, Fanger CM, Lewis RS (1997) Mitochondrial regulation of store-operated calcium signaling in T lymphocytes. J Cell Biol 137:633–648

Idevall-Hagren O, Lu A, Xie B, De Camilli P (2015) Triggered Ca2+ influx is required for extended synaptotagmin 1-induced ER-plasma membrane tethering. EMBO J 34:2291–2305

Irvine RF (1990) Quantal Ca2+ release and the control of Ca2+ entry by inositol phosphates—a possible mechanism. FEBS Lett 263:5–9

Jamieson JD, Palade GE (1967) Intracellular transport of secretory proteins in the pancreatic exocrine cell. I. Role of the peripheral elements of the Golgi complex. J Cell Biol 34:577–596

Jousset H, Frieden M, Demaurex N (2007) STIM1 knockdown reveals that store-operated Ca2+ channels located close to sarco/endoplasmic Ca2+ ATPases (SERCA) pumps silently refill the endoplasmic reticulum. J Biol Chem 282:11456–11464

Kamer KJ, Mootha VK (2014) MICU1 and MICU2 play nonredundant roles in the regulation of the mitochondrial calcium uniporter. EMBO Rep 15:299–307

Kamer KJ, Mootha VK (2015) The molecular era of the mitochondrial calcium uniporter. Nat Rev Mol Cell Biol 16:545–553

Kar P, Parekh AB (2015) Distinct spatial Ca2+ signatures selectively activate different NFAT transcription factor isoforms. Mol Cell 58:232–243

Kar P, Nelson C, Parekh AB (2011) Selective activation of the transcription factor NFAT1 by calcium microdomains near Ca2+ release-activated Ca2+ (CRAC) channels. J Biol Chem 286:14795–14803

Kar P, Samanta K, Kramer H, Morris O, Bakowski D, Parekh AB (2014) Dynamic assembly of a membrane signaling complex enables selective activation of NFAT by Orai1. Curr Biol 24:1361–1368

Kilpatrick BS, Eden ER, Schapira AH, Futter CE, Patel S (2013) Direct mobilisation of lysosomal Ca2+ triggers complex Ca2+ signals. J Cell Sci 126:60–66

Kim YJ, Guzman-Hernandez ML, Wisniewski E, Balla T (2015) Phosphatidylinositol-phosphatidic acid exchange by Nir2 at ER-PM contact sites maintains phosphoinositide signaling competence. Dev Cell 33:549–561

Krapivinsky G, Krapivinsky L, Stotz SC, Manasian Y, Clapham DE (2011) POST, partner of stromal interaction molecule 1 (STIM1), targets STIM1 to multiple transporters. Proc Natl Acad Sci U S A 108:19234–19239

Lavieu G, Orci L, Shi L, Geiling M, Ravazzola M, Wieland F, Cosson P, Rothman JE (2010) Induction of cortical endoplasmic reticulum by dimerization of a coatomer-binding peptide anchored to endoplasmic reticulum membranes. Proc Natl Acad Sci U S A 107:6876–6881

Lee HC (2011) Cyclic ADP-ribose and NAADP: fraternal twin messengers for calcium signaling. Sci China Life Sci 54:699–711

Lee MC, Miller EA, Goldberg J, Orci L, Schekman R (2004) Bi-directional protein transport between the ER and Golgi. Annu Rev Cell Dev Biol 20:87–123

Lee KP, Yuan JP, Hong JH, So I, Worley PF, Muallem S (2010) An endoplasmic reticulum/plasma membrane junction: STIM1/Orai1/TRPCs. FEBS Lett 584:2022–2027

Lefkimmiatis K, Srikanthan M, Maiellaro I, Moyer M, Curci S, Hofer AM (2009) Store-operated cyclic AMP signalling mediated by STIM1. Nat Cell Biol 11:433–442

Levine T, Loewen C (2006) Inter-organelle membrane contact sites: through a glass, darkly. Curr Opin Cell Biol 18:371–378

Lewis RS (2007) The molecular choreography of a store-operated calcium channel. Nature 446:284–287

Liou J, Kim ML, Heo WD, Jones JT, Myers JW, Ferrell JE, Meyer T (2005) STIM is a Ca2+ sensor essential for Ca2+-store-depletion-triggered Ca2+ influx. Curr Biol 15:1235–1241

Liou J, Fivaz M, Inoue T, Meyer T (2007) Live-cell imaging reveals sequential oligomerization and local plasma membrane targeting of stromal interaction molecule 1 after Ca2+ store depletion. Proc Natl Acad Sci U S A 104:9301–9306

Loewen CJ, Young BP, Tavassoli S, Levine TP (2007) Inheritance of cortical ER in yeast is required for normal septin organization. J Cell Biol 179:467–483

Luik R, Wu MM, Buchanan J, Lewis RS (2006) The elementary unit of store-operated Ca2+ entry: local activation of CRAC channels by STIM1 at ER-plasma membrane junctions. J Cell Biol 174:815–825

Luik RM, Wang B, Prakriya M, Wu MM, Lewis RS (2008) Oligomerization of STIM1 couples ER calcium depletion to CRAC channel activation. Nature 454:538–542

Lur G, Haynes LP, Prior IA, Gerasimenko OV, Feske S, Petersen OH, Burgoyne RD, Tepikin AV (2009) Ribosome-free terminals of rough ER allow formation of STIM1 puncta and segregation of STIM1 from IP(3) receptors. Curr Biol 19:1648–1653

Maiellaro I, Lefkimmiatis K, Moyer MP, Curci S, Hofer AM (2012) Termination and activation of store-operated cyclic AMP production. J Cell Mol Med 16:2715–2725

Maleth J, Choi S, Muallem S, Ahuja M (2014) Translocation between PI(4,5)P2-poor and PI(4,5)P2-rich microdomains during store depletion determines STIM1 conformation and Orai1 gating. Nat Commun 5:5843

Manford AG, Stefan CJ, Yuan HL, Macgurn JA, Emr SD (2012) ER-to-plasma membrane tethering proteins regulate cell signaling and ER morphology. Dev Cell 23:1129–1140

Manjarres IM, Rodriguez-Garcia A, Alonso MT, Garcia-Sancho J (2010) The sarco/endoplasmic reticulum Ca(2+) ATPase (SERCA) is the third element in capacitative calcium entry. Cell Calcium 47:412–418

Manjarres IM, Alonso MT, Garcia-Sancho J (2011) Calcium entry-calcium refilling (CECR) coupling between store-operated Ca2+ entry and sarco/endoplasmic reticulum Ca2+-ATPase. Cell Calcium 49:153–161

Mercer JC, Dehaven WI, Smyth JT, Wedel B, Boyles RR, Bird GS, Putney JW Jr (2006) Large store-operated calcium selective currents due to co-expression of Orai1 or Orai2 with the intracellular calcium sensor, Stim1. J Biol Chem 281:24979–24990

Michalak M, Robert Parker JM, Opas M (2002) Ca2+ signaling and calcium binding chaperones of the endoplasmic reticulum. Cell Calcium 32:269–278

Mogami H, Nakano K, Tepikin AV, Petersen OH (1997) Ca2+ flow via tunnels in polarized cells: recharging of apical Ca2+ stores by focal Ca2+ entry through basal membrane patch. Cell 88:49–55

Montalvo GB, Artalejo AR, Gilabert JA (2006) ATP from subplasmalemmal mitochondria controls Ca2+ dependent inactivation of CRAC channels. J Biol Chem 281:35616–35623

Moser von Filseck J, Copic A, Delfosse V, Vanni S, Jackson CL, Bourguet W, Drin G (2015) INTRACELLULAR TRANSPORT. Phosphatidylserine transport by ORP/Osh proteins is driven by phosphatidylinositol 4-phosphate. Science 349:432–436

Naghdi S, Waldeck-Weiermair M, Fertschai I, Poteser M, Graier WF, Malli R (2010) Mitochondrial Ca2+ uptake and not mitochondrial motility is required for STIM1-Orai1-dependent store-operated Ca2+ entry. J Cell Sci 123:2553–2564

Nunes P, Demaurex N (2014) Redox regulation of store-operated Ca2+ entry. Antioxid Redox Signal 21:915–932

Nunes P, Cornut D, Bochet V, Hasler U, Oh-Hora M, Waldburger JM, Demaurex N (2012) STIM1 juxtaposes ER to phagosomes, generating Ca(2)(+) hotspots that boost phagocytosis. Curr Biol 22:1990–1997

Okeke E, Dingsdale H, Parker T, Voronina S, Tepikin AV (2016a) Endoplasmic reticulum-plasma membrane junctions: structure, function and dynamics. J Physiol 594:2837–2847

Okeke E, Parker T, Dingsdale H, Concannon M, Awais M, Voronina S, Molgo J, Begg M, Metcalf D, Knight AE, Sutton R, Haynes L, Tepikin AV (2016b) Epithelial-mesenchymal transition, IP3 receptors and ER-PM junctions: translocation of Ca2+ signalling complexes and regulation of migration. Biochem J 473:757–767

Orci L, Ravazzola M, Le Coadic M, Shen W, Demaurex N, Cosson P (2009) STIM1-induced precortical and cortical subdomains of the endoplasmic reticulum. Proc Natl Acad Sci U S A 106:19358–19362

Orrenius S, Zhivotovsky B, Nicotera P (2003) Regulation of cell death: the calcium-apoptosis link. Nat Rev Mol Cell Biol 4:552–565

Palade GE (1955) A small particulate component of the cytoplasm. J Biophys Biochem Cytol 1:59–68

Palty R, Raveh A, Kaminsky I, Meller R, Reuveny E (2012) SARAF inactivates the store operated calcium entry machinery to prevent excess calcium refilling. Cell 149:425–438

Parekh AB (2003) Mitochondrial regulation of intracellular Ca2+ signaling: more than just simple Ca2+ buffers. News Physiol Sci 18:252–256

Parekh AB (2008) Mitochondrial regulation of store-operated CRAC channels. Cell Calcium 44:6–13

Parekh AB (2011) Decoding cytosolic Ca2+ oscillations. Trends Biochem Sci 36:78–87

Parekh AB, Muallem S (2011) Ca(2+) signalling and gene regulation. Cell Calcium 49:279

Park MK, Petersen OH, Tepikin AV (2000) The endoplasmic reticulum as one continuous Ca(2+) pool: visualization of rapid Ca(2+) movements and equilibration. EMBO J 19:5729–5739

Park C, Hoover P, Mullins F, Bachhawat P, Covington E, Raunser S, Walz T, Garcia K, Dolmetsch R, Lewis RS (2009) STIM1 clusters and activates CRAC channels via direct binding of a cytosolic domain to Orai1. Cell 136:876–890

Patron M, Checchetto V, Raffaello A, Teardo E, Reane DV, Mantoan M, Granatiero V, Szabo I, De Stefani D, Rizzuto R (2014) MICU1 and MICU2 finely tune the mitochondrial Ca2+ uniporter by exerting opposite effects on MCU activity. Mol Cell 53:726–737

Patterson RL, van Rossum DB, Gill DL (1999) Store-operated Ca2+ entry: evidence for a secretion-like coupling model. Cell 98:487–499

Perocchi F, Gohil VM, Girgis HS, Bao XR, McCombs JE, Palmer AE, Mootha VK (2010) MICU1 encodes a mitochondrial EF hand protein required for Ca2+ uptake. Nature 467:291–296

Petersen OH, Tepikin AV (2008) Polarized calcium signaling in exocrine gland cells. Annu Rev Physiol 70:273–299

Porter KR, Palade GE (1957) Studies on the endoplasmic reticulum. III. Its form and distribution in striated muscle cells. J Biophys Biochem Cytol 3:269–300

Prakriya M, Lewis RS (2001) Potentiation and inhibition of Ca(2+) release-activated Ca(2+) channels by 2-aminoethyldiphenyl borate (2-APB) occurs independently of IP(3) receptors. J Physiol 536:3–19

Prakriya M, Feske S, Gwack Y, Srikanth S, Rao A, Hogan PG (2006) Orai1 is an essential pore subunit of the CRAC channel. Nature 443:230–233

Prinz WA (2014) Bridging the gap: membrane contact sites in signaling, metabolism, and organelle dynamics. J Cell Biol 205:759–769

Putney JW Jr (1986) A model for receptor-regulated calcium entry. Cell Calcium 7:1–12

Putney JW Jr (2007) Recent breakthroughs in the molecular mechanism of capacitative calcium entry (with thoughts on how we got here). Cell Calcium 42:103–110

Putney JW Jr, Bird GS (2008) Cytoplasmic calcium oscillations and store-operated calcium influx. J Physiol 586:3055–3059

Reger JF (1961) The fine structure of neuromuscular junctions and the sarcoplasmic reticulum of extrinsic eye muscles of Fundulus heteroclitus. J Biophys Biochem Cytol 10 (4, Suppl):111–121

Ritchie MF, Samakai E, Soboloff J (2012) STIM1 is required for attenuation of PMCA-mediated Ca2+ clearance during T-cell activation. EMBO J 31:1123–1133

Rizzuto RR, Pinton P, Carrington W, Fay FS, Fogarty KE, Lifshitz LM, Tuft RA, Pozzan T (1998) Close contacts with the endoplasmic reticulum as determinants of mitochondrial Ca2+ responses. Science 280:1763–1766

Roos J, DiGregorio P, Yeromin A, Ohlsen K, Lioudyno M, Zhang S, Safrina O, Kozak J, Wagner S, Cahalan M, Velicelebi G, Stauderman K (2005) STIM1, an essential and conserved component of store-operated Ca2+ channel function. J Cell Biol 169:435–445

Rosado JA, Lopez JJ, Harper AG, Harper MT, Redondo PC, Pariente JA, Sage SO, Salido GM (2004) Two pathways for store-mediated calcium entry differentially dependent on the actin cytoskeleton in human platelets. J Biol Chem 279:29231–29235

Rosenbluth J (1962) Subsurface cisterns and their relationship to the neuronal plasma membrane. J Cell Biol 13:405–421

Rossi AE, Dirksen RT (2006) Sarcoplasmic reticulum: the dynamic calcium governor of muscle. Muscle Nerve 33:715–731

Rutkowski DT, Kaufman RJ (2004) A trip to the ER: coping with stress. Trends Cell Biol 14:20–28

Schulz TA, Choi MG, Raychaudhuri S, Mears JA, Ghirlando R, Hinshaw JE, Prinz WA (2009) Lipid-regulated sterol transfer between closely apposed membranes by oxysterol-binding protein homologues. J Cell Biol 187:889–903

Sharma S, Quintana A, Findlay GM, Mettlen M, Baust B, Jain M, Nilsson R, Rao A, Hogan PG (2013) An siRNA screen for NFAT activation identifies septins as coordinators of store-operated Ca2+ entry. Nature 499:238–242

Singaravelu K, Nelson C, Bakowski D, de Brito OM, Ng SW, Di Capite J, Powell T, Scorrano L, Parekh AB (2011) Mitofusin 2 regulates STIM1 migration from the Ca2+ store to the plasma membrane in cells with depolarized mitochondria. J Biol Chem 286:12189–12201

Smyth J, DeHaven W, Bird G, Putney JW Jr (2008) Ca2+-store-dependent and independent reversal of Stim1 localization and function. J Cell Sci 121:762–772

Smyth JT, Petranka JG, Boyles RR, Dehaven WI, Fukushima M, Johnson KL, Williams JG, Putney JW Jr (2009) Phosphorylation of STIM1 underlies suppression of store-operated calcium entry during mitosis. Nat Cell Biol 11:1465–1472

Smyth JT, Beg AM, Wu SL, Putney JW Jr, Rusan NM (2012) Phosphoregulation of STIM1 leads to exclusion of the endoplasmic reticulum from the mitotic spindle. Curr Biol 22:1487–1493

Tabas I, Ron D (2011) Integrating the mechanisms of apoptosis induced by endoplasmic reticulum stress. Nat Cell Biol 13:184–190

Takeshima H, Hoshijima M, Song LS (2015) Ca(2+) microdomains organized by junctophilins. Cell Calcium 58:349–356

Taylor CW, Rahman T, Tovey SC, Dedos SG, Taylor EJ, Velamakanni S (2009) IP3 receptors: some lessons from DT40 cells. Immunol Rev 231:23–44

Tepikin AV, Voronina SG, Gallacher DV, Petersen OH (1992) Pulsatile Ca2+ extrusion from single pancreatic acinar-cells during receptor-activated cytosolic Ca2+ spiking. J Biol Chem 267:14073–14076

Tsai FC, Meyer T (2012) Ca2+ pulses control local cycles of Lamellipodia retraction and adhesion along the front of migrating cells. Curr Biol 22:837–842

Tsai FC, Seki A, Yang HW, Hayer A, Carrasco S, Malmersjo S, Meyer T (2014) A polarized Ca2+, diacylglycerol and STIM1 signalling system regulates directed cell migration. Nat Cell Biol 16:133–144

Varnai P, Toth B, Toth DJ, Hunyady L, Balla T (2007) Visualization and manipulation of plasma membrane-endoplasmic reticulum contact sites indicates the presence of additional molecular components within the STIM1-Orai1 complex. J Biol Chem 282:29678–29690

Vig M, Peinelt C, Beck A, Koomoa DL, Rabah D, Koblan-Huberson M, Kraft S, Turner H, Fleig A, Penner R, Kinet JP (2006) CRACM1 is a plasma membrane protein essential for store-operated Ca2+ entry. Science 312:1220–1223

Voeltz GK, Rolls MM, Rapoport TA (2002) Structural organization of the endoplasmic reticulum. EMBO Rep 3:944–950

Walsh C, Barrow S, Voronina S, Chvanov M, Petersen OH, Tepikin A (2009) Modulation of calcium signalling by mitochondria. Biochim Biophys Acta 1787:1374–1382

Walsh CM, Chvanov M, Haynes LP, Petersen OH, Tepikin AV, Burgoyne RD (2010) Role of phosphoinositides in STIM1 dynamics and store-operated calcium entry. Biochem J 425:159–168

Watanabe H, Burnstock G (1976) Junctional subsurface organs in frog sympathetic ganglion cells. J Neurocytol 5:125–136

Willoughby D, Wachten S, Masada N, Cooper DM (2010) Direct demonstration of discrete Ca2+ microdomains associated with different isoforms of adenylyl cyclase. J Cell Sci 123:107–117

Willoughby D, Everett KL, Halls ML, Pacheco J, Skroblin P, Vaca L, Klussmann E, Cooper DM (2012) Direct binding between Orai1 and AC8 mediates dynamic interplay between Ca2+ and cAMP signaling. Sci Signal 5:ra29

Wu MM, Buchanan J, Luik RM, Lewis RS (2006) Ca2+ store depletion causes STIM1 to accumulate in ER regions closely associated with the plasma membrane. J Cell Biol 174:803–813

Wu MM, Covington ED, Lewis RS (2014) Single-molecule analysis of diffusion and trapping of STIM1 and Orai1 at endoplasmic reticulum-plasma membrane junctions. Mol Biol Cell 25:3672–3685

Wuytack F, Raeymaekers L, Missiaen L (2002) Molecular physiology of the SERCA and SPCA pumps. Cell Calcium 32:279–305

Xu P, Lu J, Li Z, Yu X, Chen L, Xu T (2006) Aggregation of STIM1 underneath the plasma membrane induces clustering of Orai1. Biochem Biophys Res Commun 350:969–976

Yu F, Sun L, Machaca K (2010) Constitutive recycling of the store-operated Ca2+ channel Orai1 and its internalization during meiosis. J Cell Biol 191:523–535

Yuan JP, Zeng W, Dorwart MR, Choi YJ, Worley PF, Muallem S (2009) SOAR and the polybasic STIM1 domains gate and regulate Orai channels. Nat Cell Biol 11:337–343

Zhang SL, Yu Y, Roos J, Kozak JA, Deerinck TJ, Ellisman MH, Stauderman KA, Cahalan MD (2005) STIM1 is a Ca2+ sensor that activates CRAC channels and migrates from the Ca2+ store to the plasma membrane. Nature 437:902–905

New Aspects of the Contribution of ER to SOCE Regulation: TRPC Proteins as a Link Between Plasma Membrane Ion Transport and Intracellular Ca^{2+} Stores

13

Alexis Bavencoffe, Michael Xi Zhu, and Jin-bin Tian

Abstract

Transient receptor potential canonical (TRPC) proteins were identified as molecular candidates of receptor- and/or store-operated channels because of their close homology to the *Drosophila* TRP and TRPL. Functional studies have revealed that TRPC channels play an integrated part of phospholipase C-transduced cell signaling, mediating the influx of both Ca^{2+} and Na^{+} into cells. As a consequence, the TRPC channels have diverse functional roles in different cell types, including metabotropic receptor-evoked membrane depolarization and intracellular Ca^{2+} concentration elevation. Depending on the cellular environment and the protein partners present in the channel complex, the TRPC channels display different biophysical properties and mechanisms of regulation, including but not limited to the Ca^{2+} filling state of the endoplasmic reticulum. Despite the overwhelming focus on STIM-regulated Orai channels for store-operated Ca^{2+} entry, evidence is growing for STIM-operated TRPC channel activities in various cell types, demonstrating both store-dependent and store-independent mechanisms of TRPC channel gating. The existence of physical and functional interactions between plasma membrane-localized TRPC channels and other proteins involved in sensing and regulating the intracellular Ca^{2+} store contents, such as inositol trisphosphate receptors, Junctate, and Homer, further argues for the role of TRPC proteins in linking plasma membrane ion transport with intracellular Ca^{2+} stores. The interplay among these proteins will likely define the functional significance of TRPC channel activation in different cellular contexts and under different modes of stimulations.

A. Bavencoffe • M.X. Zhu (✉) • J.-b. Tian
Department of Integrative Biology and Pharmacology, McGovern Medical School, The University of Texas Health Science Center at Houston, Houston, TX, USA
e-mail: Michael.X.Zhu@uth.tmc.edu

K. Groschner et al. (eds.), *Store-Operated Ca^{2+} Entry (SOCE) Pathways*, Advances in Experimental Medicine and Biology 993, DOI 10.1007/978-3-319-57732-6_13

Keywords
TRPC • STIM • Orai • SOCE • Isoc • SOAR domain • Trafficking

13.1 Introduction: Importance and Regulation of Ca^{2+} Homeostasis

Ca^{2+} regulates a wild range of fundamental and kinetically distinct cellular processes, ranging from secretion, contraction, transcription, and cell excitability to migration, apoptosis, differentiation, and proliferation. To initiate and regulate such multitude of responses with precision and specificity, the Ca^{2+} signal is encoded by three parameters (kinetics, amplitude, and subcellular location), which give rise to a "Ca^{2+} signature" for each cellular process (Berridge 1993; Berridge et al. 1998, 2000). The "Ca^{2+} signature" is generated through a precise control of Ca^{2+} homeostasis, involving ionic pumps, ion transporters, ion channels, and receptors. These molecules help regulate electrochemical gradients and fluxes of Ca^{2+} across plasma and intracellular membranes. In fact, among all ions, Ca^{2+} has the greatest concentration gradient across cell membranes. Extracellular Ca^{2+} concentration is about 1–2 mM, while intracellular or cytoplasmic Ca^{2+} concentration ($[Ca^{2+}]_i$) is ~100 nM in resting cells. In addition, many intracellular organelles stock Ca^{2+}. The endoplasmic reticulum (ER) is the main intracellular Ca^{2+} store with a free luminal concentration approaching 1 mM. Lysosomes, and perhaps other intracellular vesicles, also have higher free Ca^{2+} concentrations than the cytoplasm. Maintaining the Ca^{2+} gradients across membranes is crucial for all Ca^{2+}-dependent cellular processes (Berridge et al. 2003).

Elevation of $[Ca^{2+}]_i$ occurs through Ca^{2+} influx from extracellular space and/or Ca^{2+} release from intracellular stores (principally the ER). The release of Ca^{2+} from ER is mediated by inositol 1,4,5-trisphosphate (IP_3) receptors (IP_3Rs) and ryanodine receptors (RyRs) located on the ER membrane. Recent studies also revealed that two-pore channels (TPCs), P2X4 purinergic receptors, and the mucolipin family of transient receptor potential (TRP) proteins, or TRPML channels, are involved in Ca^{2+} release from endolysosomes (Calcraft et al. 2009; Dong et al. 2010; Cao et al. 2015). Similarly, multiple types of Ca^{2+}-permeable channels are present on the plasma membrane (PM), and the distinction for different forms of Ca^{2+} influx lies in the mode of activation and cell-specific distribution of these channels. In excitable cells, voltage-gated Ca^{2+} channels represent the main pathway of Ca^{2+} influx, although ligand-gated cation channels also contribute. In non-excitable cells, Ca^{2+} influx is thought to be carried out by receptor-operated channels (ROCs) and/or store-operated channels (SOCs). The former refer to Ca^{2+} entry channels activated indirectly through ligand activation of cell surface receptors coupled to G proteins and tyrosine kinases, whereas the latter describe channels that mediate Ca^{2+} influx in response to Ca^{2+} depletion from the ER.

Under physiological conditions, both receptor- and store-operated Ca^{2+} entries occur following activation of phospholipase C (PLC) by either $G_{q/11}$-coupled receptors or receptor tyrosine kinases, which in turn produces IP_3 and diacylglycerols. IP_3 activates IP_3Rs, causing Ca^{2+} release from the ER. As a consequence, the ER store is depleted, leading to SOC activation. Needless to say, one of the main functions of store-operated Ca^{2+} entry is to replenish the emptied ER stores. This is why the store-operated Ca^{2+} entry was initially designated as capacitative Ca^{2+} entry (CCE) (Putney 1986, 1991). On the other hand, it has also been recognized that Ca^{2+} entry, being store- or receptor-operated, plays important roles in prolonging Ca^{2+} elevation and contributing to spatial and temporal Ca^{2+} signaling (Berridge et al. 2003).

13.2 History of the Discovery of TRPC Channels as Candidates of Store-Operated Ca^{2+} Entry

The story of the discovery of store-operated Ca^{2+} entry dated back to 1981 when Casteels and Droogmans published that intracellular Ca^{2+} store depletion by noradrenaline in vascular smooth muscle cells increased Ca^{2+} uptake from extracellular medium, suggesting a functional link between ER Ca^{2+} release and PM Ca^{2+} influx (Casteels and Droogmans 1981). In 1986, Putney proposed the capacitative model to explain how IP_3-mediated store release triggers the novel Ca^{2+} entry pathway that responds purely to the fullness of Ca^{2+} in the internal store (Putney 1986). Although it is difficult to imagine a physiological condition at which the internal Ca^{2+} store is depleted without the concomitant activation of the PLC pathway that causes IP_3 production, or alternately the stimulation of RyRs, pure store-operated Ca^{2+} entry has been demonstrated in vitro pharmacologically in many cell types by the use of sarco-/endoplasmic reticulum Ca^{2+}-ATPase (SERCA) inhibitors, such as thapsigargin and cyclopiazonic acid (Thastrup et al. 1987; Jackson et al. 1988; Parekh and Penner 1997). In patch clamp experiments, store-operated currents, activated by SERCA inhibitors, intracellular dialysis of IP_3, or chelating cytoplasmic Ca^{2+} with EGTA or BAPTA, have been detected in multiple cell types. Among them, the most representative is the Ca^{2+} release-activated Ca^{2+} current (I_{CRAC}) first reported in T lymphocytes and mast cells (Lewis and Cahalan 1989; Hoth and Penner 1992), which is characterized by a high selectivity to Ca^{2+} ($P_{Ca}/P_{Na} = {\sim}1000$), a pronounced inward rectification with a very positive reversal potential, and an exceedingly small unitary conductance in the femtosiemen range (Hoth and Penner 1992, 1993). However, I_{CRAC} is not the only current activated by store depletion. In non-hematopoietic cells, currents possessing most properties of I_{CRAC} have been reported, but they differ from I_{CRAC} mainly by lower Ca^{2+} selectivity (e.g., $P_{Ca}/P_{Na} \sim 10$ in vascular endothelial cells, ~4 in parotid gland cells, or 1 in aortic myocytes) and larger conductance (11 pS for endothelial cells, 2.7 pS for aortic myocytes, 43 pS for pancreatic acinar cells, and 5.4 pS for pulmonary artery myocytes) (reviewed in Parekh and Putney 2005). Collectively, these have been referred to as I_{SOC}.

On the other hand, much less has been discussed for ROCs. Because receptor operation is a loosely defined term, there has been no consensus as to what constitute a truly "receptor-operated" channel. In a broad sense, receptor activation, strictly speaking here just for those receptors that stimulate PLC, triggers a cascade of cell signaling events that include phosphoinositide hydrolysis, second messenger production, Ca^{2+} store depletion, and $[Ca^{2+}]_i$ rise. As such, any channel that responds to the phosphoinositide breakdown, second messengers (IP_3, diacylglycerols, or Ca^{2+}), and/or store depletion might be called a ROC. In practice, when channel activity emerges following receptor stimulation, it is not always easy to tell which constituent(s) or step(s) of the receptor-activated pathway is responsible for activating the channel. Thus, ROCs often encompass SOCs, second messenger-activated channels, as well as Ca^{2+}-activated channels, although these concepts are not equivalent. In mast cells, a secretagogue compound 48/80 activated a 50-pS nonselective cation current, which is distinct from I_{CRAC} in its dependence on intracellular Ca^{2+}, G proteins, IP_3Rs, and in biophysical properties (Fasolato et al. 1993; Hoth et al. 1993). However, such detailed characterizations required rigorous electrophysiological manipulations, which have not been carried out for many ROCs. Pharmacological tools for ROCs and SOCs are also very rare and nonspecific, further hindering the characterization of these channels.

Because of the similarity between the pathways for insect phototransduction and PLC signaling in mammalian cells, the *Drosophila* TRP protein attracted the attention as being a candidate of SOC and/or ROC (Minke and Selinger 1991; Hardie and Minke 1992, 1993). The cloned *TRP* and *TRPL* genes were predicted to be transmembrane proteins that share sequence homology with the pore-forming subunits of voltage-gated Ca^{2+} and Na^+ channels (Montell and Rubin 1989; Phillips et al. 1992). TRP was shown to contribute to the light-induced Ca^{2+}-permeable currents in *Drosophila* photoreceptors (Hardie and Minke 1992) downstream from a rhodopsin-G_q-PLCβ pathway. Heterologous expression of TRPL in insect *Sf9* cells reconstituted a receptor-operated Ca^{2+} influx and a cation conductance (Hu et al. 1994).

These findings encouraged the search for the TRP homologue in vertebrate species. In the same month of 1995, partial sequences of TRP homologues in mouse brain and *Xenopus* oocytes (Petersen et al. 1995) and the full-length human TRPC1 sequence were reported (Wes et al. 1995; Zhu et al. 1995). Subsequently, TRPC2–TRPC7 were isolated, with functional demonstrations of their involvement in store- or receptor-operated Ca^{2+} entry (Zhu et al. 1996, 1998; Philipp et al. 1996, 1998; Boulay et al. 1997; Okada et al. 1999; Vannier et al. 1999). Some of these reports, as well as many subsequent studies, included electrophysiological data showing TRPC-mediated currents activated by receptor stimulation and/or internal Ca^{2+} store depletion (Hurst et al. 1998; Liu et al. 2000; Strübing et al. 2001). The receptor- and/or G protein (via GTPγS)-evoked TRPC currents are typically cation nonselective, with conductance values ranging from 15 to 70 pS, and outwardly rectifying. These features are different from that of I_{CRAC}; therefore, heteromultimerization of different TRPC isoforms or of TRPC with another auxiliary protein(s) had been suggested (Zhu et al. 1998; Vazquez et al. 2001). However, no successful heterologous reconstitution of a CRAC-like current has been made using TRPC in combination with any other protein, with an exception, maybe, of TRPC plus a low level of Orai1 (Liao et al. 2008). Thus, despite a large number of reports demonstrating that the knockdown or knockout of TRPC expression or dominant-negative constructs suppressed store-operated Ca^{2+} entry, the role of TRPC in store-operated Ca^{2+} entry remains controversial.

On the other hand, the more recent identification of STIM and Orai and functional reconstitution of I_{CRAC} by these proteins seem to rule out TRPCs from playing an essential role in this highly specific Ca^{2+} entry pathway (Fig. 13.1b). However, despite the overwhelming support for Orai, evidence for the involvement of TRPC channels in store-operated Ca^{2+} entry still kept on growing in recent years. Since the first edition of this book volume, new reports have appeared to support the role of TRPC1, TRPC3, TRPC4, TRPC5, and TRPC6 in store-operated Ca^{2+} entry in platelets, megakaryoblasts, primary nociceptors, endothelial cells, and myoblasts (Cioffi et al. 2012; Sundivakkam et al. 2012; Antigny et al. 2013; Lopez et al. 2013; Albarran et al. 2014; Alkhani et al. 2014; Sabourin et al. 2016). More importantly, there has been increasing appreciation that STIM1 may make the switch between store-dependent and store-independent TRPC functions (Ong et al. 2007; Zeng et al. 2008; Sundivakkam et al. 2012; Lee et al. 2014; Saul et al. 2014; Asanov et al. 2015; Desai et al. 2015) and even Orai1 may be physically associated or at least functionally coupled to a TRPC protein to alter or regulate its function (Jardin et al. 2008; Cioffi et al. 2012; Cheng et al. 2013; Saul et al. 2014; Desai et al. 2015; Xu et al. 2015). Therefore, at the level of store operation, there is significant overlap between Orai and TRPC channels, although the functional consequences of these two channel types may be quite distinct (Ong et al. 2012; Cheng et al. 2013).

13.3 Physiological Functions of TRPC Channels

TRPC channels have been implicated in many physiological functions via store- and/or receptor-operated mechanisms. Here we highlight just a few to illustrate the functional diversity of these channels. It is important to keep in mind that, first, TRPC channels are generally activated downstream from PLC stimulation and, second, channels formed by TRPCs, at least when overexpressed in heterologous systems in homomeric forms, are generally cation nonselective. This means that, when open, TRPC channels allow not only Ca^{2+} but also Na^+ influx. Often, the Na^+ influx dominates, leading to membrane depolarization and consequent activation of voltage-gated Na^+ and Ca^{2+} channels in excitable cells. In this respect, TRPC channels are similar to cation nonselective ionotropic receptors, with an exception that the activation is often triggered by metabotropic receptors. Because the channel is activated following receptor signaling, the response may show a slower kinetics, but it may last longer (Kim et al. 2003; Hartmann et al. 2008). However, the light response of the *Drosophila* photoreceptors, which occurs through G protein coupling to the activation of TRP, is extremely fast (Hardie and Raghu 2001). In non-excitable cells, the prominent Na^+ influx mediated through TRPC channels has been shown to drive Ca^{2+} influx via Na^+/Ca^{2+} exchangers working in the reverse mode (Rosker et al. 2004; Harper et al. 2013).

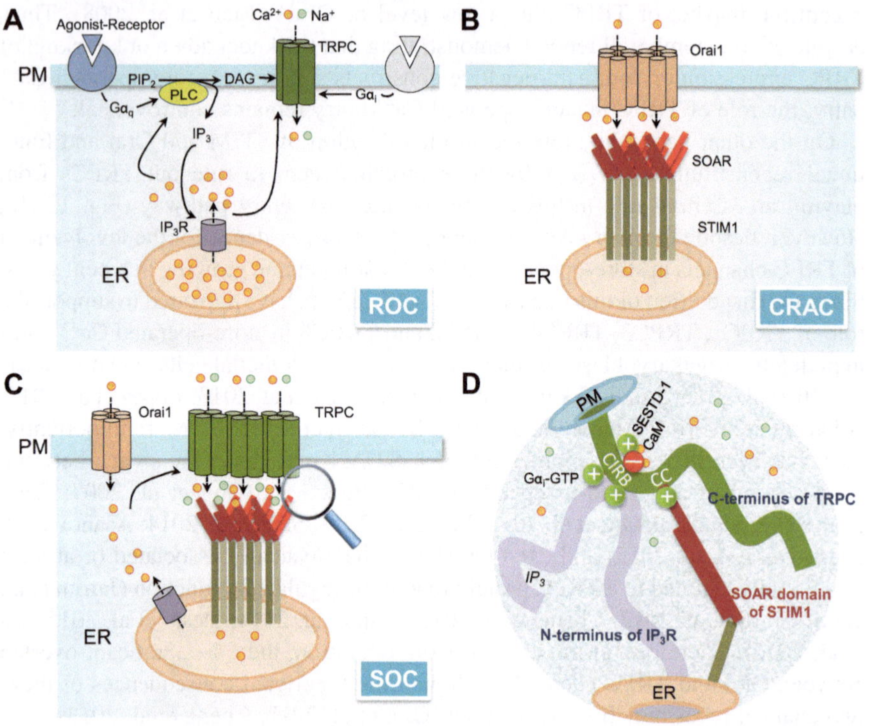

Fig. 13.1 Store-independent and store-dependent activation of TRPC channels. (**a–c**) Schematics of TRPC channels function as receptor-operated channels (ROC, **a**) or as store-operated channels (SOC, **c**). The Ca^{2+} release-activated Ca^{2+} (CRAC) channels formed by hexameric Orai1 are depicted in **b** for comparison. TRPC channels permeate the entry of both Ca^{2+} and Na^+, whereas Orai1 is highly Ca^{2+}-selective. The store-operated activation of TRPC requires STIM1; however, it is not known whether the presence of STIM1 would occlude the channel activation by other constituents of receptor stimulation, which normally involves PLC that hydrolyzes phosphatidylinositol 4,5-bisphosphate (PIP_2) to produce IP_3 and diacylglycerols (DAG). DAG directly activates some TRPC channels. IP_3 acts at IP_3Rs to release Ca^{2+} from the ER, which elevates $[Ca^{2+}]_i$. The activation of IP_3Rs can facilitate TRPC channel activation via a direct protein-protein interaction at the calmodulin- and IP_3 receptor-binding (CIRB) site of TRPC (see **d**). The elevated $[Ca^{2+}]_i$ facilitates TRPC channel opening via multiple mechanisms, including activation of PLCδ isoforms. The opening of Orai1 channels could also provide the Ca^{2+} source to facilitate TRPC (depicted in **c**). In a different mode of store operation, TRPC4 and TRPC5 can be activated downstream from $G_{i/o}$-coupled receptors via GTP-bound $G\alpha_{i/o}$ subunits in a PLC-dependent manner (Thakur et al. 2016). (**d**) An enlarged area of PM-ER junction showing the C-terminal region of TRPC where the CIRB domain interacts with the N-terminus of IP_3R, calmodulin (CaM), $G\alpha_{i/o}$-GTP (TRPC4/TRPC5), and SESTD1 (TRPC4/TRPC5) and the adjacent coiled-coil (CC) domain interacts with the STIM-Orai-activating region (SOAR) of STIM1. Positive and negative effects on TRPC activation are indicated by the "+" and "−" signs, respectively. Not shown here are Homer 1 and Junctate, which interact with both TRPC and IP_3Rs

Studies using knockout animals bearing deletion of specific TRPC genes have revealed a wide range of physiological and pathological functions associated with these channels. These include neurological and cardiovascular functions such as neurite growth (C5) (Greka et al. 2003), synaptic transmission (C1, C3, C4, C5) (Freichel et al. 2005; Yan et al. 2009; Tai et al. 2011; Stroh et al. 2012), neurodegeneration (C1, C4, C5) (Selvaraj et al. 2012; Phelan et al. 2012, 2013), fear conditioning and fear memory (C3, C4, C5) (Riccio et al. 2009, 2014; Neuner et al. 2015), motor coordination (C3) (Hartmann et al. 2008), seizure initiation (C7) (Phelan et al. 2014), endothelial permeability (C1, C3, C4, C6) (Freichel et al. 2005; Cioffi et al. 2012), vascular tone (C1, C3, C4, C6) (Freichel et al. 2005; Weissmann et al. 2006; Ding et al. 2011; Kochukov et al. 2013, 2014), and cardiac hypertrophy (C1, C4, C3, C6) (Seo et al. 2014; Camacho Londoño et al. 2015). Store-operated Ca^{2+} entry via TRPC1 is essential for saliva secretion regulated by neurotransmitters (Liu et al. 2000, 2007), which involves downstream activation of Ca^{2+}-activated K^+ channels (Liu et al. 2007) and Ca^{2+}-activated Cl^- channels (Sun et al. 2015). Ca^{2+} influx via TRPC3 is involved in regulating Ca^{2+} oscillation frequency in pancreatic acinar cells, the excess activity of which during pancreatitis is partially responsible for cell stress and damage (Kim et al. 2009). Enhanced TRPC3 activity has also been linked to renal fibrosis (Saliba et al. 2015). TRPC4 and TRPC6 are involved in electrogenic response of gastrointestinal smooth muscle cells that regulate intestinal motility (Tsvilovskyy et al. 2009). Mutations in the TRPC6 gene have been identified in patients with focal segmental glomerulosis. Many of these mutants appear to display enhanced channel activity as compared to the wild-type TRPC6 (Winn et al. 2005; Reiser et al. 2005), but loss-of-function TRPC6 mutations have also been found in some patients with this renal disease (Riehle et al. 2016). TRPC1, TRPC3, TRPC5, and TRPC6 may function together in cutaneous and auditory mechanosensation, although they do not contribute directly to the hair cell mechanotransduction (Sexton et al. 2016). TRPC7 appeared to be involved in heart failure in Dahl salt-sensitive rats (Satoh et al. 2007). All these studies point to TRPC channel functions as integrative components of the PLC signaling pathway that exert diverse physiological roles in different systems. In addition to PLC, $G_{i/o}$ proteins have also been implicated in the activation of TRPC4 and TRPC5, most likely via a direct effect of GTP-bound $G\alpha_{i/o}$ subunits on the channels (Jeon et al. 2012; Thakur et al. 2016) (Fig. 13.1a).

13.4 The Interplay Between TRPC Channels and the Internal Ca^{2+} Stores

During the first decade of the TRPC channel discovery, a significant amount of efforts was devoted to validating whether or not TRPCs form SOCs. Hence there is a rich literature on the effects of TRPC knockdown via antisense oligonucleotides or siRNA, as well as TRPC antibodies and dominant-negative constructs on store-operated Ca^{2+} entry. In a unique example, upregulation of TRPC4 and TRPC5 was found to be associated with enhanced store-operated divalent cation entry resulting from a knockdown of SERCA2 expression in cultured cardiac myocytes (Seth et al. 2004). Unlike studies specifically designed to examine how changes in TRPC expression affected store-operated Ca^{2+} entry, this observation was made without a predisposition of TRPC being involved in compensating the reduced SERCA-dependent ER Ca^{2+} uptake, and therefore, it strongly suggests a functional interplay between intracellular Ca^{2+} stores and TRPC channels.

Another piece of evidence to support the TRPC-ER link is the physical and functional interaction between TRPCs and the ER Ca^{2+} release channels, IP_3Rs. Initially, a functional interaction was demonstrated between IP_3Rs and TRPC3 expressed in HEK293 cells (Kiselyov et al. 1998), which was considered the first evidence supporting the conformational coupling hypothesis originally proposed by R. Irvine (1990) and then elaborated by M. Berridge (1995). Subsequently, a physical interaction between IP_3Rs and TRPC3 was confirmed by co-immunoprecipitation and glutathione S-transferase pull-down experiments (Boulay et al. 1999; Kiselyov et al. 1999). More interestingly, a TRPC-binding site was identified at the N-terminus of IP_3R downstream from the IP_3-binding domain, and an IP_3R-binding site was localized at the C-terminus of TRPC3, downstream from the TRP motif (Zhang et al. 2001). In addition, the IP_3R-binding site is conserved among all TRPC proteins and it overlaps with a site that binds to Ca^{2+}-calmodulin (CaM) (Zhang et al. 2001; Tang et al. 2001). Thus, the presence of a calmodulin- and IP_3 receptor-binding (CIRB) site at the C-terminus is a common feature of all TRPC proteins (Fig. 13.1d). In in vitro binding experiments, CaM inhibited IP_3R binding to the CIRB sites of TRPCs in a Ca^{2+}-dependent manner, and vice versa an IP_3R peptide representing the TRPC-binding motif displaced CaM binding to the CIRB sites. The functional correlate of this competitive binding of CaM and IP_3Rs has been examined for TRPC3 (Zhang et al. 2001) and TRPC4 (Tang et al. 2001). It was found that Ca^{2+}-CaM exerts an inhibitory effect on TRPC channels. Removal of the inhibitory CaM, either through activation of IP_3Rs by IP_3 or other means, such as pharmacologically inhibiting CaM, expressing a dominant-negative CaM mutant, or displacing CaM with a high-affinity CaM-binding peptide, activated the TRPC channels (Zhang et al. 2001; Tang et al. 2001). Importantly, the regulation of TRPC channel activity through competitive binding between CaM and IP_3Rs has also been demonstrated for an endogenous SOC, presumably composed of TRPC1, in CHO cells (Vaca and Sampieri 2002).

Interestingly, later studies from independent groups also showed that the CIRB sites of TRPC4 and TRPC5 are critical for binding by $G\alpha_i$-GTP (Jeon et al. 2012) and SESTD1 (SEC14 and spectrin domain containing 1), a Ca^{2+}-dependent phospholipid-binding protein that contains a SEC14-like domain and two spectrin repeats (Miehe et al. 2010). A few amino acids downstream from the CIRB site exists a conserved C-terminal coiled-coil domain, which may be responsible for the store-dependent activation of TRPC channels by STIM1 (Lee et al. 2014). At least for TRPC1, TRPC4, and TRPC5, single-channel activities evoked by a purified recombinant protein fragment that represents the STIM-Orai-activating region (SOAR) of STIM1 in excised inside-out patches were inhibited by CaM in a Ca^{2+}-dependent manner (Asanov et al. 2015). Therefore, the CIRB site and surrounding areas could constitute a "hotspot" of TRPC channel gating, where multiple mechanisms converge, with removing the inhibitory CaM being at least one of the major actions (Fig. 13.1d).

Also interesting is that the TRPC-IP_3R interaction may be equally important for TRPC regulation of ER Ca^{2+} release via IP_3Rs. For example, the knockout of TRPC1 in chicken B cells reduced not only store-operated Ca^{2+} entry but also IP_3R-mediated ER Ca^{2+} release (Mori et al. 2002). Additionally, a few other proteins can be involved in the TRPC-IP_3R interaction. A scaffolding protein, Homer 1, has been reported to facilitate the TRPC-IP_3R association and be involved in the functional regulation of TRPC1 and TRPC3 (Yuan et al. 2003; Kim et al. 2006). In skeletal muscles of Homer 1 knockout mice, an increased cation influx was found to be associated with reduced muscle mass and contractility, and it was corrected by silencing TRPC1 expression (Stiber et al. 2008), suggesting that the TRPC-Homer interaction is important for maintaining the low basal activity of TRPC1-containing channels in the myocytes. Another example is a trans-ER membrane protein, Junctate, which binds IP_3Rs at its cytoplasmic N-terminus. The C-terminus of Junctate is located in the ER lumen and is able to bind to Ca^{2+} at a 21:1 molar ratio with a K_D of 217 μM (Treves et al. 2004). These properties allow Junctate to sense the Ca^{2+} content in the ER lumen and transmit the information via its N-terminal interaction with IP_3Rs. Indeed, the N-terminus of Junctate was shown to associate with IP_3Rs and TRPC3 in the same complex. Junctate overexpression augmented both Ca^{2+} release from the ER store and Ca^{2+} influx from the extracellular space. Infusion of a peptide corresponding to the N-terminus of Junctate inhibited agonist-induced cation current in HEK293 cells stably expressing TRPC3. More interestingly, overexpression of the full-length Junctate strongly increased the area of the ER that is juxtaposed with the PM in the form of peripheral junctions or puncta (Treves et al. 2004, 2010). These puncta were further enriched with the overexpression of TRPC3, and they colocalized with the PM sites where agonist-induced Ca^{2+} influx was detected (Treves et al. 2010). In mouse sperm, Junctate coexists with IP_3Rs in the acrosomal crescent of the anterior head, and it interacts directly with TRPC2 and TRPC5, but not TRPC1. However, the interaction of the Junctate N-terminus with the TRPC2 C-terminus was Ca^{2+}-independent and did not appear to overlap with the CIRB site of TRPC2 (Stamboulian et al. 2005). Recently, Junctate and junctophilin-4 were both shown to be present at ER-PM junctions to interact with the Orai1-STIM1 complex and in turn regulate store Ca^{2+} release and store-operated Ca^{2+} entry (Srikanth et al. 2012; Woo et al. 2016). Thus, like STIM1, Junctate works via both TRPC and Orai channels. Collectively, these data support the view that complexes formed by IP_3Rs and TRPC channels are involved in the functional coupling between the ER Ca^{2+} stores and the PM for coordinating internal release and Ca^{2+} entry.

Finally, accumulating evidence has provided a strong support for the communication between TRPCs and the intracellular Ca^{2+} stores through STIMs. As mentioned above, STIM1 may make the switch between store-dependent and store-independent modes of TRPC channel activation, which may explain why heterologously expressed TRPCs, in the absence of STIM1 coexpression, typically exhibited store-independent activation, with the store-operated component only appearing under strictly defined conditions (Philipp et al. 1998; Zhu et al. 1998). STIM1 directly binds to TRPC1, TRPC4, and TRPC5, but not TRPC3 and TRPC6, and enhances the cation currents mediated through these channels (Worley et al. 2007; Ong et al. 2007; Yuan et al. 2007; Ambudkar et al. 2007; Liao et al. 2008, 2009; Lee et al. 2014). More amazingly, just the SOAR domain of STIM1 (aa 344–442) was able to activate TRPC1, TRPC4, and TRPC5 in inside-out membrane patches, and this occurred at a stoichiometry of two SOAR fragments per single tetrameric TRPC channel (Asanov et al. 2015). This suggests that the SOAR of STIM1 is equally important for activation of both Orai and TRPC channels, although in the full-length STIM1, the lysine doublet (KK) at positions 684 and 685 may be particularly critical for gating TRPC via an electrostatic interaction with the aspartate doublet of TRPC1 ($D_{639}D_{640}$) (Zeng et al. 2008). Apparently, as in the case between STIM1 and Orai1, the stoichiometry between STIM1 and a TRPC is crucial for store-operated channel gating (Fig. 13.1c), and competitions between Orai1 and TRPCs for STIM1 could determine whether the store-operated Ca^{2+} entry in the cell was carried by CRAC channels formed by Orai1 or SOCs formed by TRPCs (Lee et al. 2014; Saul et al. 2014; Desai et al. 2015; de Souza et al. 2015).

On the other hand, the store-operated function of TRPC channels may be dependent on Orai1 (Liao et al. 2007, 2008, 2009; Cheng et al. 2008; Lee et al. 2010). The knockdown of endogenous Orai1 expression abolished the store-operated Ca^{2+} entry and the TRPC1 currents, whereas increasing Orai1 expression enhanced these activities (Cheng et al. 2008) as well as the store-operated Ca^{2+} entry through several other TRPC channels (Liao et al. 2007, 2008, 2009). Based on the physical and functional interactions among STIM1, Orai1, and TRPCs, it has been proposed that Orai and TRPC proteins co-assemble into a channel complex (Liao et al. 2008). Indeed, knocking down Orai1 expression in endothelial cells not only reduced store-operated currents mediated by an endogenous TRPC1/TRPC4 channel but also reduced its Ca^{2+} selectivity (Cioffi et al. 2012). This mode is quite different from the common belief that Orai and TRPC form separate channels independently regulated by STIM. In more elaborated forms of STIM/Orai/TRPC interactions, these proteins were considered to coexist in a complex referred to as store-operated calcium influx complex (SOCIC), and dynamic transitions may occur between assembled channels composed of both TRPC and Orai and disassembled channels composed of either TRPC or Orai separately. Store operation, through the binding of aggregated STIM1, occurs at specialized PM domain, the lipid rafts (Vaca 2010). Nevertheless, studies have clearly shown that despite some interdependence, channels formed by Orai and TRPC proteins have distinct cellular and physiological functions (Ong et al. 2012; Cheng et al. 2013). Clearly, the interactions of TRPC, Orai, and STIM proteins and their contributions to store- and receptor-operated Ca^{2+} entry warrant further investigations.

13.5 Concluding Remarks

Active research in the past two decades has established the roles of TRPC proteins in forming channels that mediate Ca^{2+} influx in response to receptor activation and internal Ca^{2+} store depletion. Depending on the cellular environment and the protein partners present in the channel complex, the TRPC channels display different biophysical properties and mechanisms of regulation. It is generally accepted that TRPC channels are activated following the stimulation of PLC and the channels are typically cation nonselective. These channels serve functions in neurons and smooth muscles in response to metabotropic receptor signaling with membrane depolarization. In non-excitable cells, the outcome of TRPC activation is typically Ca^{2+} influx, either directly through the TRPC channels or indirectly via the action of Na^+/Ca^{2+} exchangers working in the reverse mode. TRPC channels play diverse roles in cell signaling in a wide range of cell types via both store-dependent and store-independent mechanisms. The store-operated TRPC channel activation is context dependent and requires additional protein partners, such as STIM, Orai, IP_3Rs, Junctate, and/or Homer. These proteins help to stabilize TRPC channels in defined PM microdomains, in close opposition to the ER, and facilitate the communication between the intracellular Ca^{2+} store and the channels on the PM. Thus, the regulation and physiological functions of the TRPC channels will continue to fascinate researchers in the field of Ca^{2+} signaling and reveal novel insights about important physiological and pathophysiological processes that involve the TRPC channels.

References

Albarran L, Dionisio N, Lopez E, Salido GM, Redondo PC, Rosado JA (2014) STIM1 regulates TRPC6 heteromultimerization and subcellular location. Biochem J 463:373–381

Alkhani H, Ase AR, Grant R, O'Donnell D, Groschner K, Séguéla P (2014) Contribution of TRPC3 to store-operated calcium entry and inflammatory transductions in primary nociceptors. Mol Pain 10:43

Ambudkar IS, Ong HL, Liu X, Bandyopadhyay BC, Cheng KT (2007) TRPC1: the link between functionally distinct store-operated calcium channels. Cell Calcium 42:213–223

Antigny F, Koenig S, Bernheim L, Frieden M (2013) During post-natal human myogenesis, normal myotube size requires TRPC1- and TRPC4-mediated Ca2+ entry. J Cell Sci 126:2525–2533

Asanov A, Sampieri A, Moreno C, Pacheco J, Salgado A, Sherry R, Vaca L (2015) Combined single channel and single molecule detection identifies subunit composition of STIM1-activated transient receptor potential canonical (TRPC) channels. Cell Calcium 57:1–13

Berridge MJ (1993) Inositol trisphosphate and calcium signalling. Nature 361:315–325

Berridge MJ (1995) Capacitative calcium entry. Biochem J 312:1–11

Berridge MJ, Bootman MD, Lipp P (1998) Calcium—a life and death signal. Nature 395:645–648

Berridge MJ, Lipp P, Bootman MD (2000) The versatility and universality of calcium signalling. Nat Rev Mol Cell Biol 1:11–21

Berridge MJ, Bootman MD, Roderick HL (2003) Calcium signalling: dynamics, homeostasis and remodelling. Nat Rev Mol Cell Biol 4:517–529

Boulay G, Zhu X, Peyton M, Jian M, Hurst R, Stefani E, Birnbaumer L (1997) Cloning and expression of a novel mammalian homolog of *Drosophila* transient receptor potential (Trp) involved in calcium entry secondary to activation of receptors coupled by the Gq class of G protein. J Biol Chem 272:29672–29680

Boulay G, Brown DM, Qin N, Jiang M, Dietrich A, Zhu MX, Chen Z, Birnbaumer M, Mikoshiba K, Birnbaumer L (1999) Modulation of Ca2+ entry by polypeptides of the inositol 1,4, 5-trisphosphate receptor (IP3R) that bind transient receptor potential (TRP): evidence for roles of TRP and IP3R in store depletion-activated Ca2+ entry. Proc Natl Acad Sci U S A 96:14955–14960

Calcraft PJ, Ruas M, Pan Z, Cheng X, Arredouani A, Hao X, Tang J, Rietdorf K, Teboul L, Chuang KT, Lin P, Xiao R, Wang C, Zhu Y, Lin Y, Wyatt CN, Parrington J, Ma J, Evans AM, Galione A, Zhu MX (2009) NAADP mobilizes calcium from acidic organelles through two-pore channels. Nature 459:596–600

Camacho Londoño JE, Tian Q, Hammer K, Schröder L, Camacho Londoño J, Reil JC, He T, Oberhofer M, Mannebach S, Mathar I, Philipp SE, Tabellion W, Schweda F, Dietrich A, Kaestner L, Laufs U, Birnbaumer L, Flockerzi V, Freichel M, Lipp P (2015) A background Ca2+ entry pathway mediated by TRPC1/TRPC4 is critical for development of pathological cardiac remodelling. Eur Heart J 36:2257–2266

Cao Q, Zhong XZ, Zou Y, Murrell-Lagnado R, Zhu MX, Dong XP (2015) Calcium release through P2X4 activates calmodulin to promote endolysosomal membrane fusion. J Cell Biol 209 (6):879–894

Casteels R, Droogmans G (1981) Exchange characteristics of the noradrenaline-sensitive calcium store in vascular smooth muscle cells or rabbit ear artery. J Physiol 317:263–279

Cheng KT, Liu X, Ong HL, Ambudkar IS (2008) Functional requirement for Orai1 in store operated TRPC1-STIM1 channels. J Biol Chem 283:12935–12940

Cheng KT, Ong HL, Liu X, Ambudkar IS (2013) Contribution and regulation of TRPC channels in store-operated Ca2+ entry. Curr Top Membr 71:149–179

Cioffi DL, Wu S, Chen H, Alexeyev M, St Croix CM, Pitt BR, Uhlig S, Stevens T (2012) Orai1 determines calcium selectivity of an endogenous TRPC heterotetramer channel. Circ Res 110:1435–1444

Desai PN, Zhang X, Wu S, Janoshazi A, Bolimuntha S, Putney JW, Trebak M (2015) Multiple types of calcium channels arising from alternative translation initiation of the Orai1 message. Sci Signal 8:ra74

de Souza LB, Ong HL, Liu X, Ambudkar IS (2015) Fast endocytic recycling determines TRPC1-STIM1 clustering in ER-PM junctions and plasma membrane function of the channel. Biochim Biophys Acta 1853:2709–2721

Ding Y, Winters A, Ding M, Graham S, Akopova I, Muallem S, Wang Y, Hong JH, Gryczynski Z, Yang SH, Birnbaumer L, Ma R (2011) Reactive oxygen species-mediated TRPC6 protein activation in vascular myocytes, a mechanism for vasoconstrictor-regulated vascular tone. J Biol Chem 286:31799–31809

Dong XP, Shen D, Wang X, Dawson T, Li X, Zhang Q, Cheng X, Zhang Y, Weisman LS, Delling M, Xu H (2010) PI(3,5)P2 controls membrane traffic by direct activation of mucolipin Ca2+ release channels in the endolysosome. Nat Commun 1(4):38

Fasolato C, Hoth M, Matthews G, Penner R (1993) Ca2+ and Mn2+ influx through receptor-mediated activation of nonspecific cation channels in mast cells. Proc Natl Acad Sci U S A 90:3068–3072

Freichel M, Vennekens R, Olausson J, Stolz S, Philipp SE, Weissgerber P, Flockerzi V (2005) Functional role of TRPC proteins in native systems: implications from knockout and knock-down studies. J Physiol 567:59–66

Greka A, Navarro B, Oancea E, Duggan A, Clapham DE (2003) TRPC5 is a regulator of hippocampal neurite length and growth cone morphology. Nat Neurosci 6:837–845

Hardie RC, Minke B (1992) The trp gene is essential for a light-activated Ca2+ channel in *Drosophila* photoreceptors. Neuron 8:643–651

Hardie RC, Minke B (1993) Novel Ca2+ channels underlying transduction in *Drosophila* photoreceptors: implications for phosphoinositide-mediated Ca2+ mobilization. Trends Neurosci 16:371–376

Hardie RC, Raghu P (2001) Visual transduction in *Drosophila*. Nature 413:186–193

Harper MT, Londoño JE, Quick K, Londoño JC, Flockerzi V, Philipp SE, Birnbaumer L, Freichel M, Poole AW (2013) Transient receptor potential channels function as a coincidence signal detector mediating phosphatidylserine exposure. Sci Signal 6:ra50

Hartmann J, Dragicevic E, Adelsberger H, Henning HA, Sumser M, Abramowitz J, Blum R, Dietrich A, Freichel M, Flockerzi V, Birnbaumer L, Konnerth A (2008) TRPC3 channels are required for synaptic transmission and motor coordination. Neuron 59:392–398

Hoth M, Penner R (1992) Depletion of intracellular calcium stores activates a calcium current in mast cells. Nature 355:353–356

Hoth M, Penner R (1993) Calcium release-activated calcium current in rat mast cells. J Physiol 465:359–386

Hoth M, Fasolato C, Penner R (1993) Ion channels and calcium signaling in mast cells. Ann N Y Acad Sci 707:198–209

Hu Y, Vaca L, Zhu X, Birnbaumer L, Kunze DL, Schilling WP (1994) Appearance of a novel Ca2+ influx pathway in *Sf9* insect cells following expression of the transient receptor potential-like (*trpl*) protein of *Drosophila*. Biochem Biophys Res Commun 201:1050–1056

Hurst RS, Zhu X, Boulay G, Birnbaumer L, Stefani E (1998) Ionic currents underlying HTRP3 mediated agonist-dependent Ca2+ influx in stably transfected HEK293 cells. FEBS Lett 422:333–338

Irvine RF (1990) 'Quantal' Ca2+ release and the control of Ca2+ entry by inositol phosphates—a possible mechanism. FEBS Lett 263:5–9

Jackson TR, Patterson SI, Thastrup O, Hanley MR (1988) A novel tumour promoter, thapsigargin, transiently increases cytoplasmic free Ca2+ without generation of inositol phosphates in NG115-401L neuronal cells. Biochem J 253:81–86

Jardin I, Lopez JJ, Salido GM, Rosado JA (2008) Orai1 mediates the interaction between STIM1 and hTRPC1 and regulates the mode of activation of hTRPC1-forming Ca2+ channels. J Biol Chem 283:25296–25304

Jeon JP, Hong C, Park EJ, Jeon JH, Cho NH, Kim IG, Choe H, Muallem S, Kim HJ, So I (2012) Selective Gαi subunits as novel direct activators of transient receptor potential canonical (TRPC)4 and TRPC5 channels. J Biol Chem 287:17029–17039

Kim SJ, Kim YS, Yuan JP, Petralia RS, Worley PF, Linden DJ (2003) Activation of the TRPC1 cation channel by metabotropic glutamate receptor mGluR1. Nature 426:285–291

Kim JY, Zeng W, Kiselyov K, Yuan JP, Dehoff MH, Mikoshiba K, Worley PF, Muallem S (2006) Homer 1 mediates store- and inositol 1,4,5-trisphosphate receptor-dependent translocation and retrieval of TRPC3 to the plasma membrane. J Biol Chem 281:32540–32549

Kim MS, Hong JH, Li Q, Shin DM, Abramowitz J, Birnbaumer L, Muallem S (2009) Deletion of TRPC3 in mice reduces store-operated Ca2+ influx and the severity of acute pancreatitis. Gastroenterology 137:1509–1517

Kiselyov K, Xu X, Mozhayeva G (1998) Functional interaction between InsP3 receptors and store-operated Htrp3 channels. Nature 396:478–482

Kiselyov K, Mignery GA, Zhu MX, Muallem S (1999) The N-terminal domain of the IP3 receptor gates store-operated hTrp3 channels. Mol Cell 4:423–429

Kochukov MY, Balasubramanian A, Noel RC, Marrelli SP (2013) Role of TRPC1 and TRPC3 channels in contraction and relaxation of mouse thoracic aorta. J Vasc Res 50:11–20

Kochukov MY, Balasubramanian A, Abramowitz J, Birnbaumer L, Marrelli SP (2014) Activation of endothelial transient receptor potential C3 channel is required for small conductance calcium-activated potassium channel activation and sustained endothelial hyperpolarization and vasodilation of cerebral artery. J Am Heart Assoc 3:e000913

Lee KP, Yuan JP, Hong JH, So I, Worley PF, Muallem S (2010) An endoplasmic reticulum/plasma membrane junction: STIM1/Orai1/TRPCs. FEBS Lett 584:2022–2027

Lee KP, Choi S, Hong JH, Ahuja M, Graham S, Ma R, So I, Shin DM, Muallem S, Yuan JP (2014) Molecular determinants mediating gating of transient receptor potential canonical (TRPC) channels by stromal interaction molecule 1 (STIM1). J Biol Chem 289:6372–6382

Lewis RS, Cahalan MD (1989) Mitogen-induced oscillations of cytosolic Ca2+ and transmembrane Ca2+ current in human leukemic T cells. Cell Regul 1:99–112

Liao Y, Erxleben C, Yildirim E, Abramowitz J, Armstrong DL, Birnbaumer L (2007) Orai proteins interact with TRPC channels and confer responsiveness to store depletion. Proc Natl Acad Sci U S A 104:4682–4687

Liao Y, Erxleben C, Abramowitz J, Flockerzi V, Zhu MX, Armstrong DL, Birnbaumer L (2008) Functional interactions among Orai1, TRPCs, and STIM1 suggest a STIM-regulated heteromeric Orai/TRPC model for SOCE/Icrac channels. Proc Natl Acad Sci U S A 105:2895–2900

Liao Y, Plummer NW, George MD, Abramowitz J, Zhu MX, Birnbaumer L (2009) A role for Orai in TRPC-mediated Ca2+ entry suggests that a TRPC:Orai complex may mediate store and receptor operated Ca2+ entry. Proc Natl Acad Sci U S A 106:3202–3206

Liu X, Wang W, Singh BB, Lockwich T, Jadlowiec J, O'Connell B, Wellner R, Zhu MX, Ambudkar IS (2000) Trp1, a candidate protein for the store-operated Ca2+ influx mechanism in salivary gland cells. J Biol Chem 275:3403–3411

Liu X, Cheng KT, Bandyopadhyay BC, Pani B, Dietrich A, Paria BC, Swaim WD, Beech D, Yildrim E, Singh BB, Birnbaumer L, Ambudkar IS (2007) Attenuation of store-operated Ca2+ current impairs salivary gland fluid secretion in TRPC1-/- mice. Proc Natl Acad Sci U S A 104:17542–17547

Lopez E, Berna-Erro A, Salido GM, Rosado JA, Redondo PC (2013) FKBP52 is involved in the regulation of SOCE channels in the human platelets and MEG 01 cells. Biochim Biophys Acta 1833:652–662

Miehe S, Bieberstein A, Arnould I, Ihdene O, Rütten H, Strübing C (2010) The phospholipid-binding protein SESTD1 is a novel regulator of the transient receptor potential channels TRPC4 and TRPC5. J Biol Chem 285:12426–12434

Minke B, Selinger Z (1991) Inositol lipid pathway in fly photoreceptors: excitation, calcium mobilization and retinal degeneration. Prog Retin Res 11:99–124

Montell C, Rubin GM (1989) Molecular characterization of the *Drosophila trp* locus: a putative integral membrane protein required for phototransduction. Neuron 2:1313–1323

Mori Y, Wakamori M, Miyakawa T, Hermosura M, Hara Y, Nishida M, Hirose K, Mizushima A, Kurosaki M, Mori E, Gotoh K, Okada T, Fleig A, Penner R, Iino M, Kurosaki T (2002) Transient receptor potential 1 regulates capacitative Ca2+ entry and Ca2+ release from endoplasmic reticulum in B lymphocytes. J Exp Med 195:673–681

Neuner SM, Wilmott LA, Hope KA, Hoffmann B, Chong JA, Abramowitz J, Birnbaumer L, O'Connell KM, Tryba AK, Greene AS, Savio Chan C, Kaczorowski CC (2015) TRPC3 channels critically regulate hippocampal excitability and contextual fear memory. Behav Brain Res 281:69–77

Okada T, Inoue R, Yamazaki K, Maeda A, Kurosaki T, Yamakuni T, Tanaka I, Shimizu S, Ikenaka K, Imoto K, Mori Y (1999) Molecular and functional characterization of a novel mouse transient receptor potential protein homologue TRP7. Ca2+-permeable cation channel that is constitutively activated and enhanced by stimulation of G protein-coupled receptor. J Biol Chem 274:27359–27370

Ong HL, Cheng KT, Liu X, Bandyopadhyay BC, Paria BC, Soboloff J, Pani B, Gwack Y, Srikanth S, Singh BB, Gill DL, Ambudkar IS (2007) Dynamic assembly of TRPC1-STIM1-Orai1 ternary complex is involved in store-operated calcium influx. Evidence for similarities in store-operated and calcium release-activated calcium channel components. J Biol Chem 282:9105–9116

Ong HL, Jang SI, Ambudkar IS (2012) Distinct contributions of Orai1 and TRPC1 to agonist-induced [Ca2+]i signals determine specificity of Ca2+-dependent gene expression. PLoS One 7:e47146

Parekh AB, Penner R (1997) Store depletion and calcium influx. Physiol Rev 77:901–930
Parekh AB, Putney JW Jr (2005) Store-operated calcium channels. Physiol Rev 85:757–810
Petersen CC, Berridge MJ, Borgese MF, Bennett DL (1995) Putative capacitative calcium entry channels: expression of *Drosophila* trp and evidence for the existence of vertebrate homologues. Biochem J 311:41–44
Phelan KD, Mock MM, Kretz O, Shwe UT, Kozhemyakin M, Greenfield LJ, Dietrich A, Birnbaumer L, Freichel M, Flockerzi V, Zheng F (2012) Heteromeric canonical transient receptor potential 1 and 4 channels play a critical role in epileptiform burst firing and seizure-induced neurodegeneration. Mol Pharmacol 81:384–392
Phelan KD, Shwe UT, Abramowitz J, Wu H, Rhee SW, Howell MD, Gottschall PE, Freichel M, Flockerzi V, Birnbaumer L, Zheng F (2013) Canonical transient receptor channel 5 (TRPC5) and TRPC1/4 contribute to seizure and excitotoxicity by distinct cellular mechanisms. Mol Pharmacol 83:429–438
Phelan KD, Shwe UT, Abramowitz J, Birnbaumer L, Zheng F (2014) Critical role of canonical transient receptor potential channel 7 in initiation of seizures. Proc Natl Acad Sci U S A 111:11533–11538
Philipp S, Cavalié A, Freichel M, Wissenbach U, Zimmer S, Trost C, Marquart A, Murakami M, Flockerzi V (1996) A mammalian capacitative calcium entry channel homologous to *Drosophila* TRP and TRPL. EMBO J 15:6166–6171
Philipp S, Hambrecht J, Braslavski L, Schroth G, Freichel M, Murakami M, Cavalié A, Flockerzi V (1998) A novel capacitative calcium entry channel expressed in excitable cells. EMBO J 17:4274–4282
Phillips AM, Bull A, Kelly LE (1992) Identification of a *Drosophila* gene encoding a calmodulin-binding protein with homology to the *trp* phototransduction gene. Neuron 8:631–642
Putney JW Jr (1986) A model for receptor-regulated calcium entry. Cell Calcium 7:1–12
Putney JW Jr (1991) Capacitative calcium entry revisited. Cell Calcium 11:611–624
Reiser J, Polu KR, Möller CC, Kenlan P, Altintas MM, Wei C, Faul C, Herbert S, Villegas I, Avila-Casado C, McGee M, Sugimoto H, Brown D, Kalluri R, Mundel P, Smith PL, Clapham DE, Pollak MR (2005) TRPC6 is a glomerular slit diaphragm-associated channel required for normal renal function. Nat Genet 37:739–744
Riccio A, Li Y, Moon J, Kim KS, Smith KS, Rudolph U, Gapon S, Yao GL, Tsvetkov E, Rodig SJ, Van't Veer A, Meloni EG, Carlezon WA Jr, Bolshakov VY, Clapham DE (2009) Essential role for TRPC5 in amygdala function and fear-related behavior. Cell 137:761–772
Riccio A, Li Y, Tsvetkov E, Gapon S, Yao GL, Smith KS, Engin E, Rudolph U, Bolshakov VY, Clapham DE (2014) Decreased anxiety-like behavior and Gαq/11-dependent responses in the amygdala of mice lacking TRPC4 channels. J Neurosci 34:3653–3667
Riehle M, Büscher AK, Gohlke BO, Kaßmann M, Kolatsi-Joannou M, Bräsen JH, Nagel M, Becker JU, Winyard P, Hoyer PF, Preissner R, Krautwurst D, Gollasch M, Weber S, Harteneck C (2016) TRPC6 G757D loss-of-function mutation associates with FSGS. J Am Soc Nephrol 27:2771–2783
Rosker C, Graziani A, Lukas M, Eder P, Zhu MX, Romanin C, Groschner K (2004) Ca2+ signaling by TRPC3 involves Na+ entry and local coupling to the Na+/Ca2+ exchanger. J Biol Chem 279:13696–13704
Sabourin J, Bartoli F, Antigny F, Gomez AM, Benitah JP (2016) Transient receptor potential canonical (TRPC)/Orai1-dependent store-operated Ca2+ channels: NEW TARGETS OF ALDOSTERONE IN CARDIOMYOCYTES. J Biol Chem 291:13394–13409
Saliba Y, Karam R, Smayra V, Aftimos G, Abramowitz J, Birnbaumer L, Farès N (2015) Evidence of a role for fibroblast transient receptor potential canonical 3 Ca2+ channel in renal fibrosis. J Am Soc Nephrol 26:1855–1876
Satoh S, Tanaka H, Ueda Y, Oyama J, Sugano M, Sumimoto H, Mori Y, Makino N (2007) Transient receptor potential (TRP) protein 7 acts as a G protein-activated Ca2+ channel mediating angiotensin II-induced myocardial apoptosis. Mol Cell Biochem 294:205–215
Saul S, Stanisz H, Backes CS, Schwarz EC, Hoth M (2014) How ORAI and TRP channels interfere with each other: interaction models and examples from the immune system and the skin. Eur J Pharmacol 739:49–59

Selvaraj S, Sun Y, Watt JA, Wang S, Lei S, Birnbaumer L, Singh BB (2012) Neurotoxin-induced ER stress in mouse dopaminergic neurons involves downregulation of TRPC1 and inhibition of AKT/mTOR signaling. J Clin Invest 122:1354–1367

Seo K, Rainer PP, Shalkey Hahn V, Lee DI, Jo SH, Andersen A, Liu T, Xu X, Willette RN, Lepore JJ, Marino JP Jr, Birnbaumer L, Schnackenberg CG, Kass DA (2014) Combined TRPC3 and TRPC6 blockade by selective small-molecule or genetic deletion inhibits pathological cardiac hypertrophy. Proc Natl Acad Sci U S A 111:1551–1556

Seth M, Sumbilla C, Mullen SP, Lewis D, Klein MG, Hussain A, Soboloff J, Gill DL, Inesi G (2004) Sarco(endo)plasmic reticulum Ca2+ ATPase (SERCA) gene silencing and remodeling of the Ca2+ signaling mechanism in cardiac myocytes. Proc Natl Acad Sci U S A 101:16683–16688

Sexton JE, Desmonds T, Quick K, Taylor R, Abramowitz J, Forge A, Kros CJ, Birnbaumer L, Wood JN (2016) The contribution of TRPC1, TRPC3, TRPC5 and TRPC6 to touch and hearing. Neurosci Lett 610:36–42

Srikanth S, Jew M, Kim KD, Yee MK, Abramson J, Gwack Y (2012) Junctate is a Ca2+-sensing structural component of Orai1 and stromal interaction molecule 1 (STIM1). Proc Natl Acad Sci U S A 109:8682–8687

Stamboulian S, Moutin MJ, Treves S, Pochon N, Grunwald D, Zorzato F, De Waard M, Ronjat M, Arnoult C (2005) Junctate, an inositol 1,4,5-triphosphate receptor associated protein, is present in rodent sperm and binds TRPC2 and TRPC5 but not TRPC1 channels. Dev Biol 286:326–337

Stiber JA, Zhang ZS, Burch J, Eu JP, Zhang S, Truskey GA, Seth M, Yamaguchi N, Meissner G, Shah R, Worley PF, Williams RS, Rosenberg PB (2008) Mice lacking Homer 1 exhibit a skeletal myopathy characterized by abnormal transient receptor potential channel activity. Mol Cell Biol 28:2637–2647

Stroh O, Freichel M, Kretz O, Birnbaumer L, Hartmann J, Egger V (2012) NMDA receptor-dependent synaptic activation of TRPC channels in olfactory bulb granule cells. J Neurosci 32:5737–5746

Strübing C, Krapivinsky G, Krapivinsky L, Clapham DE (2001) TRPC1 and TRPC5 form a novel cation channel in mammalian brain. Neuron 29:645–655

Sun Y, Birnbaumer L, Singh BB (2015) TRPC1 regulates calcium-activated chloride channels in salivary gland cells. J Cell Physiol 230:2848–2856

Sundivakkam PC, Freichel M, Singh V, Yuan JP, Vogel SM, Flockerzi V, Malik AB, Tiruppathi C (2012) The Ca2+ sensor stromal interaction molecule 1 (STIM1) is necessary and sufficient for the store-operated Ca2+ entry function of transient receptor potential canonical (TRPC) 1 and 4 channels in endothelial cells. Mol Pharmacol 81:510–526

Tai C, Hines DJ, Choi HB, Macvicar BA (2011) Plasma membrane insertion of TRPC5 channels contributes to the cholinergic plateau potential in hippocampal CA1 pyramidal neurons. Hippocampus 21:958–967

Tang J, Lin Y, Zhang Z, Tikunova S, Birnbaumer L, Zhu MX (2001) Identification of common binding sites for calmodulin and inositol 1,4,5-trisphosphate receptors on the carboxyl termini of trp channels. J Biol Chem 276:21303–21310

Thakur DP, Tian JB, Jeon J, Xiong J, Huang Y, Flockerzi V, Zhu MX (2016) Critical roles of Gi/o proteins and phospholipase C-δ1 in the activation of receptor-operated TRPC4 channels. Proc Natl Acad Sci U S A 113:1092–1097

Thastrup O, Linnebjerg H, Bjerrum PJ, Knudsen JB, Christensen SB (1987) The inflammatory and tumor-promoting sesquiterpene lactone, thapsigargin, activates platelets by selective mobilization of calcium as shown by protein phosphorylations. Biochim Biophys Acta 927:65–73

Treves S, Franzini-Armstrong C, Moccagatta L, Arnoult C, Grasso C, Schrum A, Ducreux S, Zhu MX, Mikoshiba K, Girard T, Smida-Rezgui S, Ronjat M, Zorzato F (2004) Junctate is a key element in calcium entry induced by activation of InsP3 receptors and/or calcium store depletion. J Cell Biol 166:537–548

Treves S, Vukcevic M, Griesser J, Armstrong CF, Zhu MX, Zorzato F (2010) Agonist-activated Ca2+ influx occurs at stable plasma membrane and endoplasmic reticulum junctions. J Cell Sci 123:4170–4181

Tsvilovskyy VV, Zholos AV, Aberle T, Philipp SE, Dietrich A, Zhu MX, Birnbaumer L, Freichel M, Flockerzi V (2009) Deletion of TRPC4 and TRPC6 in mice impairs smooth muscle contraction and intestinal motility in vivo. Gastroenterology 137:1415–1224

Vaca L (2010) SOCIC: the store-operated calcium influx complex. Cell Calcium 47:199–209

Vaca L, Sampieri A (2002) Calmodulin modulates the delay period between release of calcium from internal stores and activation of calcium influx via endogenous TRP1 channels. J Biol Chem 277:42178–42187

Vannier B, Peyton M, Boulay G, Brown D, Qin N, Jiang M, Zhu X, Birnbaumer L (1999) Mouse trp2, the homologue of the human trpc2 pseudogene, encodes mTrp2, a store depletion-activated capacitative Ca2+ entry channel. Proc Natl Acad Sci U S A 96:2060–2064

Vazquez G, Lievremont JP, St J Bird G, Putney JW Jr (2001) Human Trp3 forms both inositol trisphosphate receptor-dependent and receptor-independent store-operated cation channels in DT40 avian B lymphocytes. Proc Natl Acad Sci U S A 98:11777–11782

Weissmann N, Dietrich A, Fuchs B, Kalwa H, Ay M, Dumitrascu R, Olschewski A, Storch U, Mederos y Schnitzler M, Ghofrani HA, Schermuly RT, Pinkenburg O, Seeger W, Grimminger F, Gudermann T (2006) Classical transient receptor potential channel 6 (TRPC6) is essential for hypoxic pulmonary vasoconstriction and alveolar gas exchange. Proc Natl Acad Sci USA 103:19093–19098

Wes PD, Chevesich J, Jeromin A, Rosenberg C, Stetten G, Montell C (1995) TRPC1, a human homolog of a Drosophila store-operated channel. Proc Natl Acad Sci U S A 92:9652–9656

Winn MP, Conlon PJ, Lynn KL (2005) A mutation in the TRPC6 cation channel causes familial focal segmental glomerulosclerosis. Science 308:1801–1804

Woo JS, Srikanth S, Nishi M, Ping P, Takeshima H, Gwack Y (2016) Junctophilin-4, a component of the endoplasmic reticulum-plasma membrane junctions, regulates Ca2+ dynamics in T cells. Proc Natl Acad Sci U S A 113:2762–2767

Worley PF, Zeng W, Huang GN, Yuan JP, Kim JY, Lee MG, Muallem S (2007) TRPC channels as STIM1-regulated store-operated channels. Cell Calcium 42:205–211

Xu N, Cioffi DL, Alexeyev M, Rich TC, Stevens T (2015) Sodium entry through endothelial store-operated calcium entry channels: regulation by Orai1. Am J Physiol Cell Physiol 308:C277–C288

Yan HD, Villalobos C, Andrade R (2009) TRPC channels mediate a muscarinic receptor-induced afterdepolarization in cerebral cortex. J Neurosci 29:10038–10046

Yuan JP, Kiselyov K, Shin DM et al (2003) Homer binds TRPC family channels and is required for gating of TRPC1 by IP3 receptors. Cell 114:777–789

Yuan JP, Zeng W, Huang GN, Worley PF, Muallem S (2007) STIM1 heteromultimerizes TRPC channels to determine their function as store-operated channels. Nat Cell Biol 9:636–645

Zeng W, Yuan JP, Kim MS, Choi YJ, Huang GN, Worley PF, Muallem S (2008) STIM1 gates TRPC channels, but not Orai1, by electrostatic interaction. Mol Cell 32:439–448

Zhang Z, Tang J, Tikunova S, Johnson JD, Chen Z, Qin N, Dietrich A, Stefani E, Birnbaumer L, Zhu MX (2001) Activation of Trp3 by inositol 1,4,5-trisphosphate receptors through displacement of inhibitory calmodulin from a common binding domain. Proc Natl Acad Sci U S A 98:3168–3173

Zhu X, Chu PB, Peyton M, Birnbaumer L (1995) Molecular cloning of a widely expressed human homologue for the *Drosophila trp* gene. FEBS Lett 373:193–198

Zhu X, Jiang M, Peyton M, Boulay G, Hurst R, Stefani E, Birnbaumer L (1996) trp, a novel mammalian gene family essential for agonist-activated capacitative Ca2+ entry. Cell 85:661–671

Zhu X, Jiang M, Birnbaumer L (1998) Receptor-activated Ca2+ influx via human Trp3 stably expressed in human embryonic kidney HEK293 cells. Evidence for a non-capacitative Ca2+ entry. J Biol Chem 273:133–142

The Role of Mitochondria in the Activation/Maintenance of SOCE: Store-Operated Ca^{2+} Entry and Mitochondria

14

András Spät and Gergö Szanda

Abstract

Mitochondria extensively modify virtually all cellular Ca^{2+} transport processes, and store-operated Ca^{2+} entry (SOCE) is no exception to this rule. The interaction between SOCE and mitochondria is complex and reciprocal, substantially altering and, ultimately, fine-tuning both capacitative Ca^{2+} influx and mitochondrial function. Mitochondria, owing to their considerable Ca^{2+} accumulation ability, extensively buffer the cytosolic Ca^{2+} in their vicinity. In turn, the accumulated ion is released back into the neighboring cytosol during net Ca^{2+} efflux. Since store depletion itself and the successive SOCE are both Ca^{2+}-regulated phenomena, mitochondrial Ca^{2+} handling may have wide-ranging effects on capacitative Ca^{2+} influx at any given time. In addition, mitochondria may also produce or consume soluble factors known to affect store-operated channels. On the other hand, Ca^{2+} entering the cell during SOCE is sensed by mitochondria, and the ensuing mitochondrial Ca^{2+} uptake boosts mitochondrial energy metabolism and, if Ca^{2+} overload occurs, may even lead to apoptosis or cell death. In several cell types, mitochondria seem to be sterically excluded from the confined space that forms between the plasma membrane (PM) and endoplasmic reticulum (ER) during SOCE. This implies that high-Ca^{2+} microdomains comparable to those observed between the ER and mitochondria do not form here. In the following chapter, the above aspects of the many-sided SOCE-mitochondrion interplay will be discussed in greater detail.

A. Spät (✉)
Department of Physiology, Semmelweis University Medical School, POB 2, 1428 Budapest, Hungary

Laboratory of Molecular Physiology, Hungarian Academy of Sciences, Budapest, Hungary
e-mail: spat@eok.sote.hu

G. Szanda
Department of Physiology, Semmelweis University Medical School, POB 2, 1428 Budapest, Hungary

K. Groschner et al. (eds.), *Store-Operated Ca^{2+} Entry (SOCE) Pathways*, Advances in Experimental Medicine and Biology 993, DOI 10.1007/978-3-319-57732-6_14

Keywords
Store-operated Ca^{2+} entry • Mitochondria • Ca^{2+} uptake • Ca^{2+} release • Microdomains • NAD(P)H • Membrane potential

14.1 Introduction

Sufficient refilling of intracellular Ca^{2+} stores is required to sustain signaling during almost all Ca^{2+} release-initiated physiological processes. Impairment of capacitative or store-operated Ca^{2+} entry (SOCE) perturbs numerous cell functions and may lead to disease (Parekh 2010). Moreover, long-lasting Ca^{2+} influx and tunneling of Ca^{2+} from store-operated channels to Ca^{2+}-activable adenylyl cyclase isoforms in nearby lipid rafts and to inositol trisphosphate receptors (IP_3Rs) through the ER are required for sustained biological responses (Petersen and Verkhratsky 2007; Spät et al. 2016). Thus, factors modulating the capacitative Ca^{2+} entry machinery or simply permitting its appropriate operation may have far-reaching effects on cell function.

Mitochondria are crucial players of cellular Ca^{2+} homeostasis. By supplying adenosine triphosphate (ATP) for primary active Ca^{2+} transports, energized mitochondria are elementary factors of normal Ca^{2+} handling (Walsh et al. 2009). Mitochondria are also capable of buffering cytosolic Ca^{2+}, and thereby they extensively modulate cytosolic Ca^{2+} signals (Demaurex et al. 2009; Szanda et al. 2006; Walsh et al. 2009).

Mitochondria display high-capacity Ca^{2+} uptake by the ruthenium red (RR)-sensitive mitochondrial Ca^{2+} uniporter (MCU) which, combined with the large mitochondrial transmembrane potential ($\Delta\Psi_m$; ~180 mV, inside negative) and the ability of inorganic phosphate to chelate considerable amounts of matrix Ca^{2+}, renders mitochondria an effective Ca^{2+} "scavenger" organelle (Gunter and Pfeiffer 1990). Electrophysiological studies revealed that the MCU is a highly selective inwardly rectifying Ca^{2+} channel residing in the inner mitochondrial membrane (Kirichok et al. 2004). The opening probability of the RR-sensitive single-channel current of cardiac mitoplasts is reduced after knockout and increased after overexpression of uncoupling protein 2 (UCP2) (Bondarenko et al. 2015; Motloch et al. 2016), a protein probably participating in the control of mitochondrial Ca^{2+} uptake (Trenker et al. 2007). Importantly, the recombinant MCU protein, the pore forming 40 kDa unit of the *MCU complex* (see below), is sufficient to transport Ca^{2+} in a planar lipid bilayer on its own (Baughman et al. 2011; De Stefani et al. 2011). However, the electrophysiological characteristics of the current observed in isolated mitoplasts (mitochondria stripped of their outer membrane) require the co-expression of MCU with its accessory proteins in the *MCU complex* (Kamer and Mootha 2015). Briefly, MICU1 (mitochondrial calcium uptake 1) (Perocchi et al. 2010) and MICU2 (Plovanich et al. 2013) (or MICU3 in the central nervous system), soluble proteins of the intermembrane space, set the threshold of Ca^{2+} uptake and are responsible for the sigmoid $[Ca^{2+}]_c$ dependence of the transport rate

(Csordas et al. 2013; Kamer and Mootha 2014; Mallilankaraman et al. 2012b; Waldeck-Weiermair et al. 2015). The MICU1-MICU2 heterodimer is bound to the MCU by EMRE (essential MCU regulator), a 10 kDa protein essential for uniporter activity (Sancak et al. 2013; Tsai et al. 2016). EMRE may also act as a Ca^{2+} sensor on both sides of the inner mitochondrial membrane (IMM), functioning as a gatekeeper of the uniporter (Vais et al. 2016). MCUb inhibits channel activity (Raffaello et al. 2013). Finally, knockdown of MCUR1 (mitochondrial Ca^{2+} uniporter regulator 1) also abrogates Ca^{2+} uptake (Mallilankaraman et al. 2012a); however, whether it is a member of the MCU complex or its role is indirect is still debated (Paupe et al. 2015; Vais et al. 2015). (It may also influence mitochondrial Ca^{2+} metabolism by reducing the Ca^{2+} threshold for mitochondrial permeability transition (Chaudhuri et al. 2016).) The mitochondrial Ca^{2+} responsiveness to cytosolic Ca^{2+} signal will depend on the relative expression of these proteins in a given cell type.

Once Ca^{2+} influx lags behind Ca^{2+} efflux, mitochondria release sequestered Ca^{2+} back into the cytosol and, in turn, allow Ca^{2+} to be pumped into the ER lumen, the "original destination" of SOCE-derived Ca^{2+} (Arnaudeau et al. 2001; Malli et al. 2005).

During cell stimulation, cytosolic Ca^{2+} signal is transferred into the mitochondria, and the subsequent SOCE further augments and prolongs the mitochondrial Ca^{2+} signal (Fig. 14.1). Theoretically, mitochondrial Ca^{2+} accumulation can affect SOCE in the following ways: (1) mitochondrial Ca^{2+} buffering in the vicinity of store-operated channels could reduce Ca^{2+}-dependent SOCE inactivation, (2) mitochondrial uptake of Ca^{2+} released from internal stores may modify store depletion and could therefore affect SOCE indirectly, and (3) sequestration of Ca^{2+} entering the cell may delay store repletion and prolong SOCE. In addition, mitochondria may supply or consume soluble factors (e.g., ATP, intermediate metabolites) which influence SOCE. The influence of mitochondria on SOCE is discussed in Sect. 14.2, whereas the "mirror phenomenon," the effect of SOCE on mitochondrial functions, is dealt with in Sect. 14.3. Section 14.4 deliberates on the role of high-Ca^{2+} microdomains (HCMDs) in SOCE-induced mitochondrial Ca^{2+} uptake.

14.2 Modulation of SOCE by Mitochondrial Ca^{2+} Metabolism

14.2.1 Mitochondrial Ca^{2+} Buffering and Its Effect on SOCE

Since store-operated Ca^{2+} channels display a characteristic inactivation that is mediated by Ca^{2+} itself (Parekh 1998; Zweifach and Lewis 1995), mitochondria may reduce negative feedback by simply buffering Ca^{2+} around store-operated channels. First verification of this assumption was obtained by Hoth and co-workers in human T cells by showing that dissipation of $\Delta\Psi_m$, either with protonophore (Hoth et al. 1997) or with the blockade of the electron transport chain (Hoth et al. 2000), results in the inhibition of the sustained phase (>10 s) of

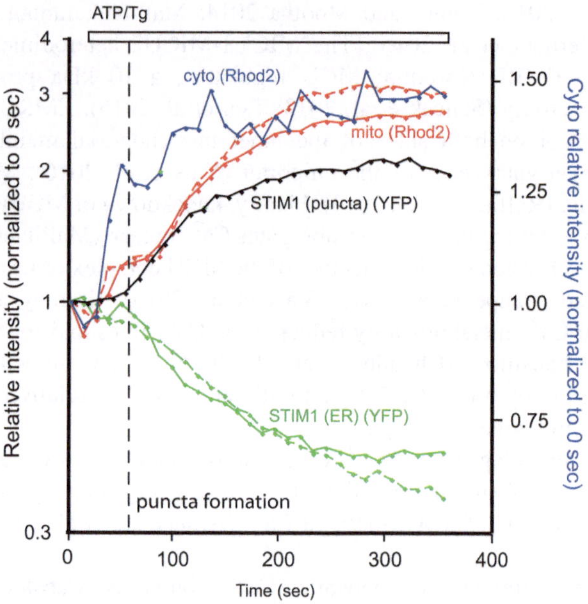

Fig. 14.1 Effect of Ca^{2+} release and the ensuing SOCE on cytosolic and mitochondrial $[Ca^{2+}]$ and the intracellular distribution of STIM1. Rhod-2 loaded COS-7 cells, transfected with YFP-STIM1 and untagged Orai1, were examined with confocal microscopy. Fifty μM ATP and 200 nM thapsigargin were added in Ca^{2+}-containing medium as indicated. The graph shows $[Ca^{2+}]_c$ (cyto, *blue trace*) measured with Rhod-2 fluorescence over the nuclear area and $[Ca^{2+}]_m$ (*red traces*) measured at two selected areas over mitochondria. STIM1 translocation was indicated either by the decreased YFP-STIM1 fluorescence in two perinuclear ER areas (*green*) or by the increased average fluorescence of several puncta at the peripheral region of the cell (*black trace*). All fluorescence intensities were normalized to 0 s. The *continuous* and *dotted* mito Rhod-2 and ER STIM1 curves show the response in two separately examined mitochondrial or ER regions. The *vertical dotted line* shows the onset of SOCE, indicated by the formation of STIM1 puncta (cf. Varnai et al. 2007). Note the small increase in $[Ca^{2+}]_m$ associated with the rapid Ca^{2+} release from the ER and a slightly delayed massive $[Ca^{2+}]_m$ elevation coinciding with STIM1 translocation to the cell periphery. Reproduced from Korzeniowski et al. (2009), with permission

SOCE. Parekh's group also analyzed this phenomenon in rat basophil leukemia (RBL) cells and confirmed the observations of Hoth's laboratory (Gilabert and Parekh 2000). The dependence of SOCE on mitochondrial function is also strengthened by data obtained in non-blood cells. Dissipation of $\Delta\Psi_m$ with the protonophore carbonyl cyanide-4-(trifluoromethoxy)phenylhydrazone (FCCP) reduced sustained SOCE in primary and immortalized rat hepatocytes (To et al. 2010) and HEK-293 cells (Mignen et al. 2005). The protonophore acted similarly in Jurkat cells, without modifying ATP and stored Ca^{2+} content of the cell (Makowska et al. 2000). It has to be emphasized that in hepatocytes, like in RBL cells, FCCP was ineffective when cytosolic Ca^{2+} was strongly buffered (Gilabert and Parekh 2000; To et al. 2010). The SOCE-promoting effect of energized mitochondria was also abolished by RR (Gilabert and Parekh 2000). (RR is the classical inhibitor of

MCU, and it also inhibits MICU1, a component of the MCU complex (see Sect. 14.1) (Perocchi et al. 2010) as well as Letm1, claimed to function as a mitochondrial Ca^{2+}/H^{+} antiporter (Jiang et al. 2009).) Altogether, Ca^{2+} uptake into mitochondria seems to be a general mechanism through which Ca^{2+} release-activated Ca^{2+} current (I_{CRAC}, the first described SOCE current) can be prolonged (Hoth et al. 1997, 2000).

14.2.2 Influence of Mitochondria on SOCE via the Modulation of Ca^{2+} Release

Depletion of the ER Ca^{2+} store induces SOCE. Mitochondrial Ca^{2+} sequestration may influence Ca^{2+} release and thereby the extent of store depletion; nevertheless pertinent observations are somewhat conflicting. Inhibition of mitochondrial Ca^{2+} uptake by means of mitochondrial depolarization in HeLa cells (Collins et al. 2000) or with RR in colon smooth muscle cells (Olson et al. 2010) reduced IP_3-mediated Ca^{2+} release. In line with these findings, energizing mitochondria shifted the IP_3-I_{CRAC} curve to the left (Gilabert et al. 2001) and thereby enhanced the sensitivity of SOCE to IP_3-induced Ca^{2+} release (Glitsch et al. 2002). On the other hand, Hajnoczky and co-workers (Hajnoczky et al. 1999) found that mitochondrial Ca^{2+} uptake attenuates, rather than promotes, IP_3-induced Ca^{2+} release in hepatocytes. The seeming contradiction between these observations may be due to the $[Ca^{2+}]$ dependence of the Ca^{2+} release processes. A significant subset of mitochondria is characteristically located close to ER Ca^{2+} channels in HCMDs and thereby effectively sequesters Ca^{2+} during Ca^{2+} release. Open probability of the ER Ca^{2+} release channels (IP_3R and ryanodine receptor) display bell-shaped dependence on $[Ca^{2+}]_c$ (Bezprozvanny et al. 1991). However, maximal open probability can be attained at strikingly differing $[Ca^{2+}]_c$ in the three isoforms of the IP_3R (reviewed in Spät et al. 2008b). Consequently, whether inhibition of mitochondrial function increases or decreases Ca^{2+} release and thereby SOCE will ultimately depend on the IP_3R isoform repertoire and resting $[Ca^{2+}]_c$ of the examined cell type.

The effect of mitochondria on agonist (IP_3)-induced store depletion might be even more complicated when the role of mitochondrial Ca^{2+} efflux is also taken into account. It was found in endothelial (Malli et al. 2003, 2005) and HeLa cells (Arnaudeau et al. 2001) that mitochondria support the maintenance of stored Ca^{2+} content during and following cell stimulation through a mechanism requiring the mitochondrial Na^{+}/Ca^{2+} exchanger. (The Na^{+}/Ca^{2+} exchanger is the main Ca^{2+} efflux transporter of mitochondria in excitable and in some non-excitable vertebrate cells (Carafoli et al. 1974).) Parekh (2008) offered a model in order to unify the contribution of mitochondrial Ca^{2+} uptake and efflux to SOCE and to store refilling. It was proposed that, by preventing Ca^{2+}-dependent slow inactivation (and by promoting store depletion in certain cell types), mitochondrial Ca^{2+} uptake first enhances SOCE. Then sequestered Ca^{2+} is released from mitochondria into the vicinity of sarcoplasmic/endoplasmic reticulum Ca^{2+}-ATPase (SERCA) pumps allowing the refilling of the ER. Thus, concerted mitochondrial Ca^{2+} uptake and efflux are able to funnel Ca^{2+} through the mitochondrial network during SOCE.

The conclusion can be drawn that the outcome of mitochondrial influence on messenger-induced store depletion cannot be predicted and therefore should be individually assessed in every cell type. It is to emphasize at this point that the effects of mitochondria on store depletion and on Ca^{2+}-dependent inactivation of store-operated channels are separate mechanisms through which mitochondria may influence SOCE. This is clearly shown by the ability of energized mitochondria to augment SOCE even after complete depletion of Ca^{2+} stores (Gilabert and Parekh 2000; Hoth et al. 1997).

14.2.3 Regulation of SOCE by Additional Mitochondrial Features

Location of mitochondria is a key element of Ca^{2+} signaling (Szanda et al. 2006; Tinel et al. 1999; Walsh et al. 2009). In Jurkat T cells, mitochondria move toward Ca^{2+} entry sites during SOCE, and the blockade of this translocation reduces the plateau of I_{CRAC} (Quintana et al. 2006). This phenomenon most probably reflects the necessity of mitochondria to buffer Ca^{2+} in the subplasmalemmal (SPL) space in order to maintain SOCE (see Sect. 14.2.1). It can be argued that the initiation of SOCE is not, or much less, sensitive to mitochondrial function in T cells (Hoth et al. 1997, 2000) simply because mitochondria are located far from store-operated channels at the beginning of the Ca^{2+} influx. It should be noted, however, that mitochondrial motility was found to be inhibited during Ca^{2+} signal in cardiac myoblasts (Yi et al. 2004) implying that movement of mitochondria toward the Ca^{2+} source is not a universal phenomenon. Furthermore, redistribution of mitochondria away from the PM fails to affect SOCE in several cell types (Frieden et al. 2004, 2005) (but see Varadi et al. 2004) strongly suggesting that mitochondria do not need to be located in the immediate vicinity of store-operated channels in order to efficiently sequester Ca^{2+}. (In fact, mitochondria may be located far from the HCMD formed between the PM and ER during SOCE and still sequester Ca^{2+} (Korzeniowski et al. 2009); this circumstance will be discussed later thoroughly.)

Soluble molecules were also proposed to connect mitochondria with SOCE. In contrast to earlier data (Gilabert and Parekh 2000; Glitsch and Parekh 2000; Hoth et al. 2000; Makowska et al. 2000), a role for ATP production in the control of SOCE was also suggested. Inhibition of the adenine nucleotide translocase enhanced the Ca^{2+}-dependent slow inactivation of I_{CRAC} in Jurkat cells (Montalvo et al. 2006). The authors concluded that ATP released from SPL mitochondria is a considerable Ca^{2+} buffer and thereby reduces Ca^{2+}-dependent slow inactivation. It is also noteworthy in this regard that mitochondrial Ca^{2+} signaling induces the formation of cAMP in the mitochondrial matrix (Di Benedetto et al. 2013; Katona et al. 2015) which, in turn, enhances ATP generation (Acin-Perez et al. 2009; Di Benedetto et al. 2013). Pyruvate is another soluble factor that was found to modulate I_{CRAC} in the physiological concentration range (Bakowski and Parekh 2007) by decreasing Ca^{2+}-dependent fast inactivation of I_{CRAC} in RBL cells. As to the possible role of free radicals in altering SOCE, nitric oxide (NO) has been shown to inhibit SOCE by activating SERCA pumps (Trepakova et al. 1999) or by suppressing mitochondrial Ca^{2+} uptake (Thyagarajan et al. 2002) (but see Glitsch et al. 2002). Modulation of the mitochondrion-SOCE relationship by free radicals or by other soluble factors may have far-reaching consequences on cellular Ca^{2+} handling as well as on human disease (Chinopoulos and Adam-Vizi 2006; Davidson and Duchen 2007; Nunes and Demaurex 2014) and therefore deserves further analysis. Additionally, ample evidence shows that impairment of SOCE due to mitochondrial malfunction interferes with cellular responses as exemplified by reduced T and mast cell activation (Chang et al. 2006; Hoth et al. 2000) and insufficient cell proliferation (Mignen et al. 2005) under such circumstances.

14.3 The Effects of SOCE on Mitochondrial Metabolism

Mitochondrial Ca^{2+} uptake activates intramitochondrial metabolic processes and provides the energy for ongoing cellular requirements. In addition to the effects of mitochondria on SOCE, the opposing action is also important since SOCE-derived Ca^{2+} significantly affects mitochondrial metabolism.

Mitochondrial metabolism is controlled by the $[Ca^{2+}]$ of the matrix. Increased $[Ca^{2+}]$ activates three dehydrogenases in mitochondrial suspension (McCormack et al. 1990) explaining the Ca^{2+}-dependent formation of mitochondrial NAD(P)H in K^+- (Pralong et al. 1992) and hormone-stimulated glomerulosa cells (Pralong et al. 1994; Rohacs et al. 1997a), in glucose-stimulated pancreatic β-cells (Pralong et al. 1994), in electrically stimulated sensory neurons (Duchen 1992), in agonist-stimulated ovarian luteal cells (Szabadkai et al. 2001), and several other cell types. The Ca^{2+}-evoked formation of reduced pyridine nucleotides results in increased production of ATP (Jouaville et al. 1999), an effect also supported by the action of Ca^{2+} on the F_1/F_o ATP synthase (Brown 1992) and the adenine nucleotide translocase (Spencer and Bygrave 1971). The significance of these metabolic effects on cell function is nicely illustrated by observations in secretory cells. Buffering of matrix Ca^{2+} results in reduced glucose-stimulated insulin release and angiotensin II-dependent aldosterone secretion (Wiederkehr et al. 2011).

Ca^{2+} entering mitochondria during SOCE stimulates the formation of mitochondrial NAD(P)H (Rohacs et al. 1997b; Szabadkai et al. 2001), but the response depends on the density of mitochondria in the SPL cytoplasm. In cell types with predominantly perinuclear location of mitochondria (e.g., in glomerulosa cells (Rohacs et al. 1997b)), the mitochondrial Ca^{2+} response to IP_3-induced Ca^{2+} release far exceeds that observed during SOCE. The opposite is true for ECV304 endothelial cells (Lawrie et al. 1996) and COS-7 cells (Korzeniowski et al. 2009) in which a relatively high fraction of the mitochondria is located in the vicinity of the PM (Lawrie et al. 1996; Yu et al. 1995).

Whereas the mitochondrial Ca^{2+} signal activates energy metabolism, excessive Ca^{2+} uptake into the mitochondria may reduce the activity of the pyruvate and oxoglutarate dehydrogenases (McCormack 1985) or may even lead to apoptosis

(Ichas and Mazat 1998) or cell death. Among other pathological states, traumatic brain injury impairs neuronal function by such Ca^{2+} overload. Downregulation of STIM2 (a sensor of ER Ca^{2+} depletion) reduced SOCE and hence also mitochondrial Ca^{2+} loading and significantly improved mitochondrial function. This observation indicates that excessive SOCE may damage mitochondria and thereby contribute to neurological disorders (Rao et al. 2015). Noteworthy, a negative feed-forward mechanism observed in human adrenocortical H295R cells may reduce the risk of mitochondrial Ca^{2+} overload and cell death. Stimulation of such cells with angiotensin II may activate p38 MAPK and novel protein kinase C isoforms (probably PKCε) prior to the onset of cytosolic Ca^{2+} signal. The activated kinases then attenuate mitochondrial Ca^{2+} signal during Ca^{2+} release and SOCE (Koncz et al. 2009; Spät et al. 2008a; Szanda et al. 2008).

14.4 The Role of High-Ca^{2+} Microdomains in the Mitochondrion-SOCE Interplay

14.4.1 [Ca^{2+}] Around SPL Mitochondria During SOCE

Mitochondrial Ca^{2+} uptake, due to the low-Ca^{2+} affinity of MCU (Gunter and Pfeiffer 1990; Kirichok et al. 2004), is generally explained by the formation of HCMD between the IP_3R or the orifice of voltage-dependent Ca^{2+} channels and the apposing mitochondrion (reviewed in Rizzuto and Pozzan 2006; Spät et al. 2008a; Szanda et al. 2006). By analogy it was presumed that the formation of HCMD is a prerequisite for mitochondrial Ca^{2+} signaling also during SOCE. In fact, several reports supported this concept.

In store-depleted pancreatic acinar cells, elevation of extracellular [Ca^{2+}] resulted in elevated mitochondrial Ca^{2+} concentration ([Ca^{2+}]$_m$) in the basolateral region only, close to the PM (Park et al. 2001). In HeLa cells, the overexpression of dynamitin (leading to the loss of the fission factor dynamin-related protein 1 (Drp-1) from the mitochondrial outer membrane) induced the shift of SPL mitochondria toward the perinuclear area. Under such conditions SOCE elicited smaller increases in [Ca^{2+}]$_c$ and [Ca^{2+}]$_m$ (Varadi et al. 2004). These latter phenomena may be attributed to reduced mitochondrial Ca^{2+} sequestration that, due to an excessive (but not measured) SPL [Ca^{2+}] elevation, attenuated SOCE. In harmony with these data, when the centripetal re-localization of mitochondria was induced with the overexpression of the fission factor human mitochondrial fission 1 protein (hFis1), SOCE-induced mitochondrial [Ca^{2+}] elevation developed significantly slower than in control cells (Feldman et al. 2010; Frieden et al. 2004).

All these studies unambiguously show that SOCE may raise [Ca^{2+}]$_m$ preferentially in SPL mitochondria, indicating that these organelles are exposed to higher [Ca^{2+}] than those located far from the cell membrane. Nevertheless, this conclusion does not mean that SPL mitochondria are exposed to [Ca^{2+}] of 10^{-5} M or more, as generally presumed for HCMD. On the contrary, quite a few observations indicate that SPL mitochondria may not be situated in the immediate vicinity of channel orifices. The lack of mitochondrial effect on the rapid inactivation of I_{CRAC} (Gilabert et al. 2001; Glitsch et al. 2002) is hardly compatible with a molecular vicinity of the mitochondria to the channel. Moreover, in quite a few cell types, mitochondria respond to moderate, even submicromolar increases in $[Ca^{2+}]_c$. For instance, in steroid-producing adrenal glomerulosa and ovarian luteal cells (Rohacs et al. 1997b; Szabadkai et al. 2001), SOCE results in a slowly developing Ca^{2+} signal that never exceeds 200 nM. In spite of these kinetic parameters, SOCE is followed by mitochondrial NAD(P)H signal.

Calcium ion permeates the voltage-dependent anion channel (VDAC, porin) in the outer mitochondrial membrane. In HeLa cells agonist-induced mitochondrial Ca^{2+} signals were enhanced by overexpression of the VDAC. In contrast, VDAC overexpression had no effect on mitochondrial Ca^{2+} uptake either during SOCE or after permeabilization and perfusion with a medium containing 5 μM Ca^{2+}. This observation indicates that perimitochondrial $[Ca^{2+}]$ exceeds 5 μM during Ca^{2+} release but does not attain this value during SOCE (Rapizzi et al. 2002). In a study on an endothelium-derived cell line, SPL $[Ca^{2+}]$ was estimated on basis of BK_{Ca} channel currents. During the SOCE phase of histamine-evoked cytosolic Ca^{2+} signals, the SPL $[Ca^{2+}]$ in the proximity of mitochondria was 0.1 μM, whereas far from mitochondria, i.e., in a region where mitochondrial Ca^{2+} buffering can be neglected, it amounted to 1.2 μM on average (Malli et al. 2003). This value is again much less than considered for HCMD.

14.4.2 Is HCMD Required for SOCE-Induced Mitochondrial Ca^{2+} Uptake?

HCMD is formed during SOCE in the cytoplasmic space between the STIM1 clusters in PM-adjacent regions of the ER and clusters of Orai1 Ca^{2+} channels in the apposing PM (Putney 2009). Exposure of mitochondria to high $[Ca^{2+}]$ requires the presence of the organelle within this well-defined space. The distance between the channels and STIM1 puncta was estimated by Varnai, Balla, and their co-workers applying a special protein engineering technique. In COS-7 cells, they targeted fluorescent proteins fused with FK506 binding protein (FKBP) or the FKBP-rapamycin binding (FRB) fragment of mTOR (mammalian target of rapamycin) to the PM and the ER, respectively. Rapamycin induced the dimerization of FKBP and FRB (provided that store depletion induced translocation of STIM1 to the peripheral ER beforehand). Orai1 appeared in the complex if the length of the linkers created a gap larger than 8 nm but smaller than 14 nm (Varnai et al. 2007). This observation is in accordance with the results of electron microscopic studies in Jurkat cells showing that STIM1-containing puncta in discrete subregions of junctional ER are located 10–25 nm from the PM (Wu et al. 2006). Other studies led to the conclusion that the distance between Orai1 and STIM1 during SOCE is smaller still (<10 nm) (Muik et al. 2008; Park et al. 2009). The conclusion can be drawn that the gap between ER and the PM at the site of SOCE is too narrow to accommodate mitochondria, at least in COS-7, HEK-293, and Jurkat cells.

We measured the relation of mitochondria, STIM1, and Orai1 in COS-7 cells. In order to enhance the active sites of SOCE, Orai1 and fluorescent protein (YFP or mRFP)-tagged STIM1 proteins were co-expressed. Ca^{2+} store depletion, induced with the purinergic agonist ATP plus the SERCA-inhibitor thapsigargin, evoked rapid increase in $[Ca^{2+}]_c$ that was associated with a moderate rise in $[Ca^{2+}]_m$ (Fig. 14.1). After a slight delay, a second phase of the Ca^{2+} response was observed, which was much more pronounced in the mitochondria than in the cytosol. The onset of this secondary Ca^{2+} rise coincided with the translocation of STIM1 from the perinuclear to peripheral ER and indicates the beginning of SOCE.

In separate experiments the cells were also transfected with mitochondria-targeted inverse pericam (i-Pericam), and the effect of Ca^{2+} depletion on the proximity of mitochondria to the site of SOCE within the $\leq$100 nm wide SPL cytoplasm was examined with total internal reflection fluorescence (TIRF) microscopy. Within 2 min of ER Ca^{2+} depletion (induced by ATP + thapsigargin in a Ca^{2+}-free medium), mRFP fluorescence became highly punctated indicating the formation of STIM1 clusters and their enrichment in close proximity to the PM. A few mitochondria showing i-Pericam fluorescence could also be located in this $\leq$100 nm SPL space. Importantly, the PM-close mitochondria showed no preferential localization close to the STIM1 puncta, but were found either in between the STIM1 patches or further away, in STIM1-free regions. In the TIRF image recorded 25 min after store depletion (Fig. 14.2), very few mitochondria could be detected

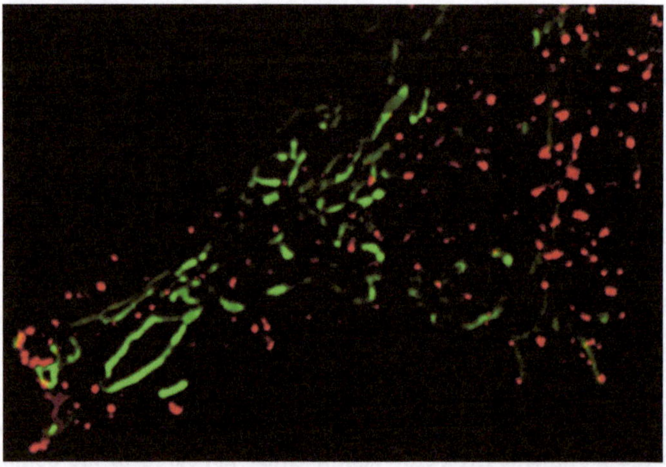

Fig. 14.2 Lack of localization of mitochondria and STIM1-labelled ER vesicles in the subplasmalemmal space of Ca^{2+}-depleted COS-7 cells. The cells were expressing mRFP-STIM1, untagged Orai1, and inverse pericam targeted to the mitochondrial matrix. They were exposed to the P2Y receptor agonist ATP and thapsigargin in Ca^{2+}-free medium for 25 min prior to acquiring this TIRF image. The punctate appearance of STIM1 (*red*) shows that STIM1 is translocated to the peripheral parts of the ER. Pericam fluorescence (*green*) is high, indicating reduced $[Ca^{2+}]_m$. (Colocalization of ER vesicles and mitochondria would bring about the merging of *red* and *green* particles into *yellow ones*.) Unpublished data of M.K. Korzeniowski, G. Szanda, T. Balla, and A. Spät

contacting any (or at least partially overlapping with) red (mRFP-STIM1) areas. Statistical analysis also excluded the colocalization of mitochondria and the STIM1 puncta. Within 10 s after the addition of Ca^{2+}, i-Pericam fluorescence rapidly decreased reflecting the SOCE-evoked increase in $[Ca^{2+}]_m$. Importantly, the rate of decrease of i-Pericam fluorescence (i.e., $[Ca^{2+}]_m$ elevation) did not differ between mitochondria far from or close to the ER. Thus, these results argue against the existence of preferential spatial positioning of mitochondria close to the STIM1-/Orai1-formed Ca^{2+} entry sites. Rather, they suggest that mitochondria are exposed to Ca^{2+} that diffuses laterally from the HCMDs formed between the PM and the SPL ER (Korzeniowski et al. 2009).

Confocal microscopic studies by Graier's group (Naghdi et al. 2010) support the above conclusion. In an endothelial cell line, they observed that more than 87% of SPL STIM1 clusters did not colocalize with SPL mitochondria. Moreover, artificial linking of mitochondria to the PM also failed to influence the intensity of SOCE. The location of mitochondria had no effect on SOCE in spite of the dependence of the latter on intact mitochondrial function. Similarly, no space for mitochondria between subplasmalemmal ER vesicles and plasma membrane, the site of SOCE, could be observed in electron microscopic studies (e.g., Dingsdale et al. 2012).

$[Ca^{2+}]$ measurements in the SPL cytosol contributed to our understanding regarding the role of HCMD in mitochondrial Ca^{2+} uptake during SOCE. Mathematical modeling predicts that $[Ca^{2+}]_c$ in the tip of PM "wrinkles" during Ca^{2+} influx may attain 0.1 mM (Brasen et al. 2010). Yet, attempts to measure $[Ca^{2+}]$ in the SPL region provided very heterogeneous data, probably reflecting significant cell-type dependence rather than only methodological differences. SOCE in HEK-293 cells raised $[Ca^{2+}]$ to merely 1 μM as estimated with PM-targeted aequorin (Nakahashi et al. 1997). Applying Fluo-4 in TIRF studies, Ca^{2+} influx often resulted in the saturation of the Ca^{2+}-sensitive dye (K_D for $Ca^{2+} = 0.35$ μM) in parotid acinar cells, but no saturation was observed in pancreatic acinar cells (Won and Yule 2006). This study suggests that $[Ca^{2+}]$ in the overall $\leq$100 nm wide SPL cytoplasm in exocrine cells does not necessarily exceed the value of a few μM. At the other extreme, in A7r5 vascular smooth muscle cells SOCE raised SPL $[Ca^{2+}]$ to 40 μM in average, as measured with PM-targeted aequorin (Marsault et al. 1997). Still higher $[Ca^{2+}]$ values, attaining 400 μM, were reported for endothelial cells in which the Ca^{2+}-sensitive protein yellow cameleon was targeted into the plasmalemmal caveolae (Isshiki et al. 2002). This Ca^{2+} response is 400-fold higher than detected with Ca^{2+}-dependent current measurements (Malli et al. 2003). The conclusion may be drawn that although in some cell types very high $[Ca^{2+}]$ may be attained around the cytosolic orifice of plasmalemmal ion channels, mitochondria may not have access to this microdomain surrounded by caveolar PM or the narrow tip of PM wrinkles.

After so many attempts of various laboratories that failed to provide conclusive results, in 2010 Pozzan's team reported experiments which successfully measured $[Ca^{2+}]$ at the outer mitochondrial membrane (OMM) (Giacomello et al. 2010). A special Ca^{2+}-sensitive probe (D1-cpV) was targeted either to the cytosol or the OMM in HeLa cells. IP_3-induced Ca^{2+} release was induced by histamine. A pixel-by-pixel analysis of fluorescence intensities revealed the formation of hot spots on the OMM, indicating the flux of Ca^{2+} through IP_3R into confined microdomains between ER and apposing mitochondria. [Ca^{2+}] in these micro-areas (about 10% of mitochondrial surface) reached values five- to tenfold higher (4–16 μM) than in the bulk cytosol. When the events in the SPL cytoplasm were examined with epifluorescence and TIRF microscopy using the same pixel-by-pixel analysis, Ca^{2+} hot spots during voltage-activated Ca^{2+} influx but not during SOCE could be revealed. The conclusion that mitochondria are excluded from the regions where store-operated Ca^{2+} channels are activated confirms our morphological observations also obtained with TIRF microscopy (Korzeniowski et al. 2009) (see above).

When the rapamycin-based dimerization technique was applied in RBL-2H3 and H9c2 cells to link ER and mitochondria together, the Ca^{2+}-sensitive fluorescent protein pericam, inserted into the interorganellar linker, indicated the formation of an IP_3-induced Ca^{2+} signal of about 4 μM at the surface of the OMM, as opposed to the 250–700 nM Ca^{2+} signal in the global cytosol. In contrast, linking the PM and the OMM, no contact points were observed for most mitochondria, showing that the dominant fraction of SPL mitochondria is unable to contact with the PM. Store-operated Ca^{2+} entry evoked by thapsigargin resulted in a relatively small pericam response, again disproving the generation of HCMD under such conditions (Csordas et al. 2010).

The TIRF-based morphological study (Korzeniowski et al. 2009), the [Ca^{2+}] measurements with the OMM-targeted Ca^{2+} sensor (Giacomello et al. 2010), as well as the experiments applying rapamycin-induced linking of the OMM with other membranes (Csordas et al. 2010) unambiguously show that there are several cell types in which SOCE-induced mitochondrial Ca^{2+} uptake occurs without the formation of HCMD. This conclusion is at variance with the general view that formation of HCMD is essential for mitochondrial Ca^{2+} uptake. It should be recalled that mitochondrial Ca^{2+} uptake at submicromolar $[Ca^{2+}]_c$ was described in various cell types (reviewed in Szanda et al. 2006). In sympathetic neurons the kinetic analysis of depolarization-induced cytosolic Ca^{2+} signals has led to the conclusion that Ca^{2+} uptake occurs already below a $[Ca^{2+}]_c$ of half μM (Colegrove et al. 2000; Pivovarova et al. 1999). The translocation of mitochondria from the SPL toward the perinuclear cytoplasm by the overexpression of the fission factor hFis1 did not modify the amplitude of the mitochondrial Ca^{2+} signal during SOCE (Frieden et al. 2004). After applying various experimental maneuvers, the plots of $[Ca^{2+}]_c$ versus $[Ca^{2+}]_m$ for single mitochondria indicated that the threshold value for mitochondrial Ca^{2+} uptake varied from 200 to 1000 nM (Collins et al. 2001). Experiments on permeabilized cells provided direct evidence in favor of mitochondrial Ca^{2+} uptake at submicromolar [Ca^{2+}] values. In three endocrine cell types, the adrenal glomerulosa cell, the ovarian luteal cells, and the insulin-producing INS-1 cell, the mitochondria respond to an elevation of perimitochondrial [Ca^{2+}] from ~60 nM to $\leq$200 nM (Pitter et al. 2002; Szabadkai et al. 2001). The threshold of net mitochondrial Ca^{2+} uptake in H295R adrenocortical cells, in HeLa cells, and in 143BmA-13 osteosarcoma cells is somewhat higher but still submicromolar (Pitter et al. 2002; Szanda et al. 2006). In all these experiments, even spermine, a potentiating factor of mitochondrial Ca^{2+} uptake (Lenzen et al. 1986), has probably been lost during cell permeabilization; therefore the threshold of net Ca^{2+} uptake in intact cells may be even lower.

The question may arise, especially in view of the low-Ca^{2+} affinity of the uptake mechanism, as to how uptake can occur from a perimitochondrial cytosol displaying only micromolar or even smaller Ca^{2+} signals. Distinct Ca^{2+} uptake mechanisms have been proposed for central and peripheral mitochondria in an endothelial cell line by Graier's group (Waldeck-Weiermair et al. 2010). They found that knockdown of UCP2/3 diminished mitochondrial Ca^{2+} uptake fueled by ER Ca^{2+} release but not that induced by SOCE. Why SOCE-induced, UCP2/3-independent Ca^{2+} uptake would not require a $[Ca^{2+}]$ of 10^{-5} M (or more) is not yet known.

Another mechanism may be considered for explaining how mitochondria can accumulate Ca^{2+} without being exposed to HCMDs during SOCE. Stimulation with Ca^{2+}-mobilizing agonists results not only in a Ca^{2+} but also in a Mg^{2+} signal and Mg^{2+}, even at physiological concentrations, and is capable of inhibiting mitochondrial Ca^{2+} uptake (Szanda et al. 2009; Tewari et al. 2014). In contrast, store-operated Ca^{2+} influx is not associated with significant elevation of $[Mg^{2+}]_c$ (Szanda et al. 2009); therefore Ca^{2+} may be accumulated by the mitochondria with higher affinity.

For the consideration of the possible role of transporters other than MCU in Ca^{2+} uptake at submicromolar $[Ca^{2+}]_c$, we refer to excellent reviews (Contreras et al. 2010; Santo-Domingo and Demaurex 2010).

14.5 Conclusions

On the whole, under a wide range of experimental conditions and in a variety of cell types, SOCE is hugely dependent on uncompromised mitochondrial metabolism and Ca^{2+} handling. Depolarization of the organelle attenuates Ca^{2+} buffering capacity and consequently hampers SOCE activation and speeds up Ca^{2+}-dependent inactivation. Additionally, mitochondria may alter the concentration of soluble factors that also influence SOCE. Premature inhibition of SOCE due to mitochondrial dysfunction may compromise normal cellular responses.

In contrast to the mitochondrial uptake of Ca^{2+} released from IP_3-sensitive stores, the formation of HCMDs may not be required for efficient mitochondrial uptake of Ca^{2+} entering through store-operated channels. The very high SPL $[Ca^{2+}]$ measured in some cell types does not seem to be a general phenomenon. The space between the STIM1-labelled ER membrane and the Orai1 in the apposing PM space, at least in quite a few cell types, is too narrow to accommodate a mitochondrion. Therefore, mitochondria may be exposed to Ca^{2+} diffusing only laterally from the HCMDs during SOCE. This means that SPL mitochondria are exposed to Ca^{2+} concentrations probably exceeding that of the global cytosolic $[Ca^{2+}]$ but which do not attain those high values estimated for the HCMDs around ER Ca^{2+} release sites.

Acknowledgment G. Szanda was supported by the János Bolyai Research Scholarship of the Hungarian Academy of Sciences.

References

Acin-Perez R, Salazar E, Kamenetsky M, Buck J, Levin LR, Manfredi G (2009) Cyclic AMP produced inside mitochondria regulates oxidative phosphorylation. Cell Metab 9:265–276

Arnaudeau S, Kelley WL, Walsh JV Jr, Demaurex N (2001) Mitochondria recycle Ca2+ to the endoplasmic reticulum and prevent the depletion of neighboring endoplasmic reticulum regions. J Biol Chem 276:29430–29439

Bakowski D, Parekh AB (2007) Regulation of store-operated calcium channels by the intermediary metabolite pyruvic acid. Curr Biol 17:1076–1081

Baughman JM, Perocchi F, Girgis HS, Plovanich M, Belcher-Timme CA, Sancak Y, Bao XR, Strittmatter L, Goldberger O, Bogorad RL, Koteliansky V, Mootha VK (2011) Integrative genomics identifies MCU as an essential component of the mitochondrial calcium uniporter. Nature 476:341–345

Bezprozvanny I, Watras J, Ehrlich BE (1991) Bell-shaped calcium-response curves of Ins(1,4,5) P3- and calcium-gated channels from endoplasmic reticulum of cerebellum. Nature 351:751–754

Bondarenko AI, Parichatikanond W, Madreiter CT, Rost R, Waldeck-Weiermair M, Malli R, Graier WF (2015) UCP2 modulates single-channel properties of a MCU-dependent Ca2+ inward current in mitochondria. Pflugers Arch 467:2509–2518

Brasen JC, Olsen LF, Hallett MB (2010) Cell surface topology creates high Ca2+ signalling microdomains. Cell Calcium 47:339–349

Brown GC (1992) Control of respiration and ATP synthesis in mammalian mitochondria and cells. Biochem J 284(Pt 1):1–13

Carafoli E, Tiozzo R, Lugli G, Crovetti F, Kratzing C (1974) The release of calcium from heart mitochondria by sodium. J Mol Cell Cardiol 6:361–371

Chang WC, Nelson C, Parekh AB (2006) Ca2+ influx through CRAC channels activates cytosolic phospholipase A2, leukotriene C4 secretion, and expression of c-fos through ERK-dependent and -independent pathways in mast cells. FASEB J 20:2381–2383

Chaudhuri D, Artiga DJ, Abiria SA, Clapham DE (2016) Mitochondrial calcium uniporter regulator 1 (MCUR1) regulates the calcium threshold for the mitochondrial permeability transition. Proc Natl Acad Sci U S A 113:E1872–E1880

Chinopoulos C, Adam-Vizi V (2006) Calcium, mitochondria and oxidative stress in neuronal pathology. Novel aspects of an enduring theme. FEBS J 273:433–450

Colegrove SL, Albrecht MA, Friel DD (2000) Dissection of mitochondrial Ca2+ uptake and release fluxes in situ after depolarization-evoked [Ca2+]i elevations in sympathetic neurons. J Gen Physiol 115:351–369

Collins TJ, Lipp P, Berridge MJ, Li WH, Bootman MD (2000) Inositol 1,4,5-trisphosphate-induced Ca2+ release is inhibited by mitochondrial depolarization. Biochem J 347:593–600

Collins TJ, Lipp P, Berridge MJ, Bootman MD (2001) Mitochondrial Ca2+ uptake depends on the spatial and temporal profile of cytosolic Ca2+ signals. J Biol Chem 276:26411–26420

Contreras L, Drago I, Zampese E, Pozzan T (2010) Mitochondria: the calcium connection. Biochim Biophys Acta 1797:607–618

Csordas G, Varnai P, Golenar T, Roy S, Purkins G, Schneider TG, Balla T, Hajnoczky G (2010) Imaging interorganelle contacts and local calcium dynamics at the ER-mitochondrial interface. Mol Cell 39:121–132

Csordas G, Golenar T, Seifert EL, Kamer KJ, Sancak Y, Perocchi F, Moffat C, Weaver D, de la Fuente Perez S, Bogorad R, Koteliansky V, Adijanto J, Mootha VK, Hajnoczky G (2013) MICU1 controls both the threshold and cooperative activation of the mitochondrial Ca(2)(+) uniporter. Cell Metab 17:976–987

Davidson SM, Duchen MR (2007) Endothelial mitochondria: contributing to vascular function and disease. Circ Res 100:1128–1141

Demaurex N, Poburko D, Frieden M (2009) Regulation of plasma membrane calcium fluxes by mitochondria. Biochim Biophys Acta 1787:1383–1394

De Stefani D, Raffaello A, Teardo E, Szabo I, Rizzuto R (2011) A forty-kilodalton protein of the inner membrane is the mitochondrial calcium uniporter. Nature 476:336–340

Di Benedetto G, Scalzotto E, Mongillo M, Pozzan T (2013) Mitochondrial Ca2+ uptake induces cyclic AMP generation in the matrix and modulates organelle ATP levels. Cell Metab 17:965–975

Dingsdale H, Haynes L, Lur G, Tepikin A (2012) The role of the ER and ER-plasma membrane junctions in the regulation of SOCE. In: Groschner K, Graier WF, Romanin C (eds) Store-operated Ca2+ entry (SOCE) pathways. Springer, Wien, pp 137–152

Duchen MR (1992) Ca2+-dependent changes in the mitochondrial energetics in single dissociated mouse sensory neurons. Biochem J 283:41–50

Feldman B, Fedida-Metula S, Nita J, Sekler I, Fishman D (2010) Coupling of mitochondria to store-operated Ca2+-signaling sustains constitutive activation of protein kinase B/Akt and augments survival of malignant melanoma cells. Cell Calcium 47:525–537

Frieden M, James D, Castelbou C, Danckaert A, Martinou JC, Demaurex N (2004) Calcium homeostasis during mitochondria fragmentation and perinuclear clustering induced by hFis1. J Biol Chem 279:22704–22714

Frieden M, Arnaudeau S, Castelbou C, Demaurex N (2005) Subplasmalemmal mitochondria modulate the activity of plasma membrane Ca2+-ATPases. J Biol Chem 280:43198–43208

Giacomello M, Drago I, Bortolozzi M, Scorzeto M, Gianelle A, Pizzo P, Pozzan T (2010) Ca2+ hot spots on the mitochondrial surface are generated by Ca2+ mobilization from stores, but not by activation of store-operated Ca2+ channels. Mol Cell 38:280–290

Gilabert JA, Parekh AB (2000) Respiring mitochondria determine the pattern of activation and inactivation of the store-operated Ca2+ current I(CRAC). EMBO J 19:6401–6407

Gilabert JA, Bakowski D, Parekh AB (2001) Energized mitochondria increase the dynamic range over which inositol 1,4,5-trisphosphate activates store-operated calcium influx. EMBO J 20:2672–2679

Glitsch MD, Parekh AB (2000) Ca2+ store dynamics determines the pattern of activation of the store-operated Ca2+ current I(CRAC) in response to InsP3 in rat basophilic leukaemia cells. J Physiol 523:283–290

Glitsch MD, Bakowski D, Parekh AB (2002) Store-operated Ca2+ entry depends on mitochondrial Ca2+ uptake. EMBO J 21:6744–6754

Gunter TE, Pfeiffer DR (1990) Mechanisms by which mitochondria transport calcium. Am J Physiol 258:C755–C786

Hajnoczky G, Hager R, Thomas AP (1999) Mitochondria suppress local feedback activation of inositol 1,4, 5-trisphosphate receptors by Ca2+. J Biol Chem 274:14157–14162

Hoth M, Fanger CM, Lewis RS (1997) Mitochondrial regulation of store-operated calcium signaling in T lymphocytes. J Cell Biol 137:633–648

Hoth M, Button DC, Lewis RS (2000) Mitochondrial control of calcium-channel gating: a mechanism for sustained signaling and transcriptional activation in T lymphocytes. Proc Natl Acad Sci U S A 97:10607–10612

Ichas F, Mazat JP (1998) From calcium signaling to cell death: two conformations for the mitochondrial permeability transition pore. Switching from low- to high-conductance state. Biochim Biophys Acta 1366:33–50

Isshiki M, Ying YS, Fujita T, Anderson RG (2002) A molecular sensor detects signal transduction from caveolae in living cells. J Biol Chem 277:43389–43398

Jiang D, Zhao L, Clapham DE (2009) Genome-wide RNAi screen identifies Letm1 as a mitochondrial Ca2+/H+ antiporter. Science 326:144–147
Jouaville LS, Pinton P, Bastianutto C, Rutter GA, Rizzuto R (1999) Regulation of mitochondrial ATP synthesis by calcium: evidence for a long-term metabolic priming. Proc Natl Acad Sci U S A 96:13807–13812
Kamer KJ, Mootha VK (2014) MICU1 and MICU2 play nonredundant roles in the regulation of the mitochondrial calcium uniporter. EMBO Rep 15:299–307
Kamer KJ, Mootha VK (2015) The molecular era of the mitochondrial calcium uniporter. Nat Rev Mol Cell Biol 16:545–553
Katona D, Rajki A, Di BG, Pozzan T, Spät A (2015) Calcium-dependent mitochondrial cAMP production enhances aldosterone secretion. Mol Cell Endocrinol 412:196–204
Kirichok Y, Krapivinsky G, Clapham DE (2004) The mitochondrial calcium uniporter is a highly selective ion channel. Nature 427:360–364
Koncz P, Szanda G, Fülöp L, Rajki A, Spät A (2009) Mitochondrial Ca2+ uptake is inhibited by a concerted action of p38 MAPK and protein kinase D. Cell Calcium 46:122–129
Korzeniowski MK, Szanda G, Balla T, Spät A (2009) Store-operated Ca2+ influx and subplasmalemmal mitochondria. Cell Calcium 46:49–55
Lawrie AM, Rizzuto R, Pozzan T, Simpson AWM (1996) A role for calcium influx in the regulation of mitochondrial calcium in endothelial cells. J Biol Chem 271:10753–10759
Lenzen S, Hickethier R, Panten U (1986) Interactions between spermine and Mg2+ on mitochondrial Ca2+ transport. J Biol Chem 261:16478–16483
Makowska A, Zablocki K, Duszynski J (2000) The role of mitochondria in the regulation of calcium influx into Jurkat cells. Eur J Biochem 267:877–884
Malli R, Frieden M, Osibow K, Zoratti C, Mayer M, Demaurex N, Graier WF (2003) Sustained Ca2+ transfer across mitochondria is essential for mitochondrial Ca2+ buffering, store-operated Ca2+ entry, and Ca2+ store refilling. J Biol Chem 278:44769–44779
Malli R, Frieden M, Trenker M, Graier WF (2005) The role of mitochondria for Ca2+ refilling of the endoplasmic reticulum. J Biol Chem 280:12114–12122
Mallilankaraman K, Cardenas C, Doonan PJ, Chandramoorthy HC, Irrinki KM, Golenar T, Csordas G, Madireddi P, Yang J, Muller M, Miller R, Kolesar JE, Molgo J, Kaufman B, Hajnoczky G, Foskett JK, Madesh M (2012a) MCUR1 is an essential component of mitochondrial Ca2+ uptake that regulates cellular metabolism. Nat Cell Biol 14:1336–1343
Mallilankaraman K, Doonan P, Cardenas C, Chandramoorthy HC, Muller M, Miller R, Hoffman NE, Gandhirajan RK, Molgo J, Birnbaum MJ, Rothberg BS, Mak DO, Foskett JK, Madesh M (2012b) MICU1 is an essential gatekeeper for MCU-mediated mitochondrial Ca2+ uptake that regulates cell survival. Cell 151:630–644
Marsault R, Murgia M, Pozzan T, Rizzuto R (1997) Domains of high Ca2+ beneath the plasma membrane of living A7r5 cells. EMBO J 16:1575–1581
McCormack JG (1985) Characterization of the effects of Ca2+ on the intramitochondrial Ca2+-sensitive enzymes from rat and within intact rat liver mitochondria. Biochem J 231:581–595
McCormack JG, Halestrap AP, Denton RM (1990) Role of calcium ions in regulation of mammalian intramitochondrial metabolism. Physiol Rev 70:391–425
Mignen O, Brink C, Enfissi A, Nadkarni A, Shuttleworth TJ, Giovannucci DR, Capiod T (2005) Carboxyamidotriazole-induced inhibition of mitochondrial calcium import blocks capacitative calcium entry and cell proliferation in HEK-293 cells. J Cell Sci 118:5615–5623
Montalvo GB, Artalejo AR, Gilabert JA (2006) ATP from subplasmalemmal mitochondria controls Ca2+-dependent inactivation of CRAC channels. J Biol Chem 281:35616–35623
Motloch LJ, Larbig R, Gebing T, Reda S, Schwaiger A, Leitner J, Wolny M, Eckardt L, Hoppe UC (2016) By regulating mitochondrial Ca2+-uptake UCP2 modulates intracellular Ca2+. PLoS One 11:e0148359
Muik M, Frischauf I, Derler I, Fahrner M, Bergsmann J, Eder P, Schindl R, Hesch C, Polzinger B, Fritsch R, Kahr H, Madl J, Gruber H, Groschner K, Romanin C (2008) Dynamic coupling of

the putative coiled-coil domain of ORAI1 with STIM1 mediates ORAI1 channel activation. J Biol Chem 283:8014–8022

Naghdi S, Waldeck-Weiermair M, Fertschai I, Poteser M, Graier WF, Malli R (2010) Mitochondrial Ca2+ uptake and not mitochondrial motility is required for STIM1-Orai1-dependent store-operated Ca2+ entry. J Cell Sci 123:2553–2564

Nakahashi Y, Nelson E, Fagan K, Gonzales E, Guillou JL, Cooper DM (1997) Construction of a full-length Ca2+-sensitive adenylyl cyclase/aequorin chimera. J Biol Chem 272:18093–18097

Nunes P, Demaurex N (2014) Redox regulation of store-operated Ca2+ entry. Antioxid Redox Signal 21:915–932

Olson ML, Chalmers S, McCarron JG (2010) Mitochondrial Ca2+ uptake increases Ca2+ release from inositol 1,4,5-trisphosphate receptor clusters in smooth muscle cells. J Biol Chem 285:2040–2050

Parekh AB (1998) Slow feedback inhibition of calcium release-activated calcium current by calcium entry. J Biol Chem 273:14925–14932

Parekh AB (2008) Mitochondrial regulation of store-operated CRAC channels. Cell Calcium 44:6–13

Parekh AB (2010) Store-operated CRAC channels: function in health and disease. Nat Rev Drug Discov 9:399–410

Park MK, Ashby MC, Erdemli G, Petersen OH, Tepikin AV (2001) Perinuclear, perigranular and sub-plasmalemmal mitochondria have distinct functions in the regulation of cellular calcium transport. EMBO J 20:1863–1874

Park CY, Hoover PJ, Mullins FM, Bachhawat P, Covington ED, Raunser S, Walz T, Garcia KC, Dolmetsch RE, Lewis RS (2009) STIM1 clusters and activates CRAC channels via direct binding of a cytosolic domain to Orai1. Cell 136:876–890

Paupe V, Prudent J, Dassa EP, Rendon OZ, Shoubridge EA (2015) CCDC90A (MCUR1) is a cytochrome c oxidase assembly factor and not a regulator of the mitochondrial calcium uniporter. Cell Metab 21:109–116

Perocchi F, Gohil VM, Girgis HS, Bao XR, McCombs JE, Palmer AE, Mootha VK (2010) MICU1 encodes a mitochondrial EF hand protein required for Ca2+ uptake. Nature 467:291–296

Petersen OH, Verkhratsky A (2007) Endoplasmic reticulum calcium tunnels integrate signalling in polarised cells. Cell Calcium 42:373–378

Pitter JG, Maechler P, Wollheim CB, Spät A (2002) Mitochondria respond to Ca2+ already in the submicromolar range: correlation with redox state. Cell Calcium 31:97–104

Pivovarova NB, Hongpaisan J, Andrews SB, Friel DD (1999) Depolarization-induced mitochondrial Ca2+ accumulation in sympathetic neurons: spatial and temporal characteristics. J Neurosci 19:6372–6384

Plovanich M, Bogorad RL, Sancak Y, Kamer KJ, Strittmatter L, Li AA, Girgis HS, Kuchimanchi S, De GJ, Speciner L, Taneja N, Oshea J, Koteliansky V, Mootha VK (2013) MICU2, a paralog of MICU1, resides within the mitochondrial uniporter complex to regulate calcium handling. PLoS One 8:e55785

Pralong WF, Hunyady L, Varnai P, Wollheim CB, Spät A (1992) Pyridine nucleotide redox state parallels production of aldosterone in potassium-stimulated adrenal glomerulosa cells. Proc Natl Acad Sci U S A 89:132–136

Pralong WF, Spät A, Wollheim CB (1994) Dynamic pacing of cell metabolism by intracellular Ca2+. J Biol Chem 269:27310–27314

Putney JW Jr (2009) Capacitative calcium entry: from concept to molecules. Immunol Rev 231:10–22

Quintana A, Schwarz EC, Schwindling C, Lipp P, Kaestner L, Hoth M (2006) Sustained activity of calcium release-activated calcium channels requires translocation of mitochondria to the plasma membrane. J Biol Chem 281:40302–40309

Raffaello A, De SD, Sabbadin D, Teardo E, Merli G, Picard A, Checchetto V, Moro S, Szabo I, Rizzuto R (2013) The mitochondrial calcium uniporter is a multimer that can include a dominant-negative pore-forming subunit. EMBO J 32:2362–2376

Rao W, Zhang L, Peng C, Hui H, Wang K, Su N, Wang L, Dai SH, Yang YF, Chen T, Luo P, Fei Z (2015) Downregulation of STIM2 improves neuronal survival after traumatic brain injury by alleviating calcium overload and mitochondrial dysfunction. Biochim Biophys Acta 1852:2402–2413

Rapizzi E, Pinton P, Szabadkai G, Wieckowski MR, Vandecasteele G, Baird G, Tuft RA, Fogarty KE, Rizzuto R (2002) Recombinant expression of the voltage-dependent anion channel enhances the transfer of Ca2+ microdomains to mitochondria. J Biol Chem 159:613–624

Rizzuto R, Pozzan T (2006) Microdomains of intracellular Ca2+: molecular determinants and functional consequences. Physiol Rev 86:369–408

Rohacs T, Nagy G, Spät A (1997a) Cytoplasmic Ca2+ signalling and reduction of mitochondrial pyridine nucleotides in adrenal glomerulosa cells in response to K+, angiotensin II and vasopressin. Biochem J 322:785–792

Rohacs T, Tory K, Dobos A, Spät A (1997b) Intracellular calcium release is more efficient than calcium influx in stimulating mitochondrial NAD(P)H formation in adrenal glomerulosa cells. Biochem J 328:525–528

Sancak Y, Markhard AL, Kitami T, Kovacs-Bogdan E, Kamer KJ, Udeshi ND, Carr SA, Chaudhuri D, Clapham DE, Li AA, Calvo SE, Goldberger O, Mootha VK (2013) EMRE is an essential component of the mitochondrial calcium uniporter complex. Science 342:1379–1382

Santo-Domingo J, Demaurex N (2010) Calcium uptake mechanisms of mitochondria. Biochim Biophys Acta 1797:907–912

Spät A, Fülöp L, Koncz P, Szanda G (2008a) When is high-Ca2+ microdomain required for mitochondrial Ca2+ uptake? Acta Physiol (Oxf) 195:139–147

Spät A, Szanda G, Csordas G, Hajnoczky G (2008b) High- and low-calcium-dependent mechanisms of mitochondrial calcium signalling. Cell Calcium 44:51–63

Spät A, Hunyady L, Szanda G (2016) Signaling interactions in the adrenal cortex. Front Endocrinol (Lausanne) 7:17

Spencer T, Bygrave FL (1971) Stimulation by calcium ions of atractyloside-sensitive adenine nucleotide translocation in rat liver mitochondria. Biochem Biophys Res Commun 43:1290–1295

Szabadkai G, Pitter JG, Spät A (2001) Cytoplasmic Ca2+ at low submicromolar concentration stimulates mitochondrial metabolism in rat luteal cells. Pflugers Arch 441:678–685

Szanda G, Koncz P, Varnai P, Spät A (2006) Mitochondrial Ca2+ uptake with and without the formation of high-Ca2+ microdomains. Cell Calcium 40:527–538

Szanda G, Koncz P, Rajki A, Spät A (2008) Participation of p38 MAPK and a novel-type protein kinase C in the control of mitochondrial Ca2+ uptake. Cell Calcium 43:250–259

Szanda G, Rajki A, Gallego-Sandin S, Garcia-Sancho J, Spät A (2009) Effect of cytosolic Mg2+ on mitochondrial Ca2+ signaling. Pflugers Arch 457:941–954

Tewari SG, Camara AK, Stowe DF, Dash RK (2014) Computational analysis of Ca2+ dynamics in isolated cardiac mitochondria predicts two distinct modes of Ca2+ uptake. J Physiol 592:1917–1930

Thyagarajan B, Malli R, Schmidt K, Graier WF, Groschner K (2002) Nitric oxide inhibits capacitative Ca2+ entry by suppression of mitochondrial Ca2+ handling. Br J Pharmacol 137:821–830

Tinel H, Cancela JM, Mogami H, Gerasimenko JV, Gerasimenko OV, Tepikin AV, Petersen OH (1999) Active mitochondria surrounding the pancreatic acinar granule region prevent spreading of inositol trisphosphate-evoked local cytosolic Ca2+ signals. EMBO J 18:4999–5008

To MS, Aromataris EC, Castro J, Roberts ML, Barritt GJ, Rychkov GY (2010) Mitochondrial uncoupler FCCP activates proton conductance but does not block store-operated Ca2+ current in liver cells. Arch Biochem Biophys 495:152–158

Trenker M, Malli R, Fertschai I, Levak-Frank S, Graier WF (2007) Uncoupling proteins 2 and 3 are fundamental for mitochondrial Ca2+ uniport. Nat Cell Biol 9:445–452

Trepakova ES, Cohen RA, Bolotina VM (1999) Nitric oxide inhibits capacitative cation influx in human platelets by promoting sarcoplasmic/endoplasmic reticulum Ca2+-ATPase-dependent refilling of Ca2+ stores. Circ Res 84:201–209

Tsai MF, Phillips CB, Ranaghan M, Tsai CW, Wu Y, Willliams C, Miller C (2016) Dual functions of a small regulatory subunit in the mitochondrial calcium uniporter complex. Elife 5:e15545

Vais H, Tanis JE, Muller M, Payne R, Mallilankaraman K, Foskett JK (2015) MCUR1, CCDC90A, is a regulator of the mitochondrial calcium uniporter. Cell Metab 22:533–535

Vais H, Mallilankaraman K, Mak DO, Hoff H, Payne R, Tanis JE, Foskett JK (2016) EMRE is a matrix Ca2+ sensor that governs gatekeeping of the mitochondrial Ca2+ uniporter. Cell Rep 14:403–410

Varadi A, Cirulli V, Rutter GA (2004) Mitochondrial localization as a determinant of capacitative Ca2+ entry in HeLa cells. Cell Calcium 36:499–508

Varnai P, Toth B, Toth DJ, Hunyady L, Balla T (2007) Visualization and manipulation of plasma membrane-endoplasmic reticulum contact sites indicates the presence of additional molecular components within the STIM1-Orai1 complex. J Biol Chem 282:29678–29690

Waldeck-Weiermair M, Malli R, Naghdi S, Trenker M, Kahn MJ, Graier WF (2010) The contribution of UCP2 and UCP3 to mitochondrial Ca2+ uptake is differentially determined by the source of supplied Ca2+. Cell Calcium 47:433–440

Waldeck-Weiermair M, Malli R, Parichatikanond W, Gottschalk B, Madreiter-Sokolowski CT, Klec C, Rost R, Graier WF (2015) Rearrangement of MICU1 multimers for activation of MCU is solely controlled by cytosolic Ca(2+). Sci Rep 5:15602

Walsh C, Barrow S, Voronina S, Chvanov M, Petersen OH, Tepikin A (2009) Modulation of calcium signalling by mitochondria. Biochim Biophys Acta 1787:1374–1382

Wiederkehr A, Szanda G, Akhmedov D, Mataki C, Heizmann CW, Schoonjans K, Pozzan T, Spät A, Wollheim CB (2011) Mitochondrial matrix calcium is an activating signal for hormone secretion. Cell Metab 13:601–611

Won JH, Yule DI (2006) Measurement of Ca2+ signaling dynamics in exocrine cells with total internal reflection microscopy. Am J Physiol Gastrointest Liver Physiol 291:G146–G155

Wu MM, Buchanan J, Luik RM, Lewis RS (2006) Ca2+ store depletion causes STIM1 to accumulate in ER regions closely associated with the plasma membrane. J Cell Biol 174:803–813

Yi M, Weaver D, Hajnoczky G (2004) Control of mitochondrial motility and distribution by the calcium signal: a homeostatic circuit. J Cell Biol 167:661–672

Yu WH, Wolfgang W, Forte M (1995) Subcellular localization of human voltage-dependent anion channel isoforms. J Biol Chem 270:13998–14006

Zweifach A, Lewis RS (1995) Slow calcium-dependent inactivation of depletion-activated calcium current. Store-dependent and -independent mechanisms. J Biol Chem 270:14445–14451

The Role of Mitochondria in the Activation/Maintenance of SOCE: Membrane Contact Sites as Signaling Hubs Sustaining Store-Operated Ca^{2+} Entry

15

Nicolas Demaurex and Daniele Guido

Abstract

Store-operated Ca^{2+} entry (SOCE) is a cell signaling pathway essential for immune and muscle function controlled by dynamic interactions between Ca^{2+}-sensing STIM proteins on the endoplasmic reticulum (ER) and Ca^{2+}-permeable ORAI channels on the plasma membrane (PM). STIM-ORAI interactions occur at membrane contact sites (MCS), evolutionarily conserved cellular structures characterized by the close apposition (10–20 nm) between the ER and target membranes that facilitate the exchange of lipids by non-vesicular transport mechanisms. STIM-ORAI interactions were considered to be restricted to ER-PM MCS, but recent evidence indicates that productive interactions take place between ER-bound STIM1 and Ca^{2+} channels located in intracellular organelles. Interactions between the ER and endosomes or lysosomes regulate the lipid homeostasis of these organelles and the propagation of Ca^{2+} signals initiated by the release of Ca^{2+} from acidic stores. Intracellular MCS also regulate the efficiency of phagocytosis, a fundamental cellular process essential for immunity and tissue homeostasis, by ensuring the coordinated opening of Ca^{2+} channels on phagocytic vacuoles and of Ca^{2+} release channels on juxtaposed ER stores. In this chapter, we review the current knowledge on the molecular composition and architecture of membrane contact sites that sustain Ca^{2+} signals at the plasma membrane and in intracellular organelles.

Keywords

Cell signaling • Ion channels • Endoplasmic reticulum • Membrane contact sites • Lipid homeostasis • Phagocytosis

N. Demaurex (✉) • D. Guido
Department of Cell Physiology and Metabolism, University of Geneva, 1 Michel-Servet, 1211 Geneva 4, Switzerland
e-mail: Nicolas.Demaurex@unige.ch

K. Groschner et al. (eds.), *Store-Operated Ca^{2+} Entry (SOCE) Pathways*, Advances in Experimental Medicine and Biology 993, DOI 10.1007/978-3-319-57732-6_15

15.1 Introduction

Cytosolic Ca^{2+} elevations regulate a vast array of vital cellular functions in all eukaryotic organisms, controlling ubiquitous processes such as cell differentiation and gene transcription and specialized functions like cardiac contraction and neurotransmission. The signal specificity is conveyed by the spatiotemporal pattern of the intracellular Ca^{2+} signals generated by the coordinated activity of channels, pumps, and exchangers, which is decoded by cell-specific Ca^{2+} effector proteins tuned to respond to distinct profiles of cytosolic Ca^{2+} elevations. Rapid elevations in the cytosolic Ca^{2+} concentration follow the activation of cell surface receptors coupled to phospholipase C (PLC) as inositol-1,4,5-trisphosphate (IP_3) generated from phosphatidylinositol-4,5-bisphosphate (PIP_2) triggers the cooperative opening of Ca^{2+} release channels located on the sarco-/endoplasmic reticulum (SR/ER), enabling the explosive release of Ca^{2+} stored within this finite intracellular compartment. Long-lasting Ca^{2+} elevations, however, require the continuous entry of Ca^{2+} across plasma membrane channels, which provides an unlimited supply of Ca^{2+} originating from the extracellular space to sustain cytosolic signals and refill the depleted stores. A Ca^{2+} entry mechanism activated by the depletion of intracellular Ca^{2+} stores termed store-operated Ca^{2+} entry (SOCE) or capacitative Ca^{2+} entry was proposed 30 years ago (Putney 1986; Putney et al. 1989) and experimentally validated at the electrophysiological level shortly thereafter (Hoth and Penner 1992; Lewis and Cahalan 1989) as underlying a Ca^{2+} release-activated Ca^{2+} current (I_{CRAC}) with very high Ca^{2+} selectivity, inward rectification, unresolvable single-channel conductance (fS), and complex regulation by both extracellular and intracellular Ca^{2+} (Parekh and Putney 2005). More than a decade elapsed before the identification of the molecular players controlling this ubiquitous signaling pathway by siRNA screens: the stromal interaction molecules (STIM) STIM1 and STIM2 (Liou et al. 2005; Roos et al. 2005), which are integral ER Ca^{2+}-sensing proteins acting as intracellular ligands to control the gating of Ca^{2+} channels on the plasma membrane belonging to the ORAI family (also known as CRACM) of which three isoforms exist in mammals (Feske et al. 2006; Vig et al. 2006; Zhang et al. 2006). Silencing of STIM or ORAI proteins strongly reduces SOCE activity, while co-expression of both molecules leads to a remarkable amplification of I_{CRAC} in cells devoid of endogenous current (Peinelt et al. 2006; Roos et al. 2005; Soboloff et al. 2006; Vig et al. 2006; Zhang et al. 2006). Genetic linkage analysis and studies in animal models established that STIM and ORAI are required for the proliferation of lymphocytes and for skeletal muscle development and function and patients with loss of function mutations in STIM1 or ORAI1 suffer from severe combined immunodeficiency and muscular dysplasia (Prakriya and Lewis 2015).

The cascade of events linking store depletion to the STIM-dependent activation of ORAI Ca^{2+} channels is now well established. Following activation of cell surface receptors coupled to PLC, STIM isoforms oligomerize as the Ca^{2+} concentration in the ER lumen decreases, exposing a cytosolic channel-activating domain (CAD) and a polybasic domain that binds PIP_2 (Liou et al. 2007; Muik et al. 2009; Yuan et al. 2009). Extended STIM oligomers then accumulate within cortical ER (cER) structures facing PIP_2-rich domains of the PM, where they trap and gate ORAI channels via their exposed CAD (McNally et al. 2013; Park et al. 2009; Zhou et al. 2010), triggering the influx of Ca^{2+} to sustain the refilling of the ER and the maintenance of long-lasting Ca^{2+} signals (Chang and Liou 2016; Prakriya and Lewis 2015). The molecular determinants that control the trafficking of STIM proteins and the opening of ORAI channels have been extensively reviewed and will not be discussed here. Instead, we will focus on the architecture of the membrane contact sites that enables productive interactions to take place between ER-bound STIM proteins and ORAI channels on target membranes.

15.2 SOCE at Plasma Membrane Contact Sites

STIM-ORAI coupling occurs at membrane contact sites (MCS), evolutionarily conserved structures that sustain non-vesicular lipid transfer at sites of close apposition (~10–30 nm) between the ER and other cellular membranes (reviewed in (Carrasco and Meyer 2011; Elbaz and Schuldiner 2011; Friedman and Voeltz 2011; Toulmay and Prinz 2011). Ca^{2+} depletion of the ER by Ca^{2+}-mobilizing agonists or by inhibitors of sarco-/endoplasmic reticulum Ca^{2+} ATPases (SERCA) such as thapsigargin, the prototypical SOCE activator, increases the amount of ER cisternae enriched in STIM molecules near the plasma membrane. These ER cisternae appear on the fluorescence microscope as punctuate structures enriched in fluorescent STIM and ORAI molecules that decorate the plasma membrane and are reported as "clusters," "junctions," or "puncta" by different research groups. Electron microscopy studies revealed that the fluorescent clusters correspond to specialized compartments derived from the ER, termed "junctional ER" or "cortical ER" due to their location at junctions forming at the cell periphery (Orci et al. 2009; Wu et al. 2006). Both the length and the number of ER structures juxtaposed to the plasma membrane increase upon store depletion (Orci et al. 2009; Wu et al. 2006) or upon enforced expression of STIM proteins, indicating that the Ca^{2+}-dependent relocation of STIM1 at the cell periphery promotes the formation and extension of MCS (Lur et al. 2009; Orci et al. 2009; Shen et al. 2011; Wu et al. 2006). The dynamic junctions generated by STIM-ORAI interactions at the ER-PM interface are reminiscent of the ER-PM junctions described half a century ago in muscle cells (Porter and Palade 1957). In striated muscle, sites of close SR-PM apposition known as diads in cardiac muscle and triads in skeletal muscle underlie the excitation-contraction coupling process that drives muscle contraction. These junctions are formed between a specialized PM invagination known as the T-tubule and the SR terminal cisternae and couple the activity of voltage-gated Ca^{2+} entry channels to the opening of ryanodine-sensitive Ca^{2+} release channels on the sarcoplasmic reticulum. At these contact sites, the proximity between the Ca^{2+} entry and Ca^{2+} release channel enables the slight influx of Ca^{2+} to trigger the rapid release of Ca^{2+} from SR stores in cardiac muscle or a conformational coupling between both channels that open the ryanodine receptors in skeletal muscle (Block et al. 1988, reviewed in Franzini-Armstrong and Jorgensen 1994). An essential difference, however, is that diads and triads are stable structures, while the junctions generated by the engagement of the SOCE machinery are transient and dynamically remodeled following STIM1 recruitment. Another difference is that the extent of plasma membrane covered by junctional ER structures is low in non-muscle cells, ranging from less than 0.25% in resting HeLa cells to 5% in cells overexpressing STIM1 and treated with thapsigargin (Orci et al. 2009; Perni et al. 2015). The scarcity of STIM-ORAI contact sites renders their detection difficult with ER or PM markers, but these structures become apparent with reporter probes enriched at ER-PM junctions via their interactions with MCS protein or lipid components (Chang et al. 2013; Jing et al. 2015; Tsai et al. 2014; Varnai et al. 2007).

The observation of focal areas of apposition between the PM and the ER in cells with replete stores and their persistence following ablation of both STIM proteins indicates that other proteins also mediate ER-PM tethering in mammalian cells (Orci et al. 2009; Shen et al. 2011). ER-PM interactions were extensively characterized in yeast, where sites of close apposition (average distance of 33 nm) between the ER and the PM cover about 50% of the PM surface, justifying the nomenclature of "cortical ER" for this cellular substructure (Pichler et al. 2001; West et al. 2011; reviewed in Toulmay and Prinz 2011). The formation and stabilization of the yeast cortical ER are mediated by six ER-resident proteins: the Ist2, a protein related to mammalian TMEM16 ion channels; the three tricalbins Tcb1p, Tcb2p, and Tcb3p that bear a cytosolic lipid-accommodating SMP (for synaptotagmin-like mitochondrial lipid-binding protein) domain as well as multiple lipids and Ca^{2+}-binding C2 domains (Creutz et al. 2004; Lee and Hong 2006); and the two vesicle-associated membrane protein-associated protein (VAP) orthologs Scs2 and Scs22 (Loewen et al. 2007). The tricalbins accumulate in the cortical ER and mediate ER-PM tethering and lipid exchange between the two membranes (Manford et al. 2012; Toulmay and Prinz 2011). Genetic deletion of all six ER-PM tethering proteins in yeast (Ist2, Scs2 and Scs22, and Tcbs 1, 2, and 3) decreases the extent of cortical ER (cER) from 40 to 5% of the PM length as the ER redistributes into internal structures (Manford et al. 2012). Of note, small regions of cER were still observed in these tethering-deficient yeast mutants, suggesting the presence of additional unknown tethering proteins. Yeast cells lacking cortical ER were still able to grow in rich medium (Manford et al. 2012) but had elevated levels of phosphatidylinositol-4-phosphate (PI4P) at the PM. This phenotype reflected the redistribution of the phosphoinositide (PI) phosphatase Sac1, which dephosphorylates PI4P on the PM in trans from the ER (Stefan et al. 2011), thereby establishing the cortical ER as a necessary platform to maintain lipid homeostasis.

The mammalian homologues of the tricalbins are the extended synaptotagmins (E-Syt1, E-Syt2, and E-Syt3) (Lee and Hong 2006; Min et al. 2007). The E-Syts associate with the ER membrane bilayer via a hairpin sequence and form homo- and heteromeric complexes that tether the ER to membranes enriched in PIP_2 (Giordano et al. 2013). Their tethering function is positively regulated by increases in cytosolic Ca^{2+} concentration via the Ca^{2+}-sensing property of E-Syt1 (Giordano et al. 2013; Idevall-Hagren et al. 2015). The cytosolic domain of purified E-Syt1 mediates efficient lipid mixing in a protein-free liposome coflotation assay (Yu et al. 2016), thereby establishing that E-Syt1 itself is a lipid transfer protein. The lipid transfer function of E-Syt1 required the SMP domain, only occurred at $Ca2^{+}$ concentrations exceeding 50 μM, and correlated with bilayer proximity in liposome-tethering assays, indicating that $Ca2^{+}$ promotes lipid transfer by tethering the two membrane bilayers. Both the Ca^{2+}-dependent tethering and the lipid transport activity of E-Syt1 were recapitulated by replacing its C2 domains with those of the synaptic vesicle fusion regulator Syt1. These in vitro findings establish E-Syts as non-vesicular lipid transfer proteins whose activity is regulated by Ca^{2+} by a membrane-tethering mechanism. Surprisingly, despite solid biochemical evidence that E-Syts mediate Ca^{2+}-dependent lipid transfer and ER-PM tethering, the combined deletion of all three E-Syt genes has no effect on mouse viability or fertility (Sclip et al. 2016; Tremblay and Moss 2016). However, genetic ablation of the three E-Syts was accompanied by an enhanced expression of STIM1, ORAI1, and TMEM110 (Tremblay and Moss 2016), suggesting that enhanced ER-PM tethering by the proteins that mediate SOCE might compensate for loss of E-Syt-mediated lipid transfer.

Despite their role in lipid transfer and in the formation and maintenance of ER-PM junctions (Giordano et al. 2013; Manford et al. 2012), E-Syts are not required for STIM-ORAI coupling, and E-Syt-depleted HeLa cells have normal SOCE despite a significant reduction in cortical ER (Giordano et al. 2013). This finding was unexpected inasmuch as STIM-ORAI interactions require the close proximity between the ER and PM membranes (Varnai et al. 2007). On the other hand, several proteins were shown to co-localize with STIM and ORAI at membrane junctions and to modulate SOCE. A non-exhaustive list includes CRACR2A (Ca^{2+} release-activated channel regulator 2A) (Srikanth et al. 2010), junctate (Guido et al. 2015; Srikanth et al. 2012), POST (partner of stromal interaction molecule 1) (Krapivinsky et al. 2011), SARAF (store-operated Ca^{2+} entry-associated regulatory factor) (Palty et al. 2012), Surf4 (Fujii et al. 2012), septins (Sharma et al. 2013), and STIMATE (STIM-activating enhancer), also known as TMEM110 (Jing et al. 2015; Quintana et al. 2015). These accessory proteins are not strictly required for the gating of ORAI channels by STIM, which can be reconstituted in yeast membrane vesicles carrying ORAI1 with bacterially expressed recombinant STIM1 (Zhou et al. 2010), but are essential for proper functioning of the SOCE circuitry in cells, as detailed in a recent review (Prakriya and Lewis 2015). Several of these proteins regulate the formation and stabilization of the membrane junctions to enable the productive interactions between STIM and ORAI. For instance, septin 4 and septin 5 are scaffolding proteins that regulate the efficiency of the recruitment of ORAI to ER-PM junctions enriched in STIM1, without affecting the translocation of STIM1 itself (Sharma et al. 2013). Septins are GTP-binding proteins recruited to negatively charged membranes, where they assemble into filamentous structures anchored to the cytoskeleton that create diffusion barriers restricting the diffusion of proteins and lipids. Septins associate with a variety of phosphoinositides at different intracellular membranes and are implicated in the regulation of endosomal trafficking (reviewed in Song et al. 2016).

15.3 SOCE at Intracellular Membrane Contact Sites

The realization that SOCE is taking place at membrane contact sites has reignited the interest into the mechanism underlying the formation of MCS, specialized junctional structures that facilitate lipid exchange, localized signaling, and non-vesicular transport between cellular structures. MCS occur when two membranes of different origins are in close juxtaposition and stabilized by tethering proteins, a situation that is distinct from the soluble N-ethylmaleimide-sensitive factor attachment receptor (SNARE)-mediated docking and fusion mechanism that drives the delivery of cargo proteins and the exchange of membrane lipids in the endocytic and secretory pathways. The minimal distance that defines MCS is still not unambiguously established, but most investigators agree on a maximal distance of 20–30 nm (Burgoyne et al. 2014; Fernandez-Busnadiego 2016). Within this distance, MCS can have different gap sizes and vary in length, with specific morphologies appearing to have distinct roles in the regulation of cellular functions (Fernandez-Busnadiego 2016). The molecular characterization of MCS is the object of intense research as these cellular structures have been linked to cell survival, organelle dynamics, autophagosome formation, lipid transport, phagocytosis, and Ca^{2+} signaling (Bravo et al. 2011; Cardenas et al. 2010; Duke et al. 2014; Elbaz and Schuldiner 2011; Guido et al. 2015; Helle et al. 2013; Nunes et al. 2012; Patel and Brailoiu 2012; Prinz 2014), and mutations in genes involved in the regulation of MCS formation have been linked to the development of several diseases (Hariri et al. 2016). MCS forming between the ER and mitochondria have been studied extensively and will not be discussed here. Instead, we will discuss the implication for Ca^{2+} signaling of MCS forming between the ER and phagosomes (Fig. 15.1) and those forming between the ER and the endo-lysosomal system (Fig. 15.2).

The ER is the organelle with the highest number of MCS with all other organelles inside the cell and has been shown to make contacts with mitochondria, phagosomes, endosomes, lysosomes, and Golgi apparatus (Carrasco and Meyer 2011; Elbaz and Schuldiner 2011; Gerasimenko et al. 2009; Helle et al. 2013; Henne et al. 2015; Levine and Loewen 2006). The revived interest in MCS follows the identification of STIM and ORAI proteins as mediators of the SOCE process and reflects the continuous methodological progress in the functional and morphological approaches aiming to probe cellular functions at the subcellular ultrastructural level. Recent fluorescence microscopy techniques, such as total internal reflection fluorescence (TIRF) microscopy (Wu et al. 2006) and super-resolution microscopy (Uno et al. 2014), combined with molecular tools such as genetically encoded Ca^{2+} indicators (GECI) (Miyawaki and Niino 2015) and fluorescent artificial tethers suitable for single-molecule localization microscopy (Csordas et al. 2010; Varnai et al. 2007) now enable the study of MCS structure and function in live cells with unprecedented resolution. In parallel, 3D-EM and cryo-electron tomography are leading an EM revival. Both fluorescence microscopy and EM techniques can be merged by the use of correlative light and electron microscopy

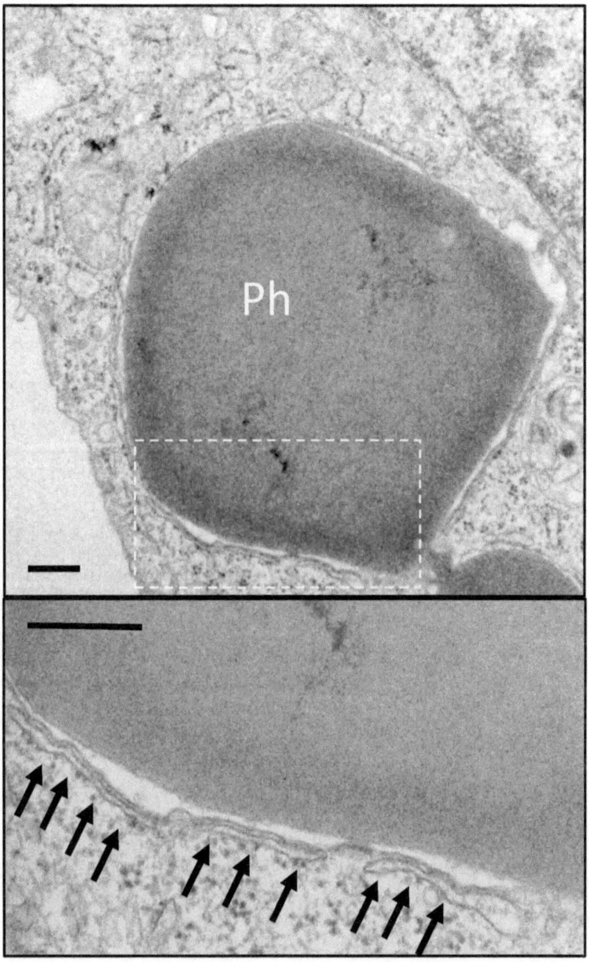

Fig. 15.1 EM micrograph of a phagosome bearing large ER-phagosome MCS. MEF cells rendered phagocytic through the expression of FCGRIIA-c-myc receptors were transfected with YFP-STIM1-4K, a STIM1 mutant that binds but fails to gate ORAI channels and makes large ER-PM contact sites (Korzeniowski et al. 2010). MEFs were exposed to red blood cells opsonized with IgG and allowed to phagocytose for 10 min before fixation and processing for EM microscopy. STIM1-4K expression stimulated the formation of large MCS (*arrows*, *inset* below) around phagosomes (*large dark object*, Ph). *Bar* = 200 nm

(CLEM), where the advantages of the two techniques are combined (van Rijnsoever et al. 2008). Notably, a new electron microscopy technique has been recently published. The researchers were able to obtain multicolor images with the full spatial electron microscopy resolution by the use of lanthanides that were locally deposited in order to concurrently label different organelles (Adams et al. 2016). In the following paragraphs, different types of MCS forming between the ER and the plasma membrane or between the ER and the membrane of other organelles will be described with an emphasis on their roles in Ca^{2+} signaling and lipid transfer. Moreover, some examples of how MCS malfunctioning is correlated to different diseases will be illustrated.

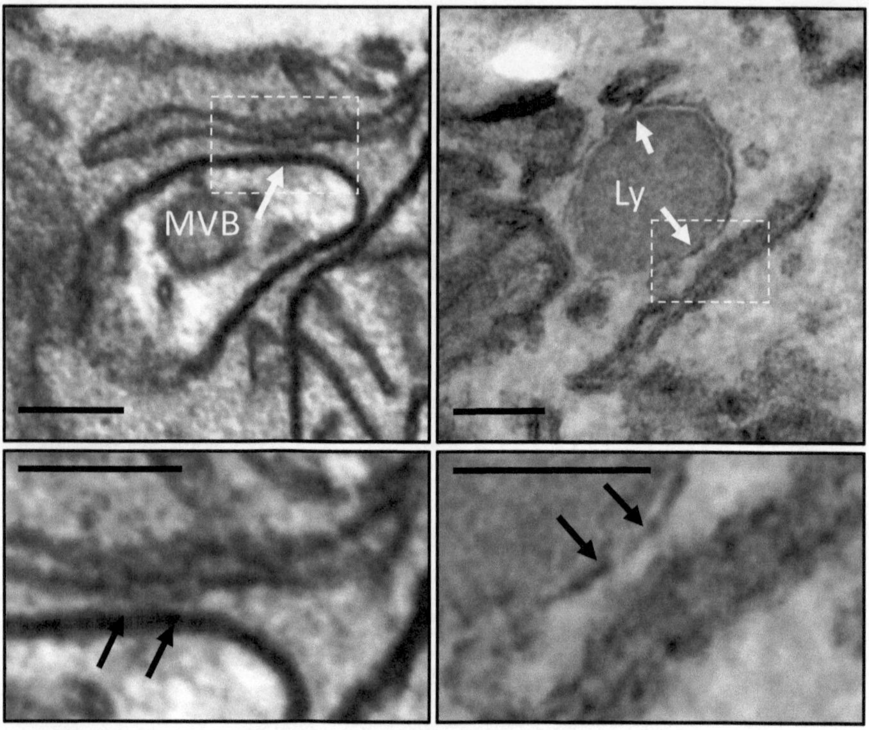

Fig. 15.2 EM micrograph of ER-endosome and ER-lysosome MCS. *Left panels* show contact sites (*white arrow* and *black arrows* in the *inset* below) between the ER and multivesicular body (MVB) endosomes in bone marrow-derived dendritic cells. *Right panels* show contact sites (*white arrows* and *black arrows* in the *inset* below) between the ER and lysosomes (Ly). Note that the *darker blurred lines* within MCS perpendicular to the membranes are characteristic electron densities believed to represent protein tethers. *Bar* = 100 nm

15.4 SOCE at ER-Phagosome Membrane Contact Sites

Phagocytosis is a fundamental cellular process essential for innate immunity and tissue homeostasis performed predominantly, but not exclusively, by the circulating and tissue-resident leukocytes, neutrophils, macrophages, and dendritic cells. Phagocytosis is associated with the occurrence of global and local Ca^{2+} signals generated by a combination of Ca^{2+} release from intracellular stores and of Ca^{2+} influx from the extracellular environment. The spatiotemporal profiles of the Ca^{2+} signals occurring during phagocytosis differ between cell types and depend on the receptor being engaged. For instance, IgG-mediated phagocytosis leads to local periphagosomal Ca^{2+} microdomains, whereas C3b-dependent phagocytosis is characterized by global Ca^{2+} elevation in both neutrophils (Dewitt and Hallett 2002; Murata et al. 1987) and macrophages (Hishikawa 1991). While phagosome formation is largely a Ca^{2+}-independent process, with neutrophils being a notable exception, Ca^{2+} elevations enhance the efficiency of phagocytosis by promoting actin dynamics and by enabling phagocytic vacuoles to gain oxidative and lytic properties via Ca^{2+}-dependent fusion with granules and lysosomes and the assembly of the NAPDH oxidase complex. The influence of Ca^{2+} on phagocytosis is exploited by some intracellular pathogens such as *Mycobacterium tuberculosis* and *Leishmania* that are able to subvert phagocytic killing by suppressing Ca^{2+} signals after being phagocytosed (Tejle et al. 2002; Malik et al. 2003). Phagosome maturation is characterized by a series of fusion events between phagosomes, endosomes, and then lysosomes, which render the novel organelle able to kill the ingested foreign particle or to degrade it in order to present antigens to cells of the adaptive immune system. Ca^{2+} signals during phagocytosis have been implicated in the generation of reactive oxygen species (ROS) and in the shedding of the actin coat that initially deforms the membrane during particle engulfment. SOCE, a mechanism that takes place at ER-PM membrane contact sites, has been implicated in the generation of pro-phagocytic Ca^{2+} signals. During phagocytosis, the molecules that mediate SOCE are detected not only at the ER-PM interface but also at ER-phagosome (ER-Ph) membrane contact sites (Fig. 15.1), where they are associated with the generation of periphagosomal Ca^{2+} microdomains that boost phagocytosis (Nunes et al. 2012). At these ER-Ph membrane contact sites, STIM1 regulates phagocytosis by promoting SOCE-dependent periphagosomal Ca^{2+} elevations, which in turn influence actin shedding. The use of either a Ca^{2+} channel blocker or a Ca^{2+} chelator significantly reduced the percentage of phagosomes associated with Ca^{2+} microdomains. Moreover, mouse embryonic fibroblasts (MEFs) overexpressing STIM1 and the phagocytic receptor FcγRIIA displayed an increase in both the number and size of ER-Ph MCS. Surprisingly, the overexpression of STIM2 does not rescue phagocytosis in *Stim1*$^{-/-}$ MEFs, suggesting a specific role for STIM1 in the generation of the periphagosomal Ca^{2+} microdomains (Nunes et al. 2012). The protein junctate, involved in the regulation of SOCE at the level of ER-PM MCS (Srikanth et al. 2012; Treves et al. 2004), has also been detected around phagosomes, suggesting that it can contribute to the formation of ER-Ph MCS. Upon junctate overexpression, MEFs ectopically expressing FcγRIIA display an increased phagocytic index, which correlated with an increased frequency of periphagosomal Ca^{2+} microdomains. In contrast to STIM1, junctate had a greater role in recruiting Ca^{2+} stores close to the phagosome membrane, favoring the generation of local Ca^{2+} microdomains via an IP_3R-dependent mechanism (Guido et al. 2015). The slight inhibition of Ca^{2+} signals observed around phagosomes exposed to Ca^{2+} channel blockers highlights the possibility of an interaction between junctate and Ca^{2+}-permeable channels resident in the phagosome membrane, a possibility supported by the reported interactions between junctate and TRPC3 channels (Treves et al. 2010). Whether Ca^{2+} is pumped inside the phagosome in an ATP-dependent manner is not known, but since a nascent phagosome is derived from the plasma membrane, some plasma membrane Ca^{2+} ATPases (PMCA) could be recruited to the phagocytic cup once phagocytic receptors are engaged. Moreover, the fusion with endosomes and lysosomes could lead to the acquisition of Ca^{2+} exchangers present in the membrane of acidic Ca^{2+} stores, as discussed in Sect. 15.5.

15.5 Ca^{2+} and Lipid Signaling at ER-Endosome/Lysosome (ER-En/Ly) Membrane Contact Sites

During the last decades, the ER has been considered the principal Ca^{2+} store in the cell, but acidic organelles are now well documented as fulfilling the same Ca^{2+}-storing role. The Ca^{2+} concentration inside these acidic organelles is estimated to be around 500 μM, which is close to the free Ca^{2+} concentration in the ER (Christensen et al. 2002; Lloyd-Evans et al. 2008). Acidic Ca^{2+} stores are implicated in several cellular functions, including receptor recycling from the plasma membrane, lipid transfer from endosomes to the ER, and fusion with phagosomes to promote the destruction of the ingested foreign particle (Eden 2016; Levin et al. 2016; Neufeld et al. 1996; Underwood et al. 1998). There is evidence showing that 80–100% of lysosomes generate MCS with the ER, a proportion much higher than any other organelle or the PM, suggesting that ER-En/Ly MCS may be an integral part of proper lysosome function (Friedman et al. 2013; Kilpatrick et al. 2013). Representative electron micrographs of ER-endosome and ER-lysosome contact sites illustrating the characteristic electron densities believed to represent protein tethers are shown in Fig. 15.2. The molecular components of ER-En/Ly MCS are still poorly characterized, but some tethers involved in these specific MCS have been recently identified. For instance, the oxysterol-binding protein-related protein (ORP1L) that is resident in late endosomes is able to interact with the vesicle-associated membrane protein-associated protein A (VAPA), an ER protein, through its diphenylalanine (FF) in an acidic tract motif (FFAT) in a cholesterol-dependent manner (Rocha et al. 2009). Both the size and the number of ER-En/Ly MCS could be increased by overexpression of an ORP1L-ΔORD mutant lacking the oxysterol-binding domain as well as by cholesterol depletion (Rocha et al. 2009). This defines ORP1L and VAPA as bona fide MCS tethers and points to a role for ER-En/Ly MCS in cholesterol sensing. ER-En/Ly MCS are likely involved in lipid homeostasis regulation, since ~30% of LDL cholesterol is transferred from the endocytic pathway to the ER via non-vesicular trafficking, and genetic ablation of hepatocyte growth factor-regulated tyrosine kinase substrate (Hrs), a key protein of the endosomal sorting complex required for transport (ESCRT), or ORP5 leads to the accumulation of cholesterol inside endosomes (Du et al. 2012). The StAR-related lipid transfer domain proteins STARD3 and STARD3NL, proteins resident in late endosomes containing the cholesterol transporter ATP-binding cassette subfamily A member 3 (ABCA3), were also reported to extend ER-En/Ly MCS (Alpy et al. 2013; van der Kant et al. 2013). VAPA also interacts with protrudin, an ER transmembrane protein able to interact with PI3P and Ras-related protein 7 (Rab7) through its FYVE (Fab-1, YGL023, Vps27, and EEA1) and Rab7-interacting domains, respectively, whose overexpression increased the number and size of ER-En/Ly MCS (Raiborg et al. 2015). However, whether these proteins trigger MCS formation or stabilize preexisting MCS is unclear. ER-En/Ly MCS are also involved in receptor endocytosis and consequently in receptor degradation. During this process, the epidermal growth factor receptor (EGFR) undergoes dephosphorylation by the protein tyrosine phosphatase 1B (PTP1B), which resides in the ER membrane. These two proteins act as tethers in the regulation of ER-En/Ly MCS (Eden et al. 2010). Some oxysterol-binding proteins localize in the ER, such as ORP5 that can interact with two endosomal proteins: the Niemann-Pick type C protein 1 (NPC1) (Du et al. 2011) and the ESCRT protein Hrs (Pridgeon et al. 2009). The relevance of ER-En/Ly membrane contact sites in the regulation of lipid metabolism and Ca^{2+} signals is highlighted in NPC1-deficient cells, which display an intracellular accumulation of cholesterol and sphingolipids (Lloyd-Evans et al. 2008) and a decreased Ca^{2+} concentration in acidic organelles. NPC1-deficient cell phenotype can be reverted by mobilizing Ca^{2+} from the ER, thereby causing an increase in cytosolic Ca^{2+} concentration (Lloyd-Evans et al. 2008). How Ca^{2+} is taken up inside acidic organelles is still not well established. Ca^{2+} ATPase activity has been reported in lysosomes isolated both from neutrophils (Klemper 1985) and fibroblasts (Lemons and Thoene 1991), and the high concentration of lysosomal protons could enable Ca^{2+} hydrogen exchangers (CAXs) to accumulate Ca^{2+} inside the organelle against its concentration gradient. Even if CAXs appear to be absent in mammalian genomes, their activity has been described in epithelial cells (Salceda and Sanchez-Chavez 2000), in PC12 cells (Mahapatra et al. 2004), and in the sheep brain cortex (Goncalves et al. 2000). This activity could reflect the combined action of sodium hydrogen exchangers driving sodium inside lysosomes to fuel the influx of Ca^{2+} across sodium Ca^{2+} exchangers, thereby mimicking an exchange between Ca^{2+} and protons. Interestingly, Melchionda and collaborators discovered the first CAX in nonplacental mammals (Melchionda et al. 2016), adding further details to this emerging topic. Whether Ca^{2+} is buffered inside acidic organelles by specific proteins is still not understood. However, there is evidence for the presence of polyphosphate in several acidic Ca^{2+} stores (Pisoni and Lindley 1992; Ruiz et al. 2004). Polyphosphate is able to interact with Ca^{2+} (Kornberg 1995) and could represent a way to buffer Ca^{2+} inside endosomes and lysosomes. Two major types of channels have been reported to be involved in the release of Ca^{2+} from acidic Ca^{2+} stores: transient receptor potential (TRP) family channels and two-pore channels (TPCs). From the TRP group, TRPM2 and TRPV2 have been localized in lysosomes (Lange et al. 2009) and early endosomes (Saito et al. 2007), respectively. Normally, both TRPMs and TRPVs are resident in the plasma membrane (Venkatachalam and Montell 2007) where they mediate Ca^{2+} influx from the extracellular environment. In the context of intracellular acidic Ca^{2+} stores, there is evidence that TRPM2 is activated by nicotinic acid adenine dinucleotide phosphate (NAADP) (Beck et al. 2006). NAADP was first reported to evoke Ca^{2+} release from acidic Ca^{2+} stores in sea urchin eggs (Churchill et al. 2002). The NAADP-dependent Ca^{2+} mobilization from lysosomes has been demonstrated by the use of glycyl-L-phenylalanine-naphthylamide (GPN). The cathepsin C-dependent hydrolysis of GPN causes the release of lysosome content via osmosis, thereby preventing NAADP-dependent Ca^{2+} release (Churchill et al. 2002; Guse and Lee 2008).

Two-pore channels are another target of NAADP. TPC1 localizes in human lysosomes and endosomes, whereas TPC2 resides predominantly in lysosomal compartments. In addition, both TCP1 and TCP2 overexpressions increase NAADP-dependent Ca^{2+} release, while TCP1 knockdown via siRNA causes a decrease in the Ca^{2+} mobilization from the acidic organelles (Brailoiu et al. 2009; Calcraft et al. 2009; Zong et al. 2009). Furthermore, there is evidence showing that Ca^{2+} release from acidic organelles, mediated by TPC activation in an NAADP-dependent manner, triggers a global Ca^{2+} elevation through Ca^{2+} release from the ER. A local Ca^{2+} event at ER-En/Ly MCS is able to induce a global cytosolic Ca^{2+} elevation via a process of Ca^{2+}-induced Ca^{2+} release (CICR), by the activation of IP_3Rs that localize in the ER membrane at the level of ER-En/Ly MCS. The ensuing depletion of ER Ca^{2+} stores leads to SOCE activation, but another cation-permeable channel, TRPML1, is directly involved in lysosomal release-related Ca^{2+} influx from the extracellular environment. TRPML1 is located on lysosomes and regulates both the release of Ca^{2+} from lysosomes and Ca^{2+} entry at the level of the plasma membrane in a SOCE-independent manner (Kilpatrick et al. 2016). TRPML1 is linked to lipid trafficking and when mutated causes mucolipidosis type IV, a lysosomal lipid storage disorder (Bargal et al. 2000). Interestingly, the use of TRPML1 agonists also reverses the lipid accumulation in other diseases like HIV-associated dementia (Bae et al. 2014) and Niemann-Pick type C disease (Shen et al. 2012). Deregulation of Ca^{2+} signaling mediated by acidic Ca^{2+} stores can lead to pancreatitis as excessive alcohol consumption leads to an increase in fatty acid ethyl esters in pancreatic acinar cells that induce the release of Ca^{2+} by IP_3Rs, thereby causing an improper activation of proteases (Gerasimenko et al. 2009).

15.6 Dynamic Regulation of Membrane Contact Sites

As discussed in the preceding parts, proteins affecting the amount, contact surface, and membrane proximity of the junctions forming at the ER-PM, ER-Ph, or ER-En/Ly interface are expected to regulate SOCE by impacting on the molecular interactions between STIM and ORAI and their assortment of associated regulatory proteins. The ER-PM gap distance correlates with the lengths of E-Syt cytosolic domains (Fernandez-Busnadiego et al. 2015) and decreases following an elevation in cytosolic Ca^{2+} (Chang et al. 2013; Fernandez-Busnadiego et al. 2015) as the shorter Ca^{2+}-sensing E-Syt1 translocates to MCS (Giordano et al. 2013; Idevall-Hagren et al. 2015). The E-Syt1-mediated decrease in ER-PM gap distance facilitates the recruitment of the phosphatidylinositol (PI) transfer protein Nir2 to ER-PM junctions (Chang et al. 2013; Chang and Liou 2016). Nir2 binds to phosphatidic acid (PA) produced by PIP_2 hydrolysis and mediates PI/PA transfer to replenish the PIP_2 consumed at sites of PLC activation (Chang et al. 2013; Kim et al. 2015). In the following part, we will discuss the potential implication of the dynamic remodeling of cortical ER structures on cytosolic Ca^{2+} signals.

Whether SOCE generates cytosolic Ca^{2+} signals depends on the fraction of entering Ca^{2+} that is captured by SERCA on the apposed ER cisternae or is extruded by Ca^{2+} pumps and exchangers on the PM. For a given number of active channels and pumps, two physical parameters influence local Ca^{2+} fluxes: (1) the dimension of cortical ER structures and (2) the distance between the ER and PM at contact sites. Electron microscopy studies indicate that STIM1-ORAI interactions take place at junctions between flat ER sheets located 10–20 nm from the PM and having a maximal diameter of 300 nm (Lur et al. 2009; Orci et al. 2009; Wu et al. 2006). Modeling of Ca^{2+} diffusion within the compartment delimited by the apposed membranes predicts that for a given number of channels, larger junctions will produce smaller cytosolic Ca^{2+} signals as the steady-state Ca^{2+} concentration at the edge of a circular microdomain decreases from 1.4 μM to ~100 nM when the radius increases from 10 to 150 nm (Hogan 2015). Large cER structures will thus favor ER refilling because a greater ER membrane surface is exposed to submicromolar Ca^{2+} concentrations that match the thermodynamics of SERCA pumps (van Breemen and Saida 1989). Geometric and thermodynamic considerations therefore predict that small MCS are associated with large $[Ca^{2+}]_{cyt}$ elevations and large MCS with efficient ER refilling. This prediction is supported by the observation that the STIM1L isoform generates smaller contacts than STIM1 but larger cytosolic Ca^{2+} elevations when expressed in *Stim1*$^{-/-}$/*Stim2*$^{-/-}$ fibroblasts (Sauc et al. 2015). Similar considerations predict that a decreased ER-PM gap will increase the Ca^{2+} concentration in the ER-PM cleft, thereby promoting the Ca^{2+}-dependent inactivation of ORAI channels. Changes in gap distance might further alter ORAI permeation and gating by modifying STIM-ORAI interactions and the accessibility of regulatory proteins in the cleft. The geometry of STIM-ORAI microdomains is dynamically regulated by organizing proteins such as E-Syts whose overexpression increases MCS length and reduces the ER-PM gap distance from 25 to 15 nm (Fernandez-Busnadiego et al. 2015). If the length and proximity of the junction determines the fate of entering Ca^{2+} and the gating of ORAI channels, these tethering proteins would endow cells with the capacity to encode Ca^{2+} signals by modulating the geometry of contact sites. This possibility is supported by a recent study reporting extension of ER-PM junctions in response to SOCE (Poteser et al. 2016). Using an elegant form of total internal reflection fluorescence (TIRF) microscopy, these authors imaged the changes in ER-PM junctional morphology with high spatiotemporal resolution by measuring the exclusion of a soluble cytosolic fluorescent protein from membrane contact sites. In this modality, organelles located within the TIRF plane (set with a minimal penetration depth of 50 nm) become visible as dark structures within the evanescent field that strictly correlate spatially with the STIM1-derived ER-PM marker MAPPER, with fluorescently tagged STIM1, and with sites of STIM-ORAI interactions measured by fluorescence resonance energy transfer (FRET). Alterations in the thickness of the evanescent fluorescent layer can then be translated into changes in contact site shape, area, and gap distance using a fluorescence density mapping function. Using this approach, the authors showed that ER-PM junctions generated by store depletion in cultured rat mast cells are highly dynamic and extend progressively when Ca^{2+} is present in the extracellular medium. Moreover, the computed gap distance was fluctuating rapidly in junctions undergoing lateral extension in a Ca^{2+}-dependent manner, while junctions forming in Ca^{2+}-free conditions were stable and failed to display lateral extension. These data indicate that the entering Ca^{2+} exerts a dynamic control on the juxtaposed cortical ER structure by promoting the lateral extension of the junctional ER. This suggests that Ca^{2+}-dependent ER-PM tethering proteins such as the Ca^{2+}-dependent E-Syts are involved in the stabilization of the STIM-ORAI junctions. This mechanism of Ca^{2+}-dependent ER-PM tethering might enhance the efficiency of SERCA-mediated ER Ca^{2+} refilling by increasing the surface of the junctional ER exposed to entering Ca^{2+}.

15.7 Conclusion

In conclusion, the recognition that the SOCE pathway is grafted on an older, evolutionarily conserved mechanism enabling the non-vesicular transfer of lipids between the membrane of the ER and target cellular membranes including, but not limited to, the plasma membrane, has led to the realization that the dynamic assembly and remodeling of the junctional ER structures bearing the SOCE regulatory STIM proteins and the Ca^{2+} sequestering SERCA pumps provide a new mechanism of regulation of cellular Ca^{2+} signals. The key questions that need to be clarified are whether the dynamic remodeling of MCS controls the efficiency of store refilling, the Ca^{2+}-dependent inactivation of ORAI channels, and the propagation of local Ca^{2+} signals and their conversion into global signals. New imaging techniques that track structural and functional changes at membrane contact sites will be required to reveal the role of membrane contact site dynamics in the regulation of physiological and pathophysiological Ca^{2+} signals.

References

Adams SR, Mackey MR, Ramachandra R, Palida Lemieux SF, Steinbach P, Bushong EA, Butko MT, Giepmans BN, Ellisman MH, Tsien RY (2016) Multicolor electron microscopy for simultaneous visualization of multiple molecular species. Cell Chem Biol 23(11):1417–1427

Alpy F, Rousseau A, Schwab Y, Legueux F, Stoll I, Wendling C, Spiegelhalter C, Kessler P, Mathelin C, Rio MC, Levine TP, Tomasetto C (2013) STARD3 or STARD3NL and VAP form a novel molecular tether between late endosomes and the ER. J Cell Sci 126(Pt 23):5500–5512

Bae M, Patel N, Xu H, Lee M, Tominaga-Yamanaka K, Nath A, Geiger J, Gorospe M, Mattson MP, Haughey NJ (2014) Activation of TRPML1 clears intraneuronal Abeta in preclinical models of HIV infection. J Neurosci 34(34):11485–11503

Bargal R, Avidan N, Ben-Asher E, Olender Z, Zeigler M, Frumkin A, Raas-Rothschild A, Glusman G, Lancet D, Bach G (2000) Identification of the gene causing mucolipidosis type IV. Nat Genet 26(1):118–123

Beck A, Kolisek M, Bagley LA, Fleig A, Penner R (2006) Nicotinic acid adenine dinucleotide phosphate and cyclic ADP-ribose regulate TRPM2 channels in T lymphocytes. FASEB J 20 (7):962–964

Block BA, Imagawa T, Campbell KP, Franzini-Armstrong C (1988) Structural evidence for direct interaction between the molecular components of the transverse tubule/sarcoplasmic reticulum junction in skeletal muscle. J Cell Biol 107(6 Pt 2):2587–2600

Brailoiu E, Churamani D, Cai X, Schrlau MG, Brailoiu GC, Gao X, Hooper R, Boulware MJ, Dun NJ, Marchant JS, Patel S (2009) Essential requirement for two-pore channel 1 in NAADP-mediated calcium signaling. J Cell Biol 186(2):201–209

Bravo R, Vicencio JM, Parra V, Troncoso R, Munoz JP, Bui M, Quiroga C, Rodriguez AE, Verdejo HE, Ferreira J, Iglewski M, Chiong M, Simmen T, Zorzano A, Hill JA, Rothermel BA, Szabadkai G, Lavandero S (2011) Increased ER–mitochondrial coupling promotes mitochondrial respiration and bioenergetics during early phases of ER stress. J Cell Sci 124 (Pt 13):2143–2152

Burgoyne T, Patel S, Eden ER (2014) Calcium signaling at ER membrane contact sites. Biochim Biophys Acta 1853(9):2012–2017

Calcraft PJ, Ruas M, Pan Z, Cheng X, Arredouani A, Hao X, Tang J, Rietdorf K, Teboul L, Chuang KT, Lin P, Xiao R, Wang C, Zhu Y, Lin Y, Wyatt CN, Parrington J, Ma J, Evans AM, Galione A, Zhu MX (2009) NAADP mobilizes calcium from acidic organelles through two-pore channels. Nature 459(7246):596–600

Cardenas C, Miller RA, Smith I, Bui T, Molgo J, Müller M, Vais H, Cheung KH, Yang J, Parker I, Thompson CB, Birnbaum MJ, Hallows KR, Foskett JK (2010) Essential regulation of cell bioenergetics by constitutive InsP3 receptor Ca2+ transfer to mitochondria. Cell 142 (2):270–283

Carrasco S, Meyer T (2011) STIM proteins and the endoplasmic reticulum-plasma membrane junctions. Annu Rev Biochem 80:973–1000

Chang CL, Liou J (2016) Homeostatic regulation of the PI(4,5)P-Ca signaling system at ER-PM junctions. Biochim Biophys Acta 1861(8 Pt B):862–873

Chang CL, Hsieh TS, Yang TT, Rothberg KG, Azizoglu DB, Volk E, Liao JC, Liou J (2013) Feedback regulation of receptor-induced Ca2+ signaling mediated by E-Syt1 and Nir2 at endoplasmic reticulum-plasma membrane junctions. Cell Rep 5(3):813–825

Christensen KA, Myers JT, Swanson JA (2002) pH-dependent regulation of lysosomal calcium in macrophages. J Cell Sci 115(Pt 3):599–607

Churchill GC, Okada Y, Thomas JM, Genazzani AA, Patel S, Galione A (2002) NAADP mobilizes Ca(2+) from reserve granules, lysosome-related organelles, in sea urchin eggs. Cell 111(5):703–708

Creutz CE, Snyder SL, Schulz TA (2004) Characterization of the yeast tricalbins: membrane-bound multi-C2-domain proteins that form complexes involved in membrane trafficking. Cell Mol Life Sci 61(10):1208–1220

Csordas G, Varnai P, Golenar T, Roy S, Purkins G, Schneider TG, Balla T, Hajnoczky G (2010) Imaging interorganelle contacts and local calcium dynamics at the ER-mitochondrial interface. Mol Cell 39(1):121–132

Dewitt S, Hallett MB (2002) Cytosolic free Ca(2+) changes and calpain activation are required for beta integrin-accelerated phagocytosis by human neutrophils. J Cell Biol 159(1):181–189

Du X, Kumar J, Ferguson C, Schulz TA, Ong YS, Hong W, Prinz WA, Parton RG, Brown AJ, Yang H (2011) A role for oxysterol-binding protein-related protein 5 in endosomal cholesterol trafficking. J Cell Biol 192(1):121–135

Du X, Kazim AS, Brown AJ, Yang H (2012) An essential role of Hrs/Vps27 in endosomal cholesterol trafficking. Cell Rep 1(1):29–35

Duke EMH, Razi M, Weston A, Guttmann P, Werner S, Henzler K, Schneider G, Tooze SA, Collinson LM (2014) Imaging endosomes and autophagosomes in whole mammalian cells using correlative cryo-fluorescence and cryo-soft X-ray microscopy (cryo-CLXM). Ultramicroscopy 143:77–87

Eden ER (2016) The formation and function of ER-endosome membrane contact sites. Biochim Biophys Acta 1861(8 Pt B):874–879

Eden ER, White IJ, Tsapara A, Futter CE (2010) Membrane contacts between endosomes and ER provide sites for PTP1B-epidermal growth factor receptor interaction. Nat Cell Biol 12 (3):267–272
Elbaz Y, Schuldiner M (2011) Staying in touch: the molecular era of organelle contact sites. Trends Biochem Sci 36(11):616–623
Fernandez-Busnadiego R (2016) Supramolecular architecture of endoplasmic reticulum-plasma membrane contact sites. Biochem Soc Trans 44(2):534–540
Fernandez-Busnadiego R, Saheki Y, De Camilli P (2015) Three-dimensional architecture of extended synaptotagmin-mediated endoplasmic reticulum-plasma membrane contact sites. Proc Natl Acad Sci U S A 112(16):E2004–E2013
Feske S, Gwack Y, Prakriya M, Srikanth S, Puppel SH, Tanasa B, Hogan PG, Lewis RS, Daly M, Rao A (2006) A mutation in Orai1 causes immune deficiency by abrogating CRAC channel function. Nature 441(7090):179–185
Franzini-Armstrong C, Jorgensen AO (1994) Structure and development of E-C coupling units in skeletal muscle. Annu Rev Physiol 56:509–534
Friedman JR, Voeltz GK (2011) The ER in 3D: a multifunctional dynamic membrane network. Trends Cell Biol 21(12):709–717
Friedman JR, DiBenedetto JR, West M, Rowland AA, Voeltz GK (2013) Endoplasmic reticulum–endosome contact increases as endosomes traffic and mature. Mol Biol Cell 24(7):1030–1040
Fujii Y, Shiota M, Ohkawa Y, Baba A, Wanibuchi H, Kinashi T, Kurosaki T, Baba Y (2012) Surf4 modulates STIM1-dependent calcium entry. Biochem Biophys Res Commun 422(4):615–620
Gerasimenko JV, Lur G, Sherwood MW, Ebisui E, Tepikin AV, Mikoshiba K, Gerasimenko OV, Petersen OH (2009) Pancreatic protease activation by alcohol metabolite depends on Ca2+ release via acid store IP3 receptors. Proc Natl Acad Sci U S A 106(26):10758–10763
Giordano F, Saheki Y, Idevall-Hagren O, Colombo SF, Pirruccello M, Milosevic I, Gracheva EO, Bagriantsev SN, Borgese N, De Camilli P (2013) PI(4,5)P(2)-dependent and Ca(2+)-regulated ER-PM interactions mediated by the extended synaptotagmins. Cell 153(7):1494–1509
Goncalves PP, Meireles SM, Neves P, Vale MGP (2000) Distinction between Ca2+ pump and Ca2 +/H+ antiport activities in synaptic vesicles of sheep brain cortex. Neurochem Int 37 (4):387–396
Guido D, Demaurex N, Nunes P (2015) Junctate boosts phagocytosis by recruiting endoplasmic reticulum Ca2+ stores near phagosomes. J Cell Sci 128(22):4074–4082
Guse AH, Lee HC (2008) NAADP: a universal Ca2+ trigger. Sci Signal 1:re10
Hariri H, Ugrankar R, Liu Y, Henne WM (2016) Inter-organelle ER-endolysosomal contact sites in metabolism and disease across evolution. Commun Integr Biol 9(3):1–14
Helle SCJ, Kanfer G, Kolar K, Lang A, Michel AH, Kornmann B (2013) Organization and function of membrane contact sites. Biochim Biophys Acta 1833(11):2526–2541
Henne WM, Liou J, Emr SD (2015) Molecular mechanisms of inter-organelle ER–PM contact sites. Curr Opin Cell Biol 35:123–130
Hishikawa T (1991) Calcium transients during Fc receptor-mediated and nonspecific phagocytosis by murine peritoneal macrophages. J Cell Biol 115(1):59–66
Hogan PG (2015) The STIM1-ORAI1 microdomain. Cell Calcium 58(4):357–367
Hoth M, Penner R (1992) Depletion of intracellular calcium stores activates a calcium current in mast cells. Nature 355(6358):353–356
Idevall-Hagren O, Lu A, Xie B, De Camilli P (2015) Triggered Ca2+ influx is required for extended synaptotagmin 1-induced ER-plasma membrane tethering. EMBO J 34 (17):2291–2305
Jing J, He L, Sun A, Quintana A, Ding Y, Ma G, Tan P, Liang X, Zheng X, Chen L, Shi X, Zhang SL, Zhong L, Huang Y, Dong MQ, Walker CL, Hogan PG, Wang Y, Zhou Y (2015) Proteomic mapping of ER-PM junctions identifies STIMATE as a regulator of Ca(2)(+) influx. Nat Cell Biol 17(10):1339–1347
Kilpatrick BS, Eden ER, Schapira AH, Futter CE, Patel S (2013) Direct mobilisation of lysosomal Ca2+ triggers complex Ca2+ signals. J Cell Sci 126(Pt 1):60–66

Kilpatrick BS, Yates E, Grimm C, Schapira AH, Patel S (2016) Endo-lysosomal TRP mucolipin-1 channels trigger global ER Ca2+ release and Ca2+ influx. J Cell Sci 129(20):3859–3867
Kim YJ, Guzman-Hernandez ML, Wisniewski E, Balla T (2015) Phosphatidylinositol-phosphatidic acid exchange by Nir2 at ER-PM contact sites maintains phosphoinositide signaling competence. Dev Cell 33(5):549–561
Klemper MS (1985) An adenosine triphosphate-dependent calcium uptake pump in human neutrophil lysosomes. J Clin Invest 76(1):303–310
Kornberg A (1995) Inorganic polyphosphate: toward making a forgotten polymer unforgettable. J Bacteriol 177(3):491–496
Korzeniowski MK, Manjarres IM, Varnai P, Balla T (2010) Activation of STIM1-Orai1 involves an intramolecular switching mechanism. Sci Signal 3:ra82.
Krapivinsky G, Krapivinsky L, Stotz SC, Manasian Y, Clapham DE (2011) POST, partner of stromal interaction molecule 1 (STIM1), targets STIM1 to multiple transporters. Proc Natl Acad Sci U S A 108(48):19234–19239
Lange I, Yamamoto S, Partida-Sanchez S, Mori Y, Fleig A, Penner R (2009) TRPM2 functions as a lysosomal Ca2+-release channel in beta cells. Sci Signal 2:ra23
Lee I, Hong W (2006) Diverse membrane-associated proteins contain a novel SMP domain. FASEB J 20(2):202–206
Lemons RM, Thoene JG (1991) Mediated calcium transport by isolated human fibroblast lysosomes. J Biol Chem 266(22):14378–14382
Levin R, Grinstein S, Canton J (2016) The life cycle of phagosomes: formation, maturation, and resolution. Immunol Rev 273(1):156–179
Levine T, Loewen C (2006) Inter-organelle membrane contact sites: through a glass, darkly. Curr Opin Cell Biol 18(4):371–378
Lewis RS, Cahalan MD (1989) Mitogen-induced oscillations of cytosolic Ca2+ and transmembrane Ca2+ current in human leukemic T cells. Cell Regul 1(1):99–112
Liou J, Kim ML, Heo WD, Jones JT, Myers JW, Ferrell JE Jr, Meyer T (2005) STIM is a Ca2+ sensor essential for Ca2+-store-depletion-triggered Ca2+ influx. Curr Biol 15(13):1235–1241
Liou J, Fivaz M, Inoue T, Meyer T (2007) Live-cell imaging reveals sequential oligomerization and local plasma membrane targeting of stromal interaction molecule 1 after Ca2+ store depletion. Proc Natl Acad Sci U S A 104(22):9301–9306
Lloyd-Evans E, Morgan AJ, He X, Smith DA, Elliot-Smith E, Sillence DJ, Churchill GC, Schuchman EH, Galione A, Platt FM (2008) Niemann-Pick disease type C1 is a sphingosine storage disease that causes deregulation of lysosomal calcium. Nat Med 14(11):1247–1255
Loewen CJ, Young BP, Tavassoli S, Levine TP (2007) Inheritance of cortical ER in yeast is required for normal septin organization. J Cell Biol 179(3):467–483
Lur G, Haynes LP, Prior IA, Gerasimenko OV, Feske S, Petersen OH, Burgoyne RD, Tepikin AV (2009) Ribosome-free terminals of rough ER allow formation of STIM1 puncta and segregation of STIM1 from IP(3) receptors. Curr Biol 19(19):1648–1653
Mahapatra NR, Mahata M, Hazra PP, McDonough PM, O'Connor DT, Mahata SK (2004) A dynamic pool of calcium in catecholamine storage vesicles. Exploration in living cells by a novel vesicle-targeted chromogranin A-aequorin chimeric photoprotein. J Biol Chem 279 (49):51107–51121
Malik ZA, Thompson CR, Hashimi S, Porter B, Iyer SS, Kusner DJ (2003) Cutting edge: mycobacterium tuberculosis blocks Ca2+ signaling and phagosome maturation in human macrophages via specific inhibition of sphingosine kinase. J Immunol 170(6):2811–2815
Manford AG, Stefan CJ, Yuan HL, Macgurn JA, Emr SD (2012) ER-to-plasma membrane tethering proteins regulate cell signaling and ER morphology. Dev Cell 23(6):1129–1140
McNally BA, Somasundaram A, Jairaman A, Yamashita M, Prakriya M (2013) The C- and N-terminal STIM1 binding sites on Orai1 are required for both trapping and gating CRAC channels. J Physiol 591(11):2833–2850
Melchionda M, Pittman JK, Mayor R, Patel S (2016) Ca2+/H+ exchange by acidic organelles regulates cell migration in vivo. J Cell Biol 212(7):803–813

Min SW, Chang WP, Sudhof TC (2007) E-Syts, a family of membranous Ca2+-sensor proteins with multiple C2 domains. Proc Natl Acad Sci U S A 104(10):3823–3828
Miyawaki A, Niino Y (2015) Molecular spies for bioimaging—fluorescent protein-based probes. Mol Cell 58(4):632–643
Muik M, Fahrner M, Derler I, Schindl R, Bergsmann J, Frischauf I, Groschner K, Romanin C (2009) A cytosolic homomerization and a modulatory domain within STIM1 C terminus determine coupling to ORAI1 channels. J Biol Chem 284(13):8421–8426
Murata T, Sullivan JA, Sawyer DW, Mandell GL (1987) Influence of type and opsonization of ingested particle on intracellular free calcium distribution and superoxide production by human neutrophils. Infect Immun 55(8):1784–1791
Neufeld EB, Cooney AM, Pitha J, Dawidowicz EA, Dwyer NK, Pentchev PG, Blanchette-Mackie EJ (1996) Intracellular trafficking of cholesterol monitored with a cyclodextrin. J Biol Chem 271(35):21604–21613
Nunes P, Cornut D, Bochet V, Hasler U, Oh-Hora M, Waldburger JM, Demaurex N (2012) STIM1 juxtaposes ER to phagosomes, generating Ca(2)(+) hotspots that boost phagocytosis. Curr Biol 22(21):1990–1997
Orci L, Ravazzola M, Le Coadic M, Shen WW, Demaurex N, Cosson P (2009) From the cover: STIM1-induced precortical and cortical subdomains of the endoplasmic reticulum. Proc Natl Acad Sci U S A 106(46):19358–19362
Palty R, Raveh A, Kaminsky I, Meller R, Reuveny E (2012) SARAF inactivates the store operated calcium entry machinery to prevent excess calcium refilling. Cell 149(2):425–438
Parekh AB, Putney JW Jr (2005) Store-operated calcium channels. Physiol Rev 85(2):757–810
Park CY, Hoover PJ, Mullins FM, Bachhawat P, Covington ED, Raunser S, Walz T, Garcia KC, Dolmetsch RE, Lewis RS (2009) STIM1 clusters and activates CRAC channels via direct binding of a cytosolic domain to Orai1. Cell 136(5):876–890
Patel S, Brailoiu E (2012) Triggering of Ca2+ signals by NAADP-gated two-pore channels: a role for membrane contact sites? Biochem Soc Trans 40(1):153–157
Peinelt C, Vig M, Koomoa DL, Beck A, Nadler MJ, Koblan-Huberson M, Lis A, Fleig A, Penner R, Kinet JP (2006) Amplification of CRAC current by STIM1 and CRACM1 (Orai1). Nat Cell Biol 8(7):771–773
Perni S, Dynes JL, Yeromin AV, Cahalan MD, Franzini-Armstrong C (2015) Nanoscale patterning of STIM1 and Orai1 during store-operated Ca2+ entry. Proc Natl Acad Sci U S A 112(40): E5533–E5542
Pichler H, Gaigg B, Hrastnik C, Achleitner G, Kohlwein SD, Zellnig G, Perktold A, Daum G (2001) A subfraction of the yeast endoplasmic reticulum associates with the plasma membrane and has a high capacity to synthesize lipids. Eur J Biochem 268(8):2351–2361
Pisoni RL, Lindley ER (1992) Incorporation of [32P]orthophosphate into long chains of inorganic polyphosphate within lysosomes of human fibroblasts. J Biol Chem 267(6):3626–3631
Porter KR, Palade GE (1957) Studies on the endoplasmic reticulum. III. Its form and distribution in striated muscle cells. J Biophys Biochem Cytol 3(2):269–300
Poteser M, Leitinger G, Pritz E, Platzer D, Frischauf I, Romanin C, Groschner K (2016) Live-cell imaging of ER-PM contact architecture by a novel TIRFM approach reveals extension of junctions in response to store-operated Ca2+-entry. Sci Rep 6:35656
Prakriya M, Lewis RS (2015) Store-operated calcium channels. Physiol Rev 95(4):1383–1436
Pridgeon JW, Webber EA, Sha D, Li L, Chin LS (2009) Proteomic analysis reveals Hrs ubiquitin-interacting motif-mediated ubiquitin signaling in multiple cellular processes. FEBS J 276 (1):118–131
Prinz WA (2014) Bridging the gap: membrane contact sites in signaling, metabolism, and organelle dynamics. J Cell Biol 205(6):759–769
Putney JW Jr (1986) A model for receptor-regulated calcium entry. Cell Calcium 7(1):1–12
Putney JW Jr, Takemura H, Hughes AR, Horstman DA, Thastrup O (1989) How do inositol phosphates regulate calcium signaling? FASEB J 3(8):1899–1905
Quintana A, Rajanikanth V, Farber-Katz S, Gudlur A, Zhang C, Jing J, Zhou Y, Rao A, Hogan PG (2015) TMEM110 regulates the maintenance and remodeling of mammalian ER-plasma

membrane junctions competent for STIM-ORAI signaling. Proc Natl Acad Sci U S A 112(51): E7083–E7092

Raiborg C, Wenzel EM, Pedersen NM, Olsvik H, Schink KO, Schultz SW, Vietri M, Nisi V, Bucci C, Brech A, Johansen T, Stenmark H (2015) Repeated ER–endosome contacts promote endosome translocation and neurite outgrowth. Nature 520(7546):234–238

Rocha N, Kuijl C, van der Kant R, Janssen L, Houben D, Janssen H, Zwart W, Neefjes J (2009) Cholesterol sensor ORP1L contacts the ER protein VAP to control Rab7–RILP–p150Glued and late endosome positioning. J Cell Biol 185(7):12091225

Roos J, DiGregorio PJ, Yeromin AV, Ohlsen K, Lioudyno M, Zhang S, Safrina O, Kozak JA, Wagner SL, Cahalan MD, Velicelebi G, Stauderman KA (2005) STIM1, an essential and conserved component of store-operated Ca2+ channel function. J Cell Biol 169(3):435–445

Ruiz FA, Lea CR, Oldfield E, Docampo R (2004) Human platelet dense granules contain polyphosphate and are similar to acidocalcisomes of bacteria and unicellular eukaryotes. J Biol Chem 279(43):44250–44257

Saito M, Hanson PI, Schlesinger P (2007) Luminal chloride-dependent activation of endosome calcium channels: patch clamp study of enlarged endosomes. J Biol Chem 282 (37):27327–27333

Salceda R, Sanchez-Chavez G (2000) Calcium uptake, release and ryanodine binding in melanosomes from retinal pigment epithelium. Cell Calcium 27(4):223–229

Sauc S, Bulla M, Nunes P, Orci L, Marchetti A, Antigny F, Bernheim L, Cosson P, Frieden M, Demaurex N (2015) STIM1L traps and gates Orai1 channels without remodeling the cortical ER. J Cell Sci 128(8):1568–1579

Sclip A, Bacaj T, Giam LR, Sudhof TC (2016) Extended Synaptotagmin (ESyt) triple knock-out mice are viable and fertile without obvious endoplasmic reticulum dysfunction. PLoS One 11: e0158295

Sharma S, Quintana A, Findlay GM, Mettlen M, Baust B, Jain M, Nilsson R, Rao A, Hogan PG (2013) An siRNA screen for NFAT activation identifies septins as coordinators of store-operated Ca2+ entry. Nature 499(7457):238–242

Shen WW, Frieden M, Demaurex N (2011) Remodelling of the endoplasmic reticulum during store-operated calcium entry. Biol Cell 103(8):365–380

Shen D, Wang X, Li X, Zhang X, Yao Z, Dibble S, Dong XP, Yu T, Lieberman AP, Showalter HD, Xu H (2012) Lipid storage disorders block lysosomal trafficking by inhibiting a TRP channel and lysosomal calcium release. Nat Commun 3:731

Soboloff J, Spassova MA, Tang XD, Hewavitharana T, Xu W, Gill DL (2006) Orai1 and STIM reconstitute store-operated calcium channel function. J Biol Chem 281(30):20661–20665

Song K, Russo G, Krauss M (2016) Septins as modulators of endo-lysosomal membrane traffic. Front Cell Dev Biol 4:124

Srikanth S, Jung HJ, Kim KD, Souda P, Whitelegge J, Gwack Y (2010) A novel EF-hand protein, CRACR2A, is a cytosolic Ca2+ sensor that stabilizes CRAC channels in T cells. Nat Cell Biol 12(5):436–446

Srikanth S, Jew M, Kim KD, Yee MK, Abramson J, Gwack Y (2012) Junctate is a Ca2+-sensing structural component of Orai1 and stromal interaction molecule 1 (STIM1). Proc Natl Acad Sci U S A 109(22):8682–8687

Stefan CJ, Manford AG, Baird D, Yamada-Hanff J, Mao Y, Emr SD (2011) Osh proteins regulate phosphoinositide metabolism at ER-plasma membrane contact sites. Cell 144(3):389–401

Tejle K, Magnusson KE, Rasmusson B (2002) Phagocytosis and phagosome maturation are regulated by calcium in J774 macrophages interacting with unopsonized prey. Biosci Rep 22 (5–6):529–540

Toulmay A, Prinz WA (2011) Lipid transfer and signaling at organelle contact sites: the tip of the iceberg. Curr Opin Cell Biol 23(4):458–463

Tremblay MG, Moss T (2016) Loss of all 3 Extended Synaptotagmins does not affect normal mouse development, viability or fertility. Cell Cycle 15(17):2360–2366

Treves S, Franzini-Armstrong C, Moccagatta L, Arnoult C, Grasso C, Schrum A, Ducreux S, Zhu MX, Mikoshiba K, Girard T, Smida-Rezgui S, Ronjat M, Zorzato F (2004) Junctate is a key element in calcium entry induced by activation of InsP3 receptors and/or calcium store depletion. J Cell Biol 166(4):537–548

Treves S, Vukcevic M, Griesser J, Armstrong CF, Zhu MX, Zorzato F (2010) Agonist-activated Ca2+ influx occurs at stable plasma membrane and endoplasmic reticulum junctions. J Cell Sci 123(Pt 23):4170–4181

Tsai FC, Seki A, Yang HW, Hayer A, Carrasco S, Malmersjo S, Meyer T (2014) A polarized Ca2+, diacylglycerol and STIM1 signalling system regulates directed cell migration. Nat Cell Biol 16 (2):133–144

Underwood KW, Jacobs NL, Howley A, Liscum L (1998) Evidence for a cholesterol transport pathway from lysosomes to endoplasmic reticulum that is independent of the plasma membrane. J Biol Chem 273(7):4266–4274

Uno SN, Kamiya M, Yoshihara T, Sugawara K, Okabe K, Tarhan MC, Fujita H, Funatsu T, Okada Y, Tobita S, Urano Y (2014) A spontaneously blinking fluorophore based on intramolecular spirocyclization for live-cell super-resolution imaging. Nat Chem 6(8):681–689

van Breemen C, Saida K (1989) Cellular mechanisms regulating [Ca2+]i smooth muscle. Annu Rev Physiol 51:315–329

van der Kant R, Zondervan I, Janssen L, Neefjes J (2013) Cholesterol-binding molecules MLN64 and ORP1L mark distinct late endosomes with transporters ABCA3 and NPC1. J Lipid Res 54 (8):2153–2165

van Rijnsoever C, Oorschot V, Klumperman J (2008) Correlative light-electron microscopy (CLEM) combining live-cell imaging and immunolabeling of ultrathin cryosections. Nat Methods 5(11):973–980

Varnai P, Toth B, Toth DJ, Hunyady L, Balla T (2007) Visualization and manipulation of plasma membrane-endoplasmic reticulum contact sites indicates the presence of additional molecular components within the STIM1-Orai1 complex. J Biol Chem 282(40):29678–29690

Venkatachalam K, Montell C (2007) TRP channels. Annu Rev Biochem 76:387–417

Vig M, Peinelt C, Beck A, Koomoa DL, Rabah D, Koblan-Huberson M, Kraft S, Turner H, Fleig A, Penner R, Kinet JP (2006) CRACM1 is a plasma membrane protein essential for store-operated Ca2+ entry. Science 312(5777):1220–1223

West M, Zurek N, Hoenger A, Voeltz GK (2011) A 3D analysis of yeast ER structure reveals how ER domains are organized by membrane curvature. J Cell Biol 193(2):333–346

Wu MM, Buchanan J, Luik RM, Lewis RS (2006) Ca2+ store depletion causes STIM1 to accumulate in ER regions closely associated with the plasma membrane. J Cell Biol 174 (6):803–813

Yu H, Liu Y, Gulbranson DR, Paine A, Rathore SS, Shen J (2016) Extended synaptotagmins are Ca2+-dependent lipid transfer proteins at membrane contact sites. Proc Natl Acad Sci U S A 113(16):4362–4367

Yuan JP, Zeng W, Dorwart MR, Choi YJ, Worley PF, Muallem S (2009) SOAR and the polybasic STIM1 domains gate and regulate Orai channels. Nat Cell Biol 11(3):337–343

Zhang SL, Yeromin AV, Zhang XH, Yu Y, Safrina O, Penna A, Roos J, Stauderman KA, Cahalan MD (2006) Genome-wide RNAi screen of Ca(2+) influx identifies genes that regulate Ca(2+) release-activated Ca(2+) channel activity. Proc Natl Acad Sci U S A 103(24):9357–9362

Zhou Y, Meraner P, Kwon HT, Machnes D, Oh-hora M, Zimmer J, Huang Y, Stura A, Rao A, Hogan PG (2010) STIM1 gates the store-operated calcium channel ORAI1 in vitro. Nat Struct Mol Biol 17(1):112–116

Zong X, Schieder M, Cuny H, Fenske S, Gruner C, Rötzer K, Griesbeck O, Harz H, Biel M, Wahl-Schott C (2009) The two-pore channel TPCN2 mediates NAADP-dependent Ca2+-release from lysosomal stores. Pflugers Arch 458(5):891–899

The Role of Mitochondria in the Activation/Maintenance of SOCE: The Contribution of Mitochondrial Ca^{2+} Uptake, Mitochondrial Motility, and Location to Store-Operated Ca^{2+} Entry

16

Roland Malli and Wolfgang F. Graier

Abstract

In most cell types, the depletion of internal Ca^{2+} stores triggers the activation of Ca^{2+} entry. This crucial phenomenon is known since the 1980s and referred to as store-operated Ca^{2+} entry (SOCE). With the discoveries of the stromal-interacting molecules (STIMs) and the Ca^{2+}-permeable Orai channels as the long-awaited molecular constituents of SOCE, the role of mitochondria in controlling the activity of this particular Ca^{2+} entry pathway is kind of buried in oblivion. However, the capability of mitochondria to locally sequester Ca^{2+} at sites of Ca^{2+} release and entry was initially supposed to rule SOCE by facilitating the Ca^{2+} depletion of the endoplasmic reticulum and removing entering Ca^{2+} from the Ca^{2+}-inhibitable channels, respectively. Moreover, the central role of these organelles in controlling the cellular energy metabolism has been linked to the activity of SOCE. Nevertheless, the exact molecular mechanisms by which mitochondria actually determine SOCE are still pretty obscure. In this essay we describe the complexity of the mitochondrial Ca^{2+} uptake machinery and its regulation, molecular components, and properties, which open new ways for scrutinizing the contribution of mitochondria to SOCE. Moreover, data concerning the variability of the morphology and cellular distribution of mitochondria as putative determinants of SOCE activation, maintenance, and termination are summarized.

Keywords

Mitochondria • Endothelial nitric oxide synthase • Ca^{2+} signaling • Store-operated Ca^{2+} entry • Mitochondrial Ca^{2+} uptake • Uncoupling protein 2 • MICU1 • MCU • Protein methylation

R. Malli • W.F. Graier (✉)
Institute of Molecular Biology and Biochemistry, Medical University of Graz, Neue Stiftingtalstrasse 6/6, 8010 Graz, Austria
e-mail: wolfgang.graier@medunigraz.at

K. Groschner et al. (eds.), *Store-Operated Ca^{2+} Entry (SOCE) Pathways*, Advances in Experimental Medicine and Biology 993, DOI 10.1007/978-3-319-57732-6_16

16.1 Introduction

Mitochondria are multifunctional organelles that fundamentally impact many different signaling pathways, thereby conferring novel complexity on the long-known cellular powerhouses (Graier et al. 2007; McBride et al. 2006). To complete this task, mitochondria communicate with other organelles and the plasma membrane in order to regulate, among other cellular signaling events, the magnitude of the entry of Ca^{2+}, nature's most versatile and ubiquitous signaling messenger (Berridge et al. 2000). It appears to be a hallmark principle in the physiology of eukaryotic cells that the particular organelle that provides energy for life is also crucially involved in the regulation of Ca^{2+} entry. Excitingly, Ca^{2+} entry has recently been demonstrated to link mitochondrial metabolic functions with cellular signal transduction. This fascinating *relay function* of the mitochondria is still largely unexplored and may be crucial for our understanding of the molecular interrelationships underlying so far unknown physiological and pathological processes. Mitochondria have been shown to contribute significantly to the activity of the so-called store-operated Ca^{2+} entry (SOCE), thereby playing multiple roles in the activation, maintenance, and termination of this vital, ubiquitous Ca^{2+} entry pathway (Demaurex et al. 2009; Parekh 2008; Parekh and Putney 2005). Similar to SOCE, the activity of endothelial nitric oxide synthase (eNOS) depends on mitochondrial Ca^{2+} uptake, while this function exceeds the organelle's role in SOCE and provides evidence for a Ca^{2+}-independent regulation of endothelial $NO^{\bullet}$ by mitochondria (Charoensin et al. 2017). However, despite some intriguing findings, the molecular mechanisms by which mitochondria apparently contribute to SOCE have not yet been clarified.

The discovery of the stromal-interacting molecule (STIM) and Orai proteins as the long-awaited constituents of the SOCE phenomenon shed light on the molecular mechanism of this ubiquitous way of Ca^{2+} entry (Frischauf et al. 2008). In view of the discovery of STIMs and Orai proteins as the key protagonists of SOCE and the recent enormous progress in the understanding of the contributors and regulation of mitochondrial Ca^{2+} uptake (Kamer et al. 2014; De Stefani and Rizzuto 2015), the role of mitochondria and their contribution to SOCE needs to be reevaluated. In this chapter we provide an overview on the molecular mechanisms, signaling pathways, and molecules that are responsible for the specific contribution of mitochondria to the activity of the stromal-interacting molecule 1 (STIM1)-/Orai1-dependent SOCE. In particular, the regulatory machinery of mitochondrial Ca^{2+} handling, mitochondrial motility, and morphology on SOCE activity will be discussed. Notably, the understanding of the actual contribution of mitochondria to SOCE is of particular interest in view of the many biological functions of the STIM1-/Orai1-dependent SOCE in lymphocytes, mast cells, platelets, endothelial cells, neurons, and cancer cells as this might lead to the development of novel anti-inflammatory, immune-modulating, antiallergic, vasoactive, neuroprotective, and anticancer drugs and therapeutic strategies.

16.1.1 Store-Operated Ca^{2+} Entry: A Ubiquitous, Enigmatic, and Vital Process

Distinctly regulated spatial and temporal fluctuations of the cellular Ca^{2+} concentration frequently control different cellular functions such diverse as proliferation and programmed cell death, in which Ca^{2+} ions serve as versatile and universal cellular messengers (Berridge et al. 2000). Cells comprise a vast amount of various Ca^{2+} binding proteins that respond to Ca^{2+} signals and subsequently trigger cell-specific signaling events yielding adequate responses (Knot et al. 2005). Essentially, two different Ca^{2+} sources, an intracellular Ca^{2+} store, which is mainly the ER, and the virtually infinite extracellular Ca^{2+} pool, are available for the generation of cellular Ca^{2+} signals (Bootman et al. 2001). Cells developed an ingenious feedback mechanism linking the ER Ca^{2+} content to the activity of Ca^{2+} entry across the plasma membrane in such a way that depletion of the ER Ca^{2+} store leads to activation of Ca^{2+} permeable ion channels in the plasma membrane. This process, mostly referred to as store-operated Ca^{2+} entry (SOCE), has been discovered in the mid-1980s (Putney 1986) and is among the most exciting and long-standing mysteries in the Ca^{2+} signaling field. SOCE was originally termed capacitative Ca^{2+} entry (CCE) and its fundamental physiological relevance was soon accepted as a process which ensures suitable ER Ca^{2+} replenishment during and/or after cell stimulation (Putney 1986, 1990, 1991), which is vital for the Ca^{2+}-dependent protein folding machinery in this organelle (Michalak et al. 2002; Osibow et al. 2006). Moreover, SOCE contributes to a variety of physiological functions such as exocytosis (Fomina and Nowycky 1999), enzyme activity (Lin et al. 2000), transcription (Dolmetsch et al. 1998), cell proliferation and migration (Potier et al. 2009), cell differentiation (Feske 2007), and apoptosis (Lampe et al. 1995). Accordingly, SOCE is considered as an important, ubiquitous physiological process (Parekh and Putney 2005).

In blood cells SOCE was firstly characterized electrophysiologically as the so-called Ca^{2+} release-activated Ca^{2+} (CRAC) current (I_{CRAC}) showing high Ca^{2+} selectivity, inward rectification, and both fast and slow Ca^{2+}-dependent inactivation (Zweifach and Lewis 1995; Parekh 1998). The measurable I_{CRAC} of blood cells is still used as the standard electrophysiological correlate of SOCE although similar but not identical currents, called CRAC-like or currents from cation-store-operated channels (cat-SOC) (Parekh and Putney 2005; McFadzean and Gibson 2002), can be recorded in other cell types perhaps pointing to the existence of various modes of store-operated Ca^{2+} entry pathways.

Since the discovery of the SOCE phenomenon, several intriguing models and mechanisms have been proposed to account for the link between ER Ca^{2+} depletion and Ca^{2+} entry (Parekh and Putney 2005; Bolotina and Csutora 2005). However, the mechanism(s) by which the ER Ca^{2+} stores communicate with the plasma membrane as well as the actual ion channel(s) responsible for SOCE remained elusive until very recently.

16.1.2 Stim and Orai: The Long-Awaited Molecular Components of SOCE

The key breakthrough in the identification of the molecular basis of SOCE was the discovery of the stromal-interacting molecule 1 (STIM1) as an essential protein for SOCE activation using RNAi (Liou et al. 2005; Roos et al. 2005). STIM1 is now known to be a Ca^{2+} sensor, detecting the luminal ER Ca^{2+} concentration via an EF-hand domain at its N-terminus, while the C-terminal part of this ER membrane spanning protein is located in the cytosol (Fig. 16.1, panel I). Upon ER Ca^{2+} depletion, STIM1 proteins oligomerize and redistribute to form clusters of STIM1 multimers at ER-plasma membrane (ER-PM) junctions (Fig. 16.1, panels I, IV, V, VI) (Liou et al. 2005; Zhang et al. 2005). This profound realignment of STIM1 proteins is an essential step in SOCE activation as thereby the message of ER Ca^{2+} depletion is transferred to ion channels at the plasma membrane. In humans STIM2 was identified (Hooper et al. 2013) which is twofold less Ca^{2+} sensitive compared to STIM1 and a weaker activator of SOCE (Soboloff et al. 2012). It has been assumed that STIM2 is important to stabilize basal cytosolic and ER Ca^{2+} levels (Brandman et al. 2007). The most prominent channel protein, which actually accomplishes SOCE, is the four-membrane spanning protein Orai1. ER Ca^{2+} depletion and the subsequent clustering of STIM1 proteins trigger the assembly of Orai1 dimers into tetramers (Penna et al. 2008). Thereby, a direct interaction of STIM1 with Orai1 proteins accomplishes the activation of SOCE (Muik et al. 2008), while additional scaffolding proteins (Varnai et al. 2007, 2009) and factors (Bolotina 2008) assist in STIM1-induced SOCE activation. Low levels of cytosolic Ca^{2+} promote STIM1 clustering upon ER Ca^{2+} depletion, whereas increased cytosolic Ca^{2+} signals prevent its assembly at ER-plasma membrane junctions (Fig. 16.1, panel VII), pointing to a negative feedback of high cytosolic Ca^{2+} to block the establishment of subplasmalemmal STIM1 clusters (Malli et al. 2008). Orai2 and Orai3 are highly conserved paralogues of Orai1, although Orai1 seems to be most important for SOCE (DeHaven et al. 2007).

It is generally believed that STIM proteins sense the Ca^{2+} depletion of the ER (Fig. 16.1, panel I), subsequently oligomerize (Fig. 16.1, panel IV), translocate to ER-plasma membrane junctions (Fig. 16.1, panel IV, V), and arrange Orai proteins (and other channels, e.g., TRPCs) into oligomers (Fig. 16.1, panel V, VI) yielding activation of these Ca^{2+}-permeable plasma membrane ion channels to establish SOCE (Fig. 16.1, panel VI). Although the concept described above is generally accepted and represents a great step in the demystification of SOCE, numerous new questions arise:

1. What is the exact mechanism of STIM1 translocation?
2. Is the translocation of STIM1 an active or passive process and which additional factors are required?
3. What is the molecular basis of the ER-plasma membrane junctions?
4. What are the mechanisms for SOCE termination?
5. What is the exact role of mitochondria in the whole scenario?

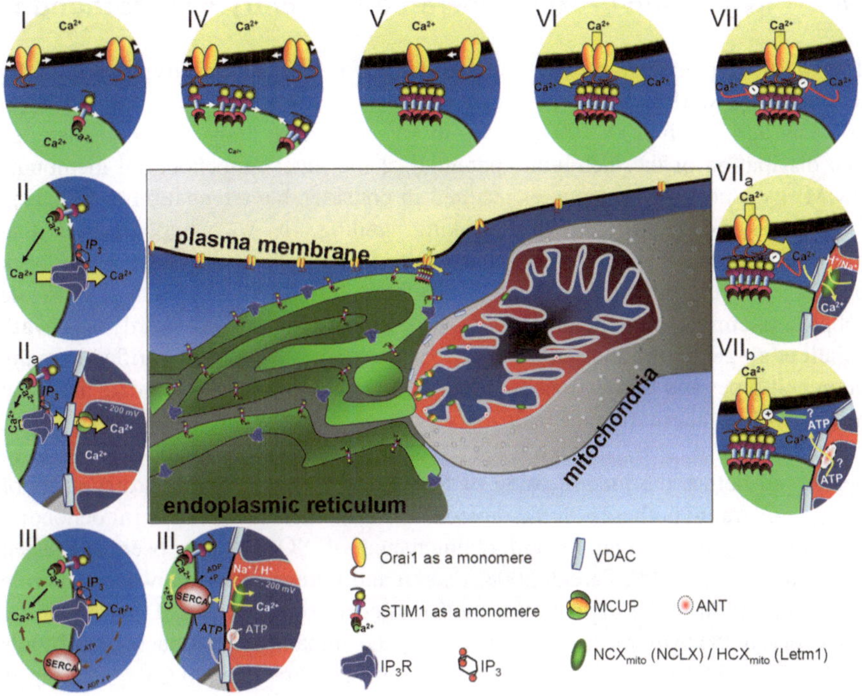

Fig. 16.1 Schematic illustration of STIM1/Orai1 signaling and possible mechanisms by which mitochondria contribute to SOCE activity. *Middle oblong drawing*: Graphic representation of a subplasmalemmal ER domain (*green*) and an adjacent mitochondrion (*gray*, *red*, *blue*). A spatial proximity between organelles and the SOCE channels, which allows protein-protein interactions as well as local control of the free Ca^{2+} concentration, other messengers, or metabolites, determines SOCE activation, maintenance, and termination. Panel **I**: Under resting conditions with a high Ca^{2+} concentration within the lumen of the ER ($[Ca^{2+}]_{ER}$), STIM1 proteins in the ER membrane and Orai1 dimers in the plasma membrane are mainly homogeneously distributed due to free lateral movements within the respective biomembranes. Panel **II**: IP_3-mediated ER Ca^{2+} release rapidly lowers $[Ca^{2+}]_{ER}$, which yields dissociation of Ca^{2+} from the EF-hand domain of STIM proteins. Panel **II$_a$**: ER Ca^{2+} release is modulated by mitochondrial Ca^{2+} uptake via the mitochondrial Ca^{2+} uniporter complex (MCUP) at ER-mitochondria junctions, which impacts on $[Ca^{2+}]_{ER}$ and, thus, on STIM1 signaling. MCUP consists of the pore-forming proteins MCU and EMRE; their main regulators MICU1, MICU2, and MCUb; and eventually the scaffolding protein MCUR1. Under conditions of enhanced PRMT1 activity, UC2/3 are essential facilitators that ensure normal MCUP activity (for details see Sect. 16.2.1.2). Panel **III**: SERCA-mediated ER Ca^{2+} replenishment counteracts IP_3-mediated ER Ca^{2+} depletion, supplying Ca^{2+} to the luminal STIM1 EF-hand domains. Panel **III$_a$**: Mitochondria are able to locally relay Ca^{2+} ions and ATP to sites of ER Ca^{2+} sequestration, thereby facilitating ER Ca^{2+} refilling and SOCE termination. Mitochondrial Ca^{2+} efflux is mainly accomplished by a Na^+/Ca^{2+} exchanger (NCX_{mito}), whereas mitochondrial ATP transport is accomplished by the adenine nucleotide translocase (ANT). Panel **IV**: Dissociation of Ca^{2+} from EF-hand domains of STIM proteins upon ER Ca^{2+} depletion triggers STIM1 oligomerization. Panel **V**: At low ER Ca^{2+} levels, STIM1 oligomers redistribute and accumulate at ER-plasma membrane junctions, which subsequently affect Orai1 dimers in the plasma membrane. Panel **VI**: Subplasmalemmal STIM1 cluster assembles Orai1 proteins to form SOCE channels at ER-plasma membrane junctions. STIM1/

16.2 Regulation of Store-Operated Ca^{2+} Entry by Mitochondria

The particular importance of mitochondria for SOCE activity was initially described by Markus Hoth, Christopher M. Fanger, and Richard S. Lewis almost 20 years ago (Hoth et al. 1997, 2000). In a very elegant study, they demonstrated that dissipation of the membrane potential of the inner mitochondrial membrane (IMM) by a chemical uncoupler resulted in considerable attenuation of SOCE in T-lymphocytes. Based on this striking finding, it was hypothesized that mitochondria buffer entering Ca^{2+} and subsequently prevent the negative feedback of Ca^{2+} on the activity of SOCE channels (Fig. 16.1, panel VII_a). While subplasmalemmal mitochondrial Ca^{2+} buffering was described shortly afterward (Malli et al. 2003a, b), the negative feedback hypothesis could be verified in many other cell types and initiated various follow-up studies aimed to elucidate the exact role as well as alternative mechanisms of mitochondria in the regulation of SOCE (Demaurex et al. 2009; Parekh 2008). Notably, most of these studies have been performed before the identification of STIM and Orai as the key protagonists of I_{CRAC}. Several hypotheses on the molecular mechanism(s) by which mitochondria contribute to the activation and maintenance of SOCE have been suggested (Demaurex et al. 2009; Parekh 2008; Parekh and Putney 2005). However, the role of mitochondria in the regulation of SOCE needed to be reassessed in view of the discovery of STIM/Orai. In a recent study, the impact of pharmacological tools (CGP 37157, FCCP, oligomycine), known to affect mitochondrial Ca^{2+} handling as well as their metabolic functions, was tested for their effects on SOCE in an endothelial cell line overexpressing STIM1 and Orai1 proteins (Naghdi et al. 2010). In this study SOCE was assessed indirectly by quantifying the raise of cytosolic and mitochondrial Ca^{2+} upon Ca^{2+} readdition after maximal ER Ca^{2+} depletion. Interestingly, only a combination of FCCP, oligomycine, and CGP37157 completely abolished mitochondrial Ca^{2+} loading by entering Ca^{2+} in STIM1/Orai1 overexpressing cells, which was associated with a pronounced impairment of the SOCE-induced cytosolic Ca^{2+} elevation (Naghdi et al. 2010). Accordingly, these findings approve the crucial importance of mitochondrial Ca^{2+} handling for the activity of STIM1-/Orai1-dependent SOCE. In line with this conclusion, more recent data clearly demonstrated the importance of mitochondrial Ca^{2+} for SOCE activity (Deak et al. 2014). In regard of SOCE activity, this work for the first time showed that mitochondrial Ca^{2+} uptake was not hampered by a toxic depolarization of the organelle but by selective diminution of the expression of either the main pore-forming protein MCU (Baughman et al. 2011; De Stefani et al. 2011) or uncoupling protein 2/3 (UCP2/3), essential facilitators of mitochondrial Ca^{2+}

Fig. 16.1 (continued) Orai1 assemblies are thought to exhibit the SOCE units. Panel **VII**: Ca^{2+} elevation at STIM1/Orai1 assemblies inhibits SOCE and impedes STIM1 clustering. Panel $\mathbf{VII_a}$: Mitochondrial buffering of entering Ca^{2+} perhaps via mitochondrial Ca^{2+} antiporter is thought to facilitate SOCE activity. Panel $\mathbf{VII_b}$: Local mitochondrial supply of ATP or other metabolites at STIM1/Orai1 SOCE units might contribute to SOCE uptake under certain conditions (Trenker et al. 2007; Waldeck-Weiermair et al. 2010b; Madreiter-Sokolowski et al. 2016b). Hence, mitofusin 2, a protein that tethers the ER to mitochondria (de Brito and Scorrano 2008), impairs STIM1 trafficking upon ER Ca^{2+} depletion independently from mitochondrial Ca^{2+} handling when mitochondria are entirely depolarized (Singaravelu et al. 2011). Nevertheless, it remains to be verified whether or not mitofusin 2 controls STIM1 functions also under physiological conditions. Notably, the idea that mitofusin 2 indeed strengthens ER-mitochondria contacts has recently been challenged by quantitative electron microscopy (Cosson et al. 2012). It was suggested that mitofusin 2 rather keeps the organelles at distance as ablation of the protein increased ER-mitochondria coupling (Filadi et al. 2015). In this study it was also demonstrated that mitofusin 2 knockout cells have reduced expression levels of MCU which explains the reduced mitochondrial Ca^{2+} signals in these cells.

In spite of the studies mentioned above, two different but certainly interrelated molecular mechanisms have been suggested by which mitochondria may influence SOCE: (1) mitochondria contribute to SOCE activity by their ability to take up (buffer) and release Ca^{2+} ions (Fig. 16.1, panel II_a, III_a, VII_a) (Demaurex et al. 2009; Parekh 2008), and (2) mitochondria regulate SOCE via small metabolites such as ATP or pyruvate, which can either facilitate ER Ca^{2+} refilling (Fig. 16.1, panel III_a) or prevent the Ca^{2+}-dependent inactivation of I_{CRAC} (Fig. 16.1, panel VII_b) (Demaurex et al. 2009; Montalvo et al. 2006). Obviously both, mitochondrial Ca^{2+} handling and their metabolic function, contribute to SOCE activity.

In addition, mitochondria have been recognized to be highly dynamic organelles that constantly rearrange their morphology and subcellular location by branching, fusion, and fission events. This plasticity of mitochondrial structures is important to maintain and adjust the functionality of these organelles (Braschi and McBride 2010). In particular, the local control of signaling events within cellular microdomains is supposed to crucially depend on the motility and morphology of mitochondria (Malli et al. 2003a, b). Hence, the versatile architectural organization of mitochondria within cells in combination with their ability to locally control the concentration of intracellular messengers and metabolites might actually contribute to the control of SOCE activity.

16.2.1 Mitochondrial Ca^{2+} Handling and Its Impact on SOCE

One distinguished feature of mitochondria that was intriguingly shown by protein-based Ca^{2+} sensors targeted to these organelles (Rizzuto et al. 1992; Nagai et al. 2001) is their active engagement in Ca^{2+} handling during physiological Ca^{2+} signaling events (Szabadkai and Duchen 2008). Importantly, mitochondrial Ca^{2+} uptake is precisely regulated because excessive mitochondrial Ca^{2+} load can lead to cell death (Hajnoczky et al. 2006). Energized mitochondria are virtually destined to sequester Ca^{2+} due to the large negative membrane potential of the IMM that provides a strong driving force for mitochondrial Ca^{2+} uptake (Duchen et al. 2008). Accordingly, it has been demonstrated that respiring, energized mitochondria determine the pattern of activation and inactivation of SOCE in eukaryotic cells due to their ability to buffer cytosolic Ca^{2+} (Gilabert et al. 2001; Gilabert and Parekh 2000). Notably, the sequestration of Ca^{2+} by mitochondria is achieved by a so far not identified mitochondrial Ca^{2+} uniporter that allows Ca^{2+} passage via the IMM (Nicholls 2005; Kirichok et al. 2004). Moreover, mitochondria contain huge amounts of Ca^{2+} binding polyphosphates in their matrix and, thus, may serve as Ca^{2+} sinks (Abramov et al. 2007).

Basically, mitochondrial Ca^{2+} sequestration can have multiple consequences for a cell. Mitochondria contain several Ca^{2+}-sensitive proteins, thus exhibiting a Ca^{2+}-sensitive target, whereas mitochondrial Ca^{2+} handling significantly impacts on cytosolic Ca^{2+} signals (Graier et al. 2007; Walsh et al. 2009). Mitochondria operate as dynamic Ca^{2+} distributing organelles whereupon mitochondrial Ca^{2+} loading is accompanied by Ca^{2+} efflux via $Na^{+}/Ca^{2+}$and H^{+}/Ca^{2+} exchangers (Demaurex et al. 2009; Gunter et al. 2000) and the NCLX (Palty et al. 2010).

Hence, during the generation of Ca^{2+} signals, a mechanism termed trans-mitochondrial Ca^{2+} flux occurs which is particularly important for the regulation of SOCE (Malli et al. 2003b, 2005; Feldman et al. 2010).

Multiple indications point to mitochondrial Ca^{2+} uptake to be fundamental for the organelle's contribution to SOCE activity. Thus, the following subsections deal with novel insights in and properties of mitochondrial Ca^{2+} uptake pathway(s).

16.2.1.1 The Mitochondrial Ca^{2+} Uptake Phenomenon Builds on Ca^{2+}-Selective Ion Channels

The phenomenon of mitochondrial Ca^{2+} uptake was already described in the early 1960s using isolated mitochondria from rat kidney (DeLuca and Engstrom 1961; Vasington and Murphy 1962). It was soon recognized that isolated mitochondria from different tissues show a similar kind of passive Ca^{2+} uptake which is driven by the electrochemical gradient and not coupled to the transport of other ions. This process was subsequently referred to as mitochondrial Ca^{2+} uniport (Fig. 16.1, panel II_a) (reviewed in Gunter and Sheu 2009). However, the mitochondrial Ca^{2+} uniport system of isolated mitochondria exhibits low affinity for Ca^{2+} with an apparent K_D of 10–15 μM (Scarpa and Graziotti 1973). As under physiological conditions raises of global cytosolic Ca^{2+} clearly remain below these levels, mitochondrial Ca^{2+} uniport was considered to be physiologically irrelevant for a rather long time (reviewed in Carafoli 2002). This view has dramatically changed in the 1990s when protein-based Ca^{2+} sensors targeted to the mitochondrial matrix allowed reliable measurements of mitochondrial Ca^{2+} signals in intact cells (Rizzuto et al. 1992). Notably, well-targeted Ca^{2+} sensors in mitochondria revealed that cytosolic Ca^{2+} signals, fueled by either intracellular Ca^{2+} release from the ER Ca^{2+} or Ca^{2+} entry, are efficiently transferred into the matrix of mitochondria (Duchen et al. 2008; Graier et al. 2007). In order to explain such findings in view of the low Ca^{2+} affinity of the mitochondrial Ca^{2+} uptake system, it was proposed that the low-affinity mitochondrial Ca^{2+} uniport system is activated by high Ca^{2+} microdomains that are probably generated predominately at sites of ER Ca^{2+} release and in some instances also at Ca^{2+} entry channels (Rizzuto et al. 1998).

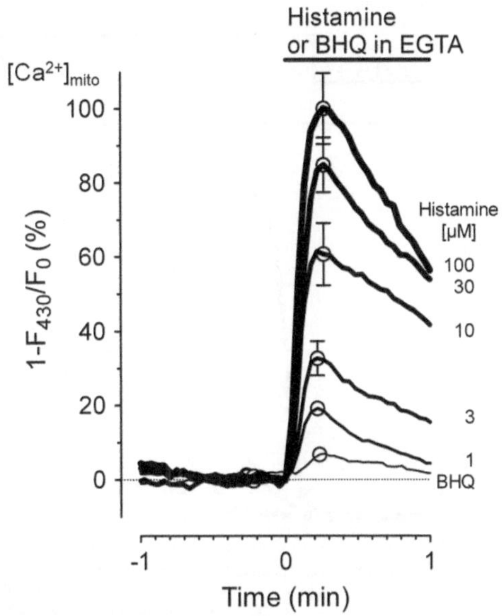

Fig. 16.2 Mitochondrial Ca^{2+} signals induced by ER Ca^{2+} mobilization. The free Ca^{2+} concentration within the mitochondrial matrix ($[Ca^{2+}]_{mito}$) was measured using cells (EAhy926) expressing mitochondrially targeted pericam, a fluorescent protein-based Ca^{2+} sensor. ER Ca^{2+} mobilization was accomplished by cell stimulation with the IP_3-generating agonist histamine at various concentrations ranging from 1 to 100 μM. Alternatively, ER Ca^{2+} depletion was induced by SERCA inhibition with 15 μM BHQ. Notably, experiments were performed in the absence of extracellular Ca^{2+} (1 mM EGTA) in order to exclude Ca^{2+} entry. Mitochondrial Ca^{2+} signals are represented as average curves in percentage of the maximal response that was achieved upon cell stimulation with 100 μM histamine of at least three independent experiments. Interestingly, although the amplitude of mitochondrial Ca^{2+} signals correlated with the strength of ER Ca^{2+} mobilization, the velocity of the Ca^{2+} transfer from the ER to mitochondria occurred independently from the strength and mode of ER Ca^{2+} release

This microdomain concept was approved by quantifying Ca^{2+} hotspots at ER-mitochondria junctions (Csordas et al. 2010; Giacomello et al. 2010) and explains how the mitochondrial Ca^{2+} regulator protein MICU1 gets activated by Ca^{2+} binding despite its rather high Ca^{2+} binding constant around 4–5 μM (Waldeck-Weiermair et al. 2015). Moreover, the promptitude of the generation of mitochondrial Ca^{2+} signals in intact endothelial cells is independent from the strength and mode of ER Ca^{2+} mobilization, further supporting the Ca^{2+} microdomain concept (Fig. 16.2). Remarkably, although it is generally accepted that mitochondria sense microdomains of high Ca^{2+} during cell stimulation, the application of mitochondrially targeted Ca^{2+} sensors also revealed mitochondrial Ca^{2+} uptake during low Ca^{2+} signals in intact cells under physiological conditions which contradicts the low-affinity mitochondrial Ca^{2+} uniport system of isolated mitochondria (Spät et al. 2008). In addition, SOCE-induced mitochondrial Ca^{2+} signals in endothelial cells are rather slow and their velocity highly correlates with

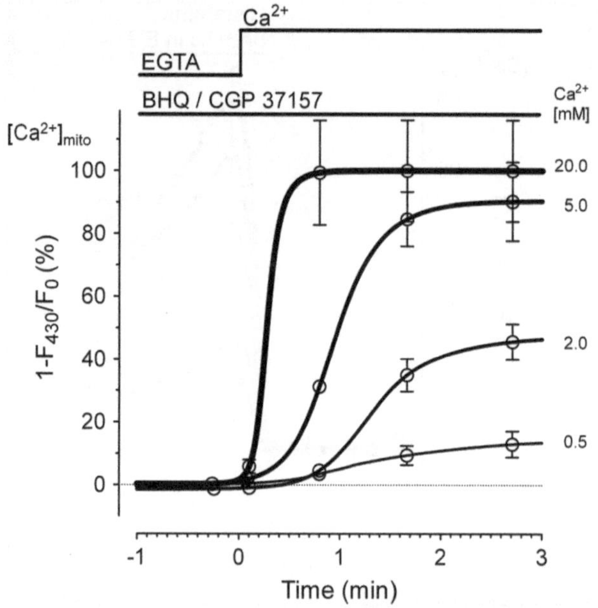

Fig. 16.3 Mitochondrial Ca^{2+} signals induced by SOCE. Mitochondrial Ca^{2+} signals exclusively fueled by SOCE were recorded in the endothelial cell line EAhy926 expressing mitochondrially targeted pericam. In order to maximally activate SOCE, cells were pretreated with 100 μM histamine and 15 μM BHQ in EGTA, which entirely depletes ER Ca^{2+} stores. Subsequently, different Ca^{2+} concentrations were added to pre-stimulated cells yielding mitochondrial Ca^{2+} signals of various strengths. Notably, both the amplitude and the velocity of mitochondrial Ca^{2+} loading correlated with the strength of SOCE

the strength of SOCE (Fig. 16.3) (Waldeck-Weiermair et al. 2010a). These observations point either to multiple mitochondrial Ca^{2+} uptake pathways with different Ca^{2+} sensitivities or a strong modulation of the Ca^{2+} sensitivity of one given Ca^{2+} uniport. Obviously, such high Ca^{2+}-sensitive uptake route(s)/mode(s) need(s) an intact cellular environment to get established and, thus, is/are not detectable using isolated mitochondria. Notably, this kind of high Ca^{2+}-sensitive mitochondrial Ca^{2+} uptake pathway(s)/mode(s) might be particularly important for the transfer of entering Ca^{2+} to the mitochondrial matrix because in many cell types mitochondria don't seem to be exposed to sites of Ca^{2+} entry (Csordas et al. 2010; Graier et al. 2007, 2008; Naghdi et al. 2010). Accordingly, SOCE-induced mitochondrial Ca^{2+} elevation and the observed feedback of mitochondrial Ca^{2+} sequestration on SOCE activity potentially depend on (a) high Ca^{2+}-sensitive mitochondrial Ca^{2+} uptake pathway(s) (Waldeck-Weiermair et al. 2010a).

In a sophisticated study using the patch clamp technique on isolated mitoplasts (i.e., swollen mitochondria lacking the outer mitochondrial membrane) of Cos-7 cells, the mitochondrial Ca^{2+} uniporter was characterized as a highly Ca^{2+}-selective inward rectifying ion channel at the inner mitochondrial membrane with quite low Ca^{2+} affinity (Kirichok et al. 2004). Importantly, using the same technique, various distinct mitochondrial Ca^{2+} channels with different occurrence, single-channel amplitude, opening time, open probability, and Ca^{2+} sensitivity were recorded in cardiac (mCa1 and mCa2 (Michels et al. 2009)), endothelial, and HeLa cells (*i*-MCC, *l*-MCC, and *xl*-MCC (Bondarenko et al. 2013)), pointing to the existence of various mitochondrial Ca^{2+} channels, at least in isolated mitoplasts. Despite the excellent characterization of the mitochondrial Ca^{2+} uniport on the electrophysiological level, the identity of the channels per se remained elusive (Malli and Graier 2010) until the groundbreaking discovery of Vamsi K. Mootha and his team (Perocchi et al. 2010).

16.2.1.2 The Mitochondrial Ca^{2+} Uniporter Complex

The first main component of the mitochondrial Ca^{2+} uniporter complex was identified by the discovery of a EF-hand-containing protein required for mitochondrial Ca^{2+} uptake in HeLa cells (mitochondrial Ca^{2+} uptake 1, MICU1) (Perocchi et al. 2010). Although this protein was known to be encoded by the *CBARA1* (alternatively referred to as *FLJ12684*) gene since several years, it was poorly characterized until then. Initially, this protein was identified as a 54-kDa autoantigen without any direct functional link to mitochondrial physiology (Aichberger et al. 2005). MICU1 is localized to mitochondria by a targeting sequence at its amino-terminus and consists of just one predicted transmembrane domain, which argues against a function of MICU1 itself as a Ca^{2+} channel in the IMM. However, by using MICU1 as a trap, several main components of the mitochondrial Ca^{2+} uniporter complex were subsequently discovered. Notably, with the mitochondrial Ca^{2+} uniporter protein (MCU) (Baughman et al. 2011; De Stefani et al. 2011) and the essential MCU regulator (EMRE) (Sancak et al. 2013), the two pore-forming proteins were found. Furthermore, MCUb, a dominant negative form of the MCU (Raffaello et al. 2013); MICU2 (Plovanich et al. 2013), which fine-tunes the function of MICU1 (Kamer and Mootha 2014; Patron et al. 2014); and MCUR1, an important regulator/scaffold protein (Mallilankaraman et al. 2012; Tomar et al. 2016), were described, thus indicating that mitochondrial Ca^{2+} uptake is achieved by a multi-protein complex under the primary control of MICU1 (Kamer et al. 2014; De Stefani and Rizzuto 2015; Elustondo et al. 2016). The regulatory role of MICU1 on MCU builds on the formation of inhibitory hexamers that disaggregate to dimers upon Ca^{2+} binding to the EF motifs and subsequently unlock MCU activity (Wang et al. 2014, 2015). In pancreatic β-cells MICU1, a different role of MICU1 has been described as an essential facilitator of MCU activity rather than a gatekeeper (Alam et al. 2012). The apparent Ca^{2+}-sensitivity of MICU1 of approx. 4–5 μM (Waldeck-Weiermair et al. 2015) perfectly matches the ion concentrations at the junction between the mitochondria and ER (Giacomello et al. 2010), thus ensuring the activation of MCU upon intracellular Ca^{2+} release by IP_3. However, activation of the protein arginine methyltransferase 1 (PRMT1) yields methylation of arginine at position 455 of MICU1 and reduces its apparent Ca^{2+} sensitivity threefold to fourfold to approx. 14 μM, thus strongly attenuating MCU activity under these conditions (Madreiter-Sokolowski et al. 2016b). Because PRMT1 activity is increased during aging and in cancer, it is suggested to protect cells against mitochondrial Ca^{2+} overload under these conditions (Madreiter-Sokolowski et al. 2016a, b).

Besides these validated components of the mitochondrial Ca^{2+} uniporter complex, several other mitochondrial proteins previously implicated in other functions were revealed to significantly contribute to mitochondrial Ca^{2+} uptake in different cell types (Hajnoczky and Csordas 2010), probably pointing to cell type-specific mitochondrial Ca^{2+} uptake machineries (Fieni et al. 2012). In this context it was demonstrated that the novel uncoupling proteins UCP2 and UCP3 are key for mitochondrial Ca^{2+} uniport in several non-excitable (cancer) cell lines (Trenker et al. 2007; Graier et al. 2008). In particular, distinct sites within the intermembrane loop 2 (IML2) of UCP2/3 appear essential for the protein's contribution to mitochondrial Ca^{2+} uptake but also to determine the Ca^{2+}-sensitivity of the respective mitochondrial sequestration machinery (Waldeck-Weiermair et al. 2010a). Hence, these two Ca^{2+} sensitivities correlated with the properties of the UCP2/3-dependent mitochondrial uptake of high (i.e., intracellularly released) and low (i.e., via the SOCE) Ca^{2+} concentrations. Moreover, it was suggested that the novel UCPs might work as sophisticated molecular switches adjusting mitochondrial Ca^{2+} uptake to Ca^{2+} signals of varying intensities (Waldeck-Weiermair et al. 2010a). However, in the endothelial cell line, tested novel UCPs exclusively contributed to mitochondrial Ca^{2+} uptake fueled by ER Ca^{2+} mobilization but not by SOCE under conditions of low endogenous expression of UCP2 and UCP3 (Waldeck-Weiermair et al. 2010a). These findings probably point to differences in the molecular composition of mitochondrial Ca^{2+} uptake sites facing either ER Ca^{2+} release or Ca^{2+} entry at the plasma membrane (see Sect. 16.2.1.3). Nonetheless, overexpression of UCP2/3 in endothelial cells also boosted mitochondrial Ca^{2+} signals fueled by SOCE, whereas cytosolic Ca^{2+} signals were not affected, indicating that an elevation of the mitochondrial Ca^{2+} uptake capacity does not necessarily augment SOCE (Trenker et al. 2007; Waldeck-Weiermair et al. 2010b). The actual molecular mechanism by which UCP2/3 contribute to mitochondrial Ca^{2+} under certain conditions was discovered recently when we demonstrated that UCP2/3 interact exclusively with methylated MICU1 via its IML2, thereby normalizing the apparent Ca^{2+} sensitivity of methylated MICU1 to that of the non-methylated MICU1 (Madreiter-Sokolowski et al. 2016b) and, thus, reestablishing normal MCU activity even under conditions of methylated MICU1. Hence, this function plays an important role in the energy metabolism of cancer cell and their protection against mitochondrial Ca^{2+} overload-triggered cell death (Madreiter-Sokolowski et al. 2016a, b).

In a genome-wide RNAi screening assay, a protein referred to as Letm1 (leucine-zipper-EF-hand-containing transmembrane region) was unexpectedly identified to function as a Ca^{2+}/H^{+} antiporter which is able to accomplish mitochondrial Ca^{2+} loading under physiological conditions (Jiang et al. 2009). Letm1 was initially characterized as a mitochondrial K^{+}/H^{+} exchanger, which controls osmotic swelling of mitochondria (Nowikovsky et al. 2004). In addition, Letm1 was supposed to contribute to the mitochondrial export machinery (Frazier et al. 2006) and the activity of OPA1, a protein that controls fusion events of the IMM and cristae remodeling (Piao et al. 2009). Although the mitochondrial Ca^{2+} transport function of Letm1 was confirmed later on as being essentially involved in the transmission of SOCE-derived Ca^{2+} into the mitochondria (Waldeck-Weiermair et al. 2011), this protein might probably exhibit a good example for a so-called moonlighting protein (Huberts and van der Klei 2010) that fulfils multiple functions related to mitochondrial (ultra)structure and ion homeostasis. Nevertheless, further studies are necessary to resolve the particular involvement of Letm1 in mitochondrial sequestration of SOCE-derived Ca^{2+} and to investigate the function of Letm1 in the interplay between SOCE and mitochondria.

16.2.1.3 Distinct Regulation of Mitochondrial Ca^{2+} Uptake Mechanism Depending on the Source of Ca^{2+}

Using siRNA-mediated knockdown of UCP2 and UCP3, it was shown that the contribution of these proteins to mitochondrial Ca^{2+} uniport is distinctly determined by the source and mode of Ca^{2+} mobilization (Waldeck-Weiermair et al. 2010a, b), while MCU, EMRE, and MICU1 are involved in any mitochondrial sequestration independently from the source of Ca^{2+} (Bondarenko et al. 2014). Interestingly, in cells with enhanced PRMT1 activity, mitochondrial Ca^{2+} uptake upon IP_3-induced Ca^{2+} release was strongly facilitated by UCP2/3, whereas the transfer of entering Ca^{2+} into mitochondria in intact endothelial cells was accomplished via an UCP2/3-independent pathway. Further results are indicative of a mitochondrial Ca^{2+} antiporter (Naghdi et al. 2010; Trenker et al. 2008) to achieve uptake of entering Ca^{2+} independently from UCP2/3-controlled mitochondrial Ca^{2+} uptake machineries in these endothelial cells (Fig. 16.1, panel VII_a). Moreover, data on the functional heterogeneity of the UCP2/3-dependent mitochondrial Ca^{2+} uniport in endothelial cells as well as recent results on the coexistence of at least two dissimilar mitochondrial Ca^{2+} channels in cardiomyocytes (Michels et al. 2009) led to the introduction of a *slot hypothesis* of mitochondrial Ca^{2+} uptake in intact cells. According to this hypothesis, limited, optimized Ca^{2+} transfer sites between the ER and mitochondria exist (Waldeck-Weiermair et al. 2010b). These mitochondrial Ca^{2+} uptake sites essentially contain MCU, EMRE, and MICU1, and if MICU1 is methylated, UCP2/3 preferentially accumulates there possibly due to a basic residue at position 169 (UCP2)/171 (UCP3) in IML2. Moreover, the number of distinct mitochondrial uptake sites for entering Ca^{2+} may be only limited by the expression of mitochondrial Ca^{2+} carriers that do not rely on UCP2/3 at moderate protein expression levels. However, upon upregulation, UCP2/3 become available for uptake of entering Ca^{2+} via the basic residue at position 164 (UCP2)/167 (UCP3) in the IML2, thus pointing to molecularly distinct sites (slots) for mitochondrial Ca^{2+} uptake depending on the source of Ca^{2+}. In summary, it was hypothesized that distinct molecular mechanisms of mitochondrial Ca^{2+} sequestration most likely meet the various demands of sufficient organelle Ca^{2+} loading from different Ca^{2+} sources in intact cells (Waldeck-Weiermair et al. 2010a, b).

According to the *slot hypothesis*, the impact of mitochondrial Ca^{2+} sequestration on ER Ca^{2+} release upon cell stimulation with an IP_3-generating agonist (Fig. 16.1, panel II, II_a), which consequently affects STIM1 oligomerization, clustering, and

SOCE activity (Fig. 16.1, panel I, IV, V, VI), is accomplished by distinct mitochondrial Ca^{2+} uptake sites (Fig. 16.1, panel II_a) that are different from those facing SOCE channels (Fig. 16.1, panel VII_a). This assumption further implicates that distinct mitochondrial Ca^{2+} uptake pathways might differentially contribute to the activation, maintenance, and termination of SOCE.

16.2.2 The Contribution of Mitochondrial Motility and Location to SOCE

Although ultimate proofs are missing, there is consensus that the predominant molecular mechanism by which mitochondria control SOCE activity is due to their ability to efficiently sequester (subplasmalemmal) Ca^{2+}. In this context it was suggested that mitochondria ought to be in close proximity to SOCE channels in order to efficiently prevent Ca^{2+}-dependent inactivation of SOCE (Bernardi 1999). This concept is further supported by the recent characterization of plasma membrane-associated membranes (PAM), which constitute an interaction platform for proteins of the plasma membrane, the ER, and, importantly, also the mitochondria (Koziel et al. 2009; Lebiedzinska et al. 2009). Using a combination of high-resolution fluorescence microscopy and patch clamp, it was indeed demonstrated that superficial mitochondria efficiently buffer subplasmalemmal Ca^{2+}. In this study, Ca^{2+}-activated K^+ channels in proximity of superficial mitochondria remained in an inactive state even upon strong cell stimulation with an IP_3-generating/Ca^{2+}-elevating agonist, while the channels got highly activated once mitochondria moved away from the channel (Malli et al. 2003a, b). These observations demonstrated that mitochondria are able to flexibly generate Ca^{2+} gradients and point to the importance of their subcellular distribution, morphology, and motility in order to contribute to the spatiotemporal pattern of subcellular Ca^{2+} signaling.

In line with these findings, it was recently reported that mitochondria actively redistribute toward the plasma membrane during Ca^{2+} entry in T-lymphocytes, which was essential for normal SOCE activity (Quintana et al. 2007). Excitingly, this result emphasizes that SOCE attracts mitochondria to relocate toward the SOCE channels, a mechanism that is feasible if one considers that Ca^{2+} controls the motility and shape of mitochondria (Fig. 16.4), probably via Miro GTPases (Saotome et al. 2008) and CaM kinase 1 alpha (Han et al. 2008), respectively.

In a recent study, mitochondria were artificially linked to the plasma membrane by expression of the designed construct mAKAP-RFP-CAAX (Naghdi et al. 2010) which consists of a mitochondrial AKAP domain, red fluorescent protein (RFP), as well as a CAAX box binding to the plasma membrane (Csordas et al. 2006; Liu et al. 2009). Expression of mAKAP-RFP-CAAX resulted in the immobilization of mitochondria at the inner site of the plasma membrane and was used to investigate the contribution of mitochondrial motility and subcellular location to SOCE in endothelial cells. Surprisingly, mitochondrial uptake of Ca^{2+} entering via the SOCE pathway was significantly reduced in cells expressing mAKAP-RFP-CAAX, while

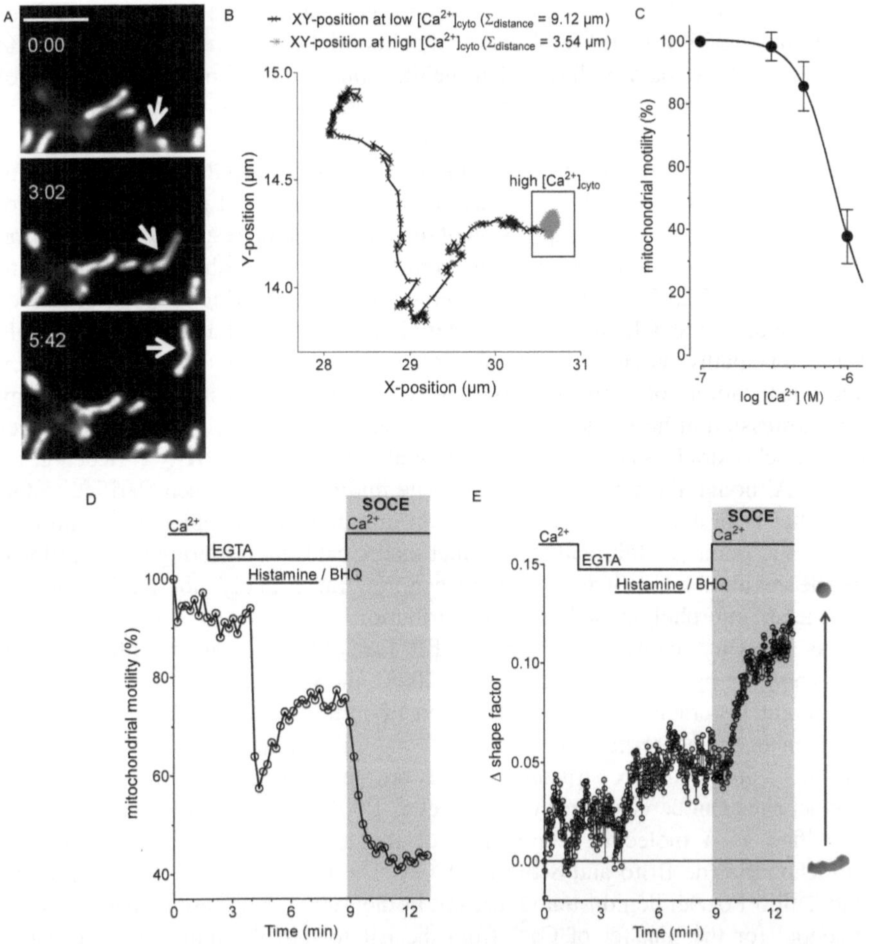

Fig. 16.4 Ca^{2+} controls mitochondrial motility and morphology. The impact of Ca^{2+} on mitochondrial motility and morphology was investigated in endothelial cells (EAhy926) transiently expressing mitochondrially targeted DsRed using a Nipkow-based array confocal laser scanning microscope. Panel **a**: Images represent a section of moving mitochondria at different time points as indicated. The *scale bar* of the *upper image* is 3 μm. Panel **b**: Analysis of mitochondrial motility by plotting the X and Y position of a single mitochondrion within 6 min prior to cell stimulation with the IP_3-generating agonist histamine (*black line*) and in the presence of the Ca^{2+} mobilizing agonist (*gray line* within the *rectangle*). Panel **c**: The Ca^{2+} concentration dependency of the inhibition of global mitochondrial dynamics in EAhy926 cells is presented (IC_{50} = 0.83 (0.68–1.0) μM). Experiments were performed with ionomycin-permeabilized (1 μM) cells. Panel **d**: The global mitochondrial motility of intact cells was analyzed during cell stimulation with 100 μM histamine and 15 μM BHQ in the absence of extracellular Ca^{2+} and upon readdition of 2 mM Ca^{2+}. Notably, the global mitochondrial motility was transiently inhibited by ER Ca^{2+} depletion due to a transient increase of the cytosolic Ca^{2+} concentration and considerably decreased upon SOCE. Panel **e**: Changes of the average shape factor (SF) that is defined as $SF = 4\pi A/P^2$ (where P is the perimeter and A the area of respective mitochondria within one given cell) over time. As indicated, cells were first stimulated with 100 μM histamine and 15 μM

the SOCE-induced cytosolic Ca^{2+} elevation was not affected. However under these conditions SOCE was still sensitive to mitochondrial depolarization with FCCP. In summary, these data indicate that neither the motility nor the location of mitochondria is of crucial importance for activation and maintenance of SOCE in the endothelial cell line used (Naghdi et al. 2010).

The importance of mitochondrial morphology and subcellular distribution for SOCE activity was further investigated by overexpression of proteins that control mitochondrial fission. Overexpression of dynamitin, which results in fragmentation and relocation of mitochondria away from the plasma membrane, was shown to attenuate SOCE in HeLa cells (Varadi et al. 2004), confirming that the vicinity of mitochondria to SOCE channels is a determinant of this particular Ca^{2+}-inhibitable Ca^{2+} entry pathway. However, in the same cell type, a pronounced fragmentation and dislocation of mitochondria away from the plasma membrane by overexpression of hFis, a key protein of the outer mitochondrial membrane inducing mitochondrial fission, only marginally affected SOCE activity (Frieden et al. 2004). Although the cause for such opposite findings has not been clarified so far, the latter is in agreement with the data on immobilized mitochondria mentioned above and point to additional facets other than local Ca^{2+} buffering that contribute to the regulation of SOCE by mitochondria, most likely not relying on the organelles' morphology and cellular distribution.

As ER Ca^{2+} depletion as well as ER Ca^{2+} refilling can be influenced by mitochondria by various means (Parekh 2008), the vicinity of mitochondria to the ER might be crucial for the contribution of mitochondria to SOCE activation, maintenance, and termination. The ER and mitochondria are distinct, highly dynamic, and complex structures which are severalfold interconnected. These contact sites can be visualized (Csordas et al. 2006), and mitofusin 2 was recently identified as a molecular component of the tethers connecting the ER with mitochondria (de Brito and Scorrano 2008). The laboratories of Rosario Rizzuto and Tullio Pozzan demonstrated already in the 1990s that these contact sites are essential for the transfer of Ca^{2+} from the ER toward the mitochondrial matrix (Rizzuto et al. 1998). Hence, mitochondria are known to impact ER Ca^{2+} release as mitochondrial Ca^{2+} withdrawal from ER Ca^{2+} release channels influences the extent of ER Ca^{2+} depletion (Fig. 16.1, panel II, II_a). This can either suppress (Hajnoczky et al. 1999) or boost (Gilabert et al. 2001) ER Ca^{2+} depletion, depending on the type of ER Ca^{2+} release channels that are in the scope of mitochondrial Ca^{2+} uptake sites. In addition, mitochondria were also shown to facilitate the transfer of Ca^{2+} to SERCA in order to enforce ER Ca^{2+} replenishment (Fig. 16.1, panel III, III_a) (Arnaudeau et al. 2001; Malli et al. 2005). Accordingly,

Fig. 16.4 (continued) BHQ in the absence of extracellular Ca^{2+} (1 mM EGTA) and upon readdition of 2 mM Ca^{2+}, which yields SOCE-induced cytosolic Ca^{2+} elevation. Notably, the ΔSF significantly increased upon SOCE. As indicated, an increase of ΔSF correlates to the transformation of flattened structures toward more spherical ones

mitochondria are able to control the ER Ca^{2+} content in a variety of ways and subsequently affect the activation, maintenance, and deactivation of SOCE (Fig. 16.1, panels II_a, III_a) (Parekh 2008). However, the contribution of mitochondrial Ca^{2+} handling to the activation status of STIM1 proteins has not been investigated so far. Nevertheless, a recent study indicates that ATP depletion, induced by inhibition of the mitochondrial metabolic function, triggers STIM1 translocation and punctae formation, while SOCE remains inactive under these conditions, thus pointing to additional aspects of mitochondrial contribution to SOCE activity (Chvanov et al. 2008).

16.3 Conclusion

While there is common agreement that mitochondria essentially contribute to activity and/or maintenance of SOCE, the actual molecular mechanisms remain still unclear. Several possibilities ranging from subplasmalemmal Ca^{2+} buffering to the generation of mediators, local ATP modulation, and regulation of STIM1 multimerization/shuttling have been presented. Though the puzzle of the molecular mechanism of mitochondrial contribution to SOCE activity/maintenance has not been entirely solved yet, it is obvious that mitochondria are multifunctional organelles that affect cellular Ca^{2+} homeostasis and, thus, the activation of SOCE by multiple means and functions.

Acknowledgments The authors thank Karin Osibow for helpful comments and proofreading. Markus Waldeck-Weiermair provided data regarding the impact of Ca^{2+} on mitochondrial motility. The laboratory of WF Graier and R Malli is supported by the Austrian Science Funds, FWF (DKplus W 1226-B18 and P 28529-B27). Microscopic equipment is part of the Nikon *Center of Excellence*, Graz, that is supported by the Austrian infrastructure program 2013/2014, Nikon Austria Inc., and BioTechMed.

References

Abramov AY, Fraley C, Diao CT, Winkfein R, Colicos MA, Duchen MR, French RJ, Pavlov E (2007) Targeted polyphosphatase expression alters mitochondrial metabolism and inhibits calcium-dependent cell death. Proc Natl Acad Sci U S A 104(46):18091–18096

Aichberger KJ, Mittermann I, Reininger R, Seiberler S, Swoboda I, Spitzauer S, Kopp T, Stingl G, Sperr WR, Valent P, Repa A, Bohle B, Kraft D, Valenta R (2005) Hom s 4, an IgE-reactive autoantigen belonging to a new subfamily of calcium-binding proteins, can induce Th cell type 1-mediated autoreactivity. J Immunol 175(2):1286–1294

Alam MR, Groschner LN, Parichatikanond W, Kuo L, Bondarenko AI, Rost R, Waldeck-Weiermair M, Malli R, Graier WF (2012) Mitochondrial Ca^{2+} uptake 1 (MICU1) and mitochondrial Ca^{2+} uniporter (MCU) contribute to metabolism-secretion coupling in clonal pancreatic β-cells. J Biol Chem 287(41):34445–34454

Arnaudeau S, Kelley WL, Walsh JV Jr, Demaurex N (2001) Mitochondria recycle Ca^{2+} to the endoplasmic reticulum and prevent the depletion of neighboring endoplasmic reticulum regions. J Biol Chem 276(31):29430–29439

Baughman JM, Perocchi F, Girgis HS, Plovanich M, Belcher-Timme CA, Sancak Y, Bao R, Strittmatter L, Goldberger O, Bogorad RL, Koteliansky V, Mootha VK (2011) Integrative genomics identifies MCU as an essential component of the mitochondrial calcium uniporter. Nature 476(7360):341–345

Bernardi P (1999) Mitochondrial transport of cations: channels, exchangers, and permeability transition. Physiol Rev 79(4):1127–1155

Berridge MJ, Lipp P, Bootman MD (2000) The versatility and universality of calcium signalling. Nat Rev Mol Cell Biol 1(1):11–21

Bolotina VM (2008) Orai, STIM1 and iPLA2beta: a view from a different perspective. J Physiol (Lond) 586(13):3035–3042

Bolotina VM, Csutora P (2005) CIF and other mysteries of the store-operated Ca^{2+}-entry pathway. Trends Biochem Sci 30(7):378–387

Bondarenko AI, Jean-Quartier C, Malli R, Graier WF (2013) Characterization of distinct single-channel properties of Ca^{2+} inward currents in mitochondria. Pflugers Arch 465(7):997–1010

Bondarenko AI, Jean-Quartier C, Parichatikanond W, Alam MR, Waldeck-Weiermair M, Malli R, Graier WF (2014) Mitochondrial Ca^{2+} uniporter (MCU)-dependent and MCU-independent Ca^{2+} channels coexist in the inner mitochondrial membrane. Pflugers Arch 466(7):1411–1420

Bootman MD, Collins TJ, Peppiatt CM, Prothero LS, MacKenzie L, De Smet P, Travers M, Tovey SC, Seo JT, Berridge MJ, Ciccolini F, Lipp P (2001) Calcium signalling—an overview. Semin Cell Dev Biol 12(1):3–10

Brandman O, Liou J, Park WS, Meyer T (2007) STIM2 is a feedback regulator that stabilizes basal cytosolic and endoplasmic reticulum Ca^{2+} levels. Cell 131(7):1327–1339

Braschi E, McBride HM (2010) Mitochondria and the culture of the Borg: understanding the integration of mitochondrial function within the reticulum, the cell, and the organism. Bioessays 32(11):958–966

Carafoli E (2002) Calcium signaling: a tale for all seasons. Proc Natl Acad Sci U S A 99 (3):1115–1122

Charoensin S, Eroglu E, Opelt M, Bischof H, Madreiter-Sokolowski CT, Kirsch A, Depaoli MR, Frank S, Schrammel A, Mayer B, Waldeck-Weiermair M, Graier WF, Malli R (2017) Intact mitochondrial Ca^{2+} uniport is essential for agonist-induced activation of endothelial nitric oxide synthase (eNOS). Free Radic Biol Med 102:248–259

Chvanov M, Walsh CM, Haynes LP, Voronina SG, Lur G, Gerasimenko OV, Barraclough R, Rudland PS, Petersen OH, Burgoyne RD, Tepikin AV (2008) ATP depletion induces translocation of STIM1 to puncta and formation of STIM1-ORAI1 clusters: translocation and re-translocation of STIM1 does not require ATP. Pflugers Arch 457(2):505–517

Cosson P, Marchetti A, Ravazzola M, Orci L (2012) Mitofusin-2 independent juxtaposition of endoplasmic reticulum and mitochondria: an ultrastructural study. PLoS One 7:e46293

Csordas G, Renken C, Varnai P, Walter L, Weaver D, Buttle KF, Balla T, Mannella CA, Hajnoczky G (2006) Structural and functional features and significance of the physical linkage between ER and mitochondria. J Cell Biol 174(7):915–921

Csordas G, Varnai P, Golenar T, Roy S, Purkins G, Schneider TG, Balla T, Hajnoczky G (2010) Imaging interorganelle contacts and local calcium dynamics at the ER-mitochondrial interface. Mol Cell 39(1):121–132

Deak AT, Blass S, Khan MJ, Groschner LN, Waldeck-Weiermair M, Hallström S, Graier WF, Malli R (2014) IP3-mediated STIM1 oligomerization requires intact mitochondrial Ca^{2+} uptake. J Cell Sci 127(Pt 13):2944–2955

de Brito OM, Scorrano L (2008) Mitofusin 2 tethers endoplasmic reticulum to mitochondria. Nature 456(7222):605–610

DeHaven WI, Smyth JT, Boyles RR, Putney JW Jr (2007) Calcium inhibition and calcium potentiation of Orai1, Orai2, and Orai3 calcium release-activated calcium channels. J Biol Chem 282(24):17548–17556

DeLuca H, Engstrom G (1961) Calcium uptake by rat kidney mitochondria. Proc Natl Acad Sci U S A 47:1744–1750

Demaurex N, Poburko D, Frieden M (2009) Regulation of plasma membrane calcium fluxes by mitochondria. Biochim Biophys Acta 1787(11):1383–1394

De Stefani D, Rizzuto R (2015) Structure and function of the mitochondrial calcium uniporter complex. Biochim Biophys Acta 1853(9):2006–2011

De Stefani D, Raffaello A, Teardo E, Szabo I, Rizzuto R (2011) A forty-kilodalton protein of the inner membrane is the mitochondrial calcium uniporter. Nature 476(7360):336–340

Dolmetsch RE, Xu K, Lewis RS (1998) Calcium oscillations increase the efficiency and specificity of gene expression. Nature 392(6679):933–936

Duchen MR, Verkhratsky A, Muallem S (2008) Mitochondria and calcium in health and disease. Cell Calcium 44(1):1–5

Elustondo PA, Nichols M, Robertson GS, Pavlov EV (2016) Mitochondrial Ca^{2+} uptake pathways. J Bioenerg Biomembr 49(1):113–119

Feldman B, Fedida-Metula S, Nita J, Sekler I, Fishman D (2010) Coupling of mitochondria to store-operated Ca^{2+}-signaling sustains constitutive activation of protein kinase B/Akt and augments survival of malignant melanoma cells. Cell Calcium 47(6):525–537

Feske S (2007) Calcium signalling in lymphocyte activation and disease. Nat Rev Immunol 7 (9):690–702

Fieni F, Lee SB, Jan YN, Kirichok Y (2012) Activity of the mitochondrial calcium uniporter varies greatly between tissues. Nat Commun 3:1317

Filadi R, Greotti E, Turacchio G, Luini A, Pozzan T, Pizzo P (2015) Mitofusin 2 ablation increases endoplasmic reticulum-mitochondria coupling. Proc Natl Acad Sci U S A 112(17):E2174–E2181

Fomina AF, Nowycky MC (1999) A current activated on depletion of intracellular Ca^{2+} stores can regulate exocytosis in adrenal chromaffin cells. J Neurosci 19(10):3711–3722

Frazier AE, Taylor RD, Mick DU, Warscheid B, Stoepel N, Meyer HE, Ryan MT, Guiard B, Rehling P (2006) Mdm38 interacts with ribosomes and is a component of the mitochondrial protein export machinery. J Cell Biol 172(4):553–564

Frieden M, James D, Castelbou C, Danckaert A, Martinou JC, Demaurex N (2004) Ca^{2+} homeostasis during mitochondrial fragmentation and perinuclear clustering induced by hFis1. J Biol Chem 279(21):22704–22714

Frischauf I, Schindl R, Derler I, Bergsmann J, Fahrner M, Romanin C (2008) The STIM/Orai coupling machinery. Channels (Austin) 2(4):261–268

Giacomello M, Drago I, Bortolozzi M, Scorzeto M, Gianelle A, Pizzo P, Pozzan T (2010) Ca^{2+} hot spots on the mitochondrial surface are generated by Ca^{2+} mobilization from stores, but not by activation of store-operated Ca^{2+} channels. Mol Cell 38(2):280–290

Gilabert JA, Parekh AB (2000) Respiring mitochondria determine the pattern of activation and inactivation of the store-operated Ca^{2+} current ICRAC. EMBO J 19(23):6401–6407

Gilabert JA, Bakowski D, Parekh AB (2001) Energized mitochondria increase the dynamic range over which inositol 1,4,5-trisphosphate activates store-operated calcium influx. EMBO J 20 (11):2672–2679

Graier WF, Frieden M, Malli R (2007) Mitochondria and Ca^{2+} signaling: old guests, new functions. Pflugers Arch 455(3):375–396

Graier W, Trenker M, Malli R (2008) Mitochondrial Ca^{2+}, the secret behind the function of uncoupling proteins 2 and 3? Cell Calcium 44(1):36–50

Gunter TE, Sheu SS (2009) Characteristics and possible functions of mitochondrial Ca^{2+} transport mechanisms. Biochim Biophys Acta 1787(11):1291–1308

Gunter TE, Buntinas L, Sparagna G, Eliseev R, Gunter K (2000) Mitochondrial calcium transport: mechanisms and functions. Cell Calcium 28(5–6):285–296

Hajnoczky G, Csordas G (2010) Calcium signalling: fishing out molecules of mitochondrial calcium transport. Curr Biol 20(20):R888–R891

Hajnoczky G, Hager R, Thomas AP (1999) Mitochondria suppress local feedback activation of inositol 1,4, 5-trisphosphate receptors by Ca^{2+}. J Biol Chem 274(20):14157–14162

Hajnoczky G, Csordas G, Das S, Garcia-Perez C, Saotome M, Sinha Roy S, Yi M (2006) Mitochondrial calcium signalling and cell death: approaches for assessing the role of mitochondrial Ca^{2+} uptake in apoptosis. Cell Calcium 40(5–6):553–560

Han XJ, Lu YF, Li S-A, Kaitsuka T, Sato Y, Tomizawa K, Nairn AC, Takei K, Matsui H, Matsushita M (2008) CaM kinase I alpha-induced phosphorylation of Drp1 regulates mitochondrial morphology. J Cell Biol 182(3):573–585

Hooper R, Samakai E, Kedra J, Soboloff J (2013) Multifaceted roles of STIM proteins. Pflugers Arch 465(10):1383–1396

Hoth M, Fanger CM, Lewis RS (1997) Mitochondrial regulation of store-operated calcium signaling in T lymphocytes. J Cell Biol 137(3):633–648

Hoth M, Button DC, Lewis RS (2000) Mitochondrial control of calcium-channel gating: a mechanism for sustained signaling and transcriptional activation in T lymphocytes. Proc Natl Acad Sci U S A 97(19):10607–10612

Huberts DHEW, van der Klei IJ (2010) Moonlighting proteins: an intriguing mode of multitasking. Biochim Biophys Acta 1803(4):520–525

Jiang D, Zhao L, Clapham DE (2009) Genome-wide RNAi screen identifies Letm1 as a mitochondrial Ca^{2+}/H^{+} antiporter. Science 326(5949):144–147

Kamer KJ, Mootha VK (2014) MICU1 and MICU2 play nonredundant roles in the regulation of the mitochondrial calcium uniporter. EMBO Rep 15(3):299–307

Kamer KJ, Sancak Y, Mootha VK (2014) The uniporter: from newly identified parts to function. Biochem Biophys Res Commun 449(4):370–372

Kirichok Y, Krapivinsky G, Clapham DE (2004) The mitochondrial calcium uniporter is a highly selective ion channel. Nature 427(6972):360–364

Knot HJ, Laher I, Sobie EA, Guatimosim S, Gomez-Viquez L, Hartmann H, Song LS, Lederer WJ, Graier WF, Malli R, Frieden M, Petersen OH (2005) Twenty years of calcium imaging: cell physiology to dye for. Mol Interv 5(2):112–127

Koziel K, Lebiedzinska M, Szabadkai G, Onopiuk M, Brutkowski W, Wierzbicka K, Wilczynski G, Pinton P, Duszynski J, Zablocki K, Wieckowski MR (2009) Plasma membrane associated membranes (PAM) from Jurkat cells contain STIM1 protein is PAM involved in the capacitative calcium entry? Int J Biochem Cell Biol 41(12):2440–2449

Lampe PA, Cornbrooks EB, Juhasz A, Johnson EM, Franklin JL (1995) Suppression of programmed neuronal death by a thapsigargin-induced Ca^{2+} influx. J Neurobiol 26(2):205–212

Lebiedzinska M, Szabadkai G, Jones AWE, Duszynski J, Wieckowski MR (2009) Interactions between the endoplasmic reticulum, mitochondria, plasma membrane and other subcellular organelles. Int J Biochem Cell Biol 41(10):1805–1816

Lin S, Fagan KA, Li KX, Shaul PW, Cooper DM, Rodman DM (2000) Sustained endothelial nitric-oxide synthase activation requires capacitative Ca^{2+} entry. J Biol Chem 275 (24):17979–17985

Liou J, Kim ML, Heo WD, Jones JT, Myers JW, Ferrell JE Jr, Meyer T (2005) STIM is a Ca^{2+} sensor essential for Ca^{2+}-store-depletion-triggered Ca^{2+} influx. Curr Biol 15(13):1235–1241

Liu X, Weaver D, Shirihai O, Hajnoczky G (2009) Mitochondrial 'kiss-and-run': interplay between mitochondrial motility and fusion-fission dynamics. EMBO J 28(20):3074–3089

Madreiter-Sokolowski CT, Gottschalk B, Parichatikanond W, Eroglu E, Klec C, Waldeck-Weiermair M, Malli R, Graier WF (2016a) Resveratrol specifically kills cancer cells by a devastating increase in the Ca^{2+} coupling between the greatly tethered endoplasmic reticulum and mitochondria. Cell Physiol Biochem 39(4):1404–1420

Madreiter-Sokolowski CT, Klec C, Parichatikanond W, Stryeck S, Gottschalk B, Pulido S, Rost R, Eroglu E, Hofmann NA, Bondarenko AI, Madl T, Waldeck-Weiermair M, Malli R, Graier WF (2016b) PRMT1-mediated methylation of MICU1 determines the UCP2/3 dependency of mitochondrial Ca^{2+} uptake in immortalized cells. Nat Commun 7:12897

Malli R, Graier WF (2010) Mitochondrial Ca^{2+} channels: great unknowns with important functions. FEBS Lett 584(10):1942–1947

Malli R, Frieden M, Osibow K, Graier WF (2003a) Mitochondria efficiently buffer subplasmalemmal Ca^{2+} elevation during agonist stimulation. J Biol Chem 278 (12):10807–10815
Malli R, Frieden M, Osibow K, Zoratti C, Mayer M, Demaurex N, Graier WF (2003b) Sustained Ca^{2+} transfer across mitochondria is essential for mitochondrial Ca^{2+} buffering, store-operated Ca^{2+} entry, and Ca^{2+} store refilling. J Biol Chem 278(45):44769–44779
Malli R, Frieden M, Trenker M, Graier WF (2005) The role of mitochondria for Ca^{2+} refilling of the endoplasmic reticulum. J Biol Chem 280(13):12114–12122
Malli R, Naghdi S, Romanin C, Graier WF (2008) Cytosolic Ca^{2+} prevents the subplasmalemmal clustering of STIM1: an intrinsic mechanism to avoid Ca^{2+} overload. J Cell Sci 121 (Pt 19):3133–3139
Mallilankaraman K, Cardenas C, Doonan PJ, Chandramoorthy HC, Irrinki KM, Golenar T, Csordas G, Madireddi P, Yang J, Müller M, Miller R, Kolesar JE, Molgo J, Kaufman B, Hajnoczky G, Foskett JK, Madesh M (2012) MCUR1 is an essential component of mitochondrial Ca^{2+} uptake that regulates cellular metabolism. Nat Cell Biol 14(12):1336–1343
McBride HM, Neuspiel M, Wasiak S (2006) Mitochondria: more than just a powerhouse. Curr Biol 16(14):R551–R560
McFadzean I, Gibson A (2002) The developing relationship between receptor-operated and store-operated calcium channels in smooth muscle. Br J Pharmacol 135(1):1–13
Michalak M, Robert Parker JM, Opas M (2002) Ca^{2+} signaling and calcium binding chaperones of the endoplasmic reticulum. Cell Calcium 32(5–6):269–278
Michels G, Khan IF, Endres-Becker J, Rottlaender D, Herzig S, Ruhparwar A, Wahlers T, Hoppe UC (2009) Regulation of the human cardiac mitochondrial Ca^{2+} uptake by 2 different voltage-gated Ca^{2+} channels. Circulation 119(18):2435–2443
Montalvo GB, Artalejo AR, Gilabert JA (2006) ATP from subplasmalemmal mitochondria controls Ca^{2+}-dependent inactivation of CRAC channels. J Biol Chem 281(47):35616–35623
Muik M, Frischauf I, Derler I, Fahrner M, Bergsmann J, Eder P, Schindl R, Hesch C, Polzinger B, Fritsch R, Kahr H, Madl J, Gruber H, Groschner K, Romanin C (2008) Dynamic coupling of the putative coiled-coil domain of ORAI1 with STIM1 mediates ORAI1 channel activation. J Biol Chem 283(12):8014–8022
Nagai T, Sawano A, Park ES, Miyawaki A (2001) Circularly permuted green fluorescent proteins engineered to sense Ca^{2+}. Proc Natl Acad Sci U S A 98(6):3197–3202
Naghdi S, Waldeck-Weiermair M, Fertschai I, Poteser M, Graier WF, Malli R (2010) Mitochondrial Ca^{2+} uptake and not mitochondrial motility is required for STIM1-Orai1-dependent store-operated Ca^{2+} entry. J Cell Sci 123(Pt 15):2553–2564
Nicholls DG (2005) Mitochondria and calcium signaling. Cell Calcium 38(3–4):311–317
Nowikovsky K, Froschauer EM, Zsurka G, Samaj J, Reipert S, Kolisek M, Wiesenberger G, Schweyen RJ (2004) The LETM1/YOL027 gene family encodes a factor of the mitochondrial K^{+} homeostasis with a potential role in the Wolf-Hirschhorn syndrome. J Biol Chem 279 (29):30307–30315
Osibow K, Frank S, Malli R, Zechner R, Graier WF (2006) Mitochondria maintain maturation and secretion of lipoprotein lipase in the endoplasmic reticulum. Biochem J 396(1):173–182
Palty R, Silverman WF, Hershfinkel M, Caporale T, Sensi SL, Parnis J, Nolte C, Fishman D, Shoshan-Barmatz V, Herrmann S, Khananshvili D, Sekler I (2010) NCLX is an essential component of mitochondrial Na^{+}/Ca^{2+} exchange. Proc Natl Acad Sci U S A 107(1):436–441
Parekh AB (1998) Slow feedback inhibition of calcium release-activated calcium current by calcium entry. J Biol Chem 273(24):14925–14932
Parekh AB (2008) Mitochondrial regulation of store-operated CRAC channels. Cell Calcium 44 (1):6–13
Parekh AB, Putney JW Jr (2005) Store-operated calcium channels. Physiol Rev 85(2):757–810
Patron M, Checchetto V, Raffaello A, Teardo E, Vecellio Reane D, Mantoan M, Granatiero V, Szabo I, De Stefani D, Rizzuto R (2014) MICU1 and MICU2 finely tune the mitochondrial Ca^{2+} uniporter by exerting opposite effects on MCU activity. Mol Cell 53(5):726–737

Penna A, Demuro A, Yeromin AV, Zhang SL, Safrina O, Parker I, Cahalan MD (2008) The CRAC channel consists of a tetramer formed by Stim-induced dimerization of Orai dimers. Nature 456 (7218):116–120
Perocchi F, Gohil VM, Girgis HS, Bao XR, McCombs JE, Palmer AE, Mootha VK (2010) MICU1 encodes a mitochondrial EF hand protein required for Ca^{2+} uptake. Nature 467 (7313):291–296
Piao L, Li Y, Kim SJ, Sohn KC, Yang KJ, Park KA, Byun HS, Won M, Hong J, Hur GM, Seok JH, Shong M, Sack R, Brazil DP, Hemmings BA, Park J (2009) Regulation of OPA1-mediated mitochondrial fusion by leucine zipper/EF-hand-containing transmembrane protein-1 plays a role in apoptosis. Cell Signal 21(5):767–777
Plovanich M, Bogorad RL, Sancak Y, Kamer KJ, Strittmatter L, Li AA, Girgis HS, Kuchimanchi S, De Groot S, Speciner L, Taneja N, OShea J, Koteliansky V, Mootha VK (2013) MICU2, a paralog of MICU1, resides within the mitochondrial uniporter complex to regulate calcium handling. PLoS One 8:e55785
Potier M, Gonzalez JC, Motiani RK, Abdullaev IF, Bisaillon JM, Singer HA, Trebak M (2009) Evidence for STIM1- and Orai1-dependent store-operated calcium influx through ICRAC in vascular smooth muscle cells: role in proliferation and migration. FASEB J 23(8):2425–2437
Putney JW Jr (1986) A model for receptor-regulated calcium entry. Cell Calcium 7(1):1–12
Putney JW Jr (1990) Capacitative calcium entry revisited. Cell Calcium 11(10):611–624
Putney JW Jr (1991) The capacitative model for receptor-activated calcium entry. Adv Pharmacol 22:251–269
Quintana A, Schwindling C, Wenning AS, Becherer U, Rettig J, Schwarz EC, Hoth M (2007) T cell activation requires mitochondrial translocation to the immunological synapse. Proc Natl Acad Sci U S A 104(36):14418–14423
Raffaello A, De Stefani D, Sabbadin D, Teardo E, Merli G, Picard A, Checchetto V, Moro S, Szabo I, Rizzuto R (2013) The mitochondrial calcium uniporter is a multimer that can include a dominant-negative pore-forming subunit. EMBO J 32(17):2362–2376
Rizzuto R, Simpson AW, Brini M, Pozzan T (1992) Rapid changes of mitochondrial Ca^{2+} revealed by specifically targeted recombinant aequorin. Nature 358(6384):325–327
Rizzuto R, Pinton P, Carrington W, Fay FS, Fogarty KE, Lifshitz LM, Tuft RA, Pozzan T (1998) Close contacts with the endoplasmic reticulum as determinants of mitochondrial Ca^{2+} responses. Science 280(5370):1763–1766
Roos J, DiGregorio PJ, Yeromin AV, Ohlsen K, Lioudyno M, Zhang S, Safrina O, Kozak JA, Wagner SL, Cahalan MD, Velicelebi G, Stauderman KA (2005) STIM1, an essential and conserved component of store-operated Ca^{2+} channel function. J Cell Biol 169(3):435–445
Sancak Y, Markhard AL, Kitami T, Kovacs-Bogdan E, Kamer KJ, Udeshi ND, Carr SA, Chaudhuri D, Clapham DE, Li AA, Calvo SE, Goldberger O, Mootha VK (2013) EMRE is an essential component of the mitochondrial calcium uniporter complex. Science 342 (6164):1379–1382
Saotome M, Safiulina D, Szabadkai G, Das S, Fransson A, Aspenstrom P, Rizzuto R, Hajnoczky G (2008) Bidirectional Ca^{2+}-dependent control of mitochondrial dynamics by the Miro GTPase. Proc Natl Acad Sci U S A 105(52):20728–20733
Scarpa A, Graziotti P (1973) Mechanisms for intracellular calcium regulation in heart. I. Stopped-flow measurements of Ca^{2+} uptake by cardiac mitochondria. J Gen Physiol 62(6):756–772
Singaravelu K, Nelson C, Bakowski D, Martins de Brito O, Ng SW, Di Capite J, Powell T, Scorrano L, Parekh AB (2011) Mitofusin 2 regulates STIM1 migration from the Ca^{2+} store to the plasma membrane in cells with depolarised mitochondria. J Biol Chem 286 (14):12189–12201
Soboloff J, Rothberg BS, Madesh M, Gill DL (2012) STIM proteins: dynamic calcium signal transducers. Nat Rev Mol Cell Biol 13(9):549–565
Spät A, Szanda G, Csordas G, Hajnoczky G (2008) High- and low-calcium-dependent mechanisms of mitochondrial calcium signalling. Cell Calcium 44(1):51–63

Szabadkai G, Duchen MR (2008) Mitochondria: the hub of cellular Ca^{2+} signaling. Physiology (Bethesda) 23:84–94

Tomar D, Dong Z, Shanmughapriya S, Koch DA, Thomas T, Hoffman NE, Timbalia SA, Goldman SJ, Breves SL, Corbally DP, Nemani N, Fairweather JP, Cutri AR, Zhang X, Song J, Jan F, Huang J, Barrero C, Rabinowitz JE, Luongo TS, Schumacher SM, Rockman ME, Dietrich A, Merali S, Caplan J, Stathopulos P, Ahima RS, Cheung JY, Houser SR, Koch WJ, Patel V, Gohil VM, Elrod JW, Rajan S, Madesh M (2016) MCUR1 is a scaffold factor for the MCU complex function and promotes mitochondrial bioenergetics. Cell Rep 15(8):1673–1685

Trenker M, Malli R, Fertschai I, Levak-Frank S, Graier WF (2007) Uncoupling proteins 2 and 3 are fundamental for mitochondrial Ca^{2+} uniport. Nat Cell Biol 9(4):445–452

Trenker M, Fertschai I, Malli R, Graier WF (2008) UCP2/3—likely to be fundamental for mitochondrial Ca^{2+} uniport. Nat Cell Biol 10(11):1237–1240

Varadi A, Cirulli V, Rutter GA (2004) Mitochondrial localization as a determinant of capacitative Ca^{2+} entry in HeLa cells. Cell Calcium 36(6):499–508

Varnai P, Toth B, Toth DJ, Hunyady L, Balla T (2007) Visualization and manipulation of plasma membrane-endoplasmic reticulum contact sites indicates the presence of additional molecular components within the STIM1-Orai1 Complex. J Biol Chem 282(40):29678–29690

Varnai P, Hunyady L, Balla T (2009) STIM and Orai: the long-awaited constituents of store-operated calcium entry. Trends Pharmacol Sci 30(3):118–128

Vasington FD, Murphy J (1962) Ca^{2+} ion uptake by rat kidney mitochondria and its dependence on respiration and phosphorylation. J Biol Chem 237:2670–2677

Waldeck-Weiermair M, Duan X, Naghdi S, Khan MJ, Trenker M, Malli R, Graier WF (2010a) Uncoupling protein 3 adjusts mitochondrial Ca^{2+} uptake to high and low Ca^{2+} signals. Cell Calcium 48(5):288–301

Waldeck-Weiermair M, Malli R, Naghdi S, Trenker M, Kahn MJ, Graier WF (2010b) The contribution of UCP2 and UCP3 to mitochondrial Ca^{2+} uptake is differentially determined by the source of supplied Ca^{2+}. Cell Calcium 47(5):433–440

Waldeck-Weiermair M, Jean-Quartier C, Rost R, Khan MJ, Vishnu N, Bondarenko AI, Imamura H, Malli R, Graier WF (2011) The leucine zipper EF hand-containing transmembrane protein 1 (LETM1) and uncoupling proteins 2 and 3 (UCP2/3) contribute to two distinct mitochondrial Ca^{2+} uptake pathways. J Biol Chem 286(32):28444–28455

Waldeck-Weiermair M, Malli R, Parichatikanond W, Gottschalk B, Madreiter-Sokolowski CT, Klec C, Rost R, Graier WF (2015) Rearrangement of MICU1 multimers for activation of MCU is solely controlled by cytosolic Ca^{2+}. Sci Rep 5:15602

Walsh C, Barrow S, Voronina S, Chvanov M, Petersen OH, Tepikin A (2009) Modulation of calcium signalling by mitochondria. Biochim Biophys Acta 1787(11):1374–1382

Wang L, Yang X, Li S, Wang Z, Liu Y, Feng J, Zhu Y, Shen Y (2014) Structural and mechanistic insights into MICU1 regulation of mitochondrial calcium uptake. EMBO J 33(6):594–604

Wang L, Yang X, Shen Y (2015) Molecular mechanism of mitochondrial calcium uptake. Cell Mol Life Sci 72(8):1489–1498

Zhang SL, Yu Y, Roos J, Kozak JA, Deerinck TJ, Ellisman MH, Stauderman KA, Cahalan MD (2005) STIM1 is a Ca^{2+} sensor that activates CRAC channels and migrates from the Ca^{2+} store to the plasma membrane. Nature 437(7060):902–905

Zweifach A, Lewis RS (1995) Slow calcium-dependent inactivation of depletion-activated calcium current. Store-dependent and -independent mechanisms. J Biol Chem 270 (24):14445–14451

Tissue Specificity: The Role of Organellar Membrane Nanojunctions in Smooth Muscle Ca^{2+} Signaling

17

Nicola Fameli, A. Mark Evans, and Cornelis van Breemen

Abstract

In this chapter we examine the importance of cytoplasmic nanojunctions—nanometer scale appositions between organellar membranes including the molecular transporters therein—to the cell signaling machinery, with specific

The online version of this chapter (doi:10.1007/978-3-319-57732-6_17) contains supplementary material, which is available to authorized users.

Animation caption: In this animation, we start from an extracellular view of one PM-SR nanospace. The PM is depicted in red. The light blue/green object is a portion of the SR. The dark blue hemispheres on it represent SERCA pumps and the yellow objects on the PM represent NCX. The white sphere within the nanospace is one Ca^{2+} undergoing three-dimensional random-walk motion. As the animation progresses, we are flying under the PM and inside the space between the PM and SR membrane and eventually out again. All the elements in this model are to scale, except for Ca^{2+}, whose radius is ten times its Bohr radius for visibility.

N. Fameli (✉)
Institute of Biophysics, Medical University of Graz, 8010 Graz, Austria

Department of Anesthesiology, Pharmacology and Therapeutics, The University of British Columbia, 2176 Health Sciences Mall, Vancouver, BC, Canada, V6T 1Z3
e-mail: nicola.fameli@ubc.ca

A.M. Evans
Centre for Integrative Physiology, College of Medicine and Veterinary Medicine, University of Edinburgh, Hugh Robson Building, George Square, Edinburgh EH8 9XD, UK

C. van Breemen (✉)
Department of Anesthesiology, Pharmacology and Therapeutics, The University of British Columbia, 2176 Health Sciences Mall, Vancouver, BC, Canada, V6T 1Z3
e-mail: breemen@interchange.ubc.ca

K. Groschner et al. (eds.), *Store-Operated Ca^{2+} Entry (SOCE) Pathways*, Advances in Experimental Medicine and Biology 993, DOI 10.1007/978-3-319-57732-6_17

reference to Ca^{2+} transport and signaling in vascular smooth muscle and endothelial cells. More specifically, we will consider the extent to which quantitative modeling may aid in the development of our understanding of these processes. Testament to the requirement for such approaches lies in the fact that recent studies have provided evermore convincing evidence in support of the view that cytoplasmic nanospaces may be as significant to the process of Ca^{2+} signaling as the Ca^{2+} transporters, release channels, and Ca^{2+}-storing organelles themselves. Moreover, the disruption and/or dysfunction of cytoplasmic nanospaces may be central to the origin of certain diseases. By way of introduction, we provide a historical perspective on the identification of smooth muscle cell plasma membrane (PM)-sarcoplasmic reticulum (SR) nanospaces and the early evidence in support of their role in the generation of asynchronous Ca^{2+} waves. We then summarize how stochastic modeling approaches can aid and guide the development of our understanding of two basic functional steps leading to healthy smooth muscle cell contraction. We furthermore outline how more sophisticated and realistic quantitative stochastic modeling may be employed not only to test working hypotheses, but also to lead in their development in a manner that informs further experimental investigation. Finally, we consider more recently defined nanospaces such as the lysosome-SR junction, by way of demonstrating the importance of quantitative stochastic modeling to our understanding of signaling mechanisms.

Keywords

Cytoplasmic nanojunctions • ER junctions • Calcium signaling • Smooth muscle • Endothelium • Lysosomes • Stochastic modeling

17.1 Introduction

Nature has designed two types of signaling systems: the "lock and key" and the "currency." Enzymes/substrates and receptors/agonists clearly belong to the former, while Ca^{2+} may be considered the analogue of money as its effects depend on where and when it is deposited. In other words, since Ca^{2+} is the universal signaling ion, the specific information contained within the Ca^{2+} signal depends on its temporal and spatial characteristics. Thereby, in smooth muscle a transient rise in the intracellular Ca^{2+} concentration ($[Ca^{2+}]_i$) can activate: calmodulin tethered to myofilaments to cause contraction, Ca^{2+}-activated K^+ channels (K_{Ca}) in the plasma membrane (PM) to promote hyperpolarization and vasodilation, refilling of the sarcoplasmic reticulum (SR) in order to maintain asynchronous Ca^{2+} oscillations, mitochondrial dehydrogenases to stimulate energy metabolism, mitochondrial cytochrome C release to cause apoptosis, and a number of as yet imperfectly defined transcription factors and ion channels to stimulate proliferation and migration. Although a quantitative, mechanistic description of how the cell accomplishes such site- and function-specific Ca^{2+} signaling is still in its infancy, we do know that it depends on the restricted diffusion of Ca^{2+} within cytoplasmic nanojunctions that are defined by membrane domains of, for example, the PM, SR, lysosomes, Golgi apparatus, and mitochondria and by the presence of the appropriate ion channels, pumps, and Ca^{2+} receptors in these domains. Indeed, as we learn in other chapters in this book (e.g., Chaps. 12 and 15), a similar type of junctional transport is also a key feature of store-operated Ca^{2+} entry (SOCE) in non-excitable cells. In fact, SOCE appears to occur principally by the tight communication between the stromal interaction molecules (STIM) clustering in endoplasmic reticulum (ER) membrane patches at the cell periphery and sets of Orai channels localized in PM domains apposing such ER patches.

Since the aforementioned functions, subject to direct regulation by Ca^{2+}, encompass much of smooth muscle physiology and pathology, it would require the volume of a book to do it justice. We therefore limit the present introduction to the biophysics of smooth muscle nanojunctions to the first of the documented cases and the one area in which an identified function has been modeled, namely, the PM-SR junction mediating refilling of the SR with Ca^{2+}. Within this example we provide a simulation of cation fluxes in the PM-SR junctions, which sustain asynchronous inositol trisphosphate (IP_3)-initiated Ca^{2+} waves, before briefly considering alternative modes of smooth muscle activation involving lysosome-SR junctions and SR Ca^{2+} release via ryanodine receptors (RyRs).

17.2 Historical Perspective: Identification of Smooth Muscle Nanospace Structure and Function

In the early 1970s, Gabella (Gabella 1971) and Somlyo (Devine et al. 1972) first described the very narrow cytoplasmic spaces located between the PM and the superficial SR, especially in the regions of caveolae, and postulated that the SR played a role in smooth muscle activation. The regular narrow width of the superficial SR lumen suggests that it is quilt-like rather than tubular in nature, as we will demonstrate below. At the PM-SR junction, the space between the two membranes can also be regarded as very thin and undulating. In 1977, one of us (CvB) noted that such a space of restricted diffusion had important functional consequences with respect to regulation of smooth muscle contractility in that the superficial SR was able to modulate the Ca^{2+} flux from the PM to the myofilaments (van Breemen 1977). Our experiments showed that the same depolarization-activated net Ca^{2+} uptake would stimulate large, rapid force development if the Ca^{2+} entered fast, but only a small, slow contraction if Ca^{2+} entry was slow, but of longer duration. This indicated that the sarcoplasmic/endoplasmic Ca^{2+}-ATPase (SERCA) pumps of the superficial SR were saturated during high rates of Ca^{2+} influx consequent to marked voltage-gated Ca^{2+} channel (VGCC) activation but provided for effective removal of local Ca^{2+} at lower rates of influx. Those experiments were also able to demonstrate that, at the same rate of Ca^{2+} entry, the rate of force development was decreased by upregulating SR Ca^{2+} uptake and increased by downregulating it. These data led to the postulation of the "Superficial Buffer Barrier" theory, which states that Ca^{2+} entry through the PM into the deeper myoplasm can be attenuated by SR Ca^{2+} uptake from a peripheral cytoplasmic space between the superficial SR and PM, in a manner characterized by restricted Ca^{2+} diffusion (van Breemen and Saida 1989). This theory has now been confirmed by a number of laboratories (Wray and Burdyga 2010), and the concept of functionally important nanojunctions between organelles is becoming central in elucidating mechanisms of smooth muscle Ca^{2+} signaling (PW Workshop 2010; Moore and Wasteneys 2012; Evans et al. 2016; van Breemen et al. 2013).

As mentioned above, there are many examples of site-specific Ca^{2+} signaling, but few have been subjected to appropriate quantitative stochastic modeling. Since we are dealing with very small spaces with volumes in the order of 10^{-18} L, containing at any time on average less than one calcium ion at resting $[Ca^{2+}]_i$, this problem does not lend itself to a deterministic quantitative treatment with bulk diffusion equations, which generally introduce too many adjustable coefficients to yield a unique solution. A stochastic approach is instead necessary, and for this reason, we will consider below two examples of stochastic nanospace modeling that have thus far been applied to smooth muscle: the PM-SR and the lysosome-SR junctions.

17.3 Mechanism of Asynchronous Ca^{2+} Waves: Experimental Evidence

In 1994 Iino first studied Ca^{2+} signaling in individual smooth muscle cells of the intact, isolated rat tail artery (Iino et al. 1994). He unexpectedly found that although the blood vessel as a whole responded to noradrenaline with tonic elevations in both Ca^{2+} concentration and force development, each individual smooth muscle cell (SMC) displayed Ca^{2+}oscillations, which were not synchronized with those of the surrounding cells. Summation of the asynchronous single cell Ca^{2+} oscillations yielded the steady-state elevations of $[Ca^{2+}]_i$ measured previously. The mechanism underlying these asynchronous oscillations is that of regenerative waves of SR Ca^{2+} release through IP_3 receptors (IP_3R) traveling along the length of the SMC. However, even though the activating Ca^{2+} is released from the SR, the oscillations are rapidly arrested by removal of extracellular Ca^{2+}, indicating the engagement of an active Ca^{2+} cycle involving both the SR and PM. In an extensive study of Ca^{2+} waves in the SMCs of the rabbit vena cava, we found that blockade of SERCA with cyclopiazonic acid (CPA) and/or receptor-operated Ca^{2+} channels (ROC) with SKF96365 abolished these Ca^{2+} oscillations; however, blockade of voltage-gated Ca^{2+} channels (VGCC) only decreased the wave frequency, but not amplitude (Lee et al. 2002). Somewhat surprisingly blockade of the Ca^{2+} influx mode of the Na^+/Ca^{2+} exchanger (NCX) with 10 μM KB-R7943 slowly, but completely, abolished the Ca^{2+} oscillations in the intact SMC, while it had little or no inhibitory effects on SERCA or RyR function. The ROC involved in this process, which in a SMC culture was identified as transient receptor potential canonical 6 (TRPC6) channel, is physiologically activated by diacylglycerol (DAG) and permeable to both Na^+ and Ca^{2+} (Poburko et al. 2007; Lemos et al. 2007). The fact that blockade of either the ROC or NCX inhibited refilling of the SR in cultured SMCs strongly suggests that the TRPC6-mediated Na^+ entry raises $[Na^+]_i$ sufficiently to switch NCX to Ca^{2+} influx mode and thereby deliver Ca^{2+} to SERCA in

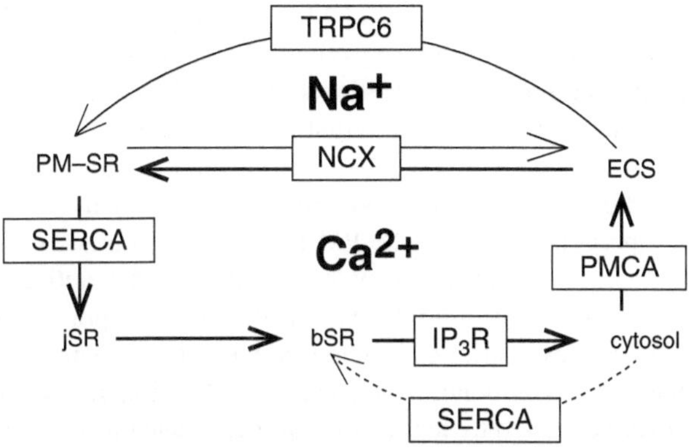

Fig. 17.1 Flow diagram for Ca^{2+} and Na^+ transport generating agonist-induced asynchronous Ca^{2+} oscillations in vascular smooth muscle. PM-SR represents the nanospace between the PM and SR at their junctions, ECS is extra-cellular space, jSR is junctional SR, bSR is bulk SR, and all other abbreviations are described in the text

the peripheral SR. Although no agonist-activated elevation of bulk $[Na^+]_i$ has been shown to be sufficient in magnitude to precipitate reverse mode NCX activity, CvB's laboratory has shown that activation of purinergic receptors in cultured SMCs causes local sub-PM $[Na^+]_i$ transients (LNats) sufficient for NCX reversal (Poburko et al. 2007). This finding is supported by a calculation of the threshold $[Na^+]_i$ for NCX reversal given by the relationship between the NCX reversal potential and the membrane potential (Blaustein and Lederer 1999).

At this point all data support the reaction scheme outlined in Fig. 17.1. SMC G-protein-linked receptor stimulation leads to activation of phospholipase C (PLC) yielding the products IP_3 and DAG. IP_3R releases Ca^{2+} from the SR to activate the myofilaments, while DAG activates TRPC6 in the PM overlying the superficial SR. Cytoplasmic Ca^{2+} is removed from the cytoplasm in large part by the PM-Ca^{2+}-ATPase (PMCA). Since the IP_3R are coactivated by both SR luminal and cytoplasmic Ca^{2+}, point release of SR Ca^{2+} synergizes with IP_3 and thus regeneratively activates adjacent IP_3Rs to cause a wave of Ca^{2+} release traveling along the entire length of the SMC. The release is periodic due to both time- and local Ca^{2+}-dependent inactivation of IP_3Rs, followed by cycles of Ca^{2+}-dependent activation subsequent to refilling at the PM-SR junctions, the speed of which determines the frequency of the Ca^{2+} oscillations. Refilling of the SR depends on Na^+ entering through ROC into the nanospace between the PM and the junctional SR, where it accumulates to a concentration sufficient to reverse NCX and thereby provide Ca^{2+} entry proximal to SERCA located in the SR membrane adjoining the PM. A population of peripheral mitochondria assists in the funneling of Ca^{2+} entry into the SR (Poburko et al. 2009). It is exactly this funneling of Ca^{2+} from the extracellular space, through the PM-SR junctions and into the SR, without raising the bulk $[Ca^{2+}]_i$, that allows the shape of the Ca^{2+} signal to be determined by SR Ca^{2+} release.

Our innovations introduced with this model are twofold: first, it specifies that refilling of the SR takes place at the PM-SR junctions and, second, that a critical step in this process is transient reversal of NCX due to generation of LNats at these same junctions. This model is admittedly complex and must therefore hold up to the following criteria: (1) removal of these junctions should abolish the asynchronous Ca^{2+} oscillations and (2) a realistic quantitative model based on experimental data should show that the above reaction scheme is indeed plausible. The first criterion has been addressed by exposing the smooth muscle tissue to calyculin A, which acts to separate the superficial SR from the PM and thereby abolishes the Ca^{2+} waves without compromising either activation of VGCCs in the PM or activation of SR Ca^{2+} release (Lee et al. 2005). Intriguingly, we have recently reported the absence of asynchronous Ca^{2+} waves in arteries isolated from elderly human patients. This lack of $[Ca^{2+}]_i$ oscillations was associated with a paucity of peripheral SR, suggesting that vascular aging may be related to degradation of membrane ultrastructure and loss of PM-SR junctions (Dai et al. 2010). In the next sections, we will explore how we addressed the second criterion.

17.4 PM-SR Junctions: Quantitative Modeling of the Active Ca^{2+} Cycle

It seems apparent from the experimental findings described thus far that only the few Ca^{2+} that populate the PM-SR nanospaces are responsible for modulating the signaling chain. This is quite remarkable as critical cellular functions pivot on the "behavior" of a handful of ions at any given time, albeit integrated over time. From this we may infer that, despite their small numbers, these ions can nonetheless cause the necessary concentration transients required to modulate transporter function. It is then evident that if it was not for the restricted volume provided by the nanojunction, there would be little capacity for such functional modulation. This is clear from more quantitative considerations. Typical nanospace volumes are of the order of 10^{-18} L, from which the simple calculation (1 ion)/(10^{-18} L)/N_A~2 μM, where N_A is Avogadro's number, tells us that an ion number change of one will be sufficient to precipitate a nanospace concentration change of the order of a few μM.

Inevitably then, at the dimensional scale of the nanojunctions, we are considering the possibility that stochastic variability plays a fundamental role in regulating intracellular Ca^{2+} signaling based on membrane-to-membrane communication. We could say that for efficient functioning of the cell signaling apparatus, the very geometry and scale of cytoplasmic nanojunctions are critical in taking advantage of the intrinsic sources of stochastic noise such as intracellular chemical reactions, transporter operation characteristics, and molecular diffusivity. Moreover, further understanding of the workings of cytoplasmic nanojunctions needs to be complemented by solid quantitative biophysical modeling.

In the following sections, we summarize our development of first-approximation quantitative stochastic models aimed at elucidating two important steps in the Ca^{2+} signaling cascade leading to contractile activation of vascular smooth muscle cells: the extracellular Ca^{2+} influx via reverse NCX during asynchronous $[Ca^{2+}]$ oscillations that serves to refill the SR store and the extracellular Na^+ influx event that allows the reversal of the NCX to occur. We will also briefly illustrate the latest directions we are taking to make our quantitative models more comprehensive and accurate, thereby giving them more predictive power.

17.4.1 Quantitative Modeling of One Mechanism for Refilling of the SR Ca^{2+} Store

One puzzling aspect of quantitatively addressing the problem of regenerative SR Ca^{2+} refilling is that there appears to be relatively few instances of PM-SR nanojunctions in a typical vascular smooth muscle cell. For example, in the inferior vena cava of the rabbit, our laboratory observed that (in 2D electron micrographs) only about 15% of the PM is apposed by portions of superficial SR (Lee et al. 2002). To address this, we built a stochastic model that would integrate experimental information of several different kinds—such as [Ca^{2+}] oscillation data from confocal microscopy, ultrastructural measurements from electron microscopy, and myographic contractile force measurements—with a simulation of the diffusive motion of ions within the cytoplasmic nanospace by a random-walk algorithm (Fameli et al. 2007). The model geometry was informed by the characterization of the PM-SR junctional spaces as observed by electron microscopy (Fig. 17.2).

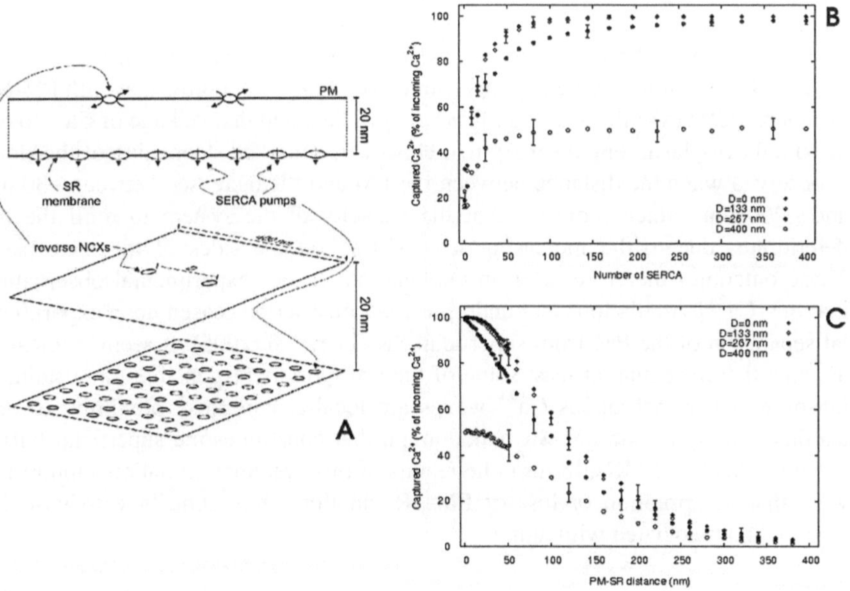

Fig. 17.2 Panel (**a**): Not-to-scale model of one PM-SR nanojunction. The dimensions are estimated from electron micrographs. Panel (**b**): Percentage of Ca^{2+} entering a PM-SR nanospace via NCX that is captured by SERCA vs number of SERCA. Panel (**c**): As in **b**, but vs distance between nanojunction membranes. The different sets of data refer to different values of the separation, *D*, between the Ca^{2+} sources (as indicated in the *insets*)

Essentially, PM-SR junctions may be represented by two planes uniformly separated by 20 nm. The lateral extension of this juxtaposition is less uniform, but it is usually a few hundred nm, and for the purposes of this approximation, we have chosen a value of 400 nm.

The diffusional trajectories of the Ca^{2+} ions were simulated by a random-walk algorithm on a cubic lattice filling the space between the two planes in Fig. 17.2a. Ions enter the junctional space through two point sources, representing NCX, and follow a 3D random walk in the PM-SR space prior to capture by a SERCA pump, implemented as a circular opening in the junctional SR with a radius of 4.5 nm, or migration beyond the nanospace by exiting through its sides to the neighboring cytoplasm. Ions that collide with regions of the SR outside the SERCA or with the PM are assumed to rebound from these surfaces and continue on with their random walk. A further assumption is that ions exiting the space at the lateral boundary of the junction are considered lost from the capture count. The main results obtained through this model are illustrated in Fig. 17.2b, c. As illustrated in panel b, data output from this quantitative model indicate that the PM-SR nanospaces within these smooth muscle cells can achieve SR Ca^{2+} capture rates sufficient to provide for good representation of the changes in Ca^{2+} observed during the repetitive Ca^{2+} waves that trigger vasoconstriction. Our model suggests that 90% or more of the incoming Ca^{2+} is captured in simulations incorporating between 100 and 200 SERCA pumps, which is entirely consistent with estimates of SERCA density from the existing literature (Inesi et al. 1990; Elmoselhi et al. 1995; Holmes et al. 2000).

The relationship between the percentage of ions captured and the distance across the PM-SR junction provided by this simulation (Fig. 17.2c) show that with PM-SR distances <30 nm refilling of the SR is very efficient and that leakage of Ca^{2+} to the proximal cytoplasm remains marginal. However, a marked steepening of the slope is observed when the distance between the PM and SR increased between ~30 nm and ~200 nm, which indicates that the capacity of the system to refill the SR deteriorates due to the increasing loss of Ca^{2+} at the sides of the nanospace. These outcomes therefore offer an explanation for our experimental observation that the $[Ca^{2+}]_i$ oscillations that underpin vasoconstriction ceased upon experimental separation of the PM from superficial SR (Lee et al. 2005). It seems logical to assume, therefore, that at least some of the transporters responsible for sustaining the observed asynchronous Ca^{2+} waves are localized in close proximity to one another and within the PM-SR nanodomain that constitutes the superficial buffer barrier in healthy SMCs. In turn, the results of this first approximation support the view that deterioration or loss of PM-SR junctions may contribute to vascular dysfunction associated with aging.

17.4.2 Quantitative Modeling of [Na^+] Transients Within the PM-SR Junction

From the above, one critical observation provided by our model is that Ca^{2+} influx for regenerative SR refilling can be provided via reversal of the NCX. For this to occur, the electrochemical potential of the NCX, E_{NCX}, and the membrane potential, V_M, must obey the relationship $E_{NCX} < V_M$ (Blaustein and Lederer 1999). This constraint implies that during activation a [Na^+] transient of the order of 30 mM or greater is necessary to cause NCX reversal. According to our previous findings, such [Na^+] spikes would have to take place in the vicinity of NCX or, in other words, in PM-SR nanospaces. Consistent with this view our laboratory has shown that in vascular SMCs, agonist stimulation elicits highly localized transient increases in [Na^+] (LNat) at the periphery of smooth muscle cells that are large enough to reverse the NCX (Poburko et al. 2007).

We therefore extended our model to incorporate the generation of LNats within the PM-SR nanospaces described in the previous section in this chapter and thereby sought to determine whether or not these would be sufficient to evoke changes in the [Na^+] within the nanospace ([Na^+]$_{ns}$) of a magnitude necessary to reverse the NCX under the conditions predicted by experimental observations (Fameli et al. 2009). Hence, we considered one Na^+ source (representing one TRPC6 channel) within the nanojunction which we positioned at the center of the PM. Given the great difference in transport capacity between the TRPC6 source and the known Na^+ sinks in the nanojunction (NCX and Na^+/K^+ATPase isoforms $\alpha_{2,3}$ (NKA$\alpha_{2,3}$)), we opted to omit any of the sinks in this first approximation of the model. As before, ions entering the nanospace follow a random walk from their source to their eventual exit through the sides of the nanojunction, with predicted changes in [Na^+]$_{ns}$ expressed as a function of distance from the source. A summary of the results obtained from this set of simulations is shown in Fig. 17.3d.

One of the most interesting outcomes of this model is that it appeared clear from the simplest implementation of the nanojunctional geometry (Fig. 17.3a) that we needed to take other ultrastructural features into account (Fig. 17.3b, c) in order to understand how [Na^+]$_{ns}$ values of the magnitude of LNats could be generated in PM-SR nanospaces. This prompted us to have a more detailed look at the electron micrographs and pay attention to some known but perhaps overlooked features, in particular to electron opaque bridging structures spanning the space between PM and SR (Devine et al. 1972; Poburko et al. 2008). Only when the junction-spanning "pillars" are incorporated in the model do our simulations reproduce [Na^+]$_{ns}$ of the same order of magnitude as the observations. This finding alone demonstrates the power of our quantitative modeling and how, when carried out in a manner informed by both cell structure and function, it may guide one to a more accurate interpretation of experimental data.

17.4.3 Stochastic Particle Simulators: Tools for Further Progress

In the previous section, we addressed the importance of almost any level of modeling to further the understanding of Ca^{2+} transport in nanojunctions, as long as the model implementation is based on firm and accurate experimental evidence. We saw how even beginning to have a grasp of what sort of numbers are involved in making the nanojunctions work can help illuminate the experimental path forward.

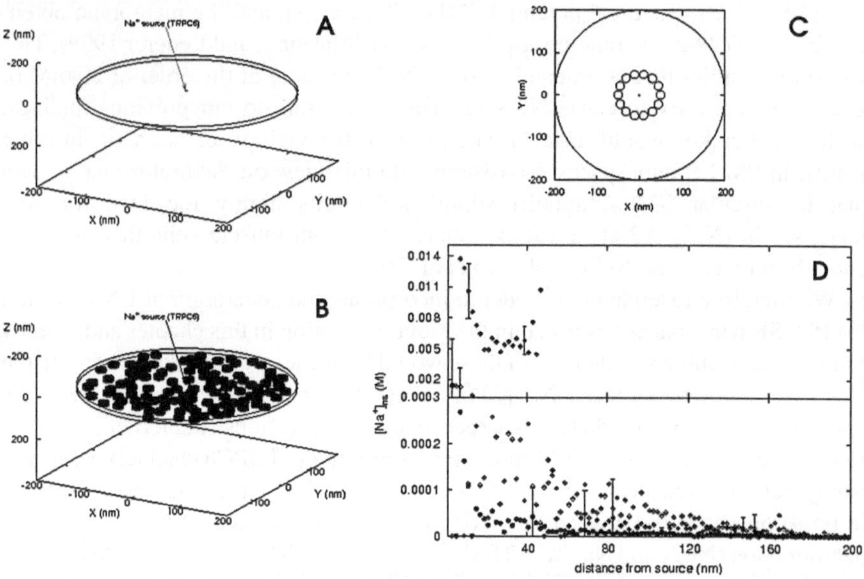

Fig. 17.3 To-scale model nanojunction used in the simulations (panels **a**, **b**, and **c**) and relative simulation data (**d**). The separation between the two surfaces (the bases of the *flat cylinder* in panels **a** and **b**) is 20 nm, as estimated by electron microscopy. Panel (**a**): bare nanojunction. Panel (**b**): nanojunction with 200 10-nm-radius randomly distributed pillars. Panel (**c**): nanojunction with 14 pillars arranged in a "porous" 50-nm-radius circular barrier around one Na^+ source. Panel (**d**): $[Na^+]_{ns}$ as a function of distance from the Na^+ source for model nanojunctions as in panels **a**, **b**, and **c** (*solid circles*, *open diamonds*, and *solid diamonds*, respectively)

We can on the other hand discern the limits of a cruder approach, in that it "only" allows us to produce arguments to corroborate conclusions gained from experiments or to establish plausible explanations for certain experimental outcomes. However, future modeling of nanojunctional biophysics must not only be required to provide plausible arguments but must also offer more detailed quantitative explanations of experimental outcomes in a manner that may provide hypotheses that can be verified by further experimentation. This can be accomplished by extending each aspect of our stochastic modeling approach, namely, the reconstruction of the ultrastructural geometry, the random-walk algorithm implementation, and the data output (both numerical and visual).

High-end stochastic particle simulators provide us with an excellent "infrastructure" in which to insert the wealth of experimental knowledge on the transport machinery of any identified intracellular nanojunction. For example, we can place within any given nanojunction the transporters, ultrastructural features, and ions with their known diffusivities and carry out realistic simulations of the diffusion of relevant molecules in a manner that predicts their impact on intracellular signaling. To achieve such goals, there are a number of key ingredients necessary for the effective development of fruitful outcomes. One is the faithful transformation of accurate ultrastructural information provided by high-resolution electron microscopy (two- and three-dimensional standard and tomographic images, respectively) into its three-dimensional "virtual" equivalent, in the guise of a computational mesh structure on which to place transporters and to perform stochastic computation as accurately as possible. At present, this process remains cumbersome and arguably the most time-consuming aspect of these studies. Another key ingredient is the provision of evermore reliable information on the localization of the transporters of interest within any given nanojunction operation. While it is relatively easy to obtain information on transporter distribution optically via immunofluorescence microscopy, accurate quantitative modeling of nanojunctional processes necessitates reliable immunogold labeling of transporters with accurate quantification of specific binding of gold particles to particular intracellular compartment elements.

Several high-quality stochastic particle simulators exist—a computational modeling scheme like that outlined in the previous section, with a geometry-building component, random-walk algorithm-computation component, and data- and visualization-output component, is what we usually refer to as stochastic particle simulators—which are able to tackle nanoscale systems. Some examples are MCell/CellBlender (Stiles et al. 1996; Stiles and Bartol 2001; MCell 2016), SmolDyn (Andrews et al. 2010; Smoldyn 2016), ChemCell (ChemCell 2016), and GridCell (Boulianne et al. 2008; GridCell 2016). Our laboratory has gained considerable experience using the first of these, which is the only one capable of handling complex geometries to reproduce biological intracellular architectures in a realistic manner while maintaining a high degree of accuracy and speed in the calculations. Ultrastructural information is reproducible by means of sophisticated 3D geometry-building software (Blender 2016), from which the mesh information can be exported and fed into MCell proper and combined with molecular and transporter information to implement random-walk computations on the given geometry inclusive of boundary conditions. MCell, in turn, outputs graphable numerical data, such as surface and volume molecular concentrations and the number of reactions. Equally significant, however, is the fact that MCell has the capacity to output visual representations of the final rendered product, which may include, for example, the given geometry of a nanojunction, the given transporters and their positions, as well as the various molecular species either present at the start of the simulation or generated by chemical reactions that are incorporated within the simulation. A further and critical aspect of working with an accurate molecular simulator such as MCell is that of the accurate provision of transporter kinetics and molecular properties, which may be adjusted as our knowledge of channel, pump, and molecular behavior develops.

To refine our SR Ca^{2+} refilling model and study Ca^{2+} transport in more detail, we subsequently developed a 3D model of a typical PM-SR nanojunction using earlier ultrastructural observations as guidance (Lee et al. 2002; Fameli et al. 2007). Within this model we placed 15 NCX working in reverse mode (Ca^{2+} influx) on the junctional PM and 200 SERCA pumps on the junctional SR membrane, as illustrated in Fig. 17.4b and in the online supplemental material.

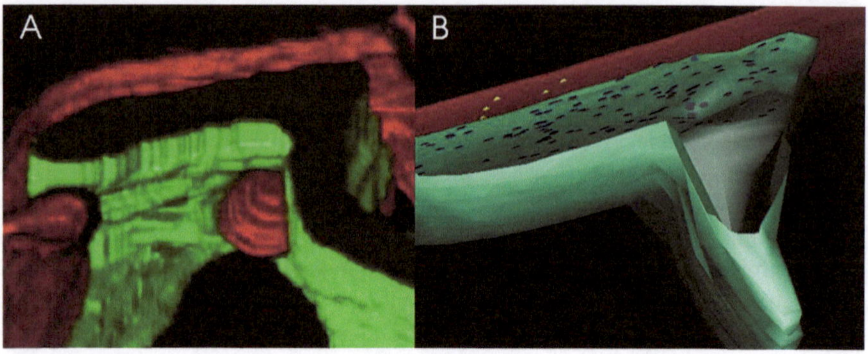

Fig. 17.4 Panel (**a**): Three-dimensional reconstruction from serial section electron micrographs of a region of a rabbit inferior vena cava smooth muscle cell, showing portions of the PM (*red*) and of superficial SR (*green*). A caveola poking through a superficial SR sheet is also visible. Panel (**b**): Three-dimensional realization of a PM (*red*)-SR (*light blue*) nanojunction inspired by the image in panel **a** and produced by DReAMM, MCell's rendering suite. The separation between PM and SR is about 20 nm. Only some of the implemented NCX (*yellow objects* on the PM) and SERCA (*blue disks* on the SR) are visible in this orientation (for further images and animations of this system, please refer to the online supplemental material associated with this article)

We ran a number of simulations in which Ca^{2+} entered the space via the set of NCX, diffused according to a random-walk algorithm and reacted with the SERCA pumps via a simplified version of a 12-state model of the latter (Läuger 1991). In the interest of studying the Ca^{2+} transport efficiency of vascular smooth muscle PM-SR nanojunctions and comparing the new results with those of our cruder models developed earlier (see previous chapters), the simulations output data on the number of Ca^{2+} that are taken up by the SERCA pumps during a time interval of the order of the Ca^{2+} oscillation period as observed by this laboratory (Lee et al. 2001). Figure 17.5 reports two typical data sets, which differ by way of the relative position of the NCX and SERCA sets. These results suggest that the communication between NCX and SERCA may vary markedly in a manner dependent on the localization pattern of the transporters. In particular, the fact that clustering of NCX produces a higher Ca^{2+} capture rate than distributing them more broadly on the junctional PM suggests that Ca^{2+} communication between the two transporters is more efficient when NCX and SERCA face each other across the nanospace in an organized way. In other words, when clusters of NCX are in "direct" communication with clusters of SERCA, transport is more efficient than when both species are scattered at random on their native membranes. In keeping with the overarching theme of this book, this intriguing possibility brings to mind the important observation that during activation of a SOCE system, first STIM and then Orai appear to cluster in the junctional ER and PM, respectively, likely to increase Ca^{2+} transport efficiency (Park et al.2009).

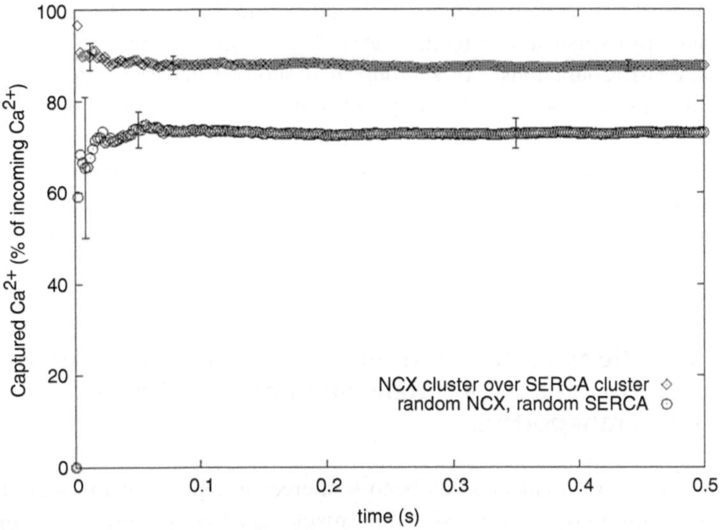

Fig. 17.5 Output from simulations run on a PM-SR nanojunction configuration as in Fig. 17.4b, with 15 NCX placed on the PM and 200 SERCA on the SR, either both at random (*dotted circles*) or in two opposing clusters near the center of their respective membranes (*dotted diamonds*)

17.4.4 PM-ER Junctions and ER Ca^{2+} Refilling in Vascular Endothelial Cells

Some functional studies in vascular endothelial cells point out that the process of ER Ca^{2+} refilling from the extracellular space may take place in a privileged manner that bypasses the bulk cytoplasm (Malli et al. 2007; Di Giuro et al. 2016). This mechanism is reminiscent of the superficial buffer-barrier effect in vascular smooth muscle (van Breemen 1977; van Breemen et al. 1986), which had been hypothesized for vascular endothelial cells (Paltauf-Doburzynska et al. 1998). Malli's study consisted of optical fluorescence measurements of cytoplasmic $[Ca^{2+}]_i$ in histamine-stimulated endothelial cells. They showed for the first time by ER Ca^{2+} release-refill-release experiments that strong attenuation of extracellular Ca^{2+} entry into the cell by membrane depolarization with high external K^+ did not change the ER Ca^{2+} releasable fraction. This suggested that the molecular mechanisms underpinning the ER Ca^{2+} refilling pathways were largely unaffected by the depolarization and likely independent of the transporters in charge of Ca^{2+} entry in the bulk cytosol. More recent experiments of NF's and CvB's groups not only confirmed the findings of Malli and co-workers but also provided strong indications that (1) the ER Ca^{2+} refilling in these cells is operated roughly in parallel by Orai1 channels and NCX1 and (2) the Ca^{2+} flow via these transporters is very likely to occur via PM-ER junctions (Di Giuro et al. 2016). The latter inference was made based on pharmacological inhibition (by 10 nM ouabain) of NKA$\alpha_{2,3}$, which have been observed to reside in PM-ER junctions in other systems. These experiments showed that interfering with the NKA in the junctions provoked an increase in the releasable fraction of ER Ca^{2+}, a finding that we can only explain assuming that the NCX portion of the ER Ca^{2+} refilling takes place via PM-ER junctions. NF's group also showed that the latter exist in the ultrastructure of endothelial cells and possess dimensional features compatible with their counterpart in vascular smooth muscle cells. Only careful protein labeling and quantitative modeling will be able to confirm the involvement of these junctions in the refilling of ER Ca^{2+} in the endothelium; however, the functional evidence reported to date already provides strong support in favor of a junctional mechanism (Di Giuro et al. 2016).

17.5 Variations on the Theme: The Spatial Organization of Intracellular Ca^{2+} Release Channels, Pumps, and Transporters

Recently, significant evidence has been gathered in support of the view that the spatial organization of the Ca^{2+} signaling machinery may provide as yet unprecedented versatility that goes beyond even that which we have thus far attempted to model. Determining factors in this respect will likely be the subcellular localization of different subtypes of Ca^{2+} release channels and Ca^{2+} pumps that may be targeted to the surface of a diverse array of Ca^{2+}-storing organelles. That this may be the case gains significant support from recent studies on pulmonary arterial smooth muscle cells.

In acutely isolated pulmonary arterial smooth muscle cells, we now know (AME, unpublished observation) that the Ca^{2+}-mobilizing messenger nicotinic acid adenine dinucleotide phosphate (NAADP) triggers bursts of Ca^{2+} release from lysosomes by a mechanism dependent upon two-pore domain channel subtype 2 (TPC2) (Calcraft et al. 2009). These Ca^{2+} bursts initiate a propagating wave by subsequently triggering Ca^{2+}-induced Ca^{2+} release from the SR via RyRs (Boittin et al. 2002). Experimental evidence from optical fluorescence studies suggests that a large proportion of lysosomes form tight clusters within the perinuclear region of pulmonary arterial smooth muscle cells and in a manner consistent with the spatially restricted nature of NAADP-dependent Ca^{2+} bursts (Kinnear et al. 2004). Moreover, these lysosomal clusters are closely associated with a subpopulation of RyRs. In fact, immunofluorescence experiments suggested that lysosome clusters may be separated from closely apposed RyRs by a narrow junction or cleft (<100 nm) that is beyond the resolution of deconvolution or confocal microscopy. The existence of these junctions was subsequently supported by transmission electron micrographs in rat pulmonary artery intact tissue (Fameli et al. 2014). Lysosome-SR junctions measured therein had mean membrane separations of 16 nm and mean lateral extensions of 262 nm. These data also suggested that if the average number of lysosomes observed per cell released their Ca^{2+} content according to the known kinetics of the putative lysosomal Ca^{2+} release channels, their combined release would be comparable in magnitude to that of the observed Ca^{2+} bursts.

Thus, lysosome-SR junctions may form a highly organized "trigger zone," or intracellular synapse, for Ca^{2+} signaling in response to NAADP. These lysosome-SR junctions may constitute a novel nanodomain with unique functional characteristics, in that the presence of such a junction offers an explanation for the fact that NAADP-dependent Ca^{2+} bursts induce propagating Ca^{2+} waves by triggering Ca^{2+}-induced Ca^{2+} release (CICR) from the SR via RyRs in an all-or-none manner, and as such these junctions may provide a "margin of safety" with respect to the initiation of vasoconstriction via discharge of lysosomal Ca^{2+} stores. This is a significant point, because the classical functions of lysosomes, such as autophagy, are executed mostly independent of lysosome-SR interactions (Gozuacik and Kimchi 2004), but are exquisitely sensitive to the luminal Ca^{2+} load of lysosomes and transmembrane Ca^{2+} fluxes. Therefore, lysosome-SR junctions must be engineered to allow for the initiation of highly localized Ca^{2+} signals that modulate lysosome-specific functions without engaging the SR while at the same time providing the capacity for initiation, as required, of propagating SR Ca^{2+} waves by way of CICR. The precise mechanisms involved appear yet more complex, because all three RyR subtypes (RyR1, RyR2, and RyR3) are expressed in arterial smooth muscle and some of these lysosome clusters may selectively couple to RyR3 (Kinnear et al. 2008). Therefore, we must ask, why RyR3? Determining factors in this respect could be the relative sensitivity of each RyR subtype to CICR, the maximum gain in response to Ca^{2+}, and the relative sensitivity of each receptor subtype to inactivation by Ca^{2+} (Chen et al. 1997; Li and Chen 2001). The threshold for activation of RyR1, RyR2, and RyR3 is similar, with channel activation at cytoplasmic Ca^{2+} concentrations above 100 nM. However, estimates of the EC50 are different, with half maximal activation at about 250 nM for RyR2 and about 400 nM for RyR3. The higher EC50 exhibited by RyR3 could be significant because this could provide for the required "margin of safety" with respect to the all-or-none amplification of Ca^{2+} bursts at the putative lysosome-SR junction; that is, the probability of false events being initiated would be lower for RyR3 than for RyR2.

Another factor that may be of significance is that, while the mean open times versus cytoplasmic Ca^{2+} concentration for RyR2 and RyR3 are comparable and increase approximately tenfold over their activation range, the mean open time for RyR1 is much lower and increases only twofold over its activation range. Furthermore, comparison of the open probability P_O versus cytoplasmic Ca^{2+} concentration curves shows that RyR3 (0–1) exhibits a higher gain in P_O than does RyR2 (0–0.9), while RyR1 (0–0.2) exhibits relatively little gain in P_O with increasing cytoplasmic Ca^{2+} concentration. Thus, once the threshold for activation is breached, RyR3 would offer greater amplification of Ca^{2+} bursts from lysosomal stores than would RyR2, while amplification via RyR1 would be marginal. There is also marked variation in the relative sensitivity of each RyR subtype to inactivation by Ca^{2+}. RyR3 exhibits the lowest sensitivity to inactivation with an IC_{50} of 3 mM, while that for RyR2 is 2 mM; in each case, channel activity may still be observed at concentrations above 10 mM. In marked contrast, RyR1 inactivation occurs within the micromolar range, and full inactivation is achieved by 1 mM; this may, in part, explain the low gain in P_O for RyR1 in response to activation by Ca^{2+}. Its sensitivity to inactivation by Ca^{2+} would therefore render RyR1 unsuitable for a role in the amplification of Ca^{2+} bursts at lysosome-SR junctions because the local Ca^{2+} concentration within this nanodomain may exceed the threshold for RyR1 inactivation. Thus, the functional properties of RyR3 make it best suited to a role in the amplification of Ca^{2+} bursts at lysosome-SR junctions (Kinnear et al. 2008).

This hypothesized functional role of lysosome-SR junctions received considerable support from a quantitative model, which we developed more recently by means of the MCell/CellBlender tools mentioned above (Fameli et al. 2014). The model includes a realistic reconstruction of the junctional geometry on the basis of our own electron microscopic observations and the implementation of both Ca^{2+} release channels on the lysosomal membrane and SERCA2 pumps on the opposing junctional SR membrane. Surface densities as well as kinetic and conductivity properties of the transporters were obtained from other published studies. Furthermore, experimentally determined open probability for the Ca^{2+} release from lysosomes allowed us to complete this model and simulate the junctional $[Ca^{2+}]$ transients from the release channel opening to their closure. Results from the model suggest that transients of sufficiently high magnitude to trigger opening of the RyR3 channels can be achieved in lysosome-SR junctions possessing the dimensional features of the observed ones. Moreover, the model also shows that only junctions with a membrane separation under approximately 30 nm would be able to support such high transients, thereby suggesting that such junctions are a necessary and sufficient condition to the generation of Ca^{2+} bursts, which may be the precursors to cell-wide Ca^{2+} waves.

Inclusion of RyR3 alone in this model cannot, however, explain the subsequent generation of a propagating Ca^{2+} wave. This is clear from the fact that RyR3 labeling declines markedly outside the perinuclear region of pulmonary arterial smooth muscle cells. Therefore, it seems unlikely that RyR3 functions to carry a propagating Ca^{2+} wave far beyond the point of initiation of CICR within the proposed lysosome-SR junction. Given this finding, it may be of significance that labeling for RyR2 increases markedly as we move from the perinuclear to the extraperinuclear region (>1.5 μM from the nucleus). This suggests that RyR2 may function to receive Ca^{2+} from RyR3 at the interface of the lysosome-SR junction and thereby allow for further propagation of the Ca^{2+} signal via CICR. Such a role would be supported by the lower EC_{50} for CICR via RyR2, which would ensure that once initiated a propagating Ca^{2+} wave would be less prone to failure (Kinnear et al. 2008).

As discussed in previous sections in this chapter, this model will only work if a SERCA pump is present to provide Ca^{2+} to the region of the SR to which RyR3 is targeted. Importantly, therefore, recently completed studies on the distribution of SERCA within these cells have provoked even greater consideration of the extent to which smooth muscle cells may utilize nanojunctions to coordinate Ca^{2+} signals. Both SERCA2a and SERCA2b were found to be expressed, and these splice variants were differentially distributed within pulmonary arterial smooth muscle cells. SERCA2a is almost entirely restricted to the perinuclear SR and thus sits proximal to RyR3 (Clark et al. 2010). Therefore, we have the necessary transporter to support uptake into the SR within the region of lysosome-SR junctions.

In marked contrast to SERCA2a, however, SERCA2b was found to be preferentially targeted to the SR proximal to the plasma membrane, i.e., to the region of the superficial buffer barrier that took center stage in our earlier discussion of Ca^{2+} signaling nanodomains. Moreover, it is this region of the SR to which RyR1 may be preferentially targeted (Clark et al. 2010). That this RyR subtype is least able to support a propagating Ca^{2+} wave may be significant too. The combination of limited support for signal propagation by RyR1 with the removal of Ca^{2+} from the cytoplasm by SERCA2b may effectively segregate the PM-SR space in these cells. That this may be the case is supported by functional data. Thus, adenylyl cyclase-coupled receptors have been shown to elicit Ca^{2+} release proximal to the PM via RyRs and thereby open Ca^{2+}-activated K^+ channels in the plasma membrane, leading to hyperpolarization followed by Ca^{2+} sequestration from the junctional space between the SR and the plasma membrane via the NCX and/or PMCA (Boittin et al. 2003).

This brings us nicely back to the importance of detailed kinetic information that may be used to formulate hypotheses that stochastic modeling may test and extend in order to inform further experimental work. SERCA2a and SERCA2b exhibit quite different kinetics, and this will undoubtedly prove to be important. SERCA2b has a higher affinity for Ca^{2+} but lower V_{max} than SERCA2a (Verboomen et al. 1994; Odermatt et al. 1996; Dode et al. 2003). SERCA2b may therefore be dominant at rest and function to maintain resting levels of Ca^{2+} in the vicinity of the contractile apparatus. However, its low V_{max} may lead to saturation of this SERCA isoform with Ca^{2+} upon release of Ca^{2+} from the central SR, via RyR3 and RyR2, in response to stimuli that elicit vasoconstriction. Thereby, SERCA2b would allow the Ca^{2+} concentration to rise in the vicinity of the contractile apparatus, until such time as vasodilation is promoted by, for example, activation of adenylyl cyclase-coupled receptors that may (1) increase the activity of SERCA2b by PKA-dependent phosphorylation (Lindemann et al. 1983; Raeymaekers et al. 1990) and facilitate the removal of Ca^{2+} from the greater cytoplasm and (2) trigger PKA-dependent Ca^{2+} release from the peripheral SR store via RyR1 to promote PM hyperpolarization and Ca^{2+} removal from PM-SR nanodomain (Clark et al. 2010). This has a physiological precedent in that it mirrors somewhat the roles of uptake 1 and uptake 2 at noradrenergic synapses, which are also dependent on the relative affinity and V_{max} of these two catecholamine transporters (Iversen 1973). These considerations also cement the view that the targeting of the appropriate ion channels, pumps, and Ca^{2+} receptors to nanojunctions can provide for the discrete regulation of different, and even opposing, cellular functions.

Importantly, the threshold for induction of CICR via RyRs may also be modulated by the luminal Ca^{2+} concentration of the SR (Beard et al. 2002; Ching et al. 2000; Gilchrist et al. 1992; Györke and Györke 1998; Tripathy and Meissner 1996), which could in turn be primed by Ca^{2+} taken up via SERCA2a during lysosomal Ca^{2+} bursts that fail to breach the threshold for CICR from the SR. Moreover lysosome-SR signaling may also occur in reverse (Morgan et al. 2013). In short, Ca^{2+} release from SR compartments could increase lysosomal Ca^{2+}

load, enhance the capacity for lysosomal Ca^{2+} release, and thus augment coupling by CICR across lysosome-SR nanojunctions.

17.6 Junctional Reorganization in Health and Disease

Intriguingly, increases in the expression of SERCA2a, RyR2, and RyR3 have been noted during the transition from a contractile to a proliferating smooth muscle phenotype (Berra-Romani et al. 2008; Magnier et al., 1992). Moreover artery culture may lead to decreases in SERCA2b expression and altered vascular reactivity (Thorne and Paul, 2003). This raises the possibility that marked reorganization of cellular nanojunctions may occur during health and disease, in a manner coupled to changes in SERCA and RyR expression at the very least. As proposed previously (van Breemen et al. 2013), such reorganization could provide for the functional diversity noted not only across different types of vascular smooth muscle but also during their differentiation. Consistent with this view, reorganization of junctional complexes within pulmonary arterial myocytes has been identified during hypoxia and may contribute to the progression. Such junctional reorganization could conceivably lead to dysfunction of enzyme systems of the lysosomes. This could contribute to pathologies associated with subclasses of lysosome storage disease such as Niemann-Pick disease type C1 (Lloyd-Evans et al. 2008; Tassoni et al. 1991) and Pompe and Gaucher disease (Jmoudiak and Futerman 2005; Noori et al. 2002), which may include hepatic portal (Tassoni et al. 1991) or pulmonary hypertension (Jmoudiak and Futerman 2005; Noori et al. 2002), dysfunctions in cholesterol trafficking (Carstea et al. 1997), and consequent increases in plasma cholesterol levels, vascular lesion formation, atherosclerosis/thrombosis, and medial degradation (Ron and Horowitz 2008; Tassoni et al. 1991).

17.7 Conclusion

Considering the complexity of the ionic mechanisms underlying the smooth muscle Ca^{2+} oscillations, it is reasonable to postulate their evolutionary advantage over a much simpler on/off switch that could, for example, be accomplished by VGCCs in the PM. Although not proven at this time, the following list of advantages appears plausible:

(a) $[Ca^{2+}]_i$ oscillations are more efficient than tonic elevations, because contraction responds to the peak rather than to the average values. This follows from the fact that the rate of smooth muscle relaxation is relatively slow, such that contraction reaches its maximum at a wave frequency of 0.5 Hz. Indeed Sneyd's group has modeled this result on the basis of a realistic model of smooth muscle contraction (Wang et al. 2010).
(b) $[Ca^{2+}]_i$ oscillations are more efficient because IP_3R-mediated Ca^{2+} release is localized near calmodulin, which is tethered to the myofilaments. Our present

model envisions that much of the activating Ca^{2+} is transferred from the junctional SR to the more central SR by a process akin to tunneling through the SR lumen before being released into the cytoplasm. We have previously reported that the peripheral cytoplasm has a significantly lower density of thick myosin filaments (Lee et al. 2002).

(c) Mitochondrial dehydrogenases respond to the frequency of $[Ca^{2+}]_{mt}$ oscillations, which would be induced by periodic Ca^{2+} release into the nanospaces between SR and mitochondria (Rizzuto et al. 2004). This mechanism ensures that energy is supplied when needed to maintain contraction.

(d) Large tonic increases in $[Ca^{2+}]_i$ might result in prolonged mitochondrial Ca^{2+} uptake leading to release of cytochrome C to induce apoptosis (Demaurex and Distelhorst 2003). This is clearly an important vascular disease mechanism to be avoided during physiological stimulation.

Although a great deal of future research is required to prove the above reasons for generating smooth muscle $[Ca^{2+}]_i$ oscillations, there can be no doubt that understanding ionic movements within nanojunctions between the PM, SR, lysosomes, and mitochondria lies at the heart of site- and function-specific Ca^{2+} signaling.

References

Andrews SS, Addy NJ, Brent R, Arkin AP (2010) Detailed simulations of cell biology with Smoldyn 2.1. PLoS Comput Biol 6(3):e1000705

Beard NA, Sakowska MM, Dulhunty AF, Laver DR (2002) Calsequestrin is an inhibitor of skeletal muscle ryanodine receptor calcium release channels. Biophys J 82(1 Pt 1):310–320

Berra-Romani R, Mazzocco-Spezzia A, Pulina MV, Golovina VA (2008) Ca2+ handling is altered when arterial myocytes progress from a contractile to a proliferative phenotype in culture. Am J Physiol Cell Physiol 295(3):C779–C790

Blaustein MP, Lederer WJ (1999) Sodium/calcium exchange: its physiological implications. Physiol Rev 79(3):763–854

Blender (2016) blender.org—Home. Available at: http://www.blender.org/

Boittin F, Galione A, Evans AM (2002) Nicotinic acid adenine dinucleotide phosphate mediates Ca2+ signals and contraction in arterial smooth muscle via a two-pool mechanism. Circ Res 91 (12):1168–1175

Boittin FX, Dipp M, Kinnear NP, Galione A, Evans AM (2003) Vasodilation by the calcium-mobilizing messenger cyclic ADP-ribose. J Biol Chem 278(11):9602–9608

Boulianne L, Al Assaad S, Dumontier M, Gross WJ (2008) GridCell: a stochastic particle-based biological system simulator. BMC Syst Biol 2:66

Calcraft PJ, Ruas M, Pan Z, Cheng X, Arredouani A, Hao X, Tang J, Rietdorf K, Teboul L, Chuang KT, Lin P, Xiao R, Wang C, Zhu Y, Lin Y, Wyatt CN, Parrington J, Ma J, Evans AM, Galione A, Zhu MX (2009) NAADP mobilizes calcium from acidic organelles through two-pore channels. Nature 459(7246):596–600

Carstea ED, Morris JA, Coleman KG, Loftus SK, Zhang D, Cummings C, Gu J, Rosenfeld MA, Pavan WJ, Krizman DB, Nagle J, Polymeropoulos MH, Sturley SL, Ioannou YA, Higgins ME, Comly M, Cooney A, Brown A, Kaneski CR, Blanchette-Mackie EJ, Dwyer NK, Neufeld EB, Chang TY, Liscum L, Strauss JF 3rd, Ohno K, Zeigler M, Carmi R, Sokol J, Markie D, O'Neill RR, van Diggelen OP, Elleder M, Patterson MC, Brady RO, Vanier MT, Pentchev PG, Tagle

DA (1997) Niemann-Pick C1 disease gene: homology to mediators of cholesterol homeostasis. Science 277(5323):228–231
ChemCell (2016) Available at: http://www.sandia.gov/
Chen SR, Li X, Ebisawa K, Zhang L (1997) Functional characterization of the recombinant type 3 Ca2+ release channel (ryanodine receptor) expressed in HEK293 cells. J Biol Chem 272 (39):24234–24246
Ching LL, Williams AJ, Sitsapesan R (2000) Evidence for Ca(2+) activation and inactivation sites on the luminal side of the cardiac ryanodine receptor complex. Circ Res 87(3):201–206
Clark JH, Kinnear NP, Kalujnaia S, Cramb G, Fleischer S, Jeyakumar LH, Wuytack F, Evans AM (2010) Identification of functionally segregated sarcoplasmic reticulum calcium stores in pulmonary arterial smooth muscle. J Biol Chem 285(18):13542–13549
Dai JM, Syyong H, Navarro-Dorado J, Redondo S, Alonso M, van Breemen C, Tejerina T (2010) A comparative study of alpha-adrenergic receptor mediated Ca(2+) signals and contraction in intact human and mouse vascular smooth muscle. Eur J Pharmacol 629(1–3):82–88
Demaurex N, Distelhorst C (2003) Cell biology. Apoptosis—the calcium connection. Science 300 (5616):65–67
Devine CE, Somlyo AV, Somlyo AP (1972) Sarcoplasmic reticulum and excitation-contraction coupling in mammalian smooth muscles. J Cell Biol 52(3):690–718
Di Giuro CML, Shrestha N, Malli R, Groschner K, van Breemen C, Fameli N (2016) Na+/Ca2+ exchangers and Orai channels jointly refill endoplasmic reticulum (ER) Ca2+ via ER nanojunctions in vascular endothelial cells. bioRxiv. Available at: https://doi.org/10.1101/084285
Dode L, Andersen JP, Leslie N, Dhitavat J, Vilsen B, Hovnanian A (2003) Dissection of the functional differences between sarco(endo)plasmic reticulum Ca2+-ATPase (SERCA) 1 and 2 isoforms and characterization of Darier disease (SERCA2) mutants by steady-state and transient kinetic analyses. J Biol Chem 278(48):47877–47889
Elmoselhi AB, Blennerhassett M, Samson SE, Grover AK (1995) Properties of the sarcoplasmic reticulum Ca(2+)-pump in coronary artery skinned smooth muscle. Mol Cell Biochem 151 (2):149–155
Evans AM, Fameli N, Ogunbayo OA, Duan J, Navarro-Dorado J (2016) From contraction to gene expression: nanojunctions of the sarco/endoplasmic reticulum deliver site- and function-specific calcium signals. Sci China Life Sci 59(8):749–763
Fameli N, van Breemen C, Kuo K (2007) A quantitative model for linking Na+/Ca2+ exchanger to SERCA during refilling of the sarcoplasmic reticulum to sustain [Ca2+] oscillations in vascular smooth muscle. Cell Calcium 42(6):565–575
Fameli N, Kuo K, van Breemen C (2009) A model for the generation of localized transient [Na+] elevations in vascular smooth muscle. Biochem Biophys Res Commun 389(3):461–465
Fameli N, Ogunbayo OA, van Breemen C, Evans AM (2014) Cytoplasmic nanojunctions between lysosomes and sarcoplasmic reticulum are required for specific calcium signaling. F1000Res 3:93
Gabella G (1971) Caveolae intracellulares and sarcoplasmic reticulum in smooth muscle. J Cell Sci 8(3):601–609
Gilchrist JS, Belcastro AN, Katz S (1992) Intraluminal Ca2+ dependence of Ca2+ and ryanodine-mediated regulation of skeletal muscle sarcoplasmic reticulum Ca2+ release. J Biol Chem 267 (29):20850–20856
Gozuacik D, Kimchi A (2004) Autophagy as a cell death and tumor suppressor mechanism. Oncogene 23(16):2891–2906
GridCell (2016) Available at: http://www.isip.ece.mcgill.ca/research/gridcell/start
Györke I, Györke S (1998) Regulation of the cardiac ryanodine receptor channel by luminal Ca2+ involves luminal Ca2+ sensing sites. Biophys J 75(6):2801–2810
Holmes ME, Samson SE, Wilson JX, Dixon SJ, Grover AK (2000) Ascorbate transport in pig coronary artery smooth muscle: Na(+) removal and oxidative stress increase loss of accumulated cellular ascorbate. J Vasc Res 37(5):390–398
Iino M, Kasai H, Yamazawa T (1994) Visualization of neural control of intracellular Ca2+ concentration in single vascular smooth muscle cells in situ. EMBO J 13(21):5026–5031

Inesi G, Sumbilla C, Kirtley ME (1990) Relationships of molecular structure and function in Ca2 (+)-transport ATPase. Physiol Rev 70(3):749–760

Iversen LL (1973) Catecholamine uptake processes. Br Med Bull 29(2):130–135

Jmoudiak M, Futerman AH (2005) Gaucher disease: pathological mechanisms and modern management. Br J Haematol 129(2):178–188

Kinnear NP, Boittin FX, Thomas JM, Galione A, Evans AM (2004) Lysosome-sarcoplasmic reticulum junctions. A trigger zone for calcium signaling by nicotinic acid adenine dinucleotide phosphate and endothelin-1. J Biol Chem 279(52):54319–54326

Kinnear NP, Wyatt CN, Clark JH, Calcraft PJ, Fleischer S, Jeyakumar LH, Nixon GF, Evans AM (2008) Lysosomes co-localize with ryanodine receptor subtype 3 to form a trigger zone for calcium signalling by NAADP in rat pulmonary arterial smooth muscle. Cell Calcium 44 (2):190–201

Läuger P (1991) Electrogenic ion pumps. Sinauer, Sunderland, MA

Lee CH, Poburko D, Sahota P, Sandhu J, Ruehlmann DO, van Breemen C (2001) The mechanism of phenylephrine-mediated [Ca(2+)](i) oscillations underlying tonic contraction in the rabbit inferior vena cava. J Physiol 534(Pt 3):641–650

Lee CH, Poburko D, Kuo KH, Seow CY, van Breemen C (2002) Ca(2+) oscillations, gradients, and homeostasis in vascular smooth muscle. Am J Physiol Heart Circ Physiol 282(5):H1571–H1583

Lee CH, Kuo KH, Dai J, Leo JM, Seow CY, Breemen Cv (2005) Calyculin-A disrupts subplasmalemmal junction and recurring Ca2+ waves in vascular smooth muscle. Cell Calcium 37(1):9–16

Lemos VS, Poburko D, Liao CH, Cole WC, van Breemen C (2007) Na+ entry via TRPC6 causes Ca2+ entry via NCX reversal in ATP stimulated smooth muscle cells. Biochem Biophys Res Commun 352(1):130–134

Li P, Chen SR (2001) Molecular basis of Ca(2)+ activation of the mouse cardiac Ca(2)+ release channel (ryanodine receptor). J Gen Physiol 118(1):33–44

Lindemann JP, Jones LR, Hathaway DR, Henry BG, Watanabe AM (1983) beta-Adrenergic stimulation of phospholamban phosphorylation and Ca2+-ATPase activity in guinea pig ventricles. J Biol Chem 258(1):464–471

Lloyd-Evans E, Morgan AJ, He X, Smith DA, Elliot-Smith E, Sillence DJ, Churchill GC, Schuchman EH, Galione A, Platt FM (2008) Niemann-Pick disease type C1 is a sphingosine storage disease that causes deregulation of lysosomal calcium. Nat Med 14(11):1247–1255

Magnier C, Papp B, Corvazier E, Bredoux R, Wuytack F, Eggermont J, Maclouf J, Enouf J (1992) Regulation of sarco-endoplasmic reticulum Ca(2+)-ATPases during platelet-derived growth factor-induced smooth muscle cell proliferation. J Biol Chem 267(22):15808–15815

Malli R, Frieden M, Hunkova M, Trenker M, Graier WF (2007) Ca2+ refilling of the endoplasmic reticulum is largely preserved albeit reduced Ca2+ entry in endothelial cells. Cell Calcium 41 (1):63–76

MCell (2016) MCell Home|Center for Quantitative Biological Simulation. Available at: http://mcell.org/

Moore ED, Wasteneys GO (2012) Nanospace biophysics. Editorial. Protoplasma 249(Suppl 1):S1

Morgan AJ, Davis LC, Wagner SK, Lewis AM, Parrington J, Churchill GC, Galione A (2013) Bidirectional Ca2+ signaling occurs between the endoplasmic reticulum and acidic organelles. J Cell Biol 200(6):789–805

Noori S, Acherman R, Siassi B, Luna C, Ebrahimi M, Pavlova Z, Ramanathan R (2002) A rare presentation of Pompe disease with massive hypertrophic cardiomyopathy at birth. J Perinat Med 30(6):517–521

Odermatt A, Kurzydlowski K, MacLennan DH (1996) The vmax of the Ca2+-ATPase of cardiac sarcoplasmic reticulum (SERCA2a) is not altered by Ca2+/calmodulin-dependent phosphorylation or by interaction with phospholamban. J Biol Chem 271(24):14206–14213

Paltauf-Doburzynska J, Posch K, Paltauf G, Graier WF (1998) Stealth ryanodine-sensitive Ca2+ release contributes to activity of capacitative Ca2+ entry and nitric oxide synthase in bovine endothelial cells. J Physiol 513(Pt 2):369–379

Park CY, Hoover PJ, Mullins FM, Bachhawat P, Covington ED, Raunser S, Walz T, Garcia KC, Dolmetsch RE, Lewis RS (2009) STIM1 clusters and activates CRAC channels via direct binding of a cytosolic domain to Orai1. Cell 136(5):876–890

Poburko D, Liao CH, Lemos VS, Lin E, Maruyama Y, Cole WC, van Breemen C (2007) Transient receptor potential channel 6-mediated, localized cytosolic [Na+] transients drive Na+/Ca2+ exchanger-mediated Ca2+ entry in purinergically stimulated aorta smooth muscle cells. Circ Res 101(10):1030–1038

Poburko D, Fameli N, Kuo KH, van Breemen C (2008) Ca2+ signaling in smooth muscle: TRPC6, NCX and LNats in nanodomains. Channels (Austin) 2(1):10–12

Poburko D, Liao CH, van Breemen C, Demaurex N (2009) Mitochondrial regulation of sarcoplasmic reticulum Ca2+ content in vascular smooth muscle cells. Circ Res 104(1):104–112

PW Workshop (2010) Nanospace biophysics. Available at: http://pwias.ubc.ca/profile/edwin-moore

Raeymaekers L, Eggermont JA, Wuytack F, Casteels R (1990) Effects of cyclic nucleotide dependent protein kinases on the endoplasmic reticulum Ca2+ pump of bovine pulmonary artery. Cell Calcium 11(4):261–268

Rizzuto R, Duchen MR, Pozzan T (2004) Flirting in little space: the ER/mitochondria Ca2+ liaison. Sci STKE 2004(215):re1

Ron I, Horowitz M (2008) Intracellular cholesterol modifies the ERAD of glucocerebrosidase in Gaucher disease patients. Mol Genet Metab 93(4):426–436

Smoldyn (2016) Available at: http://www.smoldyn.org

Stiles JR, Bartol TM (2001) Monte Carlo methods for simulating realistic synaptic microphysiology using MCell. In: De Schutter E (ed) Computational neuroscience: realistic modeling for experimentalists. CRC, Boca Raton, pp 87–127

Stiles JR, Van Helden D, Bartol TM Jr, Salpeter EE, Salpeter MM (1996) Miniature endplate current rise times less than 100 microseconds from improved dual recordings can be modeled with passive acetylcholine diffusion from a synaptic vesicle. Proc Natl Acad Sci USA 93 (12):5747–5752

Thorne GD, Paul RJ (2003) Effects of organ culture on arterial gene expression and hypoxic relaxation: role of the ryanodine receptor. Am J Physiol Cell Physiol 284(4):C999–C1005

Tripathy A, Meissner G (1996) Sarcoplasmic reticulum lumenal Ca2+ has access to cytosolic activation and inactivation sites of skeletal muscle Ca2+ release channel. Biophys J 70 (6):2600–2615

van Breemen C (1977) Calcium requirement for activation of intact aortic smooth muscle. J Physiol 272(2):317–329

van Breemen C, Saida K (1989) Cellular mechanisms regulating [Ca2+]i smooth muscle. Annu Rev Physiol 51:315–329

van Breemen C, Lukeman S, Leijten P, Yamamoto H, Loutzenhiser R (1986) The role of superficial SR in modulating force development induced by Ca entry into arterial smooth muscle. J Cardiovasc Pharmacol 8(Suppl 8):S111–S116

van Breemen C, Fameli N, Evans AM (2013) Pan-junctional sarcoplasmic reticulum in vascular smooth muscle: nanospace Ca2+ transport for site- and function-specific Ca2+ signalling. J Physiol 591(8):2043–2054

Verboomen H, Wuytack F, Van den Bosch L, Mertens L, Casteels R (1994) The functional importance of the extreme C-terminal tail in the gene 2 organellar Ca(2+)-transport ATPase (SERCA2a/b). Biochem J 303(Pt 3):979–984

Wang IY, Bai Y, Sanderson MJ, Sneyd J (2010) A mathematical analysis of agonist- and KCl-induced Ca(2+) oscillations in mouse airway smooth muscle cells. Biophys J 98 (7):1170–1181

Wray S, Burdyga T (2010) Sarcoplasmic reticulum function in smooth muscle. Physiol Rev 90 (1):113–178

Tassoni JP Jr, Fawaz KA, Johnston DE (1991) Cirrhosis and portal hypertension in a patient with adult Niemann-Pick disease. Gastroenterology 100(2):567–569

Tissue Specificity: SOCE – Implications for Ca^{2+} Handling in Endothelial Cells

18

Lothar A. Blatter

Abstract

Many cellular functions of the vascular endothelium are regulated by fine-tuned global and local, microdomain-confined changes of cytosolic free Ca^{2+} ($[Ca^{2+}]_i$). Vasoactive agonist-induced stimulation of vascular endothelial cells (VECs) typically induces Ca^{2+} release through IP_3 receptor Ca^{2+} release channels embedded in the membrane of the endoplasmic reticulum (ER) Ca^{2+} store, followed by Ca^{2+} entry from the extracellular space elicited by Ca^{2+} store depletion and referred to as capacitative or store-operated Ca^{2+} entry (SOCE). In vascular endothelial cells, SOCE is graded with the degree of store depletion and controlled locally in the subcellular microdomain where depletion occurs. SOCE provides distinct Ca^{2+} signals that selectively control specific endothelial functions: in calf pulmonary artery endothelial cells, the SOCE Ca^{2+} signal drives nitric oxide (an endothelium-derived relaxing factor of the vascular smooth muscle) production and controls activation and nuclear translocation of the transcription factor NFAT. Both cellular events are not affected by Ca^{2+} signals of comparable magnitude arising directly from Ca^{2+} release from intracellular stores, clearly indicating that SOCE regulates specific Ca^{2+}-dependent cellular tasks by a unique and exclusive mechanism. This review discusses the mechanisms of intracellular Ca^{2+} regulation in vascular endothelial cells and the role of store-operated Ca^{2+} entry for endothelium-dependent smooth muscle relaxation and nitric oxide signaling, endothelial oxidative stress response, and excitation-transcription coupling in the vascular endothelium.

L.A. Blatter (✉)
Department of Physiology and Biophysics, Rush University Medical Center, 1750 W. Harrison St., Chicago, IL 60612, USA
e-mail: Lothar_Blatter@rush.edu

K. Groschner et al. (eds.), *Store-Operated Ca^{2+} Entry (SOCE) Pathways*, Advances in Experimental Medicine and Biology 993, DOI 10.1007/978-3-319-57732-6_18

Keywords
Cytosolic calcium signaling • Excitation-transcription coupling • IP_3 receptor-induced Ca^{2+} release • Nitric oxide • Oxidative stress • SOCE • Vascular endothelium

18.1 Introduction: The Vascular Endothelium

The vascular endothelium represents the thin monolayer of cells that form the interface between the circulating blood and the underlying tissue elements of the vessel wall and the heart. While long perceived as a mere passive bystander with the sole purpose of confining luminal content within the vasculature, it has become increasingly clear that the endothelium is actively involved in a plethora of regulatory tasks including regulation of vascular permeability, vascular smooth muscle tone, blood flow and pressure, control of blood coagulation, and influencing cell and tissue development and growth, inflammatory processes, tumor development and cancer, angiogenesis, and cardiac function. Endothelial dysfunction has been implicated as a causative or at least contributing factor in cardiovascular disease processes such as atherosclerosis, hypertension, cardiac hypertrophy, and congestive heart failure. At the cellular level, many endothelial functions are regulated by the intracellular messenger Ca^{2+}. In this chapter we will review the key factors, pathways, and organelles that regulated intracellular free Ca^{2+} ($[Ca^{2+}]_i$) in vascular endothelial cells (VECs). Furthermore, we will review the role of $[Ca^{2+}]_i$, specifically store-operated Ca^{2+} entry, in important endothelial processes, including generation of the endothelium-derived relaxing factor nitric oxide (NO), response to oxidative stress, and regulation of excitation-transcription coupling in the vascular endothelium.

18.2 Intracellular Ca^{2+} Regulation in Vascular Endothelial Cells

Elevation of $[Ca^{2+}]_i$ requires mobilization of Ca^{2+} ions that occurs either through release from intracellular Ca^{2+} stores and/or Ca^{2+} entry from the extracellular space via transmembrane Ca^{2+}-conducting pathways.

18.2.1 Ca^{2+} Release from Intracellular Stores

Vascular endothelial cells are constantly exposed to circulating blood-borne humoral factors and vasoactive agonists (e.g., adenosine triphosphate (ATP), histamine, thrombin, and others) and mechanical stimuli resulting from shear stress imposed by the blood flow. Shear stress-dependent mechanical forces stimulate a number of VEC signaling events aimed to protect against atherosclerosis (Harrison 2005). The response to the stimulation by vasoactive agonists typically involves the generation of a cytosolic Ca^{2+} signal that is employed to regulate VEC functions.

Humoral agonists bind to a G-protein-coupled receptor (GPCR) which triggers the inositol 1,4,5-trisphosphate (IP_3) signaling cascade where the activation of phospholipase C (PLC) results in the formation of the second messengers diacylglycerol (DAG) and IP_3. IP_3 binds to the IP_3 receptor (IP_3R) in the membrane of the ER Ca^{2+} store. The IP_3R represents one of the two major (together with the ryanodine receptor Ca^{2+} release channel) intracellular Ca^{2+} release channels. Upon activation by IP_3, IP_3Rs liberate Ca^{2+} from the ER leading to elevations of $[Ca^{2+}]_i$ with often complex spatial and temporal characteristics. Agonist-induced elevations of $[Ca^{2+}]_i$ can occur as localized non-propagating elementary Ca^{2+} release events through opening of single channels or clusters of IP_3Rs that have been termed Ca^{2+} blips (Parker and Yao 1996) and Ca^{2+} puffs (Yao et al. 1995), respectively. We demonstrated such elementary Ca^{2+} release events in VECs (Huser and Blatter 1997; Aromolaran et al. 2007). Figure 18.1A shows examples of elementary Ca^{2+} release events elicited by ATP in a calf pulmonary artery endothelial (CPAE) cell recorded with high-resolution fluorescence confocal imaging (Fig. 18.1Aa, Ab). Ca^{2+} release events take the form of localized non-propagating Ca^{2+} transients that occurred in temporal isolation (blips) or several transients that appeared in short succession at different sites in close proximity. Together these local release transients gave rise to an elevation of $[Ca^{2+}]_i$ of wider spatial spread and longer duration due to spatial and temporal summation, reminiscent of Ca^{2+} puffs. The $[Ca^{2+}]_i$ profiles in Fig. 18.1Ac show sites where individual Ca^{2+} blips occurred (e.g., bottom trace). Detailed analysis of Ca^{2+} puffs revealed distinct step-like increases of $[Ca^{2+}]_i$ of <10 nM amplitude (inset), providing evidence of the quantal nature of Ca^{2+} release and supporting the hypothesis that Ca^{2+} puffs represent the temporal and spatial summation of individual single-channel elementary Ca^{2+} release events, i.e., Ca^{2+} blips.

Agonist stimulation can induce localized Ca^{2+} release as shown in Fig. 18.1A but also lead to propagating Ca^{2+} waves and oscillatory changes of $[Ca^{2+}]_i$ (Huser and Blatter 1997; Aromolaran et al. 2007). In response to maximal agonist stimulation, VECs typically respond by a global and transient elevation of $[Ca^{2+}]_i$ that is followed by a decline of $[Ca^{2+}]_i$ to a sustained plateau of elevated $[Ca^{2+}]_i$ that depends on extracellular Ca^{2+} and typically persists during the presence of the agonist (Fig. 18.1B, C). Inhibition of the IP_3R prevents the agonist-induced increase of $[Ca^{2+}]_i$ (Fig. 18.1D) arguing against the presence of a direct receptor-activated Ca^{2+} entry pathway; however this question has remained controversial (e.g., Girardin et al. 2010; Jousset et al. 2008).

18.2.2 Ca^{2+} Store Depletion and Ca^{2+} Influx: Store-Operated Ca^{2+} Entry

Release of Ca^{2+} from the ER leads to intracellular Ca^{2+} store depletion. As in many other cell types, in VECs Ca^{2+} store depletion results in the activation of a plasma membrane Ca^{2+} influx pathway with the primary and immediate purpose of refilling depleted stores. This Ca^{2+} influx pathway has been described first in the mid-1980s

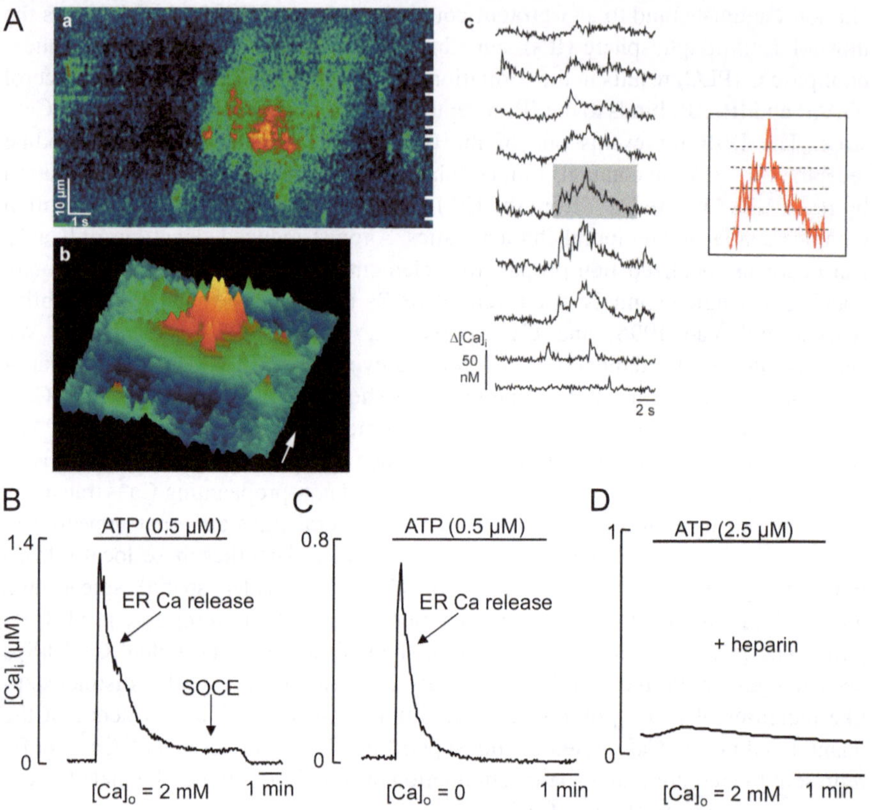

Fig. 18.1 Agonist (ATP)-induced $[Ca^{2+}]_i$ signals in CPAE cells. (**A**): Localized Ca^{2+} transients in the presence of 200 nM extracellular ATP. Confocal line scan image (**a**) and surface plot (**b**) of $[Ca^{2+}]_i$ (*blue* = 70 nM, *yellow* = 200 nM). Panel **c**: $[Ca^{2+}]_i$ profiles recorded from regions indicated by *white markings* in **Aa**. *Inset*: $[Ca^{2+}]_i$ profile from region marked by the *gray rectangle*. *Dashed lines* indicate levels of $\Delta[Ca^{2+}]_i$ ~10 nM. *Arrow*, time dimension. (**B–D**): ATP-induced Ca^{2+} transients in the presence (**B**) and absence (**C**) of extracellular Ca^{2+} and during inhibition of IP_3Rs with heparin (**D**) (panel **A** from Huser and Blatter 1997, used with permission. Panels **B** and **C** from Holda et al. 1998, used with permission)

(Putney 1986) and termed capacitative Ca^{2+} entry (CCE) or store-operated Ca^{2+} entry (SOCE) and has since been studied intensely regarding its structural features, its molecular and electrophysiological properties, its regulation, as well as its physiological and pathophysiological (CRAC channelopathies) functions (for pertinent reviews, see, e.g., Putney 1997, 2009; Berridge 1995; Holda et al. 1998; Varnai et al. 2009; Parekh and Putney 2005; Feske 2010; Smyth et al. 2010; Prakriya and Lewis 2015; Redondo and Rosado 2015).

Several studies have dealt with the demonstration of SOCE and its role in VECs (Malli et al. 2007; Abdullaev et al. 2008; Girardin et al. 2010; Madge et al. 1997; Mumtaz et al. 2010; Paltauf-Doburzynska et al. 1999; Dedkova and Blatter 2002; Holda and Blatter 1997; Huser et al. 1999; Klishin et al. 1998; Sedova and Blatter

1999; Sedova et al. 2000; Schilling et al. 1992; Gericke et al. 1993; Vaca and Kunze 1994; Dolor et al. 1992; Antigny et al. 2011). Vascular endothelial cells typically lack voltage-gated Ca^{2+} channels (Cannell and Sage 1989; Himmel et al. 1993); thus SOCE is likely the sole Ca^{2+} entry pathway in VECs (Madge et al. 1997). For example, in CPAE cells, we have shown (Sedova et al. 2000) that submaximal concentrations of the vasoactive agonist ATP caused quantal release of Ca^{2+} from the ER that was accompanied by a dose-dependent depletion of ER Ca^{2+} stores and an activity of SOCE that was graded with the amount of released Ca^{2+}. Furthermore, slow oscillations of $[Ca^{2+}]_i$ were paralleled by oscillatory activation of SOCE where activation lagged behind Ca^{2+} release and deactivation occurred on a slower time scale than activation. Activation of SOCE requires a minimal level of depletion (suggesting the existence of a threshold depletion level for activation), its activity is graded with the degree of store depletion over a broad-range ER Ca^{2+} content, and SOCE becomes fully activated before the ER is completely depleted, suggesting a safety mechanism against complete loss of stored Ca^{2+} (Fig. 18.2A). In intact CPAE cells, the Ca^{2+}-dependent, but delayed, activation of the plasma membrane Ca^{2+} ATPase (PMCA) during SOCE activity exerts a modulatory role for the net Ca^{2+} influx via SOCE (Fig. 18.2B) and cooperatively SOCE and PMCA activity control refilling of ER Ca^{2+} stores (Klishin et al. 1998). Calmodulin (CaM) and CaM kinase II (CaMKII) play an important role in the regulation of SOCE in VECs. CaMKII activates SOCE and inhibits IP_3R-mediated ER Ca^{2+} release (Aromolaran and Blatter 2005; Aromolaran et al. 2007), together resulting in conservation of stored Ca^{2+}, whereas CaM regulates the activity of PMCA and indirectly SOCE as outlined above. Several mechanisms participate in the cytosolic clearance of elevated $[Ca^{2+}]_i$. The majority of elevated $[Ca^{2+}]_i$ is sequestered by reuptake of Ca^{2+} into the ER by the sarcoplasmic/endoplasmic reticulum Ca^{2+} ATPase (SERCA) pump and extrusion across the plasma membrane by the PMCA. PMCA functions as a CaM-dependent high-affinity Ca^{2+} removal system; however under conditions of reduced PMCA activity, the low-affinity Na^+/Ca^{2+} exchange mechanism can compensate and maintain sufficient Ca^{2+} extrusion (Sedova and Blatter 1999). An important and not to be underestimated player in cytosolic Ca^{2+} sequestration are mitochondria which can take up significant amounts of Ca^{2+} in VECs (Sedova and Blatter 2000; Lawrie et al. 1996; Donnadieu and Bourguignon 1996). Mitochondrial Ca^{2+} uptake activates NO production by a mitochondria-specific nitric oxide synthase (mitoNOS). Mitochondrial NO production itself modulates mitochondrial Ca^{2+} uptake and efflux, provides a mechanism against mitochondrial Ca^{2+} overload, and represents a mechanism for fast local regulation of cell respiration, mitochondrial membrane potential, and apoptosis (Dedkova and Blatter 2005; Dedkova et al. 2004). Mitochondrial metabolism and Ca^{2+} signaling are linked to SOCE through several pathways (Parekh 2008). As discussed below NO inhibits SOCE (Dedkova and Blatter 2002); thus the mitochondrial source of NO has the potential to contribute to SOCE regulation in VECs. Furthermore, recent evidence has shown that mitochondrial Ca^{2+} uptake and extrusion play a critical role (Malli et al. 2003a, b, 2007) or are even a requirement for SOCE in VECs (Naghdi et al. 2010).

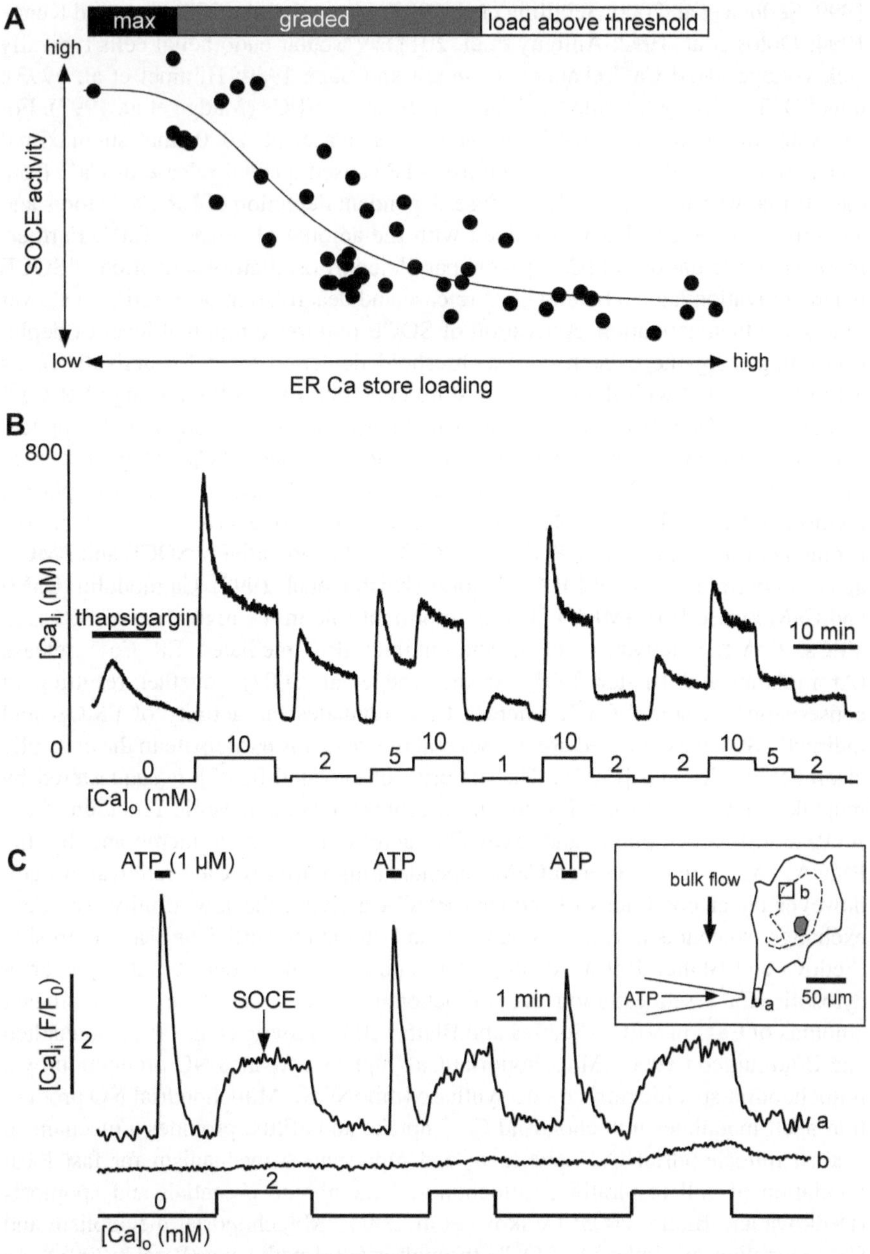

Fig. 18.2 SOCE signals in CPAE cells. (**A**) SOCE activity as a function of ER Ca^{2+} store filling (from Sedova et al. 2000, with permission). (**B**): SOCE signals induced by extracellular Ca^{2+} after permanent store depletion with the SERCA inhibitor thapsigargin. The magnitude of the SOCE signal is a function of $[Ca^{2+}]_o$ and is characterized by a rapid upstroke followed by a slower decline to an elevated plateau of $[Ca^{2+}]_i$. The latter is due to a delayed activation of PMCA (from Klishin et al. 1998, used with permission). (**C**) Focal application of ATP activates SOCE locally (site a), but not at distant sites (site b). Delayed activation of SOCE at site b is due to intra-ER redistribution of Ca^{2+} leading to ER Ca^{2+} depletion at site b (from Hüser et al. 1999, with permission)

18.2.3 SOCE: Properties and Regulation

While three decades of intensive studies of SOCE in a myriad of different tissues—mainly in electrically non-excitable cells, however, recent studies have unveiled a role for SOCE also in excitable cells such as smooth, skeletal, and cardiac muscle as well as in neurons—have generated important insight into the inner workings of SOCE, two important questions have remained enigmatic for many years, and it was not until rather recently that significant progress was made toward solving these puzzles: (1) what is the nature of the signal that conveys the information of a depleted Ca^{2+} store to the surface membrane to initiate SOCE, and (2) what is the structural and molecular identity of the SOCE pathway?

For many years the signaling mechanism relating depleted Ca^{2+} stores and activation of SOCE remained elusive despite tremendous efforts to characterize this pathway. Models that have been forwarded and debated over the years included (1) a specific diffusible messenger that is released by the depleted stores and subsequently activates Ca^{2+} entry and (2) a mechanism where conformational protein-protein interactions signal from the empty store to the surface membrane. Indeed, numerous second messengers have been implicated as a diffusible signal, including G-proteins, cGMP (cyclic guanosine monophosphate), IP_3 itself or other metabolites of the inositol phosphate pathway, elements of the P450 metabolism, various lipids (Clapham 1995), and CIF (Ca^{2+} influx factor), a phosphorylated low-molecular-weight compound proposed as a specific SOCE activating factor (Randriamampita and Tsien 1993). However none of these propositions have been proven convincingly to be the missing link for the activation of SOCE (Bolotina and Csutora 2005). The conformational-coupling models, on the other hand, put protein-protein interactions at the center of the mechanism and stress the importance of the close physical association of the surface membrane and the IP_3R ER Ca^{2+} release channel whose large cytosolic head locates closely to the surface membrane and thus to the putative surface membrane Ca^{2+} entry channel. Several studies have attempted to address this question in vascular endothelial cells. Using high-resolution confocal imaging, it could be demonstrated (Huser et al. 1999) that the local application of an agonist (ATP) caused a very localized release of Ca^{2+} from intracellular stores that was followed by localized activation of SOCE spatially restricted to the site of agonist application (Fig. 18.2C). While spatially restricted Ca^{2+} release and SOCE co-localized initially, repetitive agonist stimulation eventually led to detectable SOCE at remote sites which was due to Ca^{2+} redistribution within the ER toward the sites of release and eventually accounted for the spatial spread of SOCE activation due to gradual cell-wide store depletion. Furthermore, SOCE (but not agonist-induced ER Ca^{2+} release) was suppressed in VECs treated with cytochalasin D to disrupt the cytoskeleton (actin microfilaments) (Holda and Blatter 1997). While this effect on SOCE might be specific to VECs (indeed opposite effects were reported in different cell types, e.g., Ribeiro et al. 1997), both studies argued against a long-range cytosolic diffusible messenger and lent support to a conformational-coupling signaling mechanism. Nonetheless, all these earlier studies failed to deliver the ultimate proof for such a model in VECs.

A milestone toward resolving this conundrum was the groundbreaking work by several laboratories (Roos et al. 2005; Liou et al. 2005; Prakriya et al. 2006; Vig et al. 2006; Feske et al. 2006; Zhang et al. 2006). Their work laid the base for a new—and by now widely accepted—model for SOCE activation. In this model the ER Ca^{2+}-sensing proteins STIM1 and STIM2 (STIM, stromal interaction molecule) and the surface membrane pore-forming proteins of the Orai family (Orai1, 2, and 3) are the key players. Orai proteins are Ca^{2+} release-activated Ca^{2+}-selective ion channels. These channels have four transmembrane domains, form tetramers, and share no homology with any other ion channel. Orai proteins are now established as the molecular constituents of "calcium release-activated calcium currents" (I_{CRAC}; see below) in a number of cell types and tissues. The decreased $[Ca^{2+}]$ in the ER is sensed by STIM proteins. Subsequently, STIM proteins cluster in the ER membrane (visualized as "puncta") and relocate near the plasma membrane, where they activate Orai channels through protein-protein interaction. The requirement of STIM1 and Orai1 for the SOCE pathway (Antigny et al. 2011) and activation of I_{CRAC} has also been demonstrated in VECs (Abdullaev et al. 2008). Recent progress on the molecular biology of STIM and Orai has unveiled a large number of mutations of these proteins that result in cellular dysfunctions and have been linked to various diseases (for pertinent reviews, see, e.g., Prakriya and Lewis 2015; Feske et al. 2010; Lacruz and Feske 2015; Berna-Erro et al. 2012). Furthermore, recent advances have revealed details of the structure of the STIM and Orai proteins (Amcheslavsky et al. 2015; Prakriya 2013), have led to a better understanding of the pharmacology of SOCE (Jairaman and Prakriya 2013; Prakriya and Lewis 2015), and have solidified the notion that the STIM-Orai interaction has to be viewed as an integrated signaling complex that includes additional modulators and signaling compounds (Lopez et al. 2016; Hogan and Rao 2015; Putney 2013).

A membrane conductance directly related to store depletion, termed I_{CRAC}, was first demonstrated in mast cells (Hoth and Penner 1992). I_{CRAC} was initially described as an inwardly rectifying, highly Ca^{2+}-selective low-conductance current that could rapidly be inactivated by intracellular Ca^{2+}. Similar CRAC-like membrane currents have later been identified in VECs (Fasolato and Nilius 1998; Girardin et al. 2010). For many years I_{CRAC} has remained the best characterized SOCE Ca^{2+} current, although a growing number of reports on store-dependent conductances in different tissues and cell types with biophysical properties and modes of regulation and modulation different from the original I_{CRAC} suggested that SOCE may represent a family of membrane conductances with varying properties (Clapham 1995; Parekh and Penner 1997). More recently mammalian transient receptor potential (TRP) channels, a superfamily of nonselective cation channels, have moved into the center of focus. Among these channels, the classical or canonical TRPC family has been implicated in SOCE in a variety of cell types (e.g., Cheng et al. 2013), including VECs (Antigny et al. 2011; Groschner et al. 1998; Tiruppathi et al. 2002); however there is evidence that additional channels from other TRP families contribute to SOCE (Dragoni et al. 2015; Ma et al. 2011). Evidence has been established for a molecular association among STIM1, Orai1, and members of the TRPC family of channels (Dietrich et al. 2010; Ambudkar et al. 2007).

18.3 The Physiological Role of SOCE in the Vascular Endothelium

While the primary and immediate purpose of SOCE is the refilling of intracellular Ca^{2+} stores, it has become increasingly clear that SOCE is involved in the regulation of a range of cellular functions and tasks, in general as well as specifically in the vascular endothelium. A role for SOCE has been attributed to such diverse functions as volume regulation, secretion, phototransduction, mitogenesis and regulation of cell cycle, cell proliferation and migration, organ development, gene expression, motility, cardiovascular functions, and immune response (see reviews listed in Sect. 18.2.2). A specific role of SOCE for cell proliferation and tumorigenesis (e.g., Chen et al. 2011) has drawn increasing interest because SOCE could emerge as a potential target for cancer therapy (Xie et al. 2016; Jardin and Rosado 2016; Chen et al. 2016). In this context endothelial progenitor cells (Moccia et al. 2014) that are typically mobilized to replace dysfunctional or damaged endothelial cells (e.g., after an ischemic insult) have attracted interest as they might play a key role in sustaining tumor vascularization and growth as well as metastatization (Courjaret and Machaca 2012; Lodola et al. 2012; Sanchez-Hernandez et al. 2010; Moccia and Guerra 2016; Moccia and Poletto 2015; Fiorio Pla and Gkika 2013; Prevarskaya et al. 2011). It has been shown that SOCE plays a central role in endothelial progenitor cells (Wang et al. 2015; Zuccolo et al. 2016), lending further support to the hypothesis of SOCE as a cancer therapy target.

In the following paragraphs, we discuss several empirical studies that demonstrate a role of SOCE for three specific vascular endothelial functions: endothelial cell nitric oxide signaling, endothelial oxidative stress response, and excitation-transcription coupling.

18.3.1 SOCE and NO Signaling

The vascular endothelium plays a key role in blood flow and pressure regulation. This is achieved by the release of endothelium-derived relaxing factor (EDRF) and endothelium-derived contracting factor (EDCF) in response to neurohumoral and mechanical stimuli. Nitric oxide has been identified as the predominant (albeit not the only) EDRF (Palmer et al. 1987; Ignarro et al. 1987). NO relaxes the vascular smooth muscle (through activation of guanylate cyclase and cGMP formation) and is generated by nitric oxide synthase (endothelial isoform eNOS or NOS III) from L-arginine by a Ca^{2+}/CaM-dependent mechanism. Vasoactive agonists that raise $[Ca^{2+}]_i$ activate eNOS transiently to generate NO (Blatter et al. 1995; Nakatsubo et al. 1998; Forstermann et al. 1994). Detailed analysis of the Ca^{2+} signal and source of Ca^{2+} that activates NOS in VECs revealed (Fig. 18.3) that not ER Ca^{2+} release itself but the SOCE signal that follows Ca^{2+} release-dependent store depletion was the relevant Ca^{2+} signal for eNOS activation (Dedkova and Blatter 2002). NO generated by VECs not only affects other cell types of the vasculature, including vascular smooth muscle cells (Carrier et al. 1997), platelets (Radomski et al. 1987), and leukocytes (Kubes et al. 1991), but endothelial NO also has paracrine/autocrine

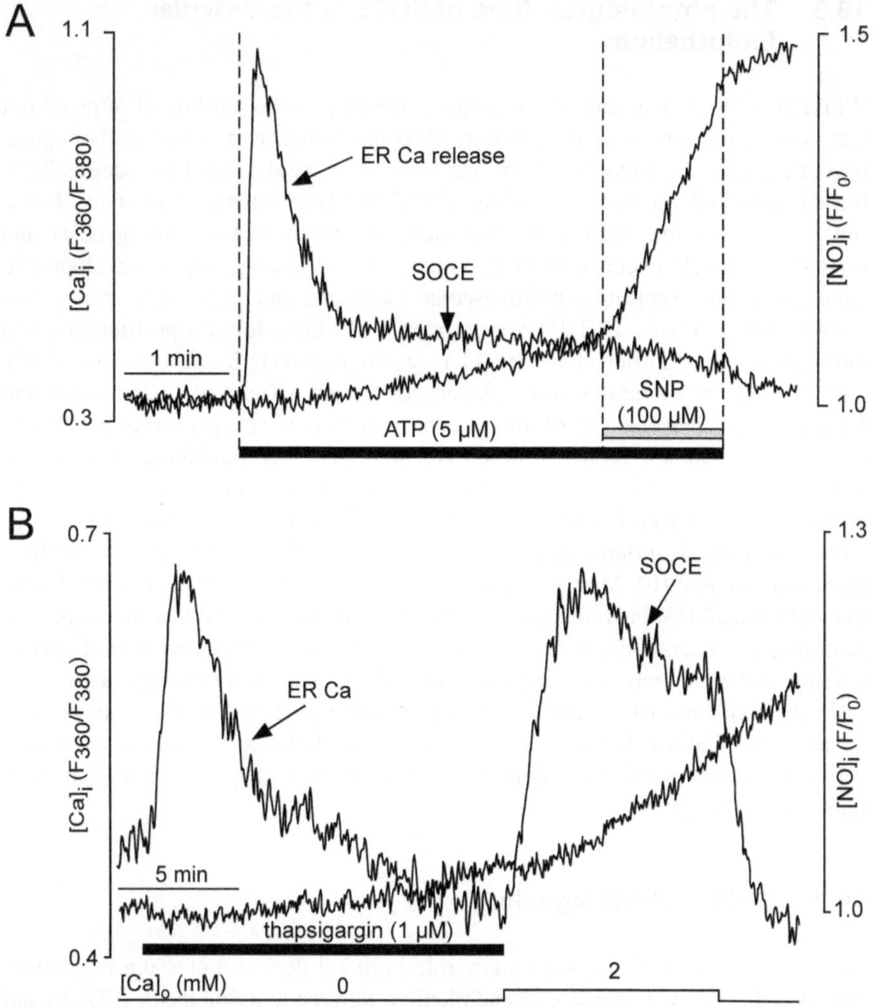

Fig. 18.3 Differential effect of ER Ca^{2+} release and SOCE on endothelial NO production. Liberation of Ca^{2+} from the ER by ATP stimulation (**a**) or SERCA inhibition with thapsigargin (**b**) has little effect on NO production. SOCE-dependent Ca^{2+} signals of similar or smaller amplitude stimulate endogenous NO formation. Exogenous NO applied in the form of sodium nitroprusside (SNP) reduces SOCE (**a**) (from Dedkova and Blatter 2002, with permission)

effects on NO and Ca^{2+} signaling in VECs. It has been shown that endogenous NO can lower basal $[Ca^{2+}]_i$ and attenuate agonist-induced Ca^{2+} responses through NO-dependent activation of PMCA but also can potentiate SOCE to sustain NO production (Chen et al. 2000). Furthermore, NO activates TRP channels in endothelial cells (Yoshida et al. 2006). This is in contrast to other studies where higher levels of NO have been shown to inhibit agonist-induced Ca^{2+} responses and Ca^{2+}

entry in endothelial cells, to reduce NOS expression levels and to suppress EDRF production (Takeuchi et al. 2004). Dedkova and Blatter (2002) showed that elevated cellular NO levels exert an autoregulatory negative feedback on $[Ca^{2+}]_i$ and NO production. This occurs through an NO/cGMP-dependent inhibition of SOCE and enhanced cytosolic Ca^{2+} clearance by accelerated reuptake of Ca^{2+} into the stores and by deactivation of SOCE as a result of enhanced filling of the stores (Dedkova and Blatter 2002). This negative feedback mechanism has a dual function, by serving as protection against excessive $[Ca^{2+}]_i$ levels and detrimentally high NO production (summarized in Yao and Huang 2003).

18.3.2 SOCE and Oxidative Stress

Under ischemic and postischemic conditions, the vascular endothelium can experience substantial degrees of oxidative stress brought upon by ischemia-related formation of reactive oxygen species (ROS) that ultimately leads to endothelial dysfunction and compromised Ca^{2+} signaling (Ward 1991; Zweier et al. 1994; Michiels et al. 1992; Zulueta et al. 1997). Several studies have found profound effects of ROS on intracellular Ca^{2+} release and SOCE in VECs (Elliott et al. 1989; Schilling and Elliott 1992; Doan et al. 1994; Elliott and Doan 1993; Graier et al. 1998; Dreher and Junod 1995; Hu and Ziegelstein 2000; Florea and Blatter 2008). Indeed, redox regulation of SOCE appears to be a common feature in various cell types (Nunes and Demaurex 2014; Bogeski et al. 2012; Prakriya and Lewis 2015). Furthermore, there is strong evidence that members of the TRPC channel family are redox-sensing channels and mediate, at least in part, cellular responses to ROS exposure and oxidative stress (Balzer et al. 1999; Poteser et al. 2006; Cioffi 2010; Kozai et al. 2014; Miller and Zhang 2011). In CPAE cells, for example (Florea and Blatter 2008), acute short-term exposure to oxidative stress had no effect on basal $[Ca^{2+}]_i$, ER Ca^{2+} release, and SOCE. In contrast, prolonged (60 min) exposure to ROS caused a profound inhibition of SOCE, whereas intracellular Ca^{2+} release initially remained unaffected. After 120 min of oxidative stress, not only was SOCE further reduced, but also Ca^{2+} release declined due to a slow depletion of the stores that resulted from SOCE inhibition. The data suggest that in VECs oxidative stress primarily affects Ca^{2+} influx in response to Ca^{2+} loss from internal stores. Vascular dysfunctions observed in hypoxia-reoxygenation situations and postischemic reperfusion injury may be explained by concomitant accumulation of large amounts of ROS (Lounsbury et al. 2000), followed by inhibition of endothelial SOCE with subsequent inhibition of NO formation and loss of an important endothelium-derived relaxing factor and paracrine mediator of blood flow regulation.

18.3.3 SOCE and Excitation-Transcription Coupling

SOCE has also been implicated in excitation-transcription coupling (ETC). A significant number of transcription factors that mediate gene and protein expression

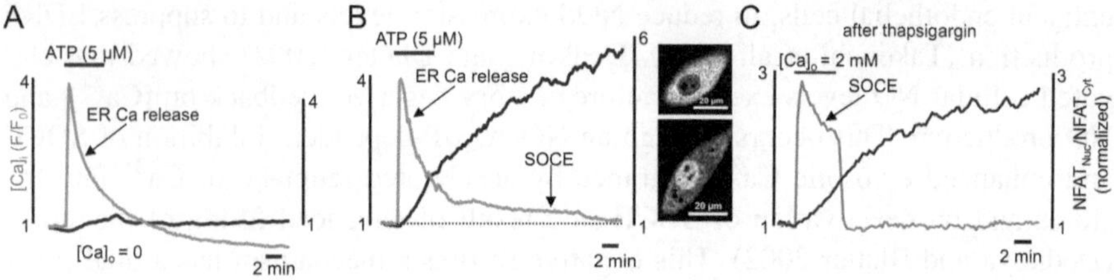

Fig. 18.4 SOCE and NFAT activation in vascular endothelial cells. (**a**) In the absence of extracellular Ca^{2+}, agonist (ATP)-induced $[Ca^{2+}]_i$ elevation (*gray trace*) fails to translocate NFAT to the nucleus (*black trace*). SOCE induced by agonist stimulation (**b**) or after store depletion with thapsigargin (**c**) is a potent stimulus for the translocation of NFAT to the nucleus in CPAE cells. *Inset* in (**b**): nuclear-cytosolic NFAT distribution of NFAT-GFP fluorescence before (*top*) and after (*bottom*) SOCE activation (reprinted from Rinne et al. 2009, with permission from Elsevier)

are Ca^{2+} dependent. Specific changes in cytosolic and nuclear Ca^{2+} signals, often in conjunction with the action of additional second messengers and signaling molecules, control the localization, translocation, and activity of these transcription factors. Vascular remodeling in response to physiological and pathological stimuli (e.g., hypertension or hypoxia) is the result of the activation of transcription factors, altered gene expression, and protein synthesis. In the vascular endothelium, hypoxia activates the transcription factor AP-1 resulting in TRPC4 expression that mediates hypoxia-induced gene expression and cell proliferation (Fantozzi et al. 2003). VECs express numerous transcription factors, including members of the NFAT family (NFAT, nuclear factor of activated T cells; Minami and Aird 2005). In the endothelium, NFAT controls gene expression during remodeling and is activated by growth factors and vasoactive agonists (Hadri et al. 2006; Boss et al. 1998; Hofer and Schweighofer 2007; Rinne et al. 2009). Activation and nuclear translocation of NFAT are mediated by dephosphorylation by the Ca^{2+}-sensitive phosphatase calcineurin (CaN). In CPAE cells, stimulation with the vasoactive agonists ATP or bradykinin in the presence of extracellular Ca^{2+} caused robust translocation of NFAT to the nucleus and activation of transcription (Rinne et al. 2009; Rinne and Blatter 2010). Influx of extracellular Ca^{2+} via SOCE, but not the Ca^{2+} release signal from the ER, was identified as the activating Ca^{2+} source (Fig. 18.4), suggesting a selective role of SOCE for gene transcription (e.g., Kar et al. 2014; Jara et al. 2014; Gwack et al. 2007).

18.4 Summary and Conclusion

A myriad of functions and tasks that are carried out in the cardiovascular system by the vascular endothelium rely on specific intracellular Ca^{2+} signals generated by exchange of Ca^{2+} ions between the cytoplasm, intracellular organelles, and the extracellular environment. Collective experimental evidence has shown that SOCE plays a center-stage role in providing the complex spatiotemporal Ca^{2+} signals required for the regulation of the diverse cellular functions of VECs and processes that may lead to cardiovascular pathologies and disease. Novel molecular and regenerative cell-based (Sanchez-Hernandez et al. 2010) therapeutic approaches that target SOCE channels and their regulatory pathways may represent future avenues for the treatment of endothelium-related cardiovascular diseases.

References

Abdullaev IF, Bisaillon JM, Potier M, Gonzalez JC, Motiani RK, Trebak M (2008) Stim1 and Orai1 mediate CRAC currents and store-operated calcium entry important for endothelial cell proliferation. Circ Res 103(11):1289–1299

Ambudkar IS, Ong HL, Liu X, Bandyopadhyay BC, Cheng KT (2007) TRPC1: the link between functionally distinct store-operated calcium channels. Cell Calcium 42(2):213–223

Amcheslavsky A, Wood ML, Yeromin AV, Parker I, Freites JA, Tobias DJ, Cahalan MD (2015) Molecular biophysics of Orai store-operated Ca2+ channels. Biophys J 108(2):237–246

Antigny F, Jousset H, Konig S, Frieden M (2011) Thapsigargin activates Ca(2)+ entry both by store-dependent, STIM1/Orai1-mediated, and store-independent, TRPC3/PLC/PKC-mediated pathways in human endothelial cells. Cell Calcium 49(2):115–127

Aromolaran AA, Blatter LA (2005) Modulation of intracellular Ca2+ release and capacitative Ca2+ entry by CaMKII inhibitors in bovine vascular endothelial cells. Am J Physiol Cell Physiol 289(6):C1426–C1436

Aromolaran AS, Zima AV, Blatter LA (2007) Role of glycolytically generated ATP for CaMKII-mediated regulation of intracellular Ca2+ signaling in bovine vascular endothelial cells. Am J Physiol Cell Physiol 293(1):C106–C118

Balzer M, Lintschinger B, Groschner K (1999) Evidence for a role of Trp proteins in the oxidative stress-induced membrane conductances of porcine aortic endothelial cells. Cardiovasc Res 42 (2):543–549

Berna-Erro A, Woodard GE, Rosado JA (2012) Orais and STIMs: physiological mechanisms and disease. J Cell Mol Med 16(3):407–424

Berridge MJ (1995) Capacitative calcium entry. Biochem J 312(Pt 1):1–11

Blatter LA, Taha Z, Mesaros S, Shacklock PS, Wier WG, Malinski T (1995) Simultaneous measurements of Ca2+ and nitric oxide in bradykinin-stimulated vascular endothelial cells. Circ Res 76(5):922–924

Bogeski I, Kilch T, Niemeyer BA (2012) ROS and SOCE: recent advances and controversies in the regulation of STIM and Orai. J Physiol 590(17):4193–4200

Bolotina VM, Csutora P (2005) CIF and other mysteries of the store-operated Ca2+-entry pathway. Trends Biochem Sci 30(7):378–387

Boss V, Wang X, Koppelman LF, Xu K, Murphy TJ (1998) Histamine induces nuclear factor of activated T cell-mediated transcription and cyclosporin A-sensitive interleukin-8 mRNA expression in human umbilical vein endothelial cells. Mol Pharmacol 54(2):264–272

Cannell MB, Sage SO (1989) Bradykinin-evoked changes in cytosolic calcium and membrane currents in cultured bovine pulmonary artery endothelial cells. J Physiol 419:555–568

Carrier GO, Fuchs LC, Winecoff AP, Giulumian AD, White RE (1997) Nitrovasodilators relax mesenteric microvessels by cGMP-induced stimulation of Ca2+-activated K channels. Am J Physiol 273(1 Pt 2):H76–H84

Chen J, Wang Y, Nakajima T, Iwasawa K, Hikiji H, Sunamoto M, Choi DK, Yoshida Y, Sakaki Y, Toyo-Oka T (2000) Autocrine action and its underlying mechanism of nitric oxide on intracellular Ca2+ homeostasis in vascular endothelial cells. J Biol Chem 275(37):28739–28749

Chen YF, Chiu WT, Chen YT, Lin PY, Huang HJ, Chou CY, Chang HC, Tang MJ, Shen MR (2011) Calcium store sensor stromal-interaction molecule 1-dependent signaling plays an important role in cervical cancer growth, migration, and angiogenesis. Proc Natl Acad Sci USA 108(37):15225–15230

Chen YF, Hsu KF, Shen MR (2016) The store-operated Ca(2+) entry-mediated signaling is important for cancer spread. Biochim Biophys Acta 1863(6 Pt B):1427–1435
Cheng KT, Ong HL, Liu X, Ambudkar IS (2013) Contribution and regulation of TRPC channels in store-operated Ca2+ entry. Curr Top Membr 71:149–179
Cioffi DL (2010) Redox regulation of endothelial canonical transient receptor potential channels. Antioxid Redox Signal 15(6):1567–1582
Clapham DE (1995) Calcium signaling. Cell 80(2):259–268
Courjaret R, Machaca K (2012) STIM and Orai in cellular proliferation and division. Front Biosci (Elite Ed) 4:331–341
Dedkova EN, Blatter LA (2002) Nitric oxide inhibits capacitative Ca2+ entry and enhances endoplasmic reticulum Ca2+ uptake in bovine vascular endothelial cells. J Physiol 539 (Pt 1):77–91
Dedkova EN, Blatter LA (2005) Modulation of mitochondrial Ca2+ by nitric oxide in cultured bovine vascular endothelial cells. Am J Physiol Cell Physiol 289(4):C836–C845
Dedkova EN, Ji X, Lipsius SL, Blatter LA (2004) Mitochondrial calcium uptake stimulates nitric oxide production in mitochondria of bovine vascular endothelial cells. Am J Physiol Cell Physiol 286(2):C406–C415
Dietrich A, Kalwa H, Gudermann T (2010) TRPC channels in vascular cell function. Thromb Haemost 103(2):262–270
Doan TN, Gentry DL, Taylor AA, Elliott SJ (1994) Hydrogen peroxide activates agonist-sensitive Ca(2+)-flux pathways in canine venous endothelial cells. Biochem J 297(Pt 1):209–215
Dolor RJ, Hurwitz LM, Mirza Z, Strauss HC, Whorton AR (1992) Regulation of extracellular calcium entry in endothelial cells: role of intracellular calcium pool. Am J Physiol 262(1 Pt 1): C171–C181
Donnadieu E, Bourguignon LY (1996) Ca2+ signaling in endothelial cells stimulated by bradykinin: Ca2+ measurement in the mitochondria and the cytosol by confocal microscopy. Cell Calcium 20(1):53–61
Dragoni S, Guerra G, Fiorio Pla A, Bertoni G, Rappa A, Poletto V, Bottino C, Aronica A, Lodola F, Cinelli MP, Laforenza U, Rosti V, Tanzi F, Munaron L, Moccia F (2015) A functional transient receptor potential vanilloid 4 (TRPV4) channel is expressed in human endothelial progenitor cells. J Cell Physiol 230(1):95–104
Dreher D, Junod AF (1995) Differential effects of superoxide, hydrogen peroxide, and hydroxyl radical on intracellular calcium in human endothelial cells. J Cell Physiol 162(1):147–153
Elliott SJ, Doan TN (1993) Oxidant stress inhibits the store-dependent Ca(2+)-influx pathway of vascular endothelial cells. Biochem J 292(Pt 2):385–393
Elliott SJ, Eskin SG, Schilling WP (1989) Effect of t-butyl-hydroperoxide on bradykinin-stimulated changes in cytosolic calcium in vascular endothelial cells. J Biol Chem 264 (7):3806–3810
Fantozzi I, Zhang S, Platoshyn O, Remillard CV, Cowling RT, Yuan JX (2003) Hypoxia increases AP-1 binding activity by enhancing capacitative Ca2+ entry in human pulmonary artery endothelial cells. Am J Physiol Lung Cell Mol Physiol 285(6):L1233–L1245
Fasolato C, Nilius B (1998) Store depletion triggers the calcium release-activated calcium current (ICRAC) in macrovascular endothelial cells: a comparison with Jurkat and embryonic kidney cell lines. Pflugers Arch 436(1):69–74
Feske S (2010) CRAC channelopathies. Pflugers Arch 460(2):417–435
Feske S, Gwack Y, Prakriya M, Srikanth S, Puppel SH, Tanasa B, Hogan PG, Lewis RS, Daly M, Rao A (2006) A mutation in Orai1 causes immune deficiency by abrogating CRAC channel function. Nature 441(7090):179–185
Feske S, Picard C, Fischer A (2010) Immunodeficiency due to mutations in ORAI1 and STIM1. Clin Immunol 135(2):169–182
Fiorio Pla A, Gkika D (2013) Emerging role of TRP channels in cell migration: from tumor vascularization to metastasis. Front Physiol 4:311

Florea SM, Blatter LA (2008) The effect of oxidative stress on Ca(2+) release and capacitative Ca(2+) entry in vascular endothelial cells. Cell Calcium 43(4):405–415

Forstermann U, Closs EI, Pollock JS, Nakane M, Schwarz P, Gath I, Kleinert H (1994) Nitric oxide synthase isozymes. Characterization, purification, molecular cloning, and functions. Hypertension 23(6 Pt 2):1121–1131

Gericke M, Droogmans G, Nilius B (1993) Thapsigargin discharges intracellular calcium stores and induces transmembrane currents in human endothelial cells. Pflugers Arch 422(6):552–557

Girardin NC, Antigny F, Frieden M (2010) Electrophysiological characterization of store-operated and agonist-induced Ca2+ entry pathways in endothelial cells. Pflugers Arch 460(1):109–120

Graier WF, Hoebel BG, Paltauf-Doburzynska J, Kostner GM (1998) Effects of superoxide anions on endothelial Ca2+ signaling pathways. Arterioscler Thromb Vasc Biol 18(9):1470–1479

Groschner K, Hingel S, Lintschinger B, Balzer M, Romanin C, Zhu X, Schreibmayer W (1998) Trp proteins form store-operated cation channels in human vascular endothelial cells. FEBS Lett 437(1-2):101–106

Gwack Y, Feske S, Srikanth S, Hogan PG, Rao A (2007) Signalling to transcription: store-operated Ca2+ entry and NFAT activation in lymphocytes. Cell Calcium 42(2):145–156

Hadri L, Pavoine C, Lipskaia L, Yacoubi S, Lompre AM (2006) Transcription of the sarcoplasmic/endoplasmic reticulum Ca2+-ATPase type 3 gene, ATP2A3, is regulated by the calcineurin/NFAT pathway in endothelial cells. Biochem J 394(Pt 1):27–33

Harrison DG (2005) The shear stress of keeping arteries clear. Nat Med 11(4):375–376

Himmel HM, Whorton AR, Strauss HC (1993) Intracellular calcium, currents, and stimulus-response coupling in endothelial cells. Hypertension 21(1):112–127

Hofer E, Schweighofer B (2007) Signal transduction induced in endothelial cells by growth factor receptors involved in angiogenesis. Thromb Haemost 97(3):355–363

Hogan PG, Rao A (2015) Store-operated calcium entry: mechanisms and modulation. Biochem Biophys Res Commun 460(1):40–49

Holda JR, Blatter LA (1997) Capacitative calcium entry is inhibited in vascular endothelial cells by disruption of cytoskeletal microfilaments. FEBS Lett 403(2):191–196

Holda JR, Klishin A, Sedova M, Huser J, Blatter LA (1998) Capacitative calcium entry. News Physiol Sci 13:157–163

Hoth M, Penner R (1992) Depletion of intracellular calcium stores activates a calcium current in mast cells. Nature 355(6358):353–356

Hu Q, Ziegelstein RC (2000) Hypoxia/reoxygenation stimulates intracellular calcium oscillations in human aortic endothelial cells. Circulation 102(20):2541–2547

Huser J, Blatter LA (1997) Elementary events of agonist-induced Ca2+ release in vascular endothelial cells. Am J Physiol 273(5 Pt 1):C1775–C1782

Huser J, Holda JR, Kockskamper J, Blatter LA (1999) Focal agonist stimulation results in spatially restricted Ca2+ release and capacitative Ca2+ entry in bovine vascular endothelial cells. J Physiol 514(Pt 1):101–109

Ignarro LJ, Buga GM, Wood KS, Byrns RE, Chaudhuri G (1987) Endothelium-derived relaxing factor produced and released from artery and vein is nitric oxide. Proc Natl Acad Sci USA 84 (24):9265–9269

Jairaman A, Prakriya M (2013) Molecular pharmacology of store-operated CRAC channels. Channels (Austin) 7(5):402–414

Jara E, Hidalgo MA, Hancke JL, Hidalgo AI, Brauchi S, Nunez L, Villalobos C, Burgos RA (2014) Delphinidin activates NFAT and induces IL-2 production through SOCE in T cells. Cell Biochem Biophys 68(3):497–509

Jardin I, Rosado JA (2016) STIM and calcium channel complexes in cancer. Biochim Biophys Acta 1863(6 Pt B):1418–1426

Jousset H, Malli R, Girardin N, Graier WF, Demaurex N, Frieden M (2008) Evidence for a receptor-activated Ca^{2+} entry pathway independent from Ca^{2+} store depletion in endothelial cells. Cell Calcium 43(1):83–94

Kar P, Samanta K, Kramer H, Morris O, Bakowski D, Parekh AB (2014) Dynamic assembly of a membrane signaling complex enables selective activation of NFAT by Orai1. Curr Biol 24 (12):1361–1368
Klishin A, Sedova M, Blatter LA (1998) Time-dependent modulation of capacitative Ca2+ entry signals by plasma membrane Ca2+ pump in endothelium. Am J Physiol 274(4 Pt 1):C1117–C1128
Kozai D, Ogawa N, Mori Y (2014) Redox regulation of transient receptor potential channels. Antioxid Redox Signal 21(6):971–986
Kubes P, Suzuki M, Granger DN (1991) Nitric oxide: an endogenous modulator of leukocyte adhesion. Proc Natl Acad Sci USA 88(11):4651–4655
Lacruz RS, Feske S (2015) Diseases caused by mutations in ORAI1 and STIM1. Ann N Y Acad Sci 1356:45–79
Lawrie AM, Rizzuto R, Pozzan T, Simpson AW (1996) A role for calcium influx in the regulation of mitochondrial calcium in endothelial cells. J Biol Chem 271(18):10753–10759
Liou J, Kim ML, Heo WD, Jones JT, Myers JW, Ferrell JE Jr, Meyer T (2005) STIM is a Ca2+ sensor essential for Ca2+-store-depletion-triggered Ca2+ influx. Curr Biol 15(13):1235–1241
Lodola F, Laforenza U, Bonetti E, Lim D, Dragoni S, Bottino C, Ong HL, Guerra G, Ganini C, Massa M, Manzoni M, Ambudkar IS, Genazzani AA, Rosti V, Pedrazzoli P, Tanzi F, Moccia F, Porta C (2012) Store-operated Ca2+ entry is remodelled and controls in vitro angiogenesis in endothelial progenitor cells isolated from tumoral patients. PLoS One 7(9): e42541
Lopez JJ, Albarran L, Gomez LJ, Smani T, Salido GM, Rosado JA (2016) Molecular modulators of store-operated calcium entry. Biochim Biophys Acta 1863(8):2037–2043
Lounsbury KM, Hu Q, Ziegelstein RC (2000) Calcium signaling and oxidant stress in the vasculature. Free Radic Biol Med 28(9):1362–1369
Ma X, Cheng KT, Wong CO, O'Neil RG, Birnbaumer L, Ambudkar IS, Yao X (2011) Heteromeric TRPV4-C1 channels contribute to store-operated Ca(2+) entry in vascular endothelial cells. Cell Calcium 50(6):502–509
Madge L, Marshall IC, Taylor CW (1997) Delayed autoregulation of the Ca2+ signals resulting from capacitative Ca2+ entry in bovine pulmonary artery endothelial cells. J Physiol 498 (Pt 2):351–369
Malli R, Frieden M, Osibow K, Graier WF (2003a) Mitochondria efficiently buffer subplasmalemmal Ca2+ elevation during agonist stimulation. J Biol Chem 278 (12):10807–10815
Malli R, Frieden M, Osibow K, Zoratti C, Mayer M, Demaurex N, Graier WF (2003b) Sustained Ca2+ transfer across mitochondria is Essential for mitochondrial Ca2+ buffering, sore-operated Ca2+ entry, and Ca2+ store refilling. J Biol Chem 278(45):44769–44779
Malli R, Frieden M, Hunkova M, Trenker M, Graier WF (2007) Ca2+ refilling of the endoplasmic reticulum is largely preserved albeit reduced Ca2+ entry in endothelial cells. Cell Calcium 41 (1):63–76
Michiels C, Arnould T, Houbion A, Remacle J (1992) Human umbilical vein endothelial cells submitted to hypoxia-reoxygenation in vitro: implication of free radicals, xanthine oxidase, and energy deficiency. J Cell Physiol 153(1):53–61
Miller BA, Zhang W (2011) TRP channels as mediators of oxidative stress. Adv Exp Med Biol 704:531–544
Minami T, Aird WC (2005) Endothelial cell gene regulation. Trends Cardiovasc Med 15 (5):174–184
Moccia F, Guerra G (2016) Ca(2+) signalling in endothelial progenitor cells: friend or foe? J Cell Physiol 231(2):314–327
Moccia F, Poletto V (2015) May the remodeling of the Ca(2)(+) toolkit in endothelial progenitor cells derived from cancer patients suggest alternative targets for anti-angiogenic treatment? Biochim Biophys Acta 1853(9):1958–1973

Moccia F, Tanzi F, Munaron L (2014) Endothelial remodelling and intracellular calcium machinery. Curr Mol Med 14(4):457–480
Mumtaz S, Burdyga G, Borisova L, Wray S, Burdyga T (2010) The mechanism of agonist induced Ca(2+) signalling in intact endothelial cells studied confocally in in situ arteries. Cell Calcium 49(1):66–77
Naghdi S, Waldeck-Weiermair M, Fertschai I, Poteser M, Graier WF, Malli R (2010) Mitochondrial Ca2+ uptake and not mitochondrial motility is required for STIM1-Orai1-dependent store-operated Ca2+ entry. J Cell Sci 123(Pt 15):2553–2564
Nakatsubo N, Kojima H, Kikuchi K, Nagoshi H, Hirata Y, Maeda D, Imai Y, Irimura T, Nagano T (1998) Direct evidence of nitric oxide production from bovine aortic endothelial cells using new fluorescence indicators: diaminofluoresceins. FEBS Lett 427(2):263–266
Nunes P, Demaurex N (2014) Redox regulation of store-operated Ca2+ entry. Antioxid Redox Signal 21(6):915–932
Palmer RM, Ferrige AG, Moncada S (1987) Nitric oxide release accounts for the biological activity of endothelium-derived relaxing factor. Nature 327(6122):524–526
Paltauf-Doburzynska J, Frieden M, Graier WF (1999) Mechanisms of Ca2+ store depletion in single endothelial cells in a Ca(2+)-free environment. Cell Calcium 25(5):345–353
Parekh AB (2008) Mitochondrial regulation of store-operated CRAC channels. Cell Calcium 44 (1):6–13
Parekh AB, Penner R (1997) Store depletion and calcium influx. Physiol Rev 77(4):901–930
Parekh AB, Putney JW Jr (2005) Store-operated calcium channels. Physiol Rev 85(2):757–810
Parker I, Yao Y (1996) Ca2+ transients associated with openings of inositol trisphosphate-gated channels in Xenopus oocytes. J Physiol 491(Pt 3):663–668
Poteser M, Graziani A, Rosker C, Eder P, Derler I, Kahr H, Zhu MX, Romanin C, Groschner K (2006) TRPC3 and TRPC4 associate to form a redox-sensitive cation channel. Evidence for expression of native TRPC3-TRPC4 heteromeric channels in endothelial cells. J Biol Chem 281(19):13588–13595
Prakriya M (2013) Store-operated Orai channels structure and function. Curr Top Membr 71:1–32
Prakriya M, Lewis RS (2015) Store-operated calcium channels. Physiol Rev 95(4):1383–1436
Prakriya M, Feske S, Gwack Y, Srikanth S, Rao A, Hogan PG (2006) Orai1 is an essential pore subunit of the CRAC channel. Nature 443(7108):230–233
Prevarskaya N, Skryma R, Shuba Y (2011) Calcium in tumour metastasis: new roles for known actors. Nat Rev Cancer 11(8):609–618
Putney JW Jr (1986) A model for receptor-regulated calcium entry. Cell Calcium 7(1):1–12
Putney JW Jr (1997) Capacitative calcium entry. R. G. Landes, Georgetown, TX
Putney JW Jr (2009) Capacitative calcium entry: from concept to molecules. Immunol Rev 231 (1):10–22
Putney JW Jr (2013) Alternative forms of the store-operated calcium entry mediators, STIM1 and Orai1. Curr Top Membr 71:109–123
Radomski MW, Palmer RM, Moncada S (1987) The anti-aggregating properties of vascular endothelium: interactions between prostacyclin and nitric oxide. Br J Pharmacol 92 (3):639–646
Randriamampita C, Tsien RY (1993) Emptying of intracellular Ca2+ stores releases a novel small messenger that stimulates Ca2+ influx. Nature 364(6440):809–814
Redondo PC, Rosado JA (2015) Store-operated calcium entry: unveiling the calcium handling signalplex. Int Rev Cell Mol Biol 316:183–226
Ribeiro CM, Reece J, Putney JW Jr (1997) Role of the cytoskeleton in calcium signaling in NIH 3T3 cells. An intact cytoskeleton is required for agonist-induced [Ca2+]i signaling, but not for capacitative calcium entry. J Biol Chem 272(42):26555–26561
Rinne A, Blatter LA (2010) A fluorescence-based assay to monitor transcriptional activity of NFAT in living cells. J Physiol 588(Pt 17):3211–3216

Rinne A, Banach K, Blatter LA (2009) Regulation of nuclear factor of activated T cells (NFAT) in vascular endothelial cells. J Mol Cell Cardiol 47(3):400–410

Roos J, DiGregorio PJ, Yeromin AV, Ohlsen K, Lioudyno M, Zhang S, Safrina O, Kozak JA, Wagner SL, Cahalan MD, Velicelebi G, Stauderman KA (2005) STIM1, an essential and conserved component of store-operated Ca2+ channel function. J Cell Biol 169(3):435–445

Sanchez-Hernandez Y, Laforenza U, Bonetti E, Fontana J, Dragoni S, Russo M, Avelino-Cruz JE, Schinelli S, Testa D, Guerra G, Rosti V, Tanzi F, Moccia F (2010) Store-operated Ca(2+) entry is expressed in human endothelial progenitor cells. Stem Cells Dev 19(12):1967–1981

Schilling WP, Elliott SJ (1992) Ca2+ signaling mechanisms of vascular endothelial cells and their role in oxidant-induced endothelial cell dysfunction. Am J Physiol 262(6 Pt 2):H1617–H1630

Schilling WP, Cabello OA, Rajan L (1992) Depletion of the inositol 1,4,5-trisphosphate-sensitive intracellular Ca2+ store in vascular endothelial cells activates the agonist-sensitive Ca(2+)-influx pathway. Biochem J 284(Pt 2):521–530

Sedova M, Blatter LA (1999) Dynamic regulation of [Ca2+]i by plasma membrane Ca(2+)-ATPase and Na+/Ca2+ exchange during capacitative Ca2+ entry in bovine vascular endothelial cells. Cell Calcium 25(5):333–343

Sedova M, Blatter LA (2000) Intracellular sodium modulates mitochondrial calcium signaling in vascular endothelial cells. J Biol Chem 275(45):35402–35407

Sedova M, Klishin A, Huser J, Blatter LA (2000) Capacitative Ca2+ entry is graded with degree of intracellular Ca2+ store depletion in bovine vascular endothelial cells. J Physiol 523 (Pt 3):549–559

Smyth JT, Hwang SY, Tomita T, DeHaven WI, Mercer JC, Putney JW (2010) Activation and regulation of store-operated calcium entry. J Cell Mol Med 14(10):2337–2349

Takeuchi K, Watanabe H, Tran QK, Ozeki M, Sumi D, Hayashi T, Iguchi A, Ignarro LJ, Ohashi K, Hayashi H (2004) Nitric oxide: inhibitory effects on endothelial cell calcium signaling, prostaglandin I2 production and nitric oxide synthase expression. Cardiovasc Res 62 (1):194–201

Tiruppathi C, Minshall RD, Paria BC, Vogel SM, Malik AB (2002) Role of Ca2+ signaling in the regulation of endothelial permeability. Vasc Pharmacol 39(4–5):173–185

Vaca L, Kunze DL (1994) Depletion of intracellular Ca2+ stores activates a Ca(2+)-selective channel in vascular endothelium. Am J Physiol 267(4 Pt 1):C920–C925

Varnai P, Hunyady L, Balla T (2009) STIM and Orai: the long-awaited constituents of store-operated calcium entry. Trends Pharmacol Sci 30(3):118–128

Vig M, Peinelt C, Beck A, Koomoa DL, Rabah D, Koblan-Huberson M, Kraft S, Turner H, Fleig A, Penner R, Kinet JP (2006) CRACM1 is a plasma membrane protein essential for store-operated Ca2+ entry. Science 312(5777):1220–1223

Wang LY, Zhang JH, Yu J, Yang J, Deng MY, Kang HL, Huang L (2015) Reduction of store-operated Ca(2+) entry correlates with endothelial progenitor cell dysfunction in atherosclerotic mice. Stem Cells Dev 24(13):1582–1590

Ward PA (1991) Mechanisms of endothelial cell injury. J Lab Clin Med 118(5):421–426

Xie J, Pan H, Yao J, Zhou Y, Han W (2016) SOCE and cancer: recent progress and new perspectives. Int J Cancer 138(9):2067–2077

Yao X, Huang Y (2003) From nitric oxide to endothelial cytosolic Ca2+: a negative feedback control. Trends Pharmacol Sci 24(6):263–266

Yao Y, Choi J, Parker I (1995) Quantal puffs of intracellular Ca2+ evoked by inositol trisphosphate in Xenopus oocytes. J Physiol 482(Pt 3):533–553

Yoshida T, Inoue R, Morii T, Takahashi N, Yamamoto S, Hara Y, Tominaga M, Shimizu S, Sato Y, Mori Y (2006) Nitric oxide activates TRP channels by cysteine S-nitrosylation. Nat Chem Biol 2(11):596–607

Zhang SL, Yeromin AV, Zhang XH, Yu Y, Safrina O, Penna A, Roos J, Stauderman KA, Cahalan MD (2006) Genome-wide RNAi screen of Ca(2+) influx identifies genes that regulate Ca(2+) release-activated Ca(2+) channel activity. Proc Natl Acad Sci USA 103(24):9357–9362

Zuccolo E, Bottino C, Diofano F, Poletto V, Codazzi AC, Mannarino S, Campanelli R, Fois G, Marseglia GL, Guerra G, Montagna D, Laforenza U, Rosti V, Massa M, Moccia F (2016) Constitutive store-operated Ca(2+) entry leads to enhanced nitric oxide production and proliferation in infantile hemangioma-derived endothelial colony-forming cells. Stem Cells Dev 25 (4):301–319

Zulueta JJ, Sawhney R, Yu FS, Cote CC, Hassoun PM (1997) Intracellular generation of reactive oxygen species in endothelial cells exposed to anoxia-reoxygenation. Am J Physiol 272(5 Pt 1): L897–L902

Zweier JL, Broderick R, Kuppusamy P, Thompson-Gorman S, Lutty GA (1994) Determination of the mechanism of free radical generation in human aortic endothelial cells exposed to anoxia and reoxygenation. J Biol Chem 269(39):24156–24162

Tissue Specificity: Store-Operated Ca^{2+} Entry in Cardiac Myocytes

19

Martin D. Bootman and Katja Rietdorf

Abstract

Calcium (Ca^{2+}) is a key regulator of cardiomyocyte contraction. The Ca^{2+} channels, pumps, and exchangers responsible for the cyclical cytosolic Ca^{2+} signals that underlie contraction are well known. In addition to those Ca^{2+} signaling components responsible for contraction, it has been proposed that cardiomyocytes express channels that promote the influx of Ca^{2+} from the extracellular milieu to the cytosol in response to depletion of intracellular Ca^{2+} stores. With non-excitable cells, this store-operated Ca^{2+} entry (SOCE) is usually easily demonstrated and is essential for prolonging cellular Ca^{2+} signaling and for refilling depleted Ca^{2+} stores. The role of SOCE in cardiomyocytes, however, is rather more elusive. While there is published evidence for increased Ca^{2+} influx into cardiomyocytes following Ca^{2+} store depletion, it has not been universally observed. Moreover, SOCE appears to be prominent in embryonic cardiomyocytes but declines with postnatal development. In contrast, there is overwhelming evidence that the molecular components of SOCE (e.g., STIM, Orai, and TRPC proteins) are expressed in cardiomyocytes from embryo to adult. Moreover, these proteins have been shown to contribute to disease conditions such as pathological hypertrophy, and reducing their expression can attenuate hypertrophic growth. It is plausible that SOCE might underlie Ca^{2+} influx into cardiomyocytes and may have important signaling functions perhaps by activating local Ca^{2+}-sensitive processes. However, the STIM, Orai, and TRPC proteins appear to cooperate with multiple protein partners in signaling complexes. It is therefore possible that some of their signaling activities are not mediated by Ca^{2+} influx signals, but by protein-protein interactions.

M.D. Bootman (✉) • K. Rietdorf (✉)
School of Life, Health and Chemical Sciences, The Open University, Walton Hall, Milton Keynes MK7 6AA, UK
e-mail: martin.bootman@open.ac.uk; katja.rietdorf@open.ac.uk

K. Groschner et al. (eds.), *Store-Operated Ca^{2+} Entry (SOCE) Pathways*, Advances in Experimental Medicine and Biology 993, DOI 10.1007/978-3-319-57732-6_19

Keywords
SOCE • Calcium • Signaling • Cardiac • Hypertrophy • Arrhythmia • Stromal interaction molecule • STIM

19.1 Store-Operated Ca^{2+} Entry in Cardiac Myocytes: Why Does It Matter?

A human heart beats over a billion times in an average life span. Each beat is coordinated by an electrical signal, an action potential, that initiates in the sinoatrial node, the pacemaking center of the heart, and then propagates through the atrial and ventricular chambers. When an action potential reaches a cardiomyocyte, it depolarizes the cell's sarcolemma (the cardiomyocyte cell membrane) and thereby triggers a brief cytosolic calcium (Ca^{2+}) signal, which in turn causes the cardiomyocyte to contract. Following depolarization, cardiomyocytes relax by resetting their cytosolic Ca^{2+} concentration through the action of Ca^{2+} transport proteins. The regular cyclical increase and decrease of cytosolic Ca^{2+} are essential for the function of the heart, and it is clearly established that alteration of Ca^{2+} signaling within cardiomyocytes contributes to cardiac dysfunction in disease situations (Kranias and Bers 2007). Many factors impinge on the generation and reversal of Ca^{2+} signals within cardiomyocytes. In particular, the content of intracellular Ca^{2+} stores and the expression of Ca^{2+} transport proteins are paramount.

Store-operated Ca^{2+} entry (SOCE) encapsulates a cellular signaling mechanism whereby depletion of intracellular Ca^{2+} stores activates channels on the surface of the cell to allow the influx of Ca^{2+}. Since cardiomyocytes have large fluxes of Ca^{2+} arising with each heartbeat, it has been argued that they do not need SOCE for Ca^{2+} store refilling or for contraction. However, even modest changes in Ca^{2+} signaling can gradually alter cardiomyocyte function (Berridge 2006). So, a source of Ca^{2+} such as SOCE, if present, has the potential to profoundly affect the heart over time. A growing body of evidence suggests that the molecular components of SOCE are expressed in cardiomyocytes and that they mediate physiological and pathological effects. Despite this evidence, it is unclear whether the cellular actions of these SOCE components are actually due to a Ca^{2+} store-dependent Ca^{2+} influx. Rather, they may have signaling roles that are distinct from regulating Ca^{2+} influx. In either case, studies of SOCE, and its molecular components, in cardiomyocytes have highlighted hitherto unknown functions of proteins in both physiology and the development of cardiac diseases.

19.2 Store-Operated Ca^{2+} Entry in Cardiac Myocytes: A Ubiquitous Mechanism for Ca^{2+} Influx into Cells?

The concept of SOCE was formulated by James Putney almost three decades ago to explain the observation that depletion of intracellular Ca^{2+} stores invariably led to increased Ca^{2+} flux across the plasma membrane (Putney 1990). All that was needed for SOCE to occur was for the intracellular Ca^{2+} stores to be depleted, whether that was caused by endogenous cellular signaling pathways or by pharmacological reagents. Around the same time, a highly Ca^{2+}-selective current, which was activated by Ca^{2+} store depletion in mast cells, was reported (Hoth and Penner 1992). This current, known as "Ca^{2+} release-activated current" (I_{CRAC}) is an electrophysiological correlate of SOCE. It took a while longer for the molecular components of SOCE to be established. The involvement of stromal interaction molecule (STIM) proteins in SOCE was elucidated from RNA interference screens, which demonstrated that loss of STIM expression significantly reduced both SOCE and I_{CRAC} (Roos et al. 2005; Liou et al. 2005). Moreover, it was also shown that Ca^{2+} store depletion caused the reversible accumulation of STIM1 within punctae that were located close to the plasma membrane, which is necessary for STIM1 to trigger Ca^{2+} influx, and that expression of a mutant STIM1 protein that could not bind Ca^{2+} caused SOCE to be constitutively active (Liou et al. 2005). These findings pinpointed STIM1 as a likely sensor of Ca^{2+} store depletion and the transducer for activation of the channels responsible for I_{CRAC}. Shortly after the identification of STIM1 as a component of SOCE, the plasma membrane channels that it activates were elucidated. These channels, denoted Orai1, Orai2, and Orai3, were also discovered using genome-wide assays (Feske et al. 2006; Vig et al. 2006; Zhang et al. 2006). Moreover, it was established that patients suffering from a condition known as severe combined immunodeficiency (SCID) had a missense mutation in Orai1 that blunted Ca^{2+} influx into their lymphocytes and that this defect could be corrected by expression of wild-type Orai1 (Feske et al. 2006).

STIM1 is a single-pass transmembrane protein with its carboxy-terminus projecting into the cytosol. While the majority of STIM1 is expressed on intracellular organelles (e.g., the endoplasmic reticulum), a significant proportion (~10–20%) of the protein is localized at the plasma membrane, where it has functions unrelated to SOCE (Thompson and Shuttleworth 2012). There has been a substantial amount of work directed toward understanding how STIM1 regulates Ca^{2+} influx into cells. In essence, STIM1 is sensitive to changes of the Ca^{2+} concentration within organelles. When organellar Ca^{2+} concentration drops below a threshold level, STIM1 triggers the opening of Ca^{2+} channels at the plasma membrane (Soboloff et al. 2012).

STIM1 reversibly binds Ca^{2+} via an "EF-hand" motif near its amino-terminus. The binding of Ca^{2+} to this site maintains STIM1 in a monomeric or dimeric form that can readily diffuse around a cell within the plane of the ER membrane. When Ca^{2+} channels located on the surface of organelles are activated or SERCA inhibitors are added to cells, there is an acute drop in organellar Ca^{2+} concentration (Liou et al. 2005). The reduction in Ca^{2+} concentration leads to the release of Ca^{2+} from the EF-hand domain of STIM1. Consequently, STIM1 undergoes molecular rearrangements wherein protein-protein interaction domains are exposed, thereby allowing STIM1 to form oligomers. The STIM1 oligomers coalesce into punctae, in association with other proteins (Jing et al. 2015), at ER regions close to the plasma membrane. The carboxy-terminal, cytosolic portion of STIM1 has coiled-coil domains that project outward upon oligomerization and enable STIM1 to span the

~15 nm distance between the ER and the plasma membrane to physically interact with, and activate, Orai channels (Zhou et al. 2017). Activation of Orai channels triggers the influx of Ca^{2+} from the extracellular milieu into a cell (Wang et al. 2010a).

The canonical transient receptor potential (TRPC) family of cation channels has also been implicated in SOCE in a variety of cell types, although there is evidence to the contrary (DeHaven et al. 2009). TRPCs are nonselective cation channels with different biophysical properties to Orai (I_{CRAC}) channels. Some members of the TRPC family are activated by lipid-derived cellular messengers (Svobodova and Groschner 2016; Beech et al. 2009). Whereas the coupling between STIM and Orai proteins invariably occurs after Ca^{2+} store depletion and is fully verified as a cause of SOCE, studies have shown that cation influx via TRPCs can be independent of Ca^{2+} store depletion and STIM1 expression (Nilius and Szallasi 2014). However, it has also been suggested that TRPCs may be activated by STIM and that STIM1, Orai, and TRPCs act as a triumvirate for Ca^{2+} influx (Ong et al. 2016). This may result from Ca^{2+} entry through a STIM1-Orai interaction that subsequently leads to the insertion of TRPC channels into the plasma membrane to prolong SOCE and downstream signaling (Cheng et al. 2011). The current arising from the interaction of STIM1, Orai, and TRPCs has been termed I_{SOC} as it is distinct from I_{CRAC} even though both currents appear to share STIM1 and Orai isoforms (Desai et al. 2015).

SOCE is essential for the refilling of intracellular Ca^{2+} stores and for the maintenance of Ca^{2+} signals such as the hormone-evoked repetitive Ca^{2+} oscillations observed in non-excitable cells (Bootman et al. 1996; Putney and Bird 2008). Intracellular Ca^{2+} stores are finite and can be rapidly depleted when exogenous agonists are applied in the absence of SOCE. Therefore, without SOCE to prolong Ca^{2+} signaling downstream, Ca^{2+}-dependent effectors may only be briefly activated. In lymphocytes, for example, Ca^{2+} influx via SOCE is essential to promote sustained cytokine production (Oh-Hora et al. 2008). However, given that most cell types express numerous Ca^{2+} channels and mechanisms for generating Ca^{2+} signals, it is likely that several Ca^{2+} influx pathways, and not solely SOCE, cooperate in the generation of cellular Ca^{2+} responses (Jones et al. 2008). In addition to prolonging Ca^{2+} oscillations, which are typically global cellular signals, SOCE has local signaling functions such as the regulation of transcription factors, enzymes, and other ion channels (Kar and Parekh 2015; Cooper 2015; Soltoff and Lannon 2013; Shuttleworth and Mignen 2003; Putney et al. 2016).

19.3 Demonstrating SOCE in Cells

The activity of SOCE in cells is most commonly demonstrated by acutely depleting intracellular Ca^{2+} stores, while cells are maintained in a Ca^{2+}-free medium, followed by re-addition of extracellular Ca^{2+} (Fig. 19.1a). An elevation of the cytosolic Ca^{2+} concentration that is concomitant with the re-addition of extracellular Ca^{2+} is typically interpreted as an influx of Ca^{2+} ions caused by activation of

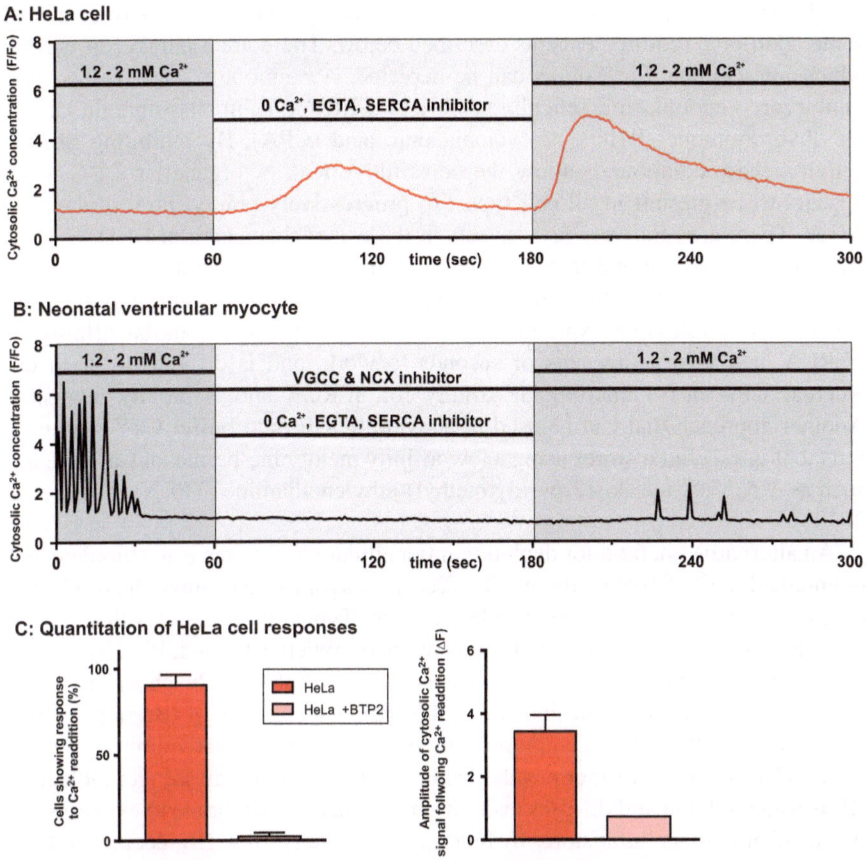

Fig. 19.1 Demonstrating SOCE following Ca^{2+} store depletion in Ca^{2+}-free medium, followed by re-addition of extracellular Ca^{2+}. Panel **a** shows the typical SOCE response of a non-excitable cell type (in this case, a HeLa cell) to the experimental protocol shown. The removal of extracellular Ca^{2+} does not alter the basal cytosolic Ca^{2+} concentration. The addition of SERCA inhibitor reveals the constitutive leak of Ca^{2+} from organelles (largely the endoplasmic reticulum in non-excitable cells), which leads to the eventual depletion of the Ca^{2+} store and activation of SOCE. Re-addition of extracellular Ca^{2+} provokes a substantial elevation of cytosolic Ca^{2+} concentration that slowly declines. The decline can be due to Ca^{2+}-dependent inhibition of SOCE or refilling of the Ca^{2+} store if a reversible SERCA inhibitor was used and washed out when extracellular Ca^{2+} was re-added. Panel **b** shows the response of a neonatal rat ventricular myocyte to the same experimental protocol as in panel **a**, with the exception that additional compounds were applied to inhibit voltage-gated Ca^{2+} channels (VGCC) and Na^+/Ca^{2+} (NCX) exchange. As is typical for cardiomyocytes, there were spontaneous Ca^{2+} oscillations evident at the start of the experiment. Panel **c** shows quantitative analysis of the response of HeLa cells to the SOCE-activating protocol shown in panel **a**. The SOCE inhibitor BTP-2 significantly decreased both the number of cells that showed a cytosolic Ca^{2+} rise following re-addition of extracellular Ca^{2+} and the amplitude of the cytosolic Ca^{2+} rise in those HeLa cells that responded

SOCE. The same protocol can be used for cardiomyocytes (Fig. 19.1b), but with some additional requirements, as described below. There are a number of ways in which intracellular Ca^{2+} stores can be depleted. A commonly used method is to inhibit sarco-endoplasmic reticulum ATPases (SERCA) using thapsigargin, di-tert-butylhydroquinone (BHQ), or cyclopiazonic acid (CPA). By inhibiting SERCA activity, these compounds allow the constitutive leak of organellar Ca^{2+}, which appears to be present in all cell types, to progressively empty intracellular Ca^{2+} stores. There are relatively few caveats in the use of thapsigargin, BHQ, or CPA. Although we have found that thapsigargin can be quite difficult to remove from plastic perfusion apparatus and microscope imaging chambers and may therefore contaminate successive experiments. Thapsigargin has a nanomolar affinity for SERCA, can take a few tens of seconds to work, and is difficult to wash out, whereas CPA has a micromolar affinity for SERCA and is rapidly reversible. Another approach that can be used to activate SOCE is to buffer Ca^{2+} within the lumen of intracellular stores using a low-affinity membrane-permeant Ca^{2+} chelator such as *N,N,N′,N′*-tetrakis(2-pyridylmethyl)-ethylenediamine (TPEN) (Wang et al. 2010b).

An alternative method for depleting intracellular Ca^{2+} stores is to stimulate cells maintained in Ca^{2+}-free medium with a receptor agonist that causes the production of a Ca^{2+}-releasing messenger inside the cells. Typically, hormones that activate phospholipase C (PLC) are used for this purpose. When activated, PLC hydrolyzes the minor membrane lipid phosphatidylinositol 4,5-bisphosphate into diacylglycerol (DAG) and inositol 1,4,5-trisphosphate (IP_3) (Berridge 2016). DAG stays within the membrane and can lead to the activation of protein kinase C, or it can be further metabolized to yield additional messenger molecules. IP_3 is water soluble and diffuses away from the plasma membrane to evoke release of Ca^{2+} from intracellular stores by binding to, and activating, IP_3 receptors (IP_3Rs; principally located on the endoplasmic reticulum, Golgi, and nuclear envelope). IP_3-generating agonists, such as hormones, neurotransmitters, and growth factors, have on many occasions been demonstrated to cause both Ca^{2+} release and Ca^{2+} influx into cells. However, while Ca^{2+} influx in response to an agonist may be due to Ca^{2+} store depletion and subsequent activation of SOCE, this is not necessarily the case since DAG and its metabolites, such as arachidonic acid, have been shown to activate cation entry that is distinct from SOCE. In particular, PLC-activating agonists have been shown to stimulate cation entry into cells via TRPCs in a Ca^{2+} store-independent manner (DeHaven et al. 2009; Svobodova and Groschner 2016; Salido et al. 2011).

In many studies, evidence for specific activation of SOCE is suggested through the use of Ca^{2+} influx inhibitors such as 2-APB, SKF96365, BTP-2, or inorganic cations (e.g., lanthanum or gadolinium) (Touchberry et al. 2011; Bootman et al. 2002, Cuddon et al. 2008) (Fig. 19.1c). However, these compounds may have questionable specificity for SOCE and have been found to affect other cellular targets (Peppiatt et al. 2003; Chen et al. 2015). An additional consideration when using cardiomyocytes to examine SOCE is Ca^{2+} influx via other mechanisms such as voltage-gated Ca^{2+} channels (VGCCs) and sodium-calcium exchange (NCX).

Both VGCCs and NCX can support spontaneous Ca^{2+} oscillation in cardiomyocytes (Fig. 19.1b) and contribute to acute cytosolic Ca^{2+} signals upon re-addition of Ca^{2+} to the extracellular medium. NCX has a key role in transporting Ca^{2+} out of cardiomyocytes after contraction has occurred. However, NCX can operate in "reverse mode" and bring Ca^{2+} into cardiomyocytes, in particular with cells that have been quiescent for a while, or have a relatively high intracellular Na^+ concentration. Both VGCCs and NCX can be blocked using specific pharmacological tools, so their putative contributions to Ca^{2+} influx signals can be negated.

IP_3Rs are a family of channels responsible for releasing Ca^{2+} from intracellular stores in many cell types, and in particular within non-excitable cells (Berridge et al. 2000). IP_3 receptors are also expressed in excitable cells, including cardiomyocytes (Lipp et al. 2000), but are often not the major Ca^{2+} release pathway. In adult cardiomyocytes, ryanodine receptors (RyRs), which are structurally similar to IP_3Rs but have approximately twice their mass and ionic conductance, are ~100-fold more abundant than IP_3Rs (Kockskamper et al. 2008) and are responsible for the Ca^{2+} signals that underlie cardiomyocyte contraction (Bers 2002). RyRs can be directly activated by caffeine or the plant alkaloid ryanodine (after which they are named). Depletion of intracellular Ca^{2+} stores by application of caffeine or ryanodine causes Ca^{2+} influx in a similar way to agonists that cause IP_3 production (Bennett et al. 1998; Hunton et al. 2004), consistent with the idea that store depletion by any means is sufficient to activate SOCE. It is therefore reasonable to postulate that cardiomyocytes, which express the molecular components of SOCE and also IP_3R and RyRs on intracellular organelles located close to the sarcolemma, have the potential to trigger SOCE following Ca^{2+} store depletion.

With regard to the heart, SOCE has been proposed to play key roles in cardiomyocyte development, homeostasis, and gene transcription. Exactly how SOCE might contribute to cardiac physiology and/or pathology is not fully clear. As discussed in the following sections, cardiomyocytes express the molecular components of SOCE pathways (e.g., STIM, Orai, and TRPC), and their expression is altered in specific cardiac disease conditions. However, the expression of these proteins does not necessarily indicate that Ca^{2+} influx via SOCE occurs in cardiomyocytes, since these proteins have been implicated in other cellular functions. Moreover, demonstrating SOCE activation in cardiomyocytes is not as straightforward as in non-excitable cells, where it is a robust cellular response (Fig. 19.1). When non-excitable cells are in a resting, unstimulated state, their cytosolic Ca^{2+} concentration is typically unaffected by the acute removal and re-addition of extracellular Ca^{2+}. However, this is not necessarily true of cardiomyocytes. Indeed, cardiomyocytes have been shown to display substantial cytosolic Ca^{2+} signals following removal and re-addition of extracellular Ca^{2+}, which were not influenced by Ca^{2+} store depletion (Zhao et al. 2015). So, the simple observation of a cytosolic Ca^{2+} signal following removal and re-addition of extracellular Ca^{2+} does not necessarily implicate SOCE. In our hands, the response of cardiomyocytes to Ca^{2+} store depletion followed by re-addition of extracellular Ca^{2+} is quite variable, with ~50% of cells showing a cytosolic Ca^{2+} increase upon re-addition of extracellular Ca^{2+} (Fig. 19.2). Although not all cells react to a typical SOCE-activating protocol, the responses that are evoked can be blocked by the same antagonists that inhibit non-excitable cell SOCE (Figs. 19.1c and 19.2b). Another factor to consider is the developmental status of cardiomyocytes. Embryonic and neonatal cardiomyocytes have a different morphology, organization, and signaling properties compared to adult cells (Bootman et al. 2011). Many of the studies proposing that SOCE is functional in cardiomyocytes have used neonatal cells, and the results do not necessarily relate to the situation in adult muscle.

Although published data is somewhat contradictory, there are reports indicating that depletion of intracellular Ca^{2+} stores within cardiomyocytes does cause Ca^{2+} influx and associated downstream responses, consistent with proposals that SOCE is an important source of Ca^{2+} to support the function of the heart. Moreover, while Ca^{2+} influx triggered by store depletion within cardiomyocytes may not always be apparent, the profound signaling functions of the molecular components of SOCE are clearly evident. We would like to direct the interested reader toward other excellent reviews and commentaries on the role of SOCE in the heart (Rosenberg 2011; Collins et al. 2013; Ruhle and Trebak 2013; Pan et al. 2014).

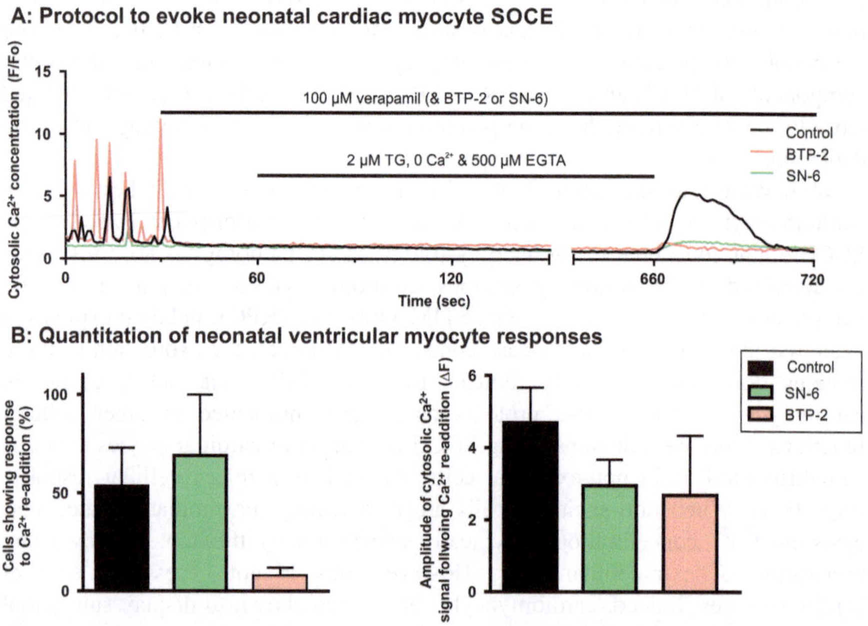

Fig. 19.2 Quantitation of SOCE responses in neonatal rat ventricular myocytes. Panel **a** illustrates the protocol used to evoke SOCE responses in neonatal rat ventricular myocytes. Panel **b** shows quantitative analysis of the response of neonatal rat ventricular myocytes to the SOCE-activating protocol shown in panel **a**. As with HeLa cells (Fig. 19.1), BTP-2 significantly ($p < 0.5$, ANOVA, calculated using GraphPad) decreased the proportion of cells showing a cytosolic response to Ca^{2+} re-addition, whereas an NCX blocker was not significantly inhibitory

19.4 Cardiac Excitation-Contraction Coupling

As mentioned earlier, Ca^{2+} is the cellular messenger that links propagating action potentials (AP) and cardiomyocyte contraction (Bootman et al. 2011). Each heartbeat involves the synchronized contraction of atrial and ventricular chambers in a process known as the "cardiac cycle" (Bers 2008). The cardiac cycle is initiated by a group of specialized pacemaking myocytes located in the right atrial wall that form the sinoatrial node (SA node). Action potentials emanate from the SA node and propagate in a coordinated manner within both the atrial chambers and ventricular chambers, thereby causing them to contract and generate the force required to pump blood around the body (Bers 2002).

As an AP sweeps over the heart, it causes cardiomyocytes to briefly depolarize and consequently activates VGCCs (specifically $Ca_V1.2$ or "L-type" VGCCs). The opening of VGCCs allows the influx of a Ca^{2+} from the extracellular milieu into a small cytoplasmic region called the "dyadic cleft," which is formed from the close apposition of the sarcolemma and the sarcoplasmic reticulum (SR, the major cardiomyocyte Ca^{2+} store). The Ca^{2+} that enters through VGCCs is significantly amplified by RyRs, which are expressed on the SR membrane facing into the dyadic cleft (Smyrnias et al. 2010). The activation of RyRs occurs via a process of "Ca^{2+}-induced Ca^{2+} release" (Roderick et al. 2003) and leads to a rapid increase of the Ca^{2+} concentration within the dyadic cleft. The Ca^{2+} signal that arises from RyR activation within the dyadic cleft is known as a "Ca^{2+} spark" (Cheng et al. 1996). When a cardiomyocyte depolarizes, Ca^{2+} sparks are simultaneously generated at thousands of dyadic clefts. Subsequently, Ca^{2+} ions diffuse from the dyadic clefts into the cytoplasm where they interact with troponin C, which promotes the interaction between actin and myosin, and thereby cause cell contraction (Berridge 2003).

Ventricular, atrial, and SA node myocytes are somewhat different in form and function, but the generation of Ca^{2+} sparks via the coupling of VGCCs to RyRs through Ca^{2+}-induced Ca^{2+} release is the same. The process leading from depolarization of a myocyte's cell membrane to its contraction is often referred to as "excitation-contraction coupling" (EC coupling). After each Ca^{2+} elevation, myocytes become refractory to further electrical stimulation for tens to hundreds of milliseconds, and during this time Ca^{2+} is returned to diastolic levels (~100 nM) by Ca^{2+}-ATPases on the sarcolemma and SR and NCX on the sarcolemma (Niggli 2011).

19.5 Cardiac Hypertrophy and Arrhythmias: Pathological Outcomes Triggered by Ca^{2+}

As described above, cardiomyocytes experience substantial cytosolic Ca^{2+} changes during EC coupling. In addition to triggering contraction, the cytosolic Ca^{2+} signals within beating cardiomyocytes regulate gene transcription, development, morphology, and metabolism. However, Ca^{2+} signals can also have pathological roles within cardiomyocytes. Two examples of the ways in which Ca^{2+} signals can provoke pathological outcomes are hypertrophy and arrhythmias.

When there is a greater hemodynamic demand, or in some disease conditions, the heart responds by initiating a growth response known as hypertrophy (Samak et al. 2016; McMullen and Jennings 2007). This is manifested as an increased thickness of the heart wall, leading to a stronger muscle. Since the heart has a modest cell proliferative capacity, hypertrophic cardiac growth occurs largely via augmented cardiomyocyte size. While cardiac hypertrophy is a helpful adaptation to cope with increased demand and in some conditions (e.g., pregnancy) can be reversible, prolonged stress on the heart promotes deleterious remodeling of cardiomyocytes, leading to a weaker, dilated heart and potentially heart failure (Umar et al. 2012; Iemitsu et al. 2001).

A number of signaling pathways have been shown to be involved in the hypertrophic growth of cardiomyocytes, and it is clearly established that Ca^{2+} signals can act as initiating events. However, not all cardiomyocyte Ca^{2+} signals trigger a hypertrophic response. In particular, the repetitive, global cytosolic Ca^{2+} signals that occur with every contraction are not necessarily involved in triggering hypertrophy (Molkentin 2006). Rather, it appears that Ca^{2+} signals arising from discrete sources, and potentially having local signaling actions, can activate hypertrophic growth. For example, IP_3 receptors located close to the nucleus (Lipp et al. 2000), or on the nuclear envelope, are able to generate Ca^{2+} signals that activate hypertrophy via the phosphatase calcineurin (CaN), which subsequently dephosphorylates the transcription factor NFAT (nuclear factor or activated T cells), thereby causing it to translocate to the nucleus and initiate gene transcription, or via calmodulin-dependent protein kinase II (CaMKII) activity and subsequent histone deacetylase (HDAC) phosphorylation (Nakayama et al. 2010; Hohendanner et al. 2014; Higazi et al. 2009). The characteristics and sources of Ca^{2+} signals that promote hypertrophy are of great interest, since their elucidation may provide targets for clinical management of deleterious cardiac remodeling. Growing evidence suggests that SOCE may provide such a local Ca^{2+} signal for hypertrophic growth of cardiomyocytes (Collins et al. 2013).

As described above, cardiomyocytes are refractory to electrical stimulation for tens to hundreds of milliseconds following an action potential. During this refractory period, Ca^{2+} is returned to diastolic levels by Ca^{2+}-ATPases on the sarcolemma and SR and NCX on the sarcolemma (Niggli 2011). In humans, sinus rhythm is typically ~60 beats per minute, but this can be altered in some disease conditions. During atrial fibrillation, for example, the atrial chambers can display activity in excess of 300 beats per minute due to spurious electrical signals, and they cannot then follow sinus rhythm (Nattel 2003). Substantial evidence has implicated spontaneous Ca^{2+} signals as a likely cause of such arrhythmic activity (Bers 2008; Heijman et al. 2012; Voigt et al. 2012). Spontaneous Ca^{2+} signals occurring during the recovery from a previous AP, or during the quiescent diastolic phase, can depolarize the sarcolemma and potentially trigger an ectopic AP or alter the refractoriness of myocytes relative to their neighbors (Johnson et al. 2012).

There are many ways in which spontaneous Ca^{2+} signals could arise in cardiomyocytes. In particular, increased Ca^{2+} content within the SR, enhanced activity of RyRs or IP_3Rs, or introduction of a Ca^{2+} source that can trigger Ca^{2+}-induced Ca^{2+} release could all potentially give rise to spontaneous Ca^{2+} signals. The activation of a Ca^{2+} influx pathway such as SOCE, albeit with a more modest current than that occurring through VGCCs, could have the potential to influence the generation of spontaneous Ca^{2+} signals and thereby by pro-arrhythmic.

19.6 Expression of SOCE Components in the Heart

19.6.1 STIM

As described above, evidence from non-excitable cells suggests that STIM1 has a major role in transducing SOCE activation by acutely sensing Ca^{2+} store depletion, forming oligomers, and associating with Orai proteins. STIM2, on the other hand, is a weaker and slower activator of Orai channels and has been suggested to be responsible for maintenance of intracellular Ca^{2+} store levels, rather than acute SOCE activation, due to the lower affinity of its EF hand for Ca^{2+} (Oh-Hora et al. 2008). Although the expression of STIM1 in cardiomyocytes is substantiated by reports from many laboratories (Correll et al. 2015; Horton et al. 2014), its expression level may be relatively low compared with other cell types. For example, the hearts of adult mice with a specific knockout of cardiomyocyte STIM1 still had ~80% of the STIM1 expression level observed in control animals, indicating that cells other than cardiomyocytes expressed the bulk of STIM1 in the heart (Collins et al. 2014).

Cardiomyocytes express three isoforms of STIM: STIM1, STIM2, and STIM1L (a splice variant of STIM1) (Zhao et al. 2015; Horton et al. 2014; Sabourin et al. 2016). It has been postulated that STIM1, and by implication SOCE, might be important in embryonic and postnatal development of cardiomyocytes. This is due to the relatively greater expression of STIM1 in early cardiomyocytes compared to adult cells (Luo et al. 2012). Moreover, STIM1, Orai1, and TRPC1 are expressed in c-kit-positive cardiac progenitor cells, where they play an essential role in cell proliferation and migration (Che et al. 2015). However, mice in which STIM1 was specifically knocked down in cardiomyocytes develop normally, with no apparent phenotypic effect until ~20 weeks of age (Collins et al. 2014). After this stage, the STIM1-deficient mice showed dilated cardiomyopathy and cardiac fibrosis, and their cardiomyocytes had elevated stress markers and altered organelle morphology. Interestingly, the overexpression of STIM1 in cardiomyocytes also has a deleterious effect. Mice in which STIM1 had been overexpressed in cardiomyocytes (to recapitulate what happens in disease situations) were unaffected until 10 weeks of age, at which point they started to show a decline in cardiac function (Correll et al. 2015). Moreover, the STIM1-overexpressing mice showed spontaneous sudden death, with complete lethality by 22 weeks of age. The exaggerated expression of STIM1 increased the activity of CaN/NFAT and

CaMKII, which would lead to hypertrophy, and also caused enhanced Ca^{2+} entry and spontaneous Ca^{2+} signals, which plausibly may have contributed to arrhythmias and sudden death. These studies highlight the importance of STIM1 for cardiomyocyte structure and function; either too little or too much STIM1 provokes pathological changes in cardiomyocytes. However, whether all the effects of reduced or enhanced STIM1 were due to changes in SOCE remains to be shown.

STIM1 expression is increased in hypertrophied cardiac myocytes (Correll et al. 2015; Luo et al. 2012; Hulot et al. 2011), and reduction of STIM1 expression can prevent the induction of hypertrophy (Luo et al. 2012; Benard et al. 2016). So, it appears that STIM1 expression is both part of the cause and the consequence of cardiomyocyte hypertrophy. Given that STIM1 can be activated by multiple cellular stressors, and not just Ca^{2+} store depletion (Soboloff et al. 2012), it likely that STIM1 participates in deleterious positive feedback signaling within cardiomyocytes. In this situation, cell stress would lead to activation of STIM1, which in turn can support unwanted phenotypic remodeling of the heart and increased expression of STIM1.

STIM1 has the potential to be signaling protein with pleiotropic effects, since it interacts with a number of proteins that have signaling and cell fate functions (Zhao et al. 2015; Benard et al. 2016; Krapivinsky et al. 2011; Hooper et al. 2013). For example, upon Ca^{2+} store depletion, STIM1 has been shown to interact with a scaffolding protein called partner of STIM1 (POST) that brings a number of associated proteins including SERCA, Na^{+}/K^{+}-ATPase, plasma membrane Ca^{2+} ATPase (PMCA), and the nuclear carrier proteins, importin-β1 and exportin-1 (Krapivinsky et al. 2011). Via its interaction with POST, STIM1 is also able to engage with these proteins. The functional consequences of STIM1 binding to the proteins that POST scaffolds are not known, but it appears that the STIM1-POST complex reduces Ca^{2+} efflux across the plasma membrane by PMCAs. STIM1 has been also been shown to negatively regulate the activity of L-type VGCCs via a direct coupling (Wang et al. 2010b). L-type VGCC peak currents in A7r5 smooth muscle cells were significantly inhibited following Ca^{2+} store depletion, suggesting that activated STIM1 can increase Ca^{2+} influx via SOCE while simultaneously reducing VGCC-mediated Ca^{2+} entry. STIM1 may therefore also play an important role in determining the activity of VGCCs in cardiomyocytes, which could impact on EC coupling and SR Ca^{2+} store filling. Interestingly, chronic reduction of L-type VGCCs (as occurs with human patients on particular medication) has been shown to cause hypertrophy, plausibly because Ca^{2+} release from RyRs is increased to compensate for the lack of Ca^{2+} influx (Goonasekera et al. 2012). The regulation of VGCCs by STIM1 could therefore be an important factor in the development of hypertrophy in cardiomyocytes.

Unlike the situation in non-excitable cells, fluorescently tagged STIM1 proteins expressed in adult cardiomyocytes were observed to be almost static and not to accumulate into punctae following depletion of Ca^{2+} from the SR (Zhao et al. 2015). Cardiomyocyte STIM1 was identified in a ~550 kDa complex, consistent with it being constitutively bound to itself and other proteins. One of its binding partners was identified as phospholamban (PLB), a negative regulator of SERCA in cardiomyocytes. By binding to PLB and disinhibiting SERCA, STIM1 was found to increase SR Ca^{2+} content and increase the frequency of spontaneous Ca^{2+} waves. However, there was no evidence for SOCE increasing SR Ca^{2+} content.

Whether the interaction of STIM1 with other proteins, such as PMCAs, PLB, and L-type VGCCs, is an important regulatory mechanism in cardiomyocytes remains to be firmly established. The fact that STIM1 knockout mice die in utero or in early life due to respiratory failure, but without an overt cardiac phenotype, may indicate that cardiac development can proceed without STIM1 although deleterious cardiomyocyte remodeling eventually occurs (Benard et al. 2016; Cahalan 2009).

19.6.2 Orai and TRPC

Cardiomyocytes express the three members of the Orai family of cation channels: Orai1, Orai2, and Orai3 (Correll et al. 2015; Horton et al. 2014; Dominguez-Rodriguez et al. 2015). Orai1-deficient mice subjected to aortic constriction to trigger pressure overload-induced hypertrophy were found to have decreased survival and a more rapid loss of cardiac function than wild-type animals. In addition, there were numerous other effects of Orai1 reduction, such as tissue fibrosis and cardiomyocyte apoptosis (Horton et al. 2014). Similar to the situation with STIM1 reduction, it appears that loss of Orai1 causes rapid deterioration of viability and cell function. Orai1 and Orai3 expression levels were found not to change during hypertrophic growth of cardiomyocytes, but the coupling of STIM1 with Orai3 became more prominent (Saliba et al. 2015). Knockdown of Orai1 (and STIM1) prevented the induction of hypertrophy in neonatal cardiac myocytes following stimulation with a PLC-stimulating agonist (Voelkers et al. 2010).

Since the roles of TRPCs in cardiomyocytes has been described many times previously (Sabourin et al. 2011; Eder and Molkentin 2011), we will only mention them briefly here. In essence, several TRPC family members are expressed in the heart (Correll et al. 2015; Dominguez-Rodriguez et al. 2015; Sabourin et al. 2011). TRPCs have been shown to form a complex with L-type VGCCs and cooperate in the initiation of beating in the developing heart (Sabourin et al. 2011). The upregulation of TRPC expression (e.g., TRPC3 and TRPC4 in adult ventricular cardiomyocytes) correlated with the appearance of increased SOCE and pro-arrhythmic spontaneous Ca^{2+} waves (Dominguez-Rodriguez et al. 2015). Moreover, TRPC3 knockout mice were less susceptible to arrhythmia induction caused by PLC-linked agonists or electrical pacing (Ju et al. 2015). Several studies have demonstrated roles for various TRPCs in hypertrophic growth of cardiomyocytes following different forms of stimulation and stress (Eder and Molkentin 2011; Makarewich et al. 2014; Seo et al. 2014).

19.6.3 Atrial Cardiomyocytes and SA Node Cells

The evidence for expression of SOCE components in the heart described above is largely derived from studies of neonatal or adult ventricular myocytes. However, it is important to point out STIM1, Orai, and TRPCs are expressed in other excitable cardiac cell types too. Atrial and SA node cells, for example, express STIM proteins, Orai channels, IP_3Rs, and TRPCs and show a cytosolic Ca^{2+} signal in response to the typical Ca^{2+} store depletion/Ca^{2+} re-addition protocol (Ju et al. 2007, 2015; Zhang et al. 2013, 2015; Liu et al. 2015). Sinoatrial node cells are critical pacemakers for cardiac beating frequency, and their pacemaking activity is known to be influenced by the rate of spontaneous Ca^{2+} sparks arising via RyRs. Ca^{2+} spark frequency is determined in part by the Ca^{2+} content of the SR, which SOCE could modulate. Application of BTP-2 or SKF96365, both of which significantly inhibited Ca^{2+} signals caused by the Ca^{2+} store depletion/Ca^{2+} re-addition protocol, reduced the frequency of sinoatrial node firing by ~15% (Liu et al. 2015). Notwithstanding issues of selectivity of the SOCE blockers used in this study, these data support the notion that that SOCE may influence cardiac beat frequency. Consistent with this, knockout of STIM1 in the SA node cells of mice was found to result in reduced Ca^{2+} influx, lesser SR Ca^{2+} content, and a reduction in heart rate (Zhang et al. 2015).

19.7 Evidence for SOCE in the Heart

As described earlier, a typical experimental protocol for establishing the activity of SOCE is to deplete intracellular Ca^{2+} stores by incubating cells maintained in Ca^{2+}-free medium with a SERCA inhibitor for several minutes, followed by re-addition of extracellular Ca^{2+}. If SOCE is active, the re-addition of extracellular Ca^{2+} would cause a rapid increase of cytosolic Ca^{2+} concentration. With non-excitable cells, this standard store depletion/Ca^{2+} re-addition protocol has generally been found to provoke robust responses. However, studies examining SOCE in cardiomyocytes using the same protocol have given quite contradictory results. In some cases, depletion of intracellular Ca^{2+} stores in neonatal and adult cardiomyocytes evoked a Ca^{2+} influx signal upon re-addition of extracellular Ca^{2+} that was visualized either as an elevation of cytosolic Ca^{2+} concentration (Uehara et al. 2002; Hunton et al. 2002, 2004; Luo et al. 2012) or refilling of the SR (Hunton et al. 2004). Studies examining the role of STIM1 in cardiomyocytes have demonstrated that Ca^{2+} influx signals observed upon re-addition of extracellular Ca^{2+} could be decreased by STIM1 shRNA or STIM1 knockout (Luo et al. 2012; Hulot et al. 2011) or increased by overexpression of STIM1 (Correll et al. 2015; Luo et al. 2012). Cytosolic Ca^{2+} signals arising from the Ca^{2+} store depletion/Ca^{2+} re-addition protocol and cation currents measured following Ca^{2+} store depletion are both increased in hypertrophied cardiomyocytes, consistent with the enhanced expression of SOCE components under these conditions (Luo et al. 2012; Hulot et al. 2011). These data support the notion that cardiomyocytes expressed a STIM1-mediated SOCE response similar to that in non-excitable cells. In contrast, a negligible response to Ca^{2+} re-addition was observed in adult rat cardiomyocytes unless they were treated with an 8-pCPT for several hours to activate EPAC and increase expression of TRPC3 and TRPC4 (Dominguez-Rodriguez et al. 2015), and in a further study, no evidence for a Ca^{2+} store-dependent Ca^{2+} influx was observed (Zhao et al. 2015).

If SOCE is a physiological mechanism for triggering Ca^{2+} influx into cardiomyocytes, it would be expected to occur in response to specific cellular stimuli. Indeed, stimulation of cardiomyocytes with thapsigargin, or PLC-activating agonists such as phenylephrine or angiotensin II, evoked persistent Ca^{2+} signals that were dependent on extracellular Ca^{2+} (Hunton et al. 2002). These Ca^{2+} influx-dependent signals were insensitive to VGCC or NCX inhibition but were prevented by reagents that block SOCE in other cell types (e.g., SKF96365). Moreover, the reagents that blocked thapsigargin-, phenylephrine-, or angiotensin II-evoked Ca^{2+} influx signals also prevented cardiomyocyte hypertrophy. Whether the mechanism of Ca^{2+} influx in this study was I_{CRAC} and I_{SOC} or another pathway is unclear, but it highlights the potential for positive inotropic and pro-hypertrophic agonists to trigger a Ca^{2+} influx signal that promotes pathological remodeling.

At present, the role of SOCE in the heart is slightly enigmatic. Published evidence suggests that Ca^{2+} influx akin to SOCE is most readily observed in embryonic cells but then declines with development until being substantially reduced, or absent, in adult cells (Uehara et al. 2002). However, deletion of STIM1, which is a key regulator of SOCE, does not affect early development of the heart, but precipitates deleterious remodeling in postnatal animals. Moreover, in some studies where STIM1, Orai, and TRPC levels have been manipulated experimentally or altered due to disease, it has been assumed that SOCE was affected without direct measurement of a store-dependent Ca^{2+} influx.

Due to space limitations, it is not possible for us to give detailed summaries of the many reports of SOCE, STIM1, Orai, and TRPC in the heart and how they affect cardiac physiology and pathology. Table 19.1 has brief notes on many of the significant findings in this topic. It is evident that SOCE, STIM1, Orai, and TRPC have been implicated in numerous processes, involving different cardiac cell types. There are some intriguing contradictions in the published literature, such as the inconsistent evidence for Ca^{2+} entry caused by Ca^{2+} store depletion in cardiomyocytes, as well as differences in the expression of STIM1, Orai, and TRPC in similar disease models.

Table 19.1 Summary of published reports showing impacts of SOCE, STIM, Orai, and TRPC on various aspects of cardiac physiology and pathology

Cell type/channels	Finding
Cardiac hypertrophy	
Neonatal and adult CM/STIM1	Robust SOCE current present in neonatal CM, marginal in adult CM from controls but prominent in cardiac hypertrophy. STIM1 expression higher in pressure overload model, protected by STIM1 silencing. STIM1 overexpression increases NFAT activity and cell size, reduced by SKF96365 (Hulot et al. 2011)
Neonatal rat CM	SOCE activated by TG, phenylephrine, angiotensin II, promoted NFAT nuclear translocation. SOCE inhibited by SKF96365 and glucosamine pretreatment (Hunton et al. 2002)
Conditional cardiac STIM1$^{-/-}$, STIM1↑ (increased expression)	STIM1$^{-/-}$ phenotype: moderate left ventricular dilatation and decreased contractility. Pressure overload-induced hypertrophy was attenuated. Increased myofibril degradation at STIM1 microdomains in the STIM1 overexpression model (Parks et al. 2016)
Neonatal and adult CM/STIM1	Robust SOCE in neonatal CM, minor in adult CM, correlating with STIM1 expression (mRNA and protein). STIM1 up- or downregulation affected SOCE Neonatal STIM1 splice (STIM1L) variant reemerges in hypertrophy (CaN/NFAT pathway). STIM1 inhibition suppressed agonist-triggered hypertrophy (Luo et al. 2012)
Neonatal CM, adult hearts/STIM1, TRPC	Endothelin increased TRPC1, SOCE, and NFAT, but not STIM1 STIM1 knockdown suppressed these effects and hypertrophy induced by aortic banding (Ohba et al. 2009)
Mouse and feline hearts/TRPCs	Hypertrophy mediated via increased TRPC currents and CaN/NFAT signaling (Makarewich et al. 2014)
Adult and neonatal CM, cardiac STIM1↑	STIM1 overexpression caused hypertrophy (via NFAT and CaMKII) activated the fetal gene program and disrupted mitochondria. Increased SOCE, SR Ca^{2+} cycling, RyR2-dependent Ca^{2+} spark activity, diastolic Ca^{2+} levels, and maximum transient amplitude. SR content unaffected. Animals died of sudden death or heart failure (Correll et al. 2015)
Cardiac STIM1$^{-/-}$ mice	Cardiac STIM1$^{-/-}$ phenotype: ER stress and mitochondrial abnormalities (>12 weeks of age), progressive decline in left ventricular function; dilated cardiomyopathy with decreased ejection fraction (contractile dysfunction) and fibrosis between 20–36 weeks of age. Death between 40 and 65 weeks of age → Cardiac STIM1 is not necessary for development, but STIM1 is essential for normal CM homeostasis (Collins et al. 2014)

(continued)

Table 19.1 (continued)

Cell type/channels	Finding
Adult rat VM/STIM1, Orai1 and Orai3	Hypertrophy: Orai1 and Orai3 protein levels unchanged, but more Orai3 and STIM1 clustering. Orai3, not Orai1, drives SOCE in hypertrophy. Orai3 also drives an AA-activated current present in hypertrophy (Saliba et al. 2015)
Neonatal CM/Orai1, TRPCs	Aldosterone increases basal Ca^{2+} and Orai1, TRPC1, TRPC4, TRPC5, and STIM1 expression and SOCE (Sabourin et al. 2016)
$Orai1^{+/-}$ mice/Orai1	Hypertrophy induced by cardiac banding caused earlier loss of cardiac function and dilated cardiomyopathy in $Orai1^{+/-}$ mice. WT mice developed the same hypertrophy level later. No differences in hypertrophy or increase in heart weight. $Orai1^{+/-}$ mice cannot compensate for the overload, leading to their earlier death (Horton et al. 2014)
Neonatal rat CM/STIM1, Orai1	PE-induced hypertrophy reduced by both STIM1 and Orai1 knockdown STIM1 knockdown → reduced diastolic Ca^{2+} and SR Ca^{2+} Orai1 knockdown → diastolic and SR Ca^{2+} not affected. Reduced cell size and CaN signaling and resulting pro-hypertrophic signaling processes. Compensatory Orai2 upregulation (Voelkers et al. 2010)
Neonatal rat CM/STIM1, Orai1	PE-induced model of hypertrophy. Hypertrophy → upregulation of CaMKIIδ, increase in hypertrophy markers ANP (atrial natriuretic peptide) and BNP (brain natriuretic peptide), CM size, increased response to Ca^{2+} re-addition and I_{CRAC} current; sensitive to CaMKIIδ siRNA, KN93, and BTP-2. Increased levels of STIM1 and Orai1 sensitive to CaMKIIδ siRNA and KN93 (Ji et al. 2016)
Cardiac fibrosis	
Rat cardiac fibroblasts/STIM1, Orai1	Angiotensin II stimulation increased STIM1 and Orai1 expression, activated SOCE and cardiac fibrosis (upregulation of fibronectin, connective tissue growth factor and smooth muscle actin, NFAT translocation). All reduced by SKF96365, STIM1, and Orai1 knockdown (Zhang et al. 2016)
Human ventricular fibroblasts/ STIM1, Orai1	Cultured fibroblasts from failing hearts: increased collagen secretion, similar collagen gene expression, similar TG-induced Ca^{2+} release and SR content, larger response to Ca^{2+} re-addition, similar STIM1, increased Orai1 expression, greater STIM-Orai co-localization. Collagen secretion is I_{CRAC} sensitive, inhibited by YM58483 (Ross et al. 2017)
Myocardial repair and metabolism	
Cardiac c-kit+ progenitor cells/ STIM1, Orai1, TRPC1	Expression of STIM1, Orai1, and TRPC1 shown. Knockdown reduced SOCE, decreased cell

(continued)

Table 19.1 (continued)

Cell type/channels	Finding
	proliferation and migration, and reduces the expression of cyclin Da, cyclin E, and/or p-Akt and cardiac repair (Che et al. 2015)
Inducible STIM1 and $STIM2^{-/-}$ mouse hearts	Increased amounts of lipid droplets in heart (and other tissues). Lipid droplet accumulation could be linked to change in metabolism similar to a starvation/fasting response. Probably mediated via cAMP (Maus et al. 2017)
Various other SOCE	
Rabbit CM	SR loading (measured via NCX current after caffeine pulse) declined with age (3, 10, and 56 days). SOCE (response to Ca^{2+} re-addition after CPA and ryanodine) became smaller over time. NCX activity modulated SOCE and SR load (decreased by SKF96365, increased by KB-R) → SOCE important for Ca^{2+} homeostasis in neonates, not essential on a beat-to-beat basis (Huang et al. 2006)
Neonatal CM/STIM1	Increased CM O-GlcNAcylation attenuated STIM1 puncta formation and SOCE and increased O-GlcNAc modification of STIM1 (Zhu-Mauldin et al. 2012)
Adult rat VM/STIM1	STIM1 not important for SOCE but for basic homeostasis via PLB binding and affecting SERCA activity (Zhao et al. 2015) Low endogenous STIM1 expression. Localized in puncta at Z-disks and along the junctional SR. No changes (localization and/or mobilization) after SR depletion, no SOCE STIM1 overexpression: no SOCE, but increased SR Ca^{2+}, increased SR leak (via PLB binding and SERCA activation). STIM1 found in large protein complex, unaltered by store depletion
Adult rat VM/TRPC3/4, EPAC	EPAC activators → increased SOCE and TRPC3 and TRPC4 protein expression and pro-arrhythmic Ca^{2+} waves. EPAC inhibitors → blunted SOCE (Dominguez-Rodriguez et al. 2015)
Human AM/TRPC1/STIM1 Orai1	TG-activated, La^{3+}-sensitive TRPC1 current in human AM. Enhanced by ET-1 and angiotensin II. Co-IP of TRPC1, STIM1, and Orai1 (Zhang et al. 2013)
HL-1/STIM1, Orai1	HL-1 cells express STIM1 and Orai1. SOCE is present, blocked by SKF96365, BTP-2, and Orai1 knockdown, which was not inhibited by VGCC or reverse mode NCX inhibitors (Touchberry et al. 2011)
HL-1/STIM1, TTCC	STIM1 knockdown reduced contractility, induced early and delayed afterdepolarizations, and increased the peak amplitude and current density of TTCC via a direct negative regulation (Nguyen et al. 2013)
SA node cells	
SA node, cardiac-specific $STIM1^{-/-}$ mouse/STIM1, Orai1	Cardiac $STIM1^{-/-}$ mice phenotype: reduced heart rate, sinus arrest, exaggerated autonomic response to cholinergic signaling. Reduced SOCE, increased LTCC activity, inhibits NCX

(continued)

Table 19.1 (continued)

Cell type/channels	Finding
	Neonatal CM: SOCE present in wild type, BTP-2 sensitive, absent in STIM1$^{-/-}$ cells. STIM1 localized at Z-lines, very limited movement (Zhang et al. 2015)
SA node/STIM1, Orai1	Isolated mouse SA node cells. SOCE present in pacemaker cells. SOCE inhibition reduces amplitude and frequency of spontaneous Ca^{2+} transients and SR content. Store depletion causes STIM1 punctae formation (Liu et al. 2015)
SA node TRPC3$^{-/-}$ mouse/TRPC3, STIM1	TRPC3$^{-/-}$ mice less susceptible to arrhythmia induction (angiotensin or pacing) Co-localization of STIM1, TRPC3, and IP_3R_2 in SA node pacemaker cells Not clearly differentiated SOCE and ROCE (Ju et al. 2015)

Abbreviations: *AM* atrial myocyte, *Ca^{2+}* calcium, *CaMKII* calmodulin-dependent protein kinase II, *CaN* calcineurin, *CM* cardiac myocyte, *CPA* cyclopiazonic acid, *DAG* diacylglycerol, *BHQ* di-tert-butylhydroquinone, *EPAC* exchange protein directly activated by cAMP, I_{CRAC} Ca^{2+} release-activated current, *HDAC* histone deacetylases, *IP_3* inositol 1,4,5-trisphosphate, *NCX* Na^+ -Ca^{2+}exchanger, *NFAT* nuclear factor of activated T cells, *PLB* phospholamban, *PLC* phospholipase C, *PMCA* plasma membrane Ca^{2+} ATPase, *POST* partner of STIM1, *RyRs* ryanodine receptors, *SCID* severe combined immunodeficiency, *SERCA* sarco-endoplasmic reticulum ATPases, *SOCE* store-operated Ca^{2+} entry, *STIM* stromal interaction molecule, *TG* thapsigargin, *TPEN* *N,N,N′,N′*-tetrakis(2-pyridylmethyl)-ethylenediamine, *TRPC* canonical transient receptor potential channel, *VGCCs* voltage-gated Ca^{2+} channels, *VM* ventricular myocyte

19.8 Summary

Despite the inconsistent reports of Ca^{2+} influx being dependent on store depletion in cardiomyocytes, there is abundant evidence that the components of SOCE are expressed in excitable cells within the heart. Moreover, their expression is linked to physiological functions such as homeostasis and pacemaking and also to pathological remodeling. The potentially complex roles of the SOCE components within cardiomyocytes are exemplified by STIM1, which is expressed at relatively high levels in neonatal cells, but not needed for cardiac development or early postnatal life. The expression of cardiomyocyte STIM1 declines in adulthood, but is reestablished during hypertrophy, and appears to contribute to the development of decompensated hypertrophy and heart failure. Given the large fluxes of cytosolic Ca^{2+} that occur during EC coupling, it is unlikely that cardiomyocytes require SOCE for refilling of the SR or to augment contraction. However, through local Ca^{2+}

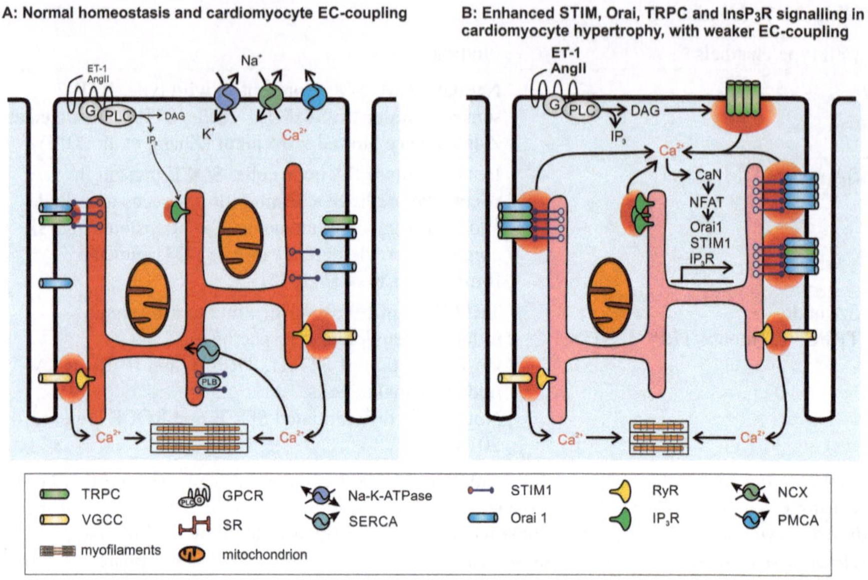

Fig. 19.3 Enhancement of EC coupling and gene transcription via local Ca^{2+} signaling by STIM1, Orai1, TRPC, and IP_3Rs in hypertrophic cardiomyocytes. Panel **a** illustrates a situation where EC coupling is the principal Ca^{2+} signaling mechanism. STIM1, Orai1, TRPC, and IP_3Rs are presumably active in this condition, albeit modestly, and make a contribution to physiological signaling activities, such as inotropy, SR refilling, metabolism, and homeostatic gene transcription. Panel **b** depicts a hypertrophic growth situation where STIM1, Orai1, TRPC, and IP_3Rs are all more highly expressed and active. In this situation, STIM1, Orai1, TRPC, and IP_3Rs could underlie pro-arrhythmic Ca^{2+} signals and enhance hypertrophic signaling pathways

signals and formation of signaling complexes, SOCE and its molecular components can have profound effects on cardiomyocyte gene transcription, function, and fate (Fig. 19.3).

References

Beech DJ, Bahnasi YM, Dedman AM, Al-Shawaf E (2009) TRPC channel lipid specificity and mechanisms of lipid regulation. Cell Calcium 45:583–588

Benard L, Oh JG, Cacheux M, Lee A, Nonnenmacher M, Matasic DS, Kohlbrenner E, Kho C, Pavoine C, Hajjar RJ, Hulot JS (2016) Cardiac Stim1 silencing impairs adaptive hypertrophy and promotes heart failure through inactivation of mTORC2/Akt signaling. Circulation 133:1458–1471. Discussion 1471

Bennett DL, Bootman MD, Berridge MJ, Cheek TR (1998) Ca2+ entry into PC12 cells initiated by ryanodine receptors or inositol 1,4,5-trisphosphate receptors. Biochem J 329(Pt 2):349–357

Berridge MJ (2003) Cardiac calcium signaling. Biochem Soc Trans 31:930–933

Berridge MJ (2006) Remodelling Ca2+ signalling systems and cardiac hypertrophy. Biochem Soc Trans 34:228–231

Berridge MJ (2016) The inositol trisphosphate/calcium signaling pathway in health and disease. Physiol Rev 96:1261–1296

Berridge MJ, Lipp P, Bootman MD (2000) The versatility and universality of calcium signaling. Nat Rev Mol Cell Biol 1:11–21

Bers DM (2002) Cardiac excitation-contraction coupling. Nature 415:198–205

Bers DM (2008) Calcium cycling and signaling in cardiac myocytes. Annu Rev Physiol 70:23–49

Bootman MD, Young KW, Young JM, Moreton RB, Berridge MJ (1996) Extracellular calcium concentration controls the frequency of intracellular calcium spiking independently of inositol 1,4,5-trisphosphate production in HeLa cells. Biochem J 314(Pt 1):347–354

Bootman MD, Collins TJ, Mackenzie L, Roderick HL, Berridge MJ, Peppiatt CM (2002) 2-aminoethoxydiphenyl borate (2-APB) is a reliable blocker of store-operated Ca2+ entry but an inconsistent inhibitor of InsP3-induced Ca2+ release. FASEB J 16:1145–1150

Bootman MD, Smyrnias I, Thul R, Coombes S, Roderick HL (2011) Atrial cardiomyocyte calcium signaling. Biochim Biophys Acta 1813:922–934

Cahalan MD (2009) STIMulating store-operated Ca(2+) entry. Nat Cell Biol 11:669–677

Che H, Li G, Sun HY, Xiao GS, Wang Y, Li GR (2015) Roles of store-operated Ca2+ channels in regulating cell cycling and migration of human cardiac c-kit+ progenitor cells. Am J Physiol Heart Circ Physiol 309:H1772–H1781

Chen KH, Liu H, Yang L, Jin MW, Li GR (2015) SKF-96365 strongly inhibits voltage-gated sodium current in rat ventricular myocytes. Pflugers Arch 467:1227–1236

Cheng H, Lederer MR, Lederer WJ, Cannell MB (1996) Calcium sparks and [Ca2+]i waves in cardiac myocytes. Am J Physiol 270:C148–C159

Cheng KT, Liu X, Ong HL, Swaim W, Ambudkar IS (2011) Local Ca(2)+ entry via Orai1 regulates plasma membrane recruitment of TRPC1 and controls cytosolic Ca(2)+ signals required for specific cell functions. PLoS Biol 9:e1001025

Collins HE, Zhu-Mauldin X, Marchase RB, Chatham JC (2013) STIM1/Orai1-mediated SOCE: current perspectives and potential roles in cardiac function and pathology. Am J Physiol Heart Circ Physiol 305:H446–H458

Collins HE, He L, Zou L, Qu J, Zhou L, Litovsky SH, Yang Q, Young ME, Marchase RB, Chatham JC (2014) Stromal interaction molecule 1 is essential for normal cardiac homeostasis through modulation of ER and mitochondrial function. Am J Physiol Heart Circ Physiol 306: H1231–H1239

Cooper DM (2015) Store-operated Ca(2)(+)-entry and adenylyl cyclase. Cell Calcium 58:368–375

Correll RN, Goonasekera SA, van Berlo JH, Burr AR, Accornero F, Zhang H, Makarewich CA, York AJ, Sargent MA, Chen X, Houser SR, Molkentin JD (2015) STIM1 elevation in the heart results in aberrant Ca(2)(+) handling and cardiomyopathy. J Mol Cell Cardiol 87:38–47

Cuddon P, Bootman MD, Richards GR, Smith AJ, Simpson PB, Roderick HL (2008) Methacholine and PDGF activate store-operated calcium entry in neuronal precursor cells via distinct calcium entry channels. Biol Res 41(2):183–195

DeHaven WI, Jones BF, Petranka JG, Smyth JT, Tomita T, Bird GS, Putney JW Jr (2009) TRPC channels function independently of STIM1 and Orai1. J Physiol 587:2275–2298

Desai PN, Zhang X, Wu S, Janoshazi A, Bolimuntha S, Putney JW Jr, Trebak M (2015) Multiple types of calcium channels arising from alternative translation initiation of the Orai1 message. Sci Signal 8:ra74

Dominguez-Rodriguez A, Ruiz-Hurtado G, Sabourin J, Gomez AM, Alvarez JL, Benitah JP (2015) Proarrhythmic effect of sustained EPAC activation on TRPC3/4 in rat ventricular cardiomyocytes. J Mol Cell Cardiol 87:74–78

Eder P, Molkentin JD (2011) TRPC channels as effectors of cardiac hypertrophy. Circ Res 108:265–272

Feske S, Gwack Y, Prakriya M, Srikanth S, Puppel SH, Tanasa B, Hogan PG, Lewis RS, Daly M, Rao A (2006) A mutation in Orai1 causes immune deficiency by abrogating CRAC channel function. Nature 441:179–185

Goonasekera SA, Hammer K, Auger-Messier M, Bodi I, Chen X, Zhang H, Reiken S, Elrod JW, Correll RN, York AJ, Sargent MA, Hofmann F, Moosmang S, Marks AR, Houser SR, Bers DM, Molkentin JD (2012) Decreased cardiac L-type Ca(2)(+) channel activity induces hypertrophy and heart failure in mice. J Clin Invest 122:280–290

Heijman J, Voigt N, Nattel S, Dobrev D (2012) Calcium handling and atrial fibrillation. Wien Med Wochenschr 162:287–291

Higazi DR, Fearnley CJ, Drawnel FM, Talasila A, Corps EM, Ritter O, McDonald F, Mikoshiba K, Bootman MD, Roderick HL (2009) Endothelin-1-stimulated InsP3-induced Ca2+ release is a nexus for hypertrophic signaling in cardiac myocytes. Mol Cell 33:472–482

Hohendanner F, McCulloch AD, Blatter LA, Michailova AP (2014) Calcium and IP3 dynamics in cardiac myocytes: experimental and computational perspectives and approaches. Front Pharmacol 5:35

Hooper R, Samakai E, Kedra J, Soboloff J (2013) Multifaceted roles of STIM proteins. Pflugers Arch 465:1383–1396

Horton JS, Buckley CL, Alvarez EM, Schorlemmer A, Stokes AJ (2014) The calcium release-activated calcium channel Orai1 represents a crucial component in hypertrophic compensation and the development of dilated cardiomyopathy. Channels 8:35–48

Hoth M, Penner R (1992) Depletion of intracellular calcium stores activates a calcium current in mast cells. Nature 355:353–356

Huang J, van Breemen C, Kuo KH, Hove-Madsen L, Tibbits GF (2006) Store-operated Ca2+ entry modulates sarcoplasmic reticulum Ca2+ loading in neonatal rabbit cardiac ventricular myocytes. Am J Physiol Cell Physiol 290:C1572–C1582

Hulot JS, Fauconnier J, Ramanujam D, Chaanine A, Aubart F, Sassi Y, Merkle S, Cazorla O, Ouille A, Dupuis M, Hadri L, Jeong D, Muhlstedt S, Schmitt J, Braun A, Benard L, Saliba Y, Laggerbauer B, Nieswandt B, Lacampagne A, Hajjar RJ, Lompre AM, Engelhardt S (2011) Critical role for stromal interaction molecule 1 in cardiac hypertrophy. Circulation 124:796–805

Hunton DL, Lucchesi PA, Pang Y, Cheng X, Dell'Italia LJ, Marchase RB (2002) Capacitative calcium entry contributes to nuclear factor of activated T-cells nuclear translocation and hypertrophy in cardiomyocytes. J Biol Chem 277:14266–14273

Hunton DL, Zou L, Pang Y, Marchase RB (2004) Adult rat cardiomyocytes exhibit capacitative calcium entry. Am J Physiol Heart Circ Physiol 286:H1124–H1132

Iemitsu M, Miyauchi T, Maeda S, Sakai S, Kobayashi T, Fujii N, Miyazaki H, Matsuda M, Yamaguchi I (2001) Physiological and pathological cardiac hypertrophy induce different molecular phenotypes in the rat. Am J Physiol Regul Integr Comp Physiol 281:R2029–R2036

Ji Y, Guo X, Zhang Z, Huang Z, Zhu J, Chen QH, Gui L (2016) CaMKIIdelta meditates phenylephrine induced cardiomyocyte hypertrophy through store-operated Ca2+ entry. Cardiovasc Pathol 27:9–17

Jing J, He L, Sun A, Quintana A, Ding Y, Ma G, Tan P, Liang X, Zheng X, Chen L, Shi X, Zhang SL, Zhong L, Huang Y, Dong MQ, Walker CL, Hogan PG, Wang Y, Zhou Y (2015) Proteomic mapping of ER-PM junctions identifies STIMATE as a regulator of Ca(2)(+) influx. Nat Cell Biol 17:1339–1347

Johnson DM, Heijman J, Bode EF, Greensmith DJ, van der Linde H, Abi-Gerges N, Eisner DA, Trafford AW, Volders PG (2012) Diastolic spontaneous calcium release from the sarcoplasmic reticulum increases beat-to-beat variability of repolarization in canine ventricular myocytes after beta-adrenergic stimulation. Circ Res 112:246–256

Jones BF, Boyles RR, Hwang SY, Bird GS, Putney JW Jr (2008) Calcium influx mechanisms underlying calcium oscillations in rat hepatocytes. Hepatology 48:1273–1281

Ju YK, Chu Y, Chaulet H, Lai D, Gervasio OL, Graham RM, Cannell MB, Allen DG (2007) Store-operated Ca2+ influx and expression of TRPC genes in mouse sinoatrial node. Circ Res 100:1605–1614

Ju YK, Lee BH, Trajanovska S, Hao G, Allen DG, Lei M, Cannell MB (2015) The involvement of TRPC3 channels in sinoatrial arrhythmias. Front Physiol 6:86

Kar P, Parekh AB (2015) Distinct spatial Ca2+ signatures selectively activate different NFAT transcription factor isoforms. Mol Cell 58:232–243

Kockskamper J, Zima AV, Roderick HL, Pieske B, Blatter LA, Bootman MD (2008) Emerging roles of inositol 1,4,5-trisphosphate signaling in cardiac myocytes. J Mol Cell Cardiol 45:128–147

Kranias EG, Bers DM (2007) Calcium and cardiomyopathies. Subcell Biochem 45:523–537

Krapivinsky G, Krapivinsky L, Stotz SC, Manasian Y, Clapham DE (2011) POST, partner of stromal interaction molecule 1 (STIM1), targets STIM1 to multiple transporters. Proc Natl Acad Sci USA 108:19234–19239

Liou J, Kim ML, Heo WD, Jones JT, Myers JW, Ferrell JE Jr, Meyer T (2005) STIM is a Ca2+ sensor essential for Ca2+-store-depletion-triggered Ca2+ influx. Curr Biol 15:1235–1241

Lipp P, Laine M, Tovey SC, Burrell KM, Berridge MJ, Li W, Bootman MD (2000) Functional InsP3 receptors that may modulate excitation-contraction coupling in the heart. Curr Biol 10:939–942

Liu J, Xin L, Benson VL, Allen DG, Ju YK (2015) Store-operated calcium entry and the localization of STIM1 and Orai1 proteins in isolated mouse sinoatrial node cells. Front Physiol 6:69

Luo X, Hojayev B, Jiang N, Wang ZV, Tandan S, Rakalin A, Rothermel BA, Gillette TG, Hill JA (2012) STIM1-dependent store-operated Ca(2)(+) entry is required for pathological cardiac hypertrophy. J Mol Cell Cardiol 52:136–147

Makarewich CA, Zhang H, Davis J, Correll RN, Trappanese DM, Hoffman NE, Troupes CD, Berretta RM, Kubo H, Madesh M, Chen X, Gao E, Molkentin JD, Houser SR (2014) Transient receptor potential channels contribute to pathological structural and functional remodeling after myocardial infarction. Circ Res 115:567–580

Maus M, Cuk M, Patel B, Lian J, Ouimet M, Kaufmann U, Yang J, Horvath R, Hornig-Do HT, Chrzanowska-Lightowlers ZM, Moore KJ, Cuervo AM, Feske S (2017) Store-operated Ca2+ entry controls induction of lipolysis and the transcriptional reprogramming to lipid metabolism. Cell Metab25:698–712

McMullen JR, Jennings GL (2007) Differences between pathological and physiological cardiac hypertrophy: novel therapeutic strategies to treat heart failure. Clin Exp Pharmacol Physiol 34:255–262

Molkentin JD (2006) Dichotomy of Ca2+ in the heart: contraction versus intracellular signaling. J Clin Invest 116:623–626

Nakayama H, Bodi I, Maillet M, DeSantiago J, Domeier TL, Mikoshiba K, Lorenz JN, Blatter LA, Bers DM, Molkentin JD (2010) The IP3 receptor regulates cardiac hypertrophy in response to select stimuli. Circ Res 107:659–666

Nattel S (2003) Atrial electrophysiology and mechanisms of atrial fibrillation. J Cardiovasc Pharmacol Ther 8(Suppl 1):S5–11

Nguyen N, Biet M, Simard E, Beliveau E, Francoeur N, Guillemette G, Dumaine R, Grandbois M, Boulay G (2013) STIM1 participates in the contractile rhythmicity of HL-1 cells by moderating T-type Ca(2+) channel activity. Biochim Biophys Acta 1833:1294–1303

Niggli E (2011) Ryanodine receptors: waking up from refractoriness. Cardiovasc Res 91:563–564

Nilius B, Szallasi A (2014) Transient receptor potential channels as drug targets: from the science of basic research to the art of medicine. Pharmacol Rev 66:676–814

Ohba T, Watanabe H, Murakami M, Sato T, Ono K, Ito H (2009) Essential role of STIM1 in the development of cardiomyocyte hypertrophy. Biochem Biophys Res Commun 389:172–176

Oh-Hora M, Yamashita M, Hogan PG, Sharma S, Lamperti E, Chung W, Prakriya M, Feske S, Rao A (2008) Dual functions for the endoplasmic reticulum calcium sensors STIM1 and STIM2 in T cell activation and tolerance. Nat Immunol 9:432–443

Ong HL, de Souza LB, Ambudkar IS (2016) Role of TRPC channels in store-operated calcium entry. Adv Exp Med Biol 898:87–109

Pan Z, Brotto M, Ma J (2014) Store-operated Ca2+ entry in muscle physiology and diseases. BMB Rep 20:69–79

Parks C, Alam MA, Sullivan R, Mancarella S (2016) STIM1-dependent Ca(2+) microdomains are required for myofilament remodeling and signaling in the heart. Sci Rep 6:25372
Peppiatt CM, Collins TJ, Mackenzie L, Conway SJ, Holmes AB, Bootman MD, Berridge MJ, Seo JT, Roderick HL (2003) 2-Aminoethoxydiphenyl borate (2-APB) antagonises inositol 1,4,5-trisphosphate-induced calcium release, inhibits calcium pumps and has a use-dependent and slowly reversible action on store-operated calcium entry channels. Cell Calcium 34:97–108
Putney JW Jr (1990) Capacitative calcium entry revisited. Cell Calcium 11:611–624
Putney JW Jr, Bird GS (2008) Cytoplasmic calcium oscillations and store-operated calcium influx. J Physiol 586:3055–3059
Putney JW Jr, Steinckwich-Besancon N, Numaga-Tomita T, Davis FM, Desai PN, D'Agostin DM, Wu S, Bird GS (2016) The functions of store-operated calcium channels. Biochi Biophys Acta. doi:10.1016/j.bbamcr.2016.11.028
Roderick HL, Berridge MJ, Bootman MD (2003) Calcium-induced calcium release. Curr Biol 13: R425
Roos J, DiGregorio PJ, Yeromin AV, Ohlsen K, Lioudyno M, Zhang S, Safrina O, Kozak JA, Wagner SL, Cahalan MD, Velicelebi G, Stauderman KA (2005) STIM1, an essential and conserved component of store-operated Ca2+ channel function. J Cell Biol 169:435–445
Rosenberg P (2011) Socking It to cardiac hypertrophy: STIM1-mediated Ca2+ entry in the cardiomyocyte. Circulation 124:766–768
Ross GR, Bajwa T Jr, Edwards S, Emelyanova L, Rizvi F, Holmuhamedov EL, Werner P, Downey FX, Tajik AJ, Jahangir A (2017) Enhanced store-operated Ca2+ influx and ORAI1 expression in ventricular fibroblasts from human failing heart. Biol Open 6:326–332
Ruhle B, Trebak M (2013) Emerging roles for native Orai Ca2+ channels in cardiovascular disease. Curr Top Membr 71:209–235
Sabourin J, Robin E, Raddatz E (2011) A key role of TRPC channels in the regulation of electromechanical activity of the developing heart. Cardiovasc Res 92:226–236
Sabourin J, Bartoli F, Antigny F, Gomez AM, Benitah JP (2016) Transient receptor potential canonical (TRPC)/Orai1-dependent store-operated Ca2+ channels: NEW TARGETS OF ALDOSTERONE IN CARDIOMYOCYTES. J Biol Chem 291:13394–13409
Saliba Y, Keck M, Marchand A, Atassi F, Ouille A, Cazorla O, Trebak M, Pavoine C, Lacampagne A, Hulot JS, Fares N, Fauconnier J, Lompre AM (2015) Emergence of Orai3 activity during cardiac hypertrophy. Cardiovasc Res 105:248–259
Salido GM, Jardin I, Rosado JA (2011) The TRPC ion channels: association with Orai1 and STIM1 proteins and participation in capacitative and non-capacitative calcium entry. Adv Exp Med Biol 704:413–433
Samak M, Fatullayev J, Sabashnikov A, Zeriouh M, Schmack B, Farag M, Popov AF, Dohmen PM, Choi YH, Wahlers T, Weymann A (2016) Cardiac hypertrophy: an introduction to molecular and cellular basis. Med Sci Monit Basic Res 22:75–79
Seo K, Rainer PP, Shalkey Hahn V, Lee DI, Jo SH, Andersen A, Liu T, Xu X, Willette RN, Lepore JJ, Marino JP Jr, Birnbaumer L, Schnackenberg CG, Kass DA (2014) Combined TRPC3 and TRPC6 blockade by selective small-molecule or genetic deletion inhibits pathological cardiac hypertrophy. Proc Natl Acad Sci USA 111:1551–1556
Shuttleworth TJ, Mignen O (2003) Calcium entry and the control of calcium oscillations. Biochem Soc Trans 31:916–919
Smyrnias I, Mair W, Harzheim D, Walker SA, Roderick HL, Bootman MD (2010) Comparison of the T-tubule system in adult rat ventricular and atrial myocytes, and its role in excitation-contraction coupling and inotropic stimulation. Cell Calcium 47:210–223
Soboloff J, Rothberg BS, Madesh M, Gill DL (2012) STIM proteins: dynamic calcium signal transducers. Nat Rev Mol Cell Biol 13:549–565
Soltoff SP, Lannon WA (2013) Activation of ERK1/2 by store-operated calcium entry in rat parotid acinar cells. PLoS One 8:e72881
Svobodova B, Groschner K (2016) Mechanisms of lipid regulation and lipid gating in TRPC channels. Cell Calcium 59:271–279

Thompson JL, Shuttleworth TJ (2012) A plasma membrane-targeted cytosolic domain of STIM1 selectively activates ARC channels, an arachidonate-regulated store-independent Orai channel. Channels 6:370–378

Touchberry CD, Elmore CJ, Nguyen TM, Andresen JJ, Zhao X, Orange M, Weisleder N, Brotto M, Claycomb WC, Wacker MJ (2011) Store-operated calcium entry is present in HL-1 cardiomyocytes and contributes to resting calcium. Biochem Biophys Res Commun 416:45–50

Uehara A, Yasukochi M, Imanaga I, Nishi M, Takeshima H (2002) Store-operated Ca2+ entry uncoupled with ryanodine receptor and junctional membrane complex in heart muscle cells. Cell Calcium 31:89–96

Umar S, Nadadur R, Iorga A, Amjedi M, Matori H, Eghbali M (2012) Cardiac structural and hemodynamic changes associated with physiological heart hypertrophy of pregnancy are reversed postpartum. J Appl Physiol 113:1253–1259

Vig M, Peinelt C, Beck A, Koomoa DL, Rabah D, Koblan-Huberson M, Kraft S, Turner H, Fleig A, Penner R, Kinet JP (2006) CRACM1 is a plasma membrane protein essential for store-operated Ca2+ entry. Science 312:1220–1223

Voelkers M, Salz M, Herzog N, Frank D, Dolatabadi N, Frey N, Gude N, Friedrich O, Koch WJ, Katus HA, Sussman MA, Most P (2010) Orai1 and Stim1 regulate normal and hypertrophic growth in cardiomyocytes. J Mol Cell Cardiol 48:1329–1334

Voigt N, Nattel S, Dobrev D (2012) Proarrhythmic atrial calcium cycling in the diseased heart. Adv Exp Med Biol 740:1175–1191

Wang Y, Deng X, Gill DL (2010a) Calcium signaling by STIM and Orai: intimate coupling details revealed. Sci Signal 3:pe42

Wang Y, Deng X, Mancarella S, Hendron E, Eguchi S, Soboloff J, Tang XD, Gill DL (2010b) The calcium store sensor, STIM1, reciprocally controls Orai and CaV1.2 channels. Science 330:105–109

Zhang SL, Yeromin AV, Zhang XH, Yu Y, Safrina O, Penna A, Roos J, Stauderman KA, Cahalan MD (2006) Genome-wide RNAi screen of Ca(2+) influx identifies genes that regulate Ca(2+) release-activated Ca(2+) channel activity. Proc Natl Acad Sci USA 103:9357–9362

Zhang YH, HJ W, Che H, Sun HY, Cheng LC, Li X, WK A, Tse HF, Li GR (2013) Functional transient receptor potential canonical type 1 channels in human atrial myocytes. Pflugers Arch 465:1439–1449

Zhang H, Sun AY, Kim JJ, Graham V, Finch EA, Nepliouev I, Zhao G, Li T, Lederer WJ, Stiber JA, Pitt GS, Bursac N, Rosenberg PB (2015) STIM1-Ca2+ signaling modulates automaticity of the mouse sinoatrial node. Proc Natl Acad Sci USA 112:E5618–E5627

Zhang B, Jiang J, Yue Z, Liu S, Ma Y, Yu N, Gao Y, Sun S, Chen S, Liu P (2016) Store-Operated Ca2+ Entry (SOCE) contributes to angiotensin II-induced cardiac fibrosis in cardiac fibroblasts. J Pharmacol Sci 132:171–180

Zhao G, Li T, Brochet DX, Rosenberg PB, Lederer WJ (2015) STIM1 enhances SR Ca2+ content through binding phospholamban in rat ventricular myocytes. Proc Natl Acad Sci USA 112: E4792–E4801

Zhou Y, Cai X, Nwokonko RM, Loktionova NA, Wang Y, Gill DL (2017) The STIM-Orai coupling interface and gating of the Orai1 channel. Cell Calcium. doi:10.1016/j.ceca.2017.01.001

Zhu-Mauldin X, Marsh SA, Zou L, Marchase RB, Chatham JC (2012) Modification of STIM1 by O-linked N-acetylglucosamine (O-GlcNAc) attenuates store-operated calcium entry in neonatal cardiomyocytes. J Biol Chem 287:39094–39106

Part III

SOCE: Molecular and Cellular (Patho)Physiology

Introduction: Overview of the Pathophysiological Implications of Store-Operated Calcium Entry in Mammalian Cells

20

Juan A. Rosado

Abstract

Since store-operated Ca^{2+} entry (SOCE) was proposed by Putney three decades ago (Putney. Cell Calcium 7:1–12, 1986), its functional role and involvement in the pathophysiology of a number of disorders has been investigated. The role of SOCE in cell physiology has been discussed in the previous chapters, and the following part is devoted to the current knowledge concerning the mechanisms underlying the development of certain diseases that involve SOCE abnormalities.

Keywords

Orai1 • STIM1 • SOCE • SCID • Stormorken syndrome • York platelet syndrome

SOCE plays an important physiological role in many cell types, including the support of cytosolic free-Ca^{2+} concentration ($[Ca^{2+}]_c$) oscillations and a number of cellular processes of great physiological relevance. Ca^{2+} signals resulting from cell stimulation with physiological concentrations of agonists usually consist of repetitive spikes of $[Ca^{2+}]_c$, referred to as Ca^{2+} oscillations (Putney 1986). The functional role of SOCE, supporting $[Ca^{2+}]_c$ oscillations, has been evidenced by pharmacological and molecular biological studies. The oscillatory pattern of Ca^{2+} signals is sensitive to treatment with 2-aminoethoxydiphenyl borate and low concentrations of gadolinium, two well-known inhibitors of SOCE (Bird and Putney 2005). Furthermore, knock-down of STIM1 or Orai1 by RNA interference has been demonstrated to result in attenuation of the frequency of $[Ca^{2+}]_c$ oscillations in response to physiological concentrations of methacholine (Wedel et al. 2007). More recently, a key role for STIM2 in the activation of SOCE and support of Ca^{2+} oscillations at low

J.A. Rosado (✉)
Cell Physiology Research Group, Department of Physiology, University of Extremadura, Cáceres, Spain
e-mail: jarosado@unex.es

K. Groschner et al. (eds.), *Store-Operated Ca^{2+} Entry (SOCE) Pathways*, Advances in Experimental Medicine and Biology 993, DOI 10.1007/978-3-319-57732-6_20

agonist concentrations has been reported. Thus, mild reductions in the endoplasmic reticulum (ER) Ca^{2+} concentration evoked by low agonist concentrations preferentially mediate SOCE via STIM2, and silencing of STIM2 impairs SOCE and Ca^{2+} oscillations without affecting STIM1-dependent Ca^{2+} signals (Thiel et al. 2013). In addition to the functional role of SOCE in the maintenance of $[Ca^{2+}]_c$ oscillations, Ca^{2+} influx via SOCE has been reported to be necessary for a number of specific cellular events, including secretion (Ma et al. 2008; Balghi et al. 2011; Pani et al. 2013; Sabourin et al. 2015), the contractile function in the skeletal muscle (Wei-Lapierre et al. 2013), platelet function (Galan et al. 2009; Gilio et al. 2010), or vascular smooth muscle cell proliferation and contraction (Dominguez-Rodriguez et al. 2012; Rodriguez-Moyano et al. 2013).

Given the importance of SOCE in cell physiology, one would expect that disorders in the mechanism of activation, maintenance, or inactivation of Ca^{2+} influx might lead to a pathophysiological phenotype. Loss- or gain-of-function gene mutations in the key components of SOCE, including STIM1 and Orai1, have been reported to underlie a number of human disorders. Probably one of the best characterized disorders directly associated with SOCE dysfunction is a hereditary form of SCID (severe combined immunodeficiency). This disorder has been attributed to a missense mutation in the *Orai1* gene, whereby a conserved arginine residue is replaced with a tryptophan at position 91, leading to complete lack of the Ca^{2+} release-activated Ca^{2+} current (I_{CRAC}) in T cells, B cells, NK cells, and fibroblasts (Feske et al. 2006). In addition to patients with the Orai1-R91W mutation, other Orai1 mutations have been found to be associated to immunodeficiency. This is the case of the lack of SOCE and CRAC channel function in patients suffering an insertion mutation in *Orai1* (A88SfsX25) that causes a frameshift beginning at residue A88 and premature termination at position 112 at the end of the first transmembrane domain, which results in abolished Orai1 expression. Similarly, missense *Orai1* mutations in the first (A103E) and third (L194P) transmembrane domains impair Orai1 expression (McCarl et al. 2009). In addition to Orai1 mutations, immunodeficiency associated to SOCE dysfunction has been attributed to an insertion mutation in STIM1 that results in a frameshift starting at position E128 of STIM1 and its premature termination at position 136 (Lacruz and Feske 2015).

In addition to immunodeficiency, *STIM1* and *Orai1* gene mutations and altered protein expression have been reported to lead to a variety of dysfunctions including myopathy, cardiac disorders, altered platelet function, and phenotype changes leading to cancer.

Concerning platelet function, a missense mutation in *STIM1* exon 7 leading to the replacement of the arginine residue at position 304 with a tryptophan (R304W), results in a gain of function that is responsible for the Stormorken syndrome in humans. The mutation affects the CC1 domain function that maintains STIM1 in an inactive state, thus causing an excessive Ca^{2+} entry into platelets and leading to premature platelet activation, constitutive phosphatidylserine exposure, thrombocytopenia, and bleeding disorders (Misceo et al. 2014; Morin et al. 2014; Nesin et al. 2014). Furthermore, patients with York platelet syndrome, a disorder characterized by thrombocytopenia, altered morphology, and delta granule deficiency, show monoallelic STIM1 mutations, i.e., either I115F or R304W resulting in gain of function (Markello et al. 2015). Both York and Stormorken syndromes are characterized by the presence of myopathy. Furthermore, as discussed in this part, several STIM1 mutations, including H72Q, D84G, or H109R/H109N, affecting the canonical EF-hand motif, have been reported to lead to tubular aggregate myopathy (Bohm et al. 2013).

Activation of SOCE as well as the level of expression of its core elements, Orai1 and STIM1, have been implicated in the development of several cancer features, including cancer cell migration, proliferation, tumor metastasis, and angiogenesis (Smani et al. 2015). In addition, a role for SOCE in apoptosis resistance in a number of cancer cell types has been proposed. An exception is prostate cancer cells, where SOCE has been proven to be pro-apoptotic, and, in these cells, Orai3 upregulation leads to altered Orai1/Orai3 expression ratio, resulting in reduced SOCE and enhanced ARC (arachidonate-regulated Ca^{2+} channel)-mediated Ca^{2+} entry, which, in turn, promotes apoptosis resistance and cancer growth (Dubois et al. 2014). The relevance of SOCE in tumor progression has been demonstrated in a number of cancer cell types (for a review, see Jardin and Rosado 2016). Undoubtedly, discerning the precise mechanism that underlies the development of the different disorders may serve as a basis for developing diagnostic and therapeutic strategies.

The chapters in this part summarize the functional role of SOCE and its core elements in different cellular models and tissues, as well as the current knowledge concerning the disorders associated to the dysfunction of the different SOCE components, including channels, Ca^{2+} sensors, and Ca^{2+} entry regulatory mechanisms. In particular, the molecular pathophysiological mechanisms of SOCE will be described in the immune system; the skeletal muscle; the cardiovascular system, including cardiac myocytes, endothelial cells, and platelets; the nervous system; and germ cells. Furthermore, the pathophysiological implications of SOCE and its core molecular components in the progression of cancer will be discussed.

Acknowledgments Supported by MINECO (Grant BFU2016-74932-C2-1-P).

References

Balghi H, Robert R, Rappaz B, Zhang X, Wohlhuter-Haddad A, Evagelidis A, Luo Y, Goepp J, Ferraro P, Romeo P, Trebak M, Wiseman PW, Thomas DY, Hanrahan JW (2011) Enhanced Ca2+ entry due to Orai1 plasma membrane insertion increases IL-8 secretion by cystic fibrosis airways. FASEB J 25:4274–4291

Bird GS, Putney JW Jr (2005) Capacitative calcium entry supports calcium oscillations in human embryonic kidney cells. J Physiol 562:697–706

Bohm J, Chevessier F, Maues De Paula A, Koch C, Attarian S, Feger C, Hantai D, Laforet P, Ghorab K, Vallat JM, Fardeau M, Figarella-Branger D, Pouget J, Romero NB, Koch M, Ebel C, Levy N, Krahn M, Eymard B, Bartoli M, Laporte J (2013) Constitutive activation of the calcium sensor STIM1 causes tubular-aggregate myopathy. Am J Hum Genet 92:271–278

Dominguez-Rodriguez A, Diaz I, Rodriguez-Moyano M, Calderon-Sanchez E, Rosado JA, Ordonez A, Smani T (2012) Urotensin-II signaling mechanism in rat coronary artery: role of STIM1 and Orai1-dependent store operated calcium influx in vasoconstriction. Arterioscler Thromb Vasc Biol 32:1325–1332

Dubois C, Vanden Abeele F, Lehen'kyi V, Gkika D, Guarmit B, Lepage G, Slomianny C, Borowiec AS, Bidaux G, Benahmed M, Shuba Y, Prevarskaya N (2014) Remodeling of channel-forming ORAI proteins determines an oncogenic switch in prostate cancer. Cancer Cell 26:19–32

Feske S, Gwack Y, Prakriya M, Srikanth S, Puppel SH, Tanasa B, Hogan PG, Lewis RS, Daly M, Rao A (2006) A mutation in Orai1 causes immune deficiency by abrogating CRAC channel function. Nature 441:179–185

Galan C, Zbidi H, Bartegi A, Salido GM, Rosado JA (2009) STIM1, Orai1 and hTRPC1 are important for thrombin- and ADP-induced aggregation in human platelets. Arch Biochem Biophys 490:137–144

Gilio K, van Kruchten R, Braun A, Berna-Erro A, Feijge MA, Stegner D, van der Meijden PE, Kuijpers MJ, Varga-Szabo D, Heemskerk JW, Nieswandt B (2010) Roles of platelet STIM1 and Orai1 in glycoprotein VI- and thrombin-dependent procoagulant activity and thrombus formation. J Biol Chem 285:23629–23638

Jardin I, Rosado JA (2016) STIM and calcium channel complexes in cancer. Biochim Biophys Acta 1863(6 Pt B):1418–1426

Lacruz RS, Feske S (2015) Diseases caused by mutations in ORAI1 and STIM1. Ann N Y Acad Sci 1356:45–79

Ma HT, Peng Z, Hiragun T, Iwaki S, Gilfillan AM, Beaven MA (2008) Canonical transient receptor potential 5 channel in conjunction with Orai1 and STIM1 allows Sr2+ entry, optimal influx of Ca2+, and degranulation in a rat mast cell line. J Immunol 180:2233–2239

Markello T, Chen D, Kwan JY, Horkayne-Szakaly I, Morrison A, Simakova O, Maric I, Lozier J, Cullinane AR, Kilo T, Meister L, Pakzad K, Bone W, Chainani S, Lee E, Links A, Boerkoel C, Fischer R, Toro C, White JG, Gahl WA, Gunay-Aygun M (2015) York platelet syndrome is a CRAC channelopathy due to gain-of-function mutations in STIM1. Mol Genet Metab 114:474–482

McCarl CA, Picard C, Khalil S, Kawasaki T, Rother J, Papolos A, Kutok J, Hivroz C, Ledeist F, Plogmann K, Ehl S, Notheis G, Albert MH, Belohradsky BH, Kirschner J, Rao A, Fischer A, Feske S (2009) ORAI1 deficiency and lack of store-operated Ca2+ entry cause immunodeficiency, myopathy, and ectodermal dysplasia. J Allergy Clin Immunol 124:1311–1318

Misceo D, Holmgren A, Louch WE, Holme PA, Mizobuchi M, Morales RJ, De Paula AM, Stray-Pedersen A, Lyle R, Dalhus B, Christensen G, Stormorken H, Tjonnfjord GE, Frengen E (2014) A dominant STIM1 mutation causes Stormorken syndrome. Hum Mutat 35:556–564

Morin G, Bruechle NO, Singh AR, Knopp C, Jedraszak G, Elbracht M, Bremond-Gignac D, Hartmann K, Sevestre H, Deutz P, Herent D, Nurnberg P, Romeo B, Konrad K, Mathieu-Dramard M, Oldenburg J, Bourges-Petit E, Shen Y, Zerres K, Ouadid-Ahidouch H, Rochette J (2014) Gain-of-function mutation in STIM1 (p.R304W) is associated with Stormorken syndrome. Hum Mutat 35:1221–1232

Nesin V, Wiley G, Kousi M, Ong EC, Lehmann T, Nicholl DJ, Suri M, Shahrizaila N, Katsanis N, Gaffney PM, Wierenga KJ, Tsiokas L (2014) Activating mutations in STIM1 and ORAI1 cause overlapping syndromes of tubular myopathy and congenital miosis. Proc Natl Acad Sci USA 111:4197–4202

Pani B, Liu X, Bollimuntha S, Cheng KT, Niesman IR, Zheng C, Achen VR, Patel HH, Ambudkar IS, Singh BB (2013) Impairment of TRPC1-STIM1 channel assembly and AQP5 translocation compromise agonist-stimulated fluid secretion in mice lacking caveolin1. J Cell Sci 126:667–675

Putney JW Jr (1986) A model for receptor-regulated calcium entry. Cell Calcium 7:1–12

Rodriguez-Moyano M, Diaz I, Dionisio N, Zhang X, Avila-Medina J, Calderon-Sanchez E, Trebak M, Rosado JA, Ordonez A, Smani T (2013) Urotensin-II promotes vascular smooth muscle cell proliferation through store operated calcium entry and EGFR transactivation. Cardiovasc Res 100:297–306

Sabourin J, Le Gal L, Saurwein L, Haefliger JA, Raddatz E, Allagnat F (2015) Store-operated Ca2+ entry mediated by Orai1 and TRPC1 participates to insulin secretion in rat beta-cells. J Biol Chem 290:30530–30539

Smani T, Shapovalov G, Skryma R, Prevarskaya N, Rosado JA (2015) Functional and physio-pathological implications of TRP channels. Biochim Biophys Acta 1853:1772–1782

Thiel M, Lis A, Penner R (2013) STIM2 drives Ca2+ oscillations through store-operated Ca2+ entry caused by mild store depletion. J Physiol 591:1433–1445

Wedel B, Boyles RR, Putney JW Jr, Bird GS (2007) Role of the store-operated calcium entry proteins Stim1 and Orai1 in muscarinic cholinergic receptor-stimulated calcium oscillations in human embryonic kidney cells. J Physiol 579:679–689

Wei-Lapierre L, Carrell EM, Boncompagni S, Protasi F, Dirksen RT (2013) Orai1-dependent calcium entry promotes skeletal muscle growth and limits fatigue. Nat Commun 4:2805

Immunological Disorders: Regulation of Ca^{2+} Signaling in T Lymphocytes

21

Sonal Srikanth, Jin Seok Woo, Zuoming Sun, and Yousang Gwack

Abstract

Engagement of T cell receptors (TCRs) with cognate antigens triggers cascades of signaling pathways in helper T cells. TCR signaling is essential for the effector function of helper T cells including proliferation, differentiation, and cytokine production. It also modulates effector T cell fate by inducing cell death, anergy (nonresponsiveness), exhaustion, and generation of regulatory T cells. One of the main axes of TCR signaling is the Ca^{2+}-calcineurin-nuclear factor of activated T cells (NFAT) signaling pathway. Stimulation of TCRs triggers depletion of intracellular Ca^{2+} store and, in turn, activates store-operated Ca^{2+} entry (SOCE) to raise the intracellular Ca^{2+} concentration. SOCE in T cells is mediated by the Ca^{2+} release-activated Ca^{2+} (CRAC) channels, which have been very well characterized in terms of their electrophysiological properties. Identification of STIM1 as a sensor to detect depletion of the endoplasmic reticulum (ER) Ca^{2+} store and Orai1 as the pore subunit of CRAC channels has dramatically advanced our understanding of the regulatory mechanism of Ca^{2+} signaling in T cells. In this review, we discuss our current understanding of Ca^{2+} signaling in T cells with specific focus on the mechanism of CRAC channel activation and regulation via protein interactions. In addition, we will discuss the role of CRAC channels in effector T cells, based on the analyses of genetically modified animal models.

S. Srikanth • J.S. Woo • Y. Gwack (✉)
Department of Physiology, David Geffen School of Medicine at UCLA, 53-266 CHS, 10833 Le Conte Avenue, Los Angeles, CA 90095, USA
e-mail: ygwack@mednet.ucla.edu

Z. Sun
Department of Molecular Immunology, Beckman Research Institute, City of Hope, Duarte, CA, USA

K. Groschner et al. (eds.), *Store-Operated Ca^{2+} Entry (SOCE) Pathways*, Advances in Experimental Medicine and Biology 993, DOI 10.1007/978-3-319-57732-6_21

Keywords
T-cell receptor signaling • CRAC channels • Orai and STIM proteins • Effector T cells • Cytokine production • Differentiation of T cells

21.1 Introduction

Ca^{2+} is ubiquitously utilized as a second messenger to regulate diverse cellular functions. However, in spite of its broad role, Ca^{2+} signaling has specificity to activate distinct downstream signaling pathways depending on the types of ligands and agonists. Each cell type, including T cells, expresses a unique blend of various Ca^{2+} channels or transporters to mediate Ca^{2+} signaling. Under resting conditions, cytoplasmic Ca^{2+} concentration ($[Ca^{2+}]$) in T cells is in the range of ~100 nM while that in the endoplasmic reticulum (ER), which serves as an intracellular Ca^{2+} store, is much higher (~400 μM). Extracellular $[Ca^{2+}]$ (e.g., blood) reaches almost 2 mM, establishing a huge $[Ca^{2+}]$ gradient between the extracellular milieu, the Ca^{2+} store, and the cytoplasm. Therefore, maintaining Ca^{2+} homeostasis in the resting state requires dynamic Ca^{2+} entry and exclusion across the membranes. After TCR stimulation, Ca^{2+} entry via store-operated Ca^{2+} (SOC) channels is a predominant mechanism to increase the intracellular Ca^{2+} concentration ($[Ca^{2+}]_i$) in T cells (Cahalan and Chandy 2009; Hogan et al. 2010; Lewis 2011; Putney 2009; Srikanth and Gwack 2013b). SOC channels were so named because they are activated by depletion of intracellular Ca^{2+} stores (Putney 1986, 2009). The Ca^{2+} release-activated Ca^{2+} (CRAC) channel is a prototype and specialized class of SOC channel in immune cells. Increased Ca^{2+} ion concentrations can affect many signaling pathways, for example, via directly binding to anionic phospholipids or activation of Ca^{2+}-sensing signaling amplifiers including calmodulin (CaM), which in turn activate a large number of protein kinases/phosphatases to affect gene expression. Because the volume of ER in T lymphocytes is much smaller than that in other cell types, SOCE via CRAC channels is particularly important for activation of the NFAT (nuclear factor of activated T cells) family of transcription factors, which require sustained levels of increased $[Ca^{2+}]_i$. In the current review, we will focus on our current understanding of the regulation of Ca^{2+} signaling in T cells and phenotypes of animal models lacking the key components of CRAC channels, which provide insights into their physiological roles in the immune system.

21.2 Signaling Pathways Activated by TCR Stimulation

Upon pathogen infection, specialized innate (e.g., dendritic cells or macrophages) and adaptive (e.g., B cells) immune cells present foreign antigens on their surface together with major histocompatibility complex (MHC) class II molecules to activate CD4+ helper T cells. Interactions between TCRs and foreign antigens presented by MHC class II molecules play an important role in T helper cell functions such as proliferation, differentiation into effector and memory cells, and massive cytokine production after being recruited into the affected sites of inflammation. In addition, interactions between self-peptides and TCRs are important for T cell development in the thymus, homeostasis, and pathological onset of autoimmune diseases (Sprent and Surh 2011). Thus, understanding of TCR signaling is crucial for development of therapy to treat patients with suboptimal immune responses (e.g., immune deficiency), lymphoproliferative disorder, and self (or microbiota)-reactive T cell-mediated autoimmunity (e.g., type I diabetes, rheumatoid arthritis, multiple sclerosis, psoriasis, and inflammatory bowel disease).

Antigen engagement of T cell receptor triggers a cascade of tyrosine phosphorylation events initiated by a co-receptor CD4 that interacts with lymphocyte-specific protein tyrosine kinase (LCK) through its intracellular domain (Fig. 21.1a). Antigen-loaded MHC class II molecules simultaneously interact with the TCR-CD3 (γ-ε, ε-δ, and ζ-ζ chains) complex and the LCK-associated co-receptor CD4. LCK then phosphorylates immunoreceptor tyrosine-based activation motifs (ITAMs) of CD3 ζ chain, which triggers the recruitment of zeta chain-associated protein kinase 70 (ZAP70). ZAP70 phosphorylates membrane-associated scaffold molecules, linker for activation of T cells (LAT) and SH2 domain-containing leukocyte protein of 76 kDa (SLP76) (Balagopalan et al. 2010; Samelson 2011; Wang et al. 2010). Together with LAT, SLP76 recruits a guanine nucleotide exchange factor (GEF), Vav1, and an enzyme, phospholipase C (PLC)-γ1. Vav1 accumulates at the immunological synapse (IS) and recruits small G proteins such as Rac1 and CDC42 (cell division control protein 42 homologue) to activate the c-Jun N-terminal kinase (JNK) and p38 MAPK (mitogen-activated protein kinase) pathways, leading to activation of AP1 transcription factors (Tybulewicz 2005). Recently, we showed that Vav1 also interacts with a large G protein, CRACR2A to activate the JNK signaling pathway (see below (Srikanth et al. 2016a)). Therefore, Vav1 acts as a signaling hub to recruit small and large G proteins to activate downstream JNK and p38 MAPK pathways. PLCγ1 recruited into the LAT/SLP76 complex hydrolyzes plasma membrane (PM)-localized phosphatidylinositol 4,5-bisphosphate (PIP_2) into second messengers, inositol trisphosphate (IP_3) and diacylglycerol (DAG), leading to activation of the Ca^{2+}-NFAT, NK-κB and ERK (extracellular signal-regulated kinase)-MAPK pathways. DAG predominantly activates NF-κB signaling pathway via activation of protein kinase C-theta (PKCθ) (Coudronniere et al. 2000; Lin et al. 2000; Sun et al. 2000). PKCθ is required to stimulate the BCM complex consisting of Bcl-10/Carma 1/MALT1, NF-κB-inducible kinase (NIK), and inhibitor of NF-κB (IκB) kinase (IKK) that eventually phosphorylates IκB (Fig. 21.1a) (Muller and Rao 2010; Smith-Garvin et al. 2009; Sun 2012). Phosphorylation of IκB leads to its degradation, allowing for nuclear translocation of NF-κB transcription factors. DAG also activates Ras-mediated signaling pathway via activation of Ras guanine nucleotide-releasing protein 1 (RasGRP1), which mediates phosphorylation-induced activation of AP-1 (Fos-Jun) transcription factors mediated by kinases, dual-specificity mitogen-activated protein kinase kinase (MEK) 1/2, and ERKs. The other product of PLCγ1 enzymatic activity, IP_3, binds to the IP_3 receptor (IP_3R) on the ER

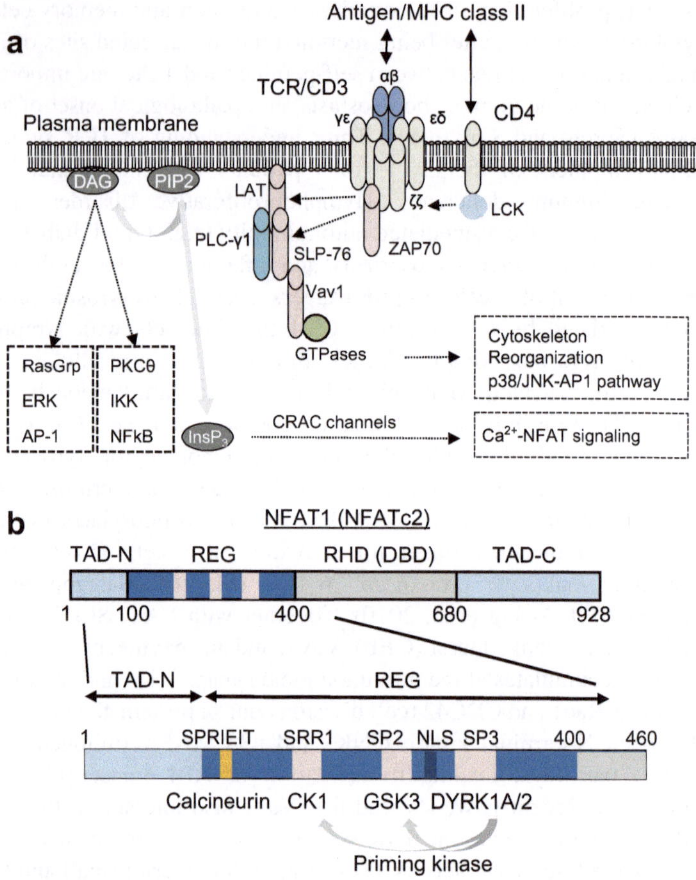

Fig. 21.1 Signaling pathways of T cell receptor stimulation. (**a**) Antigen engagement of T cell receptor induces a series of phosphorylation events and activates downstream signaling pathways. Co-receptor (e.g., CD4) ligation to MHC class II positions the protein tyrosine kinase LCK into the antigen-bound TCR/CD3 complex. LCK phosphorylates the cytoplasmic part of ζ chain to recruit ZAP70. ZAP70 phosphorylates two adaptor proteins LAT and SLP-76 that results in assembly of a signaling complex containing Vav1 and phospholipase C (PLC-γ1). Vav1 recruits downstream effector molecules including Rac1, and a Rho GTPase, CDC42 that has pleiotropic effects in cytoskeleton reorganization and p38/JNK signaling pathways. Vav1 can also recruit large Rab GTPase like CRACR2A-a to activate the JNK signaling pathway. Cytoskeleton reorganization is important for formation of the immunological synapse between antigen-presenting cells and T cells. Activated PLC-γ1 hydrolyzes PIP_2 into IP_3 and DAG. While DAG activates PKCθ-NF-κB and RasGRP-AP1 signaling pathways, IP_3 binds to the IP_3 receptor (IP_3R) on the ER membrane to empty the ER Ca^{2+} store, which induces opening of CRAC channels. Elevated $[Ca^{2+}]_i$ triggers a broad range of downstream signaling pathways including the Ca^{2+}-calmodulin/calcineurin-NFAT pathway. (**b**) Schematic of the murine NFAT1 (NFATc2) protein. The transcription activation domains that interact with transcriptional cofactors (e.g., p300) are located at the N- and C-terminus (TAD-N and TAD-C). DNA-binding domain (DBD) shows highest homology with the Rel homology domain (RHD) of Rel-family transcription factors. It also interacts with other transcription factors including AP-1 depending on the DNA sequence. The regulatory domain

membrane and releases Ca^{2+} from the ER into the cytoplasm, and this store depletion leads to activation of CRAC channels on the PM. Activation of both the Ca^{2+} and MAPK signaling pathways is essential for differentiation, and cytokine production of helper T cells and dysregulation of these pathways result in immune deficiency or autoimmune disorders in humans and mice (Constant and Bottomly 1997; Kyriakis and Avruch 2012; Zhu et al. 2010).

21.3 The Ca^{2+}-Calcineurin-NFAT Pathway in T Cells

One of the most studied Ca^{2+}-dependent signaling pathways in T cells is the calmodulin/calcineurin-NFAT pathway. Calcineurin is a Ca^{2+}-calmodulin complex-dependent serine/threonine protein phosphatase, consisting of a catalytic subunit, calcineurin A (CnAα, CnAβ, and CnAγ), and a regulatory subunit calcineurin B (CnB1 and CnB2). Upon increase of $[Ca^{2+}]_i$ via the CRAC channels, calmodulin binds Ca^{2+} and forms a complex with calcineurin, which in turn dephosphorylates the heavily phosphorylated, cytoplasmic NFAT. NFAT consists of four homologous NFAT1 (NFATc2), NFAT2 (NFATc1), NFAT3 (NFATc4), and NFAT4 (NFATc3) (Hogan et al. 2003; Macian 2005; Serfling et al. 2006; Srikanth and Gwack 2013b; Wu et al. 2007). Most of the NFAT family members are expressed in lymphocytes; however NFAT1 is predominantly expressed in naïve T cells, and its short-term dephosphorylation after stimulation depends on the levels of Ca^{2+} entry. Expression levels of a short isoform of NFAT2 (NFATc1), NFAT2/αA is also Ca^{2+}dependent (Serfling et al. 2012). NFAT2/αA plays a more positive role in T cell activation than other NFAT family members by supporting proliferation and protecting T cells against cell death upon stimulation. Dephosphorylation of NFAT exposes its nuclear localization sequence (NLS) and induces its translocation from the cytoplasm to the nucleus (Gwack et al. 2007a; Hogan et al. 2003). Nuclear NFAT forms a multimeric protein complex with itself or with other general or lineage-specific transcription factors (e.g., AP-1) to induce gene transcription involved in cytokine production, cell proliferation, growth arrest, or cell death, depending on the amplitude and duration of $[Ca^{2+}]_i$ elevation (Kim et al. 2011; Macian 2005; Macian et al. 2002).

Fig. 21.1 (continued) (REG) contains multiple phosphorylation sites to maintain cytoplasmic localization of NFAT under resting conditions and a docking site for Ca^{2+}-calmodulin-calcineurin complex (SPRIEIT motif). After increase in intracellular $[Ca^{2+}]$, Ca^{2+}-bound calmodulin forms a complex with a protein phosphatase calcineurin, which binds to the SPRIET motif and dephosphorylates the SRR (serine-rich region) and SP (Ser-Pro-X-X repeat) motifs in the regulatory domain, leading to the nuclear translocation of NFAT1 by exposing the nuclear localization sequence (NLS). SRR1, SP2, and SP3 motifs within the regulatory domain are phosphorylated by casein kinase 1 (CK1), glycogen synthase kinase 3 (GSK3), and dual-specificity tyrosine-phosphorylation-regulated kinase (DYRK) family kinases, respectively. DYRKs play a role as a priming kinase for CK1- and GSK3-mediated phosphorylation

NFAT proteins contain an N-terminal transactivation domain (TAD-N), a regulatory domain, a highly conserved DNA-binding domain (Rel homology domain, RHD) and a C-terminal transcription activation domain (TAD-C) (Fig. 21.1b). NFAT proteins all bind to a similar DNA sequence containing 5′-(A/T)GGAAA-3′ due to high conservancy within their DNA-binding domains and thus are expected to have functional redundancy (Hogan et al. 2003). TADs are least conserved, and this somehow explains distinct roles of each NFAT member because these domains can interact with different transcriptional co-activators (Mognol et al. 2016). NFATs are intrinsically localized in the cytoplasm, which is actively maintained by multiple kinases in the cytoplasm and the nucleus. The regulatory domain, which is moderately conserved among NFAT proteins, contains multiple serine-rich regions (SRRs) and Ser-Pro-X-X repeat motifs (SPs) that are phosphorylated by NFAT kinases including casein kinase 1 (CK1), glycogen synthase kinase 3 (GSK3), and dual-specificity tyrosine-phosphorylation-regulated kinase (DYRK) family (Gwack et al. 2007a; Wu et al. 2007). CK1 phosphorylates the SRR-1 region of NFAT and functions as both a maintenance kinase that keeps NFAT in the cytoplasm and an export kinase from the nucleus. GSK-3 functions as an export kinase and phosphorylates the SP-2 motif of NFAT1 and both the SP-2 and SP-3 motifs of NFAT2, and its activity is suppressed by Akt, a kinase activated in response to diverse signaling pathways in different cell types and by CD28 co-stimulatory signal (Gwack et al. 2007a). DYRKs phosphorylate the SP-3 motif of NFAT1, thereby facilitating CK1- and GSK3-mediated phosphorylation of the SRR-1 and SP-2 motifs, respectively. The substrate sites for GSK3 in NFAT are created after phosphorylation by a "priming" kinase that can be either protein kinase A (PKA) or DYRKs (Arron et al. 2006; Gwack et al. 2006). Depending on the localization, cytoplasmic DYRK2 serves as the maintenance kinase, while nuclear DYRK1A serves as the export kinase by priming phosphorylation by CK1 or GSK-3. Under resting conditions, heavily phosphorylated NFAT proteins exist in a complex with the noncoding RNA NRON (noncoding [RNA] repressor of NFAT). NRON creates a platform for RNA-protein scaffold complexes containing NFAT, NFAT kinases [e.g., CK1, GSK3, DYRK, and leucine-rich repeat kinase 2 (LRRK2)], IQ motif-containing GTPase-activating protein (IQGAP), and CaM to facilitate phosphorylation/dephosphorylation events (Liu et al. 2011; Sharma et al. 2011; Willingham et al. 2005). The regulatory domain of NFAT also contains a docking site for calcineurin, with a highly conserved consensus sequence Pro-X-Ile-X-Ile-Thr (in which X can be any amino acid) (Aramburu et al. 1999). Upon TCR stimulation-induced increase in $[Ca^{2+}]_i$, Ca^{2+}-bound CaM activates calcineurin, which dephosphorylates multiple phosphoserines in the SRR and SP motifs of NFAT regardless of its distance from the calcineurin-binding site, exposing its NLS and facilitating nuclear translocation.

In addition to the Ca^{2+}-calcineurin-NFAT pathway, increased $[Ca^{2+}]_i$ also plays a crucial role in activation of the NF-κB and ERK-MAPK signaling pathways. During formation of the BCM complex, Ca^{2+} has been implicated in phosphorylation via calmodulin kinase II (CaMKII), which is important for IκB degradation (Ishiguro et al. 2006, 2007; Oruganti et al. 2011). In addition, Ca^{2+}-activated PKCα phosphorylates NF-κB transcription factors, which is important for nuclear translocation of these factors (Liu et al. 2016). Elevated $[Ca^{2+}]_i$ also regulates the Ras-AP1 signaling pathway by binding to EF-hand motifs of RasGRP1 (Mor and Philips 2006). The Ras guanine nucleotide exchange factor (RasGEF) activity of RasGRP1 on the Golgi depends on both DAG and Ca^{2+} that eventually leads to activation of the AP-1 transcription factor. Therefore, Ca^{2+} signaling is integrated with other signaling pathways at the DNA response elements of NFAT, NF-κB, and AP-1 transcription factors, resulting in cell proliferation and survival, cytokine gene expression, differentiation, or cell death depending on the intensity of diverse signaling pathways.

21.4 Specificity of Ca^{2+} Signaling

Ca^{2+} signaling plays an important role in diverse cellular functions. However, numerous evidences suggest that Ca^{2+} can play a specialized role in activation of distinct signaling pathways depending on the amplitude, oscillation frequency, and location of its entry. In physiological conditions, unlike treatment with ionophore or a blocker of SERCA (sarcoplasmic and endoplasmic reticulum Ca^{2+} ATPase), thapsigargin, T cells show Ca^{2+} oscillations after TCR engagement that is regulated by a balance between cytoplasmic and ER Ca^{2+} concentrations and the levels of Ca^{2+} entry and exclusion (Dolmetsch and Lewis 1994; Dolmetsch et al. 1998). NFAT, AP-1, and NF-κB transcription factors were shown to be optimally activated in response to different oscillation patterns of Ca^{2+} in T cells. Transient high Ca^{2+} spikes evoked activation of JNK and NF-κB, but not NFAT, whereas prolonged low increases in $[Ca^{2+}]_i$ was sufficient for activation of NFAT, but not JNK or NF-κB (Dolmetsch et al. 1997). It was also shown that $[Ca^{2+}]$ in the microdomains near the CRAC channels is more important for nuclear translocation of NFAT than global increase of Ca^{2+} emphasizing the importance of local Ca^{2+} concentration in T cells (Kar et al. 2011). STIM1 and Orai1 proteins are crucial components of the CRAC channel in T cells (see below). Upon antigen engagement of $CD4^+$ T cells, Orai1 and STIM1 translocate into the immunological synapse, a site of contact between the TCRs and antigen-loaded MHC class II molecules, suggesting that Ca^{2+} entry via CRAC channels occurs at specific locations in T cells (Barr et al. 2008; Lioudyno et al. 2008). As expected, the site of Orai1 and STIM1 clustering was proven to be the location of Ca^{2+} entry (Luik et al. 2006).

Recent studies have shown that Ca^{2+} ions were not important for recruitment of Orai1, STIM1, and TCRs into the immunological synapse at the initial phase, but played an essential role in actin reorganization, which was crucial for the stability of the immunological synapse (Hartzell et al. 2016; Lioudyno et al. 2008). In addition, increased local $[Ca^{2+}]$ at the immunological synapse is also important for sustaining CD3 phosphorylation by neutralizing negative charges of phospholipids and generating unfolded structure of the cytoplasmic domains of CD3 chains (Shi et al. 2013). Since the immunological synapse is the site for aggregation of signaling molecules including many tyrosine and serine/threonine kinases, it is possible that clustering of Orai1 and STIM1 at the synapse is regulated by protein phosphorylation via these TCR signaling molecules, but there has been no experimental data to prove this yet. Together, these studies suggest that not only the amplitude of Ca^{2+} signaling but also its pattern and location (e.g., oscillation frequency, sustained levels, and microdomains) can provide specificity to activate downstream signaling pathways.

21.5 Components of CRAC Channels in T Cells: Orai and STIM Proteins

CRAC channels in T cells consist of pore subunits, Orai proteins, and regulatory STIM proteins. Differently from many other Ca^{2+} channels, CRAC channels are primarily gated by a direct interaction between Orai and STIM proteins. Existence of CRAC channels in immune cells was identified by electrophysiological methods based on their unique biophysical characteristics (Hoth and Penner 1992; Lewis and Cahalan 1989). Later, genome-wide RNAi screens in *Drosophila* cells identified the *Drosophila* gene *olf186-F* (named *Drosophila* Orai) and its mammalian homologues Orai1, Orai2, and Orai3 as subunits of the CRAC channels (Feske et al. 2006; Gwack et al. 2007b; Vig et al. 2006; Zhang et al. 2006). Furthermore, a missense mutation of R91W was identified in the ORAI1 gene from severe combined immune deficiency (SCID) patients that lacked functional CRAC channels and expression of wild-type Orai1 recovered CRAC currents in patient cells (Feske et al. 2006). Prior to identification of Orai1, limited RNAi screen in *Drosophila* and HeLa cells identified STIM1, a Ca^{2+}-sensing protein localized predominantly in the endoplasmic reticulum (ER) as an important regulator of CRAC channel-mediated Ca^{2+} entry (Liou et al. 2005; Roos et al. 2005; Zhang et al. 2005).

Orai1 has four transmembrane segments (TM1–TM4) with its N- and C-terminus facing the cytoplasm. The TM1 of Orai1 has been shown to line the pore, and residues in the TM1 including R91, G98, V102, and E106 are important for Ca^{2+} selectivity and gating (Cahalan and Chandy 2009; Hogan et al. 2010; Lewis 2011; McNally et al. 2012; Putney 2009; Srikanth et al. 2013; Zhang et al. 2011). The cytoplasmic N- and C-terminus of Orai1 mediates channel opening by interaction with STIM1 after store depletion. CRAC channels are also negatively regulated by excess Ca^{2+}, resulting in their Ca^{2+}-dependent inactivation (CDI) (Hoth and Penner 1992, 1993; Zweifach and Lewis 1995). In addition to channel gating, mutational studies showed that all the cytoplasmic regions of Orai1 including the N-terminus, the intracellular loop, and the C-terminus are involved in CDI (Lee et al. 2009; Mullins et al. 2009; Srikanth et al. 2010b). Thus, intracellular domains of Orai1 are important not only for channel gating but also for channel inactivation, to avoid deleterious consequences of excessive Ca^{2+} including cell death.

STIM family has two members, STIM1 and STIM2. STIM1 contains an N-terminal EF hand that detects luminal ER [Ca^{2+}], a single transmembrane

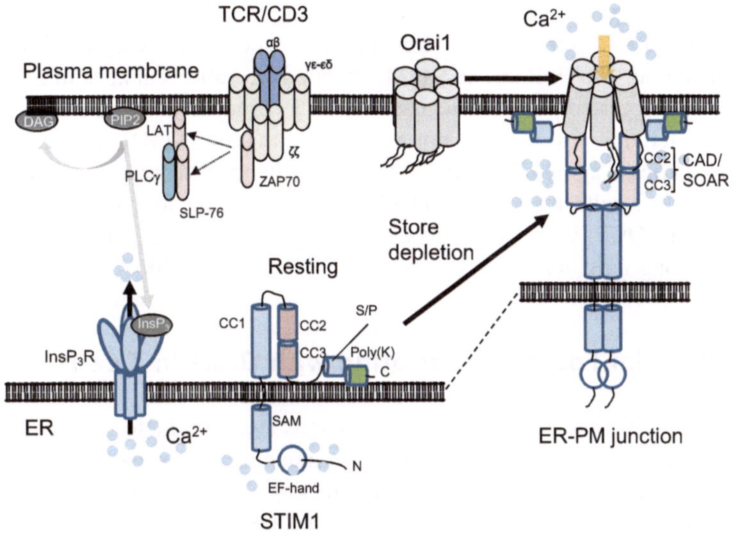

Fig. 21.2 Activation mechanism of CRAC channels in T cells. Under resting conditions, Orai1 and STIM1 are distributed at the PM and the ER membrane, respectively. Upon store depletion triggered by T cell receptor stimulation and IP_3 production via PLC-γ1, STIM1 oligomerizes by sensing ER Ca^{2+} depletion with its ER-luminal EF-hand domain and clusters at the ER-PM junctions. Upon physical interaction with the cytoplasmic, N- and C-terminus of Orai1 through the CAD/SOAR domain (coiled-coil domains 2 and 3), clustered STIM1 recruits and activates Orai1 at the ER-PM junctions. STIM1 contains an ER-luminal region comprising the EF-hand and SAM domains, a single transmembrane segment, and a cytoplasmic region. The cytoplasmic region has three coiled-coil domains (CC1, CC2, and CC3), a serine-/proline-rich domain (S/P), and a polybasic tail (poly-K) at the C-terminus that interacts with phosphoinositides after store depletion

domain, and a long C-terminal cytoplasmic region (Fig. 21.2) (reviewed in Soboloff et al. 2012). STIM1 plays a pivotal role in sensing ER [Ca^{2+}] and CRAC channel opening. Upon ER Ca^{2+} depletion, STIM1 loses bound Ca^{2+}, multimerizes, translocates to the PM-proximal ER, mediates clustering of Orai proteins, and stimulates Ca^{2+} entry (Liou et al. 2005; Roos et al. 2005; Zhang et al. 2005). STIM1 interacts with Orai1 via CRAC-activating domain (CAD)/STIM1 Orai1-activating region (SOAR) (Muik et al. 2009; Park et al. 2009; Yuan et al. 2009). The CAD/SOAR fragment of STIM1 (coiled-coil domains 2 and 3) was shown to play a pivotal role in activation of Orai1 by direct binding to its cytoplasmic N-and C-terminus. Furthermore, the stoichiometry of STIM1 binding to Orai1 can affect the fast inactivation properties of CRAC channels, indicating that STIM1 is a bona fide subunit of CRAC channels (Scrimgeour et al. 2009). These and other studies showed that CRAC channel activation involves multiple steps including STIM1 oligomerization, co-clustering of Orai1 and STIM1 at the ER-PM junctions, and gating of Orai1 (Liou et al. 2007; Muik et al. 2008, 2009; Navarro-Borelly et al. 2008; Park et al. 2009; Yuan et al. 2009). STIM2 shares similar domain structure with STIM1, but its ER-luminal EF-hand motif has lower Ca^{2+} binding affinity than STIM1, based on translocation kinetics in response to varying ER [Ca^{2+}] (Brandman et al. 2007). Because STIM2 responds to subtle changes in ER [Ca^{2+}], it has been suggested to regulate basal [Ca^{2+}] (Brandman et al. 2007). STIM2 function in activation of Orai1 is not as strong as that of STIM1, and it also shows a slower kinetics in aggregation and translocation than STIM1 (reviewed in Soboloff et al. 2012). In T cells, the role of STIM2 is not obvious due to the primary function of STIM1 in SOCE, but it seems to play a supportive role for STIM1 function in SOCE and long-term activation of NFAT (Oh-Hora et al. 2008).

21.6 Cellular Factors Associating with Orai1 in T Cells

In addition to Orai and STIM proteins, numerous auxiliary factors regulate activation and inactivation of CRAC channels by inducing conformational changes or stabilizing Orai1-STIM1 interaction. Interacting partners of Orai1 and their functions in general cell types have been recently summarized (Srikanth and Gwack 2012, 2013a; Srikanth et al. 2013). In theory, all of these interacting partners can act as regulators for Ca^{2+} signaling in T cells, but only the molecules whose functions have been validated in T cells will be discussed here.

An interacting partner of Orai1, CRAC channel regulator 2A (CRACR2A, EFCAB4B, or FLJ33805) was identified from large-scale affinity protein purification using Orai1 as bait (Srikanth et al. 2010a). CRACR2A has two splice isoforms, CRACR2A-a (~80 kDa) and CRACR2A-c (45 kDa). The short isoform CRACR2A-c is cytoplasmic and forms a ternary complex with Orai1 and STIM1 to stabilize their interaction after store depletion. Accordingly, its depletion decreases STIM1 clustering at the ER-PM junctions and, hence, SOCE. This interaction with Orai1 and STIM1 is [Ca^{2+}]$_i$-dependent, with low [Ca^{2+}]$_i$ favoring association and high [Ca^{2+}]$_i$ favoring its dissociation by sensing Ca^{2+} through its EF-hand motifs (Fig. 21.3a). The long isoform CRACR2A-a encodes a large Rab GTPase (Srikanth et al. 2016a, b; Wilson et al. 2015). CRACR2A-a is localized to the proximal Golgi area and vesicles and plays an important role in TCR signaling pathways including SOCE and JNK MAPK pathways (Srikanth et al. 2016a). While the regions for Ca^{2+} binding and interaction with Orai1 and STIM1 are conserved between the two isoforms, CRACR2A-a has an additional proline-rich domain (PRD) and a Rab GTPase domain in its C-terminus and is abundantly expressed in lymphoid organs. CRACR2A-a is unique because it clearly distinguishes itself from small Rab GTPases (~20 kDa) due to its large size (~85 kDa) and presence of multiple functional domains (Srikanth et al. 2016a; Wilson et al. 2015). The Rab GTPase domain of CRACR2A-a contains conserved guanidine-binding residues that regulate its localization in a GTP/GDP-binding and prenylation-dependent manner (Fig. 21.3b). GTP-bound and prenylated CRACR2A-a localizes within vesicles close to the *trans* Golgi network, whereas GDP-bound or unprenylated CRACR2A-a is cytosolic and rapidly degraded. Prenylation of CRACR2A-a involves geranylgeranylation at an unconventional site (CCx, x; any amino acid)

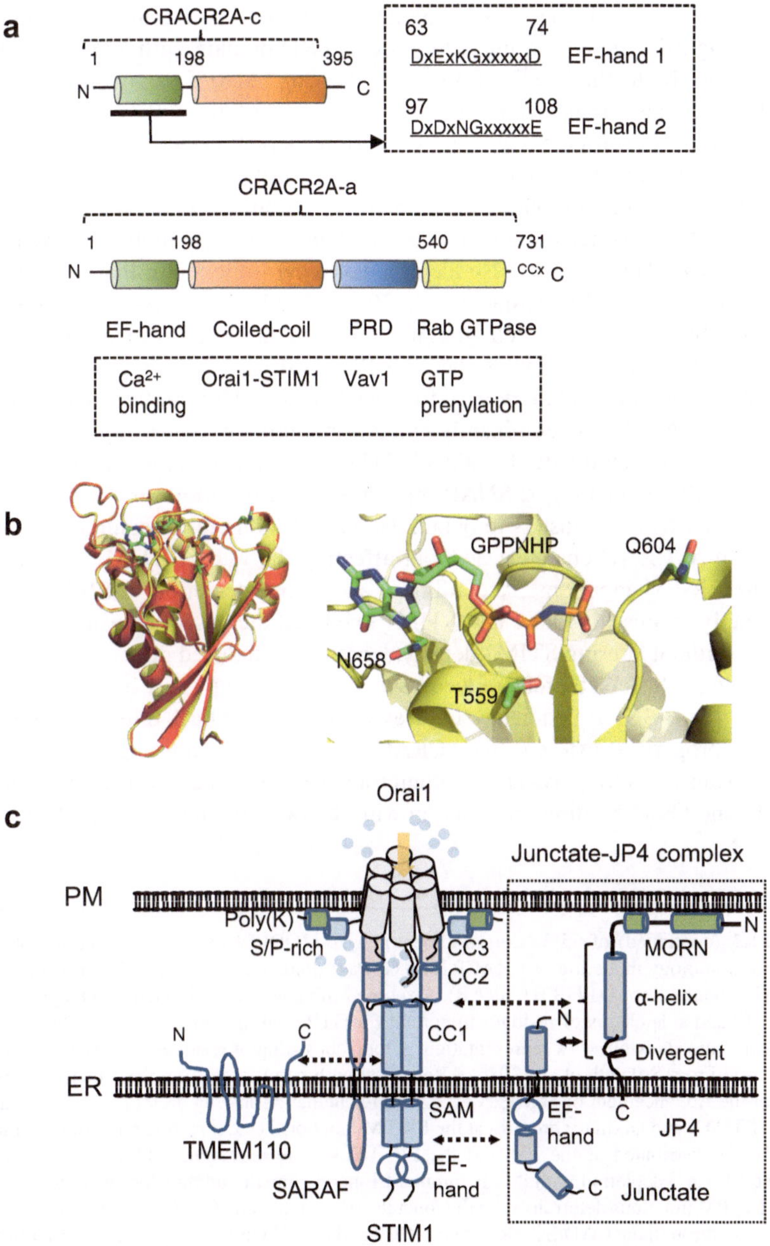

Fig. 21.3 Interacting partners of Orai1 and STIM1. (**a**) Schematic showing the domain structure of human CRACR2A-a, a member of the large Rab GTPase family. CRACR2A-a and CRACR2A-c share EF-hand motifs and a coiled-coil domain, which interact with and stabilize the Orai1-STIM1 complex. CRACR2A-a contains additional proline-rich domain (PRD), which interact with Vav1 and a Rab GTPase domain with a prenylation site at the C-terminus. GTP binding and prenylation are required for the membrane retention of CRACR2A-a. (**b**) Homology modeling of CRACR2A-a GTPase domain (*yellow*) with Rab3a (*red*). Sequence alignment between GTPase domain of CRACR2A-a and Rab3a gave a continuous alignment with sequence identity of 46% and similarity

in the C-terminus. Upon TCR stimulation CRACR2A-a translocates into the immunological synapse via interaction of its PRD domain with Vav1, to activate SOCE and both the Ca^{2+}-NFAT and the Jnk MAPK signaling pathways. CRACR2A-a also translocates into the proximity of the ER-PM junctions via vesicle trafficking after passive ER Ca^{2+} store depletion and recovers SOCE in Jurkat T cells depleted for both the isoforms, similar to CRACR2A-c. Because CRACR2A-a retains the Orai-STIM interaction domain, one can assume that it supports SOCE by interacting with both Orai1 and STIM1, similar to CRACR2A-c.

The α-SNAP protein was also identified as a cytosolic factor that interacts with both Orai1 and STIM1 (Miao et al. 2013). Depletion of α-SNAP drastically decreased SOCE in various cell types including Jurkat T cells. The original function of α-SNAP is disassembly of the SNARE (NSF attachment protein receptor) complex, a cellular machinery used for vesicle fusion. Differently from its original function, α-SNAP, a predominantly cytoplasmic protein, physically interacts with the cytosolic CAD/SOAR domain of STIM1 and the C-terminal tail of Orai1. Through this interaction, α-SNAP regulates an active molecular rearrangement within Orai1-STIM1 clusters to obtain the STIM1/Orai1 ratio required for optimal activation of CRAC channels, without affecting the rate of STIM1 translocation into the ER-PM junctions. Accordingly, after store depletion, α-SNAP-depleted cells stably expressing Orai1 and STIM1 exhibited increase in density of Orai1 in clusters without altering STIM1 density, leading to a reduced ratio of STIM1/Orai1 in individual clusters, thereby reduced SOCE. A follow-up study by the same group showed that α-SNAP deficiency induces formation of Orai1 oligomers with biophysical properties different from CRAC channels, including higher Na^+ permeability. Thus α-SNAP plays an important role in establishing a correct ratio between STIM1 and Orai1 by direct interaction with the two proteins (Li et al. 2016).

Fig. 21.3 (continued) of 65% (Clustal Omega) (*left*). MODELLER (Morreale et al. 2000) was used for homology modeling of CRACR2A-a GTPase domain to a high-resolution structure of a GPPNHP-bound Rab3a (PDB ID: 3RAB). A zoomed-in view of the GPPNHP binding site (*right*). GPPNHP and side chains of residues important for GTP binding and hydrolysis, Thr^{559}, Gln^{604}, and Asn^{658} are shown in *stick* representation. A loop consisting of residues 561–570 was removed for clarity. From Srikanth et al. (2016a). Reprinted with permission from the American Association for the Advancement of Science (AAAS). (**c**) Schematic showing protein interactions among Orai1, STIM1, and auxiliary proteins at the ER-PM junctions. Under resting conditions, Orai1 and STIM1 are distributed at the PM and the ER. Upon store depletion, STIM1 oligomerizes by sensing ER Ca^{2+} depletion with its ER-luminal EF-hand domain and translocates to form clusters at the ER-PM junctions determined by the junctate-JP4 complex in T cells. By physical interaction with Orai1 through the CAD/SOAR domain (CC2 and CC3, indicated in *red*), STIM1 recruits and activates Orai1 in the ER-PM junctions. Junctate also contains Ca^{2+}-sensing motif in the ER lumen and ER Ca^{2+} depletion increases its interaction with STIM1. SARAF and TMEM110 have been shown to translocate together with STIM1 to the ER-PM junctions. SARAF interacts with STIM1 CAD/SOAR domain to modulate slow Ca^{2+}-dependent inactivation of CRAC channels, while TMEM110 directly interacts with the CC1 region of STIM1 via its cytoplasmic C-terminus. TMEM110 is also involved in establishment of ER-PM junctions

21.7 Cellular Factors Associating with STIM1 in T Cells

Orai1 and STIM1 cluster at pre-existing junctions of the ER and the PM, a space of 10–25 nm (Varnai et al. 2007; Wu et al. 2006). In muscle cells, proteins localized to the junctions between the PM and ER/sarcoplasmic reticulum (SR) membrane form a structural foundation for Ca^{2+} dynamics essential for excitation-contraction coupling (Berridge et al. 2003; Carrasco and Meyer 2011). Various biochemical screening approaches have identified junctophilins, mitsugumins, sarcalumenin, junctin, and junctate as components of these junctions (Carrasco and Meyer 2011; Takeshima et al. 2000; Weisleder et al. 2008). Recent studies have shown that homologues and isoforms of these junctional proteins are also expressed in T cells. Srikanth et al. (2012) identified the EF-hand-containing protein, junctate as an interactor of STIM1. Junctate localization defined the sites of accumulation of CRAC channel components, since after store depletion, Orai1 and STIM1 accumulated at junctions that were already marked by junctate (Fig. 21.3c). The EF-hand motif of junctate senses ER Ca^{2+} depletion, which is important for efficient recruitment of STIM1. However, junctate alone cannot function as a membrane-tethering factor because the cytoplasmic N-terminus is very short and also does not contain any obvious PM-binding motif. In a recent study, Woo et al. (2016) identified another junctional protein, junctophilin 4 (JP4) as an interacting partner of STIM1. Junctophilin family consists of four genes JP1, JP2, JP3, and JP4 that are expressed in a tissue-specific manner and are known to form ER-PM junctions in excitable cells including skeletal muscle and cardiac and neuronal cells (Nishi et al. 2003; Takeshima et al. 2000). Junctophilins contain eight repeats of the membrane occupation and recognition nexus (MORN) motifs that bind to phospholipids in the N-terminus and a C-terminal ER membrane-spanning transmembrane segment (Garbino et al. 2009; Takeshima et al. 2000). Depletion of JP4 inhibited STIM1 recruitment into the ER-PM junctions and significantly decreased SOCE. Biochemical analyses showed a direct interaction of JP4 cytoplasmic domain with coiled-coil 1 and 2 regions of STIM1. JP4 was also shown to interact with the N-terminal cytoplasmic region of junctate. Therefore, this study demonstrates that JP4-junctate complex is localized at the ER-PM junctions in T cells and synergistically recruits STIM1 into these junctions by direct interaction. When overexpressed, STIM1 alone is sufficient to establish the ER-PM junctions using its C-terminal poly-lysine tail. However, in a physiological condition when the concentration of STIM1 is limiting or when STIM1 is unable to bind membrane phospholipids (e.g., due to low [PIP_2] in the PM), its interaction with the juncate-JP4 complex can be important for efficient assembly of a functional CRAC channel complex at the ER-PM junctions.

Transmembrane protein 110 (TMEM110 or STIM-activating enhancer) was identified as a positive regulator of SOCE using biotin-labeled protein purification and a genome-wide RNAi screen, respectively (Jing et al. 2015; Quintana et al. 2015). TMEM110 is a multi-pass ER-resident protein with its N- and C-termini facing the cytoplasm (Fig. 21.3c). Jiang et al. showed that TMEM110 interacted with the coiled-coil 1 region of STIM1 and induced its active conformation to interact with Orai1 (Jing et al. 2015). The coiled-coil 1 region of STIM1 contains an acidic amino acid motif that binds to the positively charged residues within Orai1-interacting CAD/SOAR fragment, blocking its interaction with Orai1 in an auto-inhibitory manner. Interaction of TMEM110 with the coiled-coil 1 region of STIM1 facilitated the release of this auto-inhibition. Furthermore, this study showed that depletion of TMEM110 had a modest influence on the frequency of the ER-PM junctions with 8–12% decrease in cortical ER. Another study also identified TMEM110 as an important regulator of SOCE using a siRNA screen. This study showed that siRNA-mediated depletion of TMEM110 significantly reduced the density of ER-PM junctions by >60% in HeLa/HEK293 cells, both under resting conditions and after store depletion. Importantly, artificial expansion of the junctions by overexpression of a yeast junctional protein Ist2, which is unlikely to affect STIM1 auto-inhibition, significantly rescued STIM1 translocation and SOCE. Therefore, this study concluded that TMEM110 is important for maintenance of the ER-PM junctions involved in SOCE in resting conditions and for dynamic remodeling of these junctions after store depletion. Both these studies show an important role of TMEM110 in STIM1 translocation and thereby SOCE in T cells, although further studies are required to validate whether this involves direct interaction with STIM1 or indirectly by regulating the number of ER-PM junctions.

An ER-resident protein SARAF (SOC-associated regulatory factor) was identified as an interacting partner of STIM1 and a negative regulator of SOCE, which facilitates the Ca^{2+}-dependent slow inactivation of CRAC channels (Palty et al. 2012). SARAF encodes a 339-amino acid protein containing single transmembrane segment (aa 173–195) with its N-terminus facing the ER lumen (aa 1–172) and its C-terminus facing the cytoplasm (aa 196–339) (Fig. 21.3c). SARAF contains positively charged residues, which may interact with the PM phospholipids and a serine-/proline-rich domain in its C-terminal end. Depletion of SARAF increased intracellular Ca^{2+} concentration and enhanced SOCE after store depletion, whereas its overexpression showed an opposite effect. SARAF played a negative role in SOCE with multiple modes of action: (1) interaction with the inactive form of STIM1 at the resting condition to stabilize its inactive conformation in the ER, (2) induction of Ca^{2+}-dependent inactivation of Orai1 channels after translocation to the ER-PM junctions together with STIM1, and (3) facilitating dissociation of clustered STIM1 proteins. Detailed structure-function studies identified a C-terminal inhibitory domain (CTID, aa 448–530) within STIM1 that regulates SARAF-STIM1 interaction (Jha et al. 2013). STIM1 CTID is located at the C-terminal region of the Orai1-interacting CAD/SOAR domain, and interestingly, deletion of CTID from full-length STIM1 resulted in constitutively active Orai1 channels. CTID does not bind to SARAF directly, but mediates the interaction of SARAF with the CAD/SOAR region. Therefore, this study highlights the important role of STIM1 and SARAF in Ca^{2+}-dependent inactivation of Orai1. STIM1 also negatively regulates plasma membrane Ca^{2+} ATPase (PMCA) directly or indirectly via a novel ten-transmembrane segment-containing protein, POST (partner of STIM1, TMEM20) (Krapivinsky et al. 2011; Ritchie et al. 2012). These studies suggest that both stimulation of CRAC channels and inhibition of

PMCA activity at the immunological synapse may be important for generation of sustained, local Ca^{2+} entry required for NFAT activation.

21.8 Ca^{2+} Signaling in Development of T Cells in the Thymus

In humans, loss of Orai1 or STIM1 function causes immune deficiency which is recapitulated in animal models lacking expression of these proteins. In human patients and mice lacking expression of Orai or STIM proteins, development of conventional TCRαβ+ T cells is normal. These results were unexpected because the intensity and duration of TCR signaling are important for positive and negative selection of T cells in the thymus and SOCE was expected to play a major role in these events. Although deficiency of any CRAC channel component does not influence development of conventional TCRαβ+ T cells, deficiency of both STIM1 and STIM2 impaired development of unconventional agonist-selected T cells, including regulatory T cells, invariant natural killer T (iNKT) cells, and intestinal intraepithelial lymphocytes, which are thought to require strong and sustained TCR signals during development (Oh-Hora et al. 2013). STIM1 deficiency alone dramatically reduced development of iNKT cells, while that of regulatory T cells and intestinal intraepithelial cells was normal (Oh-Hora et al. 2008). These results suggest that SOCE is important for development of agonist-selected T cells, and especially iNKT cells require high levels of $[Ca^{2+}]_i$ for their development.

Instead of CRAC channels, development of T cells depends on other modes of Ca^{2+} signaling. Recent studies on voltage-activated Ca^{2+} channels (VOCCs) provide insights into the role of Ca^{2+} signaling during T cell development and differentiation (Badou et al. 2013; Nohara et al. 2015). VOCCs are predominantly expressed in excitable cells and activated by action potentials. They are divided into L-type (Cav1.1–1.4), P/Q-type (Cav2.1), N-type (Cav2.2), R-type (Cav2.3), and T-types (Cav3.1, Cav3.2, and Cav3.3) (Christel and Lee 2012). They consist of an α1 subunit and four auxiliary subunits of α2, δ, β, and γ. The α1 subunits line the pore of the Cav channels, whereas the β subunits are regulatory proteins that are important for assembly of the channel complex, correct PM targeting, and stimulation of channel activity. Recently, it was found that deletion of only the β2 subunit profoundly influenced T cell development by decreasing the total number of thymocytes due to inhibition of the DN (double-negative)-to-DP (double-positive) transition (Badou et al. 2013; Jha et al. 2015). Protein levels of Cav1.2 and Cav1.3 were decreased in β2 subunit-deficient thymocytes, suggesting that Cav1.2/Cav1.3-β2 channels is important for thymic development of conventional T cells (Jha et al. 2015). Another study examined mice genetically deleted for expression of α1 subunit of Cav1.4 and observed subtle reduction in CD4 single-positive cells in the thymus (Omilusik et al. 2011). Both the studies using mice deficient in Cav1.4 or the regulatory β subunits identified a positive role for Cav1.2, Cav1.3, and Cav1.4 channels in T cell development.

Several questions still remain to be answered regarding the role of Ca^{2+} signaling during T cell development in the thymus. Cav channels are known to be activated by TCR stimulation, but Ca^{2+} entry via these channels is not as robust as those mediated by Orai channels. We do not understand how these moderate levels of Ca^{2+} entry mediated by Cav channels play an important role in T cell development, while the robust SOCE observed with CRAC channels is dispensable for the same. It is possible that Cav channels increase local [Ca^{2+}] to affect T cell development. Normal development of TCRαβ+ T cells in Orai1 KO or STIM1/STIM2 double KO animals suggest that CRAC channels are not important during the developmental stage. However, it is possible that other Orai channels (e.g., Orai2 or Orai3) affect T cell development in a STIM (ER Ca^{2+} store)-independent manner. In addition, the Ca^{2+} channel important for negative selection in the thymus has not been identified yet. Self-reactive T cells should have higher levels of Ca^{2+} signaling to trigger cell death (see below). Therefore, Ca^{2+} channels involved in negative selection are expected to have robust Ca^{2+} entry, but any defect in negative selection has not been identified in mice deficient for CRAC or Cav channel components. Therefore, Ca^{2+} channels that are important for each step of T cell development in the thymus remain to be uncovered in future work.

21.9 Orai and STIM Proteins in the Function of Effector T Cells

Naïve T cells undergo clonal selection, proliferation, and differentiation into effector T cells at the central or peripheral lymphoid organs (e.g., spleen or lymph nodes). For differentiation into specific effector T cells, three conditions need to be met. Signals from TCRs and co-receptor (e.g., CD28) are essential for optimal differentiation. In addition, polarizing cytokines (e.g., IL-4, IL-6, or IL-12) are essential for determination of T cell fate such as differentiation into various effector T cells including Th1, Th2, Th9, Th17, Th22, and follicular T cells. Differentiation and effector functions of these cells are controlled by signature transcription factors, for example, T-bet for Th1, GATA3 for Th2, or RORγt for Th17 cells. After differentiation at the priming sites, effector helper T cells migrate into the infected or inflamed tissues to produce cytokines that recruit and differentiate myeloid-lineage innate immune cells.

The most obvious defect in patients and mice with nonfunctional CRAC channels has been identified in various functions of effector T cells including differentiation and cytokine production. Patients with a point mutation, R91W, showed defect in cytokine production in T cells (Feske et al. 2006). In consistence, Orai1 KO or R93W (counterpart of human R91W mutation) knock-in mice also showed a reduction in cytokine production by $CD4^+$ and $CD8^+$ effector T cells (Gwack et al. 2008; McCarl et al. 2010). In addition to cytokine production, SOCE via Orai1 channels is essential for differentiation of effector T cells. Using a small molecule inhibitor of Orai1, compound 5D (*N*-[2,2,2-trichloro-1-(2-naphthylamino)ethyl]-2-furamide) identified from chemical library screen and Orai1-deficient T cells, it was shown that Orai1 plays a key role in differentiation of

Th17 cells by induction of the NFAT-RORγt pathway (Kim et al. 2014). Inhibition or deficiency of Orai1 also decreased expression of T-bet and GATA3 under Th1- and Th2-polarizing conditions, respectively, but to a much lesser degree, suggesting a higher dependence on SOCE for Th17 differentiation. Decreased Th17 differentiation in Orai1-deficient T cells could be mimicked by treatment with cyclosporine A, a blocker of calcineurin, and this defect was rescued by expression of constitutively active NFAT at least partly, suggesting that the Orai1-NFAT pathway significantly contributes to expression of RORγt and thus Th17 differentiation. Another report also validated the role of Orai1 in cytokine production in effector T cells using blockers and Orai1 knockout mice (Kaufmann et al. 2016). The major findings are consistent by demonstrating an important role of Orai1 in production of IL-17A and IFN-γ by effector T cells without affecting the function of regulatory T cells. However, there was discrepancy in the role of Orai1 in expression of T-bet and RORγt transcription factors, which could be due to differences in the residual Ca^{2+} entry between the two independent knockout mouse lines. In addition to Orai1, STIM1 deficiency also showed a pronounced reduction in SOCE and cytokine production in T cells resulting in resistance to experimental autoimmune encephalomyelitis (EAE) (Oh-Hora et al. 2008; Schuhmann et al. 2010). On the contrary, mice deficient in STIM2, another member of the STIM family, showed a mild defect in SOCE and, correspondingly, succumbed to EAE, albeit with less severe symptoms (Schuhmann et al. 2010).

The genetic manipulation of Orai and STIM proteins provides opportunities to determine the outcomes of differential SOCE levels in the functions of effector T cells. This is due to the unique, quantitative nature of Ca^{2+} signaling unlike any other signaling pathways. For example, a progressive reduction in SOCE was observed in WT, $Orai1^{+/-}$, and $Orai1^{-/-}$ T cells (Fig. 21.4a). The residual SOCE in $Orai1^{-/-}$ T cells can be blocked by the widely used CRAC channel inhibitor 2-APB (2-aminoethoxydiphenyl borate), suggesting that the residual SOCE is likely mediated by Orai2 or Orai3 proteins (Srikanth S. and Gwack Y, unpublished). Hence, SOCE in $Orai1^{-/-}$ T cells can be further reduced by expression of a dominant-negative mutant of Orai1, E106Q, which can is likely to hetero-multimerize with Orai2 or Orai3 to further decrease SOCE (Gwack et al. 2007b; Kim et al. 2011). On the contrary, expression of a mutant of Orai1 that lacks fast inactivation ($Orai1^{MutA}$) can increase SOCE in $Orai1^{-/-}$ T cells where endogenous Orai1 channels are not present (Srikanth et al. 2010b). These results provide ways to generate, in theory, five different gradients of SOCE in T cells, $Orai1^{-/-}$ T cells with expression of E106Q, $Orai1^{-/-}$, $Orai1^{+/-}$, $Orai1^{+/+}$, and $Orai1^{-/-}$ T cells with expression of MutA Orai1 (lowest to highest $[Ca^{2+}]$) (Fig. 21.4a). These genetic modifications will provide useful tools to investigate various outcomes in T cell functions responding to different levels of SOCE.

There is no data regarding a direct comparison of SOCE between Orai1 and STIM1 KO T cells, but based on the published results, it is assumed that SOCE in $STIM1^{-/-}$ T cells should be much lower than that of $Orai1^{-/-}$ T cells. SOCE in $STIM1^{-/-}STIM2^{-/-}$ (DKO) T cells is close to basal levels, which is very similar with $Orai1^{-/-}$ T cells expressing $Orai1^{E106Q}$. Progressive decrease in SOCE

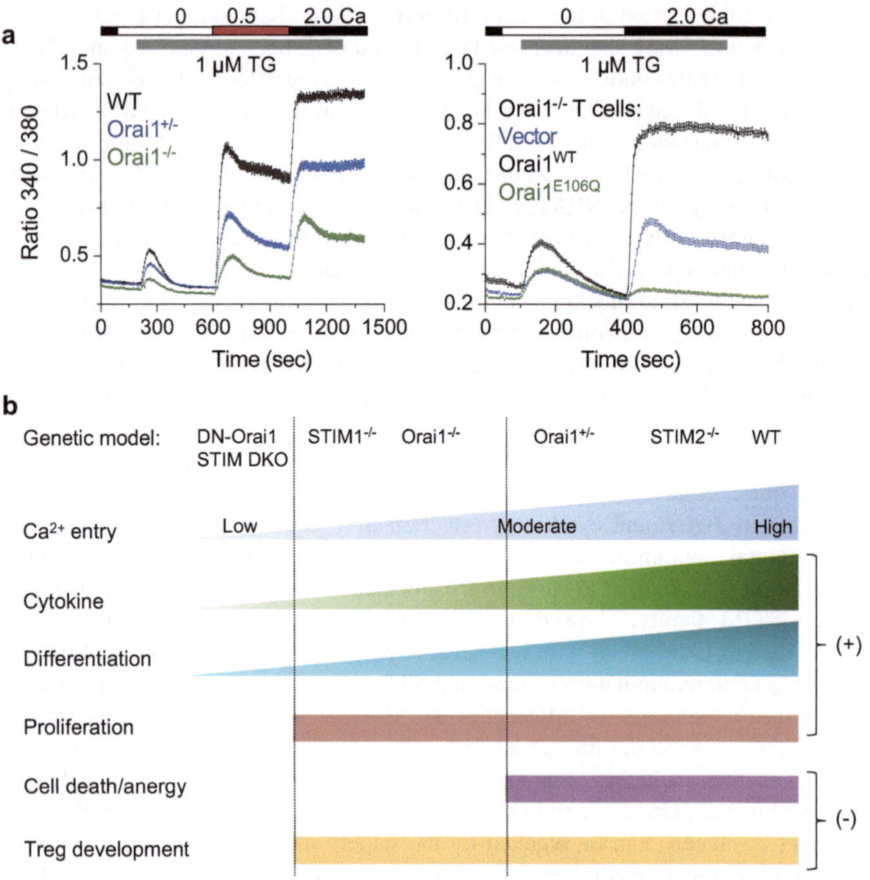

Fig. 21.4 Role of Ca^{2+} signaling in diverse aspects of T cell activation. (**a**) Gradual levels of store-operated Ca^{2+} entry generated by genetic modifications. *Left*—SOCE was measured in effector $CD4^+$ T cells from wild type (WT), Orai1 heterozygous ($Orai1^{+/-}$) and Orai1-deficient ($Orai1^{-/-}$) mice after store depletion with thapsigargin (TG) in the presence of extracellular solution containing 0.5 and 2 mM Ca^{2+}. *Right*—SOCE was measured in $Orai1^{-/-}$ $CD4^+$ T cells transduced with retroviruses expressing empty vector (vector, *blue trace*), wild type ($Orai1^{WT}$, *black trace*) or dominant-negative mutant of Orai1 ($Orai1^{E106Q}$). Data modified from article originally published in *The Journal of Immunology* (Kim DK, Srikanth S, Yee MK., Mock DC, Lawson GW, and Gwack Y. 2011. Orai1 deficiency impairs activated T cell death and enhances T cell survival J. Immunol. 187: 3620–3630. Copyright © [2011]. The American Association of Immunologists, Inc.). (**b**) Ca^{2+} requirement for cytokine production, differentiation, proliferation, cell death, and development of regulatory T cells. Cytokine production and differentiation levels (as determined by expression of signature transcription factors or surface receptors) in effector T cells gradually increase with increasing levels of intracellular Ca^{2+} concentrations. T cell proliferation does not need high Ca^{2+} entry, but requires moderate levels of elevated $[Ca^{2+}]_i$ because $Orai1^{-/-}$ or $STIM1^{-/-}$ T cells showed a normal proliferation rate after stimulation and a further reduction in $[Ca^{2+}]_i$ by overexpression of dominant-negative Orai1 (DN-Orai1) in $Orai1^{-/-}$ T cells, or deficiency of both STIM1 and STIM2 significantly inhibits proliferation. In some WT T cells, excessive $[Ca^{2+}]_i$ after TCR stimulation induces cell death and anergy, which was decreased by deficiency of Orai1. Therefore, it is likely that inhibition of CRAC channels would result in acute decrease in the immune response by decreasing cytokine production and differentiation of

observed in WT, Orai1$^{+/-}$, Orai1$^{-/-}$, or Orai1^{E106Q}-expressing cells closely correlated with cytokine expression and differentiation (as judged by the expression of key transcription factors and surface receptors) of effector T cells as depicted in Fig. 21.4b. There is a general notion that robust Ca^{2+} signaling is needed for T cell proliferation, based on the observation that absence or chelation of extracellular Ca^{2+} drastically decreases T cell proliferation. However, analysis of Orai and STIM knockout T cells reaches a different conclusion, with very low levels of SOCE as observed in STIM1- or Orai1-deficient T cells sufficient for proliferation (Kim et al. 2011; Ma et al. 2010). Similarly, double knockout of both STIM1 and STIM2, but not individual knockout, drastically decreased TCR-induced cell proliferation (Oh-Hora et al. 2008). Collectively, differentiation and cytokine production of effector T cells demand high SOCE, but the threshold levels for T cell proliferation are very low and a minimal increase in $[Ca^{2+}]_i$ is sufficient.

21.10 Ca^{2+} Signaling in Immune Suppression

Ca^{2+} signaling in T cells plays a dual role in both positive and negative immune reactions. SOCE in T cells is important for the immune suppressive mechanisms of activated T cell death, anergy, and the function of regulatory T cells. Cell death induced by TCR stimulation is critical for homeostasis of peripheral T cells after antigen clearance (Budd 2001; Krammer et al. 2007; Strasser 2005). Therefore, abrogation of T cell death leads to hypersensitive immune reaction and autoimmune disorders. Activation induced T cell death occurs through the death receptor- and mitochondria-mediated pathways. Death receptor-mediated apoptosis involves the Fas ligand (FasL)/Fas signaling pathway, primarily regulated by NFAT (Hodge et al. 1996; Macian et al. 2002; Serfling et al. 2006), while mitochondria-mediated cell death occurs due to loss of mitochondrial membrane potential (Marsden and Strasser 2003; Strasser 2005). Mitochondria-mediated cell death pathway involving the Bcl-2 family members (e.g., Bcl-2 and Bcl-X_L) and the BH3-only proteins (e.g., Bad, Bik, Bim, and Noxa) plays an important role in T cell survival and death, respectively (Budd 2001; Hildeman et al. 2002, 2007; Marrack and Kappler 2004; Strasser 2005; Strasser and Pellegrini 2004). It was shown that Orai1-deficient T cells are strongly resistant to cell death due to reduction in both death receptor- and mitochondria-mediated cell death mechanisms by decreasing expression levels of pro-apoptotic genes including FasL and Noxa as well as mitochondrial Ca^{2+} uptake (Kim et al. 2011). Similarly, STIM1 knockout T cells also displayed reduced cell death after mycobacteria infection due to reduced expression of pro-apoptotic factors, which unexpectedly resulted in hyperactive immune response and

Fig. 21.4 (continued) T cells, but in a long term, it can induce autoimmunity and hypersensitive immune responses as outcomes of suppressed cell death, anergy, and development of regulatory T cells

increased production of IFN-γ in a long term (Desvignes et al. 2015). Collectively, these results suggested that Orai- and STIM-mediated SOCE contributes to both cell death mechanisms via NFAT-mediated transcriptional regulation and accumulation of Ca^{2+} in the mitochondria.

Orai1 KO T cells show decreased cell death, but heterozygous Orai1$^{+/-}$ T cells display normal level of cell death, suggesting that intermediate levels of $[Ca^{2+}]_i$ are sufficient to induce cell death (Kim KD and Gwack Y, unpublished). Therefore, the threshold levels of $[Ca^{2+}]_i$ for activated T cell death should be between those of Orai1$^{+/-}$ and Orai1$^{-/-}$ T cells (Fig. 21.4b). However, as mentioned earlier, Orai1$^{-/-}$ T cells showed a normal rate of proliferation. One possible explanation for the difference in the threshold levels of $[Ca^{2+}]_i$ for proliferation and cell death may be caused by the differential regulation of NFAT1 and NFAT2 transcription factors. Whereas NFAT2 acted as a positive regulator for cell proliferation, NFAT1 induced cell cycle arrest and cell death (Robbs et al. 2008). Another report showed that NFAT1 and NFAT2 are preferentially activated by high and moderate Ca^{2+} levels in T cells, respectively (Srinivasan and Frauwirth 2007). Indeed, our RNA-seq analysis of Orai1$^{-/-}$ T cells indicated that expression levels of pro-apoptotic genes, major targets of NFAT1, were drastically decreased, while those of positive regulators for cell cycle progress (e.g., c-Myc, CDK4, or CDK6) that are regulated by NFAT2 were not decreased (Mognol et al. 2016; Srikanth S and Gwack Y, unpublished). Therefore, it is possible that activation of NFAT1, but not NFAT2, is inhibited in Orai1$^{-/-}$ T cells that show decreased cell death with a normal proliferation rate.

Ca^{2+} signaling also plays a pivotal role in the induction of anergy in T cells. Anergic T cells are incapable of proliferation and cytokine expression after antigen encounter (Baine et al. 2009). Technically, it is difficult to apply these standards to determine whether T cells deficient of Orai or STIM proteins undergo anergy because cytokine production in those cells are intrinsically low. However, if we judge only by proliferation after TCR stimulation, Orai1 or STIM1 KO T cells do not seem to undergo anergy because the rates of proliferation of those cells were normal (Kim et al. 2011; Ma et al. 2010). Therefore, the threshold levels of $[Ca^{2+}]_i$ for anergy induction are likely to be higher than $[Ca^{2+}]_i$ levels observed in Orai1$^{-/-}$ or STIM1$^{-/-}$ T cells. Considering the role of Ca^{2+} signaling in inhibition of excessive T cell activation, it seems reasonable that the threshold levels of Ca^{2+} required for negative immune responses including activated T cell death or anergy are higher than that for positive immune reactions.

Dysregulated Ca^{2+} signaling also induces autoimmune and lymphoproliferative disorders by defects in development of regulatory T cells. SCID patients harboring mutations in STIM1 showed autoimmune hemolytic anemia, thrombocytopenia, and enlarged spleen and lymph nodes (Picard et al. 2009). Mice lacking both STIM1 and STIM2 showed lymphoproliferative disorder in addition to SCID phenotype (Oh-Hora et al. 2008). The lymphoproliferative phenotype of STIM1/STIM2 double-knockout mice was attributed to a severe reduction in thymic Tregs, which are crucial for immune tolerance. These observations identify a strict requirement of the Ca^{2+}-NFAT signaling pathway for the development of Tregs than other T cell types. In addition to T cells, B cell-specific knockout of STIM1 and STIM2 showed decreased expression of the immune suppressive cytokine IL-10 that led to development of autoimmune diseases (Matsumoto et al. 2011). Together, these reports indicated that Ca^{2+} signaling plays an important role in both aspects of immune regulation—effector T cell activation and immune tolerance—and a block of CRAC channel activity can lead to completely opposite outcomes like immune deficiency or autoimmunity depending on the degree and duration of inhibition.

21.11 Future Directions

The CRAC channel-calcineurin-NFAT pathway has been extensively studied due to its importance in immune cell functions. Molecular understanding of this fundamental signaling pathway is crucial for development of therapy that benefits patients with immune deficiencies, autoimmune diseases, transplant rejection, and cancer. Blockers for calcineurin such as cyclosporine A and FK506 (tacrolimus) are currently used to suppress transplant rejection and acute inflammation. However, ubiquitous expression of calcineurin makes long-term treatment with the calcineurin blockers technically challenging. Identification of CRAC channel subunits provides potential new targets for drug development. Detailed structural studies targeting the pore region of Orai1 in the closed and open configuration (in the absence and presence of STIM1) will greatly help our understanding of channel regulation. In addition, identification of interacting partners of CRAC channels, particularly those predominantly expressed in the immune system, will provide new targets for therapeutic intervention of the Ca^{2+}-calcineurin-NFAT signaling pathway to balance immune reactions. Ca^{2+} is a universal second messenger; however, accumulating evidences suggest that Ca^{2+} signal has specific effects, depending on signaling patterns (e.g., amplitude and frequency of oscillation), site of accumulation (e.g., micro- or nano-domains), and cell types. Unlike other signaling pathways, it is technically possible to fine-tune the strength of Ca^{2+} signaling using genetic manipulation and expression of dominant-negative mutants. Therefore, in vivo studies using these models will provide tools to dissect outcomes of Ca^{2+} signaling under physiological conditions.

The Ca^{2+}-calcineurin-NFAT pathway is a fundamental signaling pathway for T cell physiology including proliferation, differentiation, and cytokine production as well as immune suppression. Therefore, molecular dissection of this signaling pathway will definitely provide basic knowledge on various immune phenotypes. However, translational application of inhibitors for CRAC channels needs to cross tough barriers derived from different immunological outcomes depending on the levels of Ca^{2+} entry and threshold levels for each phenotype. For example, acute inhibition of CRAC channels can have suppressive therapeutic effects by acutely shutting down effector T cell responses, but in a long term, these effects can be compromised by the lack of immune suppressive mechanisms including T cell death or development of regulatory T cells. Theoretically, we should be able to control the immunological outcomes if we understand the detailed molecular mechanisms that link the intensity and timing of Ca^{2+} signaling to T cell phenotypes.

As mentioned above, Ca^{2+} channels responsible for each step of T cell development have not been identified yet. In addition, many other non-SOC Ca^{2+} channels (e.g., Cav, purinergic receptors, and transient receptor potential channels) also exist in T cells and play important roles in effector T cell function (Badou et al. 2013; Nohara et al. 2015). Although deficiency of those channels did not decrease Ca^{2+} entry triggered by TCR stimulation as much as deficiency of Orai or STIM proteins, the immunological outcomes of inhibition of these channels were significant. These results suggest that their functions are not just supporting CRAC channels, but they may act independently of CRAC channels with the modes of activation, which are different from TCR stimulation (e.g., change in membrane potential or agonists). Therefore, identification of Ca^{2+} channels that function independent of CRAC channels in T cells is another exciting field that needs more attention.

Acknowledgments This work was supported by National Institute of Health grants AI-083432 and AI109059 (Y.G.).

References

Aramburu J, Yaffe MB, Lopez-Rodriguez C, Cantley LC, Hogan PG, Rao A (1999) Affinity-driven peptide selection of an NFAT inhibitor more selective than cyclosporin A. Science 285:2129–2133

Arron JR, Winslow MM, Polleri A, Chang CP, Wu H, Gao X, Neilson JR, Chen L, Heit JJ, Kim SK, Yamasaki N, Miyakawa T, Francke U, Graef IA, Crabtree GR (2006) NFAT dysregulation by increased dosage of DSCR1 and DYRK1A on chromosome 21. Nature 441:595–600

Badou A, Jha MK, Matza D, Flavell RA (2013) Emerging roles of L-type voltage-gated and other calcium channels in T lymphocytes. Front Immunol 4:243

Baine I, Abe BT, Macian F (2009) Regulation of T-cell tolerance by calcium/NFAT signaling. Immunol Rev 231:225–240

Balagopalan L, Coussens NP, Sherman E, Samelson LE, Sommers CL (2010) The LAT story: a tale of cooperativity, coordination, and choreography. Cold Spring Harb Perspect Biol 2: a005512

Barr VA, Bernot KM, Srikanth S, Gwack Y, Balagopalan L, Regan CK, Helman DJ, Sommers CL, Oh-Hora M, Rao A, Samelson LE (2008) Dynamic movement of the calcium sensor STIM1 and the calcium channel Orai1 in activated T-cells: puncta and distal caps. Mol Biol Cell 19:2802–2817

Berridge MJ, Bootman MD, Roderick HL (2003) Calcium signalling: dynamics, homeostasis and remodelling. Nat Rev Mol Cell Biol 4:517–529

Brandman O, Liou J, Park WS, Meyer T (2007) STIM2 is a feedback regulator that stabilizes basal cytosolic and endoplasmic reticulum Ca2+ levels. Cell 131:1327–1339

Budd RC (2001) Activation-induced cell death. Curr Opin Immunol 13:356–362

Cahalan MD, Chandy KG (2009) The functional network of ion channels in T lymphocytes. Immunol Rev 231:59–87

Carrasco S, Meyer T (2011) STIM proteins and the endoplasmic reticulum-plasma membrane junctions. Annu Rev Biochem 80:973–1000

Christel C, Lee A (2012) Ca2+-dependent modulation of voltage-gated Ca2+ channels. Biochim Biophys Acta 1820:1243–1252
Constant SL, Bottomly K (1997) Induction of Th1 and Th2 CD4+ T cell responses: the alternative approaches. Annu Rev Immunol 15:297–322
Coudronniere N, Villalba M, Englund N, Altman A (2000) NF-kappa B activation induced by T cell receptor/CD28 costimulation is mediated by protein kinase C-theta. Proc Natl Acad Sci USA 97:3394–3399
Desvignes L, Weidinger C, Shaw P, Vaeth M, Ribierre T, Liu M, Fergus T, Kozhaya L, McVoy L, Unutmaz D, Ernst JD, Feske S (2015) STIM1 controls T cell-mediated immune regulation and inflammation in chronic infection. J Clin Invest 125:2347–2362
Dolmetsch RE, Lewis RS (1994) Signaling between intracellular Ca2+ stores and depletion-activated Ca2+ channels generates [Ca2+]i oscillations in T lymphocytes. J Gen Physiol 103:365–388
Dolmetsch RE, Lewis RS, Goodnow CC, Healy JI (1997) Differential activation of transcription factors induced by Ca2+ response amplitude and duration. Nature 386:855–858
Dolmetsch RE, Xu K, Lewis RS (1998) Calcium oscillations increase the efficiency and specificity of gene expression. Nature 392:933–936
Feske S, Gwack Y, Prakriya M, Srikanth S, Puppel SH, Tanasa B, Hogan PG, Lewis RS, Daly M, Rao A (2006) A mutation in Orai1 causes immune deficiency by abrogating CRAC channel function. Nature 441:179–185
Garbino A, van Oort RJ, Dixit SS, Landstrom AP, Ackerman MJ, Wehrens XH (2009) Molecular evolution of the junctophilin gene family. Physiol Genomics 37:175–186
Gwack Y, Sharma S, Nardone J, Tanasa B, Iuga A, Srikanth S, Okamura H, Bolton D, Feske S, Hogan PG, Rao A (2006) A genome-wide Drosophila RNAi screen identifies DYRK-family kinases as regulators of NFAT. Nature 441:646–650
Gwack Y, Feske S, Srikanth S, Hogan PG, Rao A (2007a) Signalling to transcription: store-operated Ca2+ entry and NFAT activation in lymphocytes. Cell Calcium 42:145–156
Gwack Y, Srikanth S, Feske S, Cruz-Guilloty F, Oh-hora M, Neems DS, Hogan PG, Rao A (2007b) Biochemical and functional characterization of Orai proteins. J Biol Chem 282:16232–16243
Gwack Y, Srikanth S, Oh-Hora M, Hogan PG, Lamperti ED, Yamashita M, Gelinas C, Neems DS, Sasaki Y, Feske S, Prakriya M, Rajewsky K, Rao A (2008) Hair loss and defective T- and B-cell function in mice lacking ORAI1. Mol Cell Biol 28:5209–5222
Hartzell CA, Jankowska KI, Burkhardt JK, Lewis RS (2016) Calcium influx through CRAC channels controls actin organization and dynamics at the immune synapse. Elife 21:5
Hildeman DA, Zhu Y, Mitchell TC, Bouillet P, Strasser A, Kappler J, Marrack P (2002) Activated T cell death in vivo mediated by proapoptotic bcl-2 family member bim. Immunity 16:759–767
Hildeman D, Jorgensen T, Kappler J, Marrack P (2007) Apoptosis and the homeostatic control of immune responses. Curr Opin Immunol 19:516–521
Hodge MR, Ranger AM, Charles de la Brousse F, Hoey T, Grusby MJ, Glimcher LH (1996) Hyperproliferation and dysregulation of IL-4 expression in NF-ATp-deficient mice. Immunity 4:397–405
Hogan PG, Chen L, Nardone J, Rao A (2003) Transcriptional regulation by calcium, calcineurin, and NFAT. Genes Dev 17:2205–2232
Hogan PG, Lewis RS, Rao A (2010) Molecular basis of calcium signaling in lymphocytes: STIM and ORAI. Annu Rev Immunol 28:491–533
Hoth M, Penner R (1992) Depletion of intracellular calcium stores activates a calcium current in mast cells. Nature 355:353–356
Hoth M, Penner R (1993) Calcium release-activated calcium current in rat mast cells. J Physiol 465:359–386
Ishiguro K, Green T, Rapley J, Wachtel H, Giallourakis C, Landry A, Cao Z, Lu N, Takafumi A, Goto H, Daly MJ, Xavier RJ (2006) Ca2+/calmodulin-dependent protein kinase II is a modulator of CARMA1-mediated NF-kappaB activation. Mol Cell Biol 26:5497–5508

Ishiguro K, Ando T, Goto H, Xavier R (2007) Bcl10 is phosphorylated on Ser138 by Ca2+/calmodulin-dependent protein kinase II. Mol Immunol 44:2095–2100

Jha A, Ahuja M, Maleth J, Moreno CM, Yuan JP, Kim MS, Muallem S (2013) The STIM1 CTID domain determines access of SARAF to SOAR to regulate Orai1 channel function. J Cell Biol 202:71–79

Jha A, Singh AK, Weissgerber P, Freichel M, Flockerzi V, Flavell RA, Jha MK (2015) Essential roles for Cavbeta2 and Cav1 channels in thymocyte development and T cell homeostasis. Sci Signal 8:ra103

Jing J, He L, Sun A, Quintana A, Ding Y, Ma G, Tan P, Liang X, Zheng X, Chen L, Shi X, Zhang SL, Zhong L, Huang Y, Dong MQ, Walker CL, Hogan PG, Wang Y, Zhou Y (2015) Proteomic mapping of ER-PM junctions identifies STIMATE as a regulator of Ca(2)(+) influx. Nat Cell Biol 17:1339–1347

Kar P, Nelson C, Parekh AB (2011) Selective activation of the transcription factor NFAT1 by calcium microdomains near Ca2+ release-activated Ca2+ (CRAC) channels. J Biol Chem 286:14795–14803

Kaufmann U, Shaw PJ, Kozhaya L, Subramanian R, Gaida K, Unutmaz D, McBride HJ, Feske S (2016) Selective ORAI1 inhibition ameliorates autoimmune central nervous system inflammation by suppressing effector but not regulatory T cell function. J Immunol 196:573–585

Kim KD, Srikanth S, Yee MK, Mock DC, Lawson GW, Gwack Y (2011) ORAI1 deficiency impairs activated T cell death and enhances T cell survival. J Immunol 187:3620–3630

Kim KD, Srikanth S, Tan YV, Yee MK, Jew M, Damoiseaux R, Jung ME, Shimizu S, An DS, Ribalet B, Waschek JA, Gwack Y (2014) Calcium signaling via Orai1 is essential for induction of the nuclear orphan receptor pathway to drive Th17 differentiation. J Immunol 192:110–122

Krammer PH, Arnold R, Lavrik IN (2007) Life and death in peripheral T cells. Nat Rev Immunol 7:532–542

Krapivinsky G, Krapivinsky L, Stotz SC, Manasian Y, Clapham DE (2011) POST, partner of stromal interaction molecule 1 (STIM1), targets STIM1 to multiple transporters. Proc Natl Acad Sci USA 108:19234–19239

Kyriakis JM, Avruch J (2012) Mammalian MAPK signal transduction pathways activated by stress and inflammation: a 10-year update. Physiol Rev 92:689–737

Lee KP, Yuan JP, Zeng W, So I, Worley PF, Muallem S (2009) Molecular determinants of fast Ca2+-dependent inactivation and gating of the Orai channels. Proc Natl Acad Sci USA 106:14687–14692

Lewis RS (2011) Store-operated calcium channels: new perspectives on mechanism and function. Cold Spring Harb Perspect Biol 3(12):a003970

Lewis RS, Cahalan MD (1989) Mitogen-induced oscillations of cytosolic Ca2+ and transmembrane Ca2+ current in human leukemic T cells. Cell Regul 1:99–112

Li P, Miao Y, Dani A, Vig M (2016) alpha-SNAP regulates dynamic, on-site assembly and calcium selectivity of Orai1 channels. Mol Biol Cell 27(16):2542–2553

Lin X, O'Mahony A, Mu Y, Geleziunas R, Greene WC (2000) Protein kinase C-theta participates in NF-kappaB activation induced by CD3-CD28 costimulation through selective activation of IkappaB kinase beta. Mol Cell Biol 20:2933–2940

Liou J, Kim ML, Heo WD, Jones JT, Myers JW, Ferrell JE Jr, Meyer T (2005) STIM is a Ca2+ sensor essential for Ca2+-store-depletion-triggered Ca2+ influx. Curr Biol 15:1235–1241

Liou J, Fivaz M, Inoue T, Meyer T (2007) Live-cell imaging reveals sequential oligomerization and local plasma membrane targeting of stromal interaction molecule 1 after Ca2+ store depletion. Proc Natl Acad Sci USA 104:9301–9306

Lioudyno MI, Kozak JA, Penna A, Safrina O, Zhang SL, Sen D, Roos J, Stauderman KA, Cahalan MD (2008) Orai1 and STIM1 move to the immunological synapse and are up-regulated during T cell activation. Proc Natl Acad Sci USA 105:2011–2016

Liu Z, Lee J, Krummey S, Lu W, Cai H, Lenardo MJ (2011) The kinase LRRK2 is a regulator of the transcription factor NFAT that modulates the severity of inflammatory bowel disease. Nat Immunol 12:1063–1070

Liu X, Berry CT, Ruthel G, Madara JJ, MacGillivray K, Gray CM, Madge LA, McCorkell KA, Beiting DP, Hershberg U, May MJ, Freedman BD (2016) T cell receptor-induced Nuclear Factor kappaB (NF-kappaB) signaling and transcriptional activation are regulated by STIM1- and Orai1-mediated calcium entry. J Biol Chem 291:8440–8452

Luik RM, Wu MM, Buchanan J, Lewis RS (2006) The elementary unit of store-operated Ca2+ entry: local activation of CRAC channels by STIM1 at ER-plasma membrane junctions. J Cell Biol 174:815–825

Ma J, McCarl CA, Khalil S, Luthy K, Feske S (2010) T-cell-specific deletion of STIM1 and STIM2 protects mice from EAE by impairing the effector functions of Th1 and Th17 cells. Eur J Immunol 40:3028–3042

Macian F (2005) NFAT proteins: key regulators of T-cell development and function. Nat Rev Immunol 5:472–484

Macian F, Garcia-Cozar F, Im SH, Horton HF, Byrne MC, Rao A (2002) Transcriptional mechanisms underlying lymphocyte tolerance. Cell 109:719–731

Marrack P, Kappler J (2004) Control of T cell viability. Annu Rev Immunol 22:765–787

Marsden VS, Strasser A (2003) Control of apoptosis in the immune system: Bcl-2, BH3-only proteins and more. Annu Rev Immunol 21:71–105

Matsumoto M, Fujii Y, Baba A, Hikida M, Kurosaki T, Baba Y (2011) The calcium sensors STIM1 and STIM2 control B cell regulatory function through interleukin-10 production. Immunity 34:703–714

McCarl CA, Khalil S, Ma J, Oh-hora M, Yamashita M, Roether J, Kawasaki T, Jairaman A, Sasaki Y, Prakriya M, Feske S (2010) Store-operated Ca2+ entry through ORAI1 is critical for T cell-mediated autoimmunity and allograft rejection. J Immunol 185:5845–5858

McNally BA, Somasundaram A, Yamashita M, Prakriya M (2012) Gated regulation of CRAC channel ion selectivity by STIM1. Nature 482:241–245

Miao Y, Miner C, Zhang L, Hanson PI, Dani A, Vig M (2013) An essential and NSF independent role for alpha-SNAP in store-operated calcium entry. Elife 2:e00802

Mognol GP, Carneiro FR, Robbs BK, Faget DV, Viola JP (2016) Cell cycle and apoptosis regulation by NFAT transcription factors: new roles for an old player. Cell Death Dis 7:e2199

Mor A, Philips MR (2006) Compartmentalized Ras/MAPK signaling. Annu Rev Immunol 24:771–800

Morreale A, Venkatesan M, Mott HR, Owen D, Nietlispach D, Lowe PN, Laue ED (2000) Structure of Cdc42 bound to the GTPase binding domain of PAK. Nat Struct Biol 7:384–388

Muik M, Frischauf I, Derler I, Fahrner M, Bergsmann J, Eder P, Schindl R, Hesch C, Polzinger B, Fritsch R, Kahr H, Madl J, Gruber H, Groschner K, Romanin C (2008) Dynamic coupling of the putative coiled-coil domain of ORAI1 with STIM1 mediates ORAI1 channel activation. J Biol Chem 283:8014–8022

Muik M, Fahrner M, Derler I, Schindl R, Bergsmann J, Frischauf I, Groschner K, Romanin C (2009) A cytosolic homomerization and a modulatory domain within STIM1 C terminus determine coupling to ORAI1 channels. J Biol Chem 284:8421–8426

Muller MR, Rao A (2010) NFAT, immunity and cancer: a transcription factor comes of age. Nat Rev Immunol 10:645–656

Mullins FM, Park CY, Dolmetsch RE, Lewis RS (2009) STIM1 and calmodulin interact with Orai1 to induce Ca2+-dependent inactivation of CRAC channels. Proc Natl Acad Sci USA 106:15495–15500

Navarro-Borelly L, Somasundaram A, Yamashita M, Ren D, Miller RJ, Prakriya M (2008) STIM1-Orai1 interactions and Orai1 conformational changes revealed by live-cell FRET microscopy. J Physiol 586:5383–5401

Nishi M, Sakagami H, Komazaki S, Kondo H, Takeshima H (2003) Coexpression of junctophilin type 3 and type 4 in brain. Brain Res Mol Brain Res 118:102–110

Nohara LL, Stanwood SR, Omilusik KD, Jefferies WA (2015) Tweeters, woofers and horns: the complex orchestration of calcium currents in T lymphocytes. Front Immunol 6:234

Oh-Hora M, Yamashita M, Hogan PG, Sharma S, Lamperti E, Chung W, Prakriya M, Feske S, Rao A (2008) Dual functions for the endoplasmic reticulum calcium sensors STIM1 and STIM2 in T cell activation and tolerance. Nat Immunol 9:432–443

Oh-Hora M, Komatsu N, Pishyareh M, Feske S, Hori S, Taniguchi M, Rao A, Takayanagi H (2013) Agonist-selected T cell development requires strong T cell receptor signaling and store-operated calcium entry. Immunity 38(5):881–895

Omilusik K, Priatel JJ, Chen X, Wang YT, Xu H, Choi KB, Gopaul R, McIntyre-Smith A, Teh HS, Tan R, Bech-Hansen NT, Waterfield D, Fedida D, Hunt SV, Jefferies WA (2011) The Ca(v)1.4 calcium channel is a critical regulator of T cell receptor signaling and naive T cell homeostasis. Immunity 35:349–360

Oruganti SR, Edin S, Grundstrom C, Grundstrom T (2011) CaMKII targets Bcl10 in T-cell receptor induced activation of NF-kappaB. Mol Immunol 48:1448–1460

Palty R, Raveh A, Kaminsky I, Meller R, Reuveny E (2012) SARAF inactivates the store operated calcium entry machinery to prevent excess calcium refilling. Cell 149:425–438

Park CY, Hoover PJ, Mullins FM, Bachhawat P, Covington ED, Raunser S, Walz T, Garcia KC, Dolmetsch RE, Lewis RS (2009) STIM1 clusters and activates CRAC channels via direct binding of a cytosolic domain to Orai1. Cell 136:876–890

Picard C, McCarl CA, Papolos A, Khalil S, Luthy K, Hivroz C, LeDeist F, Rieux-Laucat F, Rechavi G, Rao A, Fischer A, Feske S (2009) STIM1 mutation associated with a syndrome of immunodeficiency and autoimmunity. N Engl J Med 360:1971–1980

Putney JW Jr (1986) A model for receptor-regulated calcium entry. Cell Calcium 7:1–12

Putney JW Jr (2009) Capacitative calcium entry: from concept to molecules. Immunol Rev 231:10–22

Quintana A, Rajanikanth V, Farber-Katz S, Gudlur A, Zhang C, Jing J, Zhou Y, Rao A, Hogan PG (2015) TMEM110 regulates the maintenance and remodeling of mammalian ER-plasma membrane junctions competent for STIM-ORAI signaling. Proc Natl Acad Sci USA 112: E7083–E7092

Ritchie MF, Samakai E, Soboloff J (2012) STIM1 is required for attenuation of PMCA-mediated Ca2+ clearance during T-cell activation. EMBO J 31:1123–1133

Robbs BK, Cruz AL, Werneck MB, Mognol GP, Viola JP (2008) Dual roles for NFAT transcription factor genes as oncogenes and tumor suppressors. Mol Cell Biol 28:7168–7181

Roos J, DiGregorio PJ, Yeromin AV, Ohlsen K, Lioudyno M, Zhang S, Safrina O, Kozak JA, Wagner SL, Cahalan MD, Veliçelebi G, Stauderman KA (2005) STIM1, an essential and conserved component of store-operated Ca2+ channel function. J Cell Biol 169:435–445

Samelson LE (2011) Immunoreceptor signaling. Cold Spring Harb Perspect Biol 3(12):a011510

Schuhmann MK, Stegner D, Berna-Erro A, Bittner S, Braun A, Kleinschnitz C, Stoll G, Wiendl H, Meuth SG, Nieswandt B (2010) Stromal interaction molecules 1 and 2 are key regulators of autoreactive T cell activation in murine autoimmune central nervous system inflammation. J Immunol 184:1536–1542

Scrimgeour N, Litjens T, Ma L, Barritt GJ, Rychkov GY (2009) Properties of Orai1 mediated store-operated current depend on the expression levels of STIM1 and Orai1 proteins. J Physiol 587:2903–2918

Serfling E, Klein-Hessling S, Palmetshofer A, Bopp T, Stassen M, Schmitt E (2006) NFAT transcription factors in control of peripheral T cell tolerance. Eur J Immunol 36:2837–2843

Serfling E, Avots A, Klein-Hessling S, Rudolf R, Vaeth M, Berberich-Siebelt F (2012) NFATc1/alphaA: the other face of NFAT Factors in lymphocytes. Cell Commun Signal 10:16

Sharma S, Findlay GM, Bandukwala HS, Oberdoerffer S, Baust B, Li Z, Schmidt V, Hogan PG, Sacks DB, Rao A (2011) Dephosphorylation of the nuclear factor of activated T cells (NFAT) transcription factor is regulated by an RNA-protein scaffold complex. Proc Natl Acad Sci USA 108:11381–11386

Shi X, Bi Y, Yang W, Guo X, Jiang Y, Wan C, Li L, Bai Y, Guo J, Wang Y, Chen X, Wu B, Sun H, Liu W, Wang J, Xu C (2013) Ca2+ regulates T-cell receptor activation by modulating the charge property of lipids. Nature 493:111–115

Smith-Garvin JE, Koretzky GA, Jordan MS (2009) T cell activation. Annu Rev Immunol 27:591–619
Soboloff J, Rothberg BS, Madesh M, Gill DL (2012) STIM proteins: dynamic calcium signal transducers. Nat Rev Mol Cell Biol 13:549–565
Sprent J, Surh CD (2011) Normal T cell homeostasis: the conversion of naive cells into memory-phenotype cells. Nat Immunol 12:478–484
Srikanth S, Gwack Y (2012) Orai1, STIM1, and their associating partners. J Physiol 590:4169–4177
Srikanth S, Gwack Y (2013a) Molecular regulation of the pore component of CRAC channels, Orai1. Curr Top Membr 71:181–207
Srikanth S, Gwack Y (2013b) Orai1-NFAT signalling pathway triggered by T cell receptor stimulation. Mol Cells 35:182–194
Srikanth S, Jung HJ, Kim KD, Souda P, Whitelegge J, Gwack Y (2010a) A novel EF-hand protein, CRACR2A, is a cytosolic Ca2+ sensor that stabilizes CRAC channels in T cells. Nat Cell Biol 12:436–446
Srikanth S, Jung HJ, Ribalet B, Gwack Y (2010b) The intracellular loop of Orai1 plays a central role in fast inactivation of Ca2+ release-activated Ca2+ channels. J Biol Chem 285:5066–5075
Srikanth S, Jew M, Kim KD, Yee MK, Abramson J, Gwack Y (2012) Junctate is a Ca2+-sensing structural component of Orai1 and stromal interaction molecule 1 (STIM1). Proc Natl Acad Sci USA 109:8682–8687
Srikanth S, Ribalet B, Gwack Y (2013) Regulation of CRAC channels by protein interactions and post-translational modification. Channels (Austin) 7(5):354–363
Srikanth S, Kim KD, Gao Y, Woo JS, Ghosh S, Calmettes G, Paz A, Abramson J, Jiang M, Gwack Y (2016a) A large Rab GTPase encoded by CRACR2A is a component of subsynaptic vesicles that transmit T cell activation signals. Sci Signal 9:ra31
Srikanth S, Woo JS, Gwack Y (2016b) A large Rab GTPase family in a small GTPase world. Small GTPases 24:1–6
Srinivasan M, Frauwirth KA (2007) Reciprocal NFAT1 and NFAT2 nuclear localization in CD8+ anergic T cells is regulated by suboptimal calcium signaling. J Immunol 179:3734–3741
Strasser A (2005) The role of BH3-only proteins in the immune system. Nat Rev Immunol 5:189–200
Strasser A, Pellegrini M (2004) T-lymphocyte death during shutdown of an immune response. Trends Immunol 25:610–615
Sun Z (2012) Intervention of PKC-theta as an immunosuppressive regimen. Front Immunol 3:225
Sun Z, Arendt CW, Ellmeier W, Schaeffer EM, Sunshine MJ, Gandhi L, Annes J, Petrzilka D, Kupfer A, Schwartzberg PL, Littman DR (2000) PKC-theta is required for TCR-induced NF-kappaB activation in mature but not immature T lymphocytes. Nature 404:402–407
Takeshima H, Komazaki S, Nishi M, Iino M, Kangawa K (2000) Junctophilins: a novel family of junctional membrane complex proteins. Mol Cell 6:11–22
Tybulewicz VL (2005) Vav-family proteins in T-cell signalling. Curr Opin Immunol 17:267–274
Varnai P, Toth B, Toth DJ, Hunyady L, Balla T (2007) Visualization and manipulation of plasma membrane-endoplasmic reticulum contact sites indicates the presence of additional molecular components within the STIM1-Orai1 Complex. J Biol Chem 282:29678–29690
Vig M, Peinelt C, Beck A, Koomoa DL, Rabah D, Koblan-Huberson M, Kraft S, Turner H, Fleig A, Penner R, Kinnet JP (2006) CRACM1 is a plasma membrane protein essential for store-operated Ca2+ entry. Science 312:1220–1223
Wang H, Kadlecek TA, Au-Yeung BB, Goodfellow HE, Hsu LY, Freedman TS, Weiss A (2010) ZAP-70: an essential kinase in T-cell signaling. Cold Spring Harb Perspect Biol 2:a002279
Weisleder N, Takeshima H, Ma J (2008) Immuno-proteomic approach to excitation–contraction coupling in skeletal and cardiac muscle: molecular insights revealed by the mitsugumins. Cell Calcium 43:1–8

Willingham AT, Orth AP, Batalov S, Peters EC, Wen BG, Aza-Blanc P, Hogenesch JB, Schultz PG (2005) A strategy for probing the function of noncoding RNAs finds a repressor of NFAT. Science 309:1570–1573

Wilson LA, McKeown L, Tumova S, Li J, Beech DJ (2015) Expression of a long variant of CRACR2A that belongs to the Rab GTPase protein family in endothelial cells. Biochem Biophys Res Commun 456:398–402

Woo JS, Srikanth S, Nishi M, Ping P, Takeshima H, Gwack Y (2016) Junctophilin-4, a component of the endoplasmic reticulum-plasma membrane junctions, regulates Ca2+ dynamics in T cells. Proc Natl Acad Sci USA 113:2762–2767

Wu MM, Buchanan J, Luik RM, Lewis RS (2006) Ca2+ store depletion causes STIM1 to accumulate in ER regions closely associated with the plasma membrane. J Cell Biol 174:803–813

Wu H, Peisley A, Graef IA, Crabtree GR (2007) NFAT signaling and the invention of vertebrates. Trends Cell Biol 17:251–260

Yuan JP, Zeng W, Dorwart MR, Choi YJ, Worley PF, Muallem S (2009) SOAR and the polybasic STIM1 domains gate and regulate Orai channels. Nat Cell Biol 11:337–343

Zhang SL, Yu Y, Roos J, Kozak JA, Deerinck TJ, Ellisman MH, Stauderman KA, Cahalan MD (2005) STIM1 is a Ca2+ sensor that activates CRAC channels and migrates from the Ca2+ store to the plasma membrane. Nature 437:902–905

Zhang SL, Yeromin AV, Zhang XH, Yu Y, Safrina O, Penna A, Roos J, Stauderman KA, Cahalan MD (2006) Genome-wide RNAi screen of Ca(2+) influx identifies genes that regulate Ca(2+) release-activated Ca(2+) channel activity. Proc Natl Acad Sci USA 103:9357–9362

Zhang SL, Yeromin AV, Hu J, Amcheslavsky A, Zheng H, Cahalan MD (2011) Mutations in Orai1 transmembrane segment 1 cause STIM1-independent activation of Orai1 channels at glycine 98 and channel closure at arginine 91. Proc Natl Acad Sci USA 108:17838–17843

Zhu J, Yamane H, Paul WE (2010) Differentiation of effector CD4 T cell populations. Annu Rev Immunol 28:445–489

Zweifach A, Lewis RS (1995) Rapid inactivation of depletion-activated calcium current (ICRAC) due to local calcium feedback. J Gen Physiol 105:209–226

Cardiovascular and Hemostatic Disorders: Role of STIM and Orai Proteins in Vascular Disorders

22

Jyoti Tanwar, Mohamed Trebak, and Rajender K. Motiani

Abstract

Store-operated Ca^{2+} entry (SOCE) mediated by STIM and Orai proteins is a highly regulated and ubiquitous signaling pathway that plays an important role in various cellular and physiological functions. Endoplasmic reticulum (ER) serves as the major site for intracellular Ca^{2+} storage. Stromal Interaction Molecule 1/2 (STIM1/2) sense decrease in ER Ca^{2+} levels and transmits the message to plasma membrane Ca^{2+} channels constituted by Orai family members (Orai1/2/3) resulting in Ca^{2+} influx into the cells. This increase in cytosolic Ca^{2+} in turn activates a variety of signaling cascades to regulate a plethora of cellular functions. Evidence from the literature suggests that SOCE dysregulation is associated with several pathophysiologies, including vascular disorders. Interestingly, recent studies have suggested that STIM proteins may also regulate vascular functions independent of their contribution to SOCE. In this updated book chapter, we will focus on the physiological role of STIM and Orai proteins in the vasculature (endothelial cells and vascular smooth muscle cells). We will further retrospect the literature implicating a critical role for these proteins in vascular disease.

Keywords

STIM1 • Orai1 • Orai3 • Vascular diseases • Restenosis • Hypertension • Atherosclerosis

J. Tanwar • R.K. Motiani (✉)
Systems Biology Group, CSIR-Institute of Genomics and Integrative Biology, Mathura Road, New Delhi 110020, India
e-mail: rajmotiani@igib.in

M. Trebak (✉)
Department of Cellular and Molecular Physiology, The Pennsylvania State University College of Medicine, Hershey, PA 17033, USA
e-mail: mtrebak@psu.edu

K. Groschner et al. (eds.), *Store-Operated Ca^{2+} Entry (SOCE) Pathways*, Advances in Experimental Medicine and Biology 993, DOI 10.1007/978-3-319-57732-6_22

22.1 Introduction

Vascular diseases are associated with very high morbidity and mortality rates. These disorders are largely chronic in nature, thus resulting in huge economic burden on modern societies. Vascular diseases include hypertension and a number of vascular occlusive diseases such as atherosclerosis, thrombosis, and restenosis. In most cases, the coronary vasculature, which supplies blood to the heart, is adversely affected. These vascular beds are occluded due to accumulating deposits on their wall. Although significant improvement in the treatment and health-care management has reduced the number of deaths resulting from cardiovascular disorders, they remain the number one killer in the developed world. Therefore, it is important to identify novel targetable proteins and signaling pathways that contribute to the progression of these diseases.

The role of Ca^{2+} in mediating cardiovascular diseases is well established (Wehrens and Marks 2004; Cheng and Lederer 2008; Alexopoulos and Raggi 2009). One of the most ubiquitous regulated means of Ca^{2+} influx into cells is the store-operated Ca^{2+} entry (SOCE) pathway first introduced by Putney (Putney 1986, 1990; Potier and Trebak 2008). Strong evidence supports the idea that deregulation of the SOCE pathway is an important contributor to the development of vascular disorders (Van Assche et al. 2007; Berra-Romani et al. 2008; House et al. 2008; Leung et al. 2008; Bisaillon et al. 2010; Zhang and Trebak 2011; Ruhle and Trebak 2013; Spinelli and Trebak 2016). STIM and Orai proteins were discovered in last decade as the key molecular players of the SOCE pathway (Liou et al. 2005; Roos et al. 2005; Feske et al. 2006; Vig et al. 2006; Zhang et al. 2006). STIM1 acts as an endoplasmic reticulum (ER) Ca^{2+} sensor which upon store depletion oligomerizes and interacts with plasma membrane store-operated Ca^{2+} (SOC) channels constituted by Orai1 (Liou et al. 2005; Roos et al. 2005; Feske et al. 2006; Vig et al. 2006; Zhang et al. 2006; Mignen et al. 2008; Penna et al. 2008; Park et al. 2009). Mammals have two STIM proteins (STIM1 and STIM2) and three Orai proteins (Orai1–3), all of which were shown to form channels activated by store depletion in ectopic expression system (Mercer et al. 2006; Peinelt et al. 2006; Soboloff et al. 2006). STIM1 and Orai1 have been reported to play an essential role in SOCE in the majority of cell types. However, recent studies have suggested that STIM2, Orai2, and Orai3 can also contribute to native SOCE channels (Berna-Erro et al. 2009; Motiani et al. 2010; Bandyopadhyay et al. 2011; Gruszczynska-Biegala and Kuznicki 2013; Hoth and Niemeyer 2013; Kraft 2015; Moccia et al. 2015).

Increasing evidence suggests that STIM and Orai proteins contribute to the development and progression of several pathologies. In this chapter, we will provide a brief overview of the SOCE pathway as well as STIM and Orai proteins. We will further present a detailed retrospective analysis of the current literature suggesting involvement of STIMs and Orais in vascular disorders. Finally, we will briefly discuss the possibilities of therapeutically targeting these proteins for the treatment and/or better management of vascular disorders.

22.2 SOCE and STIM/Orai Proteins

Under physiological conditions, when Gq/11 receptors or receptor tyrosine kinase that are coupled to phospholipase C (PLC) are activated by their specific agonists, PLC breaks down phosphatidylinositol 4,5 bisphosphate (PIP_2) (Michell 1975) and generates two second messengers: diacylglycerol (DAG) and inositol 1,4,5 trisphosphate (IP_3) (Berridge and Irvine 1984; Berridge 1987). DAG in turn can either directly activate ion channels on the plasma membrane (Hofmann et al. 1999; Trebak et al. 2003a, b) or can generate other second messengers such as arachidonic acid (AA), known to activate the store-independent arachidonate-regulated Ca^{2+} (ARC) channels (Shuttleworth 2009). More recent studies suggest that AA mediates its effects through Leukotriene C_4 (LTC_4) in smooth muscle cells and HEK293 cells (Gonzalez-Cobos et al. 2013; Zhang et al. 2013, 2014, 2015) (also see Sect. 22.3.3 for further details). Whereas, IP_3 binds to IP_3 receptor present on the endoplasmic reticulum (ER) resulting in release of Ca^{2+} from the ER. This release of Ca^{2+} from the ER results in depletion of intracellular Ca^{2+} stores causing activation of SOCE. Therefore, SOCE is defined as the Ca^{2+} influx into the cell via plasma membrane Ca^{2+}-permeable channels activated as a direct outcome of intracellular (primarily ER) Ca^{2+} store depletion. SOCE, originally termed capacitative Ca^{2+} entry (CCE), was first proposed and later refined by Putney (1986, 1990). The channels involved in this regulated Ca^{2+} entry are known as store-operated Ca^{2+} (SOC) channels. Ca^{2+} entry via SOC channels replenishes the stores and has also been increasingly appreciated for its role in activating and inactivating various transcription factors and regulating physiological functions (Parekh and Putney 2005; Prakriya and Lewis 2015). On the basis of his studies in parotid acinar cells, Putney proposed that in non-excitable cells, the extent of Ca^{2+} present in the intracellular stores regulates the amount of Ca^{2+}entry into the cells. Thapsigargin, a specific SERCA pump blocker that causes passive store depletion without generating second messengers, was used to demonstrate the SOCE concept pharmacologically (Takemura et al. 1989). Subsequently, Hoth and Penner (1992) showed that store depletion in RBL mast cells is associated with the development of a highly Ca^{2+} selective inward current that they termed calcium release-activated calcium current (I_{CRAC}). While SOCE was initially considered to be functional primarily in non-excitable cells, increasing evidence points toward the functional presence of SOCE and its involvement in regulating cellular functions in excitable cells such as neurons, skeletal muscle, and smooth muscle (Berna-Erro et al. 2009; Potier et al. 2009; Venkiteswaran and Hasan 2009; Gruszczynska-Biegala and Kuznicki 2013).

*St*romal *i*nteracting *m*olecule 1 (STIM1) was for long known to be outside the Ca^{2+} signaling arena as a molecule involved in binding of B-lymphocytes to stromal cells (Oritani and Kincade 1996). STIM1 gene was cloned by Parker et al. (1996) as GOK "God Only Knows" and was shown to be involved in child malignancies. It was later suggested as a tumor suppressor gene involved in skeletal muscle tumors (rhabdomyosarcomas) as its expression was downregulated during the course of skeletal muscle tumor development (Sabbioni et al. 1997). The essential role of

STIM1 and STIM2 in the SOCE pathway (as ER Ca^{2+} store sensors) came into light in 2005 when two independent studies reported their involvement in SOCE using RNAi screens (Liou et al. 2005; Roos et al. 2005).

STIM1 is involved in sensing ER Ca^{2+} store depletion during agonist stimulation, while STIM2 appears to be involved in maintaining intracellular Ca^{2+} homeostasis during basal unstimulated conditions (Brandman et al. 2007). Interestingly, studies from the Nieswandt group showed that STIM2, but not STIM1, is involved in mediating SOCE in neurons. Using transgenic mice lacking STIM2, they reported that STIM2-mediated SOCE plays a vital role in cerebral ischemia and autoimmune encephalomyelitis (Berna-Erro et al. 2009; Schuhmann et al. 2010). STIM1 has a number of discrete regions including a low affinity Ca^{2+}-binding EF-hand domain in the N-terminus facing toward the ER lumen, whereas the C-terminus of STIM1 faces the cytoplasm and contains a coiled-coil domain, a serine-proline-rich and a lysine-rich region. When ER Ca^{2+} stores are full, Ca^{2+} is bound to the EF-hand domain of STIM1 (Baba et al. 2006), but when stores are depleted, Ca^{2+} ions dissociate from the EF hand, which in turn initiates a cascade of events leading to SOC channel activation (Stathopulos et al. 2006). Upon ER Ca^{2+} store depletion, STIM1 oligomerizes and moves toward ER-plasma membrane junctions (Liou et al. 2007). In this STIM1 redistribution process, required for STIM1 to reach areas of close ER-plasma membrane junctions, microtubules play a facilitative role by optimizing the translocation of STIM1 (Smyth et al. 2007). When microtubules are disrupted or depolymerized, SOCE is inhibited (Oka et al. 2005). This oligomerization/redistribution of STIM1 is observed in the form of STIM1 clusters termed puncta. Forster Resonance Energy Transfer (FRET) and total internal reflectance fluorescence (TIRF) studies have clearly demonstrated that ER-resident STIM1 oligomerizes after store depletion, followed by its redistribution to predetermined foci in the peripheral ER (Luik et al. 2008).

Three independent labs used genome wide RNAi screens and discovered that a hypothetical protein named Orai1/CRACM1 is required for SOCE activation (Feske et al. 2006; Vig et al. 2006; Zhang et al. 2006). Feske et al. reported that Orai1 has two mammalian homologues that were named Orai2 and Orai3. Using genetic screening on lymphocyte cell lines established from a family with a severe combined immunodeficiency syndrome (SCID), the same group reported that a point mutation (arginine to tryptophan, R91W) in human Orai1 was responsible for the SCID disorder. The immunodeficient lymphocytes established by Feske et al. from SCID patients had deficient SOCE. When wild-type Orai1 was overexpressed in these immunodeficient lymphocytes, SOCE was restored (Feske et al. 2006; Parekh 2006). Subsequent mutagenesis studies showed that mutations of a glutamate (E106) residue, within the putative pore region of Orai1, by either charge neutralization by alanine or replacement of glutamate with aspartate, lead to either a dead channel or a channel with less ionic selectivity, respectively, thus establishing that Orai1 is the pore-forming unit of SOC channels (Feske et al. 2006; Peinelt et al. 2006; Yeromin et al. 2006). The ectopic co-expression of STIM1 and Orai1 by several independent groups led to large increases in SOCE and I_{CRAC} upon ER Ca^{2+} store depletion (Mercer et al. 2006; Peinelt et al. 2006; Soboloff et al. 2006).

Likewise, ectopic expression of Orai2 and Orai3 along with STIM1 can augment SOCE and I_{CRAC} (Mercer et al. 2006; DeHaven et al. 2007). These studies strongly suggested that Orai1 is the pore-forming unit of the CRAC channel and that Orai2 and Orai3 either as homomultimers or heteromultimers might function in a similar fashion in different cell types to further enhance the diversity of SOC channels. Indeed, recent literature demonstrates that Orai3 can form native functional SOC channel (Motiani et al. 2010) as well as heteromultimeric channels with Orai1 (Shuttleworth 2012; Gonzalez-Cobos et al. 2013; Thompson et al. 2013) (also see Sect. 22.3.3 for further details).

Four independent labs identified the STIM1 C-terminus region involved in direct interaction and activation of Orai1 (Kawasaki et al. 2009; Muik et al. 2009; Park et al. 2009; Yuan et al. 2009). This sequence was termed SOAR (for STIM-Orai activating region) by Yuan and colleagues and CAD (for CRAC activation domain) by Lewis and colleagues (Park et al. 2009; Yuan et al. 2009). Taken together, these studies clearly demonstrate that the C-terminus of STIM1 is responsible for the transmission of ER Ca^{2+} store depletion signal to Orai channels so that they may open for mediating SOCE.

Recent studies have identified a number of proteins that can regulate SOCE: SARAF (Palty et al. 2012), CRACR2A (Srikanth et al. 2010), STIMATE (Jing et al. 2015), Golli (Feng et al. 2006), SNAREs (Miao et al. 2013), and SEPTINS (Sharma et al. 2013). Although these proteins were reported to be playing an important role in regulating of SOCE pathway, their absolute requirement for mediating SOCE is debatable. In fact, the Hogan group showed that STIM1 and Orai1 are sufficient to form a functional SOCE pathway in vitro (Zhou et al. 2010). They used bacterially expressed recombinant STIM1 to activate human Orai1 and showed that STIM1 can activate Orai1 in the absence of other proteins. Furthermore, as mentioned earlier, studies performed by several labs have shown that co-expression of STIM1 and Orai1 alone was sufficient to cause a large increase (up to 100-fold) in I_{CRAC} and SOCE (Mercer et al. 2006; Peinelt et al. 2006; Soboloff et al. 2006), arguing that any additional endogenous cofactors are either not strictly required for SOCE activation but are involved for regulation or are present in abundant quantities in cells.

22.3 STIM and Orai Proteins in Vascular Disorders

A large numbers of studies have implicated STIM and Orai proteins in a variety of pathological conditions affecting the vascular system. Below, we will present some of the evidence suggesting critical role of STIMs and Orais in (1) thrombosis, (2) atherosclerosis, (3) restenosis, and (4) hypertension.

22.3.1 Thrombosis

Thrombosis is defined as the formation of a thrombus inside a blood vessel. When a blood vessel is injured, the body uses platelets and fibrin to form a blood clot, which in turn arrests bleeding. Thus, it is a defense mechanism used by the body to prevent blood loss. In severe cases, the thrombus can break free and travel in the lumen of the blood vessel, which in some cases lead to infarction and stroke; this process is known as embolism. Thrombosis results in narrowing of the lumen of blood vessels resulting in the obstruction of blood flow. Obstruction of coronary and cerebral arteries can lead to lethal conditions such as myocardial infarction and ischemic brain infraction resulting in heart attack and stroke, respectively.

One of the most important steps in thrombus formation is platelet activation upon stimulation with agonists like thrombin, ADP, and thromboxane A_2. Interestingly, activation of platelets via all these agonists results in elevation of intracellular Ca^{2+} levels (Rink and Sage 1990; Varga-Szabo et al. 2009). This increase in cytosolic Ca^{2+} concentration is required for platelet aggregation at the site of thrombus formation (Mazzucato et al. 2002). Strong evidence supports a role for STIM1 and Orai1 in mediating the increase in platelet Ca^{2+} concentration and thrombosis. Grosse et al. demonstrated that STIM1 is required for platelet activation. They generated a mouse cell line expressing constitutively active STIM1 by mutating an acidic amino acid (aspartate) to a neutral one (glycine) at position 84 (D84G) located in its EF hand. The platelets with EF-hand-mutated STIM1 were shown to have threefold higher basal level of intracellular Ca^{2+} in comparison to wild-type platelets. Platelets with STIM1 mutation were pre-activated, displayed higher intracellular Ca^{2+} levels, and failed to show further increase in their cytoplasmic Ca^{2+} concentrations in response to epinephrine. In contrast to wild-type platelets, the STIM1-mutated platelets were not able to adhere to fibrous collagen properly and failed to form stable three-dimensional thrombi (Grosse et al. 2007). The critical role of STIM1 in thrombus stability was further implicated by generating platelet-specific conditional STIM1 knockout mice (Ahmad et al. 2011). Using laser injury thrombosis model, Ahmad et al. explained the instability of thrombus in case of mice with STIM1 knockout platelets. They demonstrated that these platelets were inefficient in generating fibrin at the site of injury, and therefore these thrombi were less stable than those seen in WT control mice (Ahmad et al. 2011). STIM1 knockout in platelets reduces their ability to express phosphatidylserine at the plasma membrane required for thrombus stability (Gilio et al. 2010; Ahmad et al. 2011). These studies collectively suggested that STIM1 plays an important role in thrombus formation as well as its stability.

The contribution of STIM1 to platelet activation and thrombosis was studied by generating both conditional STIM1 knockout mice lacking STIM1 in platelets and STIM1 null mice ($STIM1^{-/-}$). The $STIM1^{-/-}$ mice were mostly postnatal lethal with 70% dying few hours after birth. Interestingly, the authors reported appearance of marked cyanosis in the $STIM1^{-/-}$ mice before death and suggested that it could be due to cardiopulmonary disorders (Varga-Szabo et al. 2008b). STIM1-deficient platelets showed drastic reduction in Ca^{2+} responses upon stimulation with a wide variety of agonists (thrombin, ADP, collagen-related peptide) as well as upon passive store depletion by thapsigargin. In vitro studies demonstrated that STIM1$^{-/-}$ platelets do not adhere to collagen-coated surfaces, whereas wild-type platelets both adhere and form aggregates. Furthermore, using two different in vivo injury models, these authors clearly showed that platelet aggregation was significantly delayed in STIM1$^{-/-}$ mice in comparison to wild-type mice. Moreover, platelets from STIM1$^{-/-}$ chimeras were not able to form stable thrombi, therefore suggesting a vital role for STIM1 in pathological thrombosis. Varga-Szabo et al. further demonstrated the involvement of STIM1 in ischemic brain infraction using an in vivo model of thrombus formation in brain microvasculature. The brain infarcts were significantly smaller in STIM1$^{-/-}$ chimeras in comparison to wild-type mice, and this decrease in infarct size was functionally correlated to better motor coordination and function in STIM1$^{-/-}$ chimeras (Varga-Szabo et al. 2008b).

Subsequently, Orai1, but not TRPC1, was reported to be the platelet SOC channel by the same group (Varga-Szabo et al. 2008a; Braun et al. 2009). These authors used both in vitro and in vivo approaches to demonstrate that TPRC1 neither contributes to platelet SOCE nor to their activation and function. Using TRPC1$^{-/-}$ mice, they reported that thapsigargin-, thrombin-, and collagen-related peptide (CRP)-mediated SOCE in platelets was essentially not dependent on TRPC1. Furthermore, using in vitro flow adhesion assays, no significant difference was observed in platelet aggregation between wild type and TRPC1$^{-/-}$. TRPC1$^{-/-}$ platelets were also able to form stable thrombi with kinetics similar to wild-type platelets suggesting that TRPC1 is not required for thrombosis. Using in vivo mice models, Varga-Szabo et al. (2008a) also demonstrated that TRPC1 neither contributes to physiological function of platelets (homeostasis) nor to pathological thrombosis. Braun et al. later identified Orai1 as the platelet SOC channel required for thrombus formation. The authors used platelets from Orai1$^{-/-}$ mice to demonstrate that Orai1-mediated SOCE plays a vital role in the formation of stable thrombus in vitro under high shear flow conditions. Furthermore, they reported that in vivo platelet activation and thrombi formation were significantly reduced in Orai1$^{-/-}$ chimeras that lack Orai1 expression in hematopoietic cells, in comparison to wild-type mice. These authors were also able to demonstrate that Orai1$^{-/-}$ mice were less prone to develop lethal pulmonary thromboembolism and arterial thrombi upon animal exposure to chemical insults and mechanical injuries, respectively (Braun et al. 2009). Around the same time, Feske and coworkers also reported Orai1 involvement in platelet SOCE using platelets from mice bearing the equivalent human SCID mutation in Orai1 (R93W) (Bergmeier et al. 2009). However, another group reported that both Orai1 and TRPC1 are involved in platelet SOCE and aggregation (Galan et al. 2009). The reason for the discrepancy between Galan et al. and others regarding TRPC1 involvement in platelet SOCE is not clear. Interestingly, it has been reported that platelets have two different Ca^{2+} entry mechanisms, and STIM1-Orai1 contribute to only one of them (Gilio et al. 2010). Gilio et al. showed that either STIM1 or Orai1 knockdown was not sufficient to cause inhibition of prothrombinase activity (required for thrombus formation) in the presence of thrombin. They suggested this might be due to activation of other receptor-operated Ca^{2+} entry pathways by thrombin. The use of a non-specific Ca^{2+} entry blocker, SKF96395 (at 100 μM), resulted in an almost complete inhibition of prothrombinase activity in presence of thrombin. Gilio et al. (2010) hypothesized that these receptor-activated Ca^{2+} channels could be TRPC6 and/or Orai3 as both were found highly expressed in platelets and can be blocked by SKF96395. They however ruled out the significance of STIM2 in the thrombus formation by using STIM2-deficient platelets.

Although the involvement of TRPC1 in platelet SOCE remains a contentious issue, the role of STIM1 and Orai1 in mediating platelet SOCE and thrombosis is well appreciated (Authi 2009; Bergmeier and Stefanini 2009; Varga-Szabo et al. 2009; Feske 2010; Hagedorn et al. 2010). An important observation stemming from studies on $STIM1^{-/-}$ and $Orai1^{-/-}$ chimeras is that despite the fact that STIM1 and Orai1 seem to play a role in arterial thrombosis, pulmonary thromboembolism, and ischemic brain infarction, the bleeding time in $STIM1^{-/-}$ and $Orai1^{-/-}$ chimeras was only moderately prolonged (Varga-Szabo et al. 2008b). Furthermore, human patients with mutations in STIM1 and Orai1 and abrogated SOCE function did not show prolonged bleeding times compared to normal patients (McCarl et al. 2009; Picard et al. 2009). Recently, cyclophilin-A was identified as a critical regulator of STIM1 function in platelets (Elvers et al. 2012). Authors demonstrated that cyclophilin-A silencing reduced SOCE by regulating STIM1 phosphorylation. Cyclophilin-A was further shown to mediate platelet activation and arterial thrombosis in vivo (Elvers et al. 2012). Although most of the studies have implicated an important role for ER-resident STIM1 in thrombosis, a recent study has reported an intriguing contribution of plasma membrane STIM1 to platelet aggregation and in vitro thrombus formation (Ambily et al. 2014).

Collectively, these studies suggest that while STIM1 and Orai1 are critical for the pathological thrombus formation, their contribution to physiological blood clotting is perhaps not significant. Moreover, clinical data implicates enhanced STIM1 expression and function in platelets isolated from type 2 diabetic patients in comparison to healthy individuals. It is important to note that type 2 diabetic patients are highly prone to thrombosis, and increased STIM1 function could be one of the contributing factors (Xia et al. 2015). Recent reviews have focused on the involvement of STIM1 and Orai1 as well as the possible contribution of TRPCs to platelet function and thrombosis which are listed herein (Braun et al. 2011; Lang et al. 2013; Berna-Erro et al. 2016). We look forward to future research unraveling subtle differences between platelets and leukocytes in terms of STIM1/Orai1 molecular organization and interaction that could be exploited pharmacologically. In such case, specific targeting of platelet STIM1 or Orai1 could represent an attractive target for treatment of vascular occlusive diseases caused by excessive platelet activation and thrombosis. Given the ubiquitous nature of the SOCE pathway and the major role played by STIM1/Orai1 in the immune system, the caveat of specific targeting of STIM1/Orai1 is certainly true for all diseases discussed herein where various cell types are involved, as well as any other diseases in which STIM1/Orai1 might have role.

22.3.2 Atherosclerosis

Atherosclerosis is one of the most common vascular occlusive diseases. The hallmark of atherosclerosis is chronic inflammation of the arterial wall in response to deposition of fatty materials on the blood vessel wall causing narrowing of the lumen. Diet plays a big role in the development and progression of atherosclerosis. In addition to inflammatory cells, contribution of endothelial and vascular smooth muscle cells is also well established in the development of atherosclerotic lesions. The role of STIM1 and Orai1 in atherosclerosis is starting to emerge. It has been reported that in the mice model (apo$E^{-/-}$ mice) prone to atherosclerosis, SOCE in vascular smooth muscle cells is increased before development of atherosclerotic plaques (Van Assche et al. 2007). Similarly, it has been demonstrated that both SOCE and expression of STIM1 are significantly higher in coronary smooth muscle cells of pigs fed with pro-atherosclerotic high-calorie diet in comparison to pigs raised on standard chow diet (Edwards et al. 2010). Although the expression pattern of Orai1 also increased, it was not significant. These authors further showed that in contrast to pigs fed with a high-fat diet, their exercise-trained litter mates, on same diet, had smaller SOCE and decreased expression of STIM1 at mRNA and protein levels (Edwards et al. 2010). Furthermore, coronary atherosclerosis observed in pigs fed with a high-calorie diet was attenuated when these pigs were subjected to exercise. These studies suggest that high-fat western diet increases SOCE and STIM1 expression in coronary smooth cells, which correlates with increased atherosclerotic lesions. This increase in SOCE, STIM1 expression, and coronary atherosclerosis can be attenuated by exercise (Edwards et al. 2010). One of the critical steps in the development of atherosclerotic plaque is recruitment of monocytes and neutrophils by the endothelium at the site of lesions. The recruitment of these leukocytes has been reported to occur early during the development of other vascular obstructive diseases (Welt et al. 2000). Although role for STIM1/Orai1 in monocyte recruitment is not understood, a study demonstrated that Orai1 regulates neutrophils recruitment to the inflamed vascular endothelium (Schaff et al. 2010). These authors showed that Orai1 mediates SOCE in neutrophils and is involved in neutrophils arrest and recruitment under shear flow (Schaff et al. 2010). Both monocytes and neutrophils are recruited by the endothelium in shear flow at the site of inflammatory insults; these two cell types have distinct and specific functions. Liang et al. (2016) recently reported a critical role for Orai1 in foam cells formation during atherosclerotic plaque generation. Orai1 was demonstrated to mediate macrophage SOCE in response to acute oxidized low-density lipids (ox-LDL) application. The authors showed that the Orai1-mediated SOCE could lead to calcineurin-stimulated JNK and p38 kinase activation in these cells. JNK and p38 kinase activation in turn was implicated in enhancing the scavenger receptor A expression, increasing uptake of LDL and eventually leading to higher foam cell formation (Liang et al. 2016). Further, the pathophysiological relevance of Orai1 in atherosclerosis was studied in apo$E^{-/-}$ mice. The Orai1 expression was increased in the high-fat diet induced atherosclerotic lesions of apo$E^{-/-}$ mice. Moreover, adenoviral mediated Orai1 silencing significantly decreased the atherosclerotic lesions in apo$E^{-/-}$ mice. The authors attributed this reduction in the atherosclerotic lesions upon Orai1 silencing to less immune cell infiltration to the plaque site and diminished macrophage apoptosis in the lesions. Taken together, this study strongly implicated the critical role for Orai1 in atherosclerosis by regulating both vascular inflammation and foam cell formation (Liang et al. 2016). The dysregulation of endothelial barrier function is also one of the contributing factors to vascular inflammation and subsequent atherosclerotic plaque formation. Recent literature has suggested that STIM1 plays an important role in regulating endothelial barrier function (Shinde et al. 2013; Stolwijk et al. 2016) as well as progenitor endothelial cells functions in the atherosclerotic lesions (Wang et al. 2015). We reported that STIM1 is required for thrombin-induced loss in endothelial barrier function. Interestingly, STIM1 plays a critical role in G-protein coupled receptor-activated changes in endothelial barrier function independently of Ca^{2+} and Orai1 (Stolwijk et al. 2016). STIM1 rather couples to RhoA and induces stress fiber generation leading to loss in endothelial barrier (Shinde et al. 2013). In contrast to thrombin, high-mobility group box 1 protein (HMGB1)-induced endothelial permeability was reported to require both STIM1- and Orai1-mediated SOCE (Zou et al. 2015). Zou et al. (2015) implicated involvement of Src kinase in HMGB1-induced endothelial permeability. These studies suggest that STIM1 regulates endothelial permeability, and it could connect to different cellular signaling cascades depending upon the stimuli for modulating the endothelial barrier function. Wang et al. (2015) recently reported that dysfunction of progenitor endothelial cells in atherosclerotic plaques of apo$E^{-/-}$ mice correlates with decreased SOCE, reduction in VEGF-induced Ca^{2+} oscillations, and downregulation of STIM1 as well as Orai1 expression in these cells. Taken together, all these studies suggest that STIM1 and Orai1 play a key role in the initiation and progression of atherosclerotic plaques formation.

22.3.3 Restenosis

Restenosis can be defined as reoccurrence of blood vessel narrowing leading to restricted blood flow. In order to remove the blockage from blood vessels, more than 1 million percutaneous interventions are performed every year worldwide. Quite often, these percutaneous interventions result in blood vessel injury leading to restenosis. Although the use of stents has decreased the number of restenosis cases, the rate of restenosis is still very high (Fischman et al. 1994; Serruys et al. 1994; Elezi et al. 1998). The most important pathological processes contributing to restenosis are thrombosis (see section above) and neointima formation; the latter is largely mediated by smooth muscle proliferation and migration. The resting vascular smooth muscle cells are contractile in nature and do not proliferate. However, upon vascular insults such as mechanical injuries during percutaneous interventions, the quiescent smooth muscle cells undergo phenotypic switching into a synthetic proliferative migratory phenotype (Owens et al. 2004; Kawai-Kowase and Owens 2007; House et al. 2008). This phenotypic switching is coupled to downregulation of contractile proteins and concomitant upregulation of proteins involved in smooth muscle cell proliferation and migration. Similarly, increased proliferation of synthetic smooth muscle cells is correlated with the increase in SOCE in these cells (Golovina 1999; Golovina et al. 2001). Recently, we and others have demonstrated that STIM1 and Orai1 levels are higher in synthetic vascular smooth muscle cells in comparison to quiescent smooth muscle cells from the same bed (Berra-Romani et al. 2008; Potier et al. 2009; Bisaillon et al. 2010). STIM1 and Orai1 upregulation occurred in both synthetic aortic vascular smooth muscle cells (Potier et al. 2009; Bisaillon et al. 2010) as well as mesenteric smooth muscle cells (Berra-Romani et al. 2008). Augmented STIM1 and Orai1 expression correlated with increased SOCE in synthetic smooth muscle cells (Berra-Romani et al. 2008; Potier and Trebak 2008). Using pathologically relevant primary human coronary artery smooth muscle cells, Takahashi et al. (2007) demonstrated the involvement of STIM1 in mediating SOCE and proliferation. These authors also demonstrated that STIM1 plays an important role in phosphorylating the cAMP response element-binding (CREB) transcription factor. Therefore, CREB might serve as a link between STIM1 and smooth muscle cell proliferation since it is well established that CREB activation indirectly couples to cellular proliferation (Takahashi et al. 2007).

Studies conducted in our laboratory provided the first evidence that SOCE and I_{CRAC} in vascular smooth cells are mediated by Orai1, thus solving the long-standing question of the molecular architecture of SOCE in vascular smooth muscle cells (VSMCs) (Potier et al. 2009). We also demonstrated that TRPC1/4/6 and STIM2, Orai2, and Orai3 do not contribute to SOCE in primary aortic smooth muscle cells. We showed that STIM1 and Orai1 are involved in serum-mediated proliferation and migration of vascular smooth muscle cells (Potier et al. 2009). In a subsequent study, we investigated the role of platelet-derived growth factor (PDGF), a highly potent pro-migratory and pro-proliferative factor with an established role in the development of vascular diseases (Raines 2004), in the activation of Ca^{2+} entry pathways in primary aortic smooth muscle cells. We showed that PDGF-stimulated Ca^{2+} influx occurs through SOCE and is mediated by STIM1 and Orai1, independently of TRPC1/4/6, STIM2, Orai2, and Orai3 (Bisaillon et al. 2010). Knockdown of either STIM1 or Orai1 resulted in a decrease in PDGF-mediated smooth muscle cell migration, whereas STIM2, Orai2, and Orai3 did not have a significant contribution. In vivo studies using a balloon injury model in the rat demonstrated that STIM1 and Orai1 mRNA and protein expressions were increased (12- and 15-fold, respectively, for mRNA) in smooth muscle cells isolated from injured carotid arteries in comparison to those isolated from sham control arteries (Bisaillon et al. 2010). The recent literature suggests that a number of other pro-proliferative and pro-migratory stimuli activate SOCE in SMCs (Hopson et al. 2011; Guo et al. 2012; Rodriguez-Moyano et al. 2013). Hopson et al. reported that sphingosine-1-phosphate (S1P) activates STIM1-Orai1-mediated SOCE in VSMCs. The authors also showed that S1P induced SOCE as well as STIM1/Orai1 expression was higher in proliferative VSMCs in comparison to contractile VSMCs (Hopson et al. 2011). Similarly, angiotensin II was reported to activate SOCE constituted by STIM1 and Orai1 in VSMCs (Guo et al. 2012). Guo et al. showed that angiotensin II application induces increase in STIM1 and Orai1 expression in proliferative VSMCs. Their data suggested a role for STIM1 and Orai1 in angiotensin II activated VSMC proliferation. Importantly, the authors showed that in vivo silencing of STIM1 and Orai1 results in significant decrease in angiotensin II-stimulated increase in neointima formation post-vascular injury (Guo et al. 2012). The significance of STIM1 and Orai1 in urotensin-II induced VSMC proliferation has also been reported (Rodriguez-Moyano et al. 2013). The authors reported that urotensin-II mediates its vascular remodeling effects partially through SOCE and partly by trans-activating epidermal growth factor receptor (EGFR) signaling cascade (Rodriguez-Moyano et al. 2013).

Another group has reported a similar involvement of Orai1 in mediating SOCE and proliferation of human aortic smooth muscle cells (Baryshnikov et al. 2009). They also showed that Orai1 independently of Orai2 and Orai3 mediates SOCE in human aortic smooth muscle cells. Baryshnikov et al. further demonstrated that Orai1 is functionally associated with Na^{+}-Ca^{2+} exchanger (NCX) and plasma membrane Ca^{2+}ATPase (PMCA) as knockdown of Orai1 was shown to decrease the expression of NCX and PMCA. Interestingly, they reported that although NCX colocalizes with Orai1 in plasma membrane microdomains, it does not contribute to SOCE in human aortic smooth muscle cells. These authors also confirmed the contribution of Orai1 to human vascular smooth muscle proliferation (Baryshnikov et al. 2009). In vivo studies performed using rat restenosis models have clearly demonstrated that STIM1 plays an important role in neointima formation. Using the balloon injury model, Guo et al. (2009) and Aubart et al. (2009) independently and almost simultaneously reported that in vivo knockdown of STIM1 in injured vessels resulted in substantial decrease in neointima formation. Guo and coworkers showed that STIM1 mRNA and protein levels were increased in carotid arteries 7 and 14 days post balloon angioplasty. Moreover, in vivo STIM1 knockdown with adenoviral particles into injured carotid vessel regions resulted in decreased neointima formation 14 days post injury, while rescue of STIM1 with human STIM1 plasmid led to restoration of neointima to near control levels. These authors also demonstrated the importance of STIM1 in mediating rat aortic vascular smooth muscle cell proliferation and migration in vitro along with its contribution to SOCE. They also showed that STIM1 is required for cell cycle progression as STIM1 knockdown results in cell cycle arrest in G_0/G_1 phase. The G_0/G_1 phase cell cycle arrest was concomitant with increase in expression of p21, a cyclin-dependent kinase inhibitor, and a significant reduction in phosphorylation of retinoblastoma protein (Rb) (Guo et al. 2009).

Aubart et al. (2009) published similar results showing the involvement of STIM1 in neointima formation after rat carotid angioplasty. This group used primary human coronary artery smooth muscle cells (hCASMC) for their in vitro experiments and showed that STIM1 expression was upregulated in proliferative hCASMC in comparison to quiescent hCASMC. They further demonstrated that knockdown of STIM1 results in inhibition of serum and platelet-derived growth factor-induced proliferation of hCASMC. Interestingly, hCASMC proliferation was decreased to the same levels upon incubation of hCASMC with cyclosporine A, a classical calcineurin inhibitor. Calcineurin is a Ca^{2+}-/calmodulin-dependent phosphatase that dephosphorylates various proteins including the transcription factor, nuclear factor of activated T cells (NFAT). Therefore, these authors suggested that NFAT signaling might be involved in STIM1-mediated proliferation of hCASMC by showing that STIM1 knockdown prevents nuclear translocation and activity of NFAT. It is clearly established that NFAT dephosphorylation results in its nuclear translocation and activation of pro-proliferative genes. NFAT has been reported to be involved in neointima formation, and blockade of its nuclear translocation has been suggested as an important therapeutic strategy for controlling restenosis post angioplasty (Liu et al. 2005; Yu et al. 2006). Importantly, Aubart and coworkers also demonstrated involvement of STIM1 during in vivo neointima formation, as adenoviral particles expressing shSTIM1 were able to robustly decrease neointima formation, upon balloon angioplasty, in comparison to control viral vectors encoding shRNA against the fly luciferase.

The critical role of STIM1 in vascular remodeling was further confirmed in the knockout mice with STIM1 deletion in smooth muscle cells (SM-STIM1$^{-/-}$ mice) (Mancarella et al. 2013). Work from Gill lab reported that the in vitro PDGF-stimulated VSMC proliferation and carotid artery ligation-induced neointima formation were remarkably inhibited in case of SM-STIM1$^{-/-}$ mice in comparison to littermate control mice. The authors also reported growth defects in the intestinal and vascular smooth muscle tissues in SM-STIM1$^{-/-}$ mice suggesting that STIM1 plays an important role in both smooth muscle development and injury responses (Mancarella et al. 2013). More recent data from human patients also implicates STIM1 as a biomarker for restenosis (Li et al. 2015). It was reported that the increase in serum STIM1 levels, post angioplasty in human patients, strongly correlates with restenosis. This study suggested that changes in the STIM1 levels could be evaluated as an independent risk factor for restenosis (Li et al. 2015). Taken together, all these studies provide strong evidence in support of a critical role for STIM1 in restenosis.

Our laboratory was first to demonstrate that Orai1 is an important regulator of neointima formation post-vascular injury (Zhang et al. 2011). We observed Orai1 upregulation in response to vascular injury in two independent in vivo model systems. Vascular insults due to either balloon angioplasty in rats or carotid artery ligation in mice resulted in Orai1, STIM1, and calcium-calmodulin kinase IIdelta2 (CamKIIδ2) upregulation in the medial as well as neointimal sections of the vessels (Zhang et al. 2011). Using lentiviral based Orai1 knockdown strategy in rat balloon angioplasty model, we further showed that either Orai1 or STIM1 silencing in vivo dramatically reduced neointimal hyperplasia in comparison to sh-luciferase control. Orai1 knockdown in the injured vessels led to decrease in the proliferative capabilities of the VSMCs. Further, STIM1 and CamKIIδ2 levels were diminished in the vessels treated with shOrai1. We also reported that Orai1-mediated I_{CRAC} in VSMCs regulates NFAT activation and nuclear translocation suggesting that inhibition of neointima formation by shOrai1 could be at least partially explained by decreased NFAT activation (Zhang et al. 2011).

More recent work from our laboratory suggests that Orai3 also contributes to neointima formation in rats post balloon angioplasty (Gonzalez-Cobos et al. 2013). We reported that the Orai3 expression is increased during neointima formation, and in vivo Orai3 silencing inhibits neointimal remodeling (Gonzalez-Cobos et al. 2013). Interestingly, we observed that the pathophysiological agonist thrombin activates a Ca^{2+} influx channel constituted by Orai1–Orai3 heteromultimers in VSMCs. We found that thrombin-induced Ca^{2+} influx does not depend on intracellular Ca^{2+} stores; rather it was activated by intracellular Leukotriene C_4 (LTC_4). Our data suggested that thrombin receptor stimulation in VSMCs results in LTC_4 generation, and LTC_4 activates Orai1–Orai3 heteromultimeric channel (Gonzalez-Cobos et al. 2013). We termed these channels as LTC_4-regulated Ca^{2+} (LRC) channels and the current generated upon activation of these channels as "I_{LRC}." Similar channels and currents were earlier reported in HEK cells by Shuttleworth lab (Shuttleworth 2009, 2012; Thompson et al. 2013). These authors reported that arachidonic acid (AA) activates an Orai1–Orai3 heteromultimeric channel resulting in store-independent Ca^{2+} influx (Shuttleworth 2009, 2012; Thompson et al. 2013). These channels were coined as "ARC channels" and the current was called as "I_{ARC}." Interestingly, our recent work suggests that I_{ARC} and I_{LRC} are actually the manifestation of the same channels formed by the same populations of Orai1 and Orai3 and regulated by STIM1 (Zhang et al. 2013, 2014). We further evaluated the role of LTC_4 synthase enzyme in activating I_{LRC} and its contribution to neointima formation (Zhang et al. 2015). Our data suggested that the expression of LTC_4 synthase increases in the medial and neointimal VSMCs post balloon angioplasty, and LTC_4 synthase silencing significantly decreases neointima formation. Mechanistically, both LTC_4 synthase and Orai3 mediate VSMC migration but not proliferation by regulating AKT phosphorylation independent of NFAT signaling (Zhang et al. 2015). Orai channels therefore contribute to pathological restenosis through store-dependent "I_{CRAC}" as well as store-independent "I_{LRC}" coupling them to different VSMC functions.

The evidence gathered so far in smooth muscle cells from various vascular beds by several groups clearly suggests that quiescent smooth muscle have lower STIM1, Orai1, and Orai3 expression and have an equally small Ca^{2+}influx (Zhang and Trebak 2011; Ruhle and Trebak 2013; Spinelli and Trebak 2016). The switch of smooth muscle to a more proliferative and migratory phenotype is accompanied by the upregulation of STIM1, Orai1, and Orai3 expression and the resulting increase in Ca^{2+}influx (Gonzalez-Cobos et al. 2013; Motiani et al. 2013a; Spinelli and Trebak 2016). The increase in Orais-mediated Ca^{2+} flux upon vascular insults appears to be critical for connecting to downstream effectors, which results in an increase in the proliferative and migratory abilities of smooth muscle cells (Zhang and Trebak 2011; Ruhle and Trebak 2013; Spinelli and Trebak 2016). CREB, NFAT, CamKIIδ2, and AKT are few examples of such Ca^{2+}-regulated downstream effectors, and future studies are likely to identify additional players that could be used for targeted antiproliferative and anti-migratory therapy of smooth muscle in vascular occlusive diseases. In summary, the literature strongly suggests that STIM1 and Orai channels are attractive targets for managing vascular remodeling. Indeed, a recent study suggested that sirolimus, component of certain drug-eluting stents, inhibits VSMC proliferation by inhibiting Orai1-mediated SOCE and CREB activation (Konig et al. 2013).

22.3.4 Hypertension

Hypertension is a chronic pathological condition of the systemic circulation. It is one of the major risk factors for several diseases, including heart failure, vascular occlusive diseases, myocardial infarction, renal failure, stroke, dementia, and Alzheimer's disease (Kannel 1974; de Groot et al. 2009; Nagai et al. 2010). Arterial hypertension and increased vascular tone have been associated with elevated cytosolic Ca^{2+} levels and abnormal expression of Ca^{2+} handling proteins (Wellman et al. 2001; Kitazono et al. 2002; Goulopoulou and Webb 2014). An increase in the intracellular Ca^{2+} stores has been reported in hypertensive rats in comparison to normotensive rats (Cortes et al. 1997). Giachini et al. (2009a) observed an increased expression of STIM1 and Orai1 in hypertensive rats in comparison to wild-type rats. They showed that aortas obtained from hypertensive rats have elevated spontaneous tone and force generation in response to Ca^{2+} influx. Interestingly, they observed the treatment of aortal rings with relatively high concentration of CRAC channel blockers either 100 μM 2-APB or 100 μM gadolinium resulted in decrease in the spontaneous tone of the aorta in case of hypertensive rats but not in normotensive rats. Furthermore, 2-APB and gadolinium treatment decreased the force generation in the aortal rings obtained from both groups of rats and brought it to comparable levels. However, it should be noted that the concentrations of these inhibitors used in this study are higher than the standard concentrations used to block SOCE and are likely to affect other ion channels, Ca^{2+} release channels (e.g., IP_3R), and pumps. Indeed, 100 μM 2-APB can block Ca^{2+} release via IP_3 receptors, which in turn can result in SOCE inhibition. Therefore, caution should be used in the interpretation of these results. The same authors also demonstrated that antibodies targeting STIM1 and Orai1 inhibited aortic force generation, an important contributor to hypertension, suggesting a role for STIM1 and Orai1 in force generation (Giachini et al. 2009a). This group further hypothesized that STIM1 and Orai1 can be important players involved in sexual differences associated with hypertension severity and development (Giachini et al. 2009b, 2012). In a more recent study, Giachini et al. (2012) indeed demonstrated that in comparison to female stroke-prone spontaneously hypertensive rats, male hypertensive rats expressed higher STIM1 and Orai1 protein expression as well as higher activity. Using ovariectomized female rats, they further showed that female sex hormones are involved in keeping STIM1 and Orai1 levels under check in spontaneously hypertensive rats (Giachini et al. 2012). Interestingly, it is well established that the incidence rate of hypertension is much lower in premenopausal women in comparison to men of the same age group (Safar and Smulyan 2004; Hajjar et al. 2006). These gender-dependent differences in hypertension incidence are proposed, at least partially, to be due to differences in Ca^{2+} handling protein expression, activation, and molecular mechanisms involved in cellular Ca^{2+} handling (Devynck 2002; Loukotova et al. 2002). It was reported that force generation in aortic rings obtained from female hypertensive rats is significantly less than that obtained from male hypertensive rats. Interestingly, this group also demonstrated that CRAC channel blockers as well as antibodies against STIM1 and Orai1 could abolish the differences in spontaneous contractions of aortal rings obtained from male and female hypertensive rats. Furthermore, expression of STIM1 and Orai1 was reported to be higher in hypertensive male aorta in comparison to hypertensive female and normal aorta (Giachini et al. 2009b, 2012). These studies suggest that STIM1- and Orai1-mediated SOCE could contribute to the gender differences observed in the severity of hypertension. Further in-depth studies into the differences in STIM and Orai isoform expression in hypertensive men and women will assist in better understanding of their contribution to hypertension. It might be of importance to determine the mechanisms regulating STIM and Orai isoform expression in females and whether female sex hormones such as estrogen play a role in regulating STIM and Orai expression. In this context, it is important to note that we have shown that estrogen receptor-positive breast cancer cells use mainly STIM1 and Orai3 to mediate SOCE unlike estrogen receptor-negative cells, which predominantly use STIM1 and Orai1 (Motiani et al. 2010). Further, we demonstrated that Orai3 expression is indeed regulated by estrogen (Motiani et al. 2013b). Future studies aimed at understanding the exact mechanisms involved in STIM and Orai isoform regulation and functional contribution to hypertension will likely establish them as therapeutic targets for managing hypertension.

Studies have implicated an important role of STIM2 and Orai2 in pulmonary arterial hypertension (Song et al. 2011; Fernandez et al. 2015). Song et al. demonstrated that the expression of STIM2 is higher in pulmonary artery smooth muscle cells (PASMCs) isolated from idiopathic pulmonary arterial hypertensive (PAH) patients in comparison to healthy individuals. However, these authors reported that the expression of STIM1 was not altered in PAH patients (Song et al. 2011). Further, authors showed that STIM2 knockdown in PAH-PASMCs resulted in SOCE abrogation and reduction in their proliferation. However, neither STIM2 knockdown nor its overexpression in healthy PASMCs changed their proliferation and SOCE. In a more recent study, the same group showed that the STIM2 and Orai2 expression is elevated in proliferative PASMCs in comparison to contractile PASMCs (Fernandez et al. 2015). Orai2 and STIM2 were implicated in increased SOCE observed in proliferative PASMCs. Orai2 silencing was shown to reduce SOCE whereas STIM2 overexpression was shown to increase SOCE in proliferative PASMCs (Fernandez et al. 2015). Fernandez et al. further showed that contractile PASMCs have higher voltage-dependent Ca^{2+} entry while proliferative PASMCs have enhanced SOCE. The increased proliferative capabilities of PASMCs were attributed to enhanced SOCE in these cells (Fernandez et al. 2015). Collectively, these studies implicated an important role for STIM2 and Orai2 in the PAH.

The significance of STIM1 was recently also reported in hypoxia-induced PAH (Hou et al. 2013). Hou et al. reported that hypoxia induces PASMCs proliferation and cell cycle progression. STIM1 knockdown abrogated SOCE, decreased NFAT nuclear translocation, and reduced hypoxia-stimulated PASMC proliferation and cell cycle progression suggesting that STIM1 plays a critical role in hypoxic PAH (Hou et al. 2013). Using smooth muscle-specific STIM1 knockout mice (STIM1 $SMC^{-/-}$), our group recently demonstrated an essential role of STIM1 in hypertension (Kassan et al. 2016). We showed that STIM1 expression was enhanced in the heart and vessels in WT mice upon angiotensin II infusion at week 4, which coincided with hypertension and endothelial dysfunction, and angiotensin II-induced hypertension was markedly reduced in STIM1 $SMC^{-/-}$mice (Kassan et al. 2016). Endothelial-dependent relaxation to acetylcholine was partially protected in STIM1 $SMC^{-/-}$mice subjected to angiotensin II. We further showed that STIM1 $SMC^{-/-}$mice treated with angiotensin II are protected against enhanced ER stress, oxidative stress, and fibrosis (Kassan et al. 2016). Taken together, the literature suggests that STIM and Orai proteins are critical to both pulmonary and vascular hypertension (Zhang and Trebak 2011; Pulina et al. 2013; Spinelli and Trebak 2016). Several studies discussed above implicate a very strong correlation between hypertension and enhanced STIM and/or Orai protein expression. Therefore, future studies focused on understanding the molecular mechanisms driving increase in STIM and Orai expression during hypertension would result in identification of novel targets for hypertension management.

22.4 Concluding Remarks

As discussed, since STIM and Orai proteins were identified as the molecular components of SOCE, a number of studies from several groups have identified these proteins as important contributors to vascular disorders (summarized in Table 22.1). The literature suggests that in most vascular disorders, the expression of STIM1, Orai1, or Orai3 is increased, and this increased expression results in dysregulation of cellular Ca^{2+} homeostasis eventually leading to vascular pathologies. Therefore, it is important to understand how expression of these proteins is regulated under normal physiological settings and in disease conditions. Surprisingly, there have been very few studies focused on understanding the mechanisms by which expression of STIMs and Orais is regulated (Merlet et al. 2013; Motiani et al. 2013b), and more work is required in this direction. Future studies will likely unravel the cellular and molecular mechanisms by which these proteins contribute to vascular disease. The ability of vascular cells to signal through homomultimeric or heteromultimeric Orai channels in combination with other channel molecules such as transient receptor potential canonical (TRPC) channels likely serves to further increase the diversity of Ca^{2+} signaling pathways used for the control of specific physiological and pathophysiological functions. Understanding the multiplicity of the regulatory mechanisms affecting these various Ca^{2+} channels in several cell types and the downstream signaling pathways and

Table 22.1 Expression and role of STIM and Orai in vascular diseases

Disease condition	Expression		Physiological and pathophysiological relevance	References
	STIM1/2	Orai1/2/3		
Hypertension	S1 increased	O1 increased	Increase in spontaneous aortal tone and force generation in SHR rats, which were reversed with SOCE inhibitors and antibodies against STIM1 and Orai1	Giachini et al. (2009a)
Pulmonary hypertension	S2 increased S1 no change	O2 increased	S2 is necessary for increased SOCE and proliferation in PASMC from IPAH patients	Song et al. (2011)
Hypertension			S1 is an important candidate gene for exaggerated sympathetic response to stress in SHRSP	Ferdaus et al. (2014)
Hypertension			S1 and O1 play an important role in sexual differences based vascular responses to Ca^{2+} in SHRSP	Giachini et al. (2012)
Pulmonary hypertension	S2 increased	O2 increased	Upregulation of S2 and O2 play an important role in the phenotypic transition of PASMC by increasing SOCE	Fernandez et al. (2015)
Hypoxic pulmonary hypertension	S1 increased		S1 upregulation plays a critical role in hypoxia-induced PASMC proliferation via SOC, Ca(2+), and NFAT pathway	Hou et al. (2013)
Pulmonary hypertension	S1 increased	O1 increased	PDGF upregulates S1 and O1 and hence results in increased SOCE and activates Akt/mTOR in human PASMC	Ogawa et al. (2012)
Hypertension	S1 increased		Smooth muscle S1 plays an important role in hypertension and associated cardiac fibrosis	Kassan et al. (2016)
Restenosis	Increased	Increased	STIM1 regulates proliferation of human coronary artery smooth muscle cells via CREB activation	Takahashi et al. (2007)
			STIM1 and Orai1 are upregulated in proliferative mesenteric smooth muscle cells	Berra-Romani et al. (2008)
			STIM1 and Orai1 are upregulated in rat synthetic proliferative aortic smooth muscle cells	Potier et al. (2009)

(continued)

Table 22.1 (continued)

Disease condition	Expression		Physiological and pathophysiological relevance	References
	STIM1/2	Orai1/2/3		
			STIM1 and Orai1 mediate SOCE, CRAC currents, and serum activated aortic smooth muscle cell proliferation and migration	
			Orai1 interacts with NCX and contributes to human aortic smooth muscle cell proliferation	Baryshnikov et al. (2009)
			STIM1 and Orai1 contribute to PDGF-mediated vascular smooth muscle cell migration STIM1 and Orai1 are upregulated in neointima after balloon angioplasty	Bisaillon et al. (2010)
			STIM1 contributes to neointima formation by mediating cell cycle progression of vascular smooth muscle cells	Guo et al. (2009)
			STIM1 mediates development of neointima after rat carotid angioplasty via NFAT activation	Aubart et al. (2009)
Restenosis	S1 increased	ND	Higher level of S1 in plasma determines the risk in restenosis	Li et al. (2015)
Restenosis	S1 increased	O1 increased	O1 has an important role in VSMC proliferation, migration, and neointima formation	Zhang et al. (2011)
Restenosis		O3 increased	Expression of O3 is increased in neointimal VSMC and plays crucial role in neointimal hyperplasia	Gonzalez-Cobos et al. (2013)
Restenosis		O3 increased	O3 is involved in VSMC migration (but not proliferation) and neointima formation	Zhang et al. (2015)
Restenosis	S1 increased	O1 increased	S1 and O1 are important in Ag-II-induced SOCE essential for proliferation of VSMC	Guo et al. (2012)
Atherosclerosis	Increased	Increased, but not significantly	Development of coronary atherosclerosis in pigs Exercise attenuated increase in STIM1 expression and atherosclerotic lesions	Edwards et al. (2010)

(continued)

Table 22.1 (continued)

Disease condition	Expression		Physiological and pathophysiological relevance	References
	STIM1/2	Orai1/2/3		
Atherosclerosis	ND	ND	S1 regulates HMGB1-induced endothelial cell hyperpermeability via Src activation	Zou et al. (2015)
Atherosclerosis	S1 decreased	O1 decreased	EPCs dysfunction is linked to decreased SOCE via decreased expression of S1 and O1 in atherosclerotic mice	Wang et al. (2015)
Atherosclerosis	S1 increased	O1 increased	Increased S1and O1 expression leads to increased SOCE in proliferative VSMCs induced by S1P	Hopson et al. (2011)
Atherosclerosis			Orai1 promotes atherogenesis by enhancing foam cell formation and vascular inflammation	Liang et al. (2016)
Thrombosis	S1 increased	O1 increased	Increased expression of O1 regulates calcium signaling in platelets via SGK1	Lang et al. (2013)
Thrombosis			S1 and O1 mediate GPVI-induced SOCE, procoagulant activity, and thrombus formation	Gilio et al. (2010)
Thrombosis	ND	ND	STIM1 mediates platelet activation and formation of stable 3-dimensional thrombi	Grosse et al. (2007)
			STIM1 is required for platelet aggregation and in vivo thrombus formation in brain vasculature	Varga-Szabo et al. (2008b)
			Orai1 regulates platelet activation and in vivo thrombus formation	Braun et al. (2009)

transcription factors they couple to will likely bring us closer to the effective use of these molecules as therapeutic targets for vascular and other human diseases.

Acknowledgments Mohamed Trebak is supported by R01HL123364, R01HL097111, and R21AG050072 grants from the NIH, grant NPRP8-110-3-021 from the Qatar National Research Fund (GNRF), and grant 14GRNT18880008 from the AHA. Rajender K Motiani is an INSPIRE Faculty supported by the Department of Science and Technology, India. The authors have no conflict of interests to declare.

References

Ahmad F, Boulaftali Y, Greene TK, Ouellette TD, Poncz M, Feske S, Bergmeier W (2011) Relative contributions of stromal interaction molecule 1 and CalDAG-GEFI to calcium-dependent platelet activation and thrombosis. J Thromb Haemost 9(10):2077–2086

Alexopoulos N, Raggi P (2009) Calcification in atherosclerosis. Nat Rev Cardiol 6(11):681–688

Ambily A, Kaiser WJ, Pierro C, Chamberlain EV, Li Z, Jones CI, Kassouf N, Gibbins JM, Authi KS (2014) The role of plasma membrane STIM1 and Ca(2+)entry in platelet aggregation. STIM1 binds to novel proteins in human platelets. Cell Signal 26(3):502–511

Aubart FC, Sassi Y, Coulombe A, Mougenot N, Vrignaud C, Leprince P, Lechat P, Lompre AM, Hulot JS (2009) RNA interference targeting STIM1 suppresses vascular smooth muscle cell proliferation and neointima formation in the rat. Mol Ther 17(3):455–462

Authi KS (2009) Orai1: a channel to safer antithrombotic therapy. Blood 113(9):1872–1873

Baba Y, Hayashi K, Fujii Y, Mizushima A, Watarai H, Wakamori M, Numaga T, Mori Y, Iino M, Hikida M, Kurosaki T (2006) Coupling of STIM1 to store-operated Ca2+ entry through its constitutive and inducible movement in the endoplasmic reticulum. Proc Natl Acad Sci U S A 103(45):16704–16709

Bandyopadhyay BC, Pingle SC, Ahern GP (2011) Store-operated Ca(2)+ signaling in dendritic cells occurs independently of STIM1. J Leukoc Biol 89(1):57–62

Baryshnikov SG, Pulina MV, Zulian A, Linde CI, Golovina VA (2009) Orai1, a critical component of store-operated Ca2+ entry, is functionally associated with Na+/Ca2+ exchanger and plasma membrane Ca2+ pump in proliferating human arterial myocytes. Am J Physiol Cell Physiol 297(5):C1103–C1112

Bergmeier W, Stefanini L (2009) Novel molecules in calcium signaling in platelets. J Thromb Haemost 7(Suppl 1):187–190

Bergmeier W, Oh-Hora M, McCarl CA, Roden RC, Bray PF, Feske S (2009) R93W mutation in Orai1 causes impaired calcium influx in platelets. Blood 113(3):675–678

Berna-Erro A, Braun A, Kraft R, Kleinschnitz C, Schuhmann MK, Stegner D, Wultsch T, Eilers J, Meuth SG, Stoll G, Nieswandt B (2009) STIM2 regulates capacitive Ca2+ entry in neurons and plays a key role in hypoxic neuronal cell death. Sci Signal 2(93):ra67

Berna-Erro A, Jardin I, Smani T, Rosado JA (2016) Regulation of platelet function by Orai, STIM and TRP. Adv Exp Med Biol 898:157–181

Berra-Romani R, Mazzocco-Spezzia A, Pulina MV, Golovina VA (2008) Ca2+ handling is altered when arterial myocytes progress from a contractile to a proliferative phenotype in culture. Am J Physiol Cell Physiol 295(3):C779–C790

Berridge MJ (1987) Inositol trisphosphate and diacylglycerol: two interacting second messengers. Annu Rev Biochem 56:159–193

Berridge MJ, Irvine RF (1984) Inositol trisphosphate, a novel second messenger in cellular signal transduction. Nature 312(5992):315–321

Bisaillon JM, Motiani RK, Gonzalez-Cobos JC, Potier M, Halligan KE, Alzawahra WF, Barroso M, Singer HA, Jourd'heuil D, Trebak M (2010) Essential role for STIM1/Orai1-mediated calcium influx in PDGF-induced smooth muscle migration. Am J Physiol Cell Physiol 298(5):C993–C1005

Brandman O, Liou J, Park WS, Meyer T (2007) STIM2 is a feedback regulator that stabilizes basal cytosolic and endoplasmic reticulum Ca2+ levels. Cell 131(7):1327–1339

Braun A, Varga-Szabo D, Kleinschnitz C, Pleines I, Bender M, Austinat M, Bosl M, Stoll G, Nieswandt B (2009) Orai1 (CRACM1) is the platelet SOC channel and essential for pathological thrombus formation. Blood 113(9):2056–2063

Braun A, Vogtle T, Varga-Szabo D, Nieswandt B (2011) STIM and Orai in hemostasis and thrombosis. Front Biosci (Landmark Ed) 16:2144–2160

Cheng H, Lederer WJ (2008) Calcium sparks. Physiol Rev 88(4):1491–1545

Cortes SF, Lemos VS, Stoclet JC (1997) Alterations in calcium stores in aortic myocytes from spontaneously hypertensive rats. Hypertension 29(6):1322–1328 de Groot D, Pasterkamp G, Hoefer IE (2009) Cardiovascular risk factors and collateral artery formation. Eur J Clin Invest 39(12):1036–1047

DeHaven WI, Smyth JT, Boyles RR, Putney JW Jr (2007) Calcium inhibition and calcium potentiation of Orai1, Orai2, and Orai3 calcium release-activated calcium channels. J Biol Chem 282(24):17548–17556

Devynck MA (2002) Gender and vascular smooth muscle cells: a direct influence on Ca2+ handling? J Hypertens 20(11):2139–2140

Edwards JM, Neeb ZP, Alloosh MA, Long X, Bratz IN, Peller CR, Byrd JP, Kumar S, Obukhov AG, Sturek M (2010) Exercise training decreases store-operated Ca2+entry associated with metabolic syndrome and coronary atherosclerosis. Cardiovasc Res 85(3):631–640

Elezi S, Kastrati A, Neumann FJ, Hadamitzky M, Dirschinger J, Schomig A (1998) Vessel size and long-term outcome after coronary stent placement. Circulation 98(18):1875–1880

Elvers M, Herrmann A, Seizer P, Munzer P, Beck S, Schonberger T, Borst O, Martin-Romero FJ, Lang F, May AE, Gawaz M (2012) Intracellular cyclophilin A is an important Ca(2+) regulator in platelets and critically involved in arterial thrombus formation. Blood 120(6):1317–1326

Feng JM, YK H, Xie LH, Colwell CS, Shao XM, Sun XP, Chen B, Tang H, Campagnoni AT (2006) Golli protein negatively regulates store depletion-induced calcium influx in T cells. Immunity 24(6):717–727

Ferdaus MZ, Xiao B, Ohara H, Nemoto K, Harada Y, Saar K, Hubner N, Isomura M, Nabika T (2014) Identification of Stim1 as a candidate gene for exaggerated sympathetic response to stress in the stroke-prone spontaneously hypertensive rat. PLoS One 9(4):e95091

Fernandez RA, Wan J, Song S, Smith KA, Gu Y, Tauseef M, Tang H, Makino A, Mehta D, Yuan JX (2015) Upregulated expression of STIM2, TRPC6, and Orai2 contributes to the transition of pulmonary arterial smooth muscle cells from a contractile to proliferative phenotype. Am J Physiol Cell Physiol 308(8):C581–C593

Feske S (2010) CRAC channelopathies. Pflugers Arch 460(2):417–435

Feske S, Gwack Y, Prakriya M, Srikanth S, Puppel SH, Tanasa B, Hogan PG, Lewis RS, Daly M, Rao A (2006) A mutation in Orai1 causes immune deficiency by abrogating CRAC channel function. Nature 441(7090):179–185

Fischman DL, Leon MB, Baim DS, Schatz RA, Savage MP, Penn I, Detre K, Veltri L, Ricci D, Nobuyoshi M, Cleman M, Heuser R, Almond D, Teirstein PS, Fish RD, Colombo A, Brinker J, Moses J, Shaknovich A, Hirshfeld J, Bailey S, Ellis S, Rake R, Goldberg S (1994) A randomized comparison of coronary-stent placement and balloon angioplasty in the treatment of coronary artery disease. Stent Restenosis Study Investigators. N Engl J Med 331(8):496–501

Galan C, Zbidi H, Bartegi A, Salido GM, Rosado JA (2009) STIM1, Orai1 and hTRPC1 are important for thrombin- and ADP-induced aggregation in human platelets. Arch Biochem Biophys 490(2):137–144

Giachini FR, Chiao CW, Carneiro FS, Lima VV, Carneiro ZN, Dorrance AM, Tostes RC, Webb RC (2009a) Increased activation of stromal interaction molecule-1/Orai-1 in aorta from hypertensive rats: a novel insight into vascular dysfunction. Hypertension 53(2):409–416

Giachini FR, Webb RC, Tostes RC (2009b) STIM and Orai proteins: players in sexual differences in hypertension-associated vascular dysfunction? Clin Sci (Lond) 118(6):391–396

Giachini FR, Lima VV, Filgueira FP, Dorrance AM, Carvalho MH, Fortes ZB, Webb RC, Tostes RC (2012) STIM1/Orai1 contributes to sex differences in vascular responses to calcium in spontaneously hypertensive rats. Clin Sci (Lond) 122(5):215–226

Gilio K, van Kruchten R, Braun A, Berna-Erro A, Feijge MA, Stegner D, van der Meijden PE, Kuijpers MJ, Varga-Szabo D, Heemskerk JW, Nieswandt B (2010) Roles of platelet STIM1 and Orai1 in glycoprotein VI- and thrombin-dependent procoagulant activity and thrombus formation. J Biol Chem 285(31):23629–23638

Golovina VA (1999) Cell proliferation is associated with enhanced capacitative Ca(2+) entry in human arterial myocytes. Am J Physiol 277(2 Pt 1):C343–C349

Golovina VA, Platoshyn O, Bailey CL, Wang J, Limsuwan A, Sweeney M, Rubin LJ, Yuan JX (2001) Upregulated TRP and enhanced capacitative Ca(2+) entry in human pulmonary artery myocytes during proliferation. Am J Physiol Heart Circ Physiol 280(2):H746–H755

Gonzalez-Cobos JC, Zhang X, Zhang W, Ruhle B, Motiani RK, Schindl R, Muik M, Spinelli AM, Bisaillon JM, Shinde AV, Fahrner M, Singer HA, Matrougui K, Barroso M, Romanin C, Trebak M (2013) Store-independent Orai1/3 channels activated by intracrine leukotriene C4: role in neointimal hyperplasia. Circ Res 112(7):1013–1025

Goulopoulou S, Webb RC (2014) Symphony of vascular contraction: how smooth muscle cells lose harmony to signal increased vascular resistance in hypertension. Hypertension 63(3):e33–e39

Grosse J, Braun A, Varga-Szabo D, Beyersdorf N, Schneider B, Zeitlmann L, Hanke P, Schropp P, Muhlstedt S, Zorn C, Huber M, Schmittwolf C, Jagla W, Yu P, Kerkau T, Schulze H, Nehls M, Nieswandt B (2007) An EF hand mutation in Stim1 causes premature platelet activation and bleeding in mice. J Clin Invest 117(11):3540–3550

Gruszczynska-Biegala J, Kuznicki J (2013) Native STIM2 and ORAI1 proteins form a calcium-sensitive and thapsigargin-insensitive complex in cortical neurons. J Neurochem 126 (6):727–738

Guo RW, Wang H, Gao P, Li MQ, Zeng CY, Yu Y, Chen JF, Song MB, Shi YK, Huang L (2009) An essential role for stromal interaction molecule 1 in neointima formation following arterial injury. Cardiovasc Res 81(4):660–668

Guo RW, Yang LX, Li MQ, Pan XH, Liu B, Deng YL (2012) Stim1- and Orai1-mediated store-operated calcium entry is critical for angiotensin II-induced vascular smooth muscle cell proliferation. Cardiovasc Res 93(2):360–370

Hagedorn I, Vogtle T, Nieswandt B (2010) Arterial thrombus formation. Novel mechanisms and targets. Hamostaseologie 30(3):127–135

Hajjar I, Kotchen JM, Kotchen TA (2006) Hypertension: trends in prevalence, incidence, and control. Annu Rev Public Health 27:465–490

Hofmann T, Obukhov AG, Schaefer M, Harteneck C, Gudermann T, Schultz G (1999) Direct activation of human TRPC6 and TRPC3 channels by diacylglycerol. Nature 397 (6716):259–263

Hopson KP, Truelove J, Chun J, Wang Y, Waeber C (2011) S1P activates store-operated calcium entry via receptor- and non-receptor-mediated pathways in vascular smooth muscle cells. Am J Physiol Cell Physiol 300(4):C919–C926

Hoth M, Niemeyer BA (2013) The neglected CRAC proteins: Orai2, Orai3, and STIM2. Curr Top Membr 71:237–271

Hoth M, Penner R (1992) Depletion of intracellular calcium stores activates a calcium current in mast cells. Nature 355(6358):353–356

Hou X, Chen J, Luo Y, Liu F, Xu G, Gao Y (2013) Silencing of STIM1 attenuates hypoxia-induced PASMCs proliferation via inhibition of the SOC/Ca2+/NFAT pathway. Respir Res 14:2

House SJ, Potier M, Bisaillon J, Singer HA, Trebak M (2008) The non-excitable smooth muscle: calcium signaling and phenotypic switching during vascular disease. Pflugers Arch 456 (5):769–785

Jing J, He L, Sun A, Quintana A, Ding Y, Ma G, Tan P, Liang X, Zheng X, Chen L, Shi X, Zhang SL, Zhong L, Huang Y, Dong MQ, Walker CL, Hogan PG, Wang Y, Zhou Y (2015) Proteomic mapping of ER-PM junctions identifies STIMATE as a regulator of Ca(2)(+) influx. Nat Cell Biol 17(10):1339–1347

Kannel WB (1974) Role of blood pressure in cardiovascular morbidity and mortality. Prog Cardiovasc Dis 17(1):5–24

Kassan M, Ait-Aissa K, Radwan E, Mali V, Haddox S, Gabani M, Zhang W, Belmadani S, Irani K, Trebak M, Matrougui K (2016) Essential role of smooth muscle STIM1 in hypertension and cardiovascular dysfunction. Arterioscler Thromb Vasc Biol 36(9):1900–1909

Kawai-Kowase K, Owens GK (2007) Multiple repressor pathways contribute to phenotypic switching of vascular smooth muscle cells. Am J Physiol Cell Physiol 292(1):C59–C69

Kawasaki T, Lange I, Feske S (2009) A minimal regulatory domain in the C terminus of STIM1 binds to and activates ORAI1 CRAC channels. Biochem Biophys Res Commun 385(1):49–54

Kitazono T, Ago T, Kamouchi M, Santa N, Ooboshi H, Fujishima M, Ibayashi S (2002) Increased activity of calcium channels and Rho-associated kinase in the basilar artery during chronic hypertension in vivo. J Hypertens 20(5):879–884

Konig S, Browne S, Doleschal B, Schernthaner M, Poteser M, Machler H, Wittchow E, Braune M, Muik M, Romanin C, Groschner K (2013) Inhibition of Orai1-mediated Ca(2+) entry is a key mechanism of the antiproliferative action of sirolimus in human arterial smooth muscle. Am J Physiol Heart Circ Physiol 305(11):H1646–H1657

Kraft R (2015) STIM and ORAI proteins in the nervous system. Channels (Austin) 9(5):245–252

Lang F, Munzer P, Gawaz M, Borst O (2013) Regulation of STIM1/Orai1-dependent Ca2+ signalling in platelets. Thromb Haemost 110(5):925–930

Leung FP, Yung LM, Yao X, Laher I, Huang Y (2008) Store-operated calcium entry in vascular smooth muscle. Br J Pharmacol 153(5):846–857

Li H, Jiang Z, Liu X, Yang Z (2015) Higher plasma level of STIM1, OPG are correlated with stent restenosis after PCI. Int J Clin Exp Med 8(11):21089–21097

Liang SJ, Zeng DY, Mai XY, Shang JY, Wu QQ, Yuan JN, Yu BX, Zhou P, Zhang FR, Liu YY, Lv XF, Liu J, Ou JS, Qian JS, Zhou JG (2016) Inhibition of Orai1 store-operated calcium channel prevents foam cell formation and atherosclerosis. Arterioscler Thromb Vasc Biol 36 (4):618–628

Liou J, Kim ML, Heo WD, Jones JT, Myers JW, Ferrell JE Jr, Meyer T (2005) STIM is a Ca2+ sensor essential for Ca2+-store -depletion-triggered Ca2+ influx. Curr Biol 15(13):1235–1241

Liou J, Fivaz M, Inoue T, Meyer T (2007) Live-cell imaging reveals sequential oligomerization and local plasma membrane targeting of stromal interaction molecule 1 after Ca2+ store depletion. Proc Natl Acad Sci U S A 104(22):9301–9306

Liu Z, Zhang C, Dronadula N, Li Q, Rao GN (2005) Blockade of nuclear factor of activated T cells activation signaling suppresses balloon injury-induced neointima formation in a rat carotid artery model. J Biol Chem 280(15):14700–14708

Loukotova J, Kunes J, Zicha J (2002) Gender-dependent difference in cell calcium handling in VSMC isolated from SHR: the effect of angiotensin II. J Hypertens 20(11):2213–2219

Luik RM, Wang B, Prakriya M, Wu MM, Lewis RS (2008) Oligomerization of STIM1 couples ER calcium depletion to CRAC channel activation. Nature 454(7203):538–542

Mancarella S, Potireddy S, Wang Y, Gao H, Gandhirajan RK, Autieri M, Scalia R, Cheng Z, Wang H, Madesh M, Houser SR, Gill DL (2013) Targeted STIM deletion impairs calcium homeostasis, NFAT activation, and growth of smooth muscle. FASEB J 27(3):893–906

Mazzucato M, Pradella P, Cozzi MR, De Marco L, Ruggeri ZM (2002) Sequential cytoplasmic calcium signals in a 2-stage platelet activation process induced by the glycoprotein Ibalpha mechanoreceptor. Blood 100(8):2793–2800

McCarl CA, Picard C, Khalil S, Kawasaki T, Rother J, Papolos A, Kutok J, Hivroz C, Ledeist F, Plogmann K, Ehl S, Notheis G, Albert MH, Belohradsky BH, Kirschner J, Rao A, Fischer A, Feske S (2009) ORAI1 deficiency and lack of store-operated Ca2+ entry cause immunodeficiency, myopathy, and ectodermal dysplasia. J Allergy Clin Immunol 124(6):1311–1318. e1317

Mercer JC, Dehaven WI, Smyth JT, Wedel B, Boyles RR, Bird GS, Putney JW Jr (2006) Large store-operated calcium selective currents due to co-expression of Orai1 or Orai2 with the intracellular calcium sensor, Stim1. J Biol Chem 281(34):24979–24990

Merlet E, Atassi F, Motiani RK, Mougenot N, Jacquet A, Nadaud S, Capiod T, Trebak M, Lompre AM, Marchand A (2013) miR-424/322 regulates vascular smooth muscle cell phenotype and neointimal formation in the rat. Cardiovasc Res 98(3):458–468

Miao Y, Miner C, Zhang L, Hanson PI, Dani A, Vig M (2013) An essential and NSF independent role for alpha-SNAP in store-operated calcium entry. Elife 2:e00802

Michell RH (1975) Inositol phospholipids and cell surface receptor function. Biochim Biophys Acta 415(1):81–47

Mignen O, Thompson JL, Shuttleworth TJ (2008) Orai1 subunit stoichiometry of the mammalian CRAC channel pore. J Physiol 586(2):419–425

Moccia F, Zuccolo E, Soda T, Tanzi F, Guerra G, Mapelli L, Lodola F, D'Angelo E (2015) Stim and Orai proteins in neuronal Ca(2+) signaling and excitability. Front Cell Neurosci 9:153

Motiani RK, Abdullaev IF, Trebak M (2010) A novel native store-operated calcium channel encoded by Orai3: selective requirement of Orai3 versus Orai1 in estrogen receptor-positive versus estrogen receptor-negative breast cancer cells. J Biol Chem 285(25):19173–19183

Motiani RK, Stolwijk JA, Newton RL, Zhang X, Trebak M (2013a) Emerging roles of Orai3 in pathophysiology. Channels (Austin) 7(5):392–401

Motiani RK, Zhang X, Harmon KE, Keller RS, Matrougui K, Bennett JA, Trebak M (2013b) Orai3 is an estrogen receptor alpha-regulated Ca(2)(+) channel that promotes tumorigenesis. FASEB J 27(1):63–75

Muik M, Fahrner M, Derler I, Schindl R, Bergsmann J, Frischauf I, Groschner K, Romanin C (2009) A cytosolic homomerization and a modulatory domain within STIM1 C terminus determine coupling to ORAI1 channels. J Biol Chem 284(13):8421–8426

Nagai M, Hoshide S, Kario K (2010) Hypertension and dementia. Am J Hypertens 23(2):116–124

Ogawa A, Firth AL, Smith KA, Maliakal MV, Yuan JX (2012) PDGF enhances store-operated Ca2+ entry by upregulating STIM1/Orai1 via activation of Akt/mTOR in human pulmonary arterial smooth muscle cells. Am J Physiol Cell Physiol 302(2):C405–C411

Oka T, Hori M, Ozaki H (2005) Microtubule disruption suppresses allergic response through the inhibition of calcium influx in the mast cell degranulation pathway. J Immunol 174 (8):4584–4589

Oritani K, Kincade PW (1996) Identification of stromal cell products that interact with pre-B cells. J Cell Biol 134(3):771–782

Owens GK, Kumar MS, Wamhoff BR (2004) Molecular regulation of vascular smooth muscle cell differentiation in development and disease. Physiol Rev 84(3):767–801

Palty R, Raveh A, Kaminsky I, Meller R, Reuveny E (2012) SARAF inactivates the store operated calcium entry machinery to prevent excess calcium refilling. Cell 149(2):425–438

Parekh AB (2006) Cell biology: cracking the calcium entry code. Nature 441(7090):163–165

Parekh AB, Putney JW Jr (2005) Store-operated calcium channels. Physiol Rev 85(2):757–810

Park CY, Hoover PJ, Mullins FM, Bachhawat P, Covington ED, Raunser S, Walz T, Garcia KC, Dolmetsch RE, Lewis RS (2009) STIM1 clusters and activates CRAC channels via direct binding of a cytosolic domain to Orai1. Cell 136(5):876–890

Parker NJ, Begley CG, Smith PJ, Fox RM (1996) Molecular cloning of a novel human gene (D11S4896E) at chromosomal region 11p15.5. Genomics 37(2):253–256

Peinelt C, Vig M, Koomoa DL, Beck A, Nadler MJ, Koblan-Huberson M, Lis A, Fleig A, Penner R, Kinet JP (2006) Amplification of CRAC current by STIM1 and CRACM1 (Orai1). Nat Cell Biol 8(7):771–773

Penna A, Demuro A, Yeromin AV, Zhang SL, Safrina O, Parker I, Cahalan MD (2008) The CRAC channel consists of a tetramer formed by Stim-induced dimerization of Orai dimers. Nature 456 (7218):116–120

Picard C, McCarl CA, Papolos A, Khalil S, Luthy K, Hivroz C, LeDeist F, Rieux-Laucat F, Rechavi G, Rao A, Fischer A, Feske S (2009) STIM1 mutation associated with a syndrome of immunodeficiency and autoimmunity. N Engl J Med 360(19):1971–1980

Potier M, Trebak M (2008) New developments in the signaling mechanisms of the store-operated calcium entry pathway. Pflugers Arch 457(2):405–415

Potier M, Gonzalez JC, Motiani RK, Abdullaev IF, Bisaillon JM, Singer HA, Trebak M (2009) Evidence for STIM1- and Orai1-dependent store-operated calcium influx through ICRAC in vascular smooth muscle cells: role in proliferation and migration. FASEB J 23(8):2425–2437

Prakriya M, Lewis RS (2015) Store-operated calcium channels. Physiol Rev 95(4):1383–1436

Pulina MV, Zulian A, Baryshnikov SG, Linde CI, Karashima E, Hamlyn JM, Ferrari P, Blaustein MP, Golovina VA (2013) Cross talk between plasma membrane Na(+)/Ca (2+) exchanger-1 and TRPC/Orai-containing channels: key players in arterial hypertension. Adv Exp Med Biol 961:365–374

Putney JW Jr (1986) A model for receptor-regulated calcium entry. Cell Calcium 7(1):1–12

Putney JW Jr (1990) Capacitative calcium entry revisited. Cell Calcium 11(10):611–624

Raines EW (2004) PDGF and cardiovascular disease. Cytokine Growth Factor Rev 15(4):237–254

Rink TJ, Sage SO (1990) Calcium signaling in human platelets. Annu Rev Physiol 52:431–449

Rodriguez-Moyano M, Diaz I, Dionisio N, Zhang X, Avila-Medina J, Calderon-Sanchez E, Trebak M, Rosado JA, Ordonez A, Smani T (2013) Urotensin-II promotes vascular smooth muscle cell proliferation through store-operated calcium entry and EGFR transactivation. Cardiovasc Res 100(2):297–306

Roos J, DiGregorio PJ, Yeromin AV, Ohlsen K, Lioudyno M, Zhang S, Safrina O, Kozak JA, Wagner SL, Cahalan MD, Velicelebi G, Stauderman KA (2005) STIM1, an essential and conserved component of store-operated Ca2+ channel function. J Cell Biol 169(3):435–445

Ruhle B, Trebak M (2013) Emerging roles for native Orai Ca2+ channels in cardiovascular disease. Curr Top Membr 71:209–235

Sabbioni S, Barbanti-Brodano G, Croce CM, Negrini M (1997) GOK: a gene at 11p15 involved in rhabdomyosarcoma and rhabdoid tumor development. Cancer Res 57(20):4493–4497

Safar ME, Smulyan H (2004) Hypertension in women. Am J Hypertens 17(1):82–87

Schaff UY, Dixit N, Procyk E, Yamayoshi I, Tse T, Simon SI (2010) Orai1 regulates intracellular calcium, arrest, and shape polarization during neutrophil recruitment in shear flow. Blood 115 (3):657–666

Schuhmann MK, Stegner D, Berna-Erro A, Bittner S, Braun A, Kleinschnitz C, Stoll G, Wiendl H, Meuth SG, Nieswandt B (2010) Stromal interaction molecules 1 and 2 are key regulators of autoreactive T cell activation in murine autoimmune central nervous system inflammation. J Immunol 184(3):1536–1542

Serruys PW, de Jaegere P, Kiemeneij F, Macaya C, Rutsch W, Heyndrickx G, Emanuelsson H, Marco J, Legrand V, Materne P, Belardi J, Sigwart U, Colombo A, Jacques Goy J, van den Heuvel P, Delcan J, Morel M (1994) A comparison of balloon-expandable-stent implantation with balloon angioplasty in patients with coronary artery disease. Benestent Study Group. N Engl J Med 331(8):489–495

Sharma S, Quintana A, Findlay GM, Mettlen M, Baust B, Jain M, Nilsson R, Rao A, Hogan PG (2013) An siRNA screen for NFAT activation identifies septins as coordinators of store-operated Ca2+ entry. Nature 499(7457):238–242

Shinde AV, Motiani RK, Zhang X, Abdullaev IF, Adam AP, Gonzalez-Cobos JC, Zhang W, Matrougui K, Vincent PA, Trebak M (2013) STIM1 controls endothelial barrier function independently of Orai1 and Ca2+ entry. Sci Signal 6(267):ra18

Shuttleworth TJ (2009) Arachidonic acid, ARC channels, and Orai proteins. Cell Calcium 45 (6):602–610

Shuttleworth TJ (2012) STIM and Orai proteins and the non-capacitative ARC channels. Front Biosci (Landmark Ed) 17:847–860

Smyth JT, DeHaven WI, Bird GS, Putney JW Jr (2007) Role of the microtubule cytoskeleton in the function of the store-operated Ca2+ channel activator STIM1. J Cell Sci 120(Pt 21):3762–3771

Soboloff J, Spassova MA, Tang XD, Hewavitharana T, Xu W, Gill DL (2006) Orai1 and STIM reconstitute store-operated calcium channel function. J Biol Chem 281(30):20661–20665

Song MY, Makino A, Yuan JX (2011) STIM2 contributes to enhanced store-operated Ca entry in pulmonary artery smooth muscle cells from patients with idiopathic pulmonary arterial hypertension. Pulm Circ 1(1):84–94

Spinelli AM, Trebak M (2016) Orai channel-mediated Ca2+ signals in vascular and airway smooth muscle. Am J Physiol Cell Physiol 310(6):C402–C413

Srikanth S, Jung HJ, Kim KD, Souda P, Whitelegge J, Gwack Y (2010) A novel EF-hand protein, CRACR2A, is a cytosolic Ca2+ sensor that stabilizes CRAC channels in T cells. Nat Cell Biol 12(5):436–446

Stathopulos PB, Li GY, Plevin MJ, Ames JB, Ikura M (2006) Stored Ca2+ depletion-induced oligomerization of stromal interaction molecule 1 (STIM1) via the EF-SAM region: an initiation mechanism for capacitive Ca2+ entry. J Biol Chem 281(47):35855–35862

Stolwijk JA, Zhang X, Gueguinou M, Zhang W, Matrougui K, Renken C, Trebak M (2016) Calcium signaling is dispensable for receptor-regulation of endothelial barrier function. J Biol Chem 291(44):22894–22912

Takahashi Y, Watanabe H, Murakami M, Ono K, Munehisa Y, Koyama T, Nobori K, Iijima T, Ito H (2007) Functional role of stromal interaction molecule 1 (STIM1) in vascular smooth muscle cells. Biochem Biophys Res Commun 361(4):934–940

Takemura H, Hughes AR, Thastrup O, Putney JW Jr (1989) Activation of calcium entry by the tumor promoter thapsigargin in parotid acinar cells. Evidence that an intracellular calcium pool and not an inositol phosphate regulates calcium fluxes at the plasma membrane. J Biol Chem 264(21):12266–12271

Thompson JL, Mignen O, Shuttleworth TJ (2013) The ARC channel—an endogenous store-independent Orai channel. Curr Top Membr 71:125–148

Trebak M, St J Bird G, McKay RR, Birnbaumer L, Putney JW Jr (2003a) Signaling mechanism for receptor-activated canonical transient receptor potential 3 (TRPC3) channels. J Biol Chem 278 (18):16244–16252

Trebak M, Vazquez G, Bird GS, Putney JW Jr (2003b) The TRPC3/6/7 subfamily of cation channels. Cell Calcium 33(5-6):451–461

Van Assche T, Fransen P, Guns PJ, Herman AG, Bult H (2007) Altered Ca2+ handling of smooth muscle cells in aorta of apolipoprotein E-deficient mice before development of atherosclerotic lesions. Cell Calcium 41(3):295–302

Varga-Szabo D, Authi KS, Braun A, Bender M, Ambily A, Hassock SR, Gudermann T, Dietrich A, Nieswandt B (2008a) Store-operated Ca(2+) entry in platelets occurs independently of transient receptor potential (TRP) C1. Pflugers Arch 457(2):377–387

Varga-Szabo D, Braun A, Kleinschnitz C, Bender M, Pleines I, Pham M, Renne T, Stoll G, Nieswandt B (2008b) The calcium sensor STIM1 is an essential mediator of arterial thrombosis and ischemic brain infarction. J Exp Med 205(7):1583–1591

Varga-Szabo D, Braun A, Nieswandt B (2009) Calcium signaling in platelets. J Thromb Haemost 7(7):1057–1066

Venkiteswaran G, Hasan G (2009) Intracellular Ca2+ signaling and store-operated Ca2+ entry are required in Drosophila neurons for flight. Proc Natl Acad Sci U S A 106(25):10326–10331

Vig M, Peinelt C, Beck A, Koomoa DL, Rabah D, Koblan-Huberson M, Kraft S, Turner H, Fleig A, Penner R, Kinet JP (2006) CRACM1 is a plasma membrane protein essential for store-operated Ca2+ entry. Science 312(5777):1220–1223

Wang LY, Zhang JH, Yu J, Yang J, Deng MY, Kang HL, Huang L (2015) Reduction of store-operated Ca(2+) entry correlates with endothelial progenitor cell dysfunction in atherosclerotic mice. Stem Cells Dev 24(13):1582–1590

Wehrens XH, Marks AR (2004) Novel therapeutic approaches for heart failure by normalizing calcium cycling. Nat Rev Drug Discov 3(7):565–573

Wellman GC, Cartin L, Eckman DM, Stevenson AS, Saundry CM, Lederer WJ, Nelson MT (2001) Membrane depolarization, elevated Ca(2+) entry, and gene expression in cerebral arteries of hypertensive rats. Am J Physiol Heart Circ Physiol 281(6):H2559–H2567

Welt FG, Edelman ER, Simon DI, Rogers C (2000) Neutrophil, not macrophage, infiltration precedes neointimal thickening in balloon-injured arteries. Arterioscler Thromb Vasc Biol 20(12):2553–2558

Xia W, Li Y, Wang B, Chen J, Wang X, Sun Q, Sun F, Li Z, Zhao Z (2015) Enhanced store-operated calcium entry in platelets is associated with peripheral artery disease in type 2 diabetes. Cell Physiol Biochem 37(5):1945–1955

Yeromin AV, Zhang SL, Jiang W, Yu Y, Safrina O, Cahalan MD (2006) Molecular identification of the CRAC channel by altered ion selectivity in a mutant of Orai. Nature 443(7108):226–229

Yu H, Sliedregt-Bol K, Overkleeft H, van der Marel GA, van Berkel TJ, Biessen EA (2006) Therapeutic potential of a synthetic peptide inhibitor of nuclear factor of activated T cells as antirestenotic agent. Arterioscler Thromb Vasc Biol 26(7):1531–1537

Yuan JP, Zeng W, Dorwart MR, Choi YJ, Worley PF, Muallem S (2009) SOAR and the polybasic STIM1 domains gate and regulate Orai channels. Nat Cell Biol 11(3):337–343

Zhang W, Trebak M (2011) STIM1 and Orai1: novel targets for vascular diseases? Sci China Life Sci 54(8):780–785

Zhang SL, Yeromin AV, Zhang XH, Yu Y, Safrina O, Penna A, Roos J, Stauderman KA, Cahalan MD (2006) Genome-wide RNAi screen of Ca(2+) influx identifies genes that regulate Ca(2+) release-activated Ca(2+) channel activity. Proc Natl Acad Sci U S A 103(24):9357–9362

Zhang W, Halligan KE, Zhang X, Bisaillon JM, Gonzalez-Cobos JC, Motiani RK, Hu G, Vincent PA, Zhou J, Barroso M, Singer HA, Matrougui K, Trebak M (2011) Orai1-mediated I (CRAC) is essential for neointima formation after vascular injury. Circ Res 109(5):534–542

Zhang X, Gonzalez-Cobos JC, Schindl R, Muik M, Ruhle B, Motiani RK, Bisaillon JM, Zhang W, Fahrner M, Barroso M, Matrougui K, Romanin C, Trebak M (2013) Mechanisms of STIM1 activation of store-independent leukotriene C4-regulated Ca2+ channels. Mol Cell Biol 33 (18):3715–3723

Zhang X, Zhang W, Gonzalez-Cobos JC, Jardin I, Romanin C, Matrougui K, Trebak M (2014) Complex role of STIM1 in the activation of store-independent Orai1/3 channels. J Gen Physiol 143(3):345–359

Zhang W, Zhang X, Gonzalez-Cobos JC, Stolwijk JA, Matrougui K, Trebak M (2015) Leukotriene-C4 synthase, a critical enzyme in the activation of store-independent Orai1/Orai3 channels, is required for neointimal hyperplasia. J Biol Chem 290(8):5015–5027

Zhou Y, Meraner P, Kwon HT, Machnes D, Oh-hora M, Zimmer J, Huang Y, Stura A, Rao A, Hogan PG (2010) STIM1 gates the store-operated calcium channel ORAI1 in vitro. Nat Struct Mol Biol 17(1):112–116

Zou M, Dong H, Meng X, Cai C, Li C, Cai S, Xue Y (2015) Store-operated Ca2+ entry plays a role in HMGB1-induced vascular endothelial cell hyperpermeability. PLoS One 10(4):e0123432

Cardiovascular and Hemostatic Disorders: SOCE and Ca^{2+} Handling in Platelet Dysfunction

23

Jose J. Lopez, Gines M. Salido, and Juan A. Rosado

Abstract

Among the Ca^{2+} entry mechanisms in platelets, store-operated Ca^{2+} entry (SOCE) plays a prominent role as it is necessary to achieve full activation of platelet functions and replenish intracellular Ca^{2+} stores. In platelets, as in other non-excitable cells, SOCE has been reported to involve the activation of plasma membrane channels by the ER Ca^{2+} sensor STIM1. Despite electrophysiological studies are not possible in human platelets, indirect analyses have revealed that the Ca^{2+}-permeable channels involve Orai1 and, most likely, TRPC1 subunits. A relevant role for the latter has not been found in mouse platelets. There is a body of evidence revealing a number of abnormalities in SOCE or in its molecular regulators that result in qualitative platelet disorders and, as a consequence, altered platelet responsiveness upon stimulation with multiple physiological agonists. Platelet SOCE abnormalities include STIM1 and Orai1 mutations. This chapter summarizes the current knowledge in this field, as well as the disorders associated to platelet SOCE dysfunction.

Keywords

Platelets • STIM1 • STIM2 • Orai1 • TRPC1 • SARAF • Homer1

23.1 Introduction

Platelets are anucleated cell fragments, of approximately 0.5–3.0 μm diameter, discoid, and irregularly shaped that play an essential role in hemostasis. Platelets, also called as thrombocytes, form clots, along with other plasma factors that prevent

J.J. Lopez • G.M. Salido • J.A. Rosado (✉)
Cell Physiology Research Group, Department of Physiology, University of Extremadura, Cáceres, Spain
e-mail: jarosado@unex.es

K. Groschner et al. (eds.), *Store-Operated Ca^{2+} Entry (SOCE) Pathways*, Advances in Experimental Medicine and Biology 993, DOI 10.1007/978-3-319-57732-6_23

hemorrhage. Platelets were identified by Bizzonero in 1882 (Bizzonero 1882), and the same year thrombocytopenia, a disorder characterized by a relative low platelet concentration, was reported as a cause of impaired hemostasis (Hayem 1882).

Platelets derive from larger bone marrow cells called megakaryocytes. Intravital two-photon microscopy of the bone marrow sinusoids has confirmed the generation of protrusions of megakaryocytes, known as proplatelets, that extend into the bone marrow sinusoids where cytoplasmic fragments, called preplatelets, are release and further mature into platelets within the circulation (Junt et al. 2007). More recently, an alternative mechanism for platelet generation has been described by Nishimura and coworkers. This mechanism, which might be induced by interleukin-1α and occurs independently of the megakaryopoiesis modulator, thrombopoietin, may restore circulating platelets during situations of increased platelet consumption (Nishimura et al. 2015). Despite platelets lack nucleus, they contain most intracellular organelles and subcellular structures. Platelets possess an internal membrane structure, the open canalicular system, that is continuous with, and part of, the plasma membrane and might facilitate the interaction of the plasma membrane with internal organelles. In addition, platelets exhibit a dynamic cytoskeleton, consisting of tubulin microtubules and actin filaments, which support the discoid shape of resting platelets and provide the contractile mechanism necessary for platelet function: shape changes, pseudopod formation, internal contraction, and secretion of different regulatory factors (Rosado and Sage 2000a; Fox 2001). Platelets possess a membranous system known as the dense tubular system (DTS), the analogue of the endoplasmic reticulum (ER) in other cells, which represents the major intracellular Ca^{2+} store, a reduced number of mitochondria, lysosomes, and Golgi apparatus (Ebbeling et al. 1992; Lu et al. 2013). Ultrastructural studies have revealed that platelets possess two types of secretory granules: the δ-granules and the α-granules. δ-granules store serotonin, Ca^{2+}, ATP, ADP, and pyrophosphate (Dell'Angelica et al. 2000). α-granules contain platelet factor IV, von Willebrand factor, thrombospondin, and fibrinogen. In addition, α-granules might store immunoglobulins taken from the extracellular medium by endocytosis (Harrison and Cramer 1993). Secretion of δ- and α-granules is a critical event in hemostasia. Platelet granule exocytosis has been found to be modulated by the cytosolic Ca^{2+} concentration in different ways: (1) modulation of Ca^{2+}-dependent PKC isoforms (Yoshioka et al. 2001), (2) activation of scinderin and other Ca^{2+}-dependent actin filament regulatory proteins (Rodriguez Del Castillo et al. 1992; Marcu et al. 1996), (3) induction of dense core granule secretion by activation of Munc13-4 (Shirakawa et al. 2004; Ren et al. 2010), and (4) activation of dense granule exocytosis by Rap1GAP2/synaptotagmin-like protein 1 and Rab8/synaptotagmin-like protein 4 (Neumuller et al. 2009; Hampson et al. 2013). We have recently reported that thrombin-stimulated granule secretion depends on Ca^{2+} mobilization from intracellular stores and only δ-granule secretion is partially dependent on Ca^{2+} influx from the extracellular medium through TRPC6 channels. Analyses of the kinetics of δ- and α-granule exocytosis have revealed that platelet stimulation with thrombin induces rapid release of α-granules which precedes δ-granules secretion (Lopez et al. 2015).

Cytosolic Ca^{2+} is a major signaling element in platelets. Changes in cytosolic Ca^{2+} concentration ($[Ca^{2+}]_i$) modulate a variety of functions in most cellular models, including platelets, such as granule secretion and aggregation (Rosado and Sage 2000c). In platelets, agonists increase $[Ca^{2+}]_i$ by stimulation of Ca^{2+} release from the intracellular stores, facilitation of Ca^{2+} entry from the extracellular stores, or both. The major intracellular Ca^{2+} store in platelets is the DTS. Ca^{2+} has been reported to be stored in the DTS by the sarco-/endoplasmic reticulum Ca^{2+}-ATPase (SERCA) isoform 2b, which is highly sensitive to thapsigargin but insensitive to 2,5-di-(tert-butyl)-1,4-benzohydroquinone (TBHQ) (Papp et al. 1992; Cavallini et al. 1995). In addition, a second agonist-releasable Ca^{2+} store has been described in human platelets on the base of its sensitivity to TBHQ and high concentrations of thapsigargin (Cavallini et al. 1995). This Ca^{2+} compartment, located in lysosomal-like acidic organelles (López et al. 2005), has been reported to be refilled by SERCA3 isoforms (Bobe et al. 1994). Agonists, such as ADP, release Ca^{2+} selectively from the DTS (Lopez et al. 2006a), while strong agonists like thrombin induce Ca^{2+} release from the DTS and the acidic stores through activation of protease-activated receptor (PAR)-1 and PAR-4. In addition, thrombin is able to induce selective Ca^{2+} release from the acidic stores through the activation of the glycoprotein GPIb-IX-V (Jardin et al. 2007a). The acidic store plays a relevant role in thrombin-induced platelet aggregation (Ben Amor et al. 2009).

Physiological agonists are able to activate both platelet aggregation and granule secretion by releasing the finite amount of Ca^{2+} accumulated in the intracellular stores; nevertheless, Ca^{2+} entry through plasma membrane channels is necessary to achieve full activation of platelet functions (Jardin et al. 2007b; Braun et al. 2009; Lopez et al. 2015). Different Ca^{2+} influx pathways have been described in platelets, including receptor-operated Ca^{2+} entry (ROCE) through $P2_{X1}$ receptors (MacKenzie et al. 1996; Vial et al. 2003), second messenger-operated Ca^{2+} entry (SMOCE) activated by PKC and Src tyrosine kinases (Rosado and Sage 2000d; Harper and Sage 2010), and store-operated Ca^{2+} entry (SOCE), a process controlled by the filling state of the intracellular Ca^{2+} stores (Putney 1986; Desai et al. 2015), that is a major mechanism for Ca^{2+} entry in these cells.

23.2 SOCE in Platelets

SOCE was identified three decades ago by James Putney (1986) as a mechanism for Ca^{2+} entry activated by discharge of the intracellular Ca^{2+} stores. The first demonstration of SOCE in platelets was reported by Sage and coworkers in 1989 (Sage et al. 1989). In this study, quenching of fura-2 fluorescence at the excitation wavelength of 360 nm (the isosbestic point) by Mn^{2+} was used as a marker for Ca^{2+} entry in human platelets, avoiding interference of the changes in $[Ca^{2+}]_i$ with the signal. Using this maneuver, the authors reported robust Mn^{2+} influx upon stimulation with thrombin after discharge of the internal Ca^{2+} stores. After this first description, the identification of the molecular components of SOCE in platelets has occurred in parallel with the investigations in other cell types, as SOCE in platelets share a number of features with other non-excitable cells.

SOCE is a mechanism by which discharge of receptor-releasable intracellular Ca^{2+} pools leads to the activation of a Ca^{2+} entry through plasma membrane channels. Ca^{2+} entering the cytosol then refills the intracellular Ca^{2+} compartments leading to the termination of the Ca^{2+} influx. Therefore, the simplest model to account for the activation of SOCE should include at least two components: a Ca^{2+} sensor in the lumen of the intracellular stores and a Ca^{2+}-permeable channel in the plasma membrane. The identification of these elements focused extensive investigation on this field until the characterization in 2005 of STIM1 (*ST*romal *I*nteraction *M*olecule 1) as the ER Ca^{2+} sensor (Roos et al. 2005; Zhang et al. 2005) and the identification of Orai1 as the pore-forming subunit of the CRAC (Ca^{2+} release-activated Ca^{2+}) channel in 2006 (Feske et al. 2006; Mercer et al. 2006; Peinelt et al. 2006; Prakriya et al. 2006; Soboloff et al. 2006).

As in other cells, in human platelets, the Ca^{2+} sensor of the intracellular stores was identified as the protein STIM1 (Lopez et al. 2006b). As mentioned in previous chapters, STIM1 is a Ca^{2+}-binding protein with a putative single transmembrane domain that might be located both in intracellular membranes, including the ER (Luik et al. 2006; Wu et al. 2006) and acidic Ca^{2+} stores (Zbidi et al. 2011), and in the plasma membrane (Manji et al. 2000; Lopez et al. 2006b; Spassova et al. 2006; Mignen et al. 2007). STIM1 contains different functional domains (the reader is referred to previous chapters of the book for a detailed description of STIM1 structure). Briefly, the STIM N-terminal region is located in the luminal compartment of the Ca^{2+} stores (or the extracellular medium if located in the PM) and contains the canonical, Ca^{2+}-binding, EF-hand motif (Zhang et al. 2005), as well as a "hidden," EF-hand domain, which does not bind Ca^{2+}, and the sterile-α-motif (SAM) domain involved in protein-protein interaction. The STIM1 cytosolic region comprises three conserved coiled-coil domains (CC), named CC1, CC2, and CC3; the CRAC modulatory domain (Derler et al. 2009), which includes the STIM1 homomerization domain (Yu et al. 2011); the C-terminal inhibitory domain (that regulates the interaction of STIM1 with the regulatory protein SARAF (Jha et al. 2013)); a serine-/proline-rich region; and a lysine-rich region at the end of the C-terminus, which binds to membrane phospholipids and interact with target channels (Huang et al. 2006; Jardin et al. 2013) and might be involved in enlargement of ER-plasma membrane clusters (Sauc et al. 2015). We have recently found that upon store depletion, SARAF transiently dissociates from STIM1 and interacts with the C-terminal region of Orai1, a mechanism that might be involved in the activation of SOCE (Albarran et al. 2016a). SARAF acts as a modulator of STIM1-dependent channel proteins, including the store-operated Orai1 and TRPC1 subunits (Albarran et al. 2016a, b) and the arachidonate-regulated channels (ARC; Albarran et al. 2016c), which are regulated by plasma membrane resident SARAF, whose expression in the plasma membrane is controlled by STIM1 (Albarran et al. 2016d). The CC2 and CC3 domains encompass the STIM1-Orai1-activating region (SOAR, amino acid residues 344–442) (Yuan et al. 2009), also known as OASF (Orai-activating small fragment; amino acids

233–450/474) (Muik et al. 2009), CAD (CRAC-activating domain; amino acids 342–448) (Park et al. 2009), and CCb9 (amino acids 339–444) (Kawasaki et al. 2009).

The characterization of STIM1 as the Ca^{2+} sensor in the intracellular stores allowed the identification of the Ca^{2+}-permeable channels in the plasma membrane. Electrophysiological analysis of store-mediated currents revealed the presence of two different currents: a highly Ca^{2+}-selective current (I_{CRAC}) and a heterogeneous set of nonselective currents called store-operated currents (I_{SOC}). I_{CRAC} is a non-voltage-operated current that shows a large current amplitude at negative potentials and approaches the zero current level at very positive potentials (Hoth and Penner 1992). The CRAC channels show a single channel conductance <1 pS and are highly selective for Ca^{2+} over monovalent cations, although the CRAC channels might lose their selectivity in divalent-free solutions, a situation that allows Na^+ to permeate the channels (Bakowski and Parekh 2002; Prakriya and Lewis 2002). On the other hand, the I_{SOC} currents are most likely mediated by a family of cation-permeable channels (SOC channels) that exhibit a greater conductance than CRAC channels. I_{SOC} currents have been described in a number of cell types, including vascular endothelial cells (Vaca and Kunze 1994; Fasolato and Nilius 1998), human A431 carcinoma cells (Luckhoff and Clapham 1994; Kiselyov et al. 1999), smooth muscle cells (Golovina et al. 2001; Trepakova et al. 2001), or liver cells (To et al. 2010), among others. Store-operated currents have not been recorded in platelets due to the limitation of these cells to perform electrophysiological studies; however, I_{CRAC} has been recorded in its precursor, megakaryocytes (Tolhurst et al. 2008).

The protein Orai1 has been proposed as the pore-forming subunit of the CRAC channel mediating I_{CRAC} (Feske et al. 2006; Huang et al. 2006; Luik et al. 2006), and Orai1, together with the mammalian homolog of the *Drosophila* transient receptor potential (TRP) channels, TRPC1, has been reported to form SOC channels (Desai et al. 2015; Berna-Erro et al. 2016; Ambudkar et al. 2017).

As mentioned in previous chapters of this book, the role of Orai1 in I_{CRAC} was identified by gene mapping in patients with the severe combined immune deficiency (SCID) syndrome, attributed to loss of I_{CRAC}, and whole-genome screen of *Drosophila* S2 cells (Feske et al. 2006; Vig et al. 2006; Yeromin et al. 2006). The Orai family consists of three conserved homologs: Orai1, Orai2, and Orai3 (Feske et al. 2005; Mercer et al. 2006; Zhang et al. 2006; Gwack et al. 2007; Rothberg et al. 2013). Orai1 is a protein with four putative transmembrane domains, with the N- and C-terminal domains facing the cytosol and exposing two loops (1 and 3) to the extracellular medium and loop (2) located intracellularly. Both N- and C-termini are required for the interaction with and regulation by STIM1 (Park et al. 2009; Yuan et al. 2009; Derler et al. 2013; Palty and Isacoff 2015; Palty et al. 2015). Crystallization of *Drosophila* Orai1 revealed that the CRAC channel is a hexameric complex permeable to Ca^{2+} as well as to monovalent ions in the presence of divalent cations (Hou et al. 2012). The architecture of Orai1-/TRPC1-forming SOC channels still remains unresolved. In addition to the store-operated, STIM1-activated CRAC and SOC channels, Orai1 participates in the formation of heteropentameric (3 Orai1 and 2 Orai3) Ca^{2+} channels regulated by arachidonate, a store-independent channel also regulated by STIM1 located in the plasma membrane (Thompson and Shuttleworth 2013; Zhang et al. 2014). Orai1 and STIM1 expression is mediated by the nuclear factor kappa B (NF-κB), which, in turn, is upregulated in megakaryocytes by the serum- and glucocorticoid-inducible kinase 1 (SGK1), an important process for Orai1 expression and SOCE in platelets (Lang et al. 2013). NF-κB activity is downregulated by 1,25$(OH)_2$ vitamin D3, whose formation is inhibited by klotho; thus klotho-deficient mice exhibit a reduced STIM1/Orai1 expression at the transcript and protein level, as well as attenuated platelet Ca^{2+} signaling and activation (Borst et al. 2014).

The canonical TRP (TRPC) channels were earlier identified in mammalian cells as cation-permeable channels that mediate receptor-stimulated Ca^{2+} influx (Wes et al. 1995; Zhu et al. 1995). The discovery of TRPC1 led to the identification of other six members of the TRPC channels subfamily as well as the remaining members of the TRP superfamily (28 in total) (Venkatachalam and Montell 2007). TRP channels show six transmembrane domains (S1–S6), with the N- and C-termini located in the cytosol and a reentrant loop between S5 and S6. TRPC1 forms tetrameric, polymodal, nonselective cation-permeable channels with significant Ca^{2+} permeability. Crystallographic analysis of TRPC1 is not available, but recent studies have revealed the structure of TRPV1 and TRPA1 using electron cryo-microscopy (Cao et al. 2013; Liao et al. 2013; Paulsen et al. 2015) reporting several features that might be applicable to other TRP channels. These analyses have revealed that the TRP channels show a tetrameric architecture that exhibits a fourfold symmetry around a central ion pore formed by transmembrane domains S5 and S6 and the pore loop. TRP channels contain a 25 amino acid motif, called TRP domain, encompassing the TRP box (EWKFAR) just C-terminal to S6 (Montell et al. 2002) that interacts with both the S4–S5 linker and also with an N-terminus helix located prior to S1 (Liao et al. 2013).

The first description of Orai1 in platelets was reported in 2008, where Orai1 was found to be involved a Ca^{2+} signalplex activated by Ca^{2+} store depletion that includes TRPC1, TRPC6, the type II IP_3 receptor, and SERCA3 in human platelets (Redondo et al. 2008). The expression of Orai1 in human platelets and murine megakaryocytes was confirmed by quantitative RT-PCR analysis (Tolhurst et al. 2008). Orai1 expression has also been reported in mouse platelets, and Orai1-deficient mice were found to impair platelet SOCE and thrombus formation, thus revealing the functional role of Orai1 in SOCE and platelet aggregation in mouse platelets (Braun et al. 2009). The role of Orai1 in mouse platelet function was confirmed in mice expressing an inactive Orai1 form (Orai1-R93W), which displayed a markedly attenuated SOCE, reduced integrin activation, and impaired degranulation upon activation with low agonist concentrations (Bergmeier et al. 2009). On the other hand, the first description of TRPC1 at the protein level in platelets was reported in 2000 (Rosado and Sage 2000b) where we found that TRPC1 interacts with the type II IP_3 receptor upon depletion of the intracellular Ca^{2+} stores. A previous study reported the presence of TRPC1 transcripts in the megakaryocytic cell lines MEG01, DAMI, and HEL, although the analysis was unable to detect TRPC1 mRNA in human platelets, probably as a result of rapid processing of these transcripts during platelet generation (Berg et al. 1997).

STIM1 has been found to interact with endogenously expressed TRPC1 (Lopez et al. 2006b) and Orai1 proteins (Jardin et al. 2008) in human platelets. Studies in different cell types have revealed that STIM1 undergoes a conformational change upon Ca^{2+} store depletion as a result of Ca^{2+} displacement from the STIM1 EF-hand domain. It has been hypothesized that at rest, STIM1 SOAR region is hidden from the plasma membrane channels (Fahrner et al. 2013). Discharge of the intracellular stores leads to Ca^{2+} dissociation from the EF-hand, which, in turn, changes the EF-hand-SAM conformation exposing hydrophobic domains and facilitating the formation of STIM1 dimers/oligomers (Stathopulos et al. 2006, 2009). The conformational change that occurred in the N-terminal region is transmitted to the C-terminus through the transmembrane domain, thus bringing the C-termini together (Ma et al. 2015). The CC1 region, which is clamping the remaining C-terminal portion in a tight state, releases it allowing the interaction of the SOAR region with Orai1 C- and N-termini (Derler et al. 2013; Stathopulos et al. 2013; Fahrner et al. 2014). The interaction between STIM1 and the Orai1 C-terminus involves four positively charged residues (K382, K284, K385, and K386), two aromatic amino acids (Y361 and Y362), and four hydrophobic residues (L347, L351, L373, and A376) located in the SOAR region and the amino acids L273, L276, R281, L286, and R289 located in Orai1 C-terminus, leading to what has been called the STIM1-Orai1 association pocket (SOAP) (Stathopulos et al. 2013). The interaction between STIM1 and Orai1 N-terminus remains unsolved. The activation of TRPC1 by STIM1 has been reported to require both the SOAR region, which is important for the STIM1-TRPC1 interaction (Lee et al. 2014), and also the electrostatic interaction between two negatively charged aspartates in TRPC1 (D639 and D640) with the positively charged amino acids K684 and K685 located in STIM1 polybasic domain (Zeng et al. 2008).

A recent study has reported that Orai1, TRPC1, and STIM1 might reproduce I_{SOC}, thus demonstrating that SOC channels involved the participation of these proteins (Desai et al. 2015). We have presented evidence for the existence of functional interactions between Orai1 and TRPCs under the influence of STIM1 and propose that SOC channels in human platelets are composed of heteromeric complexes that include TRPCs and Orai proteins (Jardin et al. 2008, 2009). These findings suggest that the participation of Orai1 in SOCE in human platelets might occur either by forming self-contained ion channels highly selective for Ca^{2+} and activated by STIM1, consistent with the CRAC channel hypothesis as suggested for murine megakaryocytes (Tolhurst et al. 2008), or a model in which SOC channels are formed by a combination of TRPCs and Orai proteins.

The role of the STIM1 homolog, STIM2, in SOCE in platelets is still unclear. Gilio et al. (2010) have reported that while STIM1- and Orai1-deficient mice exhibit impaired glycoprotein VI-dependent Ca^{2+} signals, PS exposure, and thrombus formation, STIM2-deficient mice show normal responses. In human platelets, STIM2 and STIM1 have been found to be expressed in acidic Ca^{2+} stores, including lysosome-related organelles and dense granules. Upon discharge of the acidic stores, using bafilomycin A1, STIM2 interacts with STIM1, SERCA, and Orai1 channels in the plasma membrane (Zbidi et al. 2011). Furthermore, treatment with thapsigargin leads to the formation of a signaling complex including STIM1, STIM2, Orai1, Orai2, and TRPC1 (Berna-Erro et al. 2012), thus suggesting a possible role for STIM2 in SOCE in human platelets.

23.3 Molecular Pathophysiology of SOCE Mechanisms in Platelets

Changes in cytosolic Ca^{2+} concentration play a key role in the regulation of platelet responsiveness, and several disorders have been found to be mediated by abnormalities in Ca^{2+} signaling, especially in SOCE. As described above, Ca^{2+} entry through plasma membrane channels is necessary to achieve full activation of platelet functions (Jardin et al. 2007b; Braun et al. 2009), and SOCE is a major mechanism for Ca^{2+} influx in these cells (Adam et al. 2016). Consequently, abnormalities in the key elements of SOCE or its molecular regulators are expected to produce platelet dysfunctions that might underlie hemostatic disorders. Therefore, identification of the molecular basis of altered cytosolic Ca^{2+} homeostasis will shed new light on the pathophysiology of cardiovascular disorders.

23.3.1 Functional Abnormalities Concerning STIM1, Orai1, and TRPC Proteins

Grosse et al. (2007) reported the first piece of evidence suggesting that STIM1 plays an important role in platelet function and mutations of human STIM1 and other proteins implicated in SOCE can be involved in inherited thrombocytopenias. Later, it was demonstrated that platelets isolated from mice lacking expression of *Stim1* gene showed both an impaired agonist-induced Ca^{2+} release from intracellular stores and SOCE activation that results in an impaired platelet function and reduced thrombus formation (Varga-Szabo et al. 2008b; Ahmad et al. 2011). Similar results were observed in Orai1-deficient mice, where platelets show attenuated SOCE and ROCE, as well as impaired agonist-induced platelet activation. This defect in platelet activation is stronger when platelets are activated via collagen receptor than when platelets are activated via G protein-coupled receptors (Braun et al. 2009; Bergmeier et al. 2013). In both STIM1- and Orai1-deficient mice platelets, SOCE inhibition causes an altered integrin activation, dense granule release, and phosphatidylserine (PS) exposure (Varga-Szabo et al. 2008b; Braun et al. 2009). However, TRPC1-deficient mice platelets showed a normal SOCE and platelet function compared to wild-type mice platelets, suggesting that Orai1 is the major SOC channels in mice platelets (Varga-Szabo et al. 2008a; Braun et al. 2009).

The functional role of STIM1 and Orai1 in the control of agonist-stimulated platelet aggregation has also been demonstrated in human platelets using a different approach. Functional knockdown of STIM1 or Orai1 by electrotransjection of cells with STIM1 (25–139) antibody or with anti-Orai1 (288–301) antibody, which impaired STIM1-Orai1complexes formation upon platelet stimulation, decreased thrombin-induced platelet aggregation and prevented ADP-stimulated aggregation as compared to cells treated with IgG of the same origin of antibodies used and containing the same concentration of preservatives. Similar results were obtained by external application of the anti-STIM1 (25–139) antibody or anti-TRPC1 antibody that bocks STIM1-TRPC1 interaction without affecting STIM1-Orai1 complexes formation, suggesting a functional role of TRPC1 in human platelet aggregation (Galan et al. 2009), in contrast to published in TRPC1-deficient mice (Varga-Szabo et al. 2008a). These discrepancies might be attributed to differences in the activation of SOCE in distinct species.

In platelets from diabetes mellitus type 2 patients, the loss of the association of STIM1 with Orai1, TRPC1, and TRPC6 results in a reduced SOCE that might be involved in the pathogenesis of the altered platelet responsiveness observed in diabetic patients (Jardin et al. 2011). As mentioned above, TRPC6 has been shown to be involved in both SOCE and non-SOCE pathways by association with different Ca^{2+}-handling proteins. Hence, depletion of intracellular Ca^{2+} stores promotes the association of TRPC6 with Orai1 and STIM1, while the non-SOCE activator OAG displaces TRPC6 from Orai1 and STIM1 and induces its association with TRPC3 (Jardin et al. 2009). Different studies suggested a functional role for TRPC6 in platelets modulating both Ca^{2+} entry and aggregation through its interaction with calmodulin and IP_3Rs (Dionisio et al. 2011) or with Orai1-STIM1-forming complexes (Jardin et al. 2011). However, the role of TRPC6 in platelet function is still unclear, and controversial studies were published using TRPC6-deficient mice (Paez Espinosa et al. 2012; Ramanathan et al. 2012; Harper et al. 2013; Albarran et al. 2014). Paez Espinosa et al. (2012) proposed that TRPC6 plays an essential role in platelet function since TRPC6-deficient mice showed an increased bleeding time and reduced thrombus formation. In contrast, two studies have showed a decrease in $[Ca^{2+}]_i$ in platelets lacking TRPC6 at resting conditions, without playing a role in the regulation of SOCE and platelet aggregation (Ramanathan et al. 2012; Albarran et al. 2014). Berna-Erro and coworkers proposed that TRPC6 regulates $[Ca^{2+}]_i$ controlling the passive Ca^{2+} leak rate from agonist-sensitive intracellular Ca^{2+} stores in resting platelets (Albarran et al. 2014).

Since the role of STIM1 and Orai1 in the mechanism of SOCE was identified, several loss-of-function mutations in both proteins have been related to different human diseases, collectively termed CRAC channelopathies, characterized by severe combined immunodeficiency (SCID)-like disease, autoimmunity, tubular aggregate myopathy, dental enamel maturation defect, and ectodermal dysplasia. Gain-of-function mutations of Orai1 are related with tubular aggregate myopathy and Stormorken-like syndrome, while gain-of-function mutations of STIM1 are associated with tubular aggregate myopathy, Stormorken syndrome, and York platelet syndrome. However, only Stormorken syndrome and York platelet syndrome are widely characterized by platelet disorders that result in bleeding diathesis and thrombocytopenia (Feske 2010; Lacruz and Feske 2015). The initial studies in genetically modified mice demonstrated that replacement of an acidic residue of aspartate by glycine at position 84 in the EF-hand motif of STIM1 induces a gain-of-function mutation of STIM1 that reduces its ability to bind Ca^{2+}, mimicking store depletion and therefore inducing a constitutive activation of STIM1 and CRAC channels that promote an activated procoagulant state of platelet due to an increased $[Ca^{2+}]_i$ in resting conditions. In vivo and in vitro functional assays demonstrated that these mutant platelets are unresponsiveness to the immunoreceptor tyrosine-based activation motif (ITAM)-containing collagen receptor complex GPVI-FcR γ-chain that causes platelet dysfunction, thrombocytopenia, and fatal bleeding disorders in mice (Grosse et al. 2007). Recently, it has been described that STIM1 (D84G) mutation also induces a constitutive activation of STIM1 and CRAC channels in human myoblast, causing tubular aggregate myopathy, but platelet function was not investigated in these patients (Bohm et al. 2013). Similar altered bleeding phenotype described above in EF-hand mutant mice has also been observed in human patients with Stormorken syndrome. This rare autosomal-dominant genetic disease was first reported in 1985, which, beside thrombocytopenia and bleeding diathesis, is characterized by congenital miosis, asplenia, headache, ichthyosis, tubular aggregate myopathy, and proximal muscle weakness (Stormorken et al. 1985, 1995; Stormorken 2002). Misceo and coworkers demonstrated that platelets isolated from Stormorken syndrome patients show an increased $[Ca^{2+}]_i$ in resting conditions and a markedly reduced both TRAP- and thapsigargin-induced SOCE compared with platelets isolated from healthy donors. As consequence of these alterations in Ca^{2+} handling, platelets from Stormorken syndrome patients are much less responsive to stimulation than those from healthy donors, and they are in a constitutively activated procoagulant state, characterized by phosphatidylserine (PS) exposure and other platelet activation markers at the plasma membrane, that promotes a decrease in lifespan of circulating platelets and a low blood platelet count, causing the mild bleeding disorder (Misceo et al. 2014), as previously Stormorken and coworkers described in 1995 (Stormorken et al. 1995). Gain-of-function mutation STIM1 (R304W) has been shown to cause the clinical phenotypes related with Stormorken syndrome (Misceo et al. 2014; Morin et al. 2014; Nesin et al. 2014). Based on structural analyses, the residue R304 is localized close to STIM1-Orai1-activating region (SOAR) in C-terminal region of STIM1. The lateral chain of the residue R304 formed a hydrogen bond with the E318 and Q314 residues, located in the first coiled-coil domain, leading the occlusion of SOAR under resting conditions (Korzeniowski et al. 2010; Cui et al. 2013; Morin et al. 2014; Lacruz and Feske 2015). Dissociation of this inhibitory helix of coiled-coil domain/SOAR interaction is required for the association of SOAR with Orai1 and the subsequent channel activation (Korzeniowski et al. 2010; Feske and Prakriya 2013). Substitution of tryptophan residue to arginine residue in position 304 increased amphipathic properties of the coiled-coil domain, avoiding the hydrogen bond with E318 and Q314 residues and leading to the release of SOAR domain that results in the constitutive activation of STIM1 and of the Orai1 channel (Shen and Demaurex 2012; Yang et al. 2012; Morin et al. 2014). Another gain-of-function mutation of STIM1, STIM1 (I115P), is mainly involved in the named York platelet syndrome (YPS) that is also associated with bleeding tendency caused by platelet abnormalities and thrombocytopenia. As described above in Stormorken syndrome, platelets are in a constitutive activated state and also show structural and functional abnormalities, including giant electron-opaque organelles and massive, multilayered target bodies and deficiency of platelet Ca^{2+} storage in delta granules. The mutation in the residue I115, located in the hidden second EF-hand domain of STIM1, induces a constitutive activation of STIM1 that increases STIM1- and Orai1-forming complexes and promotes an elevated and sustained $[Ca^{2+}]_i$. YPS patients share the STIM1 (R304W) mutation described in Stormorken syndrome patients, as well as certain symptoms such as muscle weakness with skeletal muscle atrophy (White and Gunay-Aygun 2011; White et al. 2013; Markello et al. 2015).

Although mutations in Orai1 have not been widely associated with thrombocytopenia and bleeding disorders, the loss-of-function mutation Orai1 (R93W) in mice platelets and the equivalent Orai1 (R91W) mutation in human patients, responsible for SCID, promote defects in agonist-induced Ca^{2+} influx as well as Ca^{2+}-regulated platelet functions, such as integrin activation, dense granule release, and PS exposure. However, no spontaneous bleeding or clotting disorders were observed in both cases (Bergmeier et al. 2009, 2013). The replacement of a conserved residue of arginine by tryptophan at position 91, located in the conducting pore at the cytoplasmic end of transmembrane region 1 of Orai1, avoids CRAC channel activation and Ca^{2+} influx currents through this channel (Feske et al. 2006; Thompson et al. 2009). These studies demonstrated that mutations in STIM1 or Orai1 proteins are not responsible for the clinical phenotype of Scott's syndrome, a disorder of platelet coagulant activity associated with a markedly impaired PS exposure similar to that described for Orai1 (R91W) and Orai1 (R93W) platelets (Bergmeier et al. 2009). In addition, gain-of-function mutation Orai1 (P245L) has been related with a Stormorken-like syndrome characterized by congenital miosis and tubular aggregate myopathy, but platelet dysfunction and bleeding disorders have not been described (Nesin et al. 2014). The residue P245 is located within the fourth transmembrane helix of Orai1 and plays an essential role in the stabilization of the closed state of the channel. The mutation Orai1 (P245L) stabilizes Orai1- and STIM1-forming complexes and suppresses the slow Ca^{2+}-dependent inactivation of the channel (Misceo et al. 2014; Nesin et al. 2014; Palty et al. 2015).

23.3.2 Homer1

Homer1 is a cytoplasmic adaptor family protein that contains a class II EVH1 [Ena (Enabled)/VASP (vasodilator-stimulated phosphoprotein) homology 1] domain that binds to a proline-rich PXXF motif located in different Ca^{2+}-handling protein involved in SOCE, such as TRPC1 (Yuan et al. 2003; Kim et al. 2006), STIM1 (Jardin et al. 2012), and IP_3R (Yuan et al. 2003; Kim et al. 2006). Homer1 also binds with Orai1, although this interaction is mediated by STIM1 (Jardin et al. 2012). In platelets, Homer1 has been shown to facilitate both STIM1-Orai1- and TRPC1-IP_3R-forming complexes upon cell stimulation. Impairment of Homer1 function by electrotransjection of the PPKKFR peptide, which emulates PPXXF

Homer1-binding motif, into cellular cytoplasm abolished agonist-stimulated aggregation in platelets as consequence of reduction of the maintenance of STIM1- and Orai1-forming complexes, suggesting that Homer1 regulates SOCE and, in turns, platelet function (Jardin et al. 2012).

23.4 Conclusions and Perspectives

Summarizing, SOCE is a major mechanism for Ca^{2+} influx in platelets and plays a relevant role in platelet function. In human platelets, two platelet disorders due to SOCE dysfunction have been characterized, the Stormorken and York platelet syndromes, which course with bleeding diathesis and thrombocytopenia. The mechanism underlying both syndromes includes gain-of-function mutations in STIM1. Similarly, gain-of-function mutations in Orai1 result in less characterized Stormorken-like syndromes. The precise analysis of still uncharacterized platelet dysfunctions might provide evidence for the existence of further loss- or gain-of-function mutations. Further studies should also be focused on the involvement of regulatory SOCE proteins, such as SARAF, CRACR2A, or STIMATE, as well as TRPC channels, including TRPC1 and TRPC6, in the development of platelet disorders that might be responsible for still uncharacterized pathological phenotypes.

Acknowledgments This work was supported by MINECO (Grants BFU2013-45564-C2-1-P and BFU2016-74932-C2-1-P) and Junta de Extremadura-FEDER (GR15029).

References

Adam F, Khatib AM, Lopez JJ, Vatier C, Turpin S, Muscat A, Soulet F, Aries A, Jardin I, Bobe R, Stepanian A, de Prost D, Dray C, Rosado JA, Valet P, Feve B, Siegfried G (2016) Apelin: an antithrombotic factor that inhibits platelet function. Blood 127:908–920

Ahmad F, Boulaftali Y, Greene TK, Ouellette TD, Poncz M, Feske S, Bergmeier W (2011) Relative contributions of stromal interaction molecule 1 and CalDAG-GEFI to calcium-dependent platelet activation and thrombosis. J Thromb Haemost 9:2077–2086

Albarran L, Berna-Erro A, Dionisio N, Redondo PC, Lopez E, Lopez JJ, Salido GM, Brull Sabate JM, Rosado JA (2014) TRPC6 participates in the regulation of cytosolic basal calcium concentration in murine resting platelets. Biochim Biophys Acta 1843:789–796

Albarran L, Lopez JJ, Amor NB, Martin-Cano FE, Berna-Erro A, Smani T, Salido GM, Rosado JA (2016a) Dynamic interaction of SARAF with STIM1 and Orai1 to modulate store-operated calcium entry. Sci Rep 6:24452

Albarran L, Lopez JJ, Gomez LJ, Salido GM, Rosado JA (2016b) SARAF modulates TRPC1, but not TRPC6, channel function in a STIM1-independent manner. Biochem J 473:3581–3595

Albarran L, Lopez JJ, Woodard GE, Salido GM, Rosado JA (2016c) Store-operated Ca^{2+} entry-associated regulatory factor (SARAF) plays an important role in the regulation of Arachidonate-regulated Ca2+ (ARC) channels. J Biol Chem 291:6982–6988

Albarran L, Regodon S, Salido GM, Lopez JJ, Rosado JA (2016d) Role of STIM1 in the surface expression of SARAF. Channels 11:84–88

Ambudkar I, de Souza LB, Ong HL (2017) TRPC1, Orai1, and STIM1 in SOCE: friends in tight spaces. Cell Calcium. doi:10.1016/j.ceca.2016.12.009

Bakowski D, Parekh AB (2002) Monovalent cation permeability and Ca(2+) block of the store-operated Ca(2+) current I(CRAC)in rat basophilic leukemia cells. Pflugers Arch 443:892–902

Ben Amor NB, Zbidi H, Bouaziz A, Jardin I, Hernandez-Cruz JM, Salido GM, Rosado JA, Bartegi A (2009) Acidic-store depletion is required for human platelet aggregation. Blood Coagul Fibrinolysis 20:511–516

Berg LP, Shamsher MK, El-Daher SS, Kakkar VV, Authi KS (1997) Expression of human TRPC genes in the megakaryocytic cell lines MEG01, DAMI and HEL. FEBS Lett 403:83–86

Bergmeier W, Oh-Hora M, McCarl CA, Roden RC, Bray PF, Feske S (2009) R93W mutation in Orai1 causes impaired calcium influx in platelets. Blood 113:675–678

Bergmeier W, Weidinger C, Zee I, Feske S (2013) Emerging roles of store-operated Ca(2+) entry through STIM and ORAI proteins in immunity, hemostasis and cancer. Channels (Austin) 7:379–391

Berna-Erro A, Galan C, Dionisio N, Gomez LJ, Salido GM, Rosado JA (2012) Capacitative and non-capacitative signaling complexes in human platelets. Biochim Biophys Acta 1823:1242–1251

Berna-Erro A, Jardin I, Smani T, Rosado JA (2016) Regulation of platelet function by Orai, STIM and TRP. Adv Exp Med Biol 898:157–181

Bizzonero J (1882) Über einen formbestandtheil des blutes und dessen rolle bei der thrombose und der blutgerinnung. Virchows Arch Pathol Anat Physiol Klin Med 90:261–332

Bobe R, Bredoux R, Wuytack F, Quarck R, Kovacs T, Papp B, Corvazier E, Magnier C, Enouf J (1994) The rat platelet 97-kDa Ca2+ATPase isoform is the sarcoendoplasmic reticulum Ca2 +ATPase 3 protein. J Biol Chem 269:1417–1424

Bohm J, Chevessier F, Maues De Paula A, Koch C, Attarian S, Feger C, Hantai D, Laforet P, Ghorab K, Vallat JM, Fardeau M, Figarella-Branger D, Pouget J, Romero NB, Koch M, Ebel C, Levy N, Krahn M, Eymard B, Bartoli M, Laporte J (2013) Constitutive activation of the calcium sensor STIM1 causes tubular-aggregate myopathy. Am J Hum Genet 92:271–278

Borst O, Münzer P, Schmid E, Schmidt EM, Russo A, Walker B, Yang W, Leibrock C, Szteyn K, Schmidt S, Elvers M, Faggio C, Shumilina E, Kuro-o M, Gawaz M, Lang F (2014) 1,25(OH) 2 vitamin D3-dependent inhibition of platelet Ca2+ signaling and thrombus formation in klotho-deficient mice. FASEB J 28:2108–2119

Braun A, Varga-Szabo D, Kleinschnitz C, Pleines I, Bender M, Austinat M, Bosl M, Stoll G, Nieswandt B (2009) Orai1 (CRACM1) is the platelet SOC channel and essential for pathological thrombus formation. Blood 113:2056–2063

Cao E, Liao M, Cheng Y, Julius D (2013) TRPV1 structures in distinct conformations reveal activation mechanisms. Nature 504:113–118

Cavallini L, Coassin M, Alexandre A (1995) Two classes of agonist-sensitive Ca2+ stores in platelets, as identified by their differential sensitivity to 2,5-di-(tert-butyl)-1,4-benzohydroquinone and thapsigargin. Biochem J 310(Pt 2):449–452

Cui B, Yang X, Li S, Lin Z, Wang Z, Dong C, Shen Y (2013) The inhibitory helix controls the intramolecular conformational switching of the C-terminus of STIM1. PLoS One 8:e74735

Dell'Angelica EC, Mullins C, Caplan S (2000) Lysosome-related organelles. FASEB J 14:1265–1278

Derler I, Fahrner M, Muik M, Lackner B, Schindl R, Groschner K, Romanin C (2009) A Ca2+ release-activated Ca2+ (CRAC) modulatory domain (CMD) within STIM1 mediates fast Ca2 +-dependent inactivation of ORAI1 channels. J Biol Chem 284:24933–24938

Derler I, Plenk P, Fahrner M, Muik M, Jardin I, Schindl R, Gruber HJ, Groschner K, Romanin C (2013) The extended transmembrane Orai1 N-terminal (ETON) region combines binding interface and gate for Orai1 activation by STIM1. J Biol Chem 288:29025–29034

Desai PN, Zhang X, Wu S, Janoshazi A, Bolimuntha S, Putney JW, Trebak M (2015) Multiple types of calcium channels arising from alternative translation initiation of the Orai1 message. Sci Signal 8:ra74

Dionisio N, Albarran L, Berna-Erro A, Hernandez-Cruz JM, Salido GM, Rosado JA (2011) Functional role of the calmodulin- and inositol 1,4,5-trisphosphate receptor-binding (CIRB) site of TRPC6 in human platelet activation. Cell Signal 23:1850–1856

Ebbeling L, Robertson C, McNicol A, Gerrard JM (1992) Rapid ultrastructural changes in the dense tubular system following platelet activation. Blood 80:718–723

Fahrner M, Derler I, Jardin I, Romanin C (2013) The STIM1/Orai signaling machinery. Channels (Austin) 7:330–343

Fahrner M, Muik M, Schindl R, Butorac C, Stathopulos P, Zheng L, Jardin I, Ikura M, Romanin C (2014) A coiled-coil clamp controls both conformation and clustering of stromal interaction molecule 1 (STIM1). J Biol Chem 289:33231–33244

Fasolato C, Nilius B (1998) Store depletion triggers the calcium release-activated calcium current (ICRAC) in macrovascular endothelial cells: a comparison with Jurkat and embryonic kidney cell lines. Pflugers Arch 436:69–74

Feske S (2010) CRAC channelopathies. Pflugers Arch 460:417–435

Feske S, Prakriya M (2013) Conformational dynamics of STIM1 activation. Nat Struct Mol Biol 20:918–919

Feske S, Prakriya M, Rao A, Lewis RS (2005) A severe defect in CRAC Ca2+ channel activation and altered K+ channel gating in T cells from immunodeficient patients. J Exp Med 202:651–662

Feske S, Gwack Y, Prakriya M, Srikanth S, Puppel SH, Tanasa B, Hogan PG, Lewis RS, Daly M, Rao A (2006) A mutation in Orai1 causes immune deficiency by abrogating CRAC channel function. Nature 441:179–185

Fox JE (2001) Cytoskeletal proteins and platelet signaling. Thromb Haemost 86:198–213

Galan C, Zbidi H, Bartegi A, Salido GM, Rosado JA (2009) STIM1, Orai1 and hTRPC1 are important for thrombin- and ADP-induced aggregation in human platelets. Arch Biochem Biophys 490:137–144

Gilio K, van Kruchten R, Braun A, Berna-Erro A, Feijge MA, Stegner D, van der Meijden PE, Kuijpers MJ, Varga-Szabo D, Heemskerk JW, Nieswandt B (2010) Roles of platelet STIM1 and Orai1 in glycoprotein VI- and thrombin-dependent procoagulant activity and thrombus formation. J Biol Chem 285:23629–23638

Golovina VA, Platoshyn O, Bailey CL, Wang J, Limsuwan A, Sweeney M, Rubin LJ, Yuan JX (2001) Upregulated TRP and enhanced capacitative Ca(2+) entry in human pulmonary artery myocytes during proliferation. Am J Physiol Heart Circ Physiol 280:H746–H755

Grosse J, Braun A, Varga-Szabo D, Beyersdorf N, Schneider B, Zeitlmann L, Hanke P, Schropp P, Muhlstedt S, Zorn C, Huber M, Schmittwolf C, Jagla W, Yu P, Kerkau T, Schulze H, Nehls M, Nieswandt B (2007) An EF hand mutation in Stim1 causes premature platelet activation and bleeding in mice. J Clin Invest 117:3540–3550

Gwack Y, Srikanth S, Feske S, Cruz-Guilloty F, Oh-hora M, Neems DS, Hogan PG, Rao A (2007) Biochemical and functional characterization of Orai proteins. J Biol Chem 282:16232–16243

Hampson A, O'Connor A, Smolenski A (2013) Synaptotagmin-like protein 4 and Rab8 interact and increase dense granule release in platelets. J Thromb Haemost 11:161–168

Harper MT, Sage SO (2010) Src family tyrosine kinases activate thrombin-induced non-capacitative cation entry in human platelets. Platelets 21:445–450

Harper MT, Londono JE, Quick K, Londono JC, Flockerzi V, Philipp SE, Birnbaumer L, Freichel M, Poole AW (2013) Transient receptor potential channels function as a coincidence signal detector mediating phosphatidylserine exposure. Sci Signal 6:ra50

Harrison P, Cramer EM (1993) Platelet alpha-granules. Blood Rev 7:52–62

Hayem G (1882) Sur le méchanisme de l'arrêt des hémorrhagies. C R Acad Sci 95:18

Hoth M, Penner R (1992) Depletion of intracellular calcium stores activates a calcium current in mast cells. Nature 355:353–356

Hou X, Pedi L, Diver MM, Long SB (2012) Crystal structure of the calcium release-activated calcium channel Orai. Science 338:1308–1313

Huang GN, Zeng W, Kim JY, Yuan JP, Han L, Muallem S, Worley PF (2006) STIM1 carboxyl-terminus activates native SOC, I(crac) and TRPC1 channels. Nat Cell Biol 8:1003–1010

Jardin I, Ben Amor N, Bartegi A, Pariente JA, Salido GM, Rosado JA (2007a) Differential involvement of thrombin receptors in Ca2+ release from two different intracellular stores in human platelets. Biochem J 401:167–174

Jardin I, Ben Amor N, Hernandez-Cruz JM, Salido GM, Rosado JA (2007b) Involvement of SNARE proteins in thrombin-induced platelet aggregation: evidence for the relevance of Ca2+ entry. Arch Biochem Biophys 465:16–25

Jardin I, Lopez JJ, Salido GM, Rosado JA (2008) Orai1 mediates the interaction between STIM1 and hTRPC1 and regulates the mode of activation of hTRPC1-forming Ca2+ channels. J Biol Chem 283:25296–25304

Jardin I, Gomez LJ, Salido GM, Rosado JA (2009) Dynamic interaction of hTRPC6 with the Orai1-STIM1 complex or hTRPC3 mediates its role in capacitative or non-capacitative Ca(2+) entry pathways. Biochem J 420:267–276

Jardin I, Lopez JJ, Zbidi H, Bartegi A, Salido GM, Rosado JA (2011) Attenuated store-operated divalent cation entry and association between STIM1, Orai1, hTRPC1 and hTRPC6 in platelets from type 2 diabetic patients. Blood Cells Mol Dis 46:252–260

Jardin I, Albarran L, Bermejo N, Salido GM, Rosado JA (2012) Homers regulate calcium entry and aggregation in human platelets: a role for Homers in the association between STIM1 and Orai1. Biochem J 445:29–38

Jardin I, Dionisio N, Frischauf I, Berna-Erro A, Woodard GE, Lopez JJ, Salido GM, Rosado JA (2013) The polybasic lysine-rich domain of plasma membrane-resident STIM1 is essential for the modulation of store-operated divalent cation entry by extracellular calcium. Cell Signal 25:1328–1337

Jha A, Ahuja M, Maleth J, Moreno CM, Yuan JP, Kim MS, Muallem S (2013) The STIM1 CTID domain determines access of SARAF to SOAR to regulate Orai1 channel function. J Cell Biol 202:71–79

Junt T, Schulze H, Chen Z, Massberg S, Goerge T, Krueger A, Wagner DD, Graf T, Italiano JE Jr, Shivdasani RA, von Andrian UH (2007) Dynamic visualization of thrombopoiesis within bone marrow. Science 317:1767–1770

Kawasaki T, Lange I, Feske S (2009) A minimal regulatory domain in the C terminus of STIM1 binds to and activates ORAI1 CRAC channels. Biochem Biophys Res Commun 385:49–54

Kim JY, Zeng W, Kiselyov K, Yuan JP, Dehoff MH, Mikoshiba K, Worley PF, Muallem S (2006) Homer 1 mediates store- and inositol 1,4,5-trisphosphate receptor-dependent translocation and retrieval of TRPC3 to the plasma membrane. J Biol Chem 281:32540–32549

Kiselyov KI, Semyonova SB, Mamin AG, Mozhayeva GN (1999) Miniature Ca2+ channels in excised plasma-membrane patches: activation by IP3. Pflugers Arch 437:305–314

Korzeniowski MK, Manjarres IM, Varnai P, Balla T (2010) Activation of STIM1-Orai1 involves an intramolecular switching mechanism. Sci Signal 3:ra82

Lacruz RS, Feske S (2015) Diseases caused by mutations in ORAI1 and STIM1. Ann N Y Acad Sci 1356:45–79

Lang F, Munzer P, Gawaz M, Borst O (2013) Regulation of STIM1/Orai1-dependent Ca^{2+} signalling in platelets. Thromb Haemost 110:925–930

Lee KP, Choi S, Hong JH, Ahuja M, Graham S, Ma R, So I, Shin DM, Muallem S, Yuan JP (2014) Molecular determinants mediating gating of Transient Receptor Potential Canonical (TRPC) channels by stromal interaction molecule 1 (STIM1). J Biol Chem 289:6372–6382

Liao M, Cao E, Julius D, Cheng Y (2013) Structure of the TRPV1 ion channel determined by electron cryo-microscopy. Nature 504:107–112

López JJ, Camello-Almaraz C, Pariente JA, Salido GM, Rosado JA (2005) Ca2+ accumulation into acidic organelles mediated by Ca2+- and vacuolar H+-ATPases in human platelets. Biochem J 390:243–252

Lopez JJ, Redondo PC, Salido GM, Pariente JA, Rosado JA (2006a) Two distinct Ca2+ compartments show differential sensitivity to thrombin, ADP and vasopressin in human platelets. Cell Signal 18:373–381

Lopez JJ, Salido GM, Pariente JA, Rosado JA (2006b) Interaction of STIM1 with endogenously expressed human canonical TRP1 upon depletion of intracellular Ca2+ stores. J Biol Chem 281:28254–28264

Lopez E, Bermejo N, Berna-Erro A, Alonso N, Salido GM, Redondo PC, Rosado JA (2015) Relationship between calcium mobilization and platelet alpha- and delta-granule secretion. A role for TRPC6 in thrombin-evoked delta-granule exocytosis. Arch Biochem Biophys 585:75–81

Lu W, Xu D, Tu R, Hu Z (2013) Morphology of platelet Golgi apparatus and their significance after acute cerebral infarction. Neural Regen Res 8:2134–2143

Luckhoff A, Clapham DE (1994) Calcium channels activated by depletion of internal calcium stores in A431 cells. Biophys J 67:177–182

Luik RM, Wu MM, Buchanan J, Lewis RS (2006) The elementary unit of store-operated Ca2+ entry: local activation of CRAC channels by STIM1 at ER-plasma membrane junctions. J Cell Biol 174:815–825

Ma G, Wei M, He L, Liu C, Wu B, Zhang SL, Jing J, Liang X, Senes A, Tan P, Li S, Sun A, Bi Y, Zhong L, Si H, Shen Y, Li M, Lee MS, Zhou W, Wang J, Wang Y, Zhou Y (2015) Inside-out Ca(2+) signalling prompted by STIM1 conformational switch. Nat Commun 6:7826

MacKenzie AB, Mahaut-Smith MP, Sage SO (1996) Activation of receptor-operated cation channels via P2X1 not P2T purinoceptors in human platelets. J Biol Chem 271:2879–2881

Manji SS, Parker NJ, Williams RT, van Stekelenburg L, Pearson RB, Dziadek M, Smith PJ (2000) STIM1: a novel phosphoprotein located at the cell surface. Biochim Biophys Acta 1481:147–155

Marcu MG, Zhang L, Nau-Staudt K, Trifaro JM (1996) Recombinant scinderin, an F-actin severing protein, increases calcium-induced release of serotonin from permeabilized platelets, an effect blocked by two scinderin-derived actin-binding peptides and phosphatidylinositol 4,5-bisphosphate. Blood 87:20–24

Markello T, Chen D, Kwan JY, Horkayne-Szakaly I, Morrison A, Simakova O, Maric I, Lozier J, Cullinane AR, Kilo T, Meister L, Pakzad K, Bone W, Chainani S, Lee E, Links A, Boerkoel C, Fischer R, Toro C, White JG, Gahl WA, Gunay-Aygun M (2015) York platelet syndrome is a CRAC channelopathy due to gain-of-function mutations in STIM1. Mol Genet Metab 114:474–482

Mercer JC, Dehaven WI, Smyth JT, Wedel B, Boyles RR, Bird GS, Putney JW Jr (2006) Large store-operated calcium selective currents due to co-expression of Orai1 or Orai2 with the intracellular calcium sensor, Stim1. J Biol Chem 281:24979–24990

Mignen O, Thompson JL, Shuttleworth TJ (2007) STIM1 regulates Ca2+ entry via arachidonate-regulated Ca2+-selective (ARC) channels without store depletion or translocation to the plasma membrane. J Physiol 579:703–715

Misceo D, Holmgren A, Louch WE, Holme PA, Mizobuchi M, Morales RJ, De Paula AM, Stray-Pedersen A, Lyle R, Dalhus B, Christensen G, Stormorken H, Tjonnfjord GE, Frengen E (2014) A dominant STIM1 mutation causes Stormorken syndrome. Hum Mutat 35:556–564

Montell C, Birnbaumer L, Flockerzi V, Bindels RJ, Bruford EA, Caterina MJ, Clapham DE, Harteneck C, Heller S, Julius D, Kojima I, Mori Y, Penner R, Prawitt D, Scharenberg AM, Schultz G, Shimizu N, Zhu MX (2002) A unified nomenclature for the superfamily of TRP cation channels. Mol Cell 9:229–231

Morin G, Bruechle NO, Singh AR, Knopp C, Jedraszak G, Elbracht M, Bremond-Gignac D, Hartmann K, Sevestre H, Deutz P, Herent D, Nurnberg P, Romeo B, Konrad K, Mathieu-Dramard M, Oldenburg J, Bourges-Petit E, Shen Y, Zerres K, Ouadid-Ahidouch H, Rochette J (2014) Gain-of-function mutation in STIM1 (P.R304W) is associated with Stormorken syndrome. Hum Mutat 35:1221–1232

Muik M, Fahrner M, Derler I, Schindl R, Bergsmann J, Frischauf I, Groschner K, Romanin C (2009) A cytosolic homomerization and a modulatory domain within STIM1 C terminus determine coupling to ORAI1 channels. J Biol Chem 284:8421–8426

Nesin V, Wiley G, Kousi M, Ong EC, Lehmann T, Nicholl DJ, Suri M, Shahrizaila N, Katsanis N, Gaffney PM, Wierenga KJ, Tsiokas L (2014) Activating mutations in STIM1 and ORAI1 cause overlapping syndromes of tubular myopathy and congenital miosis. Proc Natl Acad Sci U S A 111:4197–4202

Neumuller O, Hoffmeister M, Babica J, Prelle C, Gegenbauer K, Smolenski AP (2009) Synaptotagmin-like protein 1 interacts with the GTPase-activating protein Rap1GAP2 and regulates dense granule secretion in platelets. Blood 114:1396–1404

Nishimura S, Nagasaki M, Kunishima S, Sawaguchi A, Sakata A, Sakaguchi H, Ohmori T, Manabe I, Italiano JE Jr, Ryu T, Takayama N, Komuro I, Kadowaki T, Eto K, Nagai R (2015) IL-1alpha induces thrombopoiesis through megakaryocyte rupture in response to acute platelet needs. J Cell Biol 209:453–466

Paez Espinosa EV, Murad JP, Ting HJ, Khasawneh FT (2012) Mouse transient receptor potential channel 6: role in hemostasis and thrombogenesis. Biochem Biophys Res Commun 417:853–856

Palty R, Isacoff EY (2015) Cooperative binding of stromal interaction molecule 1 (STIM1) to the N and C termini of calcium release-activated calcium modulator 1 (Orai1). J Biol Chem 291 (1):334–341

Palty R, Stanley C, Isacoff EY (2015) Critical role for Orai1 C-terminal domain and TM4 in CRAC channel gating. Cell Res 25:963–980

Papp B, Enyedi A, Paszty K, Kovacs T, Sarkadi B, Gardos G, Magnier C, Wuytack F, Enouf J (1992) Simultaneous presence of two distinct endoplasmic-reticulum-type calcium-pump isoforms in human cells. Characterization by radio-immunoblotting and inhibition by 2,5-di-(t-butyl)-1,4-benzohydroquinone. Biochem J 288(Pt 1):297–302

Park CY, Hoover PJ, Mullins FM, Bachhawat P, Covington ED, Raunser S, Walz T, Garcia KC, Dolmetsch RE, Lewis RS (2009) STIM1 clusters and activates CRAC channels via direct binding of a cytosolic domain to Orai1. Cell 136:876–890

Paulsen CE, Armache JP, Gao Y, Cheng Y, Julius D (2015) Structure of the TRPA1 ion channel suggests regulatory mechanisms. Nature 520:511–517

Peinelt C, Vig M, Koomoa DL, Beck A, Nadler MJ, Koblan-Huberson M, Lis A, Fleig A, Penner R, Kinet JP (2006) Amplification of CRAC current by STIM1 and CRACM1 (Orai1). Nat Cell Biol 8:771–773

Prakriya M, Lewis RS (2002) Separation and characterization of currents through store-operated CRAC channels and Mg2+-inhibited cation (MIC) channels. J Gen Physiol 119:487–507

Prakriya M, Feske S, Gwack Y, Srikanth S, Rao A, Hogan PG (2006) Orai1 is an essential pore subunit of the CRAC channel. Nature 443:230–233

Putney JW Jr (1986) A model for receptor-regulated calcium entry. Cell Calcium 7:1–12

Ramanathan G, Gupta S, Thielmann I, Pleines I, Varga-Szabo D, May F, Mannhalter C, Dietrich A, Nieswandt B, Braun A (2012) Defective diacylglycerol-induced Ca2+ entry but normal agonist-induced activation responses in TRPC6-deficient mouse platelets. J Thromb Haemost 10:419–429

Redondo PC, Jardin I, Lopez JJ, Salido GM, Rosado JA (2008) Intracellular Ca2+ store depletion induces the formation of macromolecular complexes involving hTRPC1, hTRPC6, the type II IP3 receptor and SERCA3 in human platelets. Biochim Biophys Acta 1783:1163–1176

Ren Q, Wimmer C, Chicka MC, Ye S, Ren Y, Hughson FM, Whiteheart SW (2010) Munc13-4 is a limiting factor in the pathway required for platelet granule release and hemostasis. Blood 116:869–877

Rodriguez Del Castillo A, Vitale ML, Tchakarov L, Trifaro JM (1992) Human platelets contain scinderin, a Ca(2+)-dependent actin filament-severing protein. Thromb Haemost 67:248–251

Roos J, DiGregorio PJ, Yeromin AV, Ohlsen K, Lioudyno M, Zhang S, Safrina O, Kozak JA, Wagner SL, Cahalan MD, Velicelebi G, Stauderman KA (2005) STIM1, an essential and conserved component of store-operated Ca2+ channel function. J Cell Biol 169:435–445

Rosado JA, Sage SO (2000a) The actin cytoskeleton in store-mediated calcium entry. J Physiol 526(Pt 2):221–229

Rosado JA, Sage SO (2000b) Coupling between inositol 1,4,5-trisphosphate receptors and human transient receptor potential channel 1 when intracellular Ca2+ stores are depleted. Biochem J 350(Pt 3):631–635

Rosado JA, Sage SO (2000c) Platelet signalling: calcium. In: Gresele P, Page CP, Fuster V, Vermylen J (eds) Platelets in thrombotic and non-thrombotic disorders pathophysiology, pharmacology and therapeutics. Cambridge University Press, Cambridge, pp 260–271

Rosado JA, Sage SO (2000d) Protein kinase C activates non-capacitative calcium entry in human platelets. J Physiol 529(Pt 1):159–169

Rothberg BS, Wang Y, Gill DL (2013) Orai channel pore properties and gating by STIM: implications from the Orai crystal structure. Sci Signal 6:pe9

Sage SO, Merritt JE, Hallam TJ, Rink TJ (1989) Receptor-mediated calcium entry in fura-2-loaded human platelets stimulated with ADP and thrombin. Dual-wavelengths studies with Mn2+. Biochem J 258:923–926

Sauc S, Bulla M, Nunes P, Orci L, Marchetti A, Antigny F, Bernheim L, Cosson P, Frieden M, Demaurex N (2015) STIM1L traps and gates Orai1 channels without remodeling the cortical ER. J Cell Sci 128:1568–1579

Shen WW, Demaurex N (2012) Morphological and functional aspects of STIM1-dependent assembly and disassembly of store-operated calcium entry complexes. Biochem Soc Trans 40:112–118

Shirakawa R, Higashi T, Tabuchi A, Yoshioka A, Nishioka H, Fukuda M, Kita T, Horiuchi H (2004) Munc13-4 is a GTP-Rab27-binding protein regulating dense core granule secretion in platelets. J Biol Chem 279:10730–10737

Soboloff J, Spassova MA, Tang XD, Hewavitharana T, Xu W, Gill DL (2006) Orai1 and STIM reconstitute store-operated calcium channel function. J Biol Chem 281:20661–20665

Spassova MA, Soboloff J, He LP, Xu W, Dziadek MA, Gill DL (2006) STIM1 has a plasma membrane role in the activation of store-operated Ca(2+) channels. Proc Natl Acad Sci U S A 103:4040–4045

Stathopulos PB, Li GY, Plevin MJ, Ames JB, Ikura M (2006) Stored Ca2+ depletion-induced oligomerization of stromal interaction molecule 1 (STIM1) via the EF-SAM region: an initiation mechanism for capacitive Ca2+ entry. J Biol Chem 281:35855–35862

Stathopulos PB, Zheng L, Ikura M (2009) Stromal interaction molecule (STIM) 1 and STIM2 calcium sensing regions exhibit distinct unfolding and oligomerization kinetics. J Biol Chem 284:728–732

Stathopulos PB, Schindl R, Fahrner M, Zheng L, Gasmi-Seabrook GM, Muik M, Romanin C, Ikura M (2013) STIM1/Orai1 coiled-coil interplay in the regulation of store-operated calcium entry. Nat Commun 4:2963

Stormorken H (2002) [Stormorken's syndrome]. Tidsskr Nor Laegeforen 122:2853–2856

Stormorken H, Sjaastad O, Langslet A, Sulg I, Egge K, Diderichsen J (1985) A new syndrome: thrombocytopathia, muscle fatigue, asplenia, miosis, migraine, dyslexia and ichthyosis. Clin Genet 28:367–374

Stormorken H, Holmsen H, Sund R, Sakariassen KS, Hovig T, Jellum E, Solum O (1995) Studies on the haemostatic defect in a complicated syndrome. An inverse Scott syndrome platelet membrane abnormality? Thromb Haemost 74:1244–1251

Thompson JL, Shuttleworth TJ (2013) Molecular basis of activation of the arachidonate-regulated Ca2+ (ARC) channel, a store-independent Orai channel, by plasma membrane STIM1. J Physiol 591:3507–3523

Thompson JL, Mignen O, Shuttleworth TJ (2009) The Orai1 severe combined immune deficiency mutation and calcium release-activated Ca2+ channel function in the heterozygous condition. J Biol Chem 284:6620–6626

To MS, Aromataris EC, Castro J, Roberts ML, Barritt GJ, Rychkov GY (2010) Mitochondrial uncoupler FCCP activates proton conductance but does not block store-operated Ca(2+) current in liver cells. Arch Biochem Biophys 495:152–158

Tolhurst G, Carter RN, Amisten S, Holdich JP, Erlinge D, Mahaut-Smith MP (2008) Expression profiling and electrophysiological studies suggest a major role for Orai1 in the store-operated Ca2+ influx pathway of platelets and megakaryocytes. Platelets 19:308–313

Trepakova ES, Gericke M, Hirakawa Y, Weisbrod RM, Cohen RA, Bolotina VM (2001) Properties of a native cation channel activated by Ca2+ store depletion in vascular smooth muscle cells. J Biol Chem 276:7782–7790

Vaca L, Kunze DL (1994) Depletion of intracellular Ca2+ stores activates a Ca(2+)-selective channel in vascular endothelium. Am J Physiol 267:C920–C925

Varga-Szabo D, Authi KS, Braun A, Bender M, Ambily A, Hassock SR, Gudermann T, Dietrich A, Nieswandt B (2008a) Store-operated Ca(2+) entry in platelets occurs independently of transient receptor potential (TRP) C1. Pflugers Arch 457:377–387

Varga-Szabo D, Braun A, Kleinschnitz C, Bender M, Pleines I, Pham M, Renne T, Stoll G, Nieswandt B (2008b) The calcium sensor STIM1 is an essential mediator of arterial thrombosis and ischemic brain infarction. J Exp Med 205:1583–1591

Venkatachalam K, Montell C (2007) TRP channels. Annu Rev Biochem 76:387–417

Vial C, Pitt SJ, Roberts J, Rolf MG, Mahaut-Smith MP, Evans RJ (2003) Lack of evidence for functional ADP-activated human P2X1 receptors supports a role for ATP during hemostasis and thrombosis. Blood 102:3646–3651

Vig M, Peinelt C, Beck A, Koomoa DL, Rabah D, Koblan-Huberson M, Kraft S, Turner H, Fleig A, Penner R, Kinet JP (2006) CRACM1 is a plasma membrane protein essential for store-operated Ca2+ entry. Science 312:1220–1223

Wes PD, Chevesich J, Jeromin A, Rosenberg C, Stetten G, Montell C (1995) TRPC1, a human homolog of a Drosophila store-operated channel. Proc Natl Acad Sci U S A 92:9652–9656

White JG, Gunay-Aygun M (2011) The York platelet syndrome: a third case. Platelets 22:117–134

White JG, Pakzad K, Meister L (2013) The York platelet syndrome: a fourth case with unusual pathologic features. Platelets 24:44–50

Wu MM, Buchanan J, Luik RM, Lewis RS (2006) Ca2+ store depletion causes STIM1 to accumulate in ER regions closely associated with the plasma membrane. J Cell Biol 174:803–813

Yang X, Jin H, Cai X, Li S, Shen Y (2012) Structural and mechanistic insights into the activation of stromal interaction molecule 1 (STIM1). Proc Natl Acad Sci U S A 109:5657–5662

Yeromin AV, Zhang SL, Jiang W, Yu Y, Safrina O, Cahalan MD (2006) Molecular identification of the CRAC channel by altered ion selectivity in a mutant of Orai. Nature 443:226–229

Yoshioka A, Shirakawa R, Nishioka H, Tabuchi A, Higashi T, Ozaki H, Yamamoto A, Kita T, Horiuchi H (2001) Identification of protein kinase Calpha as an essential, but not sufficient, cytosolic factor for Ca2+-induced alpha- and dense-core granule secretion in platelets. J Biol Chem 276:39379–39385

Yu F, Sun L, Courjaret R, Machaca K (2011) Role of the STIM1 C-terminal domain in STIM1 clustering. J Biol Chem 286(10):8375–8384

Yuan JP, Kiselyov K, Shin DM, Chen J, Shcheynikov N, Kang SH, Dehoff MH, Schwarz MK, Seeburg PH, Muallem S, Worley PF (2003) Homer binds TRPC family channels and is required for gating of TRPC1 by IP3 receptors. Cell 114:777–789

Yuan JP, Zeng W, Dorwart MR, Choi YJ, Worley PF, Muallem S (2009) SOAR and the polybasic STIM1 domains gate and regulate Orai channels. Nat Cell Biol 11:337–343

Zbidi H, Jardin I, Woodard GE, Lopez JJ, Berna-Erro A, Salido GM, Rosado JA (2011) STIM1 and STIM2 are located in the acidic Ca2+ stores and associates with Orai1 upon depletion of the acidic stores in human platelets. J Biol Chem 286:12257–12270

Zeng W, Yuan JP, Kim MS, Choi YJ, Huang GN, Worley PF, Muallem S (2008) STIM1 gates TRPC channels, but not Orai1, by electrostatic interaction. Mol Cell 32:439–448

Zhang SL, Yu Y, Roos J, Kozak JA, Deerinck TJ, Ellisman MH, Stauderman KA, Cahalan MD (2005) STIM1 is a Ca2+ sensor that activates CRAC channels and migrates from the Ca2+ store to the plasma membrane. Nature 437:902–905

Zhang SL, Yeromin AV, Zhang XHF, Yu Y, Safrina O, Penna A, Roos J, Stauderman KA, Cahalan MD (2006) Genome-wide RNAi screen of Ca2+ influx identifies genes that regulate Ca2+ release-activated Ca2+ channel activity. Proc Natl Acad Sci U S A 103:9357–9362. doi:10.1073/pnas.0603161103

Zhang X, Zhang W, Gonzalez-Cobos JC, Jardin I, Romanin C, Matrougui K, Trebak M (2014) Complex role of STIM1 in the activation of store-independent Orai1/3 channels. J Gen Physiol 143:345–359

Zhu X, Chu PB, Peyton M, Birnbaumer L (1995) Molecular cloning of a widely expressed human homologue for the Drosophila trp gene. FEBS Lett 373:193–198

Cardiovascular and Hemostatic Disorders: SOCE in Cardiovascular Cells – Emerging Targets for Therapeutic Intervention

24

Klaus Groschner, Niroj Shrestha, and Nicola Fameli

Abstract

The discovery of the store-operated Ca^{2+} entry (SOCE) phenomenon is tightly associated with its recognition as a pathway of high (patho)physiological significance in the cardiovascular system. Early on, SOCE has been investigated primarily in non-excitable cell types, and the vascular endothelium received particular attention, while a role of SOCE in excitable cells, specifically cardiac myocytes and pacemakers, was initially ignored and remains largely enigmatic even to date. With the recent gain in knowledge on the molecular components of SOCE as well as their cellular organization within nanodomains, potential tissue/cell type-dependent heterogeneity of the SOCE machinery along with high specificity of linkage to downstream signaling pathways emerged for cardiovascular cells. The basis of precise decoding of cellular Ca^{2+} signals was recently uncovered to involve correct spatiotemporal organization of signaling components, and even minor disturbances in these assemblies trigger cardiovascular pathologies. With this chapter, we wish to provide an overview on current concepts of cellular organization of SOCE signaling complexes in cardiovascular cells with particular focus on the spatiotemporal aspects of coupling to downstream signaling and the potential disturbance of these mechanisms by pathogenic factors. The significance of these mechanistic concepts for the development of novel therapeutic strategies will be discussed.

K. Groschner (✉) • N. Shrestha
Institute of Biophysics, Medical University of Graz, Neue Stiftingtalstrasse 6/4, 8010 Graz, Austria
e-mail: klaus.groschner@medunigraz.at

N. Fameli
Institute of Biophysics, Medical University of Graz, Neue Stiftingtalstrasse 6/4, 8010 Graz, Austria

Department of Anesthesiology, Pharmacology, and Therapeutics, University of British Columbia, Vancouver, BC, Canada

K. Groschner et al. (eds.), *Store-Operated Ca^{2+} Entry (SOCE) Pathways*, Advances in Experimental Medicine and Biology 993, DOI 10.1007/978-3-319-57732-6_24

Keywords
STIM-Orai signaling • TRPC channels • Cardiovascular disease • ER-PM junctions

24.1 Introduction

24.1.1 SOCE Is a Critical Determinant of Cardiovascular Functions

When Orai1 was first identified as the key component of SOCE in immune cells (Feske et al. 2006; Zhang et al. 2006; Feske 2011), SOCE as a cellular phenomenon had already been characterized extensively in cardiovascular cells, and the involvement of various Ca^{2+} signaling molecules including other potential pore-forming components such as transient receptor potential canonical (TRPC) proteins had been suggested (Groschner et al. 1998). Once STIM/Orai complexes were convincingly demonstrated as the core molecular machinery possessing a highly Ca^{2+} selective conductance, which is activated upon Ca^{2+} depletion from endoplasmic reticulum (ER) (Prakriya et al. 2006), involvement of these molecules in cardiovascular SOCE was reported for a variety of cardiovascular cell preparations (Abdullaev et al. 2008; Potier et al. 2009; Völkers et al. 2012; Liu et al. 2015). This suggests that the classical, highly Ca^{2+} selective, Orai channel-mediated pathway is indeed present in cardiovascular tissues. Nonetheless, thorough electrophysiological characterization and unambiguous biophysical demonstration of native Orai conductances in cardiovascular cells is so far sparse. Moreover, expression of Orai species in healthy terminally differentiated myocytes appears rather low, and compelling evidence for a classical Orai1 conductance specifically in cardiomyocytes is still lacking. Nonetheless, expression levels of both STIM1 and Orai increase when myocytes switch to proliferative phenotypes and during pathological remodeling (Potier et al. 2009; Kassan et al. 2016; Saliba et al. 2015), and the most convincing demonstration of Orai conductances has been obtained for proliferative, synthetic vascular smooth muscle (Potier et al. 2009). Meanwhile, gene knockdown experiments in cardiac tissues provided ample evidence for a role of various molecules related to SOCE such as STIM, Orai, as well as TRP proteins in maladaptive responses (Völkers et al. 2012; Horton et al. 2014; Seo et al. 2014; Saliba et al. 2015). Since these molecules, especially STIM and TRPC, are increasingly recognized as multifunctional signaling proteins that mediate cellular responses other than SOCE (Lefkimmiatis et al. 2009; Shinde et al. 2013; Beliveau et al. 2014; Kitajima et al. 2016), the significance of SOCE in cardiovascular cells still deserves critical evaluation. Neither suppression of cellular responses by knockdown of molecules supposed to contribute to a SOCE pathway in a certain cellular setting nor the identification of cellular events that are triggered by a drop in ER Ca^{2+} levels and/or associated with enhanced Ca^{2+} entry appears sufficient to conclude a critical role of SOCE in the respective cellular function or dysfunction. It is now critical to confirm a causal relation between store-operated Ca^{2+}

conductances, identified at the electrophysiological level, and specific cardiovascular functions that are suspected to depend on SOCE. The main hurdles to complete this task are the tiny nature of the native CRAC and/or SOCE conductances in cardiovascular cells (Hunton et al. 2004) along with highly complex and prominent background conductances, especially in electrically excitable cells, as well as the lack of suitable pharmacological tools for selective targeting of SOCE. Moreover, recent evidence suggests an exquisitely efficient and specific linkage of SOCE-related Ca^{2+} handling mechanisms to downstream signaling pathways (Kar et al. 2016; Samanta and Parekh 2016). Consistently, SOCE and related Ca^{2+} entry mechanisms may be required and expressed at essentially low abundance in native cells. This scenario makes delineation of the (patho)physiological role of SOCE conductances in cardiovascular tissue a highly challenging endeavor. Importantly, SOCE channels are recently understood to serve Ca^{2+} entry by operating in a concerted and interdependent manner with other ion transport systems within the specialized and dynamic subcellular architecture of sarco-/endoplasmic reticulum (S/ER)-plasma membrane (PM) nanojunctions, intracellular regions including juxtaposed PM and S/ER membrane patches separated by 10–30 nm and extending laterally for several hundred nm, the junctional cytosol, and the local transporters (channels, pumps, exchangers). This enables spatially restricted signaling events that are well segregated from those in other cellular compartments as well as from global cytosolic Ca^{2+} changes. Despite an increasing awareness of the latter concept, investigators still widely use global cellular Ca^{2+} measurements as an experimental readout to quantify SOCE. These studies bear a high risk of misinterpretations. The advent of technologies that allow for monitoring of compartmentalized signaling processes with high spatial and temporal resolution has recently opened possibilities of more precise characterization of the functional role of individual Ca^{2+} handling units such as the SOCE machinery. In this chapter, we will cast particular emphasis on these novel aspects and the therapeutic potential of the emerging signaling concepts.

Irrespective of the above outlined difficulties in unambiguous identification of endogenous SOCE conductances and the uncertainties in determining their exact mechanistic link to cellular functions, a critical role of SOCE-related molecules in cardiovascular pathologies has been well documented by knockdown/overexpression experiments and careful pharmacological studies. Regarding the cardiovascular role of Orai1 in the sense of forming a "classical" highly Ca^{2+} selective conductance (CRAC; Hoth and Penner 1992), several investigations strongly pointed toward a specific role in proliferative cell phenotypes and adult stem cells (Abdullaev et al. 2008; Potier et al. 2009; Dragoni et al. 2014). Thus, a prominent role for SOCE apparently resides in tissue development and repair as well as in long-term adaptation and remodeling, while terminally differentiated cardiovascular cells such as cardiomyocytes and contractile smooth muscle cells may maintain and control their function predominantly by non-SOCE Ca^{2+} signaling pathways. Cardiovascular cells, which express significant SOCE, typically display proliferative and migrating phenotypes (Bisaillon et al. 2010; Potier et al. 2009), and SOCE was found to govern gene expression programs

(Kar et al. 2011, 2012b). A similar linkage to Ca^{2+}-dependent control of gene expression (Ca^{2+}/transcription coupling) has been demonstrated for both non-SOCE signaling pathways formed by cardiovascular TRPC channels (Nakayama et al. 2006; Kuwahara et al. 2006; Poteser et al. 2011). All these pathways share important structural and mechanistic features, specifically the localization of channel complexes within specialized domains of the PM, dynamic distribution of lipid domains, and physical coupling to ER-resident proteins via S/ER-PM junctions (see Part I, Chaps. 8–10). This special subcellular organization allows for an extraordinarily efficient coordination and coupling between Ca^{2+} fluxes through the plasma and the ER membrane with signaling nanospaces. In this context, ER-PM junctions facilitate SOCE, and related Ca^{2+} handling processes operate within ER-PM junctions to enable sub-plasmalemmal Ca^{2+} cycling and cellular Ca^{2+} oscillations (Thomas et al. 1996; Luo et al. 2001; Wedel et al. 2007), which are tightly linked to cell functions. SOCE-dependent oscillatory Ca^{2+} cycling is likely the basis for transcriptional regulation (Samanta and Parekh 2016) as well as pacemaking activity (Zhang et al. 2015).

In aggregate, ample evidence suggests SOCE as a determinant of the function of specialized cells in cardiovascular tissues, and this cardiovascular role of SOCE appears to involve nanojunctional signaling within ER-PM junctions and oscillatory Ca^{2+} cycling. Below, we will specifically address these aspects with a closer view on channels in endothelial cells and in cardiac pacemakers and discuss novel mechanistic concepts in terms of their potential therapeutic utilization.

24.1.2 Heterogeneity of Cardiovascular SOCE Channels

For most native tissues, including those of the cardiovascular system, heterogeneity of SOCE channels has been reported and needs consideration. Moreover, amplitude as well as spatial and temporal features of SOCE-mediated Ca^{2+} signals linked to cell functions are typically determined (shaped/tuned) by cooperating transporters and channels. This scenario makes unequivocal identification of the channel (s) involved primarily in SOCE difficult and prompts for critical evaluation by electrophysiological measurements. Heterogeneity of store-operated, Ca^{2+} permeable channels in endothelial cells was highlighted early on by electrophysiological characterization of store depletion-activated endothelial cation conductances (Fasolato and Nilius 1998; Groschner et al. 1998; Encabo et al. 1996; Freichel et al. 2001), which resolved divergent currents including both nonselective and highly Ca^{2+} selective conductances similar to the CRAC phenomenon originally described in immune cells (Hoth and Penner 1992; Lewis and Cahalan 1989). Although the molecular analysis of the underlying Ca^{2+} entry pathway in terms of signalplex composition and microdomain organization is still not comprehensive, key components of endothelial SOCE channels have been identified, and heterogeneity of endothelial Ca^{2+} channels involved in mediating agonist-induced responses of the endothelium has more recently been suggested by way of RNAi and dominant negative knockdown strategies.

24.1.3 Molecular Basis of SOCE in Endothelial Cells

For more than a decade, members of the classical/canonical subfamily of TRP cation channel proteins have been proposed as the main candidate channel structures for endothelial SOCE. Although evidence from heterologous expression studies with TRPC proteins raised substantial doubt about their direct store regulation and demonstrated rather low Ca^{2+} selectivity of these pore complexes, endothelial TRPC channels have been implicated in both nonselective (Groschner et al. 1998) and highly Ca^{2+} selective, CRAC-like endothelial conductances (Freichel et al. 2001). The identification of STIM/Orai (see also Part I) as the molecular basis of the immune cell CRAC phenomenon prompted tests for an involvement of Orai1 in endothelial SOCE. These studies yielded compelling evidence for physiologically relevant, Orai1-mediated Ca^{2+} signaling in endothelial cells (Abdullaev et al. 2008). Moreover, these investigations clearly point toward a highly specialized cellular role of Orai-mediated Ca^{2+} entry, with particular significance for the control of proliferation. By contrast, other important SOCE-dependent cell functions such as migration and barrier stability appear under the control of divergent, nonetheless related channel complexes. Such endothelial SOCE phenomena that differ in biophysical properties from classical CRAC conductances have been repeatedly observed (Groschner et al. 1994, 1998; Encabo et al. 1996) and may be based on either TRPC complexes forming a central Ca^{2+} channel structure or microdomain-organized multi-ion channel signalplexes comprised of TRPC as well as Orai and/or STIM proteins, a concept that has been put forward for other non-excitable cell types (Cheng et al. 2011). For detailed information on current concepts regarding TRPC-mediated SOCE mechanisms, see also Part I, Chaps. 8 and 9.

It is of note that classical CRAC-like endothelial current components are small and control specific cellular functions such as proliferation (Abdullaev et al. 2008). The relative contribution of Ca^{2+} channels of the TRPC family to agonist-induced Ca^{2+} entry may vary between vascular beds and was also found to undergo striking changes with cell-cell contact-dependent phenotype transitions (Graziani et al. 2010). Thus, endothelial Ca^{2+} entry mechanisms may undergo substantial reorganization during phenotype transitions leading to variable contribution of Orai and TRPC channels.

24.1.4 Molecular Basis of SOCE in Myocytes and Pacemakers

While non-excitable cells, specifically immune cells, have been the focus of SOCE research since the identification of this pathway, recognition and understanding of SOCE in excitable cells such as skeletal muscle, smooth muscle, and cardiac cells was accumulated more slowly (Casteels and Droogmans 1981; Kurebayashi and Ogawa 2001; Hunton et al. 2002; Launikonis et al. 2003; Völkers et al. 2012). As outlined above, the currents observed in cardiovascular myocytes are typically small and were detected predominantly in embryonic and proliferating cells. Consequently, SOCE has been implicated in tissue plasticity, repair, and hypertrophic growth of cardiac and vascular myocytes (Nakayama et al. 2006; Voelkers et al. 2010; Horton et al. 2014; Saliba et al. 2015). This role of SOCE is outlined and discussed in great detail in Chaps. 22–24. Genetic manipulation of STIM and Orai expression along with pharmacological studies provided evidence for a role of these two molecular key components of SOCE in cardiac as well as vascular remodeling. Of note, involvement of ER Ca^{2+} sensors of the STIM family in cardiovascular Ca^{2+} handling and in pacemaking activity has been reported even more convincingly (Shinde et al. 2013; Beliveau et al. 2014; Correll et al. 2015; Zhao et al. 2015). The latter studies demonstrate STIM as a pivotal Ca^{2+} sensor in cardiovascular cells and indicate a multifunctional role of STIM due to its ability to couple not only to Orai channels but also to other essential Ca^{2+} signaling elements such as voltage-gated channels (CaV1.2) and SERCA. To this end, it remains open if activation of STIM in myocytes triggers cytosolic Ca^{2+} elevation primarily via opening of Orai channels or via coordinated modulation of multiple Ca^{2+} handling molecules including also TRPC or NCX1 (Zhao et al. 2015). TRPC channels have been in parallel identified as essential Ca^{2+} channels in proliferating muscle tissue and proposed to function as components of pathological SOCE pathways (Nakayama et al. 2006; Kuwahara et al. 2006; Koenig et al. 2013; Seo et al. 2014; Camacho Londoño et al. 2015). Importantly, the control of cell phenotype and gene expression is closely related to the tuning of cellular Ca^{2+} oscillations (Di Capite et al. 2009; Kar et al. 2012a, 2016; Yeh et al. 2014), a mechanism that is commonly attributed to Orai channels (Parekh 2007). Considering involvement of STIM and Orai molecules in the control of functions as diverse as gene expression and pacemaking, one might consider certain heterogeneity of the underlying SOCE machinery. Electrical oscillators in biological systems typically involve both voltage-gated and Ca^{2+}-gated currents to generate cyclic electrical activity. Nonetheless, oscillatory Ca^{2+} handling and cyclic exchange of Ca^{2+} between the extracellular space and the S/ER is also a key mechanism of autonomic electrical activity in pacemakers in terms of the "Ca^{2+} clock" and a coupled intracellular and membrane clock mechanism (Lakatta et al. 2006; Yaniv et al. 2015; Sirenko et al. 2016). Importantly, Ca^{2+} and electrical oscillations in cells generally involve a rather tight coupling between Ca^{2+} transport in the ER and PM. Thus, the control of myocyte phenotype and electrical pacemaking by SOCE may share common cellular mechanisms, based on precise spatial organization of cooperating signaling molecules within S/ER-PM nanojunctions. The core molecular players in these signaling domains may be similar to those in non-excitable cells comprising STIM/Orai, and TRPC channels, all coupled intimately to ER Ca^{2+} release channels and Ca^{2+} transporters. Function and dysfunction of these signaling units is expected to depend crucially on spatial aspects and membrane architecture of the nanojunction. Current knowledge on molecular and spatial aspects of these signal transduction hubs in cardiovascular cells is provided below.

Using classical pharmacological strategies to block Orai channels and by genetic knockdown of STIM (Wedel et al. 2007; Liu et al. 2015; Zhang et al. 2015), an essential role of SOCE in Ca^{2+} oscillation as well as in sinoatrial pacemaking has been demonstrated. An example for the generation of Ca^{2+} oscillations, ostentatiously involving Orai and the cardiac ryanodine receptor subtype 2 (RyR2), is illustrated in Fig. 24.1. Similarly, extracellular Ca^{2+} dependence and sensitivity to a

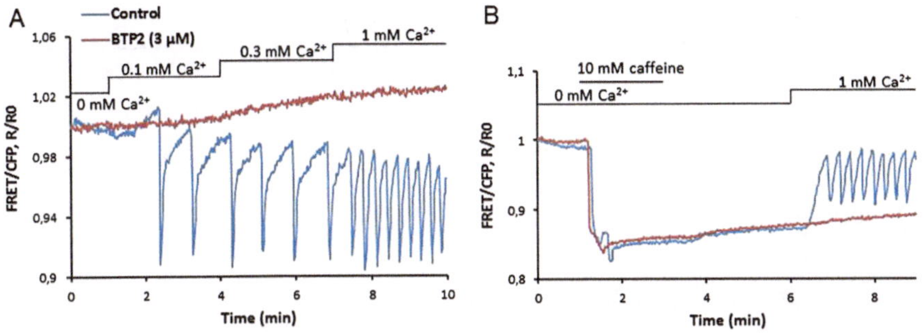

Fig. 24.1 Effect of the SOCE (Orai) blocker BTP2 on oscillations in ER Ca^{2+} levels of hRyR2_HEK293 cells. (**a**) Representative $[Ca^{2+}]_{ER}$ changes showing oscillations mediated by SOCE via STIM1-Orai1 (*blue*) and inhibition by 3 μM BTP2 (*red*). (**b**) Representative $[Ca^{2+}]_{ER}$ changes showing store depletion with 10 mM caffeine and refilling upon 1 mM extracellular Ca^{2+} addition (*blue*) and inhibition by 3 μM BTP2

blocker of Orai channels is illustrated by the experiment recording spontaneous activity in mouse pacemaker tissue (Fig. 24.2).

Figure 24.1 depicts an example of the dependence of spontaneous Ca^{2+} mobilization in RyR2 expressing HEK293 (hRyR2_HEK293) cells, measured by the ER-targeted Ca^{2+} sensor D1ER, on extracellular Ca^{2+}, and the activity of Orai channels as evidenced by the use of the pyrazole blocker BTP2. Figure 24.2 shows the dependency of firing in a pacemaker preparation on Ca^{2+} entry, most presumably via SOCE. When extracellular Ca^{2+} is removed, intracellular Ca^{2+} slowly declines, and pacemaker activity (indicated by small upward spikes on the record) ceases. With addition of extracellular Ca^{2+}, these features are reversed (compare also oscillations shown in Fig. 24.1). When SERCA is inhibited and SR Ca^{2+} is depleted, re-addition of extracellular Ca^{2+} causes a very large transient rise in intracellular Ca^{2+}, which was blocked by SKF 96365 or Gd^{3+}, both representing SOCE inhibitors (Ju et al. 2007).

Interpretation of such experimental data is complicated by the fact that all currently available pharmacological tools are insufficiently selective. Moreover, pharmacological depletion of SR stores has multiple, partly opposing effects. (1) Ca^{2+} transients are reduced and inward currents generated by Na^{+}/Ca^{2+} exchange associated with removing Ca^{2+} is reduced which tends to slow the rate of rise of the pacemaker potential and reduce the firing rate. (2) Store depletion will activate I_{SOC} and provide an inward current, which enhances pacemaker currents and increases firing rate.

The synopsis of currently available evidence suggests that SOCE channels, presumably Orai and TRPC pore complexes, operate within a well-defined environment of efficiently communicating Ca^{2+} handling molecules with spatial features set by the nanoarchitecture of ER-PM junctions. This architecture may be of similar (patho)physiological significance as expression and proper targeting of SOCE channels.

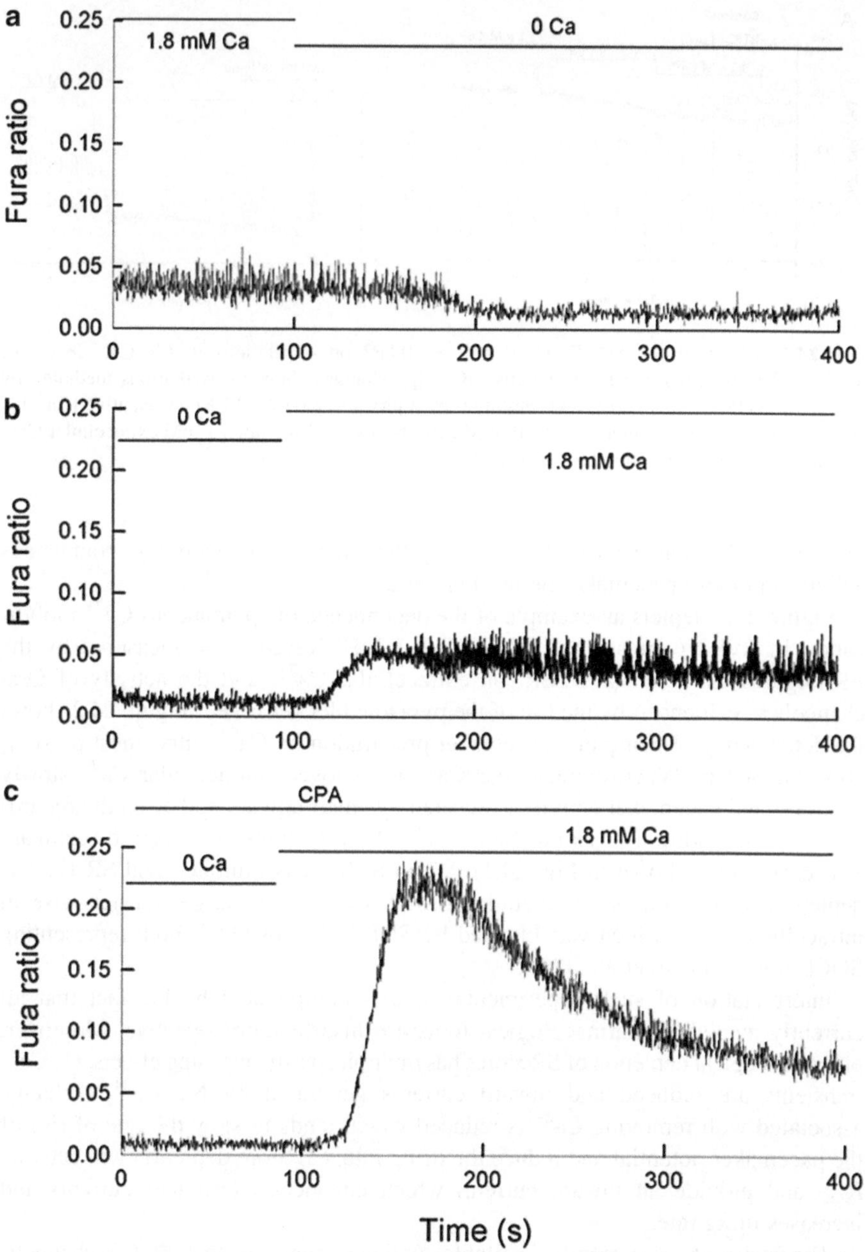

Fig. 24.2 All panels show fura-2 ratio measurements, representing intracellular Ca^{2+}, from isolated right atrial tissue including the sinoatrial node. The sinoatrial node was identified by anatomical landmarks and an adjustable shutter used to isolate the signal from this region. Panel **a**: Initial record in 1.8 mM extracellular Ca^{2+} shows spikes representing spontaneous firing from pacemaker tissue. When extracellular Ca^{2+} was reduced to zero, intracellular Ca^{2+} slowly fell and spontaneous firing ceased. Panel **b**: Replacement of extracellular Ca^{2+} increased intracellular Ca^{2+}

24.2 Spatiotemporal Aspects of Cardiovascular SOCE

SOCE channels represent a signaling element that strictly requires a specialized membrane architecture, which is formed by apposition of the superficial SR/ER membrane and the PM. SOCE competent ER-PM junctions have recently been in focus of detailed investigations of both functional (protein-interaction) and structural aspects (Schindl et al. 2009; Chang et al. 2013; Wu et al. 2014; Maleth et al. 2014; Poteser et al. 2016).

Figure 24.3 illustrates recent advances in the delineation of ER-PM junctional architecture which can be quantitatively delineated with respect to dynamics of the ER-PM junctional cleft and surface topology by a novel TIRFM method (Poteser et al. 2016) and a first application in endothelial cells. Fluorescence density mapping FDM was shown suitable to investigate the nanometer-scale architecture of junctional clefts with high temporal resolution based on the displacement of cytoplasmic fluorophore reporters by ER that approaches the PM.

Figure 24.4 gives an overview on direct and indirect interaction within the nanoarchitecture of SR/ER-PM junctions with focus on STIM-Orai or STIM-TRPC complexes and Ca^{2+} signaling by cardiovascular SOCE. Communication and signal generation is proposed to involve oscillatory Ca^{2+} cycling that is critically dependent on the physical communication between PM and SR/ER proteins as well as on spatial relations between the interaction partners in the nanodomain.

24.2.1 The Essential Role of S/ER-PM Nanojunctions

One lingering question is whether in cardiovascular cells SOCE is strictly linked to a certain population of SR/ER-PM junctions and their nanoarchitecture as it has been observed in other cell types in several studies over the past decade. In brief, pivotal experiments in Jurkat T cells revealed STIM punctae caused by ER Ca^{2+} release were co-localized with ER-PM junctions, with nanometric dimensional features as mentioned above (ER membrane to PM separation around 17 nm and extension in the 150–300 nm range) (Wu et al. 2014; Luik et al. 2006). SR/ER-PM junctions are regularly observed in cardiac myocytes (e.g., in the form of cardiac dyads), vascular smooth muscle cells (Devine et al. 1972; Poburko et al. 2004), as well as vascular endothelial cells (Di Giuro et al. 2016; see also Fig. 24.3); however, evidence that these junctions are indeed essentially required for SOCE in these cell types is still lacking.

Fig. 24.2 (continued) back to initial levels and spontaneous firing restarted. Panel **c**: When the SERCA inhibitor CPA was present, which allows the SR Ca^{2+} to be depleted, replacement of extracellular Ca^{2+} causes a very large and transient addition Ca^{2+} influx characteristic of SOCE (modified from Ju et al. 2007; see also Store-Operated Ca^{2+} Entry (SOCE) Pathways. 1st edition 2012 (eds. Groschner K, Graier WF, Romanin C; Springer Verlag Wien) Chap. 23)

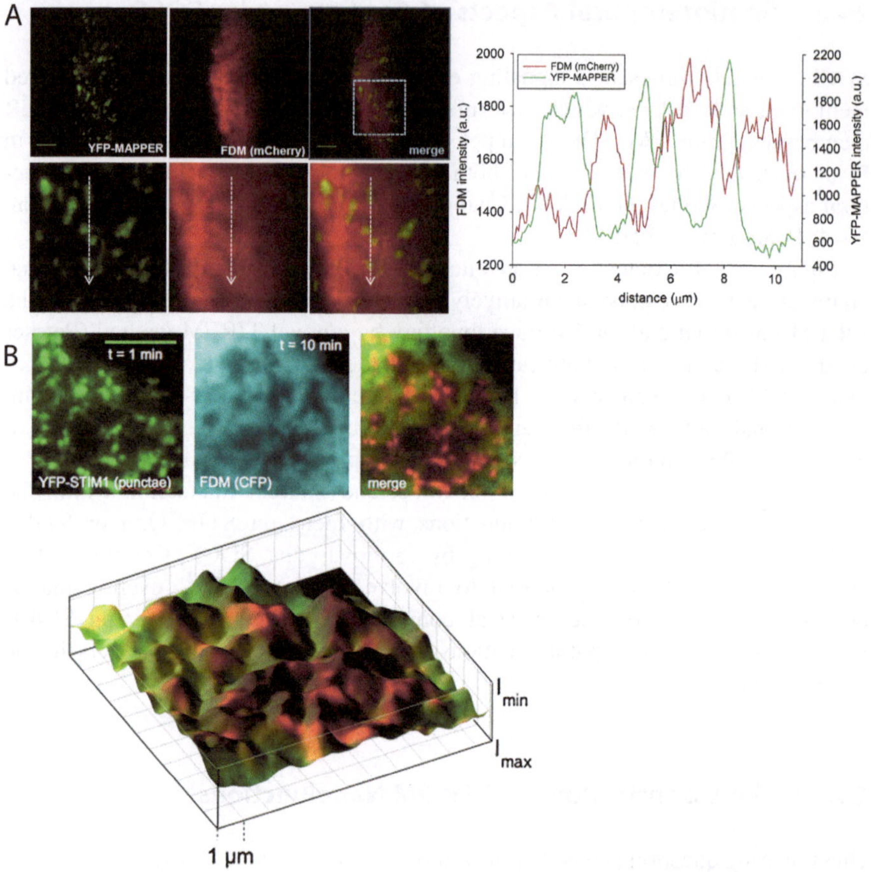

Fig. 24.3 Detection and 3D reconstruction of ER-PM junctional architecture by FDM-TIRFM. (**a**) Detection of ER-PM junctions in an EA.hy926 endothelial cell using MAPPER (Chang et al. 2013) and FDM scanning. *Top*, *left*: TIRFM micrographs of EA.hy926 cell, co-expressing YFP-MAPPER (*green*, *left*) and cytosolic mCherry (*red*, *middle*) along with the color merge (*left*). *Below*: Magnified areas indicated in *top panel*. *Right*: Line scans of YFP-MAPPER fluorescence intensity (*green*) and FDM intensity (*red*: minima correspond to ER junctional regions; see Poteser et al. 2016) as indicated by the *dotted arrow* in the images. *Scale bar* = 5 μm. (**b**) Topology of the ER membrane within the sub-plasmalemmal space of a RBL mast cell expressing YFP-STIM1 by 3D topology reconstruction from FDM data. *Top*: YFP-STIM1 punctae ($t = 1$ min after store depletion, *green*), FDM junctions (*cyan*), and color overlay (*right*). *Bottom*: 3D reconstruction of sub-plasmalemmal organelle surface within the evanescent field based on FDM and superimposed color merge image displaying localization of YFP-STIM1 clusters. I_{min} corresponds to position of the PM cytoplasmic surface. The position of STIM1 cluster (*orange*) in tight appositions (*peaks*) is shown (from Poteser et al. 2016)

Several studies suggest that the dimensional design of SR/ER-PM appositions appears critical for efficient delivery of Ca^{2+} from the extracellular space to the ER lumen, typically from Ca^{2+} ingress via a source on the PM to Ca^{2+} capture by SERCA pumps on the junctional ER membrane (described in more detail in

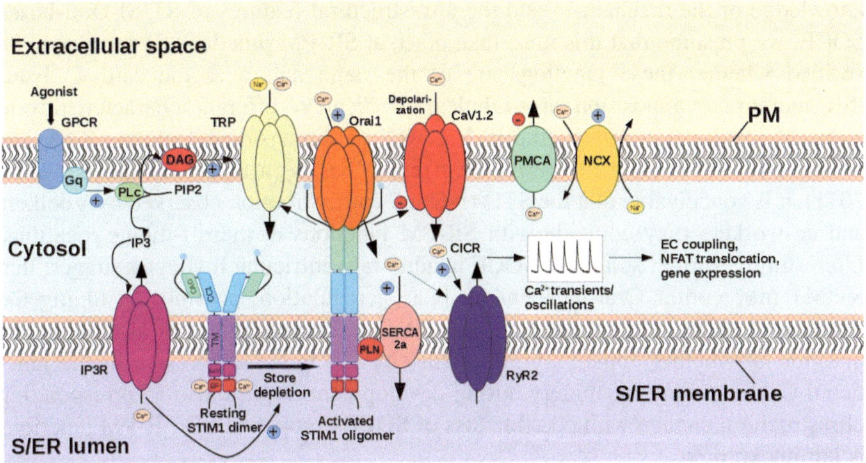

Fig. 24.4 Effectors/interaction partners of STIM1 and mechanisms associated with SOCE in cardiovascular cells. Upon stimulation, Ca^{2+} enters the cytoplasm via L-type Ca^{2+} channels (CaV1.2), which triggers Ca^{2+}-induced Ca^{2+} release (CICR) from sarcoplasmic reticulum (SR) via the ryanodine receptors (RyR2). Cytosolic Ca^{2+} is sequestered via reuptake into SR through sarco(endo)plasmic reticulum Ca^{2+}-ATPase 2a (SERCA2a) or extruded out of the cell through PM Ca^{2+} channels, Na^{+}-Ca^{2+} exchanger (NCX1), or PM Ca^{2+} ATPase (PMCA). Upon binding of agonists, such as phenylephrine and endothelin, to G protein-coupled receptors (GPCR), PLC is activated to convert phosphatidylinositol 4,5-bisphosphate (PIP_2) to diacylglycerol (DAG) and inositol 1,4,5-triphosphate (IP_3). DAG activates transient receptor potential (TRP) channels, while IP_3 stimulates IP_3 receptors (IP_3R) causing SR Ca^{2+} release. SR depletion of Ca^{2+} results in its dissociation from EF hand motifs in STIM1, leading to STIM1 oligomerization and translocation into SR-PM junctions. STIM1 oligomers interact with PM lipids through their lysine-rich domain and activate Orai1 or inhibit CaV1.2 channels via CAD domain. Ca^{2+} entry via Orai1 appears essential for S/ER refilling but also for control of oscillatory Ca^{2+} cycling, NFAT translocation, and gene expression. STIM1 also binds to phospholamban (PLN) and indirectly activates SERCA2a and inhibits PMCA. STIM1 is reported to control TRPC channels and reverse mode NCX too and is the pivotal player in ER-PM junctional signal generation

Chap. 17). This ultrastructural arrangement is most likely needed for efficient Ca^{2+} communication also in other types of SR/ER junctions, for example, in the case of lysosome-SR junctions in vascular smooth muscle, which are trigger zones for the generation of cell-wide Ca^{2+} waves (Fameli et al. 2014). In light of this knowledge about SR/ER junctions outside the realm of SOCE, it is important to explore the functional role of SR/ER-PM junctions in cardiovascular cells, as well as the possible pathophysiological consequences of deterioration of the junctional architecture.

After observation of SOCE in adult rat and neonatal rabbit ventricular myocytes (Hunton et al. 2004; Huang et al. 2006) was demonstrated before the discovery of the STIM-Orai system, Voelkers et al. (2010) showed for the first time that both Orai1 and STIM1 are prime players in neonatal rat cardiac myocyte SOCE, appear to co-localize in the perinuclear region, and can regulate both normal and hypertrophic postnatal cardiac growth. Although not emphasized in this work, given the knowledge of the mechanism and the ultrastructural features of STIM-Orai-based SOCE, we presume that this must take place at SR-PM junctions. It remains to be verified whether these junctions are of the same nature as the cardiac dyads (SR membrane apposition at t-tubules) or display different characteristics in terms of localization around the PM as well as dimensional features. Since the presence of t-tubules in the perinuclear region has been reported (Escobar et al. 2011), it is conceivable that the STIM1-Orai1 co-distribution observed by Voelkers and co-workers may coincide with SR-PM junctions at these t-tubule locations. Interestingly, further studies in SOCE in adult rat ventricular myocytes suggest that STIM1 may confer Orai-independent $[Ca^{2+}]_i$ regulation, possibly by tuning the function of SERCA2a and, in turn, the SR Ca^{2+} content (Zhao et al. 2015). The question of whether apparently important changes in the expression of the junctional Ca^{2+} handling machinery during development implies also a corresponding ultrastructural change with possible loss of SOCE-competent SR/ER-PM junctions is left unexplored.

An intriguing observation of STIM1 and Orai1 expression is reported in sino-atrial node cells (Zhang et al. 2015) in which the former protein is distributed both in sub-plasmalemmal SR, where the latter is also detected, and along the Z lines well away from the PM. In these cells, STIM1 still appears to be involved in SOCE, according to the authors, in a concerted action with Orai1. We can therefore hypothesize that this situation would again imply the presence of SR-PM junctions as well as a possible clustering of SERCA pumps to allow for Ca^{2+} SR store replenishment.

In vascular smooth muscle, it has long been established that PM-SR junctions play a fundamental role in SR Ca^{2+} refilling (van Breemen 1977; van Breemen et al. 1995; Poburko et al. 2004); however, these studies highlighted that SR Ca^{2+} refilling mainly occurs through Ca^{2+} entry via receptor-operated channels. In 2009, Potier and co-workers demonstrated that STIM1 and Orai1 were responsible for I_{CRAC} currents in vascular smooth muscle cells from rat aortas and that both were upregulated potentially in conjunction with cell proliferation and migration (Potier et al. 2009). However, the issue of privileged Ca^{2+} refilling mechanisms in SR-PM junctions of smooth muscle (vascular or otherwise) and of the specific molecular interactions within these junctions has not directly been addressed in this tissue type, although it is mentioned as the "model that best explains the activation of SOCs" in a review article (Wang et al. 2008).

In 2008, it was determined that STIM1/Orai1 can mediate SOCE also in vascular endothelial cells (Abdullaev et al. 2008). Nonetheless, it still remains open to what extent this interaction requires the existence or a specific extent/morphology of ER-PM junctions. Suspicion that the refilling of ER Ca^{2+} may take place in a privileged manner that bypasses the bulk cytoplasm arose a year earlier in a study of $[Ca^{2+}]_i$ transients arising from ER Ca^{2+} release in EA.hy926 cells (Malli et al. 2007). The hypothesis that this bypassing route may be provided by ER-PM junctions in these cells received stronger support in a recent report showing the involvement of both Orai1 and NCX in the ER Ca^{2+} refilling process (Di Giuro et al.

2016). This study also provides the only piece of evidence to date for the existence of ER-PM junctions in these cells.

While ample experimental evidence suggests that the STIM/Orai machinery supports SOCE in cardiac myocytes, vascular smooth muscle, and endothelial cells, the specific features of the SR/ER-PM junctions required to assist in ER refilling and/or in oscillatory Ca^{2+} cycling within these cardiovascular cells have yet to be carefully investigated.

24.3 SOCE and SR/ER-PM Signaling in Cardiovascular Pathologies

24.3.1 Cardiovascular SOCE as a Target of Pathogenic Factors

SOCE in cardiovascular cells displays sensitivity to an array of noxious factors. Specifically metabolic disturbances such as redox stress, hyperglycemia, hypoxia, and acidosis modify and/or disrupt SOCE pathways, and this phenomenon has been suggested as an early event in diseases such as hypertension, diabetic angiopathies, and atherosclerosis (Freichel et al. 2001; Nakayama et al. 2006; Jernigan et al. 2006; Zou et al. 2015; Wang et al. 2016b; Spinelli and Trebak 2016). In cardiac myocytes, upregulation of SOCE and enhanced expression of its molecular machinery have been reported for excessive stimulation with neurotransmitters as well as for mechanical stress and pressure overload (Collins et al. 2013). Pathological gain of SOCE function may represent a positive feed-forward element in pathological remodeling and maladaptive responses in the cardiovascular system (see also Chaps. 22–25). Of note, the evidence for modulation of SOCE by certain pathogenic factors is for a large part based on measurements of intracellular Ca^{2+} concentrations or changes in the rate of cytosolic Ca^{2+} upon elevation of extracellular Ca^{2+} (Ca^{2+} re-addition protocols). These data do not unambiguously prove that the primary target is a SOCE channel. Nonetheless, pharmacological modulation of SOCE has emerged as an attractive strategy to prevent cell dysfunction and tissue remodeling and ameliorate cardiovascular diseases.

SOCE may be involved in cardiovascular pathologies due to either loss or gain of function, and pathological alterations in SOCE may originate from altered expression of SOCE complexes, posttranslational, chemical modification of the signaling molecules or from disturbances in their junctional organization. It appears feasible to consider disruption of nanojunctional membrane architecture as a potential mechanism of SOCE pathologies. One of the most commonly accepted pathogenic metabolites reported to affect SOCE are reactive oxygen species (ROS) and reactive nitrogen species (RNS). ROS modify not only proteins but also lipids and membrane structures, and multiple mechanism of interference with SOCE may be considered. Endothelial cells in particular are constantly exposed to ROS, originating from disturbances in endothelial redox balance itself (Li and Shah 2004) or from smooth muscle, macrophages, or neutrophils (Feletou and Vanhoutte 2006). The observation of a profound redox sensitivity of endothelial Ca^{2+}

signaling gave rise to the concept of endothelial SOCE as a key player in endothelial pathophysiology (Elliott et al. 1989; Elliott and Doan 1993; Elliott and Koliwad 1995). Redox modification of Orai1 and/or TRPC-mediated Ca^{2+} entry represents a potential mechanism leading to endothelial damage at elevated levels of ROS (Florea and Blatter 2008). ROS-induced disturbances of SOCE may at a certain level cause deleterious Ca^{2+} entry along with membrane depolarization (Elliott and Koliwad 1995) resulting in cell death (Miller 2006), inability of the endothelium to communicate with blood and smooth muscle cells (Feletou and Vanhoutte 2006), or loss of endothelial barrier function (Usatyuk et al. 2006, see also Sect. 24.3.2.1).

Notably, both TRPC and Orai1 channels, representing potential components of endothelial SOCE (Groschner et al. 1998; Freichel et al. 2001; Sundivakkam et al. 2012; Abdullaev et al. 2008; Li et al. 2011), have been reported to form redox-sensitive conductances (Balzer et al. 1999; Poteser et al. 2006; Bogeski et al. 2010). For Orai1 a direct protein modification via cysteine oxidation has been demonstrated (Bogeski et al. 2010), and Orai isoforms have been recognized as differentially regulated by ROS. Orai1, but not Orai3, was found to be inhibited by oxidation due to modification of a cysteine (C195) at the extracellular portion of the protein (Bogeski et al. 2010). The intracellular signaling protein STIM1 has also been shown to sense oxidative stress via a cysteine residue (C56) located close to its ER Ca^{2+}-sensing domain (Hawkins et al. 2010). Another intracellular Ca^{2+} handling molecule that is closely related to the SOCE machinery is the IP_3-receptor, which has been demonstrated to sense redox stress (Renard-Rooney et al. 1995; Missiaen et al. 1991) and may thus provide a link between ROS and SOCE function.

A prominent, indirect mechanism by which ROS impact on SOCE is lipid modification. Oxidized lipids are well recognized as pathogenic factors in atherogenesis (Berliner and Heinecke 1996) and have been shown to augment Ca^{2+} entry into endothelial cells (Liang et al. 2010). Notably, ROS were found to activate endothelial TRPC channels by a mechanism most likely involving oxidative modification of membrane cholesterol (Poteser et al. 2006). Oxidative metabolism of cholesterol is associated with changes in membrane cholesterol content as well as membrane structure. The observed impact of ROS on channel functions might well be based on their interference with free cholesterol content in membranes and the strict dependency of SOCE on cholesterol-rich microdomain structures such as caveolae or lipid rafts (Pani and Singh 2009). Thus, metabolic disturbances that affect caveolar signaling are expected to modify or even disrupt SOCE. Alternatively, changes in free cholesterol as a consequence of oxidative metabolism may well interfere with SOCE via direct cholesterol regulation of SOCE components as recently demonstrated for Orai1 (Derler et al. 2016).

RNS were uncovered as potential regulators of SOCE, and a nitric oxide (NO)-sensitive gate has been identified in TRPC channels that is responsible for nitrosylation-mediated activation and indicates a complex regulation of TRPC channels by NO (Yoshida et al. 2006). Modulation of SOCE by NO was found strongly dependent on the expression of TRPC proteins (Thyagarajan et al. 2001). Hence, TRPC expression patterns may modify the sensitivity of cells to RNS and

ROS. Besides representing a significant source for reactive oxidative species, NO-synthase is also the origin of a more complex regulatory mechanism of SOCE. NO activates guanylate cyclase, leading to elevation of cellular cGMP, which may mediate SOCE downregulation via cGMP-dependent regulatory phosphorylation (Kwan et al. 2000; Wang et al. 2007; Xu et al. 1994).

Endothelial dysfunction associated with distorted Ca^{2+} signaling was observed in experimental hyperglycemia in animals as well as in diabetic patients (Tamareille et al. 2006; Graier et al. 1996; Chung et al. 2009; Mita et al. 2010; Paffett et al. 2007). The known association of hyperglycemia and cellular redox status suggests a redox mechanism underlying diabetes-induced impairment of endothelial SOCE and function.

Hypoxia is another pathogenic factor, which profoundly impacts on SOCE and may thereby impair cardiovascular tissue function. Hypoxia is considered a trigger event in pulmonary hypertension (Paffett et al. 2007; Fantozzi et al. 2003). Moreover, metabolic alterations associated with hypoxia and ischemia, such as changes in extracellular pH, have been identified to modify SOCE (Wakabayashi and Groschner 1996), and extracellular acidosis has been demonstrated to suppress both SOCE and SOCE-associated endothelial mediator production (Asai et al. 2009). TRPC5 has been demonstrated as pH-regulated via sensitive glutamates (E543, E595) in the extracellular portion of the molecule (Semtner et al. 2007), and recent reports indicate a general pH sensitivity of lipid-regulated TRPC channels (Huang et al. 2010).

Collectively, the reported evidence for susceptibility of SOCE to various noxious factors supports the concept that SOCE is of high significance and pivotal in cardiovascular pathology (see also Chaps. 22–25). Sensitivity of cardiovascular SOCE to metabolic disturbances may be determined by genetic variation or defects in molecular components forming endothelial SOCE channel complexes. The exact impact of variations in Orai, STIM, or TRPC genotype on cardiovascular pathophysiology remains to be elucidated.

24.3.2 Pathophysiological Consequences of Dysfunctional Cardiovascular SOCE

24.3.2.1 SOCE in Dysregulation of Vascular Tone and Inflammation

Initially, a prominent role for SOCE in endothelium-dependent control of vascular tone was suggested based on the proposed linkage between SOCE and enzymatic generation of mediators that determine vascular tone, such as NO (Freichel et al. 2001). Endothelial nitric oxide synthase (eNOS, NOSIII) and phospholipase A_2 isoforms are targeted into caveolar signaling microdomains (Michel and Vanhoutte 2010; Murakami et al. 1999; Graziani et al. 2004) and subject to regulation by cytosolic Ca^{2+} signals and caveolar Ca^{2+} entry. As many signaling components of SOCE have been found to reside in caveolae (Pani and Singh 2009), SOCE was a reasonable candidate to control activity of these endothelial enzymes. In line with this concept, knockout mice lacking the potential endothelial SOCE channel protein

TRPC4 were found to display impaired agonist-induced vasorelaxation (Freichel et al. 2001). Based on the proposed role of SOCE in formation of endothelial NO and vasoactive arachidonate metabolites, dysfunction of SOCE channels attracted particular attention as a potential origin of dysregulation of vascular tone-associated diseases, such as hypertension and coronary spasm. A classical consequence of endothelial dysfunction is strongly impaired vasomotor control due to the indispensable role of the endothelium in generation of vasoactive factors (Furchgott and Vanhoutte 1989). While control of endothelial mediator production and vascular tone appears to rely in vivo for a large part on other SOCE-independent mechanisms specifically on mechanotransduction and Piezo channels (Ranade et al. 2014), a more significant role of SOCE channels is commonly accepted for control of endothelial barrier function. Importantly, a role of Orai1 as well as a causal role of Ca^{2+} in control of barrier function has been profoundly questioned in a recent study (Stolwijk et al. 2016).

The endothelium takes an active role in preventing systemic damage by enabling the tethering, activation, and migration of leucocytes through the endothelial barrier and thereby the development of an inflammatory vascular edema. There is remarkable variation in the histological and functional properties in endothelial cells from different vascular beds, and it has been suggested that site-specific differences in vascular response to inflammatory stimuli might be due to tissue-specific representation of SOCE (Wu et al. 2005). The endothelium of the lung appears so far the best investigated tissue in terms of inflammation-associated SOCE. In pulmonary arteries, activated leucocytes have been demonstrated to induce a SOCE-dependent increase in endothelial permeability (Itagaki et al. 2010).

Inflammatory agonists, such as histamine, bradykinin, substance P, and thrombin, typically activate phospholipase C via Gq-coupled receptors and mobilize Ca^{2+} from IP_3-sensitive intracellular stores. The rise in Ca^{2+} then activates signaling pathways, which mediate cytoskeletal reorganization through myosin light chain (MLC)-dependent contraction (Moore et al. 2000; Bauer and Stevens 2002) and disassembly of VE-cadherin at the adherens junctions (Peng et al. 2005). These processes have been demonstrated as sufficient to increase endothelial permeability in the lung (Chetham et al. 1997, 1999). Rearrangement of the microfilamentous cytoskeleton represents a potential mechanism underlying SOCE-induced shape change in endothelial cells (Moore et al. 1998). Accordingly, TRPC4, a candidate SOCE channel in endothelial cells, was demonstrated to interact with several typical endothelial cytoskeletal proteins like protein 4.1, spectrin, and β-catenin (Odell et al. 2008; Cioffi et al. 2005; Graziani et al. 2010). In pulmonary arteries, involvement of the potential SOCE channel proteins TRPC1 and TRPC4 in inflammation-associated loss of endothelial barrier function (Tiruppathi et al. 2002; Paria et al. 2003) has been suggested. In $TRPC4^{-/-}$ mice, vascular permeability was reduced by about 50%, and the endothelial cells lacked thrombin-induced formation of stress fibers (Tiruppathi et al. 2002). Other studies have shown the involvement of Ca^{2+}/calmodulin-dependent protein kinase II (δ) in endothelial barrier dysfunction (Wang et al. 2010) and indicated its influence on cytoskeletal organization by affecting translocation of filamin (Wang et al. 1996) and stress fiber formation (Borbiev et al. 2003). Interestingly, CaMKII seems also to function as an upstream regulator of SOCE (Aromolaran and Blatter 2005) by inhibiting Ca^{2+} release from intracellular stores and activation of SOCE.

Ca^{2+} influx through TRPC1, commonly implicated in SOCE phenomena, has been proposed as essential for a feed-forward mechanism regulating TRPC1 expression (Paria et al. 2006) along with endothelial hyperpermeability (Yang et al. 2009). TRPC1 function may be regulated by caveolin-1, the scaffold protein that forms the structural basis of caveolae. Direct interaction of caveolin-1 with the channel may be essential for SOCE in pulmonary artery endothelial cells (Kwiatek et al. 2006).

The role of Orai1/STIM1 complexes in endothelial barrier function has more recently been in focus and investigated in detail. STIM1 was found involved in thrombin-induced endothelial Ca^{2+} signaling (Hirano et al. 2009), and an important role of STIM1 in endothelial inflammation was suggested. The key role of store depletion and STIM molecules as determinants of endothelial barrier function has been confirmed in multiple studies (Sundivakkam et al. 2012, 2013; DebRoy et al. 2014; Shinde et al. 2013). Substantial evidence has been obtained for Orai-independent, store-operated control of endothelial barrier function (DebRoy et al. 2014; Shinde et al. 2013) conferred by STIM molecules and their coupling to alternative downstream targets such as TRPC channels (Sundivakkam et al. 2012, 2013) and RhoA activation (Shinde et al. 2013). Notably, STIM/Orai-mediated SOCE in endothelial cells has been demonstrated as outlined above and found closely related to endothelial proliferation (see also Sect. 24.3.2.2; Abdullaev et al. 2008; Li et al. 2011). Although, vascular permeability in inflammation may primarily be governed via Orai-independent mechanisms in vascular cells, SOCE is unequivocally crucial in inflammatory processes due to its well-established role in the immune cells (see Chap. 21).

24.3.2.2 SOCE in Maladaptive Cardiovascular Remodeling and Disturbances of Tissue Vascularization and Vascular Repair

In both myocyte and endothelial cell gene expression, phenotype switching and proliferation is reportedly controlled by SOCE (Nakayama et al. 2006; Abdullaev et al. 2008; Bisaillon et al. 2010; Völkers et al. 2012; Correll et al. 2015). As outlined above in this chapter, the key role of SOCE in transcriptional control involves generation of frequency-coded, cellular Ca^{2+} signals of high spatial and temporal precision (Thomas et al. 1996; Samanta and Parekh 2016; Kar et al. 2016). Ca^{2+} oscillations provide information, which is decoded by multiple downstream effectors to confer transcriptional control at variable time domains (Samanta and Parekh 2016). The prominent and most intensively investigated downstream target of SOCE-controlled oscillations is the phosphatase calcineurin (CaN), which controls nuclear translocation of the transcription factor NFAT and thereby pathological remodeling processes in the cardiovascular system (Nakayama et al. 2006; Eder and Molkentin 2011). The importance of SOCE in cardiac and vascular remodeling is discussed in this book in more detail in Chaps. 22 and 25. Similar to its crucial role in maladaptive remodeling of muscle tissues, SOCE was identified as a crucial mechanism in tissue vascularization, vascular repair, as well as de novo formation of new blood vessels in terms of vasculogenesis.

Angiogenesis, the expansion of the existing vascular system, occurs during embryonic development and continues in the mature animal most commonly during wound healing (Bao et al. 2009) and tumor metastasis (Hanahan and Folkman 1996). Sprouting of endothelial cells in angiogenesis is controlled for a large part by two tyrosine kinase receptor families, the VEGF-receptors and the tie and tek kinases (Sato et al. 1995). Tie-2 is the endothelial receptor for angiopoietin-1 and angiopoietin-2, which are synthesized and secreted by the cells surrounding the vessels (Suri et al. 1996). Tie-2/angiopoietin signaling has been suggested to stimulate the production of growth factors, like PDGF and VEGF, which induce the differentiation of mesenchymal cells into pericytes or smooth muscle cells for vessel wall formation (Folkman and D'Amore 1996). The fate of vascular cells leading to endothelial maintenance, angiogenesis, or endothelial regression is considered to depend strictly on temporal changes in concentrations of these growth factors. Interestingly, VEGF-induced SOCE is inhibited by angiopoietin-1, in terms of a negative regulation that protects from growth factor-induced hyperpermeability (Jho et al. 2005).

Notably, inhibitors of SOCE have been suggested as potential tumor static agents by preventing VEGF-induced proliferation (Faehling et al. 2002) and to suppress proliferative effects of hydroxycholesterol in endothelial cells (Trevisi et al. 2010). Consistently, knockdown/silencing of STIM1 and/or Orai has been demonstrated to suppress growth factor-induced endothelial progenitor cell proliferation in hepatocytes (Shi et al. 2010) and in vascular endothelium (Abdullaev et al. 2008). In human hepatoma cells, VEGF- as well as HGF-induced SOCE and cell proliferation were reported to be dependent on TRPC6 expression that might function in cooperation with STIM1 and Orai (El Boustany et al. 2008). In renal carcinoma cells, a loss of TRPC4 and subsequently impaired Ca^{2+} signaling was found associated with diminished secretion of an angiogenesis inhibitor, thrombospondin-1, thereby enabling the angiogenic switch in tumor progression (Veliceasa et al. 2007). Of note, the endothelium is generally considered as an attractive and suitable target for anti-angiogenic tumor therapies because of its relative stable genome and comparable low tendency to develop drug resistance (Boehm et al. 1997). Recent studies indicate that angiogenesis and vasculogenesis in tumors involve modified mechanisms of Ca^{2+} handling and Ca^{2+}-dependent control of proliferation, with increased significance of SOCE specifically in endothelial progenitors (Moccia and Poletto 2015; Dragoni et al. 2014). This opens the view on more specific strategies to suppress tumor vascularization and growth by interfering with the SOCE machinery.

Vasculogenesis is a term describing the de novo formation of the vascular system that occurs in the early embryo, while a similar process takes care of vascular repair in the mature organism. Initially, vascular and hematopoietic tissues develop in parallel, involving angioblasts and hematopoietic stem cells. Growth factors secreted from endodermal cells are then believed to induce vasculogenesis in the embryo (Pardanaud et al. 1989; Ribatti et al. 2009; Cines et al. 1998). Angioblasts as well as endothelial progenitors have been identified in the adult (Asahara et al. 1997; Mead et al. 2008), indicating that vasculogenesis is not restricted to embryonic development. In fact, the delicate endothelial barrier, which governs the movement of macromolecules and cells between blood and interstitium, requires constant regeneration in order to preserve or regain tissue homeostasis and organ function in pathogenic situations, such as metabolic disorders or inflammation. An impressive example demonstrating the crucial role of endothelial regeneration in organ health is lung pathophysiology. Injury and regeneration of lung vascular endothelium have been recognized as crucial processes determining the development and progress of lung diseases ranging from pulmonary hypertension to asthma (Toya and Malik 2011). Endothelial repair is considered to involve a heterogeneous population of endothelial progenitors apparently present in both peripheral blood and within specific niches of the vessel wall. Endothelial colony-forming cells, a subtype of endothelial progenitors, show robust proliferative potential and the ability to form de novo blood vessels when transplanted into immunodeficient mice (Mead et al. 2008). These cells may be involved in neointima hyperplasia and vascular repair after injury. Endothelial repair by progenitor cells is under tight control of specific growth factors (Zhu et al. 2010), and the linked signaling pathways are typically associated with mobilization of intracellular Ca^{2+} from the ER and thus, likely to involve SOCE. SOCE has been demonstrated in endothelial progenitor cells (Sanchez-Hernandez et al. 2010), and silencing of STIM1 was shown to suppress proliferation and migration of these cells (Kuang et al. 2010; Guo et al. 2009; Shi et al. 2010). Both proliferation and migration of endothelial progenitors has been found under control of VEGF-induced SOCE and Ca^{2+} oscillations. Notably, dysfunctional SOCE was recently found to account for impaired proliferation of endothelial progenitors in a murine model of atherosclerosis (Wang et al. 2016b).

Similar to mature endothelium, the relative contribution of Ca^{2+} entry channels to growth hormone and mediator signaling may change with transition between developmental states and differentiation. Interestingly, the expression of TRPC3 was found upregulated in certain phenotypical stages during ex vivo proliferation of progenitor cell clusters (Poteser et al. 2008). Enhanced expression of TRPC3 progenitor cells was associated with promotion of VEGF-induced Ca^{2+} signaling and may represent a critical switch between growth factor-controlled Ca^{2+} signaling mechanisms during vasculogenesis (see also Store-Operated Ca^{2+} Entry (SOCE) Pathways. 1st edition 2012 (eds. Groschner K, Graier WF, Romanin C; Springer Verlag Wien) Chap. 25).

It remains to be delineated to what extent classical SOCE and STIM/Orai as the core of this Ca^{2+} entry pathway is involved in particular progenitor populations and stages of vasculogenesis. It appears likely that multiple Ca^{2+} entry mechanisms cooperate in a highly concerted manner to control proliferation and differentiation of endothelial precursors. Moreover, these mechanisms may be of variable impact at certain stages of vasculogenesis or vascular endothelial repair.

Considering endothelial progenitor cells and their responses to vascular injury as crucial for tissue homeostasis and organ function, SOCE as a mechanism critically involved in the control of proliferation, homing, and differentiation of these cells has emerged as an attractive therapeutic target for tissue regeneration.

24.3.2.3 SOCE in Arrhythmogenesis

Normal cardiac rhythm relies on a pacemaker mechanism that is currently understood to involve integration of a PM-resident electric oscillator and an intracellular SR/ER Ca^{2+} cycling mechanism. This coordinated crosstalk between plasma membrane and ER has been termed the "coupled-clock" mechanism and rests on multiple Ca^{2+}-dependent ion transport processes (Lakatta et al. 2006; Yaniv et al. 2015). For Ca^{2+} oscillations in non-excitable tissues, which have been linked to transcriptional control, a pivotal role of the phospholipase C/inositol triphosphate (IP_3) signaling has been established, and the essential contribution of SOCE and related Ca^{2+} entry mechanisms have been postulated (Wedel et al. 2007; Dupont et al. 2011; Samanta and Parekh 2016). A very similar concept appears of relevance for cardiac pacemaking (see Fig. 24.2) and generation of arrhythmicity. As for Ca^{2+} oscillations in non-excitable cells, inositide metabolism and IP_3-mediated Ca^{2+} mobilization was found to be crucial for the ER Ca^{2+} clock mechanism (Ju et al. 2012; Kapoor et al. 2015), and the importance of the crosstalk between cardiac IP_3R and RyR was uncovered. Cellular Ca^{2+} oscillations typically require a mechanism of replenishing Ca^{2+} stores via a Ca^{2+} entry mechanism to compensate for loss of Ca^{2+} into the extracellular space during cytosolic Ca^{2+} transients. The importance of "store refilling" has been clearly recognized for cardiac pacemakers (Imtiaz et al. 2010), and consequently a role for SOCE, representing a crucial process of Ca^{2+} store refilling, has been hypothesized, and the concept was corroborated by pharmacological experiments as well as by mathematical modeling approaches. The search for cardiac SOCE channels involved in the control of pacemaker activity was first focused on the TRPC family (Ju et al. 2007) and later on STIM/Orai complexes as the "core" SOCE machinery of most non-excitable tissues (Zhang et al. 2015; Liu et al. 2015). Interestingly, expression of TRPC proteins as well as STIM and Orai has been reported for sinoatrial node cells and implicated in the pacemaker activity of SOCE (Ju et al. 2007, 2015; Liu et al. 2015). Moreover, STIM has been identified as a multifunctional player in control of cardiac automaticity with an array of PM components being governed by the ER Ca^{2+} sensor (Zhang et al. 2015). These reports are in line with a concept of pacemaker Ca^{2+} cycling, which is localized in ER-PM junctions hosting a multitude of cooperating transporters and channels including Orai. Of note, TRPC3 has recently been implicated in sinoatrial as well as atrial arrhythmias (Ju et al. 2015). Although it appears unlikely that TRPC3 functions as the primary SOCE channel in pacemakers, its importance for control of pacemaker functions via GPCR-induced signals associated with SR/ER Ca^{2+} mobilization has been convincingly demonstrated (Ju et al. 2015). Moreover, TRPC3 is a second messenger (lipid)-gated cation channel that has been tightly linked to the cardiac Na^{+}-Ca^{2+} exchanger (NCX1), representing a key element of the "coupled-clock" mechanism (Doleschal et al. 2015). Importantly, functional disturbances in cardiac excitability were found associated with spatial dynamics and disruption of nanodomain communication between TRPC3 and NCX1. Collectively, these reports suggest that cardiac pacemaking and excitability involves organization of signaling complexes within ER-PM junctional domains that allow for transfer of Ca^{2+} from the ER to various Ca^{2+} transporters in the PM, which operate in a highly coordinated manner as illustrated in Fig. 24.4. In this context, SOCE may serve merely in replenishing and refilling the SR/ER Ca^{2+} levels but may as well directly support CICR via RyR2 (Fig. 24.1a). It appears important to consider in addition two more recently identified ER membrane conductances that are predicted to govern Ca^{2+} discharge from the ER, such as trimeric ER cation channels suggested to provide counter ion (K^+) fluxes (Yazawa et al. 2007) or the Ca^{2+} overload-activated TMCO1 (Wang et al. 2016a) channel, as further key players in this scenario. It remains to be clarified how this complex machinery is organized in terms of molecular architecture and to identify changes in this molecular organization that confer pacemaker dysfunction and arrhythmogenesis.

24.4 Summary, Perspectives, and Therapeutic Implications

Substantial progress has recently been made in terms of understanding molecular structures and mechanisms underlying the cardiovascular SOCE phenomenon. A prominent role of STIM1/Orai1 complexes has been demonstrated in endothelial cells, cardiac pacemaker cells, as well as pathologically remodeled smooth muscle and cardiac myocytes. As in immune cells, STIM1/Orai1 appears to constitute a Ca^{2+} entry pathway of high Ca^{2+} selectivity, which, albeit electrophysiologically barely discernible, was shown to control important cardiovascular functions. The cardiovascular SOCE mechanism is typically accompanied by activation of more prominent nonselective cation conductances and STIM-mediated alterations in voltage-gated conductances and transporters. Thus, cardiovascular SOCE is a complex phenomenon based on multiple distinct pore complexes that are expressed and organized within SR/ER-PM nanojunctions in a tissue- and phenotype-dependent manner. These cardiovascular SOCE signaling complexes are critically affected by molecules and metabolic disturbances, which induce or promote vascular injury such as reactive oxygen and nitrogen species as well as specific lipid metabolites. This is consistent with the well-recognized sensitivity of SOCE to metabolic disturbances. Genetic defects may promote sensitivity of SOCE to such pathogenic factors. It appears conceivable to hypothesize that disturbances in cardiovascular SOCE is of pathological significance. Consequently, therapeutic targeting of SOCE complexes or the underlying nanojunctional structures has emerged as a promising strategy for the prevention of diseases resulting from cardiovascular injury and maladaptive remodeling.

Acknowledgments This work was supported by the Austrian Science Funds (W1226, DKplus Metabolic and Cardiovascular Disease).

References

Abdullaev IF, Bisaillon JM, Potier M, Gonzalez JC, Motiani RK, Trebak M (2008) Stim1 and Orai1 mediate CRAC currents and store-operated calcium entry important for endothelial cell proliferation. Circ Res 103(11):1289–1299

Aromolaran AA, Blatter LA (2005) Modulation of intracellular Ca^{2+} release and capacitative Ca^{2+} entry by CaMKII inhibitors in bovine vascular endothelial cells. Am J Physiol Cell Physiol 289 (6):C1426–C1436

Asahara T, Murohara T, Sullivan A, Silver M, van der Zee R, Li T, Witzenbichler B, Schatteman G, Isner JM (1997) Isolation of putative progenitor endothelial cells for angiogenesis. Science 275(5302):964–967

Asai M, Takeuchi K, Saotome M, Urushida T, Katoh H, Satoh H, Hayashi H, Watanabe H (2009) Extracellular acidosis suppresses endothelial function by inhibiting store-operated Ca^{2+} entry via non-selective cation channels. Cardiovasc Res 83(1):97–105

Balzer M, Lintschinger B, Groschner K (1999) Evidence for a role of Trp proteins in the oxidative stress-induced membrane conductances of porcine aortic endothelial cells. Cardiovasc Res 42 (2):543–549

Bao P, Kodra A, Tomic-Canic M, Golinko MS, Ehrlich HP, Brem H (2009) The role of vascular endothelial growth factor in wound healing. J Surg Res 153(2):347–358

Bauer NN, Stevens T (2002) Putative role for a myosin motor in store-operated calcium entry. Cell Biochem Biophys 37(1):53–70

Beliveau E, Lessard V, Guillemette G (2014) STIM1 positively regulates the Ca2+ release activity of the inositol 1,4,5-trisphosphate receptor in bovine aortic endothelial cells. PLoS One 9(12): e114718

Berliner JA, Heinecke JW (1996) The role of oxidized lipoproteins in atherogenesis. Free Radic Biol Med 20(5):707–727

Bisaillon JM, Motiani RK, Gonzalez-Cobos JC, Potier M, Halligan KE, Alzawahra WF, Barroso M, Singer HA, Jourd'heuil D, Trebak M (2010) Essential role for STIM1/Orai1-mediated calcium influx in PDGF-induced smooth muscle migration. Am J Physiol Cell Physiol 298(5):C993–C1005

Boehm T, Folkman J, Browder T, O'Reilly MS (1997) Antiangiogenic therapy of experimental cancer does not induce acquired drug resistance. Nature 390(6658):404–407

Bogeski I, Kummerow C, Al-Ansary D, Schwarz EC, Koehler R, Kozai D, Takahashi N, Peinelt C, Griesemer D, Bozem M, Mori Y, Hoth M, Niemeyer BA (2010) Differential redox regulation of ORAI ion channels: a mechanism to tune cellular calcium signaling. Sci Signal 3(115):ra24

Borbiev T, Verin AD, Birukova A, Liu F, Crow MT, Garcia JG (2003) Role of CaM kinase II and ERK activation in thrombin-induced endothelial cell barrier dysfunction. Am J Physiol Lung Cell Mol Physiol 285(1):L43–L54

Camacho Londoño JE, Tian Q, Hammer K, Schröder L, Camacho Londoño J, Reil JC, He T, Oberhofer M, Mannebach S, Mathar I, Philipp SE, Tabellion W, Schweda F, Dietrich A, Kaestner L, Laufs U, Birnbaumer L, Flockerzi V, Freichel M, Lipp P (2015) A background Ca2+ entry pathway mediated by TRPC1/TRPC4 is critical for development of pathological cardiac remodelling. Eur Heart J 36(33):2257–2266

Casteels R, Droogmans G (1981) Exchange characteristics of the noradrenaline-sensitive calcium store in vascular smooth muscle cells or rabbit ear artery. J Physiol 317:263–279

Chang CL, Hsieh TS, Yang TT, Rothberg KG, Azizoglu DB, Volk E, Liao JC, Liou J (2013) Feedback regulation of receptor-induced Ca2+ signaling mediated by E-Syt1 and Nir2 at endoplasmic reticulum-plasma membrane junctions. Cell Rep 5(3):813–825

Cheng KT, Liu X, Ong HL, Swaim W, Ambudkar IS (2011) Local Ca^{2+} entry via Orai1 regulates plasma membrane recruitment of TRPC1 and controls cytosolic Ca^{2+} signals required for specific cell functions. PLoS Biol 9(3):e1001025

Chetham PM, Guldemeester HA, Mons N, Brough GH, Bridges JP, Thompson WJ, Stevens T (1997) Ca(2+)-inhibitable adenylyl cyclase and pulmonary microvascular permeability. Am J Physiol 273(1 Pt 1):L22–L30

Chetham PM, Babal P, Bridges JP, Moore TM, Stevens T (1999) Segmental regulation of pulmonary vascular permeability by store-operated Ca^{2+} entry. Am J Physiol 276(1 Pt 1): L41–L50

Chung AW, Au Yeung K, Chum E, Okon EB, van Breemen C (2009) Diabetes modulates capacitative calcium entry and expression of transient receptor potential canonical channels in human saphenous vein. Eur J Pharmacol 613(1–3):114–118

Cines DB, Pollak ES, Buck CA, Loscalzo J, Zimmerman GA, McEver RP, Pober JS, Wick TM, Konkle BA, Schwartz BS, Barnathan ES, McCrae KR, Hug BA, Schmidt AM, Stern DM (1998) Endothelial cells in physiology and in the pathophysiology of vascular disorders. Blood 91(10):3527–3561

Cioffi DL, Wu S, Alexeyev M, Goodman SR, Zhu MX, Stevens T (2005) Activation of the endothelial store-operated ISOC Ca^{2+} channel requires interaction of protein 4.1 with TRPC4. Circ Res 97(11):1164–1172

Collins HE, Zhu-Mauldin X, Marchase RB, Chatham JC (2013) STIM1/Orai1-mediated SOCE: current perspectives and potential roles in cardiac function and pathology. Am J Physiol Heart Circ Physiol 305(4):H446–H458

Correll RN, Goonasekera SA, van Berlo JH, Burr AR, Accornero F, Zhang H, Makarewich CA, York AJ, Sargent MA, Chen X, Houser SR, Molkentin JD (2015) STIM1 elevation in the heart results in aberrant Ca^{2+} handling and cardiomyopathy. J Mol Cell Cardiol 87:38–47

DebRoy A, Vogel SM, Soni D, Sundivakkam PC, Malik AB, Tiruppathi C (2014) Cooperative signaling via transcription factors NF-κB and AP1/c-Fos mediates endothelial cell STIM1 expression and hyperpermeability in response to endotoxin. J Biol Chem 289 (35):24188–24201

Derler I, Jardin I, Stathopulos PB, Muik M, Fahrner M, Zayats V, Pandey SK, Poteser M, Lackner B, Absolonova M, Schindl R, Groschner K, Ettrich R, Ikura M, Romanin C (2016) Cholesterol modulates Orai1 channel function. Sci Signal 9(412):ra10

Devine CE, Somlyo AV, Somlyo AP (1972) Sarcoplasmic reticulum and excitation-contraction coupling in mammalian smooth muscles. J Cell Biol 52:690–718

Di Capite J, Ng SW, Parekh AB (2009) Decoding of cytoplasmic Ca(2+) oscillations through the spatial signature drives gene expression. Curr Biol 19(10):853–858

Di Giuro CML, Shrestha N, Malli R, Groschner K, van Breemen C, Fameli N (2016) Na+/Ca2+ exchangers and Orai channels jointly refill endoplasmic reticulum (ER) Ca2+ via ER nanojunctions in vascular endothelial cells. bioRxiv. doi:10.1101/084285

Doleschal B, Primessnig U, Wölkart G, Wolf S, Schernthaner M, Lichtenegger M, Glasnov TN, Kappe CO, Mayer B, Antoons G, Heinzel F, Poteser M, Groschner K (2015) TRPC3 contributes to regulation of cardiac contractility and arrhythmogenesis by dynamic interaction with NCX1. Cardiovasc Res 106(1):163–173

Dragoni S, Laforenza U, Bonetti E, Reforgiato M, Poletto V, Lodola F, Bottino C, Guido D, Rappa A, Pareek S, Tomasello M, Guarrera MR, Cinelli MP, Aronica A, Guerra G, Barosi G, Tanzi F, Rosti V, Moccia F (2014) Enhanced expression of Stim, Orai, and TRPC transcripts and proteins in endothelial progenitor cells isolated from patients with primary myelofibrosis. PLoS One 9(3):e91099

Dupont G, Combettes L, Bird GS, Putney JW (2011) Calcium oscillations. Cold Spring Harb Perspect Biol 3(3). doi: 10.1101/cshperspect.a004226

Eder P, Molkentin JD (2011) TRPC channels as effectors of cardiac hypertrophy. Circ Res 108 (2):265–272

El Boustany C, Bidaux G, Enfissi A, Delcourt P, Prevarskaya N, Capiod T (2008) Capacitative calcium entry and transient receptor potential canonical 6 expression control human hepatoma cell proliferation. Hepatology 47(6):2068–2077

Elliott SJ, Doan TN (1993) Oxidant stress inhibits the store-dependent Ca(2+)-influx pathway of vascular endothelial cells. Biochem J 292(Pt 2):385–393

Elliott SJ, Koliwad SK (1995) Oxidant stress and endothelial membrane transport. Free Radic Biol Med 19(5):649–658

Elliott SJ, Eskin SG, Schilling WP (1989) Effect of t-butyl-hydroperoxide on bradykinin-stimulated changes in cytosolic calcium in vascular endothelial cells. J Biol Chem 264 (7):3806–3810

Encabo A, Romanin C, Birke FW, Kukovetz WR, Groschner K (1996) Inhibition of a store-operated Ca^{2+} entry pathway in human endothelial cells by the isoquinoline derivative LOE 908. Br J Pharmacol 119(4):702–706

Escobar M, Cardenas C, Colavita K, Petrenko NB, Franzini-Armstrong C (2011) Structural evidence for perinuclear calcium microdomains in cardiac myocytes. J Mol Cell Cardiol 50:451–459

Faehling M, Kroll J, Fohr KJ, Fellbrich G, Mayr U, Trischler G, Waltenberger J (2002) Essential role of calcium in vascular endothelial growth factor A-induced signaling: mechanism of the antiangiogenic effect of carboxyamidotriazole. FASEB J 16(13):1805–1807

Fameli N, Ogunbayo OA, van Breemen C, Evans AM (2014) Cytoplasmic nanojunctions between lysosomes and sarcoplasmic reticulum are required for specific calcium signalling. F1000Res 3:93

Fantozzi I, Zhang S, Platoshyn O, Remillard CV, Cowling RT, Yuan JX (2003) Hypoxia increases AP-1 binding activity by enhancing capacitative Ca^{2+} entry in human pulmonary artery endothelial cells. Am J Physiol Lung Cell Mol Physiol 285(6):L1233–L1245

Fasolato C, Nilius B (1998) Store depletion triggers the calcium release-activated calcium current (ICRAC) in macrovascular endothelial cells: a comparison with Jurkat and embryonic kidney cell lines. Pflugers Arch 436(1):69–74

Feletou M, Vanhoutte PM (2006) Endothelial dysfunction: a multifaceted disorder (The Wiggers Award Lecture). Am J Physiol Heart Circ Physiol 291(3):H985–H1002

Feske S (2011) Immunodeficiency due to defects in store-operated calcium entry. Ann N Y Acad Sci 1238:74–90

Feske S, Gwack Y, Prakriya M, Srikanth S, Puppel SH, Tanasa B, Hogan PG, Lewis RS, Daly M, Rao A (2006) A mutation in Orai1 causes immune deficiency by abrogating CRAC channel function. Nature 441(7090):179–185

Florea SM, Blatter LA (2008) The effect of oxidative stress on Ca^{2+} release and capacitative Ca^{2+} entry in vascular endothelial cells. Cell Calcium 43(4):405–415

Folkman J, D'Amore PA (1996) Blood vessel formation: what is its molecular basis? Cell 87 (7):1153–1155

Freichel M, Suh SH, Pfeifer A, Schweig U, Trost C, Weissgerber P, Biel M, Philipp S, Freise D, Droogmans G, Hofmann F, Flockerzi V, Nilius B (2001) Lack of an endothelial store-operated Ca^{2+} current impairs agonist-dependent vasorelaxation in TRP4-/- mice. Nat Cell Biol 3 (2):121–127

Furchgott RF, Vanhoutte PM (1989) Endothelium-derived relaxing and contracting factors. FASEB J 3(9):2007–2018

Graier WF, Simecek S, Kukovetz WR, Kostner GM (1996) High D-glucose-induced changes in endothelial Ca^{2+}/EDRF signaling are due to generation of superoxide anions. Diabetes 45 (10):1386–1395

Graziani A, Bricko V, Carmignani M, Graier WF, Groschner K (2004) Cholesterol- and caveolin-rich membrane domains are essential for phospholipase A2-dependent EDHF formation. Cardiovasc Res 64(2):234–242

Graziani A, Poteser M, Heupel WM, Schleifer H, Krenn M, Drenckhahn D, Romanin C, Baumgartner W, Groschner K (2010) Cell-cell contact formation governs Ca^{2+} signaling by TRPC4 in the vascular endothelium: evidence for a regulatory TRPC4-beta-catenin interaction. J Biol Chem 285(6):4213–4223

Groschner K, Graier WF, Kukovetz WR (1994) Histamine induces K^+, Ca^{2+}, and Cl^- currents in human vascular endothelial cells. Role of ionic currents in stimulation of nitric oxide biosynthesis. Circ Res 75(2):304–314

Groschner K, Hingel S, Lintschinger B, Balzer M, Romanin C, Zhu X, Schreibmayer W (1998) Trp proteins form store-operated cation channels in human vascular endothelial cells. FEBS Lett 437(1–2):101–106

Guo RW, Wang H, Gao P, Li MQ, Zeng CY, Yu Y, Chen JF, Song MB, Shi YK, Huang L (2009) An essential role for stromal interaction molecule 1 in neointima formation following arterial injury. Cardiovasc Res 81(4):660–668

Hanahan D, Folkman J (1996) Patterns and emerging mechanisms of the angiogenic switch during tumorigenesis. Cell 86(3):353–364

Hawkins BJ, Irrinki KM, Mallilankaraman K, Lien YC, Wang Y, Bhanumathy CD, Subbiah R, Ritchie MF, Soboloff J, Baba Y, Kurosaki T, Joseph SK, Gill DL, Madesh M (2010) S-glutathionylation activates STIM1 and alters mitochondrial homeostasis. J Cell Biol 190 (3):391–405

Hirano K, Hirano M, Hanada A (2009) Involvement of STIM1 in the proteinase-activated receptor 1-mediated Ca^{2+} influx in vascular endothelial cells. J Cell Biochem 108(2):499–507

Horton JS, Buckley CL, Alvarez EM, Schorlemmer A, Stokes AJ (2014) The calcium release-activated calcium channel Orai1 represents a crucial component in hypertrophic compensation and the development of dilated cardiomyopathy. Channels (Austin) 8(1):35–48

Hoth M, Penner R (1992) Depletion of intracellular calcium stores activates a calcium current in mast cells. Nature 355(6358):353–356

Huang J, van Breemen C, Kuo KH, Hove-Madsen L, Tibbits GF (2006) Store-operated Ca2+ entry modulates sarcoplasmic reticulum Ca2+ loading in neonatal rabbit cardiac ventricular myocytes. Am J Physiol Cell Physiol 290:C1572–C1582

Huang J, Liu CH, Hughes SA, Postma M, Schwiening CJ, Hardie RC (2010) Activation of TRP channels by protons and phosphoinositide depletion in Drosophila photoreceptors. Curr Biol 20 (3):189–197

Hunton DL, Lucchesi PA, Pang Y, Cheng X, Dell'Italia LJ, Marchase RB (2002) Capacitative calcium entry contributes to nuclear factor of activated T-cells nuclear translocation and hypertrophy in cardiomyocytes. J Biol Chem 277(16):14266–14273

Hunton DL, Zou L, Pang Y, Marchase RB (2004) Adult rat cardiomyocytes exhibit capacitative calcium entry. Am J Physiol Heart Circ Physiol 286:H1124–H1132

Imtiaz MS, von der Weid PY, Laver DR, van Helden DF (2010) SR Ca2+ store refill—a key factor in cardiac pacemaking. J Mol Cell Cardiol 49(3):412–426

Itagaki K, Zhang Q, Hauser CJ (2010) Sphingosine kinase inhibition alleviates endothelial permeability induced by thrombin and activated neutrophils. Shock 33(4):381–386

Jernigan NL, Broughton BR, Walker BR, Resta TC (2006) Impaired NO-dependent inhibition of store- and receptor-operated calcium entry in pulmonary vascular smooth muscle after chronic hypoxia. Am J Physiol Lung Cell Mol Physiol 290(3):L517–L525

Jho D, Mehta D, Ahmmed G, Gao XP, Tiruppathi C, Broman M, Malik AB (2005) Angiopoietin-1 opposes VEGF-induced increase in endothelial permeability by inhibiting TRPC1-dependent Ca^{2+} influx. Circ Res 96(12):1282–1290

Ju YK, Chu Y, Chaulet H, Lai D, Gervasio OL, Graham RM, Cannell MB, Allen DG (2007) Store-operated Ca2+ influx and expression of TRPC genes in mouse sinoatrial node. Circ Res 100 (11):1605–1614

Ju YK, Woodcock EA, Allen DG, Cannell MB (2012) Inositol 1,4,5-trisphosphate receptors and pacemaker rhythms. J Mol Cell Cardiol 53(3):375–381

Ju YK, Lee BH, Trajanovska S, Hao G, Allen DG, Lei M, Cannell MB (2015) The involvement of TRPC3 channels in sinoatrial arrhythmias. Front Physiol 6:86

Kapoor N, Tran A, Kang J, Zhang R, Philipson KD, Goldhaber JI (2015) Regulation of calcium clock-mediated pacemaking by inositol-1,4,5-trisphosphate receptors in mouse sinoatrial nodal cells. J Physiol 593(12):2649–2663

Kar P, Nelson C, Parekh AB (2011) Selective activation of the transcription factor NFAT1 by calcium microdomains near Ca2+ release-activated Ca2+ (CRAC) channels. J Biol Chem 286 (17):14795–14803

Kar P, Bakowski D, Di Capite J, Nelson C, Parekh AB (2012a) Different agonists recruit different stromal interaction molecule proteins to support cytoplasmic Ca2+ oscillations and gene expression. Proc Natl Acad Sci U S A 109(18):6969–6974

Kar P, Nelson C, Parekh AB (2012b) CRAC channels drive digital activation and provide analog control and synergy to Ca(2+)-dependent gene regulation. Curr Biol 22(3):242–247

Kar P, Mirams GR, Christian HC, Parekh AB (2016) Control of NFAT isoform activation and NFAT-dependent gene expression through two coincident and spatially segregated intracellular Ca2+ signals. Mol Cell 64(4):746–759

Kassan M, Ait-Aissa K, Radwan E, Mali V, Haddox S, Gabani M, Zhang W, Belmadani S, Irani K, Trebak M, Matrougui K (2016) Essential role of smooth muscle STIM1 in hypertension and cardiovascular dysfunction. Arterioscler Thromb Vasc Biol 36(9):1900–1909

Kitajima N, Numaga-Tomita T, Watanabe M, Kuroda T, Nishimura A, Miyano K, Yasuda S, Kuwahara K, Sato Y, Ide T, Birnbaumer L, Sumimoto H, Mori Y, Nishida M (2016) TRPC3 positively regulates reactive oxygen species driving maladaptive cardiac remodeling. Sci Rep 6:37001

Koenig S, Schernthaner M, Maechler H, Kappe CO, Glasnov TN, Hoefler G, Braune M, Wittchow E, Groschner K (2013) A TRPC3 blocker, ethyl-1-(4-(2,3,3-trichloroacrylamide) phenyl)-5-(trifluoromethyl)-1H-pyrazole-4-carboxylate (Pyr3), prevents stent-induced arterial remodeling. J Pharmacol Exp Ther 344(1):33–40

Kuang CY, Yu Y, Guo RW, Qian DH, Wang K, Den MY, Shi YK, Huang L (2010) Silencing stromal interaction molecule 1 by RNA interference inhibits the proliferation and migration of endothelial progenitor cells. Biochem Biophys Res Commun 398(2):315–320

Kurebayashi N, Ogawa Y (2001) Depletion of Ca2+ in the sarcoplasmic reticulum stimulates Ca2+ entry into mouse skeletal muscle fibres. J Physiol 533(Pt 1):185–199

Kuwahara K, Wang Y, McAnally J, Richardson JA, Bassel-Duby R, Hill JA, Olson EN (2006) TRPC6 fulfills a calcineurin signaling circuit during pathologic cardiac remodeling. J Clin Invest 116(12):3114–3126

Kwan HY, Huang Y, Yao X (2000) Store-operated calcium entry in vascular endothelial cells is inhibited by cGMP via a protein kinase G-dependent mechanism. J Biol Chem 275 (10):6758–6763

Kwiatek AM, Minshall RD, Cool DR, Skidgel RA, Malik AB, Tiruppathi C (2006) Caveolin-1 regulates store-operated Ca^{2+} influx by binding of its scaffolding domain to transient receptor potential channel-1 in endothelial cells. Mol Pharmacol 70(4):1174–1183

Lakatta EG, Vinogradova T, Lyashkov A, Sirenko S, Zhu W, Ruknudin A, Maltsev VA (2006) The integration of spontaneous intracellular Ca2+ cycling and surface membrane ion channel activation entrains normal automaticity in cells of the heart's pacemaker. Ann N Y Acad Sci 1080:178–206

Launikonis BS, Barnes M, Stephenson DG (2003) Identification of the coupling between skeletal muscle store-operated Ca2+ entry and the inositol trisphosphate receptor. Proc Natl Acad Sci U S A 100(5):2941–2944

Lefkimmiatis K, Srikanthan M, Maiellaro I, Moyer MP, Curci S, Hofer AM (2009) Store-operated cyclic AMP signalling mediated by STIM1. Nat Cell Biol 11(4):433–442

Lewis RS, Cahalan MD (1989) Mitogen-induced oscillations of cytosolic Ca^{2+} and transmembrane Ca^{2+} current in human leukemic T cells. Cell Regul 1(1):99–112

Li JM, Shah AM (2004) Endothelial cell superoxide generation: regulation and relevance for cardiovascular pathophysiology. Am J Physiol Regul Integr Comp Physiol 287(5):R1014–R1030

Li J, Cubbon RM, Wilson LA, Amer MS, McKeown L, Hou B, Majeed Y, Tumova S, Seymour VA, Taylor H, Stacey M, O'Regan D, Foster R, Porter KE, Kearney MT, Beech DJ (2011)

Orai1 and CRAC channel dependence of VEGF-activated Ca2+ entry and endothelial tube formation. Circ Res 108(10):1190–11982
Liang GH, Park S, Kim MY, Kim JA, Choi S, Suh SH (2010) Modulation of nonselective cation current by oxidized LDL and lysophosphatidylcholine and its inhibitory contribution to endothelial damage. Life Sci 86(19–20):733–739
Liu J, Xin L, Benson VL, Allen DG, Ju YK (2015) Store-operated calcium entry and the localization of STIM1 and Orai1 proteins in isolated mouse sinoatrial node cells. Front Physiol 6:69
Luik RM, Wu MM, Buchanan J, Lewis RS (2006) The elementary unit of store-operated Ca2+ entry: local activation of CRAC channels by STIM1 at ER-plasma membrane junctions. J Cell Biol 174:815–825
Luo D, Broad LM, Bird GS, Putney JW Jr (2001) Signaling pathways underlying muscarinic receptor-induced [Ca2+]i oscillations in HEK293 cells. J Biol Chem 276(8):5613–5621
Maleth J, Choi S, Muallem S, Ahuja M (2014) Translocation between PI(4,5)P2-poor and PI(4,5) P2-rich microdomains during store depletion determines STIM1 conformation and Orai1 gating. Nat Commun 5:5843
Malli R, Frieden M, Hunkova M, Trenker M, Graier WF (2007) Ca2+ refilling of the endoplasmic reticulum is largely preserved albeit reduced Ca2+ entry in endothelial cells. Cell Calcium 41:63–76
Mead LE, Prater D, Yoder MC, Ingram DA (2008) Isolation and characterization of endothelial progenitor cells from human blood. Curr Protoc Stem Cell Biol Chapter 2:Unit 2C.1
Michel T, Vanhoutte PM (2010) Cellular signaling and NO production. Pflugers Arch 459 (6):807–816
Miller BA (2006) The role of TRP channels in oxidative stress-induced cell death. J Membr Biol 209(1):31–41
Missiaen L, Taylor CW, Berridge MJ (1991) Spontaneous calcium release from inositol trisphosphate-sensitive calcium stores. Nature 352(6332):241–244
Mita M, Ito K, Taira K, Nakagawa J, Walsh MP, Shoji M (2010) Attenuation of store-operated Ca2+ entry and enhanced expression of TRPC channels in caudal artery smooth muscle from Type 2 diabetic Goto-Kakizaki rats. Clin Exp Pharmacol Physiol 37(7):670–678
Moccia F, Poletto V (2015) May the remodeling of the Ca^{2+} toolkit in endothelial progenitor cells derived from cancer patients suggest alternative targets for anti-angiogenic treatment? Biochim Biophys Acta 1853(9):1958–1973
Moore TM, Brough GH, Babal P, Kelly JJ, Li M, Stevens T (1998) Store-operated calcium entry promotes shape change in pulmonary endothelial cells expressing Trp1. Am J Physiol 275(3 Pt 1):L574–L582
Moore TM, Norwood NR, Creighton JR, Babal P, Brough GH, Shasby DM, Stevens T (2000) Receptor-dependent activation of store-operated calcium entry increases endothelial cell permeability. Am J Physiol Lung Cell Mol Physiol 279(4):L691–L698
Murakami M, Kambe T, Shimbara S, Yamamoto S, Kuwata H, Kudo I (1999) Functional association of type IIA secretory phospholipase A(2) with the glycosylphosphatidylinositol-anchored heparan sulfate proteoglycan in the cyclooxygenase-2-mediated delayed prostanoid-biosynthetic pathway. J Biol Chem 274(42):29927–29936
Nakayama H, Wilkin BJ, Bodi I, Molkentin JD (2006) Calcineurin-dependent cardiomyopathy is activated by TRPC in the adult mouse heart. FASEB J 20(10):1660–1670
Odell AF, Van Helden DF, Scott JL (2008) The spectrin cytoskeleton influences the surface expression and activation of human transient receptor potential channel 4 channels. J Biol Chem 283(7):4395–4407
Paffett ML, Naik JS, Resta TC, Walker BR (2007) Reduced store-operated Ca^{2+} entry in pulmonary endothelial cells from chronically hypoxic rats. Am J Physiol Lung Cell Mol Physiol 293 (5):L1135–L1142
Pani B, Singh BB (2009) Lipid rafts/caveolae as microdomains of calcium signaling. Cell Calcium 45(6):625–633

Pardanaud L, Yassine F, Dieterlen-Lievre F (1989) Relationship between vasculogenesis, angiogenesis and haemopoiesis during avian ontogeny. Development 105(3):473–485

Parekh AB (2007) Functional consequences of activating store-operated CRAC channels. Cell Calcium 42(2):111–121

Paria BC, Malik AB, Kwiatek AM, Rahman A, May MJ, Ghosh S, Tiruppathi C (2003) Tumor necrosis factor-alpha induces nuclear factor-kappaB-dependent TRPC1 expression in endothelial cells. J Biol Chem 278(39):37195–37203

Paria BC, Bair AM, Xue J, Yu Y, Malik AB, Tiruppathi C (2006) Ca^{2+} influx induced by protease-activated receptor-1 activates a feed-forward mechanism of TRPC1 expression via nuclear factor-kappaB activation in endothelial cells. J Biol Chem 281(30):20715–20727

Peng HH, Hodgson L, Henderson AJ, Dong C (2005) Involvement of phospholipase C signaling in melanoma cell-induced endothelial junction disassembly. Front Biosci 10:1597–1606

Poburko D, Kuo KH, Dai J, Lee CH, van Breemen C (2004) Organellar junctions promote targeted Ca2+ signaling in smooth muscle: why two membranes are better than one. Trends Pharmacol Sci 25:8–15

Poteser M, Graziani A, Rosker C, Eder P, Derler I, Kahr H, Zhu MX, Romanin C, Groschner K (2006) TRPC3 and TRPC4 associate to form a redox-sensitive cation channel. Evidence for expression of native TRPC3-TRPC4 heteromeric channels in endothelial cells. J Biol Chem 281(19):13588–13595

Poteser M, Graziani A, Eder P, Yates A, Machler H, Romanin C, Groschner K (2008) Identification of a rare subset of adipose tissue-resident progenitor cells, which express CD133 and TRPC3 as a VEGF-regulated Ca2+ entry channel. FEBS Lett 582(18):2696–2702

Poteser M, Schleifer H, Lichtenegger M, Schernthaner M, Stockner T, Kappe CO, Glasnov TN, Romanin C, Groschner K (2011) PKC-dependent coupling of calcium permeation through transient receptor potential canonical 3 (TRPC3) to calcineurin signaling in HL-1 myocytes. Proc Natl Acad Sci U S A 108(26):10556–10561

Poteser M, Leitinger G, Pritz E, Platzer D, Frischauf I, Romanin C, Groschner K (2016) Live-cell imaging of ER-PM contact architecture by a novel TIRFM approach reveals extension of junctions in response to store-operated Ca2+-entry. Sci Rep 6:35656

Potier M, Gonzalez JC, Motiani RK, Abdullaev IF, Bisaillon JM, Singer HA, Trebak M (2009) Evidence for STIM1- and Orai1-dependent store-operated calcium influx through ICRAC in vascular smooth muscle cells: role in proliferation and migration. FASEB J 23:2425–2437

Prakriya M, Feske S, Gwack Y, Srikanth S, Rao A, Hogan PG (2006) Orai1 is an essential pore subunit of the CRAC channel. Nature 443(7108):230–233

Ranade SS, Qiu Z, Woo SH, Hur SS, Murthy SE, Cahalan SM, Xu J, Mathur J, Bandell M, Coste B, Li YS, Chien S, Patapoutian A (2014) Piezo1, a mechanically activated ion channel, is required for vascular development in mice. Proc Natl Acad Sci U S A 111(28):10347–10352

Renard-Rooney DC, Joseph SK, Seitz MB, Thomas AP (1995) Effect of oxidized glutathione and temperature on inositol 1,4,5-trisphosphate binding in permeabilized hepatocytes. Biochem J 310(Pt 1):185–192

Ribatti D, Nico B, Crivellato E (2009) Morphological and molecular aspects of physiological vascular morphogenesis. Angiogenesis 12(2):101–111

Saliba Y, Keck M, Marchand A, Atassi F, Ouille A, Cazorla O, Trebak M, Pavoine C, Lacampagne A, Hulot JS, Farès N, Fauconnier J, Lompre AM (2015) Emergence of Orai3 activity during cardiac hypertrophy. Cardiovasc Res 105(3):248–259

Samanta K, Parekh AB (2016) Spatial Ca2+ profiling: decrypting the universal cytosolic Ca2+ oscillation. J Physiol. doi: 10.1113/JP272860

Sanchez-Hernandez Y, Laforenza U, Bonetti E, Fontana J, Dragoni S, Russo M, Avelino-Cruz JE, Schinelli S, Testa D, Guerra G, Rosti V, Tanzi F, Moccia F (2010) Store-operated Ca(2+) entry is expressed in human endothelial progenitor cells. Stem Cells Dev 19(12):1967–1981

Sato TN, Tozawa Y, Deutsch U, Wolburg-Buchholz K, Fujiwara Y, Gendron-Maguire M, Gridley T, Wolburg H, Risau W, Qin Y (1995) Distinct roles of the receptor tyrosine kinases Tie-1 and Tie-2 in blood vessel formation. Nature 376(6535):70–74

Schindl R, Muik M, Fahrner M, Derler I, Fritsch R, Bergsmann J, Romanin C (2009) Recent progress on STIM1 domains controlling Orai activation. Cell Calcium 46(4):227–232

Semtner M, Schaefer M, Pinkenburg O, Plant TD (2007) Potentiation of TRPC5 by protons. J Biol Chem 282(46):33868–33878

Seo K, Rainer PP, Shalkey Hahn V, Lee DI, Jo SH, Andersen A, Liu T, Xu X, Willette RN, Lepore JJ, Marino JP Jr, Birnbaumer L, Schnackenberg CG, Kass DA (2014) Combined TRPC3 and TRPC6 blockade by selective small-molecule or genetic deletion inhibits pathological cardiac hypertrophy. Proc Natl Acad Sci U S A 111(4):1551–1556

Shi Y, Song M, Guo R, Wang H, Gao P, Shi W, Huang L (2010) Knockdown of stromal interaction molecule 1 attenuates hepatocyte growth factor-induced endothelial progenitor cell proliferation. Exp Biol Med (Maywood) 235(3):317–325

Shinde AV, Motiani RK, Zhang X, Abdullaev IF, Adam AP, González-Cobos JC, Zhang W, Matrougui K, Vincent PA, Trebak M (2013) STIM1 controls endothelial barrier function independently of Orai1 and Ca2+ entry. Sci Signal 6(267):ra18

Sirenko SG, Maltsev VA, Yaniv Y, Bychkov R, Yaeger D, Vinogradova T, Spurgeon HA, Lakatta EG (2016) Electrochemical Na+ and Ca2+ gradients drive coupled-clock regulation of automaticity of isolated rabbit sinoatrial nodal pacemaker cells. Am J Physiol Heart Circ Physiol 311(1):H251–H267

Spinelli AM, Trebak M (2016) Orai channel-mediated Ca2+ signals in vascular and airway smooth muscle. Am J Physiol Cell Physiol 310(6):C402–C413

Stolwijk JA, Zhang X, Gueguinou M, Zhang W, Matrougui K, Renken C, Trebak M (2016) Calcium signaling is dispensable for receptor regulation of endothelial barrier function. J Biol Chem 291(44):22894–22912

Sundivakkam PC, Freichel M, Singh V, Yuan JP, Vogel SM, Flockerzi V, Malik AB, Tiruppathi C (2012) The Ca(2+) sensor stromal interaction molecule 1 (STIM1) is necessary and sufficient for the store-operated Ca(2+) entry function of transient receptor potential canonical (TRPC) 1 and 4 channels in endothelial cells. Mol Pharmacol 81(4):510–526

Sundivakkam PC, Natarajan V, Malik AB, Tiruppathi C (2013) Store-operated Ca2+ entry (SOCE) induced by protease-activated receptor-1 mediates STIM1 protein phosphorylation to inhibit SOCE in endothelial cells through AMP-activated protein kinase and p38β mitogen-activated protein kinase. J Biol Chem 288(23):17030–17041

Suri C, Jones PF, Patan S, Bartunkova S, Maisonpierre PC, Davis S, Sato TN, Yancopoulos GD (1996) Requisite role of angiopoietin-1, a ligand for the TIE2 receptor, during embryonic angiogenesis. Cell 87(7):1171–1180

Tamareille S, Mignen O, Capiod T, Rucker-Martin C, Feuvray D (2006) High glucose-induced apoptosis through store-operated calcium entry and calcineurin in human umbilical vein endothelial cells. Cell Calcium 39(1):47–55

Thomas AP, Bird GS, Hajnoczky G, Robb-Gaspers LD, Putney JW Jr (1996) Spatial and temporal aspects of cellular calcium signaling. FASEB J 10(13):1505–1517

Thyagarajan B, Poteser M, Romanin C, Kahr H, Zhu MX, Groschner K (2001) Expression of Trp3 determines sensitivity of capacitative Ca^{2+} entry to nitric oxide and mitochondrial Ca^{2+} handling: evidence for a role of Trp3 as a subunit of capacitative Ca^{2+} entry channels. J Biol Chem 276(51):48149–48158

Tiruppathi C, Minshall RD, Paria BC, Vogel SM, Malik AB (2002) Role of Ca^{2+} signaling in the regulation of endothelial permeability. Vascul Pharmacol 39(4–5):173–185

Toya SP, Malik AB (2011) Role of endothelial injury in disease mechanisms and contribution of progenitor cells in mediating endothelial repair. Immunobiology 217(5):569–580

Trevisi L, Bertoldo A, Agnoletto L, Poggiani C, Cusinato F, Luciani S (2010) Antiapoptotic and proliferative effects of low concentrations of 7beta-hydroxycholesterol in human endothelial cells via ERK activation. J Vasc Res 47(3):241–251

Usatyuk PV, Parinandi NL, Natarajan V (2006) Redox regulation of 4-hydroxy-2-nonenal-mediated endothelial barrier dysfunction by focal adhesion, adherens, and tight junction proteins. J Biol Chem 281(46):35554–35566 van Breemen C (1977) Calcium requirement for activation of intact aortic smooth muscle. J Physiol 272:317–329

van Breemen C, Chen Q, Laher I (1995) Superficial buffer barrier function of smooth muscle sarcoplasmic reticulum. Trends Pharmacol Sci 16:98–105

Veliceasa D, Ivanovic M, Hoepfner FT, Thumbikat P, Volpert OV, Smith ND (2007) Transient potential receptor channel 4 controls thrombospondin-1 secretion and angiogenesis in renal cell carcinoma. FEBS J 274(24):6365–6377

Voelkers M, Salz M, Herzog N, Frank D, Dolatabadi N, Frey N, Gude N, Friedrich O, Koch WJ, Katus HA, Sussman MA, Most P (2010) Orai1 and Stim1 regulate normal and hypertrophic growth in cardiomyocytes. J Mol Cell Cardiol 48:1329–1334

Völkers M, Dolatabadi N, Gude N, Most P, Sussman MA, Hassel D (2012) Orai1 deficiency leads to heart failure and skeletal myopathy in zebrafish. J Cell Sci 125(Pt 2):287–294

Wakabayashi I, Groschner K (1996) Evidence for a direct inhibitory effect of extracellular H+ on store depletion-activated Ca^{2+}entry in vascular endothelial cells. Biochem Biophys Res Commun 221(3):762–767

Wang X, Chu W, van Breemen C (1996) Potentiation of acetylcholine-induced responses in freshly isolated rabbit aortic endothelial cells. J Vasc Res 33(5):414–424

Wang X, Pluznick JL, Settles DC, Sansom SC (2007) Association of VASP with TRPC4 in PKG-mediated inhibition of the store-operated calcium response in mesangial cells. Am J Physiol Renal Physiol 293(6):F1768–F1776

Wang Y, Deng X, Hewavitharana T, Soboloff J, Gill DL (2008) Stim, ORAI and TRPC channels in the control of calcium entry signals in smooth muscle. Clin Exp Pharmacol Physiol 35:1127–1133

Wang Z, Ginnan R, Abdullaev IF, Trebak M, Vincent PA, Singer HA (2010) Calcium/Calmodulin-dependent protein kinase II delta 6 (CaMKIIdelta6) and RhoA involvement in thrombin-induced endothelial barrier dysfunction. J Biol Chem 285(28):21303–21312

Wang QC, Zheng Q, Tan H, Zhang B, Li X, Yang Y, Yu J, Liu Y, Chai H, Wang X, Sun Z, Wang JQ, Zhu S, Wang F, Yang M, Guo C, Wang H, Zheng Q, Li Y, Chen Q, Zhou A, Tang TS (2016a) TMCO1 is an ER Ca(2+) load-activated Ca(2+) channel. Cell 165(6):1454–1466

Wang YW, Zhang JH, Yu Y, Yu J, Huang L (2016b) Inhibition of store-operated calcium entry protects endothelial progenitor cells from H2O2-induced apoptosis. Biomol Ther (Seoul) 24 (4):371–379

Wedel B, Boyles RR, Putney JW Jr, Bird GS (2007) Role of the store-operated calcium entry proteins Stim1 and Orai1 in muscarinic cholinergic receptor-stimulated calcium oscillations in human embryonic kidney cells. J Physiol 579(Pt 3):679–689

Wu S, Cioffi EA, Alvarez D, Sayner SL, Chen H, Cioffi DL, King J, Creighton JR, Townsley M, Goodman SR, Stevens T (2005) Essential role of a Ca^{2+}-selective, store-operated current (ISOC) in endothelial cell permeability: determinants of the vascular leak site. Circ Res 96 (8):856–863

Wu MM, Covington ED, Lewis RS (2014) Single-molecule analysis of diffusion and trapping of STIM1 and Orai1 at endoplasmic reticulum-plasma membrane junctions. Mol Biol Cell 25:3672–3685

Xu X, Star RA, Tortorici G, Muallem S (1994) Depletion of intracellular Ca^{2+} stores activates nitric-oxide synthase to generate cGMP and regulate Ca^{2+} influx. J Biol Chem 269 (17):12645–12653

Yang LX, Guo RW, Liu B, Wang XM, Qi F, Guo CM, Shi YK, Wang H (2009) Role of TRPC1 and NF-kappaB in mediating angiotensin II-induced Ca^{2+} entry and endothelial hyperpermeability. Peptides 30(7):1368–1373

Yaniv Y, Lakatta EG, Maltsev VA (2015) From two competing oscillators to one coupled-clock pacemaker cell system. Front Physiol 6:28

Yazawa M, Ferrante C, Feng J, Mio K, Ogura T, Zhang M, Lin PH, Pan Z, Komazaki S, Kato K, Nishi M, Zhao X, Weisleder N, Sato C, Ma J, Takeshima H (2007) TRIC channels are essential for Ca2+ handling in intracellular stores. Nature 448(7149):78–82

Yeh YC, Tang MJ, Parekh AB (2014) Caveolin-1 alters the pattern of cytoplasmic Ca2+ oscillations and Ca2+-dependent gene expression by enhancing leukotriene receptor desensitization. J Biol Chem 289(25):17843–17853

Yoshida T, Inoue R, Morii T, Takahashi N, Yamamoto S, Hara Y, Tominaga M, Shimizu S, Sato Y, Mori Y (2006) Nitric oxide activates TRP channels by cysteine S-nitrosylation. Nat Chem Biol 2(11):596–607

Zhang SL, Yeromin AV, Zhang XH, Yu Y, Safrina O, Penna A, Roos J, Stauderman KA, Cahalan MD (2006) Genome-wide RNAi screen of Ca(2+) influx identifies genes that regulate Ca(2+) release-activated Ca(2+) channel activity. Proc Natl Acad Sci U S A 103(24):9357–9362

Zhang H, Sun AY, Kim JJ, Graham V, Finch EA, Nepliouev I, Zhao G, Li T, Lederer WJ, Stiber JA, Pitt GS, Bursac N, Rosenberg PB (2015) STIM1-Ca2+ signaling modulates automaticity of the mouse sinoatrial node. Proc Natl Acad Sci U S A 112:E5618–E5627

Zhao G, Li T, Brochet DX, Rosenberg PB, Lederer WJ (2015) STIM1 enhances SR Ca2+ content through binding phospholamban in rat ventricular myocytes. Proc Natl Acad Sci U S A 112: E4792–E4801

Zhu G, Huang L, Song M, Yu Z, Wu X, Zhao X, Jin J, Zhao G, Chen J, Yu S (2010) Overexpression of hepatocyte growth factor in smooth muscle cells regulates endothelial progenitor cells differentiation, migration and proliferation. Int J Cardiol 138(1):70–80

Zou M, Dong H, Meng X, Cai C, Li C, Cai S, Xue Y (2015) Store-operated Ca2+ entry plays a role in HMGB1-induced vascular endothelial cell hyperpermeability. PLoS One 10(4):e0123432

Cardiac Remodeling and Disease: SOCE and TRPC Signaling in Cardiac Pathology

25

Petra Eder

Abstract

TRPC channels have been suggested as potential candidates mediating store-operated Ca^{2+} entry (SOCE) in cardiomyocytes. There is increasing evidence that the TRPC isoforms TRPC1 and TRPC4 might fulfill the function as SOCs, in concert with or in parallel to the key players of SOCE, Orai1, and STIM1. Several other isoforms, e.g., TRPC3, TRPC6, and TRPC7, might rather associate to receptor-activated diacylglycerol (DAG)-sensitive ion channels. However, the exact activation mode has not been elucidated yet, given the characteristic of TRPC channels to heteromerize to unpredictable ion channel assemblies. Despite the incomplete information about TRPC activation, there is common agreement that they are crucial Ca^{2+} components in cardiac signaling and disease. All TRPC isoforms, TRPC1, TRPC3, TRPC4, TRPC5, TRPC6, and TRPC7, are differentially regulated in cardiac disease, and nearly all of them have been shown to impact cardiac signaling pathways that accelerate cardiac disease development. In particular, the calcineurin-nuclear factor of activated T-cell (NFAT) signaling pathway has repeatedly been linked to a TRPC-dependent Ca^{2+} influx in cardiomyocytes. Moreover, the protein kinases PKG and PKC have been found to modulate TRPC function and the hypertrophic response. Other signaling molecules, such as the serine/threonine kinase Ca^{2+}/calmodulin-dependent protein kinase II (CamKII) or the oxidative stress molecule, NADPH oxidase 2 (NOX2), have also been related to TRPC-dependent effects in the heart.

The present chapter provides a comprehensive overview of TRPC channels as Ca^{2+} entities in cardiomyocytes, their interplay with Ca^{2+} signaling pathways, and role in cardiac pathology.

P. Eder (✉)
Department of Internal Medicine I, Comprehensive Heart Failure Center Würzburg, University Hospital Würzburg, Würzburg, Germany
e-mail: eder_p@ukw.de

K. Groschner et al. (eds.), *Store-Operated Ca^{2+} Entry (SOCE) Pathways*, Advances in Experimental Medicine and Biology 993, DOI 10.1007/978-3-319-57732-6_25

Keywords
TRPC • Calcineurin • NFAT • Cardiac hypertrophy • Heart failure

25.1 Introduction

25.1.1 Ca^{2+} Regulation in Cardiomyocytes

Cardiac contraction and relaxation is based on a tightly regulated cascade of Ca^{2+} fluxes including the Ca^{2+} uptake in the cardiomyocyte, the Ca^{2+} release from the sarco/endoplasmic reticulum (SR), and the Ca^{2+} binding to the sarcomeres followed by the reuptake of Ca^{2+} in the SR and the Ca^{2+} removal out of the cell. The underlying Ca^{2+}-handling proteins in this process have been well described and include the voltage-dependent L-type Ca^{2+} channel (LTCC) allowing a Ca^{2+} entry in the cleft region of cardiomyocytes, the ryanodine receptor (RyR)-dependent Ca^{2+} release from the SR causing myocyte contraction, the sarco/endoplasmic reticulum Ca^{2+} ATPase (SERCA) responsible for the Ca^{2+} removal into the SR, and the Na/Ca exchanger (NCX) for the Ca^{2+} extrusion out of cardiomyocytes (Bers 2002).

Besides these contractile Ca^{2+} proteins, cardiomyocytes are equipped with several Ca^{2+}-handling proteins which might control signaling pathways or the Ca^{2+} homeostasis of cellular compartments and might not necessarily be part of the excitation-contraction coupling (ECC) machinery. For example, cardiomyocytes express plasmalemma Ca^{2+}-ATPases (PMCAs) with PMCA1 and PMCA4 as the most abundant isoforms (Cartwright et al. 2011). Their contribution to Ca^{2+} extrusion during ECC is less than 1% suggesting that PMCAs play a role in the local Ca^{2+} regulation affecting downstream signaling effectors such as the neuronal nitric oxide synthase (nNOS), Ras-associated factor-1, alpha1-syntrophin, Ca^{2+}-calmodulin-dependent serine protein kinase (CASK), and calcineurin (Can; Oceandy et al. 2007). In this respect, PMCAs might function as negative regulators of pathological hypertrophy limiting sub-plasmalemmal Ca^{2+} pools which are required for the activation of the calcineurin/NFAT pathway (Wu et al. 2009).

Ca^{2+} signaling might also be affected by the re-expression of Ca^{2+}-handling proteins that usually play a minor role in the healthy heart. For example, there is re-expression of T-type Ca^{2+} channels (TTCC) during pathological cardiac hypertrophy and in the post-infarction heart suggesting that they are part of the fetal gene program during pathological hypertrophy (Nuss and Houser 1993; Huang et al. 2000). The functional role of TTCCs in cardiac remodeling has been evaluated using gene-targeted mouse models. For example, pathological hypertrophy was suppressed in mice deficient for α1H ($Cav3.2^{-/-}$), a phenotype which was mechanistically coupled with less activation of the calcineurin/NFAT signaling pathway (Chiang et al. 2009). Since Ca^{2+} cycling and contractility in these mice were not impaired, α1H might be part of a membrane complex that allows local Ca^{2+} changes instead of bulk Ca^{2+}.

Store-operated Ca^{2+} entry (SOCE) has been considered as another type of Ca^{2+} influx that occurs as a result of an increased ion channel expression in the course of the reactivated fetal gene program or due to an increased ion channel activity during chronic cardiac disease development (Eder and Molkentin 2011; Hulot et al. 2011; Luo et al. 2012). In particular, TRPC channels, the Ca^{2+} release-activated channel protein Orai1, and the stromal interaction molecule STIM1 have been considered part of the molecular makeup underlying SOCE in cardiomyocytes.

25.1.2 SOCE in Cardiomyocytes

More than a decade ago, a study by Hunton et al. (2002) showed that stimulation of Gq protein coupled receptors (GqPCRs) with angiotensin II (Ang II) or phenylephrine (PE) elicited cytosolic Ca^{2+} signals in neonatal rat cardiomyocytes (NRCs) which were composed of a slowly increasing Ca^{2+} entry followed by sporadic, fast Ca^{2+} spikes. While these fast Ca^{2+} fluxes were inhibited by LTCC antagonists, the slow increase was insensitive to LTCC blockers but was rather inhibited by inhibitors of the capacitative Ca^{2+} entry. Although nonselective SOCE inhibitors were used in this study (glucosamine, SKF-96365), it was one of the first indication that cardiomyocytes allow a Ca^{2+} entry which could represent a SOCE similarly to non-excitable cells. Additional studies followed and showed that also a passive SR Ca^{2+} leak induced by the application of SERCA inhibitors results in plasma membrane Ca^{2+} fluxes (Hunton et al. 2004; Huang et al. 2006). This phenomenon was repeatedly found in neonatal cardiomyocytes and was interpreted as characteristic of an immature neonatal phenotype. Interestingly, SOCE was barely detectable in adult mouse cardiomyocytes but was described to reoccur after myocardial infarction or pressure-overload-induced cardiac hypertrophy suggesting that during the remodeling process, a re-expression of certain Ca^{2+} proteins might take place and novel Ca^{2+} fluxes are generated (Wu et al. 2010; Luo et al. 2012). Although the underlying mechanisms have not been entirely resolved, an enhanced afterload, stretch, and the chronic activation of the renin-angiotensin-aldosterone system (RAAS) might be implicated in the generation of new Ca^{2+} pathways such as SOCE. While the idea of SOCE has been accepted as critical alteration in cardiac disease, there are still uncertainties about the molecular components involved. SOCE in non-excitable cells involves the inositol 1,4,5-trisphosphate receptor (IP_3R) releasing Ca^{2+} from the endoplasmic reticulum (ER). In cardiac myocytes, IP_3Rs are 5–80 times less abundant than RyRs which might indicate a minor role for their functionality in healthy cardiomyocytes (Kockskämper et al. 2008). Interestingly, overexpression of IP_3R2 in transgenic mice results in enhanced cardiac hypertrophy following isoproterenol infusion which is mechanistically coupled with calcineurin activation (Nakayama et al. 2010). Furthermore, expression of IP_3Rs is increased in different models of heart failure which might indicate that this type of Ca^{2+} release might be relevant in pathologic situations (Go et al. 1995). It might be speculated that IP_3Rs generate local Ca^{2+} signals independently of the bulk Ca^{2+} released by RyRs which are then sensed by ion channels in the

PM. Besides, IP_3Rs are integrated in the nuclear envelope regulating nuclear Ca^{2+} and the Ca^{2+}-dependent signaling molecule serine/threonine kinase Ca^{2+}/calmodulin (CaM)-dependent protein kinase II (CaMKII), which triggers histone deacetylase 5 (HDAC5) phosphorylation and nuclear export (Wu et al. 2006).

STIM1 has been discovered as another important regulator of SOC channels in the plasma membrane. Findings from non-excitable cells have shown that the ER-Ca^{2+} release by IP_3Rs causes STIM1 to re-localize to ER sites adjacent to the plasma membrane. Subsequently, Orai1 and most likely also TRPC channels are activated and allow a Ca^{2+} entry into the cytosol (Derler et al. 2016). In the heart, the absence of STIM1 has been associated with less SR Ca^{2+} load and pressure-overload-induced cardiac hypertrophy pointing to STIM1 as critical Ca^{2+} sensor in pathological conditions (Hulot et al. 2011).

With the detection of TRPC channels in the heart, the possibility was raised that they could constitute SOCs in cardiomyocytes. Nearly all isoforms (TRPC1, TRPC3, TRPC4, TRPC5, TRPC6, TRPC7) seem to be upregulated during chronic cardiac disease in human or animal models (Eder and Molkentin 2011; Makarewich et al. 2014; Kirschmer et al. 2016). Moreover, a store-dependent activation has been found for TRPC isoforms belonging both to the diacylglycerol (DAG)-sensitive (TRPC3/6/7) and the IP_3R-sensitive subfamily (TRPC1/4/5). For example, dominant-negative (dn) TRPC3 proteins were able to inhibit a SOCE in hypertrophic cardiomyocytes (Wu et al. 2010). On the other hand, TRPC3 was activated by the application of the DAG analogue OAG rather than by passive Ca^{2+} store depletion (Makarewich et al. 2014). As another example, TRPC1 activation was associated with stretch-sensitive Ang II receptors (Seth et al. 2009) but was also shown to mediate a passive Ca^{2+} influx in concert with TRPC4 (Camacho Londono et al. 2015). Based on these findings, it might be speculated that a distinct pattern of TRPC activation does not exist or cannot easily be explored because of the complexity of ion channel heteromerization as well as the different types of cardiomyocytes (mouse and rat adult and neonatal cells) and their current neurohumoral input (cells isolated after pathological cardiac hypertrophy, myocardial infarction (MI), or non-stimulated cells) used in different studies. Also the complex formation with the classical SOCE proteins Orai1/STIM1 might influence the function of TRPC channels in the heart. For example, a recent study shows that chronic aldosterone treatment generates a SOCE which is dependent on Orai1 and TRPC1, TRPC4, and TRPC5 channels (Sabourin et al. 2016).

25.1.3 Downstream Effects of SOCE on Signaling and Cardiac Function

SOCE is linked to the stimulation of GPCR receptors induced by neurohormones such as Ang II, endothelin 1 (ET-1), or the α1-adrenergic receptor agonist phenylephrine (PE), the consecutive activation of phospholipase C (PLC), and the generation of DAG and IP_3 (Eder and Molkentin 2011). It is well known that the secretion of these hormones is a prior adaption to cardiac stress situations in order to maintain function. However, chronic secretion during chronic pathological conditions such as hypertension, aortic stenosis, and MI is connected with a variety of maladaptive effects such as fibrotic remodeling, cardiac hypertrophy, arrhythmias, necrosis, and apoptosis (Orsborne et al. 2017). Several of these adverse effects are related to alterations of the Ca^{2+} homeostasis and the transcriptional machinery. In particular, the Ca^{2+}-sensitive phosphatase CaN has evoked great interest as crucial signaling protein mediating neurohumoral input to hypertrophic remodeling (Molkentin 2013). CaN exists as a heterotrimer composed by the catalytic A subunit (CnA) and the Ca^{2+}-binding proteins calcineurin B (CnB) and calmodulin (CaM). At low Ca^{2+} concentrations, an auto-inhibitory domain blocks the activity of CaN. With increasing Ca^{2+} concentrations, the auto-inhibitory domain is released which results in a tighter interaction between CaM and CnA and the activation of CaN (Klee et al. 1998). CaN dephosphorylates the transcription factor nuclear factor of activated T-cells (NFAT) which translocates from the cytosol to the nucleus (Molkentin et al. 1998) where it regulates the expression of pro-hypertrophic genes, together with the cardiac zinc-finger transcription factor GATA4.

The Ca^{2+} source of calcineurin is still one of the biggest question marks in the scientific field. Besides alterations in amplitude or/and frequency of Ca^{2+} transients and changes in diastolic Ca^{2+} levels, Ca^{2+} sources apart from the contractile Ca^{2+} have been suggested to be linked to CaN activation in cardiomyocytes. In this respect, SOCE in cardiomyocytes has been attributed as ideal Ca^{2+} source for activating CaN, given its characteristic to allow a slowly increasing Ca^{2+} influx. In addition, it has been suggested that the activity of CaN is mechanistically linked to TRPC channels. In this context, CaN and NFAT activation was enhanced by TRPC isoforms that might function as SOCs (TRPC1, TRPC4) as well as SOC-independent TRPC isoforms (e.g., TRPC6). Interestingly, TRPC channels have been assumed to prime CaN with Ca^{2+} but are themselves downstream targets of CaN and the NFAT transcriptional activity. Thus, it has been proposed that an enhanced activation of TRPC channels after GPCR stimulation activates CaN and NFAT which in turn upregulates TRPC channels generating a forward loop from agonist stimulation and Ca^{2+} overload to hypertrophic remodeling. Indeed, nearly all TRPC isoforms seem to be upregulated in cardiac hypertrophy in humans or experimental models (see below). Moreover, NFAT binding sites have been found in TRPC1, TRPC3, and TRPC6 promoters. The mechanistic coupling between TRPC-CaN-NFAT might take place in specific subcellular platforms or Ca^{2+} microdomains. Similar to other cell types, TRPC channels might be organized in caveolae that facilitate NFAT-dependent hypertrophic remodeling (Makarewich et al. 2014). SOCE in cardiomyocytes might also indirectly affect the CaN signaling pathway. For example, STIM1 downregulation was associated with an enhanced activity of the anti-hypertrophic molecule GSK-3β. It seems that during pathological cardiac hypertrophy, STIM1 activation regulates Akt kinase activity through the activation of mTORC2 resulting in repression of the anti-hypertrophic activity of the glycogen synthase kinase *3* (GSK-3β) (Benard et al. 2016). Although not fully elucidated, this mechanism might be partially independent of the traditional function of STIM1 to govern SOCE through an interaction with Orai1.

Recently, it was suggested that the frequency of Ca^{2+} sparks is affected by the activity of STIM1 (Correll et al. 2015), as well as TRPC channels (Makarewich et al. 2014). An enhanced Ca^{2+} spark activity and thus SR Ca^{2+} leak via RyRs is usually promoted by the local activity of CamKII (Anderson et al. 2011). Given the fact that the CamKII is activated by local and high Ca^{2+} waves, a scenario is possible which implicates a SOC-mediated Ca^{2+} influx activating CamKII and thus promoting a CamKII-dependent Ca^{2+} leakage and diastolic Ca^{2+} elevation.

25.2 TRPC Channels as SOCs, Their Role in Cardiomyocyte Ca^{2+} Handling, and Physiology

The following sections give an overview of current findings on TRPC-mediated effects on cytosolic Ca^{2+}, signaling, and cardiac function. The importance of TRPC channels as SOCs as well as their functional role as receptor-operated channels (ROCs) and stretch-activated channels (SACs) will be discussed, including expressional alterations during disease, associated mechanisms, and physiological and pathophysiological effects.

25.2.1 The IP_3-sensitive Subfamily-TRPC1, 4, 5

25.2.1.1 TRPC1

There are still open questions regarding the ability of TRPC1 to integrate as subunit in TRPC tetramers (Dietrich et al. 2014). It has been supposed that TRPC1 associates with other TRPC isoforms, even across the TRPC family border, affecting ion channel properties such as the Ca^{2+} selectivity or mechanosensitivity. The integration as pore-forming subunit has also been found to increase SOCE in various cells including cardiomyocyte cultures. In a study by Ohba et al. (2007), NRCs were analyzed after chronic neurohumoral stimulation. They showed that chronic GPCR stimulation with Ang II, PE, or ET-1 resulted in a typical cardiomyocyte enlargement and enhanced hypertrophic marker gene expression. In addition, SOCE elicited by acute thapsigargin (TG) application was enhanced in cells kept under chronic neurohumoral stimulation. Among different TRPC subunits, they found an enhanced TRPC1 expression which they correlated with the enhanced SOCE activity after agonist stimulation. Indeed, downregulation of TRPC1 by a siRNA-mediated approach markedly reduced SOCE in agonist-treated NRCs. Interestingly, another recent study demonstrated that chronic aldosterone stimulation of NRCs enhanced SOCE but was abrogated by downregulating or inhibiting different TRPC isoforms and Orai1 (Sabourin et al. 2016). Based on this finding, it might be speculated that SOCE is not solely dependent on TRPC1 but rather TRPC1 together with other TRPC isoforms and even the Orai/STIM functional unit. TRPC1 was also found upregulated in the adult heart of aortic-banded rats pointing to a role of TRPC1 in pathological cardiac hypertrophy (Ohba et al. 2007). The analyses of TRPC1 knockout mice revealed convincing findings that showed a loss of the hypertrophic enlargement and functional improvement after transverse-aortic constriction of TRPC1 knockout mice (Seth et al. 2009). This cardiac phenotype was linked to a reduced TRPC-like current in cardiomyocytes of TRPC1 knockout mice after transverse aortic constriction (TAC) treatment. Further in vitro experiments suggested that a TRPC1-specific current is linked to mechanical stress (Seth et al. 2009). Stretch-dependent activation of TRPC channels might occur by sensing membrane lipid formations directly (Spassova et al. 2006) or indirectly through stretch-sensitive GPCRs. In case of TRPC1 in cardiomyocytes, it seems that mechanical stress is sensed by the Ang II receptor conferring this information to TRPC1. The opening of TRPC1-containing channels might then increase cytosolic Ca^{2+} levels and activate Ca^{2+}-dependent hypertrophic signaling pathways (Fig. 25.1). In TRPC1 knockout mice, the CaN pathway was inhibited and the activity of protein kinase B (AKT) was enhanced, which could explain the reduced hypertrophic response but better survival after TAC treatment. Another study confirmed the importance of TRPC1 in hypertrophic remodeling. However, in contrast to Seth et al., the authors of that study suggested that a double knockout TRPC1/TRPC4 was mandatory to prevent cardiac hypertrophy and fibrotic infiltration after TAC and chronic neurohumoral stimulation (Camacho Londono et al. 2015). Interestingly, cardiac protection after TAC treatment was not achieved by a double TRPC3/TRPC6 knockout. The beneficial effect of a TRPC1/TRPC4 knockout was connected with a reduced background Ca^{2+} entry in cardiomyocytes. Moreover, systolic and diastolic Ca^{2+} levels of transients were reduced in TRPC1/TRPC4 knockout cardiomyocytes both at baseline and after neurohumoral stimulation with Ang II or isoproterenol (Iso). As underlying mechanisms of the attenuated cardiac hypertrophy, both CaN and myocyte enhancer factor 2a (MEF2a) activity were reduced in TRPC1/TRPC4 knockout mice. A similar protective phenotype was revealed in triple knockout TRPC1/TRPC4/TRPC5 mice which might indicate that these three TRPC isoforms are part of the same signaling complex in cardiomyocytes. However, further studies are necessary to confirm the functionality of a TRPC1/TRPC4/TRPC5 assembly, its effects on cytosolic Ca^{2+}, and sensitivity to other stimuli such as stretch and store depletion.

25.2.1.2 TRPC4 and TRPC5

TRPC4 has been suggested as central component in cardiomyocyte Ca^{2+} signaling and cardiac disease. First evidence came from a study which analyzed the phenotype of transgenic mice which overexpressed a dominant negative (dn)-TRPC4 truncated mutant in cardiomyocytes. After TAC treatment, it became evident that these mice were protected from pathological cardiac remodeling and were prevented from a drop of cardiac function (Wu et al. 2010). This protective phenotype was associated with less CaN-NFAT signaling and less fibrotic infiltration. Assuming that dn-TRPC4 might block SOCE, a protocol was pursued to achieve a passive Ca^{2+} store release by using TG as SERCA inhibitor. Interestingly, cardiomyocytes from wildtype (WT) mice responded with a profound Ca^{2+} influx

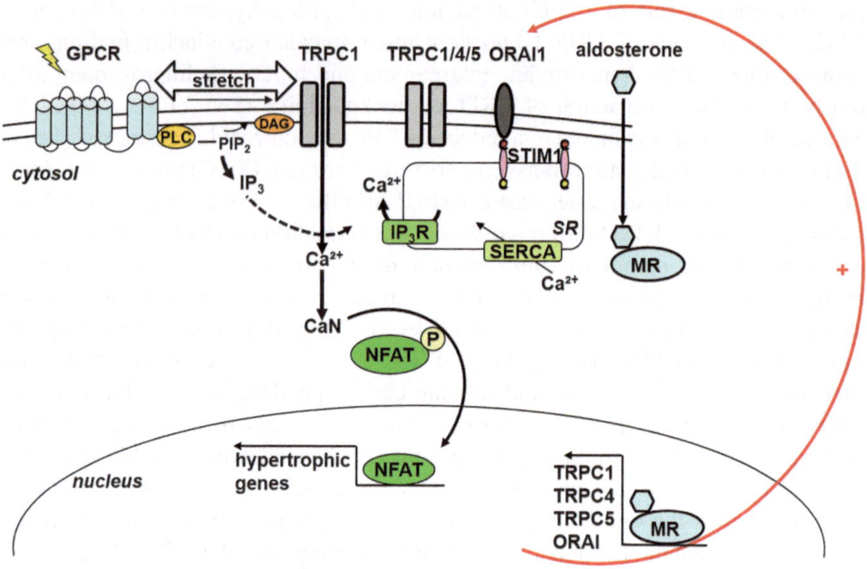

Fig. 25.1 TRPC1-dependent signaling in cardiomyocytes. TRPC1 can be activated downstream of Gq-protein-coupled receptors (GPCR) activation through endothelin 1, angiotensin II, or phenylephrine (indicated as flash) generating a background Ca^{2+} entry in cardiomyocytes. Ca^{2+} release from the endo/sarcoplasmic reticulum (SR) stores might be implicated in the activation of TRPC1. Besides, mechanical stretch through the activation of GPCRs might induce TRPC1 activation. TRPC1-dependent Ca^{2+} signals activate calcineurin (CaN) which dephosphorylates the nuclear factor of activated T-cells (NFAT). NFAT translocates to the nucleus and upregulates pro-hypertrophic genes. TRPC1 might also generate a store-operated Ca^{2+} influx in concert with TRPC4, TRPC5, and Orai1 after chronic aldosterone treatment. Aldosterone binds to mineralocorticoid receptors (MR) in the cytosol. MR might upregulate TRPC1, TRPC4, TRPC5, and Orai1 and thus increase SOCE in cardiomyocytes. *DAG* diacylglycerol, *GPCR* Gq-protein-coupled receptor, *IP_3* inositol 3-phosphate, *IP_3R* inositol 3-phosphate receptor, *Orai1* Ca^{2+} release-activated channel protein, *PIP_2* phosphatidylinositol 4,5-bisphosphate, *PLC* phospholipase C, *SERCA* sarcoplasmic/endoplasmic reticulum calcium ATPase , *STIM* stromal interaction molecule, *TRPC* transient receptor potential canonical

which was absent in control-treated mice. As pointed out at the beginning, healthy adult cardiomyocytes might not require SOCE for their proper function. A reoccurrence of SOCE in hypertrophic cardiomyocytes could therefore go along with the remodeling process from an adult to a fetal cardiac phenotype after chronic pressure overload induction. In contrast to WT animals, the SOCE in dn-C4 mice was drastically reduced. In this context, it is interesting to note that dn-TRPC3 and dn-TRPC6 mutants had similar inhibitory effects on SOCE, CaN activation, and cardiac hypertrophy. Since TRPC3 and TRPC6 more likely function as DAG-sensitive ion channels (Gudermann et al. 2004) (see also Sect. 25.2.2), this was a rather unexpected finding and reveals the difficulty to determine the actual role of individual TRPC in SOCE.

In another attempt to characterize TRPC isoforms as SOCs, a study by Houser and colleagues (Makarewich et al. 2014) focused on Ca^{2+} analyses in adult feline cardiomyocytes expressing different combinations of TRPC isoforms (TRPC4, TRPC3, TRPC6) and their respective dn-mutants (dnTRPC4 and dnTRPC6). Their findings revealed that TRPC4 was sensitive to passive Ca^{2+} store depletion while TRPC3 and TRPC6 responded to the stimulation with OAG. Interestingly, the activation of CaN-NFAT was not restricted to any of the isoforms but was similarly enhanced. This finding might be explained by the localization of TRPC channels as signaling complexes in caveolae-microdomains together with CaN and NFAT. An enhanced activity of any of these TRPC isoforms would then result in an accumulation of microdomain Ca^{2+}, which might expand to increased cytosolic Ca^{2+} levels, increased SR Ca^{2+} load, spark frequency finally resulting in CaN-NFAT, and even CamKII activation. This mechanism might accelerate cardiac hypertrophic remodeling induced by pressure overload induction or myocardial infarction. In this context, it is interesting to note that a TRPC4-dependent Ca^{2+} influx might not only affect cardiac remodeling but might affect Ca^{2+} transients and alter cardiac function. For example, dn-TRPC4 expression in transgenic mice was associated with increased Ca^{2+} transient amplitudes which remained elevated after MI, as well as after β-adrenergic stimulation suggesting an improved adrenergic responsiveness during disease when TRPC4 channels are blocked (Makarewich et al. 2014). As mentioned above, another study showed that a double knockout of TRPC1/ TRPC4 was associated with lowered systolic and diastolic levels of the Ca^{2+} transients (Camacho Londono et al. 2015). These effects seemed to be dependent on a background Ca^{2+} influx in cardiomyocytes which was clearly inhibited when TRPC1/TRPC4 were knocked out. In line with this study, we recently found that overexpression of the TRPC4 splice variants TRPC4α and TRPC4β was linked to enhanced peak Ca^{2+} levels in NRCs which enlarged even further in TRPC4β-infected cardiomyocytes after acute addition of Ang II (Kirschmer et al. 2016). Similarly, Ang II administration triggered a prominent Ca^{2+} influx in quiescent cardiomyocytes overexpressing TRPC4β but not TRPC4α. The presence of the nonselective SOCE inhibitor BTP2 blocked this Ca^{2+} influx. Interestingly, TRPC4β activity was coupled to the activation of CaN and NFAT after GPCR stimulation as established mechanistic link but did not further accelerate hypertrophy during GPCR stimulation. Instead, TRPC4β as well as TRPC4α promoted apoptosis resulting in an attenuated cell growth and impaired cell survival (Fig. 25.2). These findings show that TRPC4 channels might be able to exert dual functional roles in cardiomyocytes which might be dependent on several factors: the heteromerization pattern, the type of disease, as well as the investigated cell system.

Although TRPC5 was also shown to be upregulated in the human failing heart, its characteristics have not been pursued in further studies. As pointed out above, TRPC5 might be complexed with TRPC1 and TRPC4 and contribute to a background Ca^{2+} entry in pathological cardiac hypertrophy.

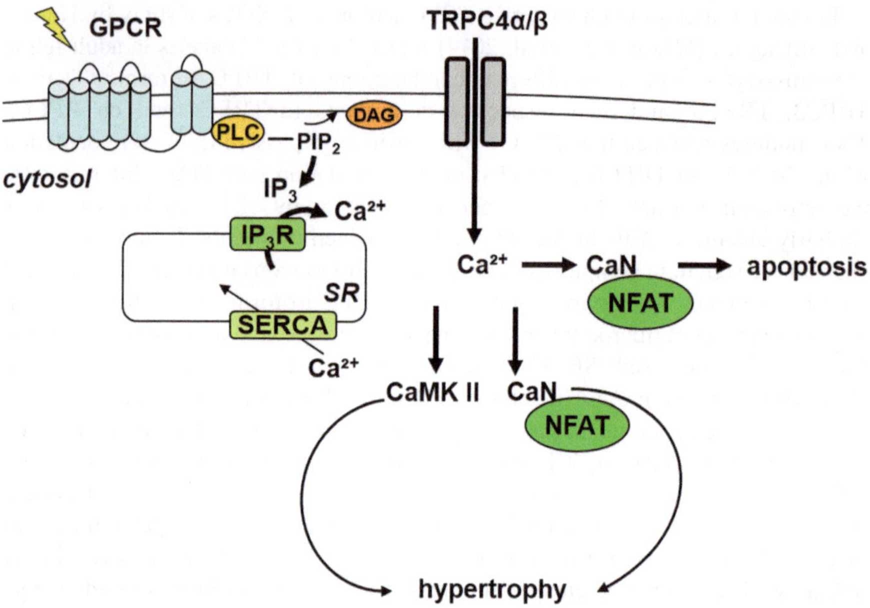

Fig. 25.2 Downstream effects of TRPC4-dependent Ca^{2+} signals in cardiomyocytes. TRPC4 can be activated downstream of Gq-protein-coupled receptor (GPCR) activation, e.g., through angiotensin II (indicated as flash) and might (similar to TRPC1) generate a background Ca^{2+} influx. This results in the activation of calcineurin (CaN) and nuclear factor of activated T-cells (NFAT) as well as the serine/threonine kinase Ca^{2+}/calmodulin-dependent protein kinase II (CamKII). Apart from promoting hypertrophy, TRPC4 activation might accelerate apoptosis and lower cardiomyocyte viability through CaN-NFAT signaling. This effect might be equally mediated by the main TRPC4 splice variants, TRPC4α and TRPC4β. *DAG* diacylglycerol, *GPCR* Gq-protein-coupled receptor, IP_3 inositol 3-phosphate, IP_3R inositol 3-phosphate receptor, PIP_2 phosphatidylinositol 4,5-bisphosphate, *PLC* phospholipase C, *SERCA* sarcoplasmic/endoplasmic reticulum calcium ATPase

25.2.2 The DAG-Sensitive Subfamily: TRPC3, TRPC6, and TRPC7

25.2.2.1 TRPC3

TRPC3 has been shown to be upregulated in different experimental models of cardiac hypertrophy, induced, for example, by chronic agonist stimulation (Brenner and Dolmetsch 2007) or pressure overload induction (Bush et al. 2006). In contrast to TRPC6, a strict DAG dependence does not necessarily exist for TRPC3. As outlined above, a dn-TRPC3 mutant expressed in cardiomyocytes reduced SOCE in cardiomyocytes isolated from pathologically hypertrophied hearts (Wu et al. 2010). Similarly, overexpression of TRPC3 in cardiomyocytes resulted in an enhanced Ca^{2+} influx induced by a concomitant stimulation with Ang II and the SERCA inhibitor cyclopiazonic acid (CPA) (Nakayama et al. 2006). TRPC3 transgenic mice develop cardiomyopathy and hypertrophy (Nakayama et al. 2006). Moreover, TRPC3 overexpression increased the sensitivity to I/R injury by enhancing apoptosis in cardiomyocytes probably through enhanced TRPC3-mediated Ca^{2+} influx and calpain cleavage (Shan et al. 2008). The apoptotic effects were inhibited by the SOCE inhibitor SKF-150606 which might indicate that TRPC3 contributes to a SOCE in the setting of I/R injury. More frequently, TRPC3 was suggested to function as DAG-sensitive ROC in cardiomyocytes. In a study by Bush et al. (2006), TRPC3 overexpression resulted in an enhanced NRC volume and sarcomeric assembly, a typical characteristic of cardiomyocyte hypertrophy. Both effects were further accelerated in cells treated with OAG and seemed to be mechanistically coupled to an enhanced activation of the CaN-NFAT signaling pathway. A SOCE-independent function was also suggested by several other studies. For example, Ca^{2+} fluxes in feline cardiomyocytes overexpressing TRPC3 were increased by the addition of OAG but not CPA (Makarewich et al. 2014). In another study, it was proposed that TRPC3 as well as TRPC6 affect hypertrophy through an interaction with the LTCC by altering the membrane potential (Onohara et al. 2006). In addition to the observation of a functional crosstalk between TRPC3 and the LTCC, it was suggested that the activity of the protein kinase C (PKC) regulates the transmission of TRPC3 activity toward the LTCC. By studying the atrial cell line HL-1, it was shown that a PKC-dependent phosphorylation of TRPC3 is necessary to complex and activate CaN and NFAT. In case of a reduced phosphorylation, TRPC3 allowed a cation influx which was conveyed to an enhanced LTCC activity generating a large Ca^{2+} influx. Interestingly, in this situation, CaN was no longer activated which might indicate that an impaired PKC-dependent phosphorylation of TRPC3 uncouples TRPC3 from CaN but retains the functional interaction with the LTCC (Poteser et al. 2011).

In more recent studies, TRPC3 was suggested as mechanosensitive ion channel regulating a stretch-induced reactive oxygen species (ROS) production in cardiomyocytes through a crosstalk with NADPH oxidase 2 (NOX2). Mechanistically, an interaction between NOX2 and TRPC3 seemed to prevent NOX2 from proteasome-dependent degradation. Reciprocally, NOX2 stabilized TRPC3 and thus enhanced its channel activity in cardiomyocytes resulting in an amplified activation of NOX2 (Kitajima et al. 2016). Thus, inhibition of TRPC3 might be beneficial to prevent cardiomyocyte hypertrophy but also attenuate oxidative stress and left ventricular dysfunction in mice with dilated cardiomyopathy (Fig. 25.3).

25.2.2.2 TRPC6

TRPC6 has been identified as key player influencing the development of heart disease not only as Ca^{2+} entity in cardiomyocytes (Kuwahara et al. 2006; Klaiber et al. 2011; Xie et al. 2012) but also in myofibroblasts (Davis et al. 2012). On the cardiomyocyte level, TRPC6 functions as a DAG- sensitive ion channel (Kuwahara et al. 2006; Xie et al. 2012) and is part of a feed-forward loop that promotes the activation of CaN and NFAT, cardiac hypertrophy and heart failure (Kuwahara et al. 2006). Consequently, overexpression of TRPC6 in transgenic mice results in a severe cardiac hypertrophy, (Kuwahara et al. 2006) with a low resistance to chronic pressure overload. The vicious circle from TRPC6 activation to CaN-NFAT signaling can be abrogated by the parallel activation of protein kinase G (PKG) through

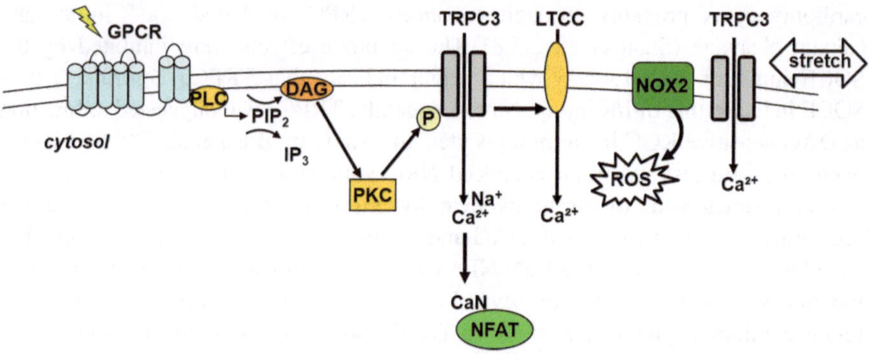

Fig. 25.3 TRPC3-dependent effects on the Ca^{2+} homeostasis in cardiomyocytes. TRPC3 can be activated downstream of the Gq-protein-coupled receptor (GPCR) signaling pathway, most likely as receptor-operated channel independently of Ca^{2+} store depletion. Protein kinase C (PKC)-dependent phosphorylation of TRPC3 results in a tight interaction with calcineurin (CaN) and the subsequent activation of CaN and nuclear factor of activated T-cells (NFAT). Inhibition of PKC disrupts this interaction and turns TRPC3 into a transcriptionally silent mode and transfers the TRPC3-mediated cation entry toward an elevated activity of the L-type Ca^{2+} channel (LTCC). Mechanical stress might also increase the activity of TRPC3 which prevents NADPH oxidase 2 (NOX2) from proteasome-dependent degradation and thus amplifies its activity as reactive oxygen species (ROS) generating enzyme. *DAG* diacylglycerol, *GPCR* Gq-protein-coupled receptor, *IP_3* inositol 3-phosphate, *IP_3R* inositol 3-phosphate receptor, *PIP_2* phosphatidylinositol 4,5-bisphosphate, *PLC* phospholipase C, *SERCA* sarcoplasmic/endoplasmic reticulum calcium ATPase

the increased secretion of cGMP by, e. g., ANP/BNP-GC-A activation (Kinoshita et al. 2010) or phosphodiesterase type 5 (PDE5) inhibition (Koitabashi et al. 2010). The interference between these pro- and anti-hypertrophic pathways is based on the phosphorylation of TRPC6 by PKG which is connected with a reduction of channel activity. A PKG-dependent inhibition of TRPC6 decreases CaN, CamKII, and ERK activation and retards ET-1 induced hypertrophic responses in cardiomyocyte cultures (Nakamura et al. 2015). The inhibitory effects are even optimized by PKG proteins that do not undergo oxidation (Nakamura et al. 2015). An interference with exocytosis of TRPC6 has also been found as mechanism to affect TRPC6 activity and TRPC6-mediated downstream effects in the heart (Fig. 25.4). In an elegant study (Xie et al. 2012), it was shown that the systemically circulating anti-aging protein Klotho blocks the phosphoinositide-3-kinase-dependent exocytosis of TRPC6 channels and thereby reduces ET-1 sensitive currents in cardiomyocytes pre-treated with isoproterenol. On the whole organ level, Klotho overexpression improved cardiac pathology and long-term survival.

25.2.2.3 TRPC7

TRPC7 is expressed in the myocardium at fairly high levels (Okada et al. 1999) and upregulated in failing hearts from Dahl salt-sensitive rats (Satoh et al. 2007). There are indications that TRPC7 is linked to Ang II-induced apoptosis in cardiomyocytes. It has also been suggested that TRPC7 associates with TRPC3 to

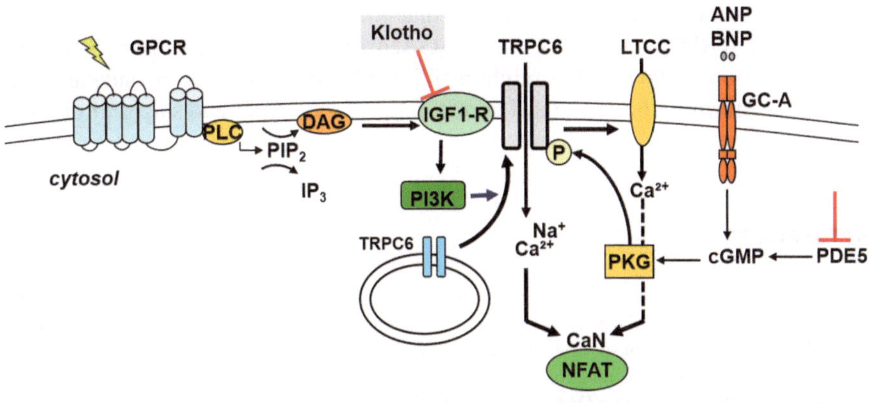

Fig. 25.4 TRPC6 as crucial Ca^{2+} entity in cardiomyocytes. TRPC6 can be activated as receptor-operated channel downstream of the Gq-protein-coupled receptor (GPCR) signaling pathway. Mechanistically linked to TRPC6 is the activation of CaN and NFAT. A TRPC6-mediated cation entry might also be coupled to an increased activity of the L-type Ca^{2+} channel (LTCC). TRPC6 channels are inhibited by protein kinase G (PKG) phosphorylation. PKG activation results from an increased cGMP secretion by guanylyl cyclase-A (GC-A) receptor stimulation with ANP/BNP or by the inhibition of cGMP degradation by phosphodiesterase type-5 inhibition (PDE5). The soluble protein Klotho blocks a phosphoinositide-3-kinase-dependent exocytosis of TRPC6 channels. *ANP* atrial natriuretic peptide, *BNP* brain natriuretic peptide, *cGMP* cyclic guanosine monophosphate, *DAG* diacylglycerol, *GPCR* Gq-protein-coupled receptor, *IGF1-R* insulin-like growth factor 1 receptor, IP_3 inositol 3-phosphate, PIP_2 phosphatidylinositol 4,5-bisphosphate, *PLC* phospholipase C

heteromeric ion channels which are activated downstream of purinergic receptor P2Y, G protein and PLCβ stimulation but not by passive Ca^{2+} store depletion (Alvarez et al. 2008). Further studies are required to thoroughly characterize the functional role of TRPC7 in the heart.

25.3 Conclusion

Over the last years, our knowledge about TRPC-dependent signaling in the heart has been tremendously advanced. A series of elegant studies has shown that TRPC channels are crucial Ca^{2+} entities that govern cardiac disease, upfront pathological cardiac hypertrophy, as well as ischemia-reperfusion injury and chronic myocardial infarction. Novel TRPC-containing signaling cascades have been discovered that might lead to a better understanding of cardiac disease development as well as management by setting the fundament for the employment of new medications. Still, cardiac-specific TRPC channel research has not been finished yet. In particular, their function as either store-operated, receptor-operated, or stretch-dependent ion channels has not been entirely resolved. There are conflicting findings regarding the precise mode of activation which might be based on the preference of TRPC subunits to heteromerize to unforeseen ion assemblies. Nevertheless, considering the current literature, there is a tendency that points to TRPC1 and TRPC4 as SOC and TRPC3, TRPC6, and TRPC7 as DAG-sensitive ion channels. To substantiate this assumption, further studies might be initiated that allow a side-by-side analysis of TRPC isoforms considering different cardiomyocyte types and stages of disease.

References

Alvarez J, Coulombe A, Cazorla O, Ugur M, Rauzier JM, Magyar J, Mathieu EL, Boulay G, Souto R, Bideaux P, Salazar G, Rassendren F, Lacampagne A, Fauconnier J, Vassort G (2008) ATP/UTP activate cation-permeable channels with TRPC3/7 properties in rat cardiomyocytes. Am J Physiol Heart Circ Physiol 295:H21–H28

Anderson ME, Brown JH, Bers DM (2011) CaMKII in myocardial hypertrophy and heart failure. J Mol Cell Cardiol 51:468–473

Benard L, Oh JG, Cacheux M, Lee A, Nonnenmacher M, Matasic DS, Kohlbrenner E, Kho C, Pavoine C, Hajjar RJ, Hulot JS (2016) Cardiac Stim1 silencing impairs adaptive hypertrophy and promotes heart failure through inactivation of mTORC2/Akt signaling. Circulation 133:1458–1471. Discussion 1471

Bers DM (2002) Cardiac excitation-contraction coupling. Nature 415:198–205

Brenner JS, Dolmetsch RE (2007) TrpC3 regulates hypertrophy-associated gene expression without affecting myocyte beating or cell size. PLoS One 2:e802

Bush EW, Hood DB, Papst PJ, Chapo JA, Minobe W, Bristow MR, Olson EN, McKinsey TA (2006) Canonical transient receptor potential channels promote cardiomyocyte hypertrophy through activation of calcineurin signaling. J Biol Chem 281:33487–33496

Camacho Londono JE, Tian Q, Hammer K, Schröder L, Camacho Londono J, Reil JC, He T, Oberhofer M, Mannebach S, Mathar I, Philipp SE, Tabellion W, Schweda F, Dietrich A, Kaestner L, Laufs U, Birnbaumer L, Flockerzi V, Freichel M, Lipp P (2015) A background Ca2+ entry pathway mediated by TRPC1/TRPC4 is critical for development of pathological cardiac remodelling. Eur Heart J 36:2257–2266

Cartwright EJ, Oceandy D, Austin C, Neyses L (2011) Ca2+ signalling in cardiovascular disease: the role of the plasma membrane calcium pumps. Sci China Life Sci 54:691–698

Chiang CS, Huang CH, Chieng H, Chang YT, Chang D, Chen JJ, Chen YC, Chen YH, Shin HS, Campbell KP, Chen CC (2009) The Ca(v)3.2 T-type Ca(2+) channel is required for pressure overload-induced cardiac hypertrophy in mice. Circ Res 104:522–530

Correll RN, Goonasekera SA, van Berlo JH, Burr AR, Accornero F, Zhang H, Makarewich CA, York AJ, Sargent MA, Chen X, Houser SR, Molkentin JD (2015) STIM1 elevation in the heart results in aberrant Ca2+ handling and cardiomyopathy. J Mol Cell Cardiol 87:38–47

Davis J, Burr AR, Davis GF, Birnbaumer L, Molkentin JD (2012) A TRPC6-dependent pathway for myofibroblast transdifferentiation and wound healing in vivo. Dev Cell 23:705–715

Derler I, Jardin I, Romanin C (2016) Molecular mechanisms of STIM/Orai communication. Am J Physiol Cell Physiol 310:C643–C662

Dietrich A, Fahlbusch M, Gudermann T (2014) Classical transient receptor potential 1 (TRPC1): channel or channel regulator? Cells 3:939–962

Eder P, Molkentin JD (2011) TRPC channels as effectors of cardiac hypertrophy. Circ Res 108:265–272

Go LO, Moschella MC, Watras J, Handa KK, Fyfe BS, Marks AR (1995) Differential regulation of two types of intracellular calcium release channels during end-stage heart failure. J Clin Invest 95:888–894

Gudermann T, Hofmann T, Mederos y Schnitzler M, Dietrich A (2004) Activation, subunit composition and physiological relevance of DAG-sensitive TRPC proteins. Novartis Found Symp 258:103–118. Discussion 118–122, 155–159, 263–266

Huang B, Qin D, Deng L, Boutjdir M, El-Sherif N (2000) Reexpression of T-type Ca2+ channel gene and current in post-infarction remodeled rat left ventricle. Cardiovasc Res 46:442–449

Huang J, van Breemen C, Kuo KH, Hove-Madsen L, Tibbits GF (2006) Store-operated Ca2+ entry modulates sarcoplasmic reticulum Ca2+ loading in neonatal rabbit cardiac ventricular myocytes. Am J Physiol Cell Physiol 290:C1572–C1582

Hulot JS, Fauconnier J, Ramanujam D, Chaanine A, Aubart F, Sassi Y, Merkle S, Cazorla O, Ouillé A, Dupuis M, Hadri L, Jeong D, Mühlstedt S, Schmitt J, Braun A, Benard L, Saliba Y, Laggerbauer B, Nieswandt B, Lacampagne A, Hajjar RJ, Lompré AM, Engelhardt S (2011) Critical role for stromal interaction molecule 1 in cardiac hypertrophy. Circulation 124:796–805

Hunton DL, Lucchesi PA, Pang Y, Cheng X, Dell'Italia LJ, Marchase RB (2002) Capacitative calcium entry contributes to nuclear factor of activated T-cells nuclear translocation and hypertrophy in cardiomyocytes. J Biol Chem 277:14266–14273

Hunton DL, Zou L, Pang Y, Marchase RB (2004) Adult rat cardiomyocytes exhibit capacitative calcium entry. Am J Physiol Heart Circ Physiol 286:H1124–H1132

Kinoshita H, Kuwahara K, Nishida M, Jian Z, Rong X, Kiyonaka S, Kuwabara Y, Kurose H, Inoue R, Mori Y, Li Y, Nakagawa Y, Usami S, Fujiwara M, Yamada Y, Minami T, Ueshima K, Nakao K (2010) Inhibition of TRPC6 channel activity contributes to the antihypertrophic effects of natriuretic peptides-guanylyl cyclase-A signaling in the heart. Circ Res 106:1849–1860

Kirschmer N, Bandleon S, von Ehrlich-Treuenstätt V, Hartmann S, Schaaf A, Lamprecht AK, Miranda-Laferte E, Langsenlehner T, Ritter O, Eder P (2016) TRPC4α and TRPC4β similarly affect neonatal cardiomyocyte survival during chronic GPCR stimulation. PLoS One 11: e0168446

Kitajima N, Numaga-Tomita T, Watanabe M, Kuroda T, Nishimura A, Miyano K, Yasuda S, Kuwahara K, Sato Y, Ide T, Birnbaumer L, Sumimoto H, Mori Y, Nishida M (2016) TRPC3 positively regulates reactive oxygen species driving maladaptive cardiac remodeling. Sci Rep 6:37001

Klaiber M, Dankworth B, Kruse M, Hartmann M, Nikolaev VO, Yang RB, Völker K, Gassner B, Oberwinkler H, Feil R, Freichel M, Groschner K, Skryabin BV, Frantz S, Birnbaumer L, Pongs O, Kuhn M (2011) A cardiac pathway of cyclic GMP-independent signaling of guanylyl cyclase A, the receptor for atrial natriuretic peptide. Proc Natl Acad Sci USA 108:18500–18505

Klee CB, Ren H, Wang X (1998) Regulation of the calmodulin-stimulated protein phosphatase, calcineurin. J Biol Chem 273:13367–13370

Kockskämper J, Zima AV, Roderick HL, Pieske B, Blatter LA, Bootman MD (2008) Emerging roles of inositol 1,4,5-trisphosphate signaling in cardiac myocytes. J Mol Cell Cardiol 45:128–147

Koitabashi N, Aiba T, Hesketh GG, Rowell J, Zhang M, Takimoto E, Tomaselli GF, Kass DA (2010) Cyclic GMP/PKG-dependent inhibition of TRPC6 channel activity and expression negatively regulates cardiomyocyte NFAT activation Novel mechanism of cardiac stress modulation by PDE5 inhibition. J Mol Cell Cardiol 48:713–724

Kuwahara K, Wang Y, McAnally J, Richardson JA, Bassel-Duby R, Hill JA, Olson EN (2006) TRPC6 fulfills a calcineurin signaling circuit during pathologic cardiac remodeling. J Clin Invest 116:3114–3126

Luo X, Hojayev B, Jiang N, Wang ZV, Tandan S, Rakalin A, Rothermel BA, Gillette TG, Hill JA (2012) STIM1-dependent store-operated Ca2+ entry is required for pathological cardiac hypertrophy. J Mol Cell Cardiol 52:136–147

Makarewich CA, Zhang H, Davis J, Correll RN, Trappanese DM, Hoffman NE, Troupes CD, Berretta RM, Kubo H, Madesh M, Chen X, Gao E, Molkentin JD, Houser SR (2014) Transient receptor potential channels contribute to pathological structural and functional remodeling after myocardial infarction. Circ Res 115:567–580

Molkentin JD (2013) Parsing good versus bad signaling pathways in the heart: role of calcineurin-nuclear factor of activated T-cells. Circ Res 113:16–19

Molkentin JD, Lu JR, Antos CL, Markham B, Richardson J, Robbins J, Grant SR, Olson EN (1998) A calcineurin-dependent transcriptional pathway for cardiac hypertrophy. Cell 93:215–228

Nakamura T, Ranek MJ, Lee DI, Shalkey Hahn V, Kim C, Eaton P, Kass DA (2015) Prevention of PKG1α oxidation augments cardioprotection in the stressed heart. J Clin Invest 125:2468–2472

Nakayama H, Wilkin BJ, Bodi I, Molkentin JD (2006) Calcineurin-dependent cardiomyopathy is activated by TRPC in the adult mouse heart. FASEB J 20:1660–1670

Nakayama H, Bodi I, Maillet M, DeSantiago J, Domeier TL, Mikoshiba K, Lorenz JN, Blatter LA, Bers DM, Molkentin JD (2010) The IP3 receptor regulates cardiac hypertrophy in response to select stimuli. Circ Res 107:659–666

Nuss HB, Houser SR (1993) T-type Ca2+ current is expressed in hypertrophied adult feline left ventricular myocytes. Circ Res 73:777–782

Oceandy D, Stanley PJ, Cartwright EJ, Neyses L (2007) The regulatory function of plasma-membrane Ca(2+)-ATPase (PMCA) in the heart. Biochem Soc Trans 35:927–930

Ohba T, Watanabe H, Murakami M, Takahashi Y, Iino K, Kuromitsu S, Mori Y, Ono K, Iijima T, Ito H (2007) Upregulation of TRPC1 in the development of cardiac hypertrophy. J Mol Cell Cardiol 42:498–507

Okada T, Inoue R, Yamazaki K, Maeda A, Kurosaki T, Yamakuni T, Tanaka I, Shimizu S, Ikenaka K, Imoto K, Mori Y (1999) Molecular and functional characterization of a novel mouse transient receptor potential protein homologue TRP7. Ca(2+)-permeable cation channel that is constitutively activated and enhanced by stimulation of G protein-coupled receptor. J Biol Chem 274:27359–27370

Onohara N, Nishida M, Inoue R, Kobayashi H, Sumimoto H, Sato Y, Mori Y, Nagao T, Kurose H (2006) TRPC3 and TRPC6 are essential for angiotensin II-induced cardiac hypertrophy. EMBO J 25:5305–5316

Orsborne C, Chaggar PS, Shaw SM, Williams SG (2017) The renin-angiotensin-aldosterone system in heart failure for the non-specialist: the past, the present and the future. Postgrad Med J 93:29–37

Poteser M, Schleifer H, Lichtenegger M, Schernthaner M, Stockner T, Kappe CO, Glasnov TN, Romanin C, Groschner K (2011) PKC-dependent coupling of calcium permeation through transient receptor potential canonical 3 (TRPC3) to calcineurin signaling in HL-1 myocytes. Proc Natl Acad Sci USA 108:10556–10561

Sabourin J, Bartoli F, Antigny F, Gomez AM, Benitah JP (2016) Transient Receptor Potential Canonical (TRPC)/Orai1-dependent Store-operated Ca2+ Channels: NEW TARGETS OF ALDOSTERONE IN CARDIOMYOCYTES. J Biol Chem 291:13394–13409

Satoh S, Tanaka H, Ueda Y, Oyama J, Sugano M, Sumimoto H, Mori Y, Makino N (2007) Transient receptor potential (TRP) protein 7 acts as a G protein-activated Ca2+ channel mediating angiotensin II-induced myocardial apoptosis. Mol Cell Biochem 294:205–215

Seth M, Zhang ZS, Mao L, Graham V, Burch J, Stiber J, Tsiokas L, Winn M, Abramowitz J, Rockman HA, Birnbaumer L, Rosenberg P (2009) TRPC1 channels are critical for hypertrophic signaling in the heart. Circ Res 105:1023–1030

Shan D, Marchase RB, Chatham JC (2008) Overexpression of TRPC3 increases apoptosis but not necrosis in response to ischemia-reperfusion in adult mouse cardiomyocytes. Am J Physiol Cell Physiol 294:C833–C841

Spassova MA, Hewavitharana T, Xu W, Soboloff J, Gill DL (2006) A common mechanism underlies stretch activation and receptor activation of TRPC6 channels. Proc Natl Acad Sci USA 103:16586–16591

Wu X, Zhang T, Bossuyt J, Li X, McKinsey TA, Dedman JR, Olson EN, Chen J, Brown JH, Bers DM (2006) Local InsP3-dependent perinuclear Ca2+ signaling in cardiac myocyte excitation-transcription coupling. J Clin Invest 116:675–682

Wu X, Chang B, Blair NS, Sargent M, York AJ, Robbins J, Shull GE, Molkentin JD (2009) Plasma membrane Ca2+-ATPase isoform 4 antagonizes cardiac hypertrophy in association with calcineurin inhibition in rodents. J Clin Invest 119:976–985
Wu X, Eder P, Chang B, Molkentin JD (2010) TRPC channels are necessary mediators of pathologic cardiac hypertrophy. Proc Natl Acad Sci USA 107:7000–7005
Xie J, Cha SK, An SW, Kuro-O M, Birnbaumer L, Huang CL (2012) Cardioprotection by Klotho through downregulation of TRPC6 channels in the mouse heart. Nat Commun 3:1238

Cardiac Remodeling and Disease: Current Understanding of STIM1/Orai1-Mediated Store-Operated Ca^{2+} Entry in Cardiac Function and Pathology

26

Fiona Bartoli and Jessica Sabourin

Abstract

For a long time, Ca^{2+} entry into cardiomyocytes was considered the sole domain of the L-type Ca^{2+} channel. Recently, STIM1/Orai1-mediated store-operated Ca^{2+} entry has been also reported to participate to Ca^{2+} influx in cardiac cells and has emerged as a key player to alter Ca^{2+} in the cardiomyocyte. In this review, we will highlight accumulated knowledge about the presence and the potential contribution of STIM1/Orai1-dependent SOCE to cardiac function and its role in the cardiac pathogenesis. Overall, even if STIM1/Orai1 proteins are present in the heart, contradictory results have been reported regarding their contribution to cardiac physiology and pathology, pointing out the necessity of further investigations, a major challenge over the coming years.

Keywords

Orai1 • STIM1 • Store-operated Ca^{2+} entry • Heart • Cardiomyocytes • Cardiac hypertrophy • Heart failure • Arrhythmias

26.1 Introduction

Heart failure (HF) is an increasingly prevalent syndrome in which abnormal cardiac function leads to inadequate supply of blood to tissues and organs for their metabolic demands. Various factors contribute to HF pathogenesis such as ischemia and myocardial infarction, hypertension, or genetic cardiomyopathies.

As a leading cardiovascular cause of death, HF has been singled out as an epidemic disease and is a staggering clinical and public health problem associated

F. Bartoli • J. Sabourin (✉)
Inserm UMR S1180, University Paris-Sud, Université Paris-Saclay, 92296 Châtenay-Malabry, France
e-mail: fiona.bartoli@u-psud.fr; jessica.sabourin@u-psud.fr

K. Groschner et al. (eds.), *Store-Operated Ca^{2+} Entry (SOCE) Pathways*, Advances in Experimental Medicine and Biology 993, DOI 10.1007/978-3-319-57732-6_26

with significant mortality and morbidity rates, particularly in patients aged 65 years and older. Despite progress in reducing HF-related mortality, hospitalizations for HF remain frequent and rates of readmissions continue to rise. As a multifactorial clinical syndrome, HF still represents an epidemic threat, highlighting the need to deeper understand the cellular mechanisms involved in the pathogenesis in order to develop innovative therapeutic strategies.

A strict spatiotemporal control of intracellular Ca^{2+} concentration is essential for many heart functions such as excitation-contraction coupling (ECC), and it is orchestrated by a wealth of ion channels that are the fundaments determining the cardiac electromechanical activity. Indeed, Ca^{2+} mishandling as a central cause of contractile dysfunction and arrhythmias is a hallmark of heart disease and has been the subject of a large body of research to understand how cardiac stress alters Ca^{2+} homeostasis in cardiomyocytes. Several groups have asserted that defects in SR Ca^{2+} uptake via the sarco-/endoplasmic reticulum Ca^{2+}-ATPase (SERCA), abnormalities in ECC such as L-type Ca^{2+} channel and ryanodine receptor (RyR) dysfunction, and/or reverse mode Na^+/Ca^{2+} exchange participate to the development of the disease (Roe et al. 2015). Currently, experimental therapies manipulating SERCA2a pump or cardiac myosin activity are in progress to improve Ca^{2+} homeostasis and cardiac function (Bernardo and Blaxall 2016).

Recently, non-voltage-gated Ca^{2+} channels such as store-operated Ca^{2+} channels (SOCs) have been reported to participate to Ca^{2+} influx in cardiac cells and have emerged as potential key players in Ca^{2+} deregulation in cardiomyocytes.

Although still under debate, the prime candidate proteins for SOCs encompass the nonselective cation channels, transient receptor potential canonical (TRPC) channels, as well as the Ca^{2+}-selective pore-forming unit of the Ca^{2+} release-activated Ca^{2+} channel (CRAC), Orai1. Stromal interaction molecule 1 (STIM1) acts as a Ca^{2+} sensor in the endo-/sarcoplasmic reticulum (ER/SR) and oligomerizes when Ca^{2+} stores are depleted. It forms clusters proximal to the plasma membrane to physically interact with and activate Orai1 and TRPC channels allowing store-operated Ca^{2+} entry (SOCE) (Prakriya 2013; Cheng et al. 2013).

In this review, we will highlight the accumulated knowledge about the presence and the contribution of STIM1/Orai1-dependent SOCE to cardiac function and its potential role in the pathogenesis of HF.

26.2 Presence of SOCE in the Heart

The first evidence of the presence of SOCE in cardiomyocytes was reported in 2002. Hunton et al. showed that in neonatal cardiomyocytes, the depletion of SR Ca^{2+} stores with either the SERCA inhibitor (thapsigargin, Tg) or inositol trisphosphate (IP_3)-generating agonists such as angiotensin (Ang II) or phenylephrine (PE) induces a sustained increase in cytoplasmic Ca^{2+} dependent on extracellular Ca^{2+}, which is abolished by nonselective SOC inhibitors, glucosamine or SKF-96365 (Hunton et al. 2002). It has been also reported that SOCE is present in cardiomyocytes at embryonic and neonatal stages and is resistant to inhibitors of voltage-dependent Ca^{2+} channels and Na^{+}/Ca^{2+} exchanger (Uehara et al. 2002). Thereafter, the presence of SOCE activated by Ang II and Tg has been shown in adult rat and mouse ventricular myocytes (Hunton et al. 2004; Kojima et al. 2012), but its amplitude is lower than in embryonic/neonatal cardiomyocytes, indicating that SOCE regulation depends on the developmental stage of the heart, which is probably linked to the developmental changes in Ca^{2+} handling between embryonic and adult hearts. In this line, we recently showed, in adult rat ventricular cardiomyocytes, low functional SOCE, which can be stimulated by Epac signaling (Dominguez-Rodriguez et al. 2015). However, the existence of SOCE was really confirmed much later by the identification of SOCE molecular components in cardiac tissue.

Indeed, several groups simultaneously identified STIM1 and Orai1 proteins in embryonic, neonatal, and adult rat ventricular cardiomyocytes (Ohba et al. 2009; Voelkers et al. 2010; Hulot et al. 2011; Volkers et al. 2012; Zhu-Mauldin et al. 2012) and asserted their requirement for SOCE. Luo et al. further showed that neonatal hearts express significant amounts of a splice variant of STIM1 (STIM1L), already described in skeletal muscle, and whose expression decreases with cardiomyocyte maturation (Luo et al. 2012). In accordance with studies on non-excitable cells, it has been shown that SR Ca^{2+} depletion leads to the formation of STIM1 puncta (Hulot et al. 2011) and its interaction with Orai1 (Zhu-Mauldin et al. 2012) in neonatal cardiomyocytes. It has been also reported that STIM1 colocalizes with key regulators of ECC such as SERCA, phospholamban, and RyR in neonatal and adult rat hearts (Zhu-Mauldin et al. 2012; Correll et al. 2015; Zhao et al. 2015). Consistent with earlier findings, we also recently reported a physical interaction between STIM1 and Orai1 elicited by Ca^{2+} store depletion in neonatal cardiomyocytes (Sabourin et al. 2016). In this in vitro model, one study suggested, using Ca^{2+} imaging and Sr^{2+} permeability assays, that SOCE is highly selective for Ca^{2+}, reflecting that SOCE is mediated by STIM1/Orai1 proteins rather than TRPC channels (Zhu-Mauldin et al. 2012). By contrast, with the whole-cell patch-clamp mode, a method that precisely determines ion selectivity of channels, we recorded a mixed nature of I_{SOC} carried by a dynamic assembly of TRPC-STIM1-Orai1 (Sabourin et al. 2016).

Interestingly, Hulot et al. found two separate STIM1-dependent currents in adult rat cardiomyocytes (Hulot et al. 2011): one with double rectifying current rapidly activated by Ca^{2+} store depletion and the other fully active even when Ca^{2+} stores are replete, with predominantly inward rectifying current reflecting Orai1-activated currents. This work therefore joins a growing number of studies that show a more rapid kinetics of I_{SOC} activation in cardiac myocytes than in non-excitable cells. This more rapid activation and store-independent current may be explained, at least, by the constitutive association of STIM1 and Orai1 at the membrane, as observed in sinoatrial node cells (Zhang et al. 2015). In line with the idea, STIM1 appears to form pre-constituted "punctate" structures without the need of Ca^{2+} store depletion in neonatal and adult cardiomyocytes (Parks et al. 2016). Furthermore, in skeletal muscle, STIM1L has been demonstrated to colocalize with Orai1 channels and to interact with actin to form permanent clusters, allowing the immediate activation of SOCE (Darbellay et al. 2011). Similar experiments with STIM1L may provide new data regarding the kinetics of SOCE activation in cardiomyocytes.

The presence of both STIM1 and Orai1-mediating SOCE tightly coupled to SR Ca^{2+} release has also been found in HL-1 cells, an immortal atrial cardiomyocyte cell line (Touchberry et al. 2011). Nonetheless, the existence of cardiac STIM1-/Orai1-dependent SOCE machinery raises the need to understand its underlying physiological role in the heart.

26.3 Physiological Roles of SOCE in the Heart

Firstly, it has been observed that siRNA-mediated knockdown (KD) of STIM1 impacts the cytosolic and SR Ca^{2+} handling leading to lower diastolic Ca^{2+} levels and a decline in SR Ca^{2+} content in neonatal cardiomyocytes, while silencing Orai1 expression did not appreciably alter these parameters (Voelkers et al. 2010). In addition, the frequency of spontaneous Ca^{2+} transients is significantly decreased in STIM1 and Orai1-KD cardiomyocytes. These authors also showed that KD of Orai1, but not of STIM1, causes a significant decrease in neonatal cardiomyocytes' size associated with a reduction of calcineurin (CnA) and Ca^{2+}/calmodulin-dependent kinase II (CaMKII) activity under resting conditions. By contrast, at adult stages, genetic manipulation of STIM1 in vitro and in vivo using gain- and loss-of-function strategies greatly influenced cardiac size and function (Hulot et al. 2011). In HL-1 cells, the KD of Orai1 also decreases SOCE and both cytosolic and SR Ca^{2+} levels (Touchberry et al. 2011).

It has been also shown that STIM1 participates to the contractile rhythmicity of HL-1 cells by moderating T-type Ca^{2+} channel expression and activity (Nguyen et al. 2013). Of note, STIM1 and Orai1 limit the surface expression and activity of the L-type Ca^{2+} channels (LTCCs) via physical interaction in A7r5 vascular smooth muscle cells and $Ca_v1.2$ expressing HEK-293 cells (Wang et al. 2010). We and others also demonstrated in embryonic and adult cardiac cells that TRPC channels form macromolecular complexes with the LTCCs, in particular in caveolae (Sabourin et al. 2011; Makarewich et al. 2014). To date, the interaction between STIM1/Orai1 and LTCCs in adult cardiomyocytes remains to be determined.

In our recent study, we showed a diastolic Ca^{2+} overload induced by chronic aldosterone treatment in electrically paced neonatal cardiomyocytes that was totally prevented when SOCE was blocked by BTP2 (*N*-[4-[3,5-bis(trifluoromethyl)-1H-pyrazol-1-yl]phenyl]-4-methyl-1,2,3-thiadiazole-5-carboxamide) or when Orai1 was specifically blocked by Synta66 (*N*-(2′,5′-dimethoxy[1,1′-biphenyl]-4-yl)-3-fluoro-4-pyridinecarboxamide). This phenomenon was also partially prevented in the presence of the serum- and glucocorticoid-regulated kinase 1 (SGK1) inhibitor GSK, suggesting that Orai1-mediated SOCE plays a role in the regulation of cardiac diastolic Ca^{2+} homeostasis, via an SGK1-dependent mechanism (Sabourin et al. 2016).

Recent in vivo studies have proved the contribution of Orai1 and Orai3 activity on the regulation of myocardial electromechanical stability (Wolkowicz et al. 2011; Wang et al. 2012) and of contractile development machinery (Horton et al. 2014). Collins et al. have demonstrated that STIM1 is essential for normal cardiac homeostasis through modulation of ER and mitochondrial functions (Collins et al. 2014).

Altogether, we and others proposed that STIM1/Orai1-mediated SOCE, predominantly expressed in developing cardiomyocytes, plays a role in maintaining diastolic Ca^{2+} homeostasis over time under unstressed conditions and that SOCE is crucial for normal postnatal cardiac growth.

26.4 Pathological Roles of SOCE in the Heart

Despite an uncertain role of SOCE in normal cardiac physiology, notably in adults, several studies reported strong evidence that deregulation of SOCE pathway is an important contributor to the development and progression of cardiac pathologies such as cardiac hypertrophy, atrial and ventricular arrhythmias, and dilated cardiomyopathy leading to heart failure.

26.4.1 Cardiac Hypertrophy

Cardiac hypertrophy is the primary mechanism by which the heart responds to cardiac stressors such as myocardial infarction or hypertension-induced pressure overload, to preserve the pump function. Hypertrophy from different pathologic stimuli is associated with sustained increases in intracellular Ca^{2+} which activate signaling pathways such as CnA/NFAT (calcineurin/nuclear factor of activated T cells) and CaMKII/HDACs (Ca^{2+}/calmodulin-dependent protein kinase II/histone deacetylases) signaling leading to hypertrophic transcription programs, pathological growth, and cardiac remodeling, in fine altering cardiac function. Several studies proposed a key role for STIM1/Orai1-mediated SOCE in altered Ca^{2+} signaling underlying the development of cardiac hypertrophy, by changing the fetal gene program governed by CnA/NFAT signaling.

The first study on neonatal cardiomyocytes showed that a treatment with IP_3-generating agonists such as PE or Ang II for 48 h produced an increased intracellular Ca^{2+} concentration and an increased cell area, and led to the nuclear translocation of NFAT. All these parameters were prevented in the presence of nonselective SOC inhibitors, glucosamine or SKF-96365, and, to a lesser extent, in the presence of LTCC inhibitor (Hunton et al. 2002). Subsequently, similar results were found on neonatal cardiomyocytes where KD of STIM1 and Orai1 suppressed enhanced SOCE and increased NFAT activation and cell size induced by 48 h treatment with endothelin 1 (ET-1) or PE (Ohba et al. 2009; Voelkers et al. 2010; Hulot et al. 2011). Conversely, neonatal cardiomyocytes overexpressing STIM1 are significantly larger and display enhanced NFAT activity, which are prevented in the presence of SKF-96365 (Hulot et al. 2011). Furthermore, both STIM1 and Orai1

KD completely abrogated PE-mediated hypertrophic neonatal myocyte growth, by inhibiting CaMKII and ERK1/2 (extracellular signal-regulated kinases 1/2) signaling pathway, while only Orai1 KD prevented PE-mediated CnA-dependent prohypertrophic signaling (Voelkers et al. 2010).

Later, STIM1L has been identified as predominant at neonatal stages, but its expression decreases in adults and only reemerges with hypertrophic agonist- or afterload-induced cardiac stress. Indeed, mice subjected to transverse aortic constriction (TAC) during 3 weeks to trigger ventricular hypertrophy exhibit an increase in STIM1L mRNA and protein levels associated with robust SOCE compared to sham animals, consistent with the fact that reactivation of STIM1L expression increases SOCE during hypertrophic process (Luo et al. 2012). A similar induction of STIM1L was observed on isolated adult cardiomyocytes stimulated with PE (Luo et al. 2012). More recently, it has been shown that treatment of human embryonic stem cell-derived cardiomyocytes (hESC-CM) with PE during 48 h causes a marked hypertrophy, accompanied with an upregulation of Orai1 protein. The development of hypertrophy is suspended when Orai1 expression is silenced using siRNA or using dominant-negative construct of Orai1^{G98A} or by nitric oxide (NO) and cyclic guanosine monophosphate (cGMP) via protein kinase G (PKG) activation. Nevertheless, these anti-hypertrophic effects are lost when Orai1 is mutated on serine 34, suggesting that NO, cGMP, and PKG inhibit hypertrophy of hESC-CM cells via PKG-mediated phosphorylation of Orai1 channels on serine 34 (Wang et al. 2015).

All these in vitro data clearly demonstrate that STIM1-/Orai1-dependent Ca^{2+} entry is instrumental for pathological cardiac hypertrophy development. However, its contribution in adults and in an integrated in vivo model remains limited.

Hulot and colleagues were the first in 2011 to demonstrate an overexpression of STIM1 protein and enhanced SOC current in the in vivo model of pressure overload-induced left ventricular hypertrophy in rat. By contrast, STIM1 gene silencing results in less SOCE and protects the heart from the development of hypertrophy by reducing the CnA/NFATc3 signaling cascade (Hulot et al. 2011). A more recent report also found that STIM1 deletion protects the heart from pressure overload-induced cardiac hypertrophy in mice (Parks et al. 2016). For Orai1, only one study observed an upregulation of Orai1 expression at mRNA and protein levels in mice after pressure overload-induced cardiac hypertrophy, without further functional investigation (Volkers et al. 2012).

Overall, these studies strongly support a role for STIM1-mediated SOCE as playing a key role in regulating cardiac hypertrophy and further suggest that increased STIM1 protein that occurs in response to hypertrophic stimuli is responsible for fetal gene program activation, a characteristic of stressed cardiomyocytes. However, additional studies are clearly necessary to determine the relative importance and the role of Orai1 in mediating this response.

26.4.2 Dilated Cardiomyopathy and Heart Failure

Cardiac hypertrophy with normal cardiac function tends to progress over time to depressed cardiac function, described as "hypertrophy decompensation," clinically resulting in the syndrome of HF. The patients with HF are highly exposed to sudden cardiac death (SCD) due to electrical and mechanical disorders of the heart. The decompensation phase is associated with progressive ventricular dilatation and systolic dysfunction, and one of the processes involved in is the loss of cardiomyocytes and their replacement by fibrous tissue leading to a decrease in cardiac contractile performance. Deregulation of Ca^{2+} homeostasis is important in this pathological process. Recent studies tried to show the involvement of SOCE in the transition from cardiac hypertrophy to HF.

Importantly, it has been described that Orai1-/STIM1-deficient patients or Orai1-/STIM1-deficient mice exhibit immunodeficiencies, muscular hypotonia, ectodermal dysplasia, autoimmunity, and lymphoproliferative diseases, but no obvious cardiac muscle-related phenotypes and no prejudice to the cardiovascular function (Feske 2010), consistent with the idea that the SOCE pathway is not crucial during cardiac muscle development and/or contractility under normal physiological conditions. Alteration of heart function related to defect in SOCE might appear with aging, stressful stimuli, or enduring exercise, but could not be revealed due to early death of patients or deficient mice.

However, in a zebrafish model, the inactivation of Orai1 results in the development of HF and the progressive loss of skeletal muscle myofiber integrity. Indeed, 48 h post fertilization, Orai1-deficient embryos spontaneously develop severe HF associated with bradycardia, decreased blood circulation, and blood congestion at the inflow tract and have reduced skeletal muscle force generation without affecting early cardiogenesis and cardiomyocyte differentiation (Volkers et al. 2012). Orai1 appears to be essential for sarcomere growth and function by regulating the localization of calsarcins, a family of striated muscle-specific proteins, at the sarcomeric Z-disk domain. The absence of a comparable cardiac deficit in human patients may be attributed to the presence of Orai3 in mammals.

Similarly, the global Orai1-deficient mice submitted to TAC during 8 weeks (Horton et al. 2014) show a significantly reduced survival rate, a much earlier loss of cardiac function and greater dilatation of the left ventricle compared to WT mice, suggesting that Orai1 deficiency accelerates the pathology and leads rapidly to dilated cardiomyopathy and HF. The maximum rate of hypertrophy is similarly reached in both groups, but Orai1-deficient mice are not able to compensate the overload, leading to the development of dilated cardiomyopathy. Thus, Orai1 may play an important role in the progression of this pathology (Horton et al. 2014). The authors explained these differences by significant apoptosis without major differences in cellular hypertrophy and hypertrophic markers and fibrosis. However, the difference in amount of apoptosis observed in WT and $Orai1^{+/-}$ mice is not clearly established. Of interest, these clinical symptoms seem contradictory to in vitro studies where KD of Orai1 is associated with protection from hypertrophic phenotype. Therefore, cardiac protective or deleterious effect of Orai1 is still under debate. Moreover, the question remains whether those effects are related to systemic (including inflammatory responses or vegetative nervous system) or cardiac-specific alteration and to examine the cardiac phenotype of this mouse model under physiological conditions.

Regarding STIM1, Correll et al. developed a mouse model overexpressing STIM1 in the heart to mimic the STIM1 upregulation observed during cardiac disease (Correll et al. 2015). As expected, STIM1 transgenic (Tg) mice show enhanced SOCE. Under basal condition, STIM1 Tg mice exhibit SCD at 6 weeks of age, and when they reached 12 weeks of age, they develop HF with hypertrophy, with induction of the fetal gene program, mitochondrial alterations, loss of ventricular function, and pulmonary edema. At cellular level, cardiac myocytes isolated from STIM1 Tg mice show spontaneous Ca^{2+} transients and Ca^{2+} waves prevented by SKF-96365, increased LTCC current and Ca^{2+} spark frequency. Moreover, STIM1 Tg mice submitted to pressure overload or neurohormonal stimulation display accelerated hypertrophic phenotype (Correll et al. 2015). Thus, this study supports the idea that increased STIM1 expression is a maladaptive alteration in the heart.

Collins and colleagues developed a constitutive cardiomyocyte-restricted STIM1-KO mouse which exhibits increased ER stress with ER dilation, mitochondrial disorganization with increased mitochondrial fission at 12 weeks of age (Collins et al. 2014). A progressive decline in cardiac function was observed at 20 weeks of age, associated with the presence of inflammatory infiltrate and fibrosis, and was progressively worsened at 36 weeks with a marked left ventricular dilatation (Collins et al. 2014). Another group demonstrated that the inducible cardiac-specific STIM1-KO mouse model, under unstressed conditions, exhibits left ventricular dilatation associated with reduced cardiac contractility, which is consistent with previous studies (Parks et al. 2016). They further revealed that STIM1 deletion causes impaired cell contractility. They speculated that the STIM1-dependent Ca^{2+} signals determine restricted Ca^{2+} microdomains that drive myofilament disassembly and cytoskeletal remodeling (Parks et al. 2016). In line with these results, after 5 weeks of AAV9 (adeno-associated virus serotype 9) injection encoding shRNA directed against STIM1, mice developed a mild cardiac dilation and dysfunction with progressive decline in fractional shortening under physiological conditions (Benard et al. 2016). Nevertheless, STIM1 silencing promotes the rapid transition from cardiac hypertrophy to heart failure (Benard et al. 2016) in contrast to previous observations (Hulot et al. 2011; Parks et al. 2016). Thus, the presence of STIM1 appears to be instrumental for the persistence of adaptive cardiac hypertrophy. They also supported a novel model whereby STIM1 activation following hypertrophic stimulation is essential to modulate Akt kinase activity through activation of mTORC2, repressing the anti-hypertrophic and anti-apoptotic GSK-3β activity (Benard et al. 2016).

Overall, these studies strongly support a key role of STIM1-mediated SOCE in regulating cardiac hypertrophy and HF. However, further investigations are clearly necessary with Orai1 overexpressing or silencing murine models in a cardiac-specific manner, which may provide the best way to understand the contribution of Orai1 in mediating cardiac hypertrophy and HF.

26.4.3 Arrhythmias

Approximately 50% of deaths among patients with HF are classified as SCD, mainly caused by lethal arrhythmias, primary from ventricular tachycardia (VT) degenerating to ventricular fibrillation, in which aberrant Ca^{2+} fluxes are a recurrent theme. Mechanisms underlying these arrhythmias are multifactorial, and recently it has been proposed that SOC channels and especially Orai channel family could be involved in the arrhythmogenic process.

Recently, STIM1/Orai1-mediated SOCE has been proposed to influence pace-making activity of the sinoatrial node by linking the Ca^{2+} and the membrane clocks (Zhang et al. 2015; Liu et al. 2015).

In ventricular cardiomyocytes, the fluctuation of intracellular Ca^{2+} leads to rhythmic contraction, and Nguyen et al. hypothesized that STIM1 should participate in the regulation of the contractile activity of cardiomyocyte-derived HL-1 cells. Indeed, STIM1 KD perturbs the rate of cell contraction and induces irregular spontaneous Ca^{2+} oscillations and arrhythmogenic events, such as early or delayed afterdepolarizations (Nguyen et al. 2013). STIM1 silencing also increases the current density of T-type Ca^{2+} channel suggesting that STIM1 is a negative regulator of these voltage-dependent Ca^{2+} channels and maintains a constant cardiac rhythm by preventing Ca^{2+} overload-induced arrhythmias (Nguyen et al. 2013). In vivo and in vitro STIM1 overexpression in the adult heart and isolated ventricular myocytes generates spontaneous Ca^{2+} transients and leads to the development of arrhythmogenic Ca^{2+} waves and cytosolic and SR Ca^{2+} overloads, which are potential triggers of SCD (Zhao et al. 2015; Correll et al. 2015).

Several studies focused on the relation between 2-APB, an activator of Orai1 and Orai3 at high concentration, and the initiation of arrhythmias in both atrial and ventricular myocytes. 2-APB administration causes sporadic or tachycardic ectopy in superfused rat left atria and induces automatic activity in nonautomatic cardiac muscle, suppressed by SKF-96365 (Wolkowicz et al. 2011). The 2-APB also provokes high-frequency ventricular electrical activity and heart mechanical collapse on Langendorff-perfused rat heart, and the SKF-96365 reverses this type of ventricular fibrillation (Wang et al. 2012).

These data suggest that STIM1/Orai-mediated SOCE may be an important regulator of myocardial electromechanical stability and could be involved in the initiation of atrial and ventricular arrhythmias.

26.5 Conclusion

Over the past few years, collectively, we have significantly improved the basic molecular insight of STIM-/Orai1-mediated SOCE and highlighted the role of these proteins in the regulation of basal Ca^{2+} homeostasis, but also in the development and the progression of different cardiac diseases. Hence, STIM1/Orai1-dependent machinery may be a novel target to consider for pharmacological intervention in the treatment of heart disease.

However, many questions remain regarding the contribution of STIM1 and Orai1 proteins to cardiac physiology and pathology. Understanding the cellular and integrated mechanisms by which STIM1 and Orai1 proteins exert their physiological roles will be a major challenge over the coming years. More particularly, it is important to notice that all studies mentioned in this review employed artificial and nonphysiological conditions to stimulate SOCE (depletion of stores with irreversible SERCA pump inhibitors in Ca^{2+}-free medium for example). Cardiomyocytes in vivo are never exposed to these harsh conditions; therefore, it is important to determine what physiological conditions activate SOCE in cardiomyocytes. Further studies are also needed to understand the relative importance of SOCs compared to voltage-gated Ca^{2+} channels signaling in the heart and to examine a potential functional interaction between LTCCs and STIM/Orai1 pathway as observed with TRPC-/LTCC-formed macromolecular complexes at embryonic and adult stages (Sabourin et al. 2011; Makarewich et al. 2014). Moreover, we and others showed that Orai1 channels form multimeric complexes with TRPC channels which expand the complexity of Orai1-mediated Ca^{2+} signaling pathways in cardiac physiopathology.

In addition to SOCE, there is also growing evidence of store-independent Ca^{2+} entry (SICE) primarily characterized as an endogenous arachidonic acid-regulated Ca^{2+} (ARC) channel carried by Orai1/Orai3 pentamers, but our knowledge of Orai3 function in the heart is limited. Lastly, STIM2 and Orai2, two underestimated proteins, have never been investigated despite their presence in the heart.

Despite promising findings, future studies are needed to include SOCE in the current model of cardiomyocyte Ca^{2+} signaling.

Acknowledgments We thank Jean-Pierre Benitah and Fabrice Antigny for their careful reading. We thank the Attractivity grant 2014 from the University of Paris-Sud, Agence Nationale de la Recherche (ANR-13-BSV1-0023-01 and ANR-15-CE14-0005), and CORDDIM (Région Ile de France) for their support.

References

Benard L, Oh JG, Cacheux M, Lee A, Nonnenmacher M, Matasic DS, Kohlbrenner E, Kho C, Pavoine C, Hajjar RJ, Hulot JS (2016) Cardiac Stim1 silencing impairs adaptive hypertrophy and promotes heart failure through inactivation of mTORC2/Akt signaling. Circulation 133:1458–1471. Discussion 1471

Bernardo BC, Blaxall BC (2016) From bench to bedside: new approaches to therapeutic discovery for heart failure. Heart Lung Circ 25:425–434

Cheng KT, Ong HL, Liu X, Ambudkar IS (2013) Contribution and regulation of TRPC channels in store-operated Ca2+ entry. Curr Top Membr 71:149–179

Collins HE, He L, Zou L, Qu J, Zhou L, Litovsky SH, Yang Q, Young ME, Marchase RB, Chatham JC (2014) Stromal interaction molecule 1 is essential for normal cardiac homeostasis through modulation of ER and mitochondrial function. Am J Physiol Heart Circ Physiol 306: H1231–H1239

Correll RN, Goonasekera SA, van Berlo JH, Burr AR, Accornero F, Zhang H, Makarewich CA, York AJ, Sargent MA, Chen X, Houser SR, Molkentin JD (2015) STIM1 elevation in the heart results in aberrant Ca(2)(+) handling and cardiomyopathy. J Mol Cell Cardiol 87:38–47

Darbellay BS, Arnaudeau S, Bader CR, Konig S, Bernheim L (2011) STIM1L is a new actin-binding splice variant involved in fast repetitive Ca2+ release. J Cell Biol 194:335–346

Dominguez-Rodriguez A, Ruiz-Hurtado G, Sabourin J, Gomez AM, Alvarez JL, Benitah JP (2015) Proarrhythmic effect of sustained EPAC activation on TRPC3/4 in rat ventricular cardiomyocytes. J Mol Cell Cardiol 87:74–78

Feske S (2010) CRAC channelopathies. Pflugers Arch 460:417–435

Horton JS, Buckley CL, Alvarez EM, Schorlemmer A, Stokes AJ (2014) The calcium release-activated calcium channel Orai1 represents a crucial component in hypertrophic compensation and the development of dilated cardiomyopathy. Channels (Austin) 8:35–48

Hulot JS, Fauconnier J, Ramanujam D, Chaanine A, Aubart F, Sassi Y, Merkle S, Cazorla O, Ouille A, Dupuis M, Hadri L, Jeong D, Muhlstedt S, Schmitt J, Braun A, Benard L, Saliba Y, Laggerbauer B, Nieswandt B, Lacampagne A, Hajjar RJ, Lompre AM, Engelhardt S (2011) Critical role for stromal interaction molecule 1 in cardiac hypertrophy. Circulation 124:796–805

Hunton DL, Lucchesi PA, Pang Y, Cheng X, Dell'Italia LJ, Marchase RB (2002) Capacitative calcium entry contributes to nuclear factor of activated T-cells nuclear translocation and hypertrophy in cardiomyocytes. J Biol Chem 277:14266–14273

Hunton DL, Zou L, Pang Y, Marchase RB (2004) Adult rat cardiomyocytes exhibit capacitative calcium entry. Am J Physiol Heart Circ Physiol 286:H1124–H1132

Kojima A, Kitagawa H, Omatsu-Kanbe M, Matsuura H, Nosaka S (2012) Presence of store-operated Ca2+ entry in C57BL/6J mouse ventricular myocytes and its suppression by sevoflurane. Br J Anaesth 109:352–360

Liu J, Xin L, Benson VL, Allen DG, Ju YK (2015) Store-operated calcium entry and the localization of STIM1 and Orai1 proteins in isolated mouse sinoatrial node cells. Front Physiol 6:69

Luo X, Hojayev B, Jiang N, Wang ZV, Tandan S, Rakalin A, Rothermel BA, Gillette TG, Hill JA (2012) STIM1-dependent store-operated Ca(2)(+) entry is required for pathological cardiac hypertrophy. J Mol Cell Cardiol 52:136–147

Makarewich CA, Zhang H, Davis J, Correll RN, Trappanese DM, Hoffman NE, Troupes CD, Berretta RM, Kubo H, Madesh M, Chen X, Gao E, Molkentin JD, Houser SR (2014) Transient receptor potential channels contribute to pathological structural and functional remodeling after myocardial infarction. Circ Res 115:567–580

Nguyen N, Biet M, Simard E, Beliveau E, Francoeur N, Guillemette G, Dumaine R, Grandbois M, Boulay G (2013) STIM1 participates in the contractile rhythmicity of HL-1 cells by moderating T-type Ca(2+) channel activity. Biochim Biophys Acta 1833:1294–1303

Ohba T, Watanabe H, Murakami M, Sato T, Ono K, Ito H (2009) Essential role of STIM1 in the development of cardiomyocyte hypertrophy. Biochem Biophys Res Commun 389:172–176

Parks C, Alam MA, Sullivan R, Mancarella S (2016) STIM1-dependent Ca(2+) microdomains are required for myofilament remodeling and signaling in the heart. Sci Rep 6:25372

Prakriya M (2013) Store-operated Orai channels: structure and function. Curr Top Membr 71:1–32

Roe AT, Frisk M, Louch WE (2015) Targeting cardiomyocyte Ca2+ homeostasis in heart failure. Curr Pharm Des 21:431–448

Sabourin J, Robin E, Raddatz E (2011) A key role of TRPC channels in the regulation of electromechanical activity of the developing heart. Cardiovasc Res 92:226–236

Sabourin J, Bartoli F, Antigny F, Gomez AM, Benitah JP (2016) Transient receptor potential canonical (TRPC)/Orai1-dependent store-operated Ca2+ channels: NEW TARGETS OF ALDOSTERONE IN CARDIOMYOCYTES. J Biol Chem 291:13394–13409

Touchberry CD, Elmore CJ, Nguyen TM, Andresen JJ, Zhao X, Orange M, Weisleder N, Brotto M, Claycomb WC, Wacker MJ (2011) Store-operated calcium entry is present in HL-1 cardiomyocytes and contributes to resting calcium. Biochem Biophys Res Commun 416:45–50

Uehara A, Yasukochi M, Imanaga I, Nishi M, Takeshima H (2002) Store-operated Ca2+ entry uncoupled with ryanodine receptor and junctional membrane complex in heart muscle cells. Cell Calcium 31:89–96

Voelkers M, Salz M, Herzog N, Frank D, Dolatabadi N, Frey N, Gude N, Friedrich O, Koch WJ, Katus HA, Sussman MA, Most P (2010) Orai1 and Stim1 regulate normal and hypertrophic growth in cardiomyocytes. J Mol Cell Cardiol 48:1329–1334

Volkers M, Dolatabadi N, Gude N, Most P, Sussman MA, Hassel D (2012) Orai1 deficiency leads to heart failure and skeletal myopathy in zebrafish. J Cell Sci 125:287–294

Wang Y, Deng X, Mancarella S, Hendron E, Eguchi S, Soboloff J, Tang XD, Gill DL (2010) The calcium store sensor, STIM1, reciprocally controls Orai and CaV1.2 channels. Science 330:105–109

Wang P, Umeda PK, Sharifov OF, Halloran BA, Tabengwa E, Grenett HE, Urthaler F, Wolkowicz PE (2012) Evidence that 2-aminoethoxydiphenyl borate provokes fibrillation in perfused rat hearts via voltage-independent calcium channels. Eur J Pharmacol 681:60–67

Wang Y, Li ZC, Zhang P, Poon E, Kong CW, Boheler KR, Huang Y, Li RA, Yao X (2015) Nitric oxide-cGMP-PKG pathway acts on Orai1 to inhibit the hypertrophy of human embryonic stem cell-derived cardiomyocytes. Stem Cells 33:2973–2984

Wolkowicz PE, Huang J, Umeda PK, Sharifov OF, Tabengwa E, Halloran BA, Urthaler F, Grenett HE (2011) Pharmacological evidence for Orai channel activation as a source of cardiac abnormal automaticity. Eur J Pharmacol 668:208–216

Zhang H, Sun AY, Kim JJ, Graham V, Finch EA, Nepliouev I, Zhao G, Li T, Lederer WJ, Stiber JA, Pitt GS, Bursac N, Rosenberg PB (2015) STIM1-Ca2+ signaling modulates automaticity of the mouse sinoatrial node. Proc Natl Acad Sci U S A 112:E5618–E5627

Zhao G, Li T, Brochet DX, Rosenberg PB, Lederer WJ (2015) STIM1 enhances SR Ca2+ content through binding phospholamban in rat ventricular myocytes. Proc Natl Acad Sci U S A 112: E4792–E4801

Zhu-Mauldin X, Marsh SA, Zou L, Marchase RB, Chatham JC (2012) Modification of STIM1 by O-linked N-acetylglucosamine (O-GlcNAc) attenuates store-operated calcium entry in neonatal cardiomyocytes. J Biol Chem 287:39094–39106

Neurological and Motor Disorders: Neuronal Store-Operated Ca^{2+} Signaling – An Overview and Its Function

27

Sunitha Bollimuntha, Biswaranjan Pani, and Brij B. Singh

Abstract

Calcium (Ca^{2+}) is a ubiquitous second messenger that performs significant physiological task such as neurosecretion, exocytosis, neuronal growth/differentiation, and the development and/or maintenance of neural circuits. An important regulatory aspect of neuronal Ca^{2+} homeostasis is store-operated Ca^{2+} entry (SOCE) which, in recent years, has gained much attention for influencing a variety of nerve cell responses. Essentially, activation of SOCE ensues following the activation of the plasma membrane (PM) store-operated Ca^{2+} channels (SOCC) triggered by the depletion of endoplasmic reticulum (ER) Ca^{2+} stores. In addition to the TRPC (transient receptor potential canonical) and the Orai family of ion channels, STIM (stromal interacting molecule) proteins have been baptized as key molecular regulators of SOCE. Functional significance of the TRPC channels in neurons has been elaborately studied; however, information on Orai and STIM components of SOCE, although seems imminent, is currently limited. Importantly, perturbations in SOCE have been implicated in a spectrum of neuropathological conditions. Hence, understanding the precise involvement of SOCC in neurodegeneration would presumably unveil avenues for plausible therapeutic interventions. We thus review the role of SOCE-regulated neuronal Ca^{2+} signaling in selecting neurodegenerative conditions.

Keywords

TRPC • Calcium • Oxidative and ER stress • Neurodegenerative diseases

S. Bollimuntha (✉) • B. Pani • B.B. Singh (✉)
Department of Biomedical Sciences, School of Medicine and Health Sciences, University of North Dakota, Grand Forks, ND 58202, USA
e-mail: sunitha.bollimuntha@med.und.edu; brij.singh@med.und.edu

K. Groschner et al. (eds.), *Store-Operated Ca^{2+} Entry (SOCE) Pathways*, Advances in Experimental Medicine and Biology 993, DOI 10.1007/978-3-319-57732-6_27

27.1 Introduction

27.1.1 Molecular Determents of Ca^{2+} Homeostasis

Calcium (Ca^{2+}) plays an important role in influencing virtually every aspect of cell's life (Berridge et al. 2000; Putney 2011; Berridge 2012). Most of the cellular signaling events are orchestrated by an increase in the level of cytosolic Ca^{2+} ($[Ca^{2+}]_{cyt}$), which under basal condition is maintained at about 100 nM (Berridge et al. 2003; Clapham 2007; Meldolesi et al. 1988; Miller 1988). The elevations in $[Ca^{2+}]_{cyt}$ are largely regulated by a combination of intracellular Ca^{2+} release and extracellular Ca^{2+} influx events (Putney and Bird 2008). Intracellular compartments such as the endoplasmic reticulum (ER) and mitochondria, where the level of stored Ca^{2+} ranges from 0.1 to 1.0 mM, serve as a primary source of internal Ca^{2+} release (Birnbaumer 2009). On the other hand, Ca^{2+} channels located at the plasma membrane (PM) bring about an elevation in the cytosolic Ca^{2+} level by transporting Ca^{2+} across the membrane from the extracellular milieu, where the levels of Ca^{2+} are generally maintained at about 1–2 mM (Berridge et al. 2003; Clapham 2007). Following an increase in $[Ca^{2+}]_{cyt}$, the Ca^{2+} pumps located in the PM, the ER, or the Golgi complex work in concert to help in attaining a basal level of $[Ca^{2+}]_{cyt}$, either by extrusion of Ca^{2+} to the extracellular environment or by its sequestration into the organellar lumen. Ca^{2+} is intracellularly compartmentalized by ATP-driven Ca^{2+} pumps such as sarco-endoplasmic reticulum Ca^{2+} ATPase (SERCA) and secretory pathway Ca^{2+} ATPase (SPCA) localized at the ER and Golgi complex, respectively (Guerini et al. 2005; Strehler and Treiman 2004). Yet another important organelle, which helps in sequestering Ca^{2+}, is the mitochondrion. Ca^{2+} readily diffuses through large pores in the mitochondrial outer membrane but crosses through channels and transporters across the inner membrane. One such highly Ca^{2+}-selective channel identified was the MiCa (Kirichok et al. 2004). Extrusion of $[Ca^{2+}]_{cyt}$ is achieved by PM-associated pumps such as PMCA and exchangers like Na^+/Ca^{2+} (NCX) and Na^+/Ca^{2+}-K^+ (NCKX). The NCX and NCKX exchange one Ca^{2+} for three Na^+ (NCX) or cotransport one K^+ with one Ca^{2+} in exchange for four Na^+ (NCKX) (Guerini et al. 2005; Hilgemann et al. 2006). On the other hand, Ca^{2+} ATPases lower intracytoplasmic Ca^{2+} levels by exchanging protons for two (SERCA) or one (PMCA) Ca^{2+} per ATP hydrolyzed. These ATPases strive to strike a balance and maintain a low steady-state $[Ca^{2+}]_{cyt}$ (Brini and Carafoli 2009; Gouaux and Mackinnon 2005). In essence, this regulated flux of Ca^{2+} in and out of the cytosol results in $[Ca^{2+}]_{cyt}$ oscillations that range from about 100 nM in resting to about 1000 nM in stimulated state. This oscillatory pattern of $[Ca^{2+}]_{cyt}$ is physiologically vital for the regulation of numerous signaling processes (Berridge 1997; Parekh 2011). The cellular responses mediated by Ca^{2+} are broadly categorized as fast or short-term responses (operates in seconds to minutes) and slow or long-term responses (that take from an hour up to several days). The fast responses such as muscle contraction, neurotransmission, and neurosecretion require rapid and localized Ca^{2+} signals with reasonably short span of Ca^{2+} spikes (Putney and Bird 2008), whereas long-term cellular responses such as gene regulation, proliferation, and differentiation rely on relatively slow and sustained Ca^{2+} signals for efficient execution (Berridge et al. 2000; Gwack et al. 2007). Hence, the onset of Ca^{2+} signals and the regulation of associated physiological events are spatiotemporally regulated (Berridge 1997; Berridge et al. 2003; Berridge and Dupont 1994).

27.1.2 Neuronal Store-Operated Ca^{2+} Entry

In general, TRP channels are defined as nonselective cation-permeable channels since, in addition to permeating Ca^{2+}, they also allow Na^+ and Mg^{2+} to flow through (Zhu et al. 1996). The TRPC (TRP canonical) subfamily constitutes a total of seven TRPC proteins from TRPC1 through TRPC7 (Montell 2005b; Venkatachalam and Montell 2007). Physiological activation of TRPC channels is brought about by a PLC-mediated signaling event following the activation of PM-associated G protein-coupled receptors (GPCRs) or receptor tyrosine kinases (RTKs), which in turn lead to the hydrolysis of phosphatidylinositol 4,5-bisphosphate (PIP_2) into membrane-bound diacylglycerol (DAG) and soluble IP_3. IP_3 binds to IP3R, which functions as ligand-gated channel and evokes Ca^{2+} release from the ER into the cytoplasm through IP_3R expressed on the ER membrane leading to a depletion of ER Ca^{2+} stores. This very store depletion triggers the activation of PM-associated Ca^{2+} channels, and hence the phenomenon is termed store-operated Ca^{2+} entry (SOCE) (as indicated in Fig. 27.1). Recent advances have identified that TRPC and Orai channels could promote SOCE and are thus referred to as store-operated Ca^{2+} channels or SOCC (Barritt 1999; Berridge 1995; Parekh and Penner 1997; Putney 2013). Activated TRPC channels mediate transmembrane fluxes of Ca^{2+} down their electrochemical gradients, thereby raising intracellular Ca^{2+} levels. A part of the Ca^{2+} is pumped back into the ER by SERCA pumps to maintain a steady-state Ca^{2+} concentration of the ER, thus preventing it from exhausting all the Ca^{2+} stores, which partially explains the significance of SOCE. Alternatively, an increase in cytoplasmic Ca^{2+} levels also regulates a plethora of critical cellular functions including, but not limited to, secretion, proliferation, differentiation, apoptosis, and motility (Beck et al. 2008; Bollimuntha et al. 2005a; Cai et al. 2006; Fabian et al. 2008; Pani et al. 2006; Singh et al. 2001; Selvaraj et al. 2012). More recently, the ER Ca^{2+} sensor STIM1 (Liou et al. 2005; Roos et al. 2005) and the PM Ca^{2+} channel Orai1 (Feske et al. 2006; Vig et al. 2006) have been identified as molecular components of SOCE. As depicted in Fig. 27.1, STIM1 has also been shown to activate TRPC channels albeit in a way distinct from that of activating Orai1 channels (Huang et al. 2006; Yuan et al. 2009; Zeng et al. 2008). Interestingly, TRPC proteins have been reported to associate with STIM1 as well as Orai1 and exist as ternary complexes (Cheng et al. 2008; Liao et al. 2007, 2008; Ong et al. 2007). Thus, this heterotypic association of PM-SOCC and their differential activation by STIM1 reveals yet another level of diversity in the regulation of SOCE.

SOCE has also been identified in many types of excitable cells and is important in maintaining Ca^{2+} homeostasis (Majewski and Kuznicki 2015; Sun et al. 2014a).

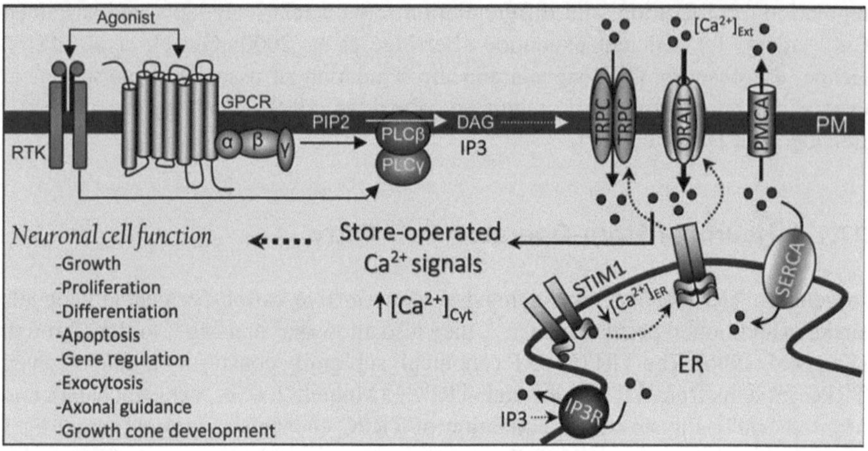

Fig. 27.1 Store-operated Ca^{2+} entry and neuronal function. This illustration depicts the current understanding of SOCE regulated by STIM1-mediated activation of PM-SOCC (TRPC/Orai channels). Agonist-induced activation of PM receptors (GPCR/RTK) results in the generation of the diffusible cellular messenger IP_3 following the PLC-mediated hydrolysis of PIP_2. IP_3 binds to its receptor (IP_3R) in the ER depleting the ER Ca^{2+} stores. This leads to STIM1 oligomerization and recruitment of the STIM1 clusters to ER-PM juxtaposed sites where STIM1 physically gates the TRPC and Orai1 channels to bring about Ca^{2+} entry. This raises the $[Ca^{2+}]_{cyt}$ which not only aids in the SERCA pump-mediated ER store refilling but also promotes the regulation of several cellular functions. Additionally, diacylglycerol (DAG), the membrane-associated lipid messenger, also has the ability to activate and select TRPC channels independent of the ER store and presumably STIM1 as well

For a long time, neuronal cells have been thought not to have a SOCE pathway (Friel and Tsien 1992), since these cells have the voltage-gated Ca^{2+} channels and the receptor-activated Ca^{2+} channels such as glutamate and AMPA receptors as their prime Ca^{2+} influx components. However, many recent studies have provided evidence of functional SOCE in a variety of excitable cell types (Akbari et al. 2004; Emptage et al. 2001; Selvaraj et al. 2012). The impact of Ca^{2+} release from neuronal ER stores, and subsequent activation of SOCE channels on physiological function, has been best studied using various *Drosophila* mutant phenotypes (Banerjee et al. 2004; Montell 2005a; Venkiteswaran and Hasan 2009). These studies support the fundamental concept that IP_3-mediated Ca^{2+} release, followed by Ca^{2+} entry, is required in specific classes of neurons for phototransduction or other neuronal functions. This is further supported by a recent study which confirmed the functional importance of type 1 IP_3R function in mouse neuronal cells (van de Leemput et al. 2007). The existence of SOCE in neuronal cells was originally demonstrated in bovine adrenal chromaffin cells, in the PC12 cell line (Cheek and Thastrup 1989; Clementi et al. 1992), and later in other neuronal cells including the Aplysia bag cell neurons (Philipp et al. 1998; Pizzo et al. 2001). Importantly, the relevance of SOCE channels in the development and plasticity of the nervous system was initially demonstrated by Mu-ming Poo's group

(Nishiyama et al. 2000). These results were further verified by Emptage et al. (2001) who showed that store depletion was essential in the opening of the SOCE channels in hippocampal synaptic boutons, and this entry was essential in modulating the rate of neurotransmitter release along with modulating the synaptic plasticity. This underscores SOCE as a critical regulator of neuronal (patho) physiology.

Ca^{2+} entry via the G protein-coupled mechanism has been implicated in the shaping of action potentials, synaptic transmission, and sensory transduction (Congar et al. 1997; Linden 1994). Additionally, changes in cytosolic Ca^{2+} are also known to regulate the motility of many cellular structures, including the axonal growth cones (Gomez et al. 2001) and dendritic filopodia of developing neurons (Lohmann et al. 2005). Thus, it can be anticipated that TRPC channels may have a significant role in regulating these fundamental neuronal processes. Indeed, Ca^{2+} entry through the TRPC channel has been shown to play a critical role in basic fibroblast growth factor (bFGF)-induced cortical neural stem cell (NSC) proliferation (Fiorio Pla et al. 2005). It has been shown that activation of TRPC1 (C1), but not TRPC3 (C3), regulates cell proliferation in neural stem cells. In contrast, Ca^{2+} influx through C1 and C3 controlled the switch between proliferation and differentiation in immortalized hippocampal H19-7 cells (Wu et al. 2004). Similarly, in mammals, C3 and C6, but not C1, have been shown to be associated with BDNF-mediated neuronal growth (Jia et al. 2007). These distinct physiological outcomes can be partially explained by the differences in spatial localization and functional activation of individual TRPC channels in diverse neuronal populations. Additionally, the differential distribution of membrane receptors (GPCRs/RTKs), concentration, and availability of the desired agonist can further add to the complex regulation of a distinct set of TRPC channels. Importantly, both Orai and STIM proteins are expressed in skeletal and brain tissues, albeit with different expression patterns (Kraft 2015), and STIM2-deficient mice were protected from cerebral damage after ischemic stroke (Berna-Erro et al. 2009). However, another study showed that loss of STIM2 induced a massive neuronal loss in the hippocampal region (Sun et al. 2014b), suggesting that STIM2 could have differential role on different cells.

27.2 Ca^{2+} and Neuropathophysiology

Neuronal cell injury is mediated via both increase and decrease of cytosolic Ca^{2+} concentration (Sattler and Tymianski 2000). Changes in intracellular Ca^{2+} concentration stimulate a number of intracellular events and could either trigger or inhibit the cell death process (Berridge et al. 2000; Putney 2003). Importantly, disturbances in Ca^{2+} homeostasis have been implicated in many neurodegenerative diseases such as PD, AD, and HD (Albers and Beal 2000; Bollimuntha et al. 2005b; O'Bryant et al. 2009; Zuccato and Cattaneo 2007, 2009; Sun et al. 2014a). It is not surprising that disturbances in Ca^{2+} signaling pathways underlie neuronal loss, since many factors involved in neuronal function are dependent on Ca^{2+} signaling (Berridge et al. 2000; Putney 2003). However, the cellular mechanism (s) underlying neurodegeneration, due to alterations in Ca^{2+} homeostasis, remains to be elucidated (Bezprozvanny and Mattson 2008). In addition, several factors including generation of free radicals, impairment of mitochondrial function, ER stress, and apoptosis have been proposed to be regulated by alterations in cytosolic Ca^{2+}. Increased cytosolic Ca^{2+} leads to inappropriate activation of Ca^{2+}-dependent processes, which stay inactive at low Ca^{2+} levels, causing metabolic derangements leading to neuronal death (Berridge et al. 2000; Bezprozvanny and Mattson 2008; Putney 2003). In contrast, decrease in ER Ca^{2+} can induce ER stress, which can activate cell death cascades (Ermak and Davies 2002) suggesting that controlled Ca^{2+} influx from external media is critical for neuronal function and its survival (Selvaraj et al. 2012). Since TRPCs are essential for replenishing and maintaining ER Ca^{2+}, chronic depletion of ER Ca^{2+}, as would occur in the absence of TRPC function, could influence ER-dependent processes such as protein folding and trafficking, ER stress response, and apoptosis. In the following sections, we discuss the findings that link TRP channels and SOCE with the neurodegenerative condition (Fig. 27.2).

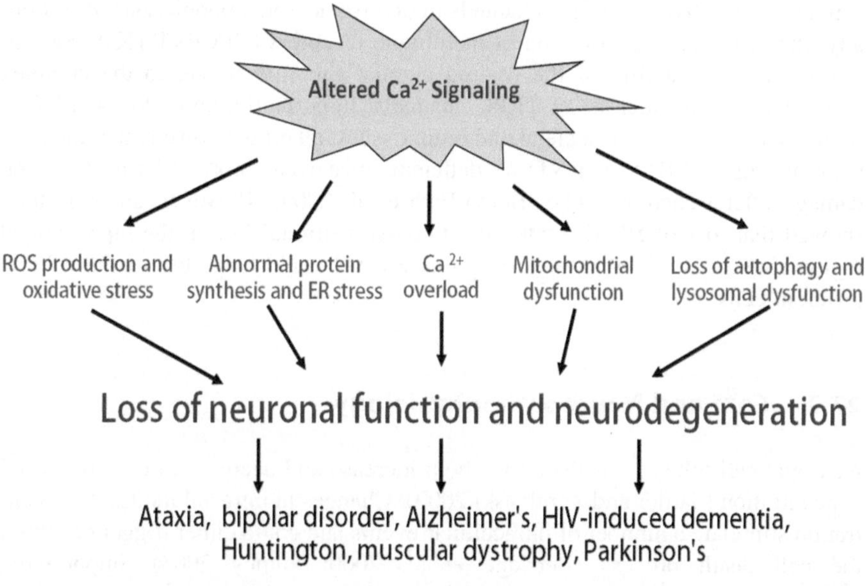

Fig. 27.2 Consequences of abnormal Ca^{2+} signaling. This illustration depicts the current understanding of the consequences of abnormal Ca^{2+} signaling. Altered Ca^{2+} signaling leads to issues such as oxidative stress, ROS production, Ca^{2+} overload, unfolded protein response and ER stress, mitochondrial and lysosomal dysfunction, and inhibition of autophagy. This leads to loss of neuronal function and induction of neuronal diseases

27.2.1 SOCE Potentiates Alzheimer's Disease

Alzheimer's disease (AD) is a progressive and irreversible neurodegenerative disorder marked by neuronal atrophy, synapse degeneration, amyloid plaques, and neurofibrillary tangles. This disease can be divided into two major forms, one that has an early onset, the familial Alzheimer's disease (FAD), and the other that has a late onset, the sporadic Alzheimer's disease (SAD). FAD accounts for less than 10% of all the AD, however, have provided clues to understand the molecular and cellular basis of this genetically complex disease. The pathogenesis of FAD has been attributed to missense mutations in one of the three principle genes encoding amyloid precursor protein (APP), presenilin 1 (PS1), and presenilin 2 (PS2). APP is a plasma membrane protein which is processed in succession by β- and γ-secretase yielding Aβ (40 and 42) peptides based on the site of cleavage. Aβ40 is the soluble form and is secreted out, whereas Aβ42, formed from a missense mutation of APP, is an insoluble form and is retained within the PM. This then leads to the formation of Aβ aggregates in specific regions of the brain, thereby causing neurodegeneration (De Strooper and Annaert 2000). PS1 and PS2 are integral membrane proteins localized in the ER and are implicated in the γ-secretase activity of APP. Interestingly, presenilins are also suggested to function as ER Ca^{2+} leak channels (Tu et al. 2006). Accumulating evidence over the last decade has shown a strong link between pathogenesis of AD and impaired Ca^{2+} homeostasis (Bezprozvanny and Mattson 2008). Since Ca^{2+} signaling depends on both the release of Ca^{2+} from the ER and the subsequent influx across the PM, dysregulation in any of these pathways can create an imbalance in the overall cellular Ca^{2+} homeostasis. Information obtained from mutant PS1 and PS2 transgenic mice as well as from primary cortical neurons expressing various presenilin mutations has shown that in AD, the basal ER Ca^{2+} stores are larger which results in an increased dynamic flow of Ca^{2+} into the cytosol (LaFerla 2002; Stutzmann et al. 2006). The proposed mechanism for this effect is reasoned due to the disruption of Ca^{2+} leak function of mutant presenilin creating an imbalance between Ca^{2+} leak and Ca^{2+} reuptake. Contrary to this "Ca^{2+} overload" hypothesis, there are strong evidences showing no increase in Ca^{2+} stores; rather Ca^{2+} stores in PS1 and PS2 mutant phenotypes are reduced (Cheung et al. 2008; Giacomello et al. 2005; Lessard et al. 2005). It has also been shown that FAD PS1 and PS2 mutants interact with IP_3R rendering it sensitive to lower IP_3 concentrations, thus affecting the receptor's gating and resulting in exaggerated $[Ca^{2+}]_{cyt}$ (Cheung et al. 2008). These discrepancies could be attributed to the type of mutations on the presenilin and the cell type used, which would dictate the eventual phenotypic disparity. In another set of studies, the increase in Ca^{2+} influx is also attributed to the increased expression of ryanodine receptors (RyR) in AD-affected cells (Stutzmann et al. 2006). However, irrespective of the content of Ca^{2+} stores, SOCE was abrogated with most of the presenilin mutation. One of the logical interpretations could be that since the size of the ER store always remains high, the threshold levels of store depletion required to trigger SOCE are not attained, and hence a decrease in SOCE is observed. On the other hand, when PS1 function was abrogated by knockdown or dominant negative strategy, there was a marked potentiation of SOCE suggesting that PS1 in general negatively regulates SOCE and their mutations create a gain-of-function phenotype that would further augment the inhibition. It is of note that the elevated $[Ca^{2+}]_{cyt}$ acts proximal to the disease phenotype resulting in the enhanced secretase activity and thus capable of modulating Aβ peptide levels (Yoo et al. 2000). However, the addition of Aβ peptides to the cells expressing PS1 mutations did not show any significant increase in SOCE. Interestingly, a recent study showed that mouse embryonic fibroblast lacking presenilins had increased levels of STIM1 and decreased levels of STIM2 expression with a marked potentiation of SOCE. Expression of FAD mutants in these cells attenuated SOCE without altering the STIM expression (Bojarski et al. 2009). In contrast, a recent report showed that STIM2 is essential for the SOCE in hippocampal neurons and downregulation of STIM2 leads to massive loss of neurons. Consistent with this, samples of AD patients also showed a decrease in STIM2, but not STIM1 (Sun et al. 2014a). However, still the molecular identity of the Ca^{2+} channel that could contribute to these results (Orai vs. TRPC) is not yet established. In addition, STIM2 has been shown to actually inhibit SOCE in some cases and contribute to basal Ca^{2+} regulation, rather than agonist-mediated Ca^{2+} entry. Thus, in this context, the question of how STIM2 functions differently in neuronal vs. nonneuronal cells needs further clarification.

27.2.2 Loss of SOCE Can Induce Parkinson's Disease

Parkinson's disease (PD) is the second most common progressive neurodegenerative disorder caused by the degeneration of a particular group of neurons in the substantia nigra pars compacta, which control motor regulation of the limbic system. As is the case with Alzheimer's disease, this also has two forms—the less prevalent familial PD and the more prevalent sporadic PD. The pathological hallmarks of the disease are pronounced degeneration of nigrostriatal dopaminergic neurons and the presence of cytoplasmic inclusions called Lewy bodies (LBs) formed by the aggregation of several proteins including α-synuclein, parkin, ubiquitin, and neurofilaments. Not all cases of PD exhibit LB pathogenesis and hence have to be diagnosed by clinical symptoms. Even though the actual cause or etiology of sporadic PD is not clearly understood, the most common modes of neuronal degeneration, as understood from gene mutations in familial PD, appear to be due to mitochondrial dysfunction, oxidative stress, and impairment of the ubiquitin proteasome pathway. These effects are similar to those induced by 1-methyl, 4-phenyl, 1, 2, 3, 4-tetrapyridine (MPTP) neurotoxin, and hence it is commonly used as toxin-based model to study PD (Dauer and Przedborski 2003). MPP+, the metabolite of MPTP, is preferentially taken up by dopaminergic neurons through dopamine transporter, which then gets accumulated in the mitochondria and inhibits complex I of the respiratory chain. As a result, there is a loss of mitochondrial membrane potential, depletion of ATP, and increased free radical production creating oxidative stress within the cell (Przedborski et al. 2004). The free radicals thus released from the mitochondria disturb the intracellular Ca^{2+} homeostasis, possibly by compromising the function of Ca^{2+} signaling components of the ER and PM (Mattson 2007). With regard to ER store depletion, the cell has its own way of replenishing the stores through the activation of SOCE channels. But following MPP+ treatment, TRPC1 expression is drastically decreased. As a result, the ER fails to maintain the normal Ca^{2+} levels required for its proper functioning, which eventually leads to ER stress (Bollimuntha et al. 2005b; Selvaraj et al. 2009, 2012). The mitochondrion on the other hand takes up the excess Ca^{2+} present in the cytoplasm (released from the ER) and loses its membrane potential due to Ca^{2+} overload. This in turn exacerbates the free radical production, thus creating a vicious cycle of oxidative stress, which initiates an intrinsic apoptotic cascade. Our findings also suggest that overexpression of TRPC1 combats some of the negative effects of MPP^+-induced oxidative stress and mitochondrial dysfunction, thus providing the cells an opportunity to recover from the stressed state. In addition, autophagy has been also shown to work in conjunction or independently to help the cell cope with ER stress by removal of misfolded proteins, and autophagy has been shown to modulate by TRPC1 and Ca^{2+}, suggesting that these could also contribute to the survival of these neurons (Sukumaran et al. 2016). In addition, dopamine and other neurotransmitter levels regulate mood disorders such as schizophrenia, anxiety, bipolar disorders, and others, which could also be regulated by Ca^{2+} signaling; however, the role of Ca^{2+} channels in this regard is still not well defined.

27.2.3 A Functional Look at SOCE in Ataxia

Cerebellar ataxias represent a heterogeneous group of neurodegenerative disorders. To date, 26 different genetic subtypes of spinocerebellar ataxias (SCA) have been identified. Spinocerebellar ataxia type 1 (SCA1) is an inheritable autosomal dominant neurodegenerative disease caused by expansions of a CAG trinucleotide repeats encoding polyglutamine tract, within the SCA1-encoded ATXN1 (ataxin 1) protein. The pathogenesis involves a gain-of-function mutation in ATXN1, which mainly affects the Purkinje cells of the cerebellum and brainstem neurons (Vig et al. 2001). Although the genetic diversity and affected cellular pathways of hereditary ataxias are broad, one common theme among these genes is their effects on maintaining Ca^{2+} balance primarily in the cerebellum (Mark et al. 2017). Moreover, it has also been shown that the polyglutamine mutation may affect the gene expression pattern and was found that certain neuronal genes involved in signal transduction and Ca^{2+} homeostasis were downregulated in a sequential pattern in SCA1 mice (Lin et al. 2000). Among other genes, this study identified IP_3R, SERCA2, inositol polyphosphate 5-phosphatase type 1 (5-phosphatase), and TRPC3 genes important for Ca^{2+} homeostasis. Importantly, IP_3R and SERCA2 were downregulated in the first 2 weeks of age, whereas in the subsequent 3–4 weeks, phosphatase and TRPC3 were downregulated. These results are striking given the fact that IP_3R, SERCA2, and TRPC3 constitute the molecular components of SOCE. This clearly infers that in SCA1 the ER Ca^{2+} homeostasis can be compromised under sustained stress conditions, which can lead to a variety of cellular dysfunctions. Additionally, spinocerebellar ataxia14 (SCA14), one of the subtypes of spinocerebellar ataxias, is caused by mutations in the protein kinase C gamma (PKCγ) gene (Adachi et al. 2008). PKCγ has been found to negatively regulate TRPC3 activity by phosphorylating TRPC3, but in the case of SCA14, it fails to phosphorylate TRPC3 resulting in sustained Ca^{2+} influx following stimulation. Hence, the PKCγ kinase activity is seemingly indispensable for TRPC3 phosphorylation, and feedback regulation of Ca^{2+} influx, which when fails, as in the case of SCA14, might cause imbalance in cellular Ca^{2+} homeostasis and account for disease progression. One of the intriguing features is the differences in the expression and mechanism of action of TRPC3 in two different subtypes of SCAs—rather opposing functions. Furthermore, in an attempt to identify crucial gene products implicated in cerebral ataxia, phenotype-driven dominant mutagenesis screens were performed, and an ataxic mouse mutant by the name moonwalker (Mwk) mice was identified. These mice exhibited a gain-of-function mutation (T635A) of TRPC3 that rendered the channel constitutively active due to the lack of negative feedback regulation by PKCγ-mediated phosphorylation of the threonine residue (Becker et al. 2009). As a result of impaired TRPC3 channel function, diminished dendritic arborization and progressive loss of Purkinje neurons were observed. The disparity in the mode of expression and functionality of TRPC3 in various forms of SCA, which eventually results in degeneration of cerebellar neurons and ataxic phenotype, can be reconciled from the fact that TRPC3 might not be the sole cause of the disease, rather a component of secondary effect in the dysfunctional signaling cascade. Importantly, to establish TRPC3-dependent mechanisms, it has been recently shown that activated Ca^{2+} signaling is coupled to lipid metabolism and the regulation of Purkinje cell development in the Mwk cerebellum (Dulneva et al. 2015), confirming the role of TRPC3 in ataxia.

27.2.4 SOCE in Muscular Dystrophy

Duchenne muscular dystrophy (DMD) is a progressive neuromuscular disease characterized by muscle degeneration. The pathology of this disease is attributed to the deficiency of dystrophin, which is located at the cytoplasmic face of the sarcolemma. In normal skeletal muscle cells, dystrophin acts as a bridge that connects the extracellular matrix protein laminin-2 to the cytoskeletal f-actin filaments. Hence, in DMD the lack of dystrophin and consequent disorganization of the cytoskeleton can make the PM vulnerable to mechanical damage and can result in functional deregulation of several PM ion channels (Gailly 2002; Vandebrouck et al. 2002). Store-operated Ca^{2+} channels, TRPC1 and TRPC4 in particular, are among those whose disruption can lead to increased Ca^{2+} influx resulting in cellular Ca^{2+} overload, and their knockdown resulted in a tenfold decrease in Ca^{2+} influx. Hence, the imbalance in intracellular Ca^{2+} levels and the resulting Ca^{2+}-dependent activation of proteases are suggested to be the probable reasons for muscle degeneration in DMD (Vandebrouck et al. 2002). In another interesting study, forced expression of minidystrophin at the plasma membrane resulted in regaining normal Ca^{2+} homeostasis in dystrophin-deficient Sol8 myotubes. It is suggested that under normal conditions minidystrophin negatively regulates store-operated Ca^{2+} channels resulting in reduced store-dependent Ca^{2+} influx and hence overexpression of minidystrophin in dystrophin-lacking myotubes restored the function of store-dependent channels, resulting in less Ca^{2+} influx and subsequent prevention of Ca^{2+} mishandling (Vandebrouck et al. 2006). Importantly, it has recently been shown that Ca^{2+} influx across an unstable sarcolemma due to increased activity of a STIM1-Orai1 complex is a disease determinant in muscular dystrophy, and, hence, SOCE represents a potential therapeutic target (Goonasekera et al. 2014).

27.2.5 SOCE Induces HIV-Associated Dementia

Human immunodeficiency virus (HIV) infection-induced cognitive impairment has been suggested to precipitate in memory loss and dementia. Thus, HIV-associated dementia (HAD) forms a neuropathological manifestation of the virus infection (Masliah et al. 1996; Mattson et al. 2005). Activation of immune cells by HIV results in the secretion of soluble factors that destabilize neuronal Ca^{2+} homeostasis, encourage oxidative stress, and result in neural damage, which is thought to underlie the cognitive-motor dysfunction that develops in many HIV-infected patients. The HIV-1 transactivator protein (Tat) has been reported to be involved in cognitive impairment (Li et al. 2004) and promotes neuronal death by inducing mitochondrial dysfunction and oxidative stress (Norman et al. 2007; Self et al. 2004). Additionally, evidence also suggests a role of HIV-1 Tat in altering cellular Ca^{2+} homeostasis (Bonavia et al. 2001; Self et al. 2004), presumably interfering with molecular components of Ca^{2+} stores (Haughey et al. 1999; Norman et al. 2008). Tat, via Ca^{2+} dysregulation, has been shown to promote calpain-1 cleavage of p35 to p25, which hyperactivates CDK5 resulting in abnormal phosphorylation of downstream targets such as Tau, collapsin response mediator protein-2 (CRMP2), doublecortin (DCX), and MEF2. Thus, it is conceivable that SOCE might affect Tat toxicity in a way analogous to the protective role of TRPC1 in MPP^+-induced PD (Bollimuntha et al. 2005b; Selvaraj et al. 2009). Indeed, two recent interesting studies from the same group have implicated the involvement of TRPC channels in Tat toxicity and HAD (Yao et al. 2009a, b). The first study reports that the HIV-1 Tat induces significant cell death in rat midbrain neurons. Treatment of the neurons with the chemokine CCL2, a member of the C-C subfamily of chemokines, alternatively known as monocyte chemoattractant protein-1 (MCP-1), substantially alleviates Tat toxicity and improves neuronal survival. The premise that CCL2 engages PLCβ-mediated IP_3 production and that IP_3-mediated store depletion can activate TRPC channels, thus contributing to neuroprotection, prompted this group to further investigate the involvement of TRPC channels. Interestingly, SKF-96365 and 2-APB, the nonspecific SOCE blockers, significantly blocked the CCL2-induced neuroprotection against Tat toxicity. Further, using RNAi the authors found that TRPC1 and TRPC5 channels were involved in mediating the protective effects of CCL2, preferably via an ERK/CREB signaling axis (Yao et al. 2009b). In the similar context of HIV-1 Tat-induced neurotoxicity, this group demonstrated the involvement of TRPC channels in platelet-derived growth factor (PDGF)-mediated protection of the rat midbrain neurons. In addition to the neuroprotective effects of PDGF in vitro, its pretreatment in vivo also rescued loss of dopaminergic neurons following Tat exposure. This study revealed a prominent role of TRPC5 and TRPC6 channels in engaging a Pyk2/ERK/CREB pathway to elicit PDGF-mediated neuroprotection (Yao et al. 2009b). In another study, pretreatment of cells with either the PI3K inhibitor or a TRPC blocker resulted in the failure of PDGF to inactivate GSK3β, thereby suggesting the intersection of PI3K and TRPC signaling at GSK3β (Peng et al. 2012).

27.2.6 SOCE and Oxidative Stress

The steady-state generation and proper compartmentalization of free radicals to specific cellular microdomains transduce physiological redox signals to carry out anabolic activities such as proliferation, migration, transcription, and neurotransmission. However, in specific pathological condition, these oxidants can leave their spatial confinements, thereby leading to oxidative stress (Durackova 2010; Kamata and Hirata 1999). Oxidant-induced toxicity and cognitive impairment have been reported in a variety of neurodegenerative diseases including AD, PD, ALS, HD, stroke, and prion pathogenesis (Bezprozvanny and Hayden 2004; Cassarino et al. 1997; Dauer and Przedborski 2003; Mattson 2007; Patten et al. 2010). A prime commonality in free radical production and associated neurodegeneration is mitochondrial dysfunction, and there appears to be a positive correlation between them in regulating disease progression (Lin and Beal 2006). Changes in the levels of cytosolic Ca^{2+} influence numerous physiological processes (Berridge 1998; Berridge et al. 2000). Free radicals can influence the physiological activity of a cell by altering cytosolic Ca^{2+}, either by mobilizing Ca^{2+} from internal stores or by modulating Ca^{2+} influx across the PM. Interestingly, perturbation in Ca^{2+} homeostasis has also been described to result in the generation of free radicals, leading to oxidant-induced neuronal programmed cell death (Bezprozvanny and Mattson 2008; Chakraborti et al. 1999; Ermak and Davies 2002; Mattson 2007). Thus, it is conceivable that the components of neuronal SOCE will have a prominent influence on regulating the redox status of a cell. Alternatively, there also lies the possibility that the SOCE channels can themselves be regulated by cellular redox fluctuations. From a neuronal perspective, evidence on oxidative stress-regulated SOCE is limited; however, in non-excitable cells the same has been more elaborately studied.

Several members of TRPs and more recently the Orai1 and STIM2 components of SOCE have been identified to be redox responsive (Aarts and Tymianski 2005;

Berna-Erro et al. 2009; Birnbaumer 2009; Bogeski et al. 2010; Miller 2006). One of the earliest evidences that oxidants have the potential to activate SOCC comes from a study demonstrating H_2O_2 to induce Ca^{2+} influx in canine endothelial cells. Interestingly, this H_2O_2-induced Ca^{2+} influx was dose dependent and sensitive to the SOCE blocker SKF-96365 (Doan et al. 1994). Additionally, in human oocytes a similar H_2O_2 dose-dependent increase in Ca^{2+} influx was observed which was sensitive to 2-APB, yet another SOCE blocker (Martin-Romero et al. 2008). Further, in porcine aortic endothelial cells (PAECs), induction of oxidative stress activates nonselective cation entry. In the initial study, treatment of the endothelial cells with *tert*-butyl hydroperoxide elicited cation conductance which was suppressed by dominantly interfering with the function of TRPC3 channels (Balzer et al. 1999). In a subsequent study, the same group showed that the stress-induced cation conductance in PAECs was due to the heteromeric assembly of TRPC3/TRPC4 channels. TRPC1 in these cells weakly interacted with TRPC3; however, its involvement in heteromeric channel assembly and in redox sensing remains to be determined (Poteser et al. 2006). Interestingly, amyloid-β aggregates have been shown to induce oxidative stress (Bezprozvanny and Mattson 2008), and since TRPC3 has a redox-sensing ability, it can potentially exacerbate AD. Similarly, TRPC5 has also been demonstrated to be activated by NO via S-nitrosylation (Yoshida et al. 2006) and thus possess the ability to participate in neurodegenerative conditions. In addition to members of the TRPC subfamily, the TRPMs have also been identified as redox sensitive. In particular, TRPM2 and TRPM7 have been associated with oxidative insult-induced cell death. TRPM2 is known to be activated by ROS including H_2O_2, and oxidant-induced Ca^{2+} influx via TRPM2 channels results in cell death (Hara et al. 2002). H_2O_2 and amyloid β-peptide have been shown to induce cell death in rat striatal neurons expressing endogenous TRPM2, and inhibiting the function of TRPM2 improves cell survival following stress (McNulty and Fonfria 2005). A study also describes PARP (poly ADP ribose polymerase) as a mediator between oxidative stress and TRPM2 activation and that PARP inhibitors tend to suppress TRPM2 function and protect cells from oxidative injury (Fonfria et al. 2004; Miller 2004; Yang et al. 2006). Similarly, H_2O_2 also activates TRPM7 conductance, and RNAi-mediated inhibition of the channel reduces oxidant-induced Ca^{2+} influx, suppresses ROS production, and protects neurons from anoxic cell death (Aarts and Tymianski 2005; Yamamoto et al. 2009). Interestingly, in a human endothelial cell line, it was recently shown that DHA, a polyunsaturated fatty acid, attenuates H_2O_2-induced oxidative stress by reducing Ca^{2+} influx. DHA apparently alters the lipid composition of the plasma membrane raft domains, and in doing so, it displaces the caveolar raft-localized TRPC1 to non-raft compartment, thus rendering the channel nonfunctional (Ye et al. 2010). In cultured hippocampal neurons, stress-induced exacerbation in a TRP channel-related Ca^{2+} influx leads to neuronal death, and pretreatments of the cells with the antioxidant alpha-tocopherol protect the neurons by inhibiting the TRP-mediated intracellular Ca^{2+} overload (Crouzin et al. 2007). Additionally, in *Drosophila* photoreceptors, anoxia activates TRP and TRP-like (TRPL) currents, and genetic depletion of these channels is shown to protect the photoreceptor cells from anoxic stress (Agam et al. 2000). From the findings discussed above, it follows that oxidative stress activates SOCE, and in a pathological situation, oxidant-induced exaggeration in SOCE can potentially amplify cell injury. Thus, components of SOCE emerge as attractive targets in controlling oxidative stress-induced cell death. A reduction in SOCE in glutamate-resistant HT22 cells protected against oxidative death, suggesting that dysregulated Ca^{2+} entry through Orai1 mediates the detrimental Ca^{2+} entry in programmed cell death induced by GSH depletion (Henke et al. 2013). However, ROS has also been shown to regulate SOCE, which could be due to direct effects on the core SOCE machinery, including STIM proteins and their partner channels, or indirectly by influencing the status of ER Ca^{2+} stores. STIM1 was shown to be S-glutathionylated at C56 amino acid during oxidative stress induced by bacterial lipopolysaccharide (LPS) or butathionine sulfoximine-induced GSH depletion, which lowered the affinity of STIM1 for Ca^{2+} and facilitated oligomerization, leading to store-independent activation of STIM1 (Nunes and Demaurex 2014). Similarly, Orai1 was also recently demonstrated to act as a direct redox sensor mainly by virtue of a reactive cysteine at position C195, and exposure to H_2O_2 inhibited Ca^{2+} influx in normal Orai1, but not in Orai1 mutated at position 195 (Nunes and Demaurex 2014). These results suggest that ROS regulates the Ca^{2+} channels and its modulators that modify Ca^{2+} entry and could promote degeneration.

27.3 SOCE, Membrane Rafts, and Neurodegeneration

It is being widely accepted that the PM is not just a uniform stretch of the protein-lipid bilayer, rather it is spatially organized into multiple microdomains that constitute unique signaling nodes to efficiently relay external signals into the cell's interior. Regions of the PM that are rich in cholesterol and sphingolipids fall into a broader class of membrane microdomain termed as membrane "lipid rafts" that have the ability to cluster specific signaling molecules required to bring about a specific cellular response. Involvement of membrane rafts in the progression of neurodegeneration has been described in a variety of dementing conditions including AD, PD, ALS, and prion pathogenesis. In AD, the generation of amyloid β-peptide by the processing of amyloid precursor protein (APP) hallmarks disease initiation (De Strooper and Annaert 2000). Interestingly, in cultured neurons it had been shown that acute extraction of membrane cholesterol by MβCD or by chronic statin treatments to inhibit cholesterol synthesis reduces the accumulation of amyloid β-peptide (Ehehalt et al. 2003). Similarly, molecules involved in PD pathogenesis such as parkin, PINK1, α-synuclein, and *LRRK2* have been shown to be associated with lipid raft microdomains (Hatano et al. 2007; Park et al. 2009; Parkin et al. 1999). Importantly, most TRPCs have been identified to be expressed in lipid rafts, and Ca^{2+} entry was dependent on the integrity of lipid rafts (Pani and Singh 2009). However, information on the membrane raft perspective of SOCE regulation in disease progression is still lacking. Since components of the SOCC localize to membrane rafts and work in concert with multiple regulators, it would be worthwhile to investigate their physical state in various neurodegenerative conditions such as attenuation of TRPC1-SOCE in PD and exaggeration of SOCE in neuronal oxidative stress and AD. This will not only provide mechanistic insight into the altered Ca^{2+} homeostasis but will also presumably shed light on the onset and progression of the disease.

27.4 Concluding Remarks

SOCE is a unique aspect of neuronal Ca^{2+} signaling wherein the intracellular Ca^{2+} stores communicate with the plasma membrane channels to orchestrate cellular Ca^{2+} homeostasis. This process is essential for maintaining intracellular Ca^{2+} stores as well as influencing vital physiological processes of the excitable cells. TRPC and Orai channels have been recently identified as SOCE channels, and there has been intense focus on TRPC channels as recent data from various labs indicate a possible role of TRPC channels in neuronal function. However, conclusive data regarding the exact physiological function of most of these channels in neuronal cells is still lacking. Studies using single- or multiple-gene knockouts will be essential to validate the diverse roles of these channels in neuronal function. Additionally, as TRPC proteins cross talk with Orai and are activated by STIM1, studies focused on delineating the function of the "TRPC-Orai-STIM" complex in neuronal context will be critical in understanding their plausible role in neuronal pathophysiology.

Acknowledgments We duly acknowledge the grant support from the National Institutes of Health (DE017102; DE024300; GM113123).

References

Aarts MM, Tymianski M (2005) TRPMs and neuronal cell death. Pflugers Arch 451:243–249

Adachi N, Kobayashi T, Takahashi H, Kawasaki T, Shirai Y, Ueyama T, Matsuda T, Seki T, Sakai N, Saito N (2008) Enzymological analysis of mutant protein kinase C gamma causing spinocerebellar ataxia type 14 and dysfunction in Ca2+ homeostasis. J Biol Chem 283:19854–19863

Agam K, von Campenhausen M, Levy S, Ben-Ami HC, Cook B, Kirschfeld K, Minke B (2000) Metabolic stress reversibly activates the Drosophila light-sensitive channels TRP and TRPL in vivo. J Neurosci 20:5748–5755

Akbari Y, Hitt BD, Murphy MP, Dagher NN, Tseng BP, Green KN, Golde TE, LaFerla FM (2004) Presenilin regulates capacitative calcium entry dependently and independently of gamma-secretase activity. Biochem Biophys Res Commun 322:1145–1152

Albers DS, Beal MF (2000) Mitochondrial dysfunction and oxidative stress in aging and neurodegenerative disease. J Neural Transm Suppl 59:133–154

Balzer M, Lintschinger B, Groschner K (1999) Evidence for a role of Trp proteins in the oxidative stress-induced membrane conductances of porcine aortic endothelial cells. Cardiovasc Res 42:543–549

Banerjee S, Lee J, Venkatesh K, Wu CF, Hasan G (2004) Loss of flight and associated neuronal rhythmicity in inositol 1,4,5-trisphosphate receptor mutants of Drosophila. J Neurosci 24:7869–7878

Barritt GJ (1999) Receptor-activated Ca2+ inflow in animal cells: a variety of pathways tailored to meet different intracellular Ca2+ signalling requirements. Biochem J 337(Pt 2):153–169

Beck B, Lehen'kyi V, Roudbaraki M, Flourakis M, Charveron M, Bordat P, Polakowska R, Prevarskaya N, Skryma R (2008) TRPC channels determine human keratinocyte differentiation: new insight into basal cell carcinoma. Cell Calcium 43:492–505

Becker EB, Oliver PK, Glitsch MD, Banks GT, Achilli F, Hardy A, Nolan PM, Fisher EM, Davies KE (2009) A point mutation in TRPC3 causes abnormal Purkinje cell development and cerebellar ataxia in moonwalker mice. Proc Natl Acad Sci U S A 106:6706–6711

Berna-Erro A, Braun A, Kraft R, Kleinschnitz C, Schuhmann MK, Stegner D, Wultsch T, Eilers J, Meuth SG, Stoll G, Nieswandt B (2009) STIM2 regulates capacitive Ca2+ entry in neurons and plays a key role in hypoxic neuronal cell death. Sci Signal 2(93):ra67

Berridge MJ (1995) Capacitative calcium entry. Biochem J 312(Pt 1):1–11

Berridge MJ (1997) The AM and FM of calcium signalling. Nature 386:759–760

Berridge MJ (1998) Neuronal calcium signaling. Neuron 21:13–26

Berridge MJ (2012) Calcium signalling remodelling and disease. Biochem Soc Trans 40:297–309

Berridge MJ, Dupont G (1994) Spatial and temporal signalling by calcium. Curr Opin Cell Biol 6:267–274

Berridge MJ, Lipp P, Bootman MD (2000) The versatility and universality of calcium signalling. Nat Rev Mol Cell Biol 1:11–21

Berridge MJ, Bootman MD, Roderick HL (2003) Calcium signalling: dynamics, homeostasis and remodeling. Nat Rev Mol Cell Biol 4:517–729

Bezprozvanny I, Hayden MR (2004) Deranged neuronal calcium signaling and Huntington disease. Biochem Biophys Res Commun 322:1310–1317

Bezprozvanny I, Mattson MP (2008) Neuronal calcium mishandling and the pathogenesis of Alzheimer's disease. Trends Neurosci 31:454–463

Birnbaumer L (2009) The TRPC class of ion channels: a critical review of their roles in slow, sustained increases in intracellular Ca(2+) concentrations. Annu Rev Pharmacol Toxicol 49:395–426

Bogeski I, Kummerow C, Al-Ansary D, Schwarz EC, Koehler R, Kozai D, Takahashi N, Peinelt C, Griesemer D, Bozem M, Mori Y, Hoth M, Niemeyer BA (2010) Differential redox regulation of ORAI ion channels: a mechanism to tune cellular calcium signaling. Sci Signal 3(115):ra24

Bojarski L, Pomorski P, Szybinska A, Drab M, Skibinska-Kijek A, Gruszczynska-Biegala J, Kuznicki J (2009) Presenilin-dependent expression of STIM proteins and dysregulation of capacitative Ca2+ entry in familial Alzheimer's disease. Biochim Biophys Acta 1793:1050–1057

Bollimuntha S, Cornatzer E, Singh BB (2005a) Plasma membrane localization and function of TRPC1 is dependent on its interaction with beta-tubulin in retinal epithelium cells. Vis Neurosci 22:163–170

Bollimuntha S, Singh BB, Shavali S, Sharma SK, Ebadi M (2005b) TRPC1-mediated inhibition of 1-methyl-4-phenylpyridinium ion neurotoxicity in human SH-SY5Y neuroblastoma cells. J Biol Chem 280:2132–2140

Bonavia R, Bajetto A, Barbero S, Albini A, Noonan DM, Schettini G (2001) HIV-1 Tat causes apoptotic death and calcium homeostasis alterations in rat neurons. Biochem Biophys Res Commun 288:301–308

Brini M, Carafoli E (2009) Calcium pumps in health and disease. Physiol Rev 89:1341–1378

Cai S, Fatherazi S, Presland RB, Belton CM, Roberts FA, Goodwin PC, Schubert MM, Izutsu KT (2006) Evidence that TRPC1 contributes to calcium-induced differentiation of human keratinocytes. Pflugers Arch 452:43–52

Cassarino DS, Fall CP, Swerdlow RH, Smith TS, Halvorsen EM, Miller SW, Parks JP, Parker WD Jr, Bennett JP Jr (1997) Elevated reactive oxygen species and antioxidant enzyme activities in animal and cellular models of Parkinson's disease. Biochim Biophys Acta 1362:77–86

Chakraborti T, Das S, Mondal M, Roychoudhury S, Chakraborti S (1999) Oxidant, mitochondria and calcium: an overview. Cell Signal 11:77–85

Cheek TR, Thastrup O (1989) Internal Ca2+ mobilization and secretion in bovine adrenal chromaffin cells. Cell Calcium 10:213–221
Cheng KT, Liu X, Ong HL, Ambudkar IS (2008) Functional requirement for Orai1 in store-operated TRPC1-STIM1 channels. J Biol Chem 283:12935–12940
Cheung KH, Shineman D, Muller M, Cardenas C, Mei L, Yang J, Tomita T, Iwatsubo T, Lee VM, Foskett JK (2008) Mechanism of Ca2+ disruption in Alzheimer's disease by presenilin regulation of InsP3 receptor channel gating. Neuron 58:871–883
Clapham DE (2007) Calcium signaling. Cell 131:1047–1058
Clementi E, Scheer H, Zacchetti D, Fasolato C, Pozzan T, Meldolesi J (1992) Receptor-activated Ca2+ influx. Two independently regulated mechanisms of influx stimulation coexist in neurosecretory PC12 cells. J Biol Chem 267:2164–2172
Congar P, Leinekugel X, Ben-Ari Y, Crepel V (1997) A long-lasting calcium-activated nonselective cationic current is generated by synaptic stimulation or exogenous activation of group I metabotropic glutamate receptors in CA1 pyramidal neurons. J Neurosci 17:5366–5379
Crouzin N, de Jesus Ferreira MC, Cohen-Solal C, Aimar RF, Vignes M, Guiramand J (2007) Alpha-tocopherol-mediated long-lasting protection against oxidative damage involves an attenuation of calcium entry through TRP-like channels in cultured hippocampal neurons. Free Radic Biol Med 42:1326–1337
Dauer W, Przedborski S (2003) Parkinson's disease: mechanisms and models. Neuron 39:889–909
De Strooper B, Annaert W (2000) Proteolytic processing and cell biological functions of the amyloid precursor protein. J Cell Sci 113(Pt 11):1857–1870
Doan TN, Gentry DL, Taylor AA, Elliott SJ (1994) Hydrogen peroxide activates agonist-sensitive Ca(2+)-flux pathways in canine venous endothelial cells. Biochem J 297(Pt 1):209–215
Dulneva A, Lee S, Oliver PL, Di Gleria K, Kessler BM, Davies KE, Becker EB (2015) The mutant Moonwalker TRPC3 channel links calcium signaling to lipid metabolism in the developing cerebellum. Hum Mol Genet 24:4114–4125
Durackova Z (2010) Some current insights into oxidative stress. Physiol Res 59:459–469
Ehehalt R, Keller P, Haass C, Thiele C, Simons K (2003) Amyloidogenic processing of the Alzheimer beta-amyloid precursor protein depends on lipid rafts. J Cell Biol 160:113–123
Emptage NJ, Reid CA, Fine A (2001) Calcium stores in hippocampal synaptic boutons mediate short-term plasticity, store-operated Ca2+ entry, and spontaneous transmitter release. Neuron 29:197–208
Ermak G, Davies KJ (2002) Calcium and oxidative stress: from cell signaling to cell death. Mol Immunol 38:713–721
Fabian A, Fortmann T, Dieterich P, Riethmuller C, Schon P, Mally S, Nilius B, Schwab A (2008) TRPC1 channels regulate directionality of migrating cells. Pflugers Arch 457:475–484
Feske S, Gwack Y, Prakriya M, Srikanth S, Puppel SH, Tanasa B, Hogan PG, Lewis RS, Daly M, Rao A (2006) A mutation in Orai1 causes immune deficiency by abrogating CRAC channel function. Nature 441:179–185
Fiorio Pla A, Maric D, Brazer SC, Giacobini P, Liu X, Chang YH, Ambudkar IS, Barker JL (2005) Canonical transient receptor potential 1 plays a role in basic fibroblast growth factor (bFGF)/FGF receptor-1-induced Ca2+ entry and embryonic rat neural stem cell proliferation. J Neurosci 25:2687–2701
Fonfria E, Marshall IC, Benham CD, Boyfield I, Brown JD, Hill K, Hughes JP, Skaper SD, McNulty S (2004) TRPM2 channel opening in response to oxidative stress is dependent on activation of poly(ADP-ribose) polymerase. Br J Pharmacol 143:186–192
Friel DD, Tsien RW (1992) A caffeine- and ryanodine-sensitive Ca2+ store in bullfrog sympathetic neurones modulates effects of Ca2+ entry on [Ca2+]i. J Physiol 450:217–246
Gailly P (2002) New aspects of calcium signaling in skeletal muscle cells: implications in Duchenne muscular dystrophy. Biochim Biophys Acta 1600:38–44
Giacomello M, Barbiero L, Zatti G, Squitti R, Binetti G, Pozzan T, Fasolato C, Ghidoni R, Pizzo P (2005) Reduction of Ca2+ stores and capacitative Ca2+ entry is associated with the familial

Alzheimer's disease presenilin-2 T122R mutation and anticipates the onset of dementia. Neurobiol Dis 18:638–648

Gomez TM, Robles E, Poo M, Spitzer NC (2001) Filopodial calcium transients promote substrate-dependent growth cone turning. Science 291:1983–1987

Goonasekera SA, Davis J, Kwong JQ, Accornero F, Wei-LaPierre L, Sargent MA, Dirksen RT, Molkentin JD (2014) Enhanced Ca^{2+} influx from STIM1-Orai1 induces muscle pathology in mouse models of muscular dystrophy. Hum Mol Genet 23:3706–3715

Gouaux E, Mackinnon R (2005) Principles of selective ion transport in channels and pumps. Science 310:1461–1465

Guerini D, Coletto L, Carafoli E (2005) Exporting calcium from cells. Cell Calcium 38:281–289

Gwack Y, Feske S, Srikanth S, Hogan PG, Rao A (2007) Signalling to transcription: store-operated Ca2+ entry and NFAT activation in lymphocytes. Cell Calcium 42:145–156

Hara Y, Wakamori M, Ishii M, Maeno E, Nishida M, Yoshida T, Yamada H, Shimizu S, Mori E, Kudoh J, Shimizu N, Kurose H, Okada Y, Imoto K, Mori Y (2002) LTRPC2 Ca2+-permeable channel activated by changes in redox status confers susceptibility to cell death. Mol Cell 9:163–173

Hatano T, Kubo S, Imai S, Maeda M, Ishikawa K, Mizuno Y, Hattori N (2007) Leucine-rich repeat kinase 2 associates with lipid rafts. Hum Mol Genet 16:678–690

Haughey NJ, Holden CP, Nath A, Geiger JD (1999) Involvement of inositol 1,4,5-trisphosphate-regulated stores of intracellular calcium in calcium dysregulation and neuron cell death caused by HIV-1 protein tat. J Neurochem 73:1363–1374

Henke N, Albrecht P, Bouchachia I, Ryazantseva M, Knoll K, Lewerenz J, Kaznacheyeva E, Maher P, Methner A (2013) The plasma membrane channel ORAI1 mediates detrimental calcium influx caused by endogenous oxidative stress. Cell Death Dis 4:e470

Hilgemann DW, Yaradanakul A, Wang Y, Fuster D (2006) Molecular control of cardiac sodium homeostasis in health and disease. J Cardiovasc Electrophysiol 17(Suppl 1):S47–S56

Huang GN, Zeng W, Kim JY, Yuan JP, Han L, Muallem S, Worley PF (2006) STIM1 carboxyl-terminus activates native SOC, I(crac) and TRPC1 channels. Nat Cell Biol 8:1003–1010

Jia Y, Zhou J, Tai Y, Wang Y (2007) TRPC channels promote cerebellar granule neuron survival. Nat Neurosci 10:559–567

Kamata H, Hirata H (1999) Redox regulation of cellular signalling. Cell Signal 11:1–14

Kirichok Y, Krapivinsky G, Clapham DE (2004) The mitochondrial calcium uniporter is a highly selective ion channel. Nature 427:360–364

Kraft R (2015) STIM and ORAI proteins in the nervous system. Channels (Austin) 9:245–252

LaFerla FM (2002) Calcium dyshomeostasis and intracellular signalling in Alzheimer's disease. Nat Rev Neurosci 3:862–872

Lessard CB, Lussier MP, Cayouette S, Bourque G, Boulay G (2005) The overexpression of presenilin2 and Alzheimer's-disease-linked presenilin2 variants influences TRPC6-enhanced Ca2+ entry into HEK293 cells. Cell Signal 17:437–445

Li ST, Matsushita M, Moriwaki A, Saheki Y, Lu YF, Tomizawa K, Wu HY, Terada H, Matsui H (2004) HIV-1 Tat inhibits long-term potentiation and attenuates spatial learning [corrected]. Ann Neurol 55:362–371

Liao Y, Erxleben C, Yildirim E, Abramowitz J, Armstrong DL, Birnbaumer L (2007) Orai proteins interact with TRPC channels and confer responsiveness to store depletion. Proc Natl Acad Sci U S A 104:4682–4687

Liao Y, Erxleben C, Abramowitz J, Flockerzi V, Zhu MX, Armstrong DL, Birnbaumer L (2008) Functional interactions among Orai1, TRPCs, and STIM1 suggest a STIM-regulated heteromeric Orai/TRPC model for SOCE/Icrac channels. Proc Natl Acad Sci U S A 105:2895–2900

Lin MT, Beal MF (2006) Mitochondrial dysfunction and oxidative stress in neurodegenerative diseases. Nature 443:787–795

Lin X, Antalffy B, Kang D, Orr HT, Zoghbi HY (2000) Polyglutamine expansion down-regulates specific neuronal genes before pathologic changes in SCA1. Nat Neurosci 3:157–163

Linden R (1994) The survival of developing neurons: a review of afferent control. Neuroscience 58:671–682
Liou J, Kim ML, Heo WD, Jones JT, Myers JW, Ferrell JE Jr, Meyer T (2005) STIM is a Ca2+ sensor essential for Ca2+-store-depletion-triggered Ca2+ influx. Curr Biol 15:1235–1241
Lohmann C, Finski A, Bonhoeffer T (2005) Local calcium transients regulate the spontaneous motility of dendritic filopodia. Nat Neurosci 8:305–312
Majewski L, Kuznicki J (2015) SOCE in neurons: signaling or just refilling? Biochim Biophys Acta 1853:1940–1952
Mark MD, Schwitalla JC, Groemmke M, Herlitze S (2017) Keeping our calcium in balance to maintain our balance. Biochem Biophys Res Commun 483(4):1040–1050
Martin-Romero FJ, Ortiz-de-Galisteo JR, Lara-Laranjeira J, Dominguez-Arroyo JA, Gonzalez-Carrera E, Alvarez IS (2008) Store-operated calcium entry in human oocytes and sensitivity to oxidative stress. Biol Reprod 78:307–315
Masliah E, Ge N, Mucke L (1996) Pathogenesis of HIV-1 associated neurodegeneration. Crit Rev Neurobiol 10:57–67
Mattson MP (2007) Calcium and neurodegeneration. Aging Cell 6:337–350
Mattson MP, Haughey NJ, Nath A (2005) Cell death in HIV dementia. Cell Death Differ 12(Suppl 1):893–904
McNulty S, Fonfria E (2005) The role of TRPM channels in cell death. Pflugers Arch 451:235–242
Meldolesi J, Volpe P, Pozzan T (1988) The intracellular distribution of calcium. Trends Neurosci 11:449–452
Miller RJ (1988) Calcium signalling in neurons. Trends Neurosci 11:415–419
Miller BA (2004) Inhibition of TRPM2 function by PARP inhibitors protects cells from oxidative stress-induced death. Br J Pharmacol 143:515–516
Miller BA (2006) The role of TRP channels in oxidative stress-induced cell death. J Membr Biol 209:31–41
Montell C (2005a) TRP channels in Drosophila photoreceptor cells. J Physiol 567:45–51
Montell C (2005b) The TRP superfamily of cation channels. Sci STKE 2005:re3
Nishiyama M, Hong K, Mikoshiba K, Poo MM, Kato K (2000) Calcium stores regulate the polarity and input specificity of synaptic modification. Nature 408:584–588
Norman JP, Perry SW, Kasischke KA, Volsky DJ, Gelbard HA (2007) HIV-1 trans activator of transcription protein elicits mitochondrial hyperpolarization and respiratory deficit, with dysregulation of complex IV and nicotinamide adenine dinucleotide homeostasis in cortical neurons. J Immunol 178:869–876
Norman JP, Perry SW, Reynolds HM, Kiebala M, De Mesy Bentley KL, Trejo M, Volsky DJ, Maggirwar SB, Dewhurst S, Masliah E, Gelbard HA (2008) HIV-1 Tat activates neuronal ryanodine receptors with rapid induction of the unfolded protein response and mitochondrial hyperpolarization. PLoS One 3:e3731
Nunes P, Demaurex N (2014) Redox regulation of store-operated Ca2+ entry. Antioxid Redox Signal 21(6):915–932
O'Bryant SE, Hobson V, Hall JR, Waring SC, Chan W, Massman P, Lacritz L, Cullum CM, Diaz-Arrastia R (2009) Brain-derived neurotrophic factor levels in Alzheimer's disease. J Alzheimers Dis 17:337–341
Ong HL, Cheng KT, Liu X, Bandyopadhyay BC, Paria BC, Soboloff J, Pani B, Gwack Y, Srikanth S, Singh BB, Gill DL, Ambudkar IS (2007) Dynamic assembly of TRPC1-STIM1-Orai1 ternary complex is involved in store-operated calcium influx. Evidence for similarities in store-operated and calcium release-activated calcium channel components. J Biol Chem 282:9105–9116
Pani B, Singh BB (2009) Lipid rafts/caveolae as microdomains of calcium signaling. Cell Calcium 45:625–633
Pani B, Cornatzer E, Cornatzer W, Shin DM, Pittelkow MR, Hovnanian A, Ambudkar IS, Singh BB (2006) Up-regulation of transient receptor potential canonical 1 (TRPC1) following sarco

(endo)plasmic reticulum Ca2+ ATPase 2 gene silencing promotes cell survival: a potential role for TRPC1 in Darier's disease. Mol Biol Cell 17:4446–4458
Parekh AB (2011) Decoding cytosolic Ca(2+) oscillations. Trends Biochem Sci 36(2):78–87
Parekh AB, Penner R (1997) Store depletion and calcium influx. Physiol Rev 77:901–930
Park JY, Kim KS, Lee SB, Ryu JS, Chung KC, Choo YK, Jou I, Kim J, Park SM (2009) On the mechanism of internalization of alpha-synuclein into microglia: roles of ganglioside GM1 and lipid raft. J Neurochem 110:400–411
Parkin ET, Turner AJ, Hooper NM (1999) Amyloid precursor protein, although partially detergent-insoluble in mouse cerebral cortex, behaves as an atypical lipid raft protein. Biochem J 344(Pt 1):23–30
Patten DA, Germain M, Kelly MA, Slack RS (2010) Reactive oxygen species: stuck in the middle of neurodegeneration. J Alzheimers Dis 20(Suppl 2):S357–S367
Peng F, Yao H, Akturk HK, Buch S (2012) Platelet-derived growth factor CC-mediated neuroprotection against HIV Tat involves TRPC-mediated inactivation of GSK 3beta. PLoS One 7(10):e47572
Philipp S, Hambrecht J, Braslavski L, Schroth G, Freichel M, Murakami M, Cavalie A, Flockerzi V (1998) A novel capacitative calcium entry channel expressed in excitable cells. EMBO J 17:4274–4282
Pizzo P, Burgo A, Pozzan T, Fasolato C (2001) Role of capacitative calcium entry on glutamate-induced calcium influx in type-I rat cortical astrocytes. J Neurochem 79:98–109
Poteser M, Graziani A, Rosker C, Eder P, Derler I, Kahr H, Zhu MX, Romanin C, Groschner K (2006) TRPC3 and TRPC4 associate to form a redox-sensitive cation channel. Evidence for expression of native TRPC3-TRPC4 heteromeric channels in endothelial cells. J Biol Chem 281:13588–13595
Przedborski S, Tieu K, Perier C, Vila M (2004) MPTP as a mitochondrial neurotoxic model of Parkinson's disease. J Bioenerg Biomembr 36:375–379
Putney JW Jr (2003) Capacitative calcium entry in the nervous system. Cell Calcium 34:339–344
Putney JW Jr (2011) The physiological function of store-operated calcium entry. Neurochem Res 36:1157–1165
Putney JW Jr (2013) Alternative forms of the store-operated calcium entry mediators, STIM1 and Orai1. Curr Top Membr 71:109–123
Putney JW, Bird GS (2008) Cytoplasmic calcium oscillations and store-operated calcium influx. J Physiol 586:3055–3059
Roos J, DiGregorio PJ, Yeromin AV, Ohlsen K, Lioudyno M, Zhang S, Safrina O, Kozak JA, Wagner SL, Cahalan MD, Velicelebi G, Stauderman KA (2005) STIM1, an essential and conserved component of store-operated Ca2+ channel function. J Cell Biol 169:435–445
Sattler R, Tymianski M (2000) Molecular mechanisms of calcium-dependent excitotoxicity. J Mol Med 78:3–13
Self RL, Mulholland PJ, Nath A, Harris BR, Prendergast MA (2004) The human immunodeficiency virus type-1 transcription factor Tat produces elevations in intracellular Ca2+ that require function of an N-methyl-D-aspartate receptor polyamine-sensitive site. Brain Res 995:39–45
Selvaraj S, Watt JA, Singh BB (2009) TRPC1 inhibits apoptotic cell degeneration induced by dopaminergic neurotoxin MPTP/MPP(+). Cell Calcium 46:209–218
Selvaraj S, Sun Y, Watt JA, Song S, Lei S, Birnbaumer L, Singh BB (2012) Toxin-induced ER stress in mice involves down regulation of TRPC1 and inhibition of the AKT/mTOR signaling. J Clin Invest 122:1354–1367
Singh BB, Zheng C, Liu X, Lockwich T, Liao D, Zhu MX, Birnbaumer L, Ambudkar IS (2001) Trp1-dependent enhancement of salivary gland fluid secretion: role of store-operated calcium entry. FASEB J 15:1652–1654
Strehler EE, Treiman M (2004) Calcium pumps of plasma membrane and cell interior. Curr Mol Med 4:323–335

Stutzmann GE, Smith I, Caccamo A, Oddo S, Laferla FM, Parker I (2006) Enhanced ryanodine receptor recruitment contributes to Ca2+ disruptions in young, adult, and aged Alzheimer's disease mice. J Neurosci 26:5180–5189
Sukumaran P, Schaar A, Sun Y, Singh BB (2016) Functional role of TRP channels in modulating ER stress and autophagy. Cell Calcium 60(2):123–132
Sun Y, Sukumaran P, Singh BB (2014a) Physiological function and characterization of TRPCs in non-excitable and excitable cells. Cells 3:455–475
Sun S, Zhang H, Liu J, Popugaeva E, Xu NJ, Feske S, White CL, Bezprozvanny I (2014b) Reduced synaptic STIM2 expression and impaired store-operated calcium entry cause destabilization of mature spines in mutant presenilin mice. Neuron 82:79–93
Tu H, Nelson O, Bezprozvanny A, Wang Z, Lee SF, Hao YH, Serneels L, De Strooper B, Yu G, Bezprozvanny I (2006) Presenilins form ER Ca2+ leak channels, a function disrupted by familial Alzheimer's disease-linked mutations. Cell 126:981–993
Vandebrouck C, Martin D, Colson-Van Schoor M, Debaix H, Gailly P (2002) Involvement of TRPC in the abnormal calcium influx observed in dystrophic (mdx) mouse skeletal muscle fibers. J Cell Biol 158:1089–1096
Vandebrouck A, Ducret T, Basset O, Sebille S, Raymond G, Ruegg U, Gailly P, Cognard C, Constantin B (2006) Regulation of store-operated calcium entries and mitochondrial uptake by minidystrophin expression in cultured myotubes. FASEB J 20:136–138
van de Leemput J, Chandran J, Knight MA, Holtzclaw LA, Scholz S, Cookson MR, Houlden H, Gwinn-Hardy K, Fung HC, Lin X, Hernandez D, Simon-Sanchez J, Wood NW, Giunti P, Rafferty I, Hardy J, Storey E, Gardner RJ, Forrest SM, Fisher EM, Russell JT, Cai H, Singleton AB (2007) Deletion at ITPR1 underlies ataxia in mice and spinocerebellar ataxia 15 in humans. PLoS Genet 3:e108
Venkatachalam K, Montell C (2007) TRP channels. Annu Rev Biochem 76:387–417
Venkiteswaran G, Hasan G (2009) Intracellular Ca2+ signaling and store-operated Ca2+ entry are required in Drosophila neurons for flight. Proc Natl Acad Sci U S A 106:10326–10331
Vig PJ, Subramony SH, McDaniel DO (2001) Calcium homeostasis and spinocerebellar ataxia-1 (SCA-1). Brain Res Bull 56:221–225
Vig M, Peinelt C, Beck A, Koomoa DL, Rabah D, Koblan-Huberson M, Kraft S, Turner H, Fleig A, Penner R, Kinet JP (2006) CRACM1 is a plasma membrane protein essential for store-operated Ca2+ entry. Science 312:1220–1223
Wu X, Zagranichnaya TK, Gurda GT, Eves EM, Villereal ML (2004) A TRPC1/TRPC3-mediated increase in store-operated calcium entry is required for differentiation of H19-7 hippocampal neuronal cells. J Biol Chem 279:43392–43402
Yamamoto S, Shimizu S, Mori Y (2009) Involvement of TRPM2 channel in amplification of reactive oxygen species-induced signaling and chronic inflammation. Nippon Yakurigaku Zasshi 134:122–126
Yang KT, Chang WL, Yang PC, Chien CL, Lai MS, Su MJ, Wu ML (2006) Activation of the transient receptor potential M2 channel and poly(ADP-ribose) polymerase is involved in oxidative stress-induced cardiomyocyte death. Cell Death Differ 13:1815–1826
Yao H, Peng F, Dhillon N, Callen S, Bokhari S, Stehno-Bittel L, Ahmad SO, Wang JQ, Buch S (2009a) Involvement of TRPC channels in CCL2-mediated neuroprotection against tat toxicity. J Neurosci 29:1657–1669
Yao H, Peng F, Fan Y, Zhu X, Hu G, Buch SJ (2009b) TRPC channel-mediated neuroprotection by PDGF involves Pyk2/ERK/CREB pathway. Cell Death Differ 16:1681–1693
Ye S, Tan L, Ma J, Shi Q, Li J (2010) Polyunsaturated docosahexaenoic acid suppresses oxidative stress induced endothelial cell calcium influx by altering lipid composition in membrane caveolar rafts. Prostaglandins Leukot Essent Fatty Acids 83:37–43
Yoo AS, Cheng I, Chung S, Grenfell TZ, Lee H, Pack-Chung E, Handler M, Shen J, Xia W, Tesco G, Saunders AJ, Ding K, Frosch MP, Tanzi RE, Kim TW (2000) Presenilin-mediated modulation of capacitative calcium entry. Neuron 27:561–572

Yoshida T, Inoue R, Morii T, Takahashi N, Yamamoto S, Hara Y, Tominaga M, Shimizu S, Sato Y, Mori Y (2006) Nitric oxide activates TRP channels by cysteine S-nitrosylation. Nat Chem Biol 2:596–607

Yuan JP, Zeng W, Dorwart MR, Choi YJ, Worley PF, Muallem S (2009) SOAR and the polybasic STIM1 domains gate and regulate Orai channels. Nat Cell Biol 11:337–343

Zeng W, Yuan JP, Kim MS, Choi YJ, Huang GN, Worley PF, Muallem S (2008) STIM1 gates TRPC channels, but not Orai1, by electrostatic interaction. Mol Cell 32:439–448

Zhu X, Jiang M, Peyton M, Boulay G, Hurst R, Stefani E, Birnbaumer L (1996) Trp, a novel mammalian gene family essential for agonist-activated capacitative Ca2+ entry. Cell 85:661–671

Zuccato C, Cattaneo E (2007) Role of brain-derived neurotrophic factor in Huntington's disease. Prog Neurobiol 81:294–330

Zuccato C, Cattaneo E (2009) Brain-derived neurotrophic factor in neurodegenerative diseases. Nat Rev Neurol 5:311–322

Neurological and Motor Disorders: TRPC in the Skeletal Muscle

28

Sophie Saüc and Maud Frieden

Abstract

Transient receptor potential canonical (TRPC) channels belong to the large family of TRPs that are mostly nonselective cation channels with a great variety of gating mechanisms. TRPC are composed of seven members that can all be activated downstream of agonist-induced phospholipase C stimulation, but some members are also stretch-activated and/or are part of the store-operated Ca^{2+} entry (SOCE) pathway. Skeletal muscles generate contraction via an explosive increase of cytosolic Ca^{2+} concentration resulting almost exclusively from sarcoplasmic reticulum Ca^{2+} channel opening. Even if neglected for a long time, it is now commonly accepted that Ca^{2+} entry via SOCE and other routes is essential to sustain contractions of the skeletal muscle. In addition, Ca^{2+} influx is required during muscle regeneration, and alteration of the influx is associated with myopathies. In this chapter, we review the implication of TRPC channels at different stages of muscle regeneration, in adult muscle fibers, and discuss their implication in myopathies.

Keywords

TRPC • Skeletal muscle • Calcium entry • Muscle regeneration • Myopathies

28.1 Introduction

The skeletal muscle has a cell architecture dedicated to allow synchronized contractions along the entire fiber that can have very large dimensions (from micrometers to centimeters). To achieve this synchronism, cytosolic Ca^{2+} has to

S. Saüc • M. Frieden (✉)
Department of Cell Physiology and Metabolism, University of Geneva, 1 rue Michel Servet, 1211 Geneva, Switzerland
e-mail: sophie.Sauc@unige.ch; maud.frieden@unige.ch

K. Groschner et al. (eds.), *Store-Operated Ca^{2+} Entry (SOCE) Pathways*, Advances in Experimental Medicine and Biology 993, DOI 10.1007/978-3-319-57732-6_28

rise throughout the whole fiber in a uniform manner. A unique feature of striated muscles (cardiac and skeletal) is the presence of transverse tubules (t-tubules) that are deep plasma membrane (PM) invaginations perpendicular to the long axis of the fiber occurring at intervals of less than 2 μm along the entire length of the cells (Flucher and Franzini-Armstrong 1996). As well, the organization of the sarcoplasmic reticulum (SR) is very peculiar and consists of longitudinal parts surrounding the myofibrils (actin and myosin filaments) and terminal enlargements that come in close apposition to the t-tubules called the terminal cisternae. The SR membrane next to the t-tubules is the junctional SR (Sorrentino 2011), and this arrangement forms the triad where the excitation-contraction (EC) coupling takes place leading to the synchronous Ca^{2+} elevation required to trigger efficient contraction (Fig. 28.1). EC coupling starts by the arrival of the action potential down the t-tubules, which is sensed by the dihydropyridine receptor (DHPR), the voltage sensor of the system consisting of the Cav1.1 L-type Ca^{2+} channel. The action potential leads to a conformational change of DHPR that is transmitted to the

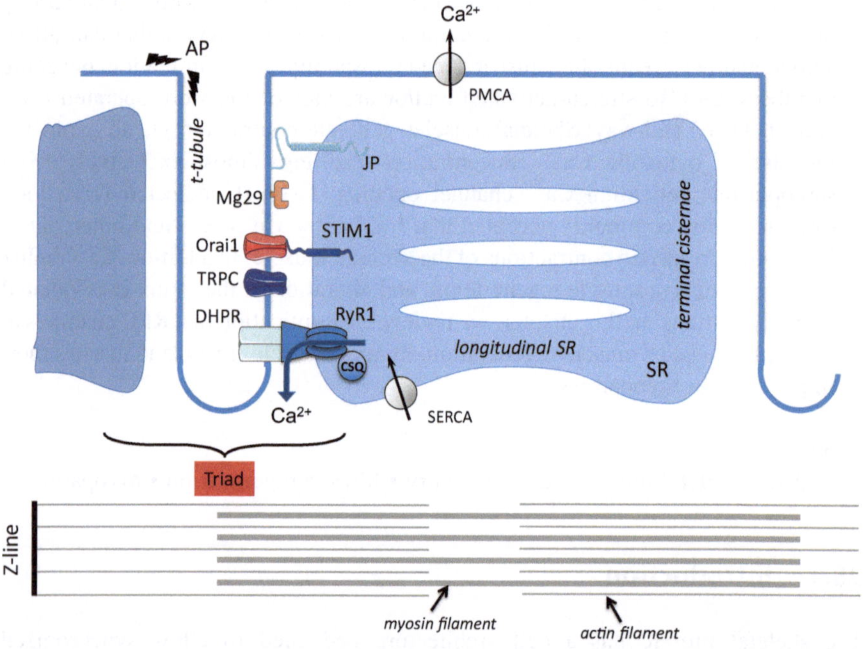

Fig. 28.1 Schematic representation of the spatial organization of the triad and the contractile apparatus. Several triadic proteins are represented: dihydropyridine receptor (DHPR) and mitsugumin 29 (Mg29) on the t-tubule and ryanodine receptor 1 (RyR1) and junctophilin (JP) on the terminal cisternae of the SR. STIM1 and Orai1 are depicted at the triad according to most reports investigating their localization. TRPC is represented on the t-tubule, even if depending on the TRPC and the stage of differentiation, they can be localized at the sarcolemma or in intracellular compartments. Calsequestrin (CSQ) and the Ca^{2+} pumps (SERCA and PMCA) are also depicted. Part of a sarcomere is represented, with one Z line and the actin and myosin filaments

ryanodine receptor 1 (RyR1) located on the junctional SR. The simultaneous openings of RyR1 allow an explosive Ca^{2+} release and the subsequent muscle contraction that ended by the decrease of cytosolic Ca^{2+} concentration. This is due to highly efficient Ca^{2+} repumping in the SR by the sarco-endoplasmic reticulum Ca^{2+} ATPase (SERCA) and to a lesser extends to Ca^{2+} extrusion via plasma membrane Ca^{2+} ATPase (PMCA) and Na^{+}/Ca^{2+} exchanger. The very brief skeletal muscle action potential (2–5 ms) precludes the opening of DHPR, and thus the Ca^{2+} needed for contraction comes almost exclusively from the SR (Dirksen 2009). Hence, for a very long time, it was assumed that external Ca^{2+} entry was for no need in skeletal muscle contraction, which was supported by the observation that several muscle twitches can occur in the absence of external Ca^{2+} likely due to the efficient recycling of Ca^{2+} back to the SR. In this context, the interest for store-operated Ca^{2+} entry (SOCE) in the skeletal muscle was minimal. The first report of SOCE in the skeletal muscle came in 2001 and the seminal paper of Kurebayashi (Kurebayashi and Ogawa 2001). It showed the requirement of Ca^{2+} entry in order to allow store refilling and sustained force development in the extensor digitorum longus (EDL) muscle upon repetitive electrical stimulations. Subsequent studies further highlighted the presence of Ca^{2+} entry pathways independent of DHPR in the skeletal muscle and their physiological relevance. The interest for SOCE gained tremendously after the identification of STIM/Orai and the finding that impaired SOCE has deleterious impact on muscle function. Besides the severe defect in immune function, loss-of-function mutations in STIM1 or Orai1 proteins lead to hypotonia in most patients harboring such mutations. The defect is congenital and characterized by muscle weakness and enhanced fatigability (McCarl et al. 2009; Lacruz and Feske 2015). In animal models as well, the absence of SOCE in the skeletal muscle results in mice more prone to fatigue that display a reduced muscle mass and force (Stiber et al. 2008a; Li et al. 2012; Wei-Lapierre et al. 2013). Interestingly, gain of function mutations in STIM1 or Orai1 results as well in muscular weakness, mainly associated with tubular aggregate myopathy syndrome (Bohm et al. 2014).

A particular feature of skeletal muscle SOCE is its kinetic of activation. In most cells, the machinery leading to SOCE, meaning the STIM1 aggregation, translocation to the PM, unfolding, and eventually Orai1 opening, takes tens of seconds to be completed (Prakriya and Lewis 2015). In the skeletal muscle, however, the process is very fast, less than a second according to some studies (Launikonis and Rios 2007; Launikonis et al. 2009; Edwards et al. 2010). Several aspects can account for this rapidity like the architecture of the SR that is permanently close to t-tubules at the triad, and thus STIM1 and Orai1 are close to each other even before store depletion. It was also proposed that the presence of an alternative splice variant of STIM1, called STIM1L (long isoform) that is highly expressed in the skeletal muscle, participated to the fast SOCE. Indeed, STIM1L was shown to colocalize with Orai1 at rest, and the onset of SOCE was significantly delayed in the absence of the long isoform (Darbellay et al. 2011). In addition to SOCE, mainly two other Ca^{2+} entry pathways were described in the skeletal muscle: the stretch-activated Ca^{2+} entry [SACE (Franco and Lansman 1990)] and the excitation-coupling Ca^{2+} entry (ECCE

(Cherednichenko et al. 2004)). ECCE is activated by repetitive electrical stimulations but is store independent and does not require STIM1 or Orai1. The channel(s) responsible for ECCE remains to be determined (Lyfenko and Dirksen 2008), but DHPR was postulated to be involved in this process (Bannister et al. 2009). SACE was shown to be an important player during muscle differentiation and, in adult, is likely responsible for abnormal Ca^{2+} entry in myopathies such as Duchenne muscular dystrophy. Several members of the transient receptor potential (TRP) family are activated by stretch (Maroto et al. 2005; Spassova et al. 2006) and thus are candidates for SACE in the skeletal muscle. Actually, TRPC channels were first described in the skeletal muscle about 15 years ago and were shown to be part of the SOCE pathway and likely involved as SACE channels (Vandebrouck et al. 2002). TRPC channels belong to the large family of TRP channels that comprises six subfamilies called TRPC (canonical), TRPV (vanilloid), TRPM (melastatin), TRPA (ankyrin), TRPP (polycystic), and TRPML (mucolipin). All have six transmembrane domains and assemble as tetramers. They are mainly nonselective cation channels with various Ca^{2+} selectivity ranging from high Ca^{2+} selectivity (TRPV5 and TRPV6) to Ca^{2+} impermeant (TRPM4 and TRPM5) (Nilius and Owsianik 2011). The mammalian TRPC family is composed of seven members (six in human, TRPC2 being a pseudogene) that are activated downstream to the phospholipase C (PLC) pathway. They contain three to four ankyrin repeats in the N-terminal side and a conserved TRP domain on the C-terminal side. According to their sequence homology, they are further subdivided into four categories, TRPC1, TRPC2, TRPC3/6/7, and TRPC4/5. Second messengers like diacylglycerol (DAG) directly activate TRPC3/6/7 (Hofmann et al. 1999; Dietrich et al. 2005). TRPC1 and TRPC4 were shown to be part of SOCE in several cell types but not all, while it is more debated for the other TRPCs (Ong et al. 2016). However, STIM1 was reported to directly bind TRPC1/4/5 and indirectly TRPC3/6, conferring them a SOCE function (Huang et al. 2006; Yuan et al. 2007; Zeng et al. 2008). As already mentioned, several TRP channels are mechanosensitive (Pedersen and Nilius 2007), and in particular TRPC1 was shown to be stretch activated (Maroto et al. 2005). Importantly, TRPC can form heteromers with other TRPCs or with TRP from other families, leading to change in their own properties both in terms of unitary conductance but also in their mode of activation. Consequently, it is frequently problematic to assign a particular function to specific members of the TRPC family.

28.1.1 Muscle Differentiation

The muscle has the ability to regenerate during the entire life, even if this capacity declines with age. The regeneration takes place following everyday life injuries but also in case of myopathies. There, the fibers are more fragile and get repeatedly damaged, which imposes their frequent repair. Hence, the regenerative capability of the skeletal muscle is more rapidly exhausted leading to a dramatic decline of muscle performance and ambulatory impairment (Allen et al. 2016). Muscle regeneration is possible thanks to the presence of a pool of satellite cells that are localized between the plasma membrane and the basal lamina of muscle fibers. These cells are muscle stem cells that are normally quiescent but get activated following an injury or a strong mechanical stress (Hawke and Garry 2001). The activation leads to cell proliferation as myoblasts followed by a cell cycle arrest and the beginning of the differentiation process. Eventually the cells fused together or with existing fibers in order to regenerate the lesion. The sequence of events taking place during regeneration is very well controlled and, at least for the initial steps, can be recapitulated in vitro. Myoblast differentiation requires the activation of mainly three families of transcription factors: the myogenic regulatory factor (MRF), myocyte enhancer factor 2 (MEF2), and nuclear factor of activated T-cells (NFAT). Ca^{2+} signals are of crucial importance for the activation of NFAT but also for MEF2 (Bentzinger et al. 2012). In myoblasts, the Ca^{2+} signals are mostly due to Ca^{2+} release from inositol-1,4,5-trisphosphate receptor (IP_3R) accompanied by Ca^{2+} entry through diverse PM channels not gated by voltage. Rapidly after the initiation of differentiation, the RyR1 and DHPR are expressed and participate to Ca^{2+} signals, even though initially the EC coupling takes place at the PM rather than at the triad, whose structure is established later on during the differentiation (Flucher and Franzini-Armstrong 1996). Impairment of SOCE due to alteration of the STIM1/Orai1 machinery led to defect of differentiation (Stiber et al. 2008a; Darbellay et al. 2009). As well, several TRPCs were shown to be required for a proper differentiation program, and we will discuss their involvement in Sect. 28.3, but first we will consider the role of TRPC in adult fibers and myotubes.

28.2 Role of TRPC in Myotubes and Adult Fibers

In 2002, the first paper describing TRPC channels in mouse skeletal muscle was published in which TRPC1/2/3/4 and TRPC6 were detected both at the transcript and protein levels, with TRPC2 and TRPC3 being found in intracellular compartments, and TRPC1/4 and TRPC6 at the PM (Vandebrouck et al. 2002). Few years later, the group of Brinkmeier investigated the expression profile of all TRPs in mouse tissues using standard and real-time RT-PCR and in situ hybridization techniques (Kunert-Keil et al. 2006). Among the TRPCs, the authors found that TRPC3 was the most expressed, followed by TRPC1 and TRPC6. This study also highlighted that the expression profile of the channels might differ according to the mouse strain considered (C57Bl/10, NOD, and Balb/c). A follow-up study focusing on the skeletal muscle reported that TRPC3 and TRPC6 are mainly expressed (both at mRNA and protein levels; Kruger et al. 2008), while in human skeletal muscle, the transcripts of TRPC1/3/4/5 and TRPC6 were found (Riccio et al. 2002). A similar expression profile except for TRPC5 was reported in human myotubes (Antigny et al. 2013). Surprisingly, the vast majority of studies investigating the role of TRPC in the skeletal muscle focused on TRPC1, TRPC3, and TRPC4, whereas almost nothing is known regarding the function of TRPC5 and specially TRPC6 that is abundant in the muscle.

In adult mouse muscle fibers, several reports demonstrated that TRPC1 is localized at the PM. Such results come either from electrical measurements of the channel activity (Vandebrouck et al. 2002, 2007; Ducret et al. 2006; Zanou et al. 2010) or from immunostaining of endogenous TRPC1 (Vandebrouck et al. 2002; Stiber et al. 2008b). On the contrary, in 2009, Berbey reported that TRPC1 was mainly localized on the longitudinal part of the SR (Berbey et al. 2009). The endogenous TRPC1 and the overexpressed YFP-TRPC1 gave similar intracellular localization, and it was thus postulated that TRPC1 worked as a SR leak channel. This study was however debated 1 year later, as the group of Gailly showed, using TRPC1$^{-/-}$ mice, that the antibody used in the Berbey's report was not specific (Tajeddine et al. 2010), and thus raised questions about the localization of TRPC1. In other tissues, TRPC1 was shown to be intracellularly located (Nesin and Tsiokas 2014), and in the skeletal muscle, other members like TRPC3 were found in intracellular compartments (Vandebrouck et al. 2002). Nevertheless, it appears that in the muscle at least an important fraction of TRPC1 is at the PM and plays a role in Ca^{2+} entry.

It is generally accepted that TRPC1 can function as mechanosensitive channel. In the skeletal muscle, several papers using patch-clamp recordings or Ca^{2+} imaging experiments identified TRPC1 as being activated by mechanical stretch of the PM (Formigli et al. 2005; Ducret et al. 2006; Louis et al. 2008). A dysregulation of this process is part of the alteration observed in dystrophies (see Sect. 28.4). Using the TRPC1$^{-/-}$ mouse model, Zanou published a comprehensive study on the role of TRPC1 in the skeletal muscle (Zanou et al. 2010). Interestingly, using single-channel recordings, the authors claimed that TRPC1 was neither activated by mechanical stretch nor by store depletion. Using Mn^{2+} quench technique to assess Ca^{2+} entry, they further confirmed no alteration of SOCE in TRPC1$^{-/-}$ muscle. The size of the muscles was smaller in TRPC1$^{-/-}$ animals showing an involvement of TRPC1 in myogenesis that is in line with other observations (see Sect. 28.3). The knockout animals performed like wild type when they were subject to free running, while they did not resist to fatigue upon strenuous exercise. The decrease fatigue resistance was further confirmed on isolated soleus muscle (Zanou et al. 2010). The susceptibility to fatigue of TRPC1$^{-/-}$ animals is an interesting finding for two reasons: first because this phenotype is reminiscent from the STIM1- and/or Orai1-deficient animals, which would suggest that these molecules are working together in the same signaling pathway, like it was shown in salivary glands (Ambudkar 2014). The second aspect is precisely the fact that the group of Gailly did not see TRPC1 as being part of SOCE (Zanou et al. 2010). Hence, it could be that STIM1/Orai1 are working independently from TRPC1 in the muscle but that both Ca^{2+} entry pathways are important in preventing muscle fatigue. More studies are required to solve the molecular arrangements of these channels. As well, the mode of activation of TRPC1 in adult remains to be determined as neither store depletion nor mechanical stretch appeared to open the channel in adult muscle fibers (Zanou et al. 2010), which is different from what was reported on immature fibers (see Sect. 28.3) and in other cellular systems (Maroto et al. 2005). In murine myotubes, TRPC1 was reported to form stable complex with both dystrophin and α1-syntrophin, two key molecules found at the costameres (the molecular complex that bridges the myofilaments to the PM) and being involved in the membrane stability (Vandebrouck et al. 2007). TRPC4 was later on shown to be part of the same complex (Sabourin et al. 2009). The overexpression of TRPC1/TRPC4 enhanced SOCE in myotubes, and the downregulation of α1-syntrophin increased Ca^{2+} influx dependent on TRPC1 and TRPC4 (Sabourin et al. 2009). Hence, it emerged that the proper function of these channels is controlled, at least partially, by their anchoring to a macromolecular complex at the costameres. It is likely that not all TRPCs are part of macromolecular complex and that other pools of the channels might be localized at other places like the triad, but this remains to be determined.

The first paper describing TRPC3 in the muscle showed an intracellular localization of the channel in myotubes (Vandebrouck et al. 2002). Later on, it was reported that TRPC3 was more expressed in slow than in fast muscle fibers and that it participated to Ca^{2+} influx, a finding in favor of a PM localization (Rosenberg et al. 2004). Interestingly, the expression of TRPC3 increased after free running exercise only in slow fibers that are the ones required preferentially in such training. The overexpression of TRPC3 enhanced NFAT activity that in turn appeared to control the expression level of TRPC3. It was thus proposed that neuromuscular activity in slow fibers promotes TRPC3 expression that by activating the calcineurin-NFAT pathway participates to the maintenance of the slow oxidative phenotype. Overexpression of the channel however was not associated with SOCE enhancement, neither it altered the EC coupling. This is different to what was reported in myotubes, where siTRPC3 decreased EC coupling (Woo et al. 2008) (see Sect. 28.3). Hence, it is likely that during the maturation of the fibers, TRPC3 modulates EC coupling, while in adult fibers, the function of TRPC3 is modified. For instance, Lanner et al. (2009) revealed that TRPC3-induced Ca^{2+} entry is an important regulator of insulin-induced glucose uptake in adult skeletal muscle. This glucose uptake depends on the proper translocation of the glucose transporter 4 (GLUT4) from internal vesicles toward the PM. Indeed, GLUT4 and TRPC3 colocalized at the level of t-tubules, the preferential site of glucose uptake in skeletal muscles (Lauritzen et al. 2006). It emerged however that TRPC3 was not directly activated following insulin receptor stimulation but was activated by DAG and modulated the response to insulin. The conclusion raised by the authors was that a pool of TRPC3 is permanently expressed at the PM and that a vesicular pool associated with GLUT4 translocates to the PM in response to insulin.

Finally, it was reported that TRPC1 and TRPC3 associated in the skeletal muscle. Woo determined that the interaction site resides in the ankyrin repeat region of TRPC3 and that disruption of this interaction led to a decreased basal Ca^{2+} level in myotubes (Woo et al. 2014). This is in line with another paper showing an interaction between TRPC1 and TRPC3 in myotubes by co-immunoprecipitation (Cheung et al. 2011). The authors did not investigate the functional role of the heteromers but interestingly found that after few days of differentiation, these TRPCs are significantly more expressed in myotubes compared to the unfused cells, called reserve cells (Cheung et al. 2011). This differential expression profile would be very interesting to understand in the context of muscle regeneration and the role of TRPCs in myotubes versus reserve cells.

28.3 Role of TRPC in Muscle Differentiation

The expression level of TRPC1 increased during differentiation with a peak at around 24 h followed by a decline (Louis et al. 2008; Formigli et al. 2009). This transient elevation is accompanied by an increase of basal Ca^{2+} entry assessed by Mn^{2+} quench that was blocked by GsMTx4 (Louis et al. 2008), a spider venom toxin specific of SACE channels (Suchyna et al. 2000). This elevated basal Ca^{2+} entry was also reduced by siTRPC1, which as well reduced SOCE in myoblasts (Louis et al. 2008). Consistently, decreasing the expression level of TRPC1 significantly impaired muscle differentiation. The same study reported a defect of myoblast fusion related to an impairment of migration and alignment of the cells prior fusion. As well, the activation of calpain, a family of Ca^{2+}-dependent cysteine proteases implicated in fusion (Leloup et al. 2006), was impaired in siTRPC1-treated cells. The lack of migration and alignment was further confirmed using cultured $TRPC1^{-/-}$ myoblasts but also in $TRPC1^{-/-}$ mice model where the regeneration of the muscle after cardiotoxin-induced muscle lesion was evaluated. The delayed muscle regeneration in $TRPC1^{-/-}$ mice correlated with a decreased expression of the transcription factors Myf5, MyoD, and myogenin. In addition, the Akt/mTOR signaling pathway was severely downregulated in regenerating muscles from $TRPC1^{-/-}$ mice (Zanou et al. 2012). The group of Maecci also reported the role of TRPC1 in sphingosine 1-phosphate (S1P)-stimulated myogenesis in C2C12, a mouse myoblast cell line (Formigli et al. 2009). S1P is a bioactive lipid with numerous biological actions including pro-myogenic effects (Squecco et al. 2006) and was shown to activate SACE channels (Formigli et al. 2005). In line, Formigli et al. reported that TRPC1 were activated by stretch and stimulated by S1P. The channels were found in lipid rafts, and cholesterol depletion altered TRPC1 localization and drastically reduced S1P-induced current. Furthermore, siTRPC1 had a strong impact on differentiation, as almost no fusion event took place after 2–3 days of differentiation (Formigli et al. 2009). Connexins, and in particular Cx43, are implicated during myogenesis (Araya et al. 2005; Squecco et al. 2006). In C2C12, S1P-activated TRPC1 led to Ca^{2+} elevation that together with calpain activation favored the expression of Cx43 (Meacci et al. 2010). All these stimulated pathways in turn promoted muscle differentiation and fusion. Interestingly, the overexpression of TRPC1 while enhancing SOCE in myotubes delayed muscle differentiation leading to the formation of small myotubes (Olah et al. 2011). To note, the overexpression of TRPC1 led to a decrease in STIM1 expression, which could represent a compensatory mechanism that attenuated Ca^{2+} overload and might explain the delayed differentiation (Olah et al. 2011). This finding is however in contradiction with data obtained on human muscle differentiation, where the overexpression of TRPC1 accelerated the process of myogenesis, together with an increase of SOCE (Antigny et al. 2013). In line with the data obtained with mouse cells, the invalidation of TRPC1 in human myoblasts affected differentiation, mainly by altering the fusion process leading to myotubes of smaller size (Antigny et al. 2013). In a technically challenging paper, Liu measured the Ca^{2+} signals in satellite cells in situ, meaning in their normal environment between the PM and the basal lamina (Liu and Schneider 2014). They assessed satellite cell Ca^{2+} elevation in response to fibroblast growth factor 2 (FGF2) that was associated with an increased number of MyoD-positive cells and of NFATc2/c3 nuclear translocation, both hallmarks of satellite cell activation. In parallel, the authors reported a strong expression of TRPC1 in satellite cells and an important decrease of the Ca^{2+} response by SKF-96365, a non-specific TRPC channel blocker, which led them postulating that TRPC1 is implicated in FGF2-induced Ca^{2+} signals. This is an attractive hypothesis but that would need to be confirmed with more specific tools than SKF-96365. Recently, a study investigating the process of muscle reloading after disuse clearly showed that TRPC1 is necessary for the recovering phase (Xia et al. 2016). Indeed, during the unloading period, TRPC1 expression decreased (Zhang et al. 2014) and upon reloading increased again to reach the control level after 28 days. In case of siRNA treatment against TRPC1 during the reloading period, the muscle regeneration was significantly delayed. The authors further showed that inhibition of the calcineurin-NFAT pathway downregulated TRPC1 expression, which would imply a link between calcineurin and TRPC1 during muscle regrowth (Xia et al. 2016). Simulated microgravity that mimics muscle disuse resulted also in downregulation of TRPC1 expression. In addition, myoblast proliferation slowed down under this condition, an effect recapitulated by SKF-96365, which could involve TRPC1 as an important regulator of cell cycle (Benavides Damm et al. 2013). Altogether, these studies implicated TRPC1 as a channel necessary for the proper muscle differentiation, both in human and murine models. The channel is likely activated by stretch but also upon store depletion, at least during myogenesis, and participates to the establishment of the differentiation program, from satellite cell activation to myoblast fusion.

The expression level of TRPC3 is relatively low in myoblasts and rapidly increases after the initiation of differentiation. It then declines and stabilizes in myotubes at level above the one found in myoblasts (Cheung et al. 2011; Lee et al. 2006; Woo et al. 2008). However, Lee did not report a major alteration of the differentiation process in the absence of TRPC3 (Lee et al. 2006). In line, SOCE was not altered by TRPC3 silencing, but as TRPC1 was upregulated in this condition, a compensatory mechanism could have taken place leading to an apparent normal SOCE (Lee et al. 2006). In other cellular systems, the implication of TRPC3 in SOCE is a controversial issue that might arise from the cell type considered but also be related to the expression level of TRPC3 (Lichtenegger and Groschner 2014). After few days of differentiation, the expression level of several triadic proteins like triadin, calsequestrin (CSQ), and junctophilin 1 (JP1) was upregulated in myotube knockdown (KD) for TRPC3, suggesting an implication of TRPC3 in the control of triadic protein expression. It was reported that in myotubes, while not in adult fibers (see Sect. 28.2), the EC coupling assessed by KCl pushes was decreased in TRPC3 KD cells. Neither the DHPR/RyR1 levels nor the SR Ca^{2+} content was affected, but the activity of RyR1 appeared diminished. A direct interaction between RyR1 and TRPC3 could however not be demonstrated, while TRPC3 directly interacted with TRPC1, JP2, calmodulin (CaM), Homer, calreticulin, and mitsugumin 29 (Mg29), which might constitute a link between TRPC3 and RyR1 (Woo et al. 2008). The interaction site between Mg29 and TRPC3 was determined to be on the N-terminal side of Mg29, and the disruption of the interaction between both proteins resulted in reduced Ca^{2+} transients upon KCl stimulation (Woo et al. 2015). Hence, it appeared that in myotubes, TRPC3 modulated the efficiency of EC coupling, likely through interaction with other proteins like Mg29 or RyR1. In parallel study, the same group investigated the role of TRPC3 independently from its involvement in EC coupling, by generating TRPC3 and DHPR double KD myoblast cell line. Upon differentiation, the phenotype was dramatic with the induction of apoptosis associated with an increased SR Ca^{2+} content and a higher basal cytosolic Ca^{2+} concentration (Woo et al. 2010). As such important effect on the differentiation was not observed, neither in TRPC3 KD nor DHPR KD cells, it was proposed that both Ca^{2+} entry channels constituted a redundant pathway, at least during the course of myogenesis (Woo et al. 2010). The authors thus attributed a role for TRPC3 during differentiation, which probably needs further studies to be confirmed. It also remains to be determined whether the partners of TRPC3 and its functional interaction with RyR1 are taking place in adult tissue or whether it is an intermediate state occurring during the differentiation process only. According to the lack of impact of TRPC3 invalidation in EC coupling in adult fibers (Rosenberg et al. 2004; Lanner et al. 2009), one can postulate that the interaction partners of TRPC3 indeed change during the process of differentiation. Overall, a clear picture regarding the role of TRPC3 in muscle development is hard to define, due to the multiple direct or indirect partners of TRPC3, and because the genetic KD of TRPC3 alters the expression level of several other proteins involved in Ca^{2+} homeostasis. One should mention that TRPC3 knockout mice displayed a locomotor defect, but this defect was linked to alterations at the level of the Purkinje cells of the cerebellum that express high amount of TRPC3 and that strongly contributes to the movement coordination (Hartmann et al. 2008). Finally, in human, muscle differentiation was altered upon invalidation of TRPC4, associated with a reduction of SOCE. Furthermore, MEF2 expression level was decreased, and the fusion process was impaired, leading to the formation of smaller-sized myotubes. A very similar phenotype was observed upon TRPC1 invalidation, highlighting the role of these channels in the late fusion events in human myogenesis (Antigny et al. 2013).

28.4 TRPC and Muscle Diseases

In spite of the absence of identified mutations in TRPC coding genes linked to inherited muscle diseases, likely due to overlapping function between different TRPCs (Nilius and Owsianik 2010), numerous lines of evidence have revealed the implication of these channels in the pathogenesis of several muscle diseases and especially in muscular dystrophies. Muscular dystrophies are inherited diseases caused by gene mutations of proteins forming the membrane anchored dystrophin-associated protein complex (DAPC) linking the extracellular matrix to the cytoskeleton. The most studied dystrophy is the Duchenne muscular dystrophy (DMD) due to mutation in the dystrophin gene itself and characterized by progressive muscle degeneration associated with shorter life span, for review (Vallejo-Illarramendi et al. 2014). One hypothesis to explain muscle wasting in DMD is the activation of Ca^{2+}-dependent proteases such as calpains and reactive oxygen species (ROS) production because of increased cytoplasmic and mitochondrial Ca^{2+} concentration, respectively (Gailly et al. 2007; Spencer et al. 1995; Allen et al. 2016). Calpain activation as well as ROS production would lead to cycle of muscle fiber necrosis followed by regeneration, exhausting the pool of satellite cells and subsequently decreasing the amount of muscle fibers. Despite some studies with muscle fibers from *mdx* mice (the animal model of DMD) or differentiated human DMD myotubes reporting normal cytoplasmic Ca^{2+} level at rest (Gailly et al. 1993; Head 1993; Pressmar et al. 1994; Collet et al. 1999), the general consensus is that dystrophin-deficient myofibers or myotubes, when submitted to contractions, maintain persistent elevated intracellular Ca^{2+} concentration (Turner et al. 1988; Bakker et al. 1993; Imbert et al. 1995). Whether Ca^{2+} entry would be sufficient to induce a complete dystrophic phenotype was proved definitively in 2009 by Millay and co-workers (Millay et al. 2009). In their study, they used a constitutively active form of TRPC3 overexpressed specifically in muscles which induced a muscular dystrophy syndrome with small muscle fibers, central nucleation, and fibrosis, all hallmarks of dystrophic fibers, as well as mRNA expression profile similar to those of $Scgd^{-/-}$ mice (δ-sarcoglycan deleted mouse model for muscular dystrophy) (Millay et al. 2009). In addition, overexpression of a dominant-negative form of TRPC6 (dnTRPC6) in this TRPC3 overexpression background reverted the dystrophic phenotype both in *mdx* and in $Scgd^{-/-}$ mouse, likely due to formation of hetero-tetramers between endogenous TRPC and dnTRPC6. Such inactive hetero-tetramers by inhibiting TRPC-dependent Ca^{2+} overload in two different mouse models for dystrophy suggested that TRPC dysregulation caused the disease (Millay et al. 2009). TRPC were selected as responsible of dystrophy, as previous work has highlighted their dysregulation in *mdx* mice (Vandebrouck et al. 2002). Actually, as the Ca^{2+} defect in DMD and *mdx* fibers was revealed only when they contracted, one of the first candidates that was investigated for deregulated Ca^{2+} entry was stretch-activated Ca^{2+} channels (SAC) (Yeung et al. 2005). Since 1994, it was known that SAC channels had higher activity in *mdx* fibers than in WT (Franco-Obregon and Lansman 1994). These results were confirmed a decade later when streptomycin and GsMTx4 (both blockers of SAC) were applied to *mdx* fibers in order to prevent the rise of resting intracellular Ca^{2+} after tetanic stimulations. In this study, Yeung et al. (2005) also showed that supplementation of drinking water of *mdx* mice with streptomycin decreased the number of muscle fibers with central nuclei. In parallel, the group of Gailly demonstrated that SAC and SOCE channels expressed in muscle fibers shared similar characteristics as unitary conductance (between 7 and 8 pS with 110 mM Ca^{2+} in the pipette), current-voltage relationship, and drug sensitivity (to Gd^{3+}, SKF-96365, 2-aminoethoxydiphenyl borate, and GsMTx4) (Ducret et al. 2006). These results reinforced the previous discovery 3 years before that TRPC channels are part of the pathway leading to Ca^{2+} overload in *mdx* muscle fibers (Vandebrouck et al. 2002). In addition, Vandebrouck et al. (2002) showed that the occurrence of the basal, the Tg-induced, and the caffeine-induced higher Ca^{2+} current seen in *mdx*-derived muscle fibers was significantly reduced when TRPC1 and TRPC4 were silenced, without affecting their other characteristics (open probability, open and close time constant, activation by store depletion). That was the first study suggesting that TRPC1 and TRPC4 might have a role in DMD etiology, but the link between DAPC alteration and TRPC1 and TRPC4 channel overactivation remained unclear until Constantin's works (Vandebrouck et al. 2007; Sabourin et al. 2009). Co-immunoprecipitation experiments of muscle extracts from WT mice revealed that TRPC1, TRPC4, and dystrophin were part of the same complex through their shared interaction with α1-syntrophin, a signaling protein of the DAPC. In *mdx* fibers, this interaction was decreased due to reduction of the amount of α1-syntrophin at the PM. In addition, GST pulldown assay allowed them to identify the PDZ (PSD95-disk large-zonula occludens protein) domain of α1-syntrophin as the interacting domain with both TRPC1 and TRPC4. Finally, functional studies confirmed that TRPC1, TRPC4, and α1-syntrophin are working together. Whereas *mdx*-derived myotubes displayed two times higher SOCE when measured with Mn^{2+} quench assay, these myotubes transfected with recombinant α1-syntrophin showed a reduction of the Ca^{2+} influx rate. On the contrary, myotubes with functional mini-dystrophin (a functional shorter form of the dystrophin easier to transfect) when transfected with siRNA against α1-syntrophin showed enhanced Ca^{2+}, but this effect was reversed by additional silencing of either TRPC1 or TRPC4 (Vandebrouck et al. 2007; Sabourin et al. 2009). So the increased Ca^{2+} entry typical of dystrophic muscle cells would be due at least in part to defective TRPC1 and TRPC4 regulation because of the destabilized DAPC leading to decreased amount of α1-syntrophin signaling at the PM. In addition to α1-syntrophin, other scaffolding proteins such as Homer 1 (Stiber et al. 2008b) and caveolin-1 (Gervasio et al. 2008) have been shown to interact with TRPC1. Stiber et al. in 2008 while studying *Homer* $1^{-/-}$ mice detected an increased whole-cell basal current blocked by GsMTx4 and an increased basal Ca^{2+} entry, measured by Ca^{2+} imaging lost after TRPC1 silencing (Stiber et al. 2008b). The *Homer* $1^{-/-}$ mouse displayed dystrophic phenotype with diminished cross-sectional area of muscle fibers associated with reduced contractile force. The authors ruled out a possible alteration of EC coupling and neuromuscular junction formation defect as the leading cause of reduced skeletal muscle contractility but attributed it to TRPC1-dependent Ca^{2+} influx dysregulation. They showed partial colocalization in adult muscle between Homer 1 and TRPC1 as well as co-immunoprecipitation of the two proteins. They also detected a significant decrease in membrane stiffness probably due to increase of calpain activity assessed by upregulation of calpastatin (an endogenous calpain inhibitor that is upregulated in case of excessive calpain activity). The authors proposed Homer 1 as a regulatory protein of the mechanosensitive TRPC1 channel. In the absence of Homer 1, the opening probability of TRPC1 would increase inducing Ca^{2+} overload and subsequent muscle necrosis (Stiber et al. 2008b). Finally, the last pathway identified in regulating TRPC1 activity in dystrophic fibers was the ROS-Src1-caveolin-3 pathway (Gervasio et al. 2008). Caveolin-3 is the muscle-specific isoform of caveolin proteins necessary for the formation of caveolae which are membrane invaginations rich in particular lipids and implicated in cell signaling. Src kinase is known to be activated by ROS and interacts with TRPC3 (Kawasaki et al. 2006) and caveolin-3 (Li et al. 1996). Gervasio showed that the expression of caveolin-3, TRPC1, and Src kinase all increased in *mdx* mice (Gervasio et al. 2008). In addition, co-immunoprecipitation, FRET, and colocalization experiments showed that TRPC1 and caveolin interacted in C2 myoblast cell line. The authors demonstrated that application of H_2O_2 induced Src activation, which in turn promoted on the one hand TRPC1 and caveolin interaction and on the other hand TRPC1-caveolin-dependent Ca^{2+} influx both in C2 myoblasts and isolated *mdx* muscle fibers. Moreover, inhibition of Src by the ROS scavenger Tiron decreased SAC activity and thus Ca^{2+} influx in *mdx* muscle fibers preventing stretch-induced muscle damage. Thus, ROS production would be a consequence but also a cause of the Ca^{2+} overload, via activation of TRPC1, which would accelerate muscle damage of the dystrophic muscle. However, whether or not TRPC expression is altered in dystrophic muscle fibers is not proved unequivocally. Whereas Kruger reported no major variations or even a decrease of the mRNA level of the different TRPC in *mdx* compared to WT mice skeletal muscle, Matsumura detected higher TRPC1 expression in the diaphragm of *mdx* mice which correlates with the more pronounced wasting of respiratory muscles compared to limb muscles (Kruger et al. 2008; Matsumura et al. 2011). The discrepancy between these results might come from the heterogeneity between age, muscle type, gender, or even species (human compared to mice). Hence, it is likely that in muscle fibers, a macromolecular complex encompassing dystrophin and other scaffolding proteins (caveolin-3, Homer 1, α1-syntrophin), localized at the costameres, regulates the opening of SAC channels (which can also work as SOCE channels). This complex controls Ca^{2+} influx that occurs upon contractions, especially eccentric contractions that are the one implying stretching of the sarcolemma together with shortening. Indeed, eccentric contractions are the most deleterious exercise for dystrophic fibers likely because the DAPC complex is altered and not capable of maintaining TRPC channels in a closed conformation when fibers contract (Fig. 28.2).

In addition to dystrophy, TRPCs are implicated in the etiology of another muscle disease: myasthenia gravis (MG). MG is a rare autoimmune disease characterized by fatigability and muscle weakness as a result of neuromuscular junction defect mainly due to autoantibody directed against acetylcholine receptors. However,

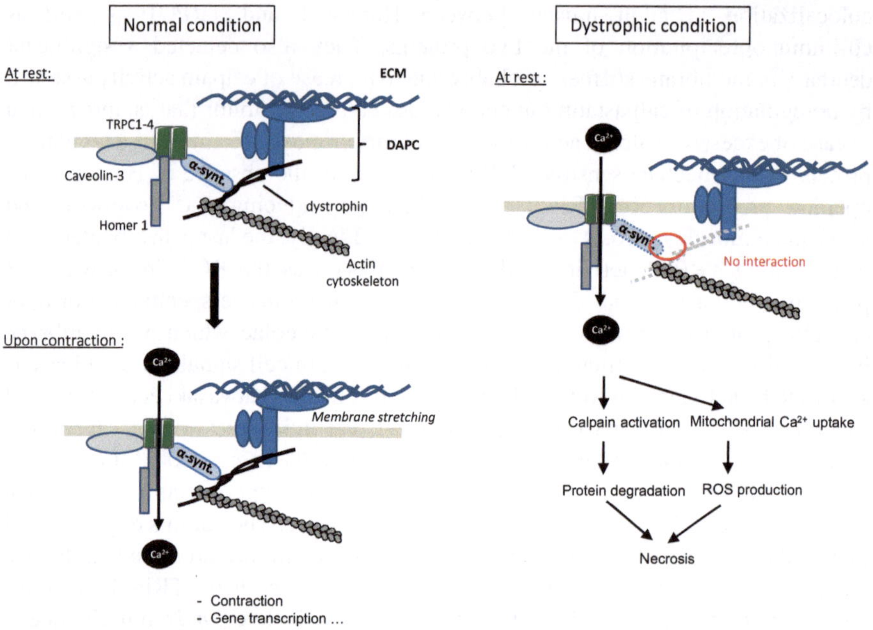

Fig. 28.2 Hypothetical model for Ca^{2+} overload in muscle dystrophic fibers. In normal condition, the dystrophin-associated protein complex (DAPC) stabilizes the sarcolemma during contraction by linking the actin cytoskeleton to the extracellular matrix (ECM). Several scaffolding proteins such as α1-syntrophin (α-synt.), Homer 1, and caveolin-3 are associated with this complex, and their interaction with TRPC1–4 channel heteromers allows the maintenance of the channels in a close conformation. Upon contractions, the stretching of the membrane likely modifies these interactions, which allows TRPC opening and Ca^{2+} entry. In dystrophic fibers, the absence of dystrophin or other proteins of the DAPC weakens the interactions between TRPC and its DAPC-associated regulators. The channels are no more regulated, and their opening probability increases inducing Ca^{2+} overload and activation of the downstream pathways leading to muscle necrosis

some MG patients presented anti-RyR1 antibodies inducing EC coupling deficiency (Romi et al. 2007). As TRPC3 was shown to interact with RyR1 and might have a role in EC function (Woo et al. 2008; see Sect. 28.3), the presence of anti-TRPC3 antibodies was tested and detected (in addition to anti-RyR1) in the serum, of several MG patients (Takamori 2008). The authors suspected the more severe MG phenotype of those patients to be due to the combined TRPC3 and RyR1 antibodies in their serum, thus highlighting the importance of TRPC3 in proper muscle function.

28.5 Conclusion

The interest for Ca^{2+} entry in skeletal muscle function is rather recent, as for a long time it was assumed that the internal Ca^{2+} recycling was efficient enough to avoid decrease of SR Ca^{2+} content upon contractions. It is now accepted that the skeletal muscle required Ca^{2+} influx for their proper function, and thus the interest for the different routes allowing Ca^{2+} entry is growing. Besides the prototypic SOCE players STIM1 and Orai1, the TRPCs, which are non-voltage-gated cation channels, are also implicated in several functions of the skeletal muscle. These channels are polymodal in their mode of activation, which confers them a great variety of responses but complicates the study of their involvement in specific functions. However, it emerges that skeletal muscle differentiation requires Ca^{2+} entry and TRPC channels, especially TRPC1. TRPC3 is also implicated while to a lesser degree, and TRPC4 was only reported to be important for human myogenesis. In adult tissue, the same three TRPCs are involved in different physiological processes, and interestingly, their mode of activation and/or their binding partners appeared to change during the time of differentiation. In myopathies like DMD, the absence of dystrophin led to an alteration of TRPC1/TRPC4 regulation resulting in unregulated Ca^{2+} entry that is deleterious for the muscle fibers. Hence, TRPC could be potential therapeutic targets to slow down the progression of this disease.

References

Allen DG, Whitehead NP, Froehner SC (2016) Absence of dystrophin disrupts skeletal muscle signaling: roles of Ca2+, reactive oxygen species, and nitric oxide in the development of muscular dystrophy. Physiol Rev 96:253–305

Ambudkar IS (2014) Ca(2+) signaling and regulation of fluid secretion in salivary gland acinar cells. Cell Calcium 55:297–305

Antigny F, Koenig S, Bernheim L, Frieden M (2013) During post-natal human myogenesis, normal myotube size requires TRPC1- and TRPC4-mediated Ca(2+) entry. J Cell Sci 126:2525–2533

Araya R, Eckardt D, Maxeiner S, Kruger O, Theis M, Willecke K, Saez JC (2005) Expression of connexins during differentiation and regeneration of skeletal muscle: functional relevance of connexin43. J Cell Sci 118:27–37

Bakker AJ, Head SI, Williams DA, Stephenson DG (1993) Ca2+ levels in myotubes grown from the skeletal muscle of dystrophic (mdx) and normal mice. J Physiol 460:1–13

Bannister RA, Pessah IN, Beam KG (2009) The skeletal L-type Ca(2+) current is a major contributor to excitation-coupled Ca(2+) entry. J Gen Physiol 133:79–91

Benavides Damm T, Richard S, Tanner S, Wyss F, Egli M, Franco-Obregon A (2013) Calcium-dependent deceleration of the cell cycle in muscle cells by simulated microgravity. FASEB J 27:2045–2054

Bentzinger CF, Wang YX, Rudnicki MA (2012) Building muscle: molecular regulation of myogenesis. Cold Spring Harb Perspect Biol 4(2). doi:10.1101/cshperspect.a008342

Berbey C, Weiss N, Legrand C, Allard B (2009) Transient receptor potential canonical type 1 (TRPC1) operates as a sarcoplasmic reticulum calcium leak channel in skeletal muscle. J Biol Chem 284:36387–36394

Bohm J, Chevessier F, Koch C, Peche GA, Mora M, Morandi L, Pasanisi B, Moroni I, Tasca G, Fattori F, Ricci E, Penisson-Besnier I, Nadaj-Pakleza A, Fardeau M, Joshi PR, Deschauer M, Romero NB, Eymard B, Laporte J (2014) Clinical, histological and genetic characterisation of patients with tubular aggregate myopathy caused by mutations in STIM1. J Med Genet 51:824–833

Cherednichenko G, Hurne AM, Fessenden JD, Lee EH, Allen PD, Beam KG, Pessah IN (2004) Conformational activation of Ca2+ entry by depolarization of skeletal myotubes. Proc Natl Acad Sci U S A 101:15793–15798

Cheung KK, Yeung SS, Au SW, Lam LS, Dai ZQ, Li YH, Yeung EW (2011) Expression and association of TRPC1 with TRPC3 during skeletal myogenesis. Muscle Nerve 44:358–365

Collet C, Allard B, Tourneur Y, Jacquemond V (1999) Intracellular calcium signals measured with indo-1 in isolated skeletal muscle fibres from control and mdx mice. J Physiol 520 (Pt 2):417–429

Darbellay B, Arnaudeau S, Konig S, Jousset H, Bader C, Demaurex N, Bernheim L (2009) STIM1- and Orai1-dependent store-operated calcium entry regulates human myoblast differentiation. J Biol Chem 284:5370–5380

Darbellay B, Arnaudeau S, Bader CR, Konig S, Bernheim L (2011) STIM1L is a new actin-binding splice variant involved in fast repetitive Ca2+ release. J Cell Biol 194:335–346

Dietrich A, Kalwa H, Rost BR, Gudermann T (2005) The diacylgylcerol-sensitive TRPC3/6/7 subfamily of cation channels: functional characterization and physiological relevance. Pflugers Arch 451:72–80

Dirksen RT (2009) Checking your SOCCs and feet: the molecular mechanisms of Ca2+ entry in skeletal muscle. J Physiol 587:3139–3147

Ducret T, Vandebrouck C, Cao ML, Lebacq J, Gailly P (2006) Functional role of store-operated and stretch-activated channels in murine adult skeletal muscle fibres. J Physiol 575:913–924

Edwards JN, Murphy RM, Cully TR, von Wegner F, Friedrich O, Launikonis BS (2010) Ultra-rapid activation and deactivation of store-operated Ca(2+) entry in skeletal muscle. Cell Calcium 47:458–467

Flucher BE, Franzini-Armstrong C (1996) Formation of junctions involved in excitation-contraction coupling in skeletal and cardiac muscle. Proc Natl Acad Sci U S A 93:8101–8106

Formigli L, Meacci E, Sassoli C, Chellini F, Giannini R, Quercioli F, Tiribilli B, Squecco R, Bruni P, Francini F, Zecchi-Orlandini S (2005) Sphingosine 1-phosphate induces cytoskeletal reorganization in C2C12 myoblasts: physiological relevance for stress fibres in the modulation of ion current through stretch-activated channels. J Cell Sci 118:1161–1171

Formigli L, Sassoli C, Squecco R, Bini F, Martinesi M, Chellini F, Luciani G, Sbrana F, Zecchi-Orlandi S, Francini F, Maecci E (2009) Regulation of transient receptor potential canonical channel 1 (TRPC1) by sphingosine 1-phosphate in C2C12 myoblasts and its relevance for a role of mechanotransduction in skeletal muscle differentiation. J Cell Sci 122:1322–1333

Franco A Jr, Lansman JB (1990) Calcium entry through stretch-inactivated ion channels in mdx myotubes. Nature 344:670–673

Franco-Obregon A Jr, Lansman JB (1994) Mechanosensitive ion channels in skeletal muscle from normal and dystrophic mice. J Physiol 481(Pt 2):299–309

Gailly P, Boland B, Himpens B, Casteels R, Gillis JM (1993) Critical evaluation of cytosolic calcium determination in resting muscle fibres from normal and dystrophic (mdx) mice. Cell Calcium 14:473–483

Gailly P, De Backer F, Van Schoor M, Gillis JM (2007) In situ measurements of calpain activity in isolated muscle fibres from normal and dystrophin-lacking mdx mice. J Physiol 582:1261–1275

Gervasio OL, Whitehead NP, Yeung EW, Phillips WD, Allen DG (2008) TRPC1 binds to caveolin-3 and is regulated by Src kinase—role in Duchenne muscular dystrophy. J Cell Sci 121:2246–2255

Hartmann J, Dragicevic E, Adelsberger H, Henning HA, Sumser M, Abramowitz J, Blum R, Dietrich A, Freichel V, Birnbaumer L, Konnerth A (2008) TRPC3 channels are required for synaptic transmission and motor coordination. Neuron 59:392–398

Hawke TJ, Garry DJ (2001) Myogenic satellite cells: physiology to molecular biology. J Appl Physiol 91:534–551

Head SI (1993) Membrane potential, resting calcium and calcium transients in isolated muscle fibres from normal and dystrophic mice. J Physiol 469:11–19

Hofmann T, Obukhov AG, Schaefer M, Harteneck C, Gudermann T, Schultz G (1999) Direct activation of human TRPC6 and TRPC3 channels by diacylglycerol. Nature 397:259–263

Huang GN, Zeng W, Kim JY, Yuan JP, Han L, Muallem S, Worley PF (2006) STIM1 carboxyl-terminus activates native SOC, I(crac) and TRPC1 channels. Nat Cell Biol 8:1003–1010

Imbert N, Cognard C, Duport G, Guillou C, Raymond G (1995) Abnormal calcium homeostasis in Duchenne muscular dystrophy myotubes contracting in vitro. Cell Calcium 18:177–186

Kawasaki BT, Liao Y, Birnbaumer L (2006) Role of Src in C3 transient receptor potential channel function and evidence for a heterogeneous makeup of receptor- and store-operated Ca2+ entry channels. Proc Natl Acad Sci U S A 103:335–340

Kruger J, Kunert-Keil C, Bisping F, Brinkmeier H (2008) Transient receptor potential cation channels in normal and dystrophic mdx muscle. Neuromuscul Disord 18:501–513

Kunert-Keil C, Bisping F, Kruger J, Brinkmeier H (2006) Tissue-specific expression of TRP channel genes in the mouse and its variation in three different mouse strains. BMC Genomics 7:159

Kurebayashi N, Ogawa Y (2001) Depletion of Ca2+ in the sarcoplasmic reticulum stimulates Ca2+ entry into mouse skeletal muscle fibres. J Physiol 533:185–199

Lacruz RS, Feske S (2015) Diseases caused by mutations in ORAI1 and STIM1. Ann N Y Acad Sci 1356:45–79

Lanner JT, Bruton JD, Assefaw-Redda Y, Andronache Z, Zhang SJ, Severa D, Zhang ZB, Melzer W, Zhang SL, Katz A, Westerblad H (2009) Knockdown of TRPC3 with siRNA coupled to carbon nanotubes results in decreased insulin-mediated glucose uptake in adult skeletal muscle cells. FASEB J 23:1728–1738

Launikonis BS, Rios E (2007) Store-operated Ca2+ entry during intracellular Ca2+ release in mammalian skeletal muscle. J Physiol 583:81–97

Launikonis BS, Stephenson DG, Friedrich O (2009) Rapid Ca2+ flux through the transverse tubular membrane, activated by individual action potentials in mammalian skeletal muscle. J Physiol 587:2299–2312

Lauritzen HP, Ploug T, Prats C, Tavare JM, Galbo H (2006) Imaging of insulin signaling in skeletal muscle of living mice shows major role of T-tubules. Diabetes 55:1300–1306

Lee EH, Cherednichenko G, Pessah IN, Allen PD (2006) Functional coupling between TRPC3 and RyR1 regulates the expressions of key triadic proteins. J Biol Chem 281:10042–10048

Leloup L, Mazeres G, Daury L, Cottin P, Brustis JJ (2006) Involvement of calpains in growth factor-mediated migration. Int J Biochem Cell Biol 38:2049–2063

Li S, Couet J, Lisanti MP (1996) Src tyrosine kinases, Galpha subunits, and H-Ras share a common membrane-anchored scaffolding protein, caveolin. Caveolin binding negatively regulates the auto-activation of Src tyrosine kinases. J Biol Chem 271:29182–29190

Li T, Finch EA, Graham V, Zhang ZS, Ding JD, Burch J, Oh-hora M, Rosenberg P (2012) STIM1-Ca(2+) signaling is required for the hypertrophic growth of skeletal muscle in mice. Mol Cell Biol 32:3009–3017

Lichtenegger M, Groschner K (2014) TRPC3: a multifunctional signaling molecule. Handb Exp Pharmacol 222:67–84

Liu Y, Schneider MF (2014) FGF2 activates TRPC and Ca(2+) signaling leading to satellite cell activation. Front Physiol 5:38

Louis M, Zanou N, Van Schoor M, Gailly P (2008) TRPC1 regulates skeletal myoblast migration and differentiation. J Cell Sci 121:3951–3959

Lyfenko AD, Dirksen RT (2008) Differential dependence of store-operated and excitation-coupled Ca2+ entry in skeletal muscle on STIM1 and Orai1. J Physiol 586:4815–4824

Maroto R, Raso A, Wood TG, Kurosky A, Martinac B, Hamill OP (2005) TRPC1 forms the stretch-activated cation channel in vertebrate cells. Nat Cell Biol 7:179–185

Matsumura CY, Taniguti AP, Pertille A, Santo Neto H, Marques MJ (2011) Stretch-activated calcium channel protein TRPC1 is correlated with the different degrees of the dystrophic phenotype in mdx mice. Am J Physiol Cell Physiol 301:C1344–C1350

McCarl CA, Picard C, Khalil S, Kawasaki T, Rother J, Papolos A, Kutok J, Hivroz C, Ledeist F, Plogmann K, Ehl S, Notheis G, Albert MH, Belohradsky BH, Kirschner J, Rao A, Fischer A, Feske S (2009) ORAI1 deficiency and lack of store-operated Ca2+ entry cause immunodeficiency, myopathy, and ectodermal dysplasia. J Allergy Clin Immunol 124:1311–1318

Meacci E, Bini F, Sassoli C, Martinesi M, Squecco R, Chellini F, Zecchi-Orlandini S, Francini F, Formigli L (2010) Functional interaction between TRPC1 channel and connexin-43 protein: a novel pathway underlying S1P action on skeletal myogenesis. Cell Mol Life Sci 67:4269–4285

Millay DP, Goonasekera SA, Sargent MA, Maillet M, Aronow BJ, Molkentin JD (2009) Calcium influx is sufficient to induce muscular dystrophy through a TRPC-dependent mechanism. Proc Natl Acad Sci U S A 106:19023–19028

Nesin V, Tsiokas L (2014) Trpc1. Handb Exp Pharmacol 222:15–51

Nilius B, Owsianik G (2010) Transient receptor potential channelopathies. Pflugers Arch 460:437–450

Nilius B, Owsianik G (2011) The transient receptor potential family of ion channels. Genome Biol 12:218

Olah T, Fodor J, Ruzsnavszky O, Vincze J, Berbey C, Allard B, Csernoch L (2011) Overexpression of transient receptor potential canonical type 1 (TRPC1) alters both store operated calcium entry and depolarization-evoked calcium signals in C2C12 cells. Cell Calcium 49:415–425

Ong HL, de Souza LB, Ambudkar IS (2016) Role of TRPC Channels in Store-Operated Calcium Entry. Adv Exp Med Biol 898:87–109

Pedersen SF, Nilius B (2007) Transient receptor potential channels in mechanosensing and cell volume regulation. Methods Enzymol 428:183–207

Prakriya M, Lewis RS (2015) Store-operated calcium channels. Physiol Rev 95:1383–1436

Pressmar J, Brinkmeier H, Seewald MJ, Naumann T, Rudel R (1994) Intracellular Ca2+ concentrations are not elevated in resting cultured muscle from Duchenne (DMD) patients and in MDX mouse muscle fibres. Pflugers Arch 426:499–505

Riccio A, Medhurst AD, Mattei C, Kelsell RE, Calver AR, Randall AD, Benham CD, Pangalos MN (2002) mRNA distribution analysis of human TRPC family in CNS and peripheral tissues. Brain Res Mol Brain Res 109:95–104

Romi F, Aarli JA, Gilhus NE (2007) Myasthenia gravis patients with ryanodine receptor antibodies have distinctive clinical features. Eur J Neurol 14:617–620

Rosenberg P, Hawkins A, Stiber J, Shelton JM, Hutcheson K, Bassel-Duby R, Shin DM, Yan Z, Williams RS (2004) TRPC3 channels confer cellular memory of recent neuromuscular activity. Proc Natl Acad Sci U S A 101:9387–9392

Sabourin J, Lamiche C, Vandebrouck A, Magaud C, Rivet J, Cognard C, Bourmeyster N, Constantin B (2009) Regulatiqn of TRPC1 and TRPC4 cation channels requires an alpha1-syntrophin-dependent complex in skeletal mouse myotubes. J Biol Chem 284:36248–36261

Sorrentino V (2011) Sarcoplasmic reticulum: structural determinants and protein dynamics. Int J Biochem Cell Biol 43:1075–1078

Spassova MA, Hewavitharana T, Xu W, Soboloff J, Gill DL (2006) A common mechanism underlies stretch activation and receptor activation of TRPC6 channels. Proc Natl Acad Sci U S A 103:16586–16591

Spencer MJ, Croall DE, Tidball JG (1995) Calpains are activated in necrotic fibers from mdx dystrophic mice. J Biol Chem 270:10909–10914

Squecco R, Sassoli C, Nuti F, Martinesi M, Chellini F, Nosi D, Zecchi-Orlandini S, Francini F, Formigli L, Maecci E (2006) Sphingosine 1-phosphate induces myoblast differentiation through Cx43 protein expression: a role for a gap junction-dependent and -independent function. Mol Biol Cell 17:4896–4910

Stiber J, Hawkins A, Zhang ZS, Wang S, Burch J, Graham V, Ward CC, Seth M, Finch E, Malouf N, Williams RD, JP E, Rosenberg P (2008a) STIM1 signalling controls store-operated calcium entry required for development and contractile function in skeletal muscle. Nat Cell Biol 10:688–697

Stiber JA, Zhang ZS, Burch J, JP E, Zhang S, Truskey GA, Seth M, Yamaguchi N, Meissner G, Shah R, Worley PF, Williams RS, Rosenberg PB (2008b) Mice lacking Homer 1 exhibit a skeletal myopathy characterized by abnormal transient receptor potential channel activity. Mol Cell Biol 28:2637–2647

Suchyna TM, Johnson JH, Hamer K, Leykam JF, Gage DA, Clemo HF, Baumgarten CM, Sachs F (2000) Identification of a peptide toxin from Grammostola spatulata spider venom that blocks cation-selective stretch-activated channels. J Gen Physiol 115:583–598

Tajeddine N, Zanou N, Van Schoor M, Lebacq J, Gailly P (2010) TRPC1: subcellular localization? J Biol Chem 285:le1. Author reply le2

Takamori M (2008) Autoantibodies against TRPC3 and ryanodine receptor in myasthenia gravis. J Neuroimmunol 200:142–144

Turner PR, Westwood T, Regen CM, Steinhardt RA (1988) Increased protein degradation results from elevated free calcium levels found in muscle from mdx mice. Nature 335:735–738

Vallejo-Illarramendi A, Toral-Ojeda I, Aldanondo G, Lopez de Munain A (2014) Dysregulation of calcium homeostasis in muscular dystrophies. Expert Rev Mol Med 16:e16

Vandebrouck C, Martin D, Colson-Van Schoor M, Debaix H, Gailly P (2002) Involvement of TRPC in the abnormal calcium influx observed in dystrophic (mdx) mouse skeletal muscle fibers. J Cell Biol 158:1089–1096

Vandebrouck A, Sabourin J, Rivet J, Balghi H, Sebille S, Kitzis A, Raymond G, Cognard C, Bourmeyster N, Constantin B (2007) Regulation of capacitative calcium entries by alpha1-syntrophin: association of TRPC1 with dystrophin complex and the PDZ domain of alpha1-syntrophin. FASEB J 21:608–617

Wei-Lapierre L, Carrell EM, Boncompagni S, Protasi F, Dirksen RT (2013) Orai1-dependent calcium entry promotes skeletal muscle growth and limits fatigue. Nat Commun 4:2805

Woo JS, Kim DH, Allen PD, Lee EH (2008) TRPC3-interacting triadic proteins in skeletal muscle. Biochem J 411:399–405

Woo JS, Cho CH, Kim DH, Lee EH (2010) TRPC3 cation channel plays an important role in proliferation and differentiation of skeletal muscle myoblasts. Exp Mol Med 42:614–627

Woo JS, Lee KJ, Huang M, Cho CH, Lee EH (2014) Heteromeric TRPC3 with TRPC1 formed via its ankyrin repeats regulates the resting cytosolic Ca2+ levels in skeletal muscle. Biochem Biophys Res Commun 446:454–459

Woo JS, Hwang JH, Huang M, Ahn MK, Cho CH, Ma J, Lee EH (2015) Interaction between mitsugumin 29 and TRPC3 participates in regulating Ca(2+) transients in skeletal muscle. Biochem Biophys Res Commun 464:133–139

Xia L, Cheung KK, Yeung SS, Yeung EW (2016) The involvement of transient receptor potential canonical type 1 in skeletal muscle regrowth after unloading-induced atrophy. J Physiol 594:3111–3126

Yeung EW, Whitehead NP, Suchyna TM, Gottlieb PA, Sachs F, Allen DG (2005) Effects of stretch-activated channel blockers on $[Ca2+]_i$ and muscle damage in the mdx mouse. J Physiol 562:367–380

Yuan JP, Zeng W, Huang GN, Worley PF, Muallem S (2007) STIM1 heteromultimerizes TRPC channels to determine their function as store-operated channels. Nat Cell Biol 9:636–645

Zanou N, Shapovalov G, Louis M, Tajeddine N, Gallo C, Van Schoor M, Anguish I, Cao ML, Schakman O, Dietrich A, Lebacq J, Ruegg U, Roulet E, Birnbaumer L, Gailly P (2010) Role of TRPC1 channel in skeletal muscle function. Am J Physiol Cell Physiol 298:C149–C162

Zanou N, Schakman O, Louis P, Ruegg UT, Dietrich A, Birnbaumer L, Gailly P (2012) Trpc1 ion channel modulates phosphatidylinositol 3-kinase/Akt pathway during myoblast differentiation and muscle regeneration. J Biol Chem 287:14524–14534

Zeng W, Yuan JP, Kim MS, Choi YJ, Huang GN, Worley PF, Muallem S (2008) STIM1 gates TRPC channels, but not Orai1, by electrostatic interaction. Mol Cell 32:439–448

Zhang BT, Yeung SS, Cheung KK, Chai ZY, Yeung EW (2014) Adaptive responses of TRPC1 and TRPC3 during skeletal muscle atrophy and regrowth. Muscle Nerve 49:691–699 [illegible] ZS, Burch [illegible], Zhang [illegible], Tucker GM, Stelt [illegible] Yan [illegible] Stubb [illegible], Wesley [illegible], Williams RS, [illegible] (2007) [illegible] skeletal [illegible] calcineurin signaling [illegible] potential receptor channel [illegible] [illegible] 2[illegible]

Sa[illegible] Johnson [illegible], Hanley [illegible], [illegible] Gee [illegible], Clapham [illegible] Bennett [illegible] Sabatini [illegible] (2007) Identification of a [illegible] from [illegible] [illegible] skeletal muscle. J Cell Physiol [illegible]

[illegible] Vandebrouck [illegible], Van [illegible] (2010) TRPC [illegible] J Biol Chem [illegible]

[illegible] (2011) [illegible] TRPC [illegible] muscle [illegible]

[illegible] (2001) [illegible] potential [illegible] regulation [illegible] Nature [illegible]

Vandebrouck [illegible] (2002) [illegible] calcium [illegible] muscle [illegible]

Vandebrouck [illegible] (2005) [illegible] TRPC [illegible]

Vandebrouck [illegible], Sabourin [illegible] (2007) [illegible] Brain [illegible]

[illegible] (2007) [illegible] growth and [illegible]

[illegible] TRPC [illegible]

[illegible] (2007) TRPC [illegible] channel [illegible] muscle [illegible] Exp Mol Med [illegible]

[illegible] (2010) [illegible] TRPC [illegible] skeletal muscle [illegible] Biochem Biophys Res Commun [illegible]

[illegible] (2008) [illegible] TRPC3 [illegible] skeletal muscle [illegible]

[illegible] (2011) [illegible] skeletal muscle [illegible]

[illegible] (2007) [illegible] muscle [illegible] mdx mouse [illegible]

[illegible] TRPC [illegible] channels [illegible] store-operated channels [illegible]

Zanou N, [illegible], Louis M, [illegible] (2010) Role of TRPC1 channel in skeletal muscle function. Am J Physiol Cell Physiol [illegible]

Zanou N, [illegible] (2012) [illegible] TRPC [illegible] muscle [illegible] J Biol Chem [illegible]

[illegible] (2011) TRPC [illegible]

Zhang [illegible] (2011) [illegible] TRPC [illegible]

Fertility: Store-Operated Ca^{2+} Entry in Germ Cells – Role in Egg Activation

29

Zoltan Machaty, Chunmin Wang, Kiho Lee, and Lu Zhang

Abstract

At the time of fertilization, the sperm activates the egg and induces embryonic development by triggering an elevation in the egg's intracellular free Ca^{2+} concentration. In mammals the initial Ca^{2+} rise is followed by a series of repetitive Ca^{2+} transients (known as oscillations) that last for several hours. Although the source of Ca^{2+} during the signaling process is primarily the egg's smooth endoplasmic reticulum, the oscillations stop in the absence of extracellular Ca^{2+} indicating that a Ca^{2+} influx across the plasma membrane is essential to sustain them. Depletion of the intracellular stores using specific inhibitors generates a Ca^{2+} entry across the plasma membrane of eggs of various species, and a continuous influx of Ca^{2+} has been linked to the sperm-induced Ca^{2+} oscillations in the mouse; these data indicate that store-operated Ca^{2+} entry (SOCE) operates in eggs and may be the mechanism that maintains the long-lasting Ca^{2+} signal at fertilization. Recent findings suggest that the signaling proteins STIM1 and Orai1 are present in eggs; they are responsible for mediating SOCE, and their functions are essential for proper Ca^{2+} signaling at fertilization to support normal embryo development.

Keywords

SOCE • STIM • Orai • Egg • Activation • Fertilization • Calcium influx

Z. Machaty (✉) • L. Zhang
Department of Animal Sciences, Lilly Hall of Life Sciences, Purdue University, West Lafayette, IN, USA
e-mail: zmachaty@purdue.edu

C. Wang
Vivere Health, Houston, TX, USA

K. Lee
Department of Animal and Poultry Sciences, Virginia Tech, Blacksburg, VA, USA

K. Groschner et al. (eds.), *Store-Operated Ca^{2+} Entry (SOCE) Pathways*, Advances in Experimental Medicine and Biology 993, DOI 10.1007/978-3-319-57732-6_29

29.1 Introduction

The female gamete is produced by meiosis which consists of two rounds of nuclear divisions (Whitaker 1996). During the final stage of their production, immature oocytes undergo maturation, and mature oocytes (known as eggs) are released from the ovary. At this point, their development is stopped at a species-specific stage of the cell cycle; ovulated mammalian eggs are arrested at the metaphase stage of the second meiotic division. Under physiological conditions, it is the fertilizing sperm that restarts the cell cycle machinery; by fusing to the egg, it stimulates a series of events in the egg cytoplasm that are collectively known as activation. A number of evidences indicate that the key trigger for egg activation is a Ca^{2+} signal: an elevation in the egg's intracellular Ca^{2+} level is responsible for all additional downstream events associated with fertilization. Results from sea urchins, fishes, and frogs indicate that in these animals the sperm generates a single Ca^{2+} transient that leads to egg activation. In others including mammals, the sperm-induced Ca^{2+} signal takes the form of long-lasting oscillations in the intracellular free Ca^{2+} level (reviewed by Stricker 1999). It was also observed that the external Ca^{2+} concentration has a profound effect on the sperm-induced Ca^{2+} signal (Igusa and Miyazaki 1983).

29.2 Relevance of the Sperm-Induced Ca^{2+} Oscillations

Depending on the species, the events of egg activation include a transient elevation in the intracellular free Ca^{2+} concentration of the egg, the release of the content of the cortical granules that results in a block to polyspermy, resumption of meiosis, changes in intracellular pH, recruitment of maternal mRNAs, formation of the maternal and paternal pronuclei, initiation of DNA synthesis, and cleavage (Schultz and Kopf 1995). The oscillating signal seems to be important to ensure complete activation of the egg. By applying repetitive electroporation to induce different numbers of Ca^{2+} transients in eggs, it was demonstrated that each Ca^{2+} rise pushes progressively forward the activation process (Ducibella et al. 2002). Cortical granule exocytosis takes place in a stepwise manner with each consequential Ca^{2+} rise triggering the release of the content of additional granules. Furthermore, early events such as resumption of meiosis are stimulated by a single elevation in the intracellular Ca^{2+} level, while others require more transients. The meiotic arrest in mammalian eggs is maintained by the M-phase-promoting factor (MPF) composed of a regulatory subunit, cyclin B1, and a catalytic subunit, cyclin-dependent kinase 1. For the resumption of meiosis (i.e., proceeding from metaphase II to anaphase II), MPF activity in the ooplasm must drop. This is achieved by destroying cyclin B1, and it is now known that the destruction of cyclin B1 is stimulated by the sperm-induced Ca^{2+} signal (Jones 2007). In freshly ovulated mouse eggs, a treatment with ethanol (ethanol induces a single Ca^{2+} rise in eggs) induced meiotic resumption because one Ca^{2+} spike was sufficient to downregulate MPF activity. However, the eggs were not able to exit meiosis, but instead they got rearrested at a so-called third metaphase stage, which indicated that MPF activity in the ooplasm was restored (Kubiak 1989). In addition, the oscillatory Ca^{2+} signal is also important for the proper stimulation of events that occur later during development. Although a static, sustained Ca^{2+} signal may push the egg through the first mitotic cell cycle, postimplantation development in the resultant embryos is poor (Ozil et al. 2005).

29.3 How Are the Ca^{2+} Oscillations Generated?

It was discovered a long time ago that the Ca^{2+} signal upon fertilization is produced when Ca^{2+} is released from the intracellular Ca^{2+} stores. In sea urchin eggs, it was shown that during fertilization there is an increased turnover of phospholipids and an elevation in inositol 1,4,5-trisphosphate (IP_3) levels (Turner et al. 1986). This led to the conclusion that the Ca^{2+} rise is triggered when IP_3 binds to its receptor on the surface of the endoplasmic reticulum (ER). However, it was unclear how the sperm triggers IP_3 production following interaction of the gametes. The existence of a sperm-derived factor with a role in activation has long been suspected. It was indicated by the finding that injection of a soluble sperm extract into hamster and mouse eggs induced long-lasting Ca^{2+} oscillations similar to those measured during fertilization (Swann 1990). The fact that the sperm extract showed high phospholipase C (PLC) activity and caused a large increase in IP_3 levels in sea urchin egg homogenates implied that the active factor may be a PLC (Jones et al. 1998). However, none of the known PLC isoforms induced repetitive Ca^{2+} transients when injected at physiological levels into the egg cytoplasm. The hunt for the elusive sperm factor eventually led to the discovery of a novel phospholipase C isoform in mammalian sperm. In the mouse, PLCζ (PLCzeta) was found to be expressed exclusively in the testes, and when its complementary RNA (cRNA) was microinjected into eggs, an oscillating Ca^{2+} signal was generated that was indistinguishable from that measured during fertilization (Saunders et al. 2002). Although a number of other sperm factors have been proposed, it is now widely accepted that PLCζ is the key molecule essential for the generation of the oscillatory Ca^{2+} signal during mammalian fertilization.

There is no complete agreement on what makes the signal oscillate. According to the Ca^{2+}-dependent IP_3 production model, it is the regenerative production of IP_3 that creates Ca^{2+} oscillations. When PLCζ is introduced into the egg, it stimulates IP_3 production at resting Ca^{2+} levels. This leads to Ca^{2+} release from the ER; increased Ca^{2+} concentration will further stimulate PLCζ leading to elevated IP_3 production, which in turn results in further Ca^{2+} release during the rising phase of Ca^{2+} transients (Dupont and Dumollard 2004). Another popular model explains the train of spikes with the biphasic response of the IP_3 receptor/channel to cytosolic Ca^{2+}. In this scenario, a stable IP_3 level delivers a constant stimulation to the IP_3 receptor, which opens at low Ca^{2+} levels but closes when Ca^{2+} concentration is high in the cytoplasm (Bezprozvanny et al. 1991). It was also suggested that in mammalian eggs two mechanisms may exist that are responsible for the generation of the repetitive Ca^{2+} signal (Swann and Yu 2008).

While most of the attention has been focused on the rising phase of the Ca^{2+} transients, little is known about the events that take place after the Ca^{2+} is released into the cytosol. Prolonged high Ca^{2+} concentration in the cytoplasm can be detrimental so Ca^{2+} must be removed. Part of the Ca^{2+} is pumped back into the stores by sarcoplasmic/endoplasmic reticulum calcium ATPase (SERCA) pumps or into the mitochondria, while some Ca^{2+} is extruded from the cell by means of plasma membrane Ca^{2+} ATPase (PMCA) pumps and the Na^+/Ca^{2+} exchanger (reviewed by Dumollard et al. 2002). At the same time, an influx of Ca^{2+} accompanies each transient during the sperm-induced Ca^{2+} oscillations (McGuinness et al. 1996).

29.4 Store-Operated Ca^{2+} Entry in Oocytes/Eggs

It has been known for quite some time that the female gamete contains the machinery that is able to generate a Ca^{2+} influx across the plasma membrane. The fertilizing sperm triggered recurring hyperpolarizing responses in hamster eggs that were the consequence of Ca^{2+}-activated K^+ conductance across the plasma membrane due to periodic release of Ca^{2+} from the intracellular stores (Igusa and Miyazaki 1983). The hyperpolarizing responses were abolished in the absence of extracellular Ca^{2+}, while the frequency of the responses was dependent on the Ca^{2+} concentration in the holding medium. This clearly indicated a Ca^{2+} flux into the eggs and suggested a possible linkage between Ca^{2+} transport and Ca^{2+} release. The extracellular Ca^{2+} was later shown to be a prerequisite for the maintenance of the sperm-induced Ca^{2+} oscillations, and a model was proposed according to which the Ca^{2+} influx is responsible for reloading the Ca^{2+} stores (Miyazaki 1991).

A Ca^{2+} influx generated after an IP_3-induced Ca^{2+} signal was also detected in immature *Xenopus* oocytes. The oocytes isolated from *Xenopus* ovaries provided a convenient system for studying Ca^{2+} signaling and became popular model cells primarily because they have numerous Ca^{2+}-dependent chloride (Cl^-) channels in their plasma membrane facilitating swift detection of cytosolic free Ca^{2+} (Barish 1983). In these cells, agonist activation of several different drug receptors linked to IP_3 or the direct injection of IP_3 evoked transient inward Cl^- currents when the driving force for Ca^{2+} was increased by creating more negative membrane potentials (Parker et al. 1985; Parker and Miledi 1987). Removing external Ca^{2+} or injecting the Ca^{2+}-chelating agent EGTA into the oocytes abolished the Cl^- currents so it was concluded that IP_3 enhanced the permeability of the plasma membrane to Ca^{2+}, so raising the electrical driving force amplified the Ca^{2+} influx resulting in the opening of Ca^{2+}-dependent Cl^- channels.

A clear indication that an influx of Ca^{2+} is generated in eggs because of the release of Ca^{2+} from the intracellular stores came from experiments in the mouse. In unfertilized eggs, an influx pathway could be activated with the SERCA pump inhibitor thapsigargin, indicating the coupling of the Ca^{2+} entry mechanism to store depletion (Kline and Kline 1992). In addition, the sperm-induced Ca^{2+} transients were suppressed by thapsigargin or a decline in extracellular Ca^{2+}. Others also found that the frequency of the oscillations during fertilization depended on the extracellular Ca^{2+} concentration (Shiina et al. 1993). These data implied that during fertilization the Ca^{2+} permeability of the plasma membrane increased and the enhanced Ca^{2+} influx was responsible for maintaining the Ca^{2+} oscillations by refilling the intracellular stores.

During the years, the store-operated Ca^{2+} entry (SOCE) mechanism in immature *Xenopus* oocytes was dissected and analyzed in much detail. Injection of a non-metabolizable form of IP_3 evoked periodic release of Ca^{2+} from the intracellular store, and more importantly, the Ca^{2+} release activated a Ca^{2+} influx pathway which could modify the oscillatory pattern and refill the IP_3 pools (Parekh et al. 1993). Others also found that depletion of the intracellular stores in *Xenopus* oocytes using thapsigargin or a non-metabolizable IP_3 analogue stimulated a Ca^{2+} entry (Lupu-Meiri et al. 1993; Yao and Parker 1993). Interestingly, SOCE inactivated during oocyte maturation in *Xenopus* as the oocytes became capable of fertilization. The inactivation took place at the time of germinal vesicle breakdown and coincided with the activation of MPF (Machaca and Haun 2002), but the exact mechanism of inactivation became clear only after the identification of the components of the SOCE cascade (discussed below).

The onset of SOCE under physiological conditions and its potential relevance was demonstrated when a continuous Ca^{2+} influx associated with fertilization was detected in mouse eggs using the manganese-quench technique (McGuinness et al. 1996). The influx was controlled by the intracellular stores as indicated by the facts that it initiated upon store depletion during the first sperm-induced Ca^{2+} spike, whereas in the interspike period as the stores began to refill, the rate of Ca^{2+} entry became attenuated. The Ca^{2+} influx seemed to set the frequency of the oscillations by controlling the time necessary to refill the ER thereby setting the stage for the next Ca^{2+} transient. Finally, SOCE was also demonstrated in eggs of other species. Emptying the ER of pig and human eggs using thapsigargin in Ca^{2+}-free medium promoted an increase in cytoplasmic Ca^{2+} levels after Ca^{2+} was added back to the medium (Machaty et al. 2002; Martin-Romero et al. 2008). In the pig, a divalent cation influx was also generated after an IP_3-induced release of Ca^{2+} from the intracellular stores. The increases were blocked by inhibitors of store-operated Ca^{2+} channels further strengthening the idea that it was the emptying of the stores that triggered the Ca^{2+} entry.

29.5 Defining the SOCE Pathway

Although the mechanism of SOCE was clearly shown in eggs of various species and their function during fertilization was also demonstrated, the components as well as the regulation of the pathway have remained obscure. Evidences indicated that eggs of various species contain voltage-gated Ca^{2+} channels that mediated a Ca^{2+}-dependent action potential (Miyazaki and Igusa 1981). However, results of electrophysiological studies suggested that the channels that transported Ca^{2+} across the plasma membrane during fertilization probably belonged to another family of Ca^{2+} channels (Miyazaki and Igusa 1982). It was then suggested that in *Xenopus* oocytes the store depletion-induced Ca^{2+} influx was modulated by a mechanism that involved a phosphatase and a diffusible messenger (Parekh et al. 1993). Store emptying was believed to release a diffusible molecule that moved from the stores to the plasma membrane and activated the Ca^{2+} channels. Others found that the SOCE pathway created by a treatment with thapsigargin was both potentiated and inhibited by Ca^{2+} itself (Petersen and Berridge 1994). It was hypothesized that Ca^{2+} entering the cytoplasm stimulated Ca^{2+} influx and high cytosolic Ca^{2+} concentrations eventually inactivated SOCE. In addition, low-level activation of protein kinase C (PKC) stimulated Ca^{2+} influx, while higher-level PKC activity inhibited it. Others also reported that full-length PKC overexpression or PKC stimulation increased the frequency of sperm-induced Ca^{2+} oscillations in mouse eggs (Halet et al. 2004) and constitutively active PKC constructs caused a persistent rise in cytosolic free Ca^{2+} levels after a Ca^{2+} influx was stimulated by store depletion (Madgwick et al. 2005) arguing that PKC positively regulated SOCE.

Membrane fusion was also implicated in the activation of store-operated Ca^{2+} channels. When protein components of the exocytotic apparatus were blocked using specific toxins, SOCE was inhibited suggesting that membrane trafficking played a role in the activation of the Ca^{2+} influx (Yao et al. 1999). Downregulation of Rho A, a small GTPase protein, increased the amplitude of the current through the store-operated Ca^{2+} channels, and because Rho A was known to regulate membrane trafficking, it provided further support for the idea that the exocytosis of membrane channels may regulate store-operated Ca^{2+} channels in *Xenopus* oocytes. Another line of experiments suggested the potential involvement of the Ca^{2+}/calmodulin-dependent protein kinase II (CaMKII) pathway. The idea was particularly appealing because CaMKII plays an important role in decoding intracellular Ca^{2+} signals, and it was argued that CaMKII-mediated regulation of SOCE would provide a mechanism for Ca^{2+} to control its own entry. Unfortunately, the relevant findings here are difficult to reconcile. In one study, the amplitude of Ca^{2+}-evoked Cl^- transients triggered by store depletion increased when CaMKII activity in the oocytes was inhibited; this was interpreted as an indication that Ca^{2+} entering the cell after store depletion activated CaMKII which through a negative feedback mechanism inhibited SOCE (Matifat et al. 1997). Results from another work indicated however, that a rise in the cytoplasmic free Ca^{2+} level enhanced Ca^{2+} entry and this increase was mediated by CaMKII (Machaca 2003). Finally, heterologous expression of TRP enhanced SOCE in *Xenopus* oocytes (Petersen et al. 1995), and it was demonstrated that *Xenopus* oocytes and porcine eggs contain endogenous TRP channels (Bobanovic et al. 1999; Machaty et al. 2002). Consequently, TRP was considered a candidate channel for the SOCE pathway. Antisense oligonucleotides designed against the human TRP1 protein inhibited Ca^{2+} influx evoked by thapsigargin which implied that SOCE was under the control of TRP proteins at least in *Xenopus* oocytes (Tomita et al. 1998) although additional data questioned this hypothesis (Brereton et al. 2000).

The breakthrough in identifying the components of the SOCE pathway that operate in female gametes came when in somatic cells the stromal interaction molecule 1 (STIM1) protein was shown to be the sensor of Ca^{2+} content of the ER lumen that links store depletion to Ca^{2+} influx (Liou et al. 2005; Roos et al. 2005). Our laboratory was the first to investigate the existence of STIM1 in eggs and demonstrated its presence both at the transcript and protein levels using the pig as a model (Koh et al. 2009). We designed oligonucleotide primers based on a porcine EST sequence that showed high similarity with STIM1 of other species (human, mouse, and bovine). Using the primers for PCR cloning revealed the existence of STIM1 transcripts in pig eggs. With immunocytochemical analysis, we showed that like in other cell types, the protein was predominantly located in the cytosol and rearranged in clusters near the plasma membrane after a thapsigargin-induced Ca^{2+} release. Exogenously expressed YFP-STIM1 (STIM1 conjugated to the yellow fluorescent protein) behaved in a very similar manner: first, it localized in the cytoplasm, and a thapsigargin-evoked Ca^{2+} store depletion triggered its redistribution close to the plasma membrane. These results supported the hypothesis that Ca^{2+} store depletion in eggs induces STIM1 translocation to stimulate Ca^{2+} release-activated Ca^{2+} channel activity in the plasma membrane.

The essential role of STIM1 in the regulation of SOCE in eggs was demonstrated using the technique of RNA interference. siRNA was produced to target a sequence that lies in the SAM domain-encoding region of STIM1. The siRNA was injected into eggs 36 h after the beginning of maturation; by this time, most of the eggs extruded their first polar body, and removing the surrounding cumulus cells to facilitate microinjection did not perturb the progression of meiosis. The injected eggs were incubated for 15 h; this was followed by monitoring the changes in the intracellular free Ca^{2+} concentration after store depletion and Ca^{2+} add-back. Knockdown of STIM1 in porcine eggs completely suppressed thapsigargin-induced Ca^{2+} influx. STIM1 overexpression on the other hand moderately enhanced the activity of the store-operated Ca^{2+} channels. The protein was overexpressed by microinjection into eggs of an mRNA encoding YFP-STIM1. The exogenously expressed STIM1 was demonstrated in the injected eggs by western blot analysis 15 h after microinjection. Adding Ca^{2+} to the external medium after a thapsigargin-induced store depletion led to higher intracellular Ca^{2+} levels in YFP-STIM1 mRNA-injected eggs compared to the controls, and the influx was completely blocked by La^{3+}, an inhibitor of store-operated Ca^{2+} channels. This was in agreement with some earlier findings where overexpression of STIM1 caused a modest increase in SOCE in somatic cell types (Roos et al. 2005; Peinelt et al. 2006).

These data indicated that STIM1 was crucial for the generation of SOCE in pig eggs. Next, we wanted to know whether it had any role in sustaining the long-lasting Ca^{2+} signal associated with mammalian fertilization. In order to investigate this, STIM1 levels were suppressed in eggs prior to fertilization. To allow for maximum downregulation of STIM1, we decided to inject siRNA into oocytes before rather than after meiotic maturation. As the presence of cumulus cells is necessary for proper maturation, the siRNA was injected into cumulus-enclosed oocytes. The microinjected oocytes were then incubated in maturation medium for

44 h. During this time, the oocytes reached the metaphase stage of the second meiotic division when they could be fertilized; this time was also sufficient (as established in HeLa cells) to completely downregulate STIM1 expression after a treatment with siRNA (Liou et al. 2005). At the end of the maturation period, the cumulus cells together with the zona pellucida were removed; removal of the zona makes interaction between the gametes easier. The cells were then loaded with fura-2 and fertilized, and the changes in their intracellular free Ca^{2+} levels were monitored. Control eggs injected with non-silencing siRNA showed Ca^{2+} oscillations that lasted for several hours (Fig. 29.1). Zona-free eggs become polyspermic, and the frequency of the sperm-induced Ca^{2+} signal rises with increasing number of spermatozoa fusing with the egg (Faure et al. 1999); in our case, we typically registered 1–3 spikes per hour. However, the eggs injected with STIM1 siRNA showed only a single Ca^{2+} transient; additional Ca^{2+} rises were not detected in eggs that lacked STIM1 (Lee et al. 2012).

Downregulation of STIM1 prior to fertilization was detrimental to embryo development as well. The percentage of eggs undergoing cleavage in the control groups (eggs injected with saline or scrambled siRNA) was approximately 70%, while cleavage in eggs with downregulated STIM1 dropped markedly, to around 25%. Similar tendencies were observed in blastocyst formation. Only 0.8% of the STIM1 siRNA-injected eggs reached the blastocyst stage 7 days after fertilization; blastocyst formation in the control groups was around 15%. The decrease in embryo

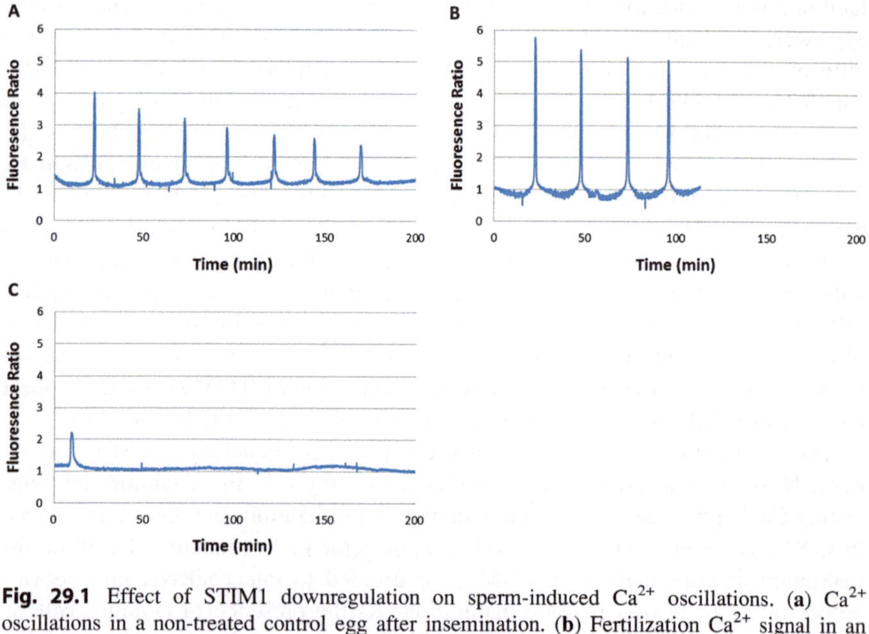

Fig. 29.1 Effect of STIM1 downregulation on sperm-induced Ca^{2+} oscillations. (**a**) Ca^{2+} oscillations in a non-treated control egg after insemination. (**b**) Fertilization Ca^{2+} signal in an egg injected with scrambled siRNA. (**c**) Sperm-induced Ca^{2+} signal in an egg injected with STIM1 siRNA. Note the lack of repetitive oscillations in these eggs

development was clearly due to the absence of STIM1 because when siRNA against STIM1 was delivered into eggs 6 h after fertilization, embryo cleavage was not affected: the proportion of two-cell embryos after injection of saline or STIM1 siRNA was around 70% in both groups (Lee et al. 2010). In mouse eggs, STIM1 was also implicated in the process of Ca^{2+} signaling during fertilization. In resting cells, STIM1 was distributed in discrete clusters in the egg cortex and co-localized with the ER marker calreticulin (Gómez-Fernández et al. 2009). Thapsigargin or fertilization induced a redistribution of the protein in larger areas close to the surface of the egg. Taken together, the results presented indicate that STIM1 plays a critical role during fertilization by sustaining the Ca^{2+} oscillations through the stimulation of the SOCE cascade. They also imply that normal STIM1 function has an effect on subsequent embryo development.

To better understand the mechanism underlying the Ca^{2+} signals that accompany fertilization, we investigated the presence of Orai1, the channel component of the SOCE pathway (Feske et al. 2006; Vig et al. 2006b), in pig eggs. Messenger RNA was isolated from pig eggs and used as a template; subsequent RT-PCR using the appropriate oligonucleotide primers clearly indicated the existence of Orai1 transcripts in the eggs (Wang et al. 2010a). The presence of the Orai1 protein was demonstrated by western blot analysis. For the assay 400 eggs were pooled; a rabbit anti-Orai1 antibody and an appropriate secondary antibody identified the protein with the expected size. The intracellular localization of Orai1 was determined by means of immunocytochemistry. As in other cell types, Orai1 localized primarily at the plasma membrane of the eggs. Next, the complete Orai1 open reading frame was cloned and tagged with the green fluorescent protein to generate a plasmid encoding the GFP-Orai1 fusion protein. The plasmid was transcribed in vitro, and the resultant mRNA was microinjected into the eggs. Like the endogenous protein, GFP-tagged Orai1 localized in the cell cortex implying a role for Orai1 as a membrane channel.

We then investigated whether Orai1 is involved in generating the Ca^{2+} influx after depletion of the intracellular Ca^{2+} stores. GFP-tagged Orai1 was expressed in pig eggs by microinjecting GFP-Orai1 mRNA; control cells were microinjected with the carrier medium only. The intracellular stores of the eggs were then depleted in Ca^{2+}-free medium. Subsequent addition of Ca^{2+} to the cells induced an elevation in the cytosolic free Ca^{2+} concentration in both Orai1-overexpressing and control eggs, an indication of SOCE. The amplitude of the Ca^{2+} rise was markedly lower in Orai1-overexpressing eggs compared to the control. In addition, the Ca^{2+} influx was also slower in GFP-Orai1 mRNA-injected eggs: the time between Ca^{2+} add-back and the peak values were 827.5 ± 122.8 and 97.5 ± 67.5 s, respectively ($p = 0.002$; Wang et al. 2010b). These findings indicated that elevated levels of Orai1 in eggs had a negative effect on SOCE activity. This is in accordance with other studies showing that in somatic cell types, expression of Orai1 alone decreased SOCE and only when co-expressed with STIM1 did Orai1 enhance Ca^{2+} influx after store depletion (Mercer et al. 2006; Peinelt et al. 2006). Finally, downregulation of Orai1 in the eggs abolished the Ca^{2+} oscillations associated with fertilization (Fig. 29.2). Eggs microinjected with Orai1 siRNA at the beginning of maturation and cultured for 44 h displayed only the initial Ca^{2+} spike induced by the fertilizing sperm, and additional transients were not detected (Wang et al. 2012).

In the immature *Xenopus* oocyte, SOCE was also mediated by an interplay between ER-resident STIM1 and Orai1 channels in the plasma membrane (Yu et al. 2009). Depletion of Ca^{2+} stores led to co-clustering of STIM1 and Orai1 at the plasma membrane that was associated with the generation of large store-operated Ca^{2+} currents. These experiments also revealed how SOCE is inactivated during *Xenopus* oocyte maturation, a phenomenon that has been mentioned earlier. Uncoupling of store depletion from Ca^{2+} entry in the maturing

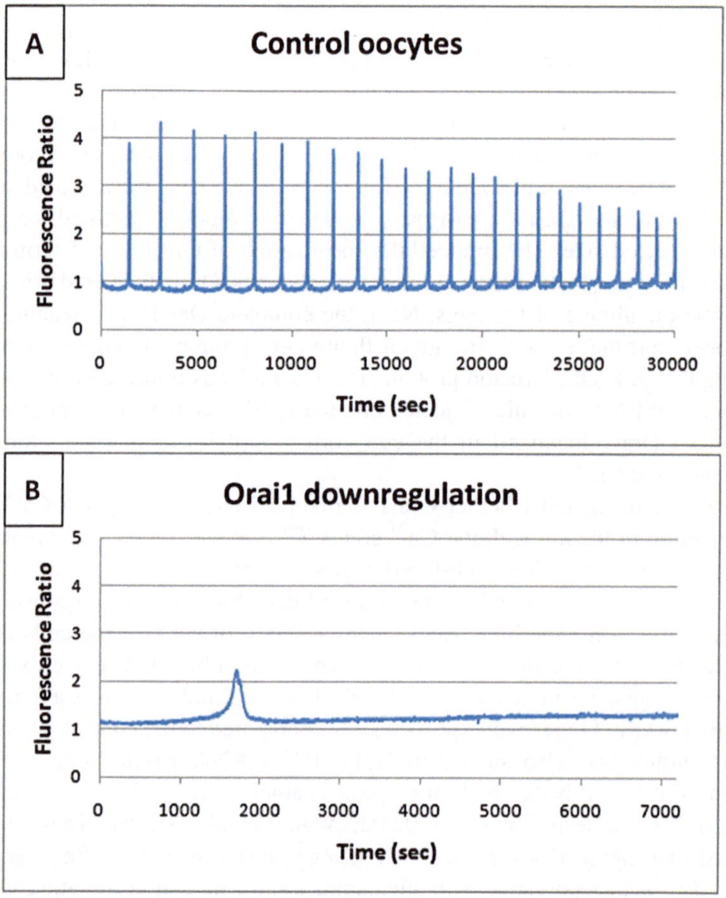

Fig. 29.2 Effects of Orai1 downregulation on the sperm-induced Ca^{2+} signal. (**a**) Control (scrambled siRNA-injected) eggs showed normal Ca^{2+} oscillations. (**b**) Downregulation of Orai1 caused the premature termination of the Ca^{2+} signal. Each trace is a representative response collected from a single egg

Xenopus oocyte was found to be the result of internalization of Orai1 into an intracellular vesicular compartment and the inability of STIM1 to cluster in response to store depletion. Blocking exocytosis by expressing a dominant negative SNAP-25 mutant in oocytes increased the presence of Orai1 in the intracellular region, indicating that Orai1 continuously recycles between the cell membrane and an endosomal compartment. It has recently been demonstrated that recycling of Orai1 between the endosome and the plasma membrane was mediated through a Rho- and Rab5-dependent endocytic pathway and that store depletion shifted endosomal Orai1 to the cell membrane (Yu et al. 2010). These data are consistent with the earlier findings that blocking exocytosis in *Xenopus* oocytes inhibited SOCE and that downregulation of Rho A, a small GTPase involved in membrane trafficking, increased the amplitude of the current through the store-operated Ca^{2+} channels (Yao et al. 1999).

When we compared the abundance of Orai1 transcripts in oocytes at various developmental stages, we found that Orai1 expression in immature oocytes was significantly higher compared to that after maturation (Wang et al. 2010a). This is somewhat different from that seen in *Xenopus* oocytes where total cellular Orai1 levels did not change during maturation although the protein was removed from the plasma membrane (Yu et al. 2009). Nevertheless, it indicates a similar tendency: that the number of Orai1 channels available to mediate a transmembrane Ca^{2+} current decreases during oocyte maturation. It is also consistent with the finding that Orai1 overexpression decreases SOCE and reveals an interesting feature of Orai1 function in female gametes. Oocytes acquire the ability to activate and form an embryo during maturation; this in mammals means that they become capable of generating repetitive Ca^{2+} signals in response to fertilization (Jones et al. 1995). The process is associated with the remodeling of Ca^{2+} signaling pathways and involves changes such as reorganization of the ER, increase in the number of IP_3 receptors, elevation in the concentration of stored Ca^{2+}, and redistribution of Ca^{2+}-binding proteins in the ER lumen (Carroll et al. 1996; Machaca 2007). As the increase in the intracellular free Ca^{2+} concentration after Ca^{2+} add-back in thapsigargin-treated oocytes was distinctly lower before compared to that measured after maturation (Fig.29.3; Wang et al. 2012), it now appears that a decrease in Orai1 in the plasma membrane also contributes to the ability of eggs to mount the

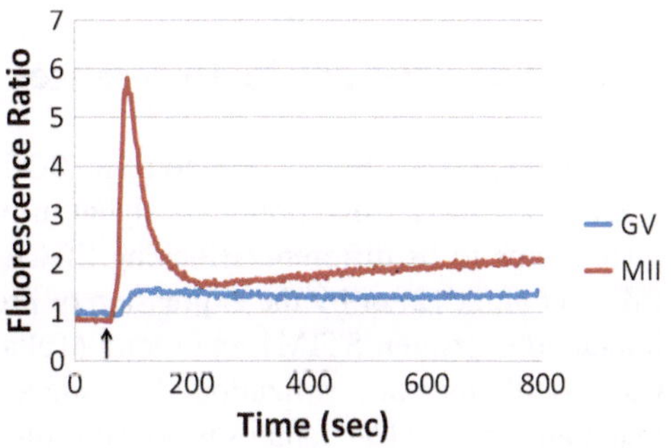

Fig. 29.3 Store-operated Ca^{2+} entry in oocytes before and after maturation. The Ca^{2+} influx stimulated by store depletion was markedly lower in immature oocytes compared to those that completed maturation. The *arrow* indicates the time of Ca^{2+} add-back after a treatment with thapsigargin

proper Ca^{2+} signal during fertilization. In *Xenopus* where fertilization evokes only a single Ca^{2+} transient, this serves as a safety mechanism: SOCE inactivates via Orai1 internalization to prevent premature activation (Yu et al. 2010), whereas in mammals a reduced Orai1 expression sets the appropriate Orai1 to STIM1 ratio so that an optimized SOCE will be able to sustain the repetitive Ca^{2+} oscillations.

SOCE is activated when after store depletion STIM1 oligomerizes and moves to ER sites that are closely juxtaposed to the plasma membrane and its carboxy-terminal region physically interacts with Orai1 (Luik et al. 2006; Prakriya et al. 2006; Vig et al. 2006a). In order to further confirm that it is SOCE mediated by an interaction between STIM1 and Orai1 that operates during fertilization, we have employed Förster resonance energy transfer (FRET) analysis. FRET is a distance-dependent proximity probe, which allows for the detection of energy transfer between two light-sensitive molecules if they are sufficiently close to each other (Gandhi and Isacoff 2005). The two constructs for the analysis, STIM1 tagged with the cyan fluorescent protein (STIM1-CFP) and Orai1 fused to the yellow fluorescent protein (Orai1-YFP), were the generous gift of Dr. Murali Prakriya at Northwestern University. cDNA of the constructs were transcribed in vitro to generate mRNAs; the mRNAs were injected into pig eggs, and after expression of the tagged proteins, the eggs were fertilized. In some eggs FRET, in others changes in the cytosolic Ca^{2+} levels, were then monitored. In principle, when the two proteins, STIM1 and Orai1, get close to each other, i.e., the distance between the donor CFP and the acceptor YFP becomes less than 10 nm, CFP can transfer energy to YFP resulting in an increase in the intensity of the fluorescence emitted by YFP.

The fertilizing sperm induced Ca^{2+} oscillations in the eggs, whose frequency was higher in cells that were injected with mRNAs compared to the control non-injected eggs. This is understandable if we consider that in these cells more STIM1 and Orai1 proteins were available to generate Ca^{2+} entry. In addition, we detected repetitive increases in FRET in the mRNA-injected eggs. This suggests that STIM1 repeatedly interacted with Orai1 to generate a Ca^{2+} influx whenever the stores became empty. Importantly, the frequency of the FRET signal was similar to that of the Ca^{2+} oscillations, indicating that they are related and they take place in synchrony in the fertilized egg (our unpublished observation). This suggests that during the sperm-induced Ca^{2+} oscillations, STIM1 interacts with Orai1 in a repetitive manner.

29.6 Species-Specific Differences

The findings described above point at SOCE as the mechanism responsible for maintaining the fertilization Ca^{2+} signal in pig eggs. Interestingly, the situation in mice seems to be different. Blocking SOCE in mouse eggs by specific inhibitors (Miao et al. 2012) or by the expression of protein fragments that interfere with the interaction between STIM1 and Orai1 (Takahashi et al. 2013) did not prevent the Ca^{2+} oscillations at fertilization. This suggests that in mouse eggs the Ca^{2+} entry that maintains the Ca^{2+} spikes is not controlled by the filling status of the stores. As mentioned earlier, the Ca^{2+} influx in mouse eggs was reported to be under the control of PKC. Phorbol esters are effective activators of PKC, and when applied to mouse eggs, they triggered low-amplitude Ca^{2+} transients and, also, a number of activation events downstream of the Ca^{2+} signal (Cuthbertson and Cobbold 1985; Colonna et al. 1989). Stimulation of PKC also led to Ca^{2+} influx and caused Ca^{2+} oscillations (Yu et al. 2008). The TRP channel became the center of attention as it can serve as a Ca^{2+} influx channel in various cell types (Ramsey et al. 2006), they are expressed in eggs (Petersen et al. 1995; Machaty et al. 2002), and because certain TRP isoforms are in fact regulated by PKC (Hardie 2007). An ion channel current activated by TRP agonists was then identified in mouse eggs; the current was absent in eggs of transgenic animals lacking TRPV3 channels. However, although the TRP agonists caused Ca^{2+} entry and subsequent activation in wild-type eggs, TRPV3$^{-/-}$ eggs displayed normal Ca^{2+} oscillations upon fertilization (Carvacho et al. 2013) indicating that these channels are not essential for normal fertilization. Subsequently, the T-type channel $Ca_V3.2$ was suggested as part of the mechanism that mediates Ca^{2+} influx. Mouse eggs lacking the α1 subunit of this channel were unable to express a voltage-activated Ca^{2+} current, accumulated abnormally low levels of Ca^{2+} in their endoplasmic reticulum during maturation, and displayed impaired Ca^{2+} oscillations at fertilization (Bernhardt et al. 2015). This indicates the involvement of $Ca_V3.2$ channels in Ca^{2+} homeostasis and signaling, but the fact that mice having nonfunctional $Ca_V3.2$ channels are still fertile (although they show impaired fertility) implies that additional influx mechanisms must also be involved. Recent research identified the transient receptor potential melastanin 7 (TRPM7)-like channel as a potential candidate. Such channels are expressed in mouse eggs, and their pharmacological block prevents spontaneous Ca^{2+} influxes in immature oocytes, and it also disrupts the repetitive Ca^{2+} signal after sperm-egg fusion (Carvacho et al. 2016; Bernhardt et al. 2017). In chicken lymphocytes, TRPM7, via its intrinsic kinase domain, has been shown to modulate SOCE (Faouzi et al. 2017); thus, it is going to be interesting to learn the complex interactions that mediate Ca^{2+} entry at fertilization in mice and potentially in other species as well.

29.7 Conclusions

The results discussed above strongly imply that SOCE mediated by an interaction between STIM1 and Orai1 is essential to sustain the long-lasting Ca^{2+} oscillations during mammalian fertilization, at least in some species. However, Ca^{2+} oscillations are not always mediated by store-operated Ca^{2+} channels but may involve membrane channels opened by lipid second messengers (Shuttleworth 1999) or other channels such as TRPs that are associated with both store- and receptor-operated Ca^{2+} entries (Grimaldi et al. 2003). As mentioned above, female gametes also contain TRP channels, and certain TRP channels may be activated by STIM1 (Huang et al. 2006; Yuan et al. 2009). This necessitates further investigation regarding the exact mechanism that maintains the Ca^{2+} oscillations at fertilization in other species including humans. In addition, there is increasing evidence that intracellular Ca^{2+} signals control a number of behavioral responses in sperm as well including capacitation, acrosome reaction, hyperactivated flagellar motility, and chemotaxis (reviewed by Florman et al. 2008). These events require the presence of extracellular Ca^{2+}, and in some cases Ca^{2+} may be provided through a store-operated influx mechanism. A better understanding of the molecular basis of the various Ca^{2+} entry mechanisms and the characterization of their interaction during fertilization are expected to make the fertilization process more effective during the application of various assisted reproductive technologies. In addition, characterization of the pharmacology of the components has the potential to lead to the development of new contraceptive methods.

References

Barish ME (1983) A transient calcium-dependent chloride current in the immature Xenopus oocyte. J Physiol 342:309–325

Bernhardt ML, Zhang Y, Erxleben CF, Padilla-Banks E, McDonough CE, Miao YL, Armstrong DL, Williams CJ (2015) CaV3.2 T-type channels mediate Ca^{2+} entry during oocyte maturation and following fertilization. J Cell Sci 128:4442–4452

Bernhardt ML, Padilla-Banks E, Stein P, Zhang Y, Williams CJ (2017) Store-operated Ca^{2+} entry is not required for fertilization-induced Ca^{2+} signaling in mouse eggs. Cell Calcium. doi:10.1016/j.ceca.2017.02.004

Bezprozvanny I, Watras J, Ehrlich BE (1991) Bell-shaped calcium-response curves of Ins(1,4,5)P_3- and calcium-gated channels from endoplasmic reticulum of cerebellum. Nature 351:751–754

Bobanovic LK, Laine M, Petersen CC, Bennett DL, Berridge MJ, Lipp P, Ripley SJ, Bootman MD (1999) Molecular cloning and immunolocalization of a novel vertebrate trp homologue from Xenopus. Biochem J 340:593–599

Brereton HM, Harland ML, Auld AM, Barritt GJ (2000) Evidence that the TRP-1protein is unlikely to account for store-operated Ca^{2+} inflow in Xenopus laevis oocytes. Mol Cell Biochem 214:63–74

Carroll J, Jones KT, Whittingham DG (1996) Ca^{2+} release and the development of Ca^{2+} release mechanisms during oocyte maturation: a prelude to fertilization. Rev Reprod (3):137–143

Carvacho I, Lee HC, Fissore RA, Clapham DE (2013) TRPV3 channels mediate strontium-induced mouse-egg activation. Cell Rep 5:1375–1386

Carvacho I, Ardestani G, Lee HC, McGarvey K, Fissore RA, Lykke-Hartmann K (2016) TRPM7-like channels are functionally expressed in oocytes and modulate post-fertilization embryo development in mouse. Sci Rep 6:34236

Colonna R, Tatone C, Malgaroli A, Eusebi F, Mangia F (1989) Effects of protein kinase C stimulation and free Ca^{2+} rise in mammalian egg activation. Gamete Res 24:171–183

Cuthbertson KS, Cobbold PH (1985) Phorbol ester and sperm activate mouse oocytes by inducing sustained oscillations in cell Ca^{2+}. Nature 316:541–542

Ducibella T, Huneau D, Angelichio E, Xu Z, Schultz RM, Kopf GS, Fissore R, Madoux S, Ozil JP (2002) Egg-to-embryo transition is driven by differential responses to Ca^{2+} oscillation number. Dev Biol 250:280–291

Dumollard R, Carroll J, Dupont G, Sardet C (2002) Calcium wave pacemakers in eggs. J Cell Sci 115:3557–3564

Dupont G, Dumollard R (2004) Simulation of calcium waves in ascidian eggs: insights into the origin of the pacemaker sites and the possible nature of the sperm factor. J Cell Sci 117:4313–4323

Faouzi M, Kilch T, Horgen FD, Fleig A, Penner R (2017) The TRPM7 channel kinase regulates store-operated calcium entry. J Physiol. doi:10.1113/JP274006

Faure JE, Myles DG, Primakoff P (1999) The frequency of calcium oscillations in mouse eggs at fertilization is modulated by the number of fused sperm. Dev Biol 213:370–377

Feske S, Gwack Y, Prakriya M, Srikanth S, Puppel SH, Tanasa B, Hogan PG, Lewis RS, Daly M, Rao A (2006) A mutation in Orai1 causes immune deficiency by abrogating CRAC channel function. Nature 441:179–185

Florman HM, Jungnickel MK, Sutton KA (2008) Regulating the acrosome reaction. Int J Dev Biol 52:503–510

Gandhi CS, Isacoff EY (2005) Shedding light on membrane proteins. Trends Neurosci 28:472–479

Gómez-Fernández C, Pozo-Guisado E, Ganan-Parra M, Perianes MJ, Alvarez IS, Martin-Romero FJ (2009) Relocalization of STIM1 in mouse oocytes at fertilization: early involvement of store-operated calcium entry. Reproduction 138:211–221

Grimaldi M, Maratos M, Verma A (2003) Transient receptor potential channel activation causes a novel form of $[Ca^{2+}]_i$ oscillations and is not involved in capacitative Ca^{2+} entry in glial cells. J Neurosci 23:4737–4745

Halet G, Tunwell R, Parkinson SJ, Carroll J (2004) Conventional PKCs regulate the temporal pattern of Ca^{2+} oscillations at fertilization in mouse eggs. J Cell Biol 164:1033–1044

Hardie RC (2007) TRP channels and lipids: from Drosophila to mammalian physiology. J Physiol 578:9–24

Huang GN, Zeng W, Kim JY, Yuan JP, Han L, Muallem S, Worley PF (2006) STIM1 carboxyl-terminus activates native SOC, ICRAC and TRPC1 channels. Nat Cell Biol 8:1003–1010

Igusa Y, Miyazaki S (1983) Effects of altered extracellular and intracellular calcium concentration on hyperpolarizing responses of the hamster egg. J Physiol 340:611–632

Jones KT (2007) Intracellular calcium in the fertilization and development of mammalian eggs. Clin Exp Pharmacol Physiol 34:1084–1089

Jones KT, Carroll J, Whittingham DG (1995) Ionomycin, thapsigargin, ryanodine and sperm sensitive calcium release increase during meiotic maturation of mouse oocytes. J Biol Chem 270:6671–6677

Jones KT, Cruttwell C, Parrington J, Swann K (1998) A mammalian sperm cytosolic phospholipase C activity generates inositol trisphosphate and causes Ca^{2+} release in sea urchin egg homogenates. FEBS Lett 437:297–300

Kline D, Kline JT (1992) Thapsigargin activates a calcium influx pathway in the unfertilized mouse egg and suppresses repetitive calcium transients in the fertilized egg. J Biol Chem 267:17624–17630

Koh S, Lee K, Wang C, Cabot RA, Machaty Z (2009) STIM1 regulates store-operated Ca^{2+} entry in oocytes. Dev Biol 330:368–376

Kubiak JZ (1989) Mouse oocytes gradually develop the capacity for activation during the metaphase II arrest. Dev Biol 136:537–545

Lee K, Wang C, Chaille JM, Machaty Z (2010) STIM1 is required for normal fertilization in the pig. In: Proceedings of the annual meeting of the Society for the Study of Reproduction, Milwaukee, Wisconsin, p 38

Lee K, Wang C, Machaty Z (2012) STIM1 is required for Ca^{2+} signaling during mammalian fertilization. Dev Biol 367:154–162

Liou J, Kim ML, Heo WD, Jones JT, Myers JW, Ferrell JE Jr, Meyer T (2005) STIM is a Ca^{2+} sensor essential for Ca^{2+}-store-depletion-triggered Ca^{2+} influx. Curr Biol 15:1235–1241

Luik RM, MM W, Buchanan J, Lewis RS (2006) The elementary unit of store-operated Ca^{2+} entry: local activation of CRAC channels by STIM1 at ER-plasma membrane junctions. J Cell Biol 174:815–825

Lupu-Meiri M, Beit-Or A, Christensen SB, Oron Y (1993) Calcium entry in Xenopus oocytes: effects of inositol trisphosphate, thapsigargin and DMSO. Cell Calcium 14:101–110

Machaca K (2003) Ca^{2+}-calmodulin-dependent protein kinase II potentiates store-operated Ca^{2+} current. J Biol Chem 278:33730–33737

Machaca K (2007) Ca^{2+} signaling differentiation during oocyte maturation. J Cell Physiol 213:331–340

Machaca K, Haun S (2002) Induction of maturation-promoting factor during Xenopus oocyte maturation uncouples Ca^{2+} store depletion from store-operated Ca^{2+} entry. J Cell Biol 156:75–85

Machaty Z, Ramsoondar JJ, Bonk AJ, Bondioli KR, Prather RS (2002) Capacitative calcium entry mechanism in porcine oocytes. Biol Reprod 66:667–674

Madgwick S, Levasseur M, Jones KT (2005) Calmodulin-dependent protein kinase II, and not protein kinase C, is sufficient for triggering cell-cycle resumption in mammalian eggs. J Cell Sci 118:3849–3859

Martin-Romero FJ, Ortiz-de-Galisteo JR, Lara-Laranjeira J, Dominguez-Arroyo JA, Gonzalez-Carrera E, Alvarez IS (2008) Store-operated calcium entry in human oocytes and sensitivity to oxidative stress. Biol Reprod 78:307–315

Matifat F, Fournier F, Lorca T, Capony JP, Brûlé G, Collin T (1997) Involvement of the Ca^{2+}/calmodulin-dependent protein kinase II pathway in the Ca^{2+}-mediated regulation of the capacitative Ca^{2+} entry in Xenopus oocytes. Biochem J 322:267–272

McGuinness OM, Moreton RB, Johnson MH, Berridge MJ (1996) A direct measurement of increased divalent cation influx in fertilised mouse oocytes. Development 122:2199–2206

Mercer JC, Dehaven WI, Smyth JT, Wedel B, Boyles RR, Bird GS, Putney JW Jr (2006) Large store-operated calcium selective currents due to co-expression of Orai1 or Orai2 with the intracellular calcium sensor, Stim1. J Biol Chem 281:24979–24990

Miao YL, Stein P, Jefferson WN, Padilla-Banks E, Williams CJ (2012) Calcium influx-mediated signaling is required for complete mouse egg activation. Proc Natl Acad Sci U S A 109:4169–4174

Miyazaki S (1991) Repetitive calcium transients in hamster oocytes. Cell Calcium 12:205–216

Miyazaki S, Igusa Y (1981) Ca^{2+}-dependent action potential and Ca^{2+}-induced fertilization potential in golden hamster eggs. In: Ohnishi ST, Endo M (eds) The mechanisms of gated calcium transport across biological membranes. Academic, New York, pp 305–311

Miyazaki S, Igusa Y (1982) Ca-mediated activation of a K current at fertilization of golden hamster eggs. Proc Natl Acad Sci U S A 79:931–935

Ozil JP, Markoulaki S, Toth S, Matson S, Banrezes B, Knott JG, Schultz RM, Huneau D, Ducibella T (2005) Egg activation events are regulated by the duration of a sustained $[Ca^{2+}]_{cyt}$ signal in the mouse. Dev Biol 282:39–54

Parekh AB, Terlau H, Stühmer W (1993) Depletion of $InsP_3$ stores activates a Ca^{2+} and K^+ current by means of a phosphatase and a diffusible messenger. Nature 364:814–818

Parker I, Miledi R (1987) Inositol trisphosphate activates a voltage-dependent calcium influx in Xenopus oocytes. Proc R Soc Lond B Biol Sci 231:27–36

Parker I, Gundersen CB, Miledi R (1985) A transient inward current elicited by hyperpolarization during serotonin activation in Xenopus oocytes. Proc R Soc Lond B Biol Sci 223:279–292

Peinelt C, Vig M, Koomoa DL, Beck A, Nadler MJ, Koblan-Huberson M, Lis A, Fleig A, Penner R, Kinet JP (2006) Amplification of CRAC current by STIM1 and CRACM1 (Orai1). Nat Cell Biol 8:771–773

Petersen CC, Berridge MJ (1994) The regulation of capacitative calcium entry by calcium and protein kinase C in Xenopus oocytes. J Biol Chem 269:32246–32253

Petersen CC, Berridge MJ, Borgese MF, Bennett DL (1995) Putative capacitative calcium entry channels: expression of Drosophila trp and evidence for the existence of vertebrate homologues. Biochem J 311:41–44

Prakriya M, Feske S, Gwack Y, Srikanth S, Rao A, Hogan PG (2006) Orai1 is an essential pore subunit of the CRAC channel. Nature 443:230–233

Ramsey IS, Delling M, Clapham DE (2006) An introduction to TRP channels. Annu Rev Physiol 68:619–647

Roos J, DiGregorio PJ, Yeromin AV, Ohlsen K, Lioudyno M, Zhang S, Safrina O, Kozak JA, Wagner SL, Cahalan MD, Veliçelebi G, Stauderman KA (2005) STIM1, an essential and conserved component of store-operated Ca^{2+} channel function. J Cell Biol 169:435–445

Saunders CM, Larman MG, Parrington J, Cox LJ, Royse J, Blayney LM, Swann K, Lai FA (2002) PLCzeta: a sperm-specific trigger of Ca^{2+} oscillations in eggs and embryo development. Development 129:3533–3544

Schultz RM, Kopf GS (1995) Molecular basis of mammalian egg activation. Curr Top Dev Biol 30:21–62

Shiina Y, Kaneda M, Matsuyama K, Tanaka K, Hiroi M, Doi K (1993) Role of the extracellular Ca^{2+} on the intracellular Ca^{2+} changes in fertilized and activated mouse oocytes. J Reprod Fertil 97:143–150

Shuttleworth TJ (1999) What drives calcium entry during $[Ca^{2+}]_i$ oscillations?—challenging the capacitative model. Cell Calcium 25:237–246

Stricker SA (1999) Comparative biology of calcium signaling during fertilization and egg activation in animals. Dev Biol 211:157–176

Swann K (1990) A cytosolic sperm factor stimulates repetitive calcium increases and mimics fertilization in hamster eggs. Development 110:1295–1302

Swann K, Yu Y (2008) The dynamics of calcium oscillations that activate mammalian eggs. Int J Dev Biol 52:585–594

Takahashi T, Kikuchi T, Kidokoro Y, Shirakawa H (2013) Ca^{2+} influx-dependent refilling of intracellular Ca2+ stores determines the frequency of Ca2+ oscillations in fertilized mouse eggs. Biochem Biophys Res Commun 430:60–65

Tomita Y, Kaneko S, Funayama M, Kondo H, Satoh M, Akaike A (1998) Intracellular Ca^{2+} store-operated influx of Ca^{2+} through TRP-R, a rat homolog of TRP, expressed in Xenopus oocytes. Neurosci Lett 248:195–198

Turner PR, Jaffe LA, Fein A (1986) Regulation of cortical vesicle exocytosis in sea urchin eggs by inositol 1,4,5-trisphosphate and GTP-binding protein. J Cell Biol 102:70–76

Vig M, Beck A, Billingsley JM, Lis A, Parvez S, Peinelt C, Koomoa DL, Soboloff J, Gill DL, Fleig A, Kinet JP, Penner R (2006a) CRACM1 multimers form the ion-selective pore of the CRAC channel. Curr Biol 16:2073–2079

Vig M, Peinelt C, Beck A, Koomoa DL, Rabah D, Koblan-Huberson M, Kraft S, Turner H, Fleig A, Penner R, Kinet JP (2006b) CRACM1 is a plasma membrane protein essential for store-operated Ca^{2+} entry. Science 312:1220–1223

Wang C, Lee K, Machaty Z (2010a) Characterization of Orai1 in porcine oocytes. Reprod Domest Anim 45(suppl 3):73

Wang C, Lee K, Machaty Z (2010b) The role of Orai1 in store-operated calcium entry in porcine oocytes. In: Proceedings of the annual meeting of the Society for the Study of Reproduction, Milwaukee, Wisconsin, p 126

Wang C, Lee K, Gajdócsi E, Bali Papp Á, Machaty Z (2012) Orai1 mediates store-operated Ca^{2+} entry during fertilization in mammalian oocytes. Dev Biol 365:414–423

Whitaker M (1996) Control of meiotic arrest. Rev Reprod 1:127–135

Yao Y, Parker I (1993) Inositol trisphosphate-mediated Ca^{2+} influx into Xenopus oocytes triggers Ca^{2+} liberation from intracellular stores. J Physiol 468:275–295

Yao Y, Ferrer-Montiel AV, Montal M, Tsien RY (1999) Activation of store-operated Ca^{2+} current in Xenopus oocytes requires SNAP-25 but not a diffusible messenger. Cell 98:475–485

Yu Y, Halet G, Lai FA, Swann K (2008) Regulation of diacylglycerol production and protein kinase C stimulation during sperm- and PLCζ-mediated mouse egg activation. Biol Cell 100:633–643

Yu F, Sun L, Machaca K (2009) Orai1 internalization and STIM1 clustering inhibition modulate SOCE inactivation during meiosis. Proc Natl Acad Sci U S A 106:17401–17406

Yu F, Sun L, Machaca K (2010) Constitutive recycling of the store-operated Ca^{2+} channel Orai1 and its internalization during meiosis. J Cell Biol 191:523–535

Yuan JP, Zeng W, Dorwart MR, Choi YJ, Worley PF, Muallem S (2009) SOAR and the polybasic STIM1 domains gate and regulate Orai channels. Nat Cell Biol 11:337–343

Metabolic Disorders and Cancer: Hepatocyte Store-Operated Ca^{2+} Channels in Nonalcoholic Fatty Liver Disease

30

Eunüs S. Ali, Grigori Y. Rychkov, and Greg J. Barritt

Abstract

In steatotic hepatocytes, intracellular Ca^{2+} homeostasis is substantially altered compared to normal. Decreased Ca^{2+} in the endoplasmic reticulum (ER) can lead to ER stress, an important mediator of the progression of liver steatosis to nonalcoholic steatohepatitis, type 2 diabetes, and hepatocellular carcinoma. Store-operated Ca^{2+} channels (SOCs) in hepatocytes are composed principally of Orai1 and STIM1 proteins. Their main role is the maintenance of adequate Ca^{2+} in the lumen of the ER. In steatotic hepatocytes, store-operated Ca^{2+} entry (SOCE) is substantially inhibited. This inhibition is associated with a decrease in Ca^{2+} in the ER. Lipid-induced inhibition of SOCE is mediated by protein kinase C (PKC) and may involve the phosphorylation and subsequent inhibition of Orai1. Experimental inhibition of SOCE enhances lipid accumulation in normal hepatocytes incubated in the presence of exogenous fatty acids. The antidiabetic drug exendin-4 reverses the lipid-induced inhibition of SOCE and decreases liver lipid with rapid onset. It is proposed that lipid-induced inhibition of SOCE in the plasma membrane and of SERCA2b in the ER membrane leads to a persistent decrease in ER Ca^{2+}, ER stress, and the ER stress response, which in turn enhances (amplifies) lipid accumulation. A low level of persistent SOCE due to chronic ER Ca^{2+} depletion in steatotic hepatocytes may contribute to an

E.S. Ali • G.J. Barritt (✉)
Department of Medical Biochemistry, School of Medicine, Flinders University, GPO Box 2100, Adelaide, SA 5001, Australia

Centre for Neuroscience, School of Medicine, Flinders University, GPO Box 2100, Adelaide, SA 5001, Australia
e-mail: eunus.ali_csl@yahoo.com.au; greg.barritt@flinders.edu.au

G.Y. Rychkov
School of Medicine, The University of Adelaide, Adelaide, SA, Australia

South Australian Health and Medical Research Institute, Adelaide, SA 5005, Australia
e-mail: grigori.rychkov@adelaide.edu.au

K. Groschner et al. (eds.), *Store-Operated Ca^{2+} Entry (SOCE) Pathways*, Advances in Experimental Medicine and Biology 993, DOI 10.1007/978-3-319-57732-6_30

elevated cytoplasmic-free Ca^{2+} concentration leading to the activation of calcium-calmodulin kinase II (CaMKII), decreased lipid removal by autophagy, and insulin resistance. It is concluded that lipid-induced inhibition of SOCE plays an important role in the progression of liver steatosis to insulin insensitivity and hepatocellular carcinoma.

Keywords
Store-operated Ca^{2+} entry • Liver • Steatosis • Exendin-4 • GLP-1 • Intracellular Ca^{2+} • Cyclic AMP

30.1 Introduction

The liver plays a central role in regulating whole-body carbohydrate, lipid, and protein homeostasis, as well as performing specific functions such as the synthesis and transport of bile acids and the detoxification of endogenous and exogenous metabolites (Boyer 2002; Leite and Nathanson 2001). Hepatocytes (parenchymal cells) are the predominant cell type comprising about 70% of the total liver cells (about 90% of liver volume). Hepatocytes are principally responsible for, or contribute to, all major functions of the liver. Other cell types include endothelial, smooth muscle, biliary epithelial (cholangiocytes), hepatic stellate, Kupffer (macrophages), and oval (fat storing) cells (Boyer 2002; Dagli et al. 1998; Leite and Nathanson 2001; Young and Heath 2000). These have specific functions including mediating the flow of blood and bile, defense mechanisms, and liver regeneration. In this review, we focus on the role of store-operated Ca^{2+} channels (SOCs) in hepatocytes.

SOCs are the main pathway of Ca^{2+} entry to hepatocytes under normal physiological conditions (Aromataris et al. 2008; Barritt et al. 2008; Jones et al. 2008; Rychkov et al. 2001; Wilson et al. 2015). Their role is the maintenance of an appropriate concentration of Ca^{2+} in the lumen of the endoplasmic reticulum (ER) and, via Ca^{2+} transfer from the ER to mitochondria, of Ca^{2+} in the mitochondrial matrix (Aromataris et al. 2008; Barritt et al. 2008; Wilson et al. 2015). Ca^{2+} entry through SOCs may also contribute directly to local and global increases in Ca^{2+} induced by hormones and to the maintenance of the cytoplasmic-free Ca^{2+} concentration ($[Ca^{2+}]_{cyt}$) in non-stimulated (resting) hepatocytes (Barritt et al. 2008). During the past 10 years, work in a number of laboratories has shown that a central feature of liver steatosis and many other liver diseases is ER stress and the ER stress response (Bozaykut et al. 2016; Han and Kaufman 2016; Pagliassotti et al. 2016). The accumulation of lipid droplets in hepatocytes and resulting steatosis are central components of nonalcoholic fatty liver disease (NAFLD). This condition can progress to nonalcoholic steatosis (NASH), insulin resistance, and ultimately to hepatocellular carcinoma (HCC) (Cabre et al. 2016; Karin and Dhar 2016; Ma et al. 2016; Wong et al. 2016).

Recently, it has been shown that a decrease in Ca^{2+} uptake to the ER, which leads to substantially decreased $[Ca^{2+}]_{ER}$, is central to the development of ER stress in steatotic hepatocytes (Arruda and Hotamisligil 2015; Fu et al. 2011; Park et al. 2010). The decrease in Ca^{2+} uptake to the ER is due to lipid-induced inhibition of both SERCA2b in the ER and store-operated Ca^{2+} entry (SOCE) in the plasma membrane (Ali et al. 2016; Arruda and Hotamisligil 2015; Fu et al. 2011; Park et al. 2010; Wilson et al. 2015). Lipid-induced inhibition of SOCE in steatotic hepatocytes is associated with decreased $[Ca^{2+}]_{ER}$ and an exacerbation of lipid accumulation (Wilson et al. 2015). Lipid-induced inhibition of SOCE can be reversed by the glucagon-like peptide-1 (GLP-1) receptor agonist exendin-4, an antidiabetic drug (Ali et al. 2016). Other studies have provided evidence which suggests that liver steatosis is associated with an elevated resting hepatocyte $[Ca^{2+}]_{cyt}$ (Ozcan et al. 2013; Ozcan and Tabas 2016; Park and Lee 2014; Park et al. 2014). It is proposed that in steatotic hepatocytes chronically elevated $[Ca^{2+}]_{cyt}$ inhibits the autophagic pathway and activates calmodulin-dependent protein kinase II (CaMKII). Disruption of the autophagic pathway contributes to lipid accumulation, while activated CaMKII contributes to the development of liver insulin resistance (Ozcan et al. 2013; Ozcan and Tabas 2016). Administration of inhibitors of voltage-operated Ca^{2+} channels (VOCCs) in an animal model of obesity was found to increase lipid droplet autophagy and decrease lipid in steatotic livers (Park and Lee 2014; Park et al. 2014).

In this review, we aim to summarize recent results on lipid-induced inhibition of SOCE in steatotic hepatocytes and to discuss the role this inhibition of SOCE is proposed to play in the development of ER stress and in the downstream consequences of chronic ER stress. Since SOCE can potentially regulate $[Ca^{2+}]_{cyt}$ both directly and indirectly (via Ca^{2+} in the ER) and there appear to be some different observations, assumptions, and conclusions about the effects of steatosis on $[Ca^{2+}]_{cyt}$ in hepatocytes, this aspect of the functions of SOCE in hepatocytes is also addressed.

30.2 Roles of Store-Operated Ca^{2+} Channels in Normal Hepatocytes

Increases in $[Ca^{2+}]_{cyt}$ in hepatocytes mediate the hormonal regulation of liver metabolism and the regulation of bile flow and many other functions of the liver. Hormones which use changes in $[Ca^{2+}]_{cyt}$ as an intracellular messenger include adrenaline, glucagon, vasopressin, insulin, several other hormones, and cytokines (Aromataris et al. 2006; Barritt et al. 2008; Bartlett et al. 2014; Exton et al. 1981; Rieusset 2017; Wilson et al. 2015). In addition, the maintenance of adequate concentrations of free Ca^{2+} in the lumen of the ER and in the mitochondrial matrix is essential for normal lipid, carbohydrate, and protein synthesis (Arruda and Hotamisligil 2015; Fu et al. 2011; Park et al. 2010; Rieusset 2017).

Hormone-induced increases in $[Ca^{2+}]_{cyt}$ are principally generated by the activation of phospholipase C, formation of inositol 1,4,5-trisphosphate (IP_3), and the subsequent IP_3-induced release of Ca^{2+} from the ER (Barritt et al. 2008; Bartlett et al. 2014; Burgess et al. 1984; Derler et al. 2016a; Fahrner et al. 2013; Prakriya and Lewis 2015). At normal physiological concentrations of a given hormone, the hormone-induced increase in IP_3 causes oscillations in $[Ca^{2+}]_{cyt}$. These transmit a Ca^{2+} signal, coded by amplitude and frequency, to Ca^{2+}-sensitive target enzymes and proteins (Woods et al. 1986, 1987) (reviewed in Barritt et al. 2008; Gaspers and Thomas 2005). These oscillations in $[Ca^{2+}]_{cyt}$ are principally due to the release of Ca^{2+} from the ER through IP_3 receptors and the subsequent reuptake of Ca^{2+} to the ER via the ER (Ca^{2+} + Mg^{2+})ATPase2b (SERCA2b) Ca^{2+} pumps (Jones et al. 2008; Parekh 2011). The major components of the hepatocyte intracellular Ca^{2+} signaling system are shown in a highly schematic form in Fig. 30.1.

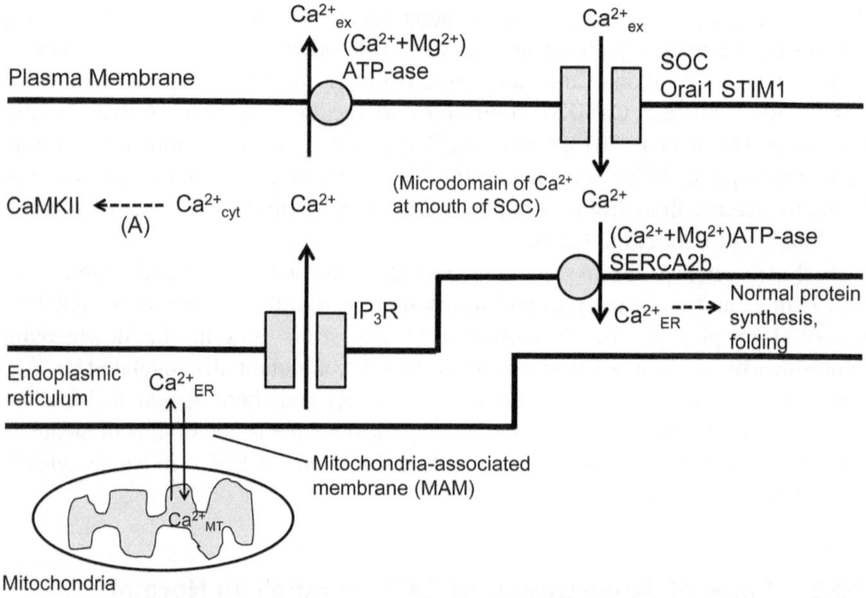

Fig. 30.1 A highly simplified scheme showing the role of store-operated Ca^{2+} channels in maintaining an adequate concentration of free Ca^{2+} in the lumen of the endoplasmic reticulum, in the mitochondrial matrix, and indirectly in the cytoplasmic space of normal hepatocytes. The binding of hormones to receptors in the plasma membrane activates phospholipase C (not shown) leading to the generation of IP_3 which releases Ca^{2+} from the ER via IP_3 receptors (IP_3R) and results in a decrease in $[Ca^{2+}]_{ER}$. This generates a cytoplasmic Ca^{2+} signal which activates downstream Ca^{2+}-dependent enzymes such as calcium-calmodulin kinase II (CAMII kinase). During hormonal Ca^{2+} signaling, some Ca^{2+} is transferred from the cytoplasmic space (and from the ER via the cytoplasmic space) to the extracellular space via the plasma membrane (Ca^{2+} + Mg^{2+})ATPase. To replenish the ER Ca^{2+} store, the decrease in ER $[Ca^{2+}]_{ER}$ activates SOCE leading to Ca^{2+} uptake to the ER via SOCs and SERCA2b. Some ER Ca^{2+} is also transferred to the matrix of mitochondria associated with the ER through mitochondria-associated membranes (MAM) of the ER. It is thought that, under normal conditions, refilling of the ER from SOCE involves only a modest increase in $[Ca^{2+}]_{cyt}$ in the vicinity of the mouth of the SOC due to rapid uptake of Ca^{2+} to the ER via SERCA2b and also by nearby mitochondria (not shown)

During this cytoplasmic Ca^{2+} signaling, some Ca^{2+} is transported from the cytoplasmic space to the extracellular space by the plasma membrane (Ca^{2+} + Mg^{2+})-ATPase and possibly also by the (Na^+ – Ca^{2+}) exchange system (reviewed in Barritt et al. 2008). Ca^{2+} entry through store-operated Ca^{2+} channels (SOCE) in the plasma membrane is required to maintain sufficient amounts of Ca^{2+} within the hepatocyte and the hormone-initiated $[Ca^{2+}]_{cyt}$ signal over periods of minutes or hours. SOCE is also needed to replenish the ER Ca^{2+} stores after the hormonal signal is turned off and to maintain appropriate concentrations of Ca^{2+} in the lumen of the ER and in the mitochondrial matrix over time (Jones et al. 2008; Prakriya and Lewis 2015; Wilson et al. 2015).

Measurement of both mRNA and protein expression has shown that in primary rat hepatocytes Orai1 is the predominantly expressed isoform, with some expression of Orai3 and little or none of Orai2 (Jones et al. 2008; Wilson et al. 2015). The major isoform of STIM expressed in rat hepatocytes is STIM1, with smaller amounts of STIM2 (Jones et al. 2008; Wilson et al. 2015). In H4IIE rat liver cells, a similar pattern of expression of Orai and STIM has been observed (Wilson et al. 2015). In rat and mouse hepatocytes, the SOCE channel pore is composed predominantly of the Orai1 (channel pore) and STIM1 (activator) polypeptides (Aromataris et al. 2008; Jones et al. 2008; Litjens et al. 2007). Studies with human hepatocytes suggest that transient receptor potential canonical 6 (TRPC6) may also make a contribution to SOCE (El Boustany et al. 2008). As in other animal cell types, Orai1 is activated by a decrease in Ca^{2+} in the lumen of the ER at the cortical endoplasmic reticulum-plasma membrane (ER-PM) junctions. This alters the conformation of STIM1 leading to the binding of STIM1 to Orai1 and opening of the channel pore (Prakriya and Lewis 2015; Scrimgeour et al. 2014). Indirect evidence suggests that in hepatocytes, and possibly also in some other cell types, a decrease in Ca^{2+} in a small subregion of the ER, rather than in the bulk of the ER, is all that is needed to activate SOCE (Castro et al. 2009; Gregory et al. 1999). Other experiments have suggested that in hepatocytes Ca^{2+} release from the ER in this subregion is mediated principally by type 1 IP_3 receptors (IP_3R) (Gregory et al. 1999).

It is thought that in many types of animal cells, most of the Ca^{2+} which enters the cell through SOCs is immediately transported to the ER by SERCA pumps and to the mitochondrial matrix through the mitochondrial Ca^{2+} uniporter. As a result, the increase in $[Ca^{2+}]_{cyt}$ due to SOCE is likely to be limited to a small region or microdomain in the vicinity of the SOC pore (Courjaret and Machaca 2014; Golovina 2005; Jousset et al. 2007; Vaca 2010). In some situations, this increase in $[Ca^{2+}]_{cyt}$ due to SOCE has important signaling roles such as the regulation of adenylate cyclase (Chiono et al. 1995; Cooper 2015). In hepatocytes it is also likely that under normal conditions most Ca^{2+} which enters the cell through SOCs is transported to the ER by SERCA pumps and to the mitochondrial matrix through the mitochondrial Ca^{2+} uniporter. However, we are not aware of any studies on hepatocytes which have used confocal fluorescence imaging or total internal reflection fluorescence (TIRF) to investigate the increase in $[Ca^{2+}]_{cyt}$ underneath the plasma membrane during SOCE. Some studies using epifluorescence have reported little or no increase in $[Ca^{2+}]_{cyt}$ during SOCE (Barritt 1999; Chan et al. 2004; Fernando et al. 1998; To et al. 2010) (Fig. 30.1).

The description above of the role and mechanisms of activation of SOCs in hepatocytes is likely a very much simplified view of what is a much more complicated set of intracellular signaling pathways. Relatively little is known of how the components of the hepatocyte SOC network and subsequent changes in intracellular Ca^{2+} operate in the dimensions of both time and intracellular space. For example, different extracellular signals such as insulin, glucagon, adrenaline, and adenosine triphosphate (ATP) will act for different time periods and their signals will be influenced by other extracellular and intracellular conditions, such as the amount of intracellular lipid and level of reactive oxygen species (ROS).

Several other Ca^{2+}-permeable channels in the plasma membrane of hepatocytes also mediate Ca^{2+} entry (Barritt et al. 2008; El Boustany et al. 2008; Jones et al. 2008; Kheradpezhouh et al. 2014; Rychkov and Barritt 2011). These are likely activated in response to specific extracellular and intracellular signals, some of which are only generated in pathological conditions. These other plasma membrane Ca^{2+} entry pathways include TRPM2 nonselective cation channels, which are activated by ROS, a number of other TRP channels, and stretch-activated Ca^{2+} permeable channels (Barritt et al. 2008; Jones et al. 2008; Kheradpezhouh et al. 2014; Rychkov and Barritt 2011). Most studies have failed to find any evidence for the expression of voltage-operated Ca^{2+} channels (VOCCs) in hepatocytes (Auld et al. 2000; Brereton et al. 1997; Graf and Häussinger 1996; Sawanobori et al. 1989). However, VOCCs are expressed in other cell types present in the liver (Bataller et al. 2001; Hijioka et al. 1992; Lobeck et al. 2016; Pinzani et al. 1992).

30.3 Liver Steatosis and Its Progression to Insulin Resistance and Hepatocellular Carcinoma

Liver steatosis in subjects with NAFLD is a major risk factor for the development of hepatic insulin resistance and type 2 diabetes. Up to one-third of the population of many countries exhibits NAFLD (Liu et al. 2016; Samuel and Shulman 2016; Tilg et al. 2017). Under some conditions NAFLD can progress to NASH, cirrhosis, fibrosis, and hepatocellular carcinoma (Cabre et al. 2016; Karin and Dhar 2016; Ma et al. 2016; Wong et al. 2016). Hepatocellular carcinoma is the sixth most common cancer worldwide and the second leading cause of cancer-related deaths. Now that good progress has been made in the treatment of hepatitis B and C viral infections, NAFLD is the major risk factor for the development of hepatocellular carcinoma (Cabre et al. 2016; Karin and Dhar 2016; Ma et al. 2016; Wong et al. 2016). It is proposed that the progression of liver steatosis to hepatocellular carcinoma, which takes place over a time span of years, is mediated in part by increases in ROS, ER stress and inflammation, and the provision of an environment that enhances the formation of tumor promoters (Cabre et al. 2016; Font-Burgada et al. 2016; Karin and Dhar 2016). In addition, acetyl-CoA derived from free fatty acids provides a major fuel source for tumor cell metabolism (Röhrig and Schulze 2016).

In the steatotic liver di- and triacylglycerol, and esterified cholesterol, derived from an available excess of free fatty acids and cholesterol, accumulate in lipid droplets in the cytoplasmic space (Dong and Czaja 2011; Liu and Czaja 2013; Mashek et al. 2015; Walther and Farese 2012). Some lipid droplets are also present in normal (non-steatotic) hepatocytes. These provide a reservoir of stored lipids for cellular metabolism. Lipids within lipid droplets can be hydrolyzed by lipases in the cytoplasmic space leading to the release of fatty acids for β-oxidation (Dong and Czaja 2011; Liu and Czaja 2013; Mashek et al. 2015; Walther and Farese 2012). In addition, an alternative pathway for the release of free fatty acids involves autophagy and lysosomal degradation (Dong and Czaja 2011; Liu and Czaja 2013; Martinez-Lopez and Singh 2015).

Mature lipid droplets are cytoplasmic organelles comprised principally of a hydrophobic core of triacylglycerol and esterified cholesterol surrounded by a monolayer of phospholipid and protein (Dong and Czaja 2011; Liu and Czaja 2013; Mashek et al. 2015; Walther and Farese 2012). Lipid droplets in steatotic hepatocytes can be detected and quantitated using agents such as Nile Red and fluorescent microscopy and can be observed in greater detail by electron microscopy (Sharma et al. 2016; Steenks et al. 2010; Wilson et al. 2015) (Fig. 30.2). There are several nutritional and genetic animal models of liver steatosis and various cellular models in which hepatocytes are incubated in the presence of saturated

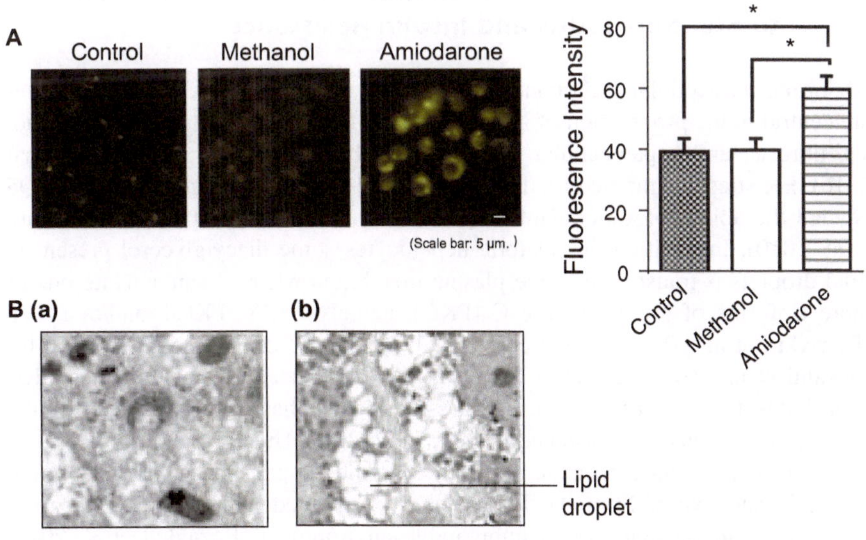

Fig. 30.2 Lipid droplets in liver cells incubated in the presence of an inhibitor of fatty acid oxidation observed using Nile Red and fluorescence microscopy, and in the genetically obese Zucker rat observed in electron micrographs. (**A**) H4IIE rat liver cells incubated in the presence of an inhibitor of fatty acid oxidation, amiodarone, or vehicle, methanol. Lipid droplets in the cytoplasmic space were observed using Nile Red and fluorescence microscopy (from Wilson et al. 2015). (**B**) Electron micrographs of liver sections from lean (**a**) and obese (**b**) Zucker rats showing the presence of lipid droplets (*arrow*) in obese livers (**b**) (from Steenks et al. 2010)

and/or unsaturated fatty acids, or amiodarone, an inhibitor of fatty acid oxidation (Lau et al. 2016; Nakamura et al. 2009; Wilson et al. 2015) (Fig. 30.2A).

Since the accumulation of lipid in liver in NAFLD is a major risk factor for development of type 2 diabetes and hepatocellular carcinoma, reduction of liver lipid offers a strategy for preventing the onset of these diseases (Crane and McGowan 2016; Khera et al. 2016; Kim 2016; Nyenwe et al. 2011). Diet, exercise, and weight loss presently offer the best approach to reduce liver lipid (Crane and McGowan 2016; Khera et al. 2016; Kim 2016; Nyenwe et al. 2011). However, these strategies are only effective for a relatively small number of NAFLD patients (Khera et al. 2016; Kim 2016; Scherer and Dufour 2016). Moreover, there are few approved pharmacological therapies for reduction of hepatic lipids in patients with NAFLD and NASH (Musso et al. 2016; Scherer and Dufour 2016). However, a number of antidiabetic drugs have been shown to induce weight loss and liver fat in NAFLD. These are presently being evaluated for their effectiveness and safety in decreasing liver steatosis (Crane and McGowan 2016; Khera et al. 2016; Kim 2016). Examples are exendin-4 and other slowly metabolized analogues of the incretin hormone GLP-1 (Campbell and Drucker 2013).

30.4 Roles of Abnormal Intracellular Ca^{2+} Concentrations, Endoplasmic Reticulum Stress, and Protein Kinase C in the Development of Steatosis and Its Progression to Steatohepatitis and Insulin Resistance

Abnormal intracellular Ca^{2+} concentrations and chronic ER stress in hepatocytes are central to the progression of NAFLD to NASH, insulin resistance, liver cirrhosis, fibrosis, and hepatocellular carcinoma (Cabre et al. 2016; Karin and Dhar 2016). ER stress is initiated by decreased $[Ca^{2+}]_{ER}$ and by the generation of ROS beyond the ability of antioxidant enzymes to remove them (Fu et al. 2011; Park et al. 2010). In addition, in steatotic hepatocytes some diacylglycerol present in lipid droplets is transferred to the plasma membrane where it can activate one or more isoforms of protein kinase C (PKC), namely PKCε, PKCδ, and/or PKCβ (Bozaykut et al. 2016; Cantley et al. 2013; Gao et al. 2016; Greene et al. 2010; Mekahli et al. 2011; Pagliassotti et al. 2016; Perry et al. 2014). Activated PKC contributes to the initiation of ER stress and also has important roles in the development of insulin resistance (Greene et al. 2010; Huang et al. 2009).

ER stress in hepatocytes arises when the enzymes responsible for normal protein synthesis and export from the ER become overloaded with newly synthesized proteins or are subjected to an unfavorable environment (Bozaykut et al. 2016; Han and Kaufman 2016; Ozcan and Tabas 2016; Pagliassotti et al. 2016; Wang and Kaufman 2014). Enzymes in the lumen of the ER and transmembrane ER proteins are responsible for the synthesis of lipids and proteins, the correct folding of newly synthesized proteins, and the export of proteins from the ER lumen to the cytoplasmic space or to vesicles destined for exocytosis. Under abnormal conditions, such as initiated by a decrease in $[Ca^{2+}]_{ER}$ and/or an increase in ROS, protein folding in the ER lumen is disrupted and misfolded proteins accumulate in the lumen. Initially, compensatory mechanisms, involving inositol-requiring protein-1 (IRE1) and activating transcription factor 6 (ATF6), attempt to adapt the ER and the cell to ER stress through the unfolded protein response (UPR), which is also called the ER stress response (Bozaykut et al. 2016; Han and Kaufman 2016; Ozcan and Tabas 2016; Pagliassotti et al. 2016; Wang and Kaufman 2014). The UPR also involves an increase in the synthesis of chaperones and other proteins responsible for correct folding of newly synthesized proteins. However, under conditions of chronic and severe ER stress, these compensatory mechanisms are overwhelmed. Instead of the activation of protective compensatory pathways, the UPR activates RNA-like ER kinase (PERK) and the apoptotic pathway. Chronic ER stress leads to enhanced lipid synthesis, disruption of insulin-induced Akt phosphorylation, and the development of hepatic insulin resistance (Bozaykut et al. 2016; Fu et al. 2012; Gao et al. 2016; Ozcan et al. 2013; Ozcan and Tabas 2016; Rieusset et al. 2016; Samuel and Shulman 2012; Sun and Lazar 2013). Several groups have shown that strategies which reduce ER stress can decrease lipid and reduce insulin resistance in steatotic hepatocytes (Engin and Hotamisligil 2010; Kammoun et al. 2009; Kars et al. 2010; Ozcan et al. 2006; Samuel and Shulman 2012; Xiao et al. 2011).

30.5 Lipid-Induced Inhibition of SERCA2b in Steatotic Hepatocytes

In the livers of obese mice the mRNA and protein expression of SERCA2b, the main isoform of SERCA present in liver, are greatly reduced compared to that in normal liver (Egnatchik et al. 2014; Park et al. 2010; Zhang et al. 2014). Restoration of SERCA2b expression in the livers of obese and diabetic mice was found to alleviate ER stress and restore blood glucose homeostasis to near normal values (Egnatchik et al. 2014; Park et al. 2010; Zhang et al. 2014). Moreover, overexpression of SERCA2b was found to decrease the expression of stearoyl-CoA desaturase-1, diacylglycerol acyltransferase 2, fatty acid synthase, and acetyl-Co carboxylase 2, enzymes involved in lipid synthesis, and to decrease liver triglyceride (Park et al. 2010). Fu et al. (2011) undertook a proteomic analysis of ER proteins from lean and obese mouse livers and a quantitative analysis of lipids extracted from the ER of these livers. They observed an increase in the ratio of phosphatidylcholine/phosphatidylethanolamine in ER membranes isolated from the livers of obese mice. This was associated with decreased SERCA activity measured in liver microsomes (Fu et al. 2011). Taken together, these results indicate that in steatotic hepatocytes the activity of SERCA2b is decreased compared to that in normal hepatocytes. This is due to the decreased expression of this Ca^{2+} transporter and to inhibition of the activity by the elevated phosphatidylcholine/phosphatidyl ethanolamine ratio in the ER membrane. It was proposed that these changes lead to decreased $[Ca^{2+}]_{ER}$, an essential component of the pathway in development of ER stress in hepatocytes in steatotic livers (Arruda and Hotamisligil 2015; Fu et al. 2011; Kang et al. 2016; Mekahli et al. 2011; Park et al. 2010; Wei et al. 2009).

The importance of decreased $[Ca^{2+}]_{ER}$ in the progression of fatty liver to steatosis and insulin resistance has been confirmed by studies which have shown that increased expression of SERCA2b, as described above, or allosteric activation of SERCA2b restores $[Ca^{2+}]_{ER}$, reduces ER stress, and decreases hepatic steatosis and insulin resistance in vivo (Kang et al. 2016; Park et al. 2010). Kang et al. (2016) employed CDN1163, a novel allosteric activator of SERCA which lowered blood glucose concentrations, improved glucose tolerance, and decreased liver steatosis when administered to genetically obese mice exhibiting insulin resistance and type 2 diabetes.

30.6 Lipid-Induced Inhibition of Store-Operated Ca^{2+} Entry in Steatotic Hepatocytes

30.6.1 Effects of Lipid on Store-Operated Ca^{2+} Entry in Steatotic Hepatocytes

The presence of lipid droplets in rat liver cells pre-incubated with palmitate and in steatotic hepatocytes isolated from the livers of genetically obese Zucker rats was found to be associated with a substantial inhibition of SOCE and reduction of $[Ca^{2+}]_{ER}$ (Fig. 30.3B(a) cf. Figs. 30.3A(a) and 30.3C, D ((b), vehicle cf. (a), vehicle)) (Wilson et al. 2015). The lipid-induced inhibition of SOCE was also associated with a substantial reduction in the ability of the hormones ATP and phenylephrine to induce an increase in $[Ca^{2+}]_{cyt}$ in hepatocytes incubated in the presence of $[Ca^{2+}]_{ext}$. SOCE was measured using the Ca^{2+} sensor fura-2 and fluorescence imaging and by patch clamp recording (Wilson et al. 2015). The experiments using fura-2 employed the "Ca^{2+} add-back protocol" in which hepatocytes loaded with fura-2 were initially incubated in the absence of $[Ca^{2+}]_{ext}$, the SERCA inhibitor 2,5-di-(*tert*-butyl)-1,4-benzohydro-quinone (BHQ) was added to release Ca^{2+} from the ER, and then $[Ca^{2+}]_{ext}$ added to initiate Ca^{2+} entry through SOCs. In patch clamp experiments, activation of SOCE was initiated using IP_3 in the pipette solution or ATP in the bath. Ca^{2+} in the ER was measured indirectly by the increase in $[Ca^{2+}]_{cyt}$ induced by BHQ.

It was argued that the lipid-induced inhibition of SOCE contributes to the observed decrease in $[Ca^{2+}]_{ER}$ in steatotic hepatocytes (Fu et al. 2011; Park et al. 2010; Wilson et al. 2015). However, it is unclear whether, in normal hepatocytes, the inhibition of SOCE alone would cause a decrease in $[Ca^{2+}]_{ER}$. The pharmacological inhibition of SOCE or knockdown of Orai1 or STIM1 using siRNA results in some reduction of hormone-generated oscillations in $[Ca^{2+}]_{cyt}$ (Jones et al. 2008). This suggests that inhibition of SOCE does lead to a decrease in $[Ca^{2+}]_{ER}$ in hepatocytes. Moreover, experiments with HEK cells have provided a more direct evidence that inhibition of SOCE decreases $[Ca^{2+}]_{ER}$ (Manjarres et al. 2011; Srivats et al. 2016). However, the results of studies with HeLa cells in which SOCE was inhibited by knockdown of STIM1 suggest that decreased SOCE has minimal effects on $[Ca^{2+}]_{ER}$ (Jousset et al. 2007).

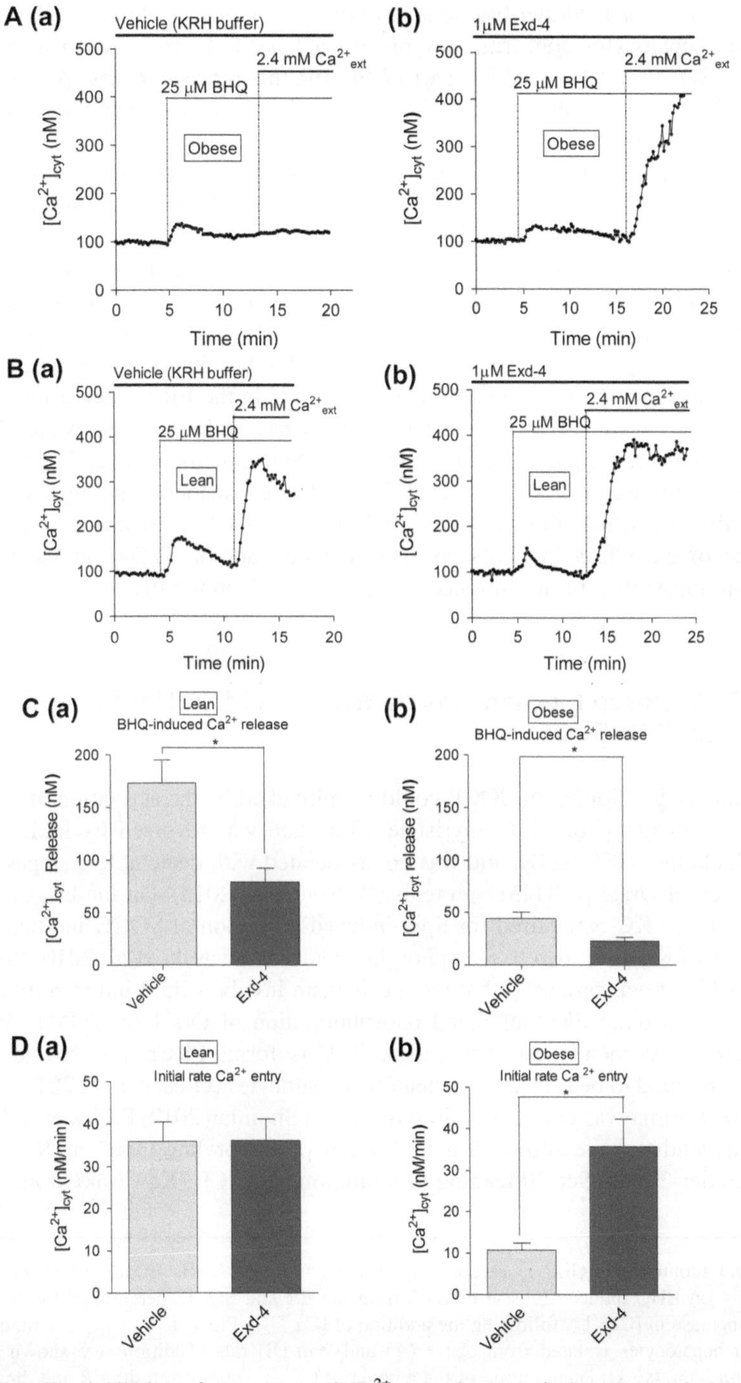

Fig. 30.3 The inhibition of store-operated Ca^{2+} entry in steatotic hepatocytes isolated from genetically obese Zucker rats and reversal of this inhibition by the glucagon-like peptide-1

Another unresolved question related to the lipid-induced inhibition of SOCE in steatotic hepatocytes concerns how much Ca^{2+} needs to be released in order to activate SOCE and from which part of the ER this release occurs. As described above, lipid-induced inhibition of SOCE is associated with a substantial decrease in $[Ca^{2+}]_{ER}$ (Wilson et al. 2015). It is possible that the observed lipid-induced inhibition of SOCE in steatotic hepatocytes is due simply to the movement of only a few molecules of STIM1 to the ER-PM junction and hence limited activation of Orai1 following the experimental application of BHQ, IP_3, or ATP. As mentioned above, previous studies on the role of ER Ca^{2+} release in the activation of hepatocyte SOCs have provided indirect evidence that the activation of SOCE involves a small subregion of the ER rather than the whole ER (Castro et al. 2009; Gregory et al. 1999). It was argued that Ca^{2+} stored in this ER subregion may not be altered in steatotic hepatocytes even though the bulk of Ca^{2+} in the ER is decreased in these cells (Wilson et al. 2015). Moreover, as described in Sect. 30.8, exendin-4 can completely reverse lipid-induced inhibition of SOCE without reversing the lipid-induced decrease in $[Ca^{2+}]_{ER}$ (Ali et al. 2016). These considerations suggest that the mechanism by which lipid induces inhibition of SOCE in steatotic cells in the absence of exendin-4 involves changes in Orai1 and/or STIM1 at the ER-PM junction, rather than to lipid-induced changes in Ca^{2+} in the ER.

30.6.2 Proposed Mechanisms by Which Lipid, in Lipid Droplets, Inhibits SOCE

Lipid-induced inhibition of SOCE could be mimicked by the activation of PKC by the phorbol ester phorbol 12-myristate 13-acetate was reversed by inhibitors of PKC, including GF10923X, and was not associated with detectable changes in the expression of Orai1 or STIM1 proteins (Wilson et al. 2015). On the basis that the activation of PKC is required for lipid-induced inhibition of SOCE and the observation that PKC is known to phosphorylate Orai1 (Kawasaki et al. 2010; Srikanth et al. 2013), it was proposed that the mechanism involves lipid-induced inhibition of SOCE involving PKC-mediated phosphorylation of Orai1 or STIM1 (Wilson et al. 2015). As mentioned above, three PKC isoforms PKCε, PKCδ, and PKCβ have been found to be activated in steatotic hepatocytes (Greene et al. 2010; Huang et al. 2009; Jornayvaz et al. 2011; Jornayvaz and Shulman 2012; Puljak et al. 2005). Previous studies have shown that PKCβ can phosphorylate Orai1 at N-terminal residues Ser-27 and Ser-30 leading to inhibition of SOCE (Kawasaki et al. 2010).

Fig. 30.3 (continued) (GLP-1) receptor agonist, exendin-4. (**A, B**) Effects of steatosis and exendin-4 on BHQ-induced release of Ca^{2+} from the ER and SOCE, determined by the initial rate of increase in $[Ca^{2+}]_{cyt}$ following the addition of $[Ca^{2+}]_{ext}$. Plots of $[Ca^{2+}]_{cyt}$ as a function of time for hepatocytes isolated from obese (**A**) and lean (**B**) rats. Additions are shown by the *horizontal bars*. (**C, D**) Quantitation of the amount of Ca^{2+} released from the ER and the rate of SOCE (from Ali et al. 2016)

No PKC-mediated phosphorylation of STIM1 had yet been reported (Srikanth et al. 2013). These considerations suggest that in steatotic hepatocytes PKC phosphorylates Orai1 rather than STIM1. Further experiments are required to test whether this PKCβ isoform is involved in the inhibition of SOCE, and whether Orai1 is the target.

Other possible mechanisms for lipid-induced inhibition of SOCE include the direct interaction of membrane cholesterol and/or fatty acids with Orai1 and/or STIM1. It has been shown that Orai1 is inhibited by an increase in cholesterol in the plasma membrane (Derler et al. 2016b). Moreover, polyunsaturated fatty acids can inhibit the coupling of STIM1 to Orai1, possibly by disrupting the oligomerization of STIM1 (Holowka et al. 2014). While these mechanisms cannot be excluded, they are considered unlikely since they do not necessarily involve PKC and on the basis that Orai1 rather than STIM1 is the likely target of inhibition. The proposed mechanism by which lipid in lipid droplets inhibits SOCE in steatotic hepatocytes is summarized in Fig. 30.4 (steps 6, 7, 8, and 1). This involves the movement of diacylglycerol (DAG) from lipid droplets to the plasma membrane, the activation of PKCβ (or PKCε or PKCδ) at the plasma membrane, phosphorylation of the Orai1 polypeptide, and the inhibition of the Orai1 Ca^{2+}-selective channel.

30.7 Inhibition of Store-Operated Ca^{2+} Entry Enhances Further Lipid Accumulation in the Presence of a Supply of Exogenous Fatty Acids

Pharmacological inhibition of SOCE or knockdown of Orai1 and STIM1 using siRNA was found to cause a twofold enhancement of lipid droplet accumulation in normal liver cells or hepatocytes incubated in the presence of exogenous fatty acids (Wilson et al. 2015). Since the main function of SOCE is thought to be to replenish the ER Ca^{2+} stores following Ca^{2+} release induced by IP_3 (Carrasco and Meyer 2011; Prakriya and Lewis 2015), it was suggested that this observation indicates that in steatotic liver cells inhibited SOCE acts in concert with inhibition of SERCA2b to create a low $[Ca^{2+}]_{ER}$ which then leads to an enhancement of cytoplasmic lipid droplet accumulation. Hence it was proposed that the lipid-induced inhibition of SOCE and the associated low $[Ca^{2+}]_{ER}$ act as a positive feedback loop to enhance further lipid accumulation (Wilson et al. 2015). Such an amplification cycle for the effects of ER stress on lipid synthesis has previously been suggested (Dara et al. 2011).

The enhanced lipid accumulation in this experimental model is likely initiated by a substantial decrease in $[Ca^{2+}]_{ER}$ due to lipid-induced inhibition of both SOCE and SERCA2b, leading to ER stress and the ER stress response. However, the mechanism(s) by which ER stress and the ER stress response lead to an increase in the formation of lipid droplets are complex and are not well understood. The mechanism(s) could involve enhanced lipogenesis, decreased lipolysis, and fatty acid oxidation and/or long-chain fatty acid toxicity (Dara et al. 2011; Wang and Kaufman 2014). There is some evidence that a likely mechanism involves enhanced

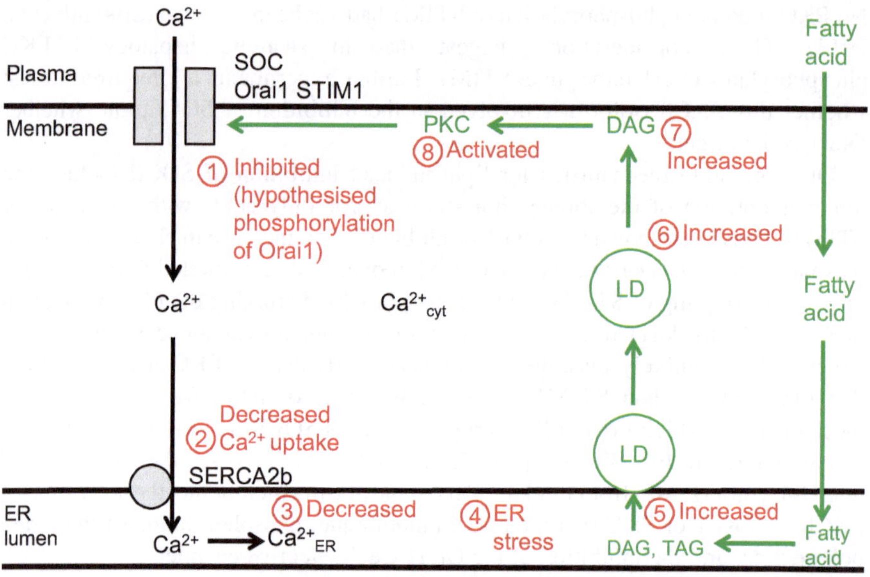

Fig. 30.4 Schematic representation of the proposed mechanism by which lipid-induced inhibition of store-operated Ca^{2+} entry leads to enhanced lipid accumulation in hepatocytes in the presence of a supply of exogenous fatty acids. The scheme shows, in *black*, the role of SOCs, composed of the Orai1 and STIM1 proteins, in maintaining a normal concentration of Ca^{2+} in the lumen of the ER. In steatotic hepatocytes supplied with exogenous fatty acids (shown in *green*), the fatty acids enter the ER lumen and are esterified to diacylglycerol (DAG) and triacylglycerol (TAG). The DAG and TAG are formed into lipid droplets which bud from the surface of the ER and reside in the cytoplasmic space. Some lipid droplets diffuse to the plasma membrane where DAG is transferred to the membrane and activates PKC, which in turn is hypothesized to phosphorylate and inhibit Orai1. The numbers and steps indicated in *red* show the sequence of events through which it is proposed that lipid-induced inhibition of SOCE amplifies the accumulation of lipid droplets in the cytoplasmic space. These are: (1) lipid-induced inhibition of SOCE; (2), decreased Ca^{2+} uptake to ER due to decreased supply of Ca^{2+} from SOCE and inhibited SERCA2b activity; (3) decreased $[Ca^{2+}]_{ER}$; (4) maintained or enhanced ER stress and the ER stress response; (5) increased synthesis of DAG and TAG and lipid droplets; and (6) increased number of lipid droplets in the cytoplasmic space. This in turn maintains: (7) increased DAG in the plasma membrane, (8) enhanced activation of PKC and (1) inhibition of SOCE

lipid synthesis (Fu et al. 2011; Samuel and Shulman 2012; Wang and Kaufman 2014). The pathway by which exogenous fatty acids increase the formation of lipid droplets in the cytoplasmic space of hepatocytes involves the esterification of fatty acids to di- and triacylglycerol, the esterification of cholesterol, and the synthesis of requisite proteins in the lumen of the ER (Mashek et al. 2015; Walther and Farese 2012; Wang and Kaufman 2014). These are then packaged into a budding lipid droplet, which initially forms within the membrane bilayer of the ER, developing into an independent organelle located in the cytoplasmic space. These processes take place at locations where the ER membrane interacts with mitochondria and peroxisomes. That part of the ER which interacts with mitochondria is called the mitochondria-associated ER membrane.

A hypothesized mechanism by which inhibition of SOCE enhances the formation of lipid droplets is shown in Fig. 30.4 (steps 1–8). It is based, in part, on evidence that decreased $[Ca^{2+}]_{ER}$ and ER stress enhance lipid synthesis and fatty acid esterification (Fu et al. 2011; Wang and Kaufman 2014) but changes in Ca^{2+} in the mitochondrial matrix may also play a role. Thus, Arruda et al. (2014) showed in a mouse model of obesity that obesity (and presumably liver steatosis) is associated with an increase in mitochondria-associated ER membranes and an increase in the transfer of Ca^{2+} from ER to mitochondria, resulting in mitochondrial Ca^{2+} overload, increased generation of ROS, and decreased oxidative phosphorylation, including β-oxidation of fatty acids.

In the light of the overall idea that inhibition of SOCE may be involved in enhancing lipid accumulation in hepatocytes, it is interesting to note that in studies with *Drosophila* Baumbach et al. found STIM to be an important regulator of fat storage (Baumbach et al. 2014; Zhang and Thomas 2016). Knockdown of STIM in fat storage tissues substantially increased the amount of fat stored. Further experiments led to the conclusion that the actions of STIM in maintaining normal levels of fat in fat storage tissues are mediated indirectly by sNPF (short neuropeptide F), a functional homologue of the mammalian orexigenic neuropeptide Y, in the *Drosophila* brain. sNPF, in turn, regulates *Drosophila* food intake.

Based on the results of many previous studies which have established that changes in intracellular Ca^{2+} regulate lipid metabolism, and on the recent observation, described above, indicating a role for STIM in fat storage in *Drosophila*, Maus et al. (2017) investigated the effects of inhibition of SOCE on lipid accumulation in various animal tissues and cells. Employing a tamoxifen-inducible mouse model in which the *STIM1* or *STIM2* genes can be deleted to abolish SOCE, they showed that inhibition of SOCE leads to a large increase in lipid in skeletal and heart muscle and in liver. Moreover, they also found that cultured fibroblasts obtained from patients with loss-of-function mutations in *Orai1* or *STIM1*, which abolished SOCE, exhibit an increase in lipid accumulation compared with controls. Further experiments provided evidence that in fibroblasts and NIH3T3L-1 cells, SOCE is required to maintain mitochondrial fatty acid oxidation and lipolysis. The idea that SOCE provides a supply of intracellular Ca^{2+}, which activates adenylate cyclase, leading to the generation of cyclic AMP, activation of PKA, activation of hormone sensitive lipase, and enhanced lipolysis is consistent with these results (Maus et al. 2017).

30.8 Strategies for the Pharmacological Restoration of Ca^{2+} Entry Through SOCE in Steatotic Hepatocytes

As described in Sect. 30.3, NAFLD is a major risk factor for the development of hepatic insulin resistance, type 2 diabetes, and hepatocellular carcinoma (Liu et al. 2016; Samuel and Shulman 2016; Tilg et al. 2017). This implies that in NAFLD subjects reduction of liver fat is a strategy to prevent later development of type

2 diabetes and hepatocellular carcinoma in these subjects (Crane and McGowan 2016; Khera et al. 2016; Kim 2016; Nyenwe et al. 2011). Limited pharmacological strategies are presently available for the reduction of liver fat (Crane and McGowan 2016; Khera et al. 2016; Kim 2016; Musso et al. 2016; Scherer and Dufour 2016). However, some antidiabetic drugs, including GLP-1 agonists, have been shown to reduce weight and liver fat in obese subjects (Crane and McGowan 2016). On the basis that SOCE provides Ca^{2+} to maintain normal $[Ca^{2+}]_{ER}$, it is a reasonable theoretical prediction that pharmacological activation of SOCE in steatotic hepatocytes would lead to a greater supply of Ca^{2+} to the lumen of the ER, an increase in $[Ca^{2+}]_{ER}$, and a reduction of ER stress. However, at present, there is little experimental evidence to indicate whether this prediction is accurate. As mentioned above, there is presently only indirect evidence that, in the absence of any changes in SERCA2b activity in normal hepatocytes, a decrease in SOCE leads to a decrease in $[Ca^{2+}]_{ER}$. Moreover, there is also little evidence that an increase in SOCE leads to an increase in $[Ca^{2+}]_{ER}$ in these cells. It is also not known whether the activation of SOCE in steatotic hepatocytes, in which SERCA2b is also inhibited, would lead to an increase in $[Ca^{2+}]_{ER}$.

In an attempt to find pharmacological agents which activate SOCE in steatotic hepatocytes, Ali et al. (2016) found that the GLP-1 receptor agonist exendin-4 completely reverses the lipid-induced inhibition of SOCE (Fig. 30.3A, D). Half-maximal effect exendin-4 on lipid-induced inhibition of SOCE was observed at 100 nM exendin-4, a concentration comparable to that for the actions of exendin-4 on other targets (Lee et al. 2012, 2014). Exendin-4 also inhibited the accumulation of lipid in liver cells incubated in the presence of exogenous fatty acids and in the presence of a pharmacological inhibitor of SOCE (Ali et al. 2016). Under the conditions employed, exendin-4 did not reverse the lipid-induced inhibition of ER Ca^{2+} release (Fig. 30.3A, C). This inability of exendin-4 to increase $[Ca^{2+}]_{ER}$ under these conditions is possibly because, as discussed above, one of the mechanisms by which lipid inhibits SERCA2b involves a decrease in the expression of this protein (Park et al. 2010). In the experiments with steatotic hepatocytes from obese Zucker rats, exendin-4 was applied for a period of only 15 min before the measurement of SOCE. This is unlikely be sufficient time for SERCA2b to increase from low to basal protein levels.

The action of exendin-4 in reversing lipid-induced inhibition of SOCE was rapid in onset (about 70% recovery in 10 min), mimicked by GLP-1 and by dibutyryl cyclic AMP, and associated with the formation of cyclic AMP (Ali et al. 2016). Of particular interest was the decrease in cytoplasmic lipid induced by exendin-4 (Fig. 30.5) which also exhibited a very rapid onset (halftime about 6.5 min). Rp-cyclic AMPS, an inhibitor of protein kinase A (PKA) (Beebe et al. 1987), prevented the ability of exendin-4 to reverse the lipid-induced inhibition of SOCE (Ali et al. 2016), indicating that the increase in cyclic AMP induced by exendin-4 is a necessary step in this mechanism. It was proposed that the mechanism by which exendin-4 reverses lipid-induced inhibition of SOCE involves GLP-1 receptors, an increase in cyclic AMP, activation of PKA, enhanced lipolysis, decreased DAG, decreased PKC activity, and de-inhibition of SOCE (Fig. 30.6)

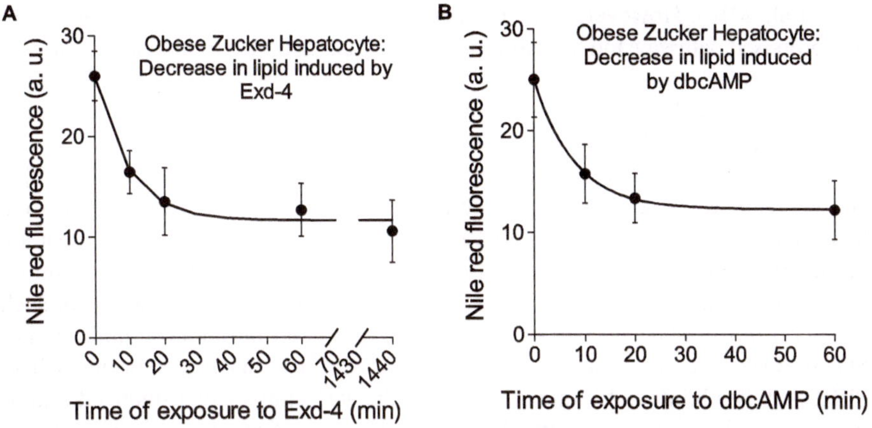

Fig. 30.5 The glucagon-like peptide-1 (GLP-1) receptor agonist exendin-4 and dibutyryl cyclic AMP each cause a decrease in lipid in steatotic hepatocytes isolated from genetically obese Zucker rats with similar time courses and a rapid onset. Cytoplasmic lipid in hepatocytes isolated from obese Zucker rats incubated in the presence of exendin-4 (Exd-4) (**a**) or cyclic AMP (**b**), added at $t = 0$ min, was measured using Nile Red and fluorescence imaging (from Ali et al. 2016)

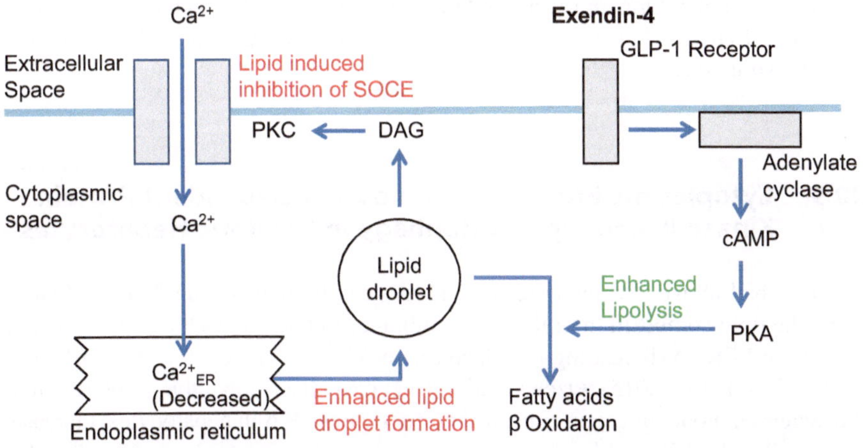

Fig. 30.6 A schematic representation of the proposed mechanism by which the GLP-1 receptor agonist exendin-4 reverses the lipid-induced inhibition of store-operated Ca^{2+} entry in hepatocytes. It is proposed that in steatotic hepatocytes, lipid droplets in the cytoplasmic space increase DAG in the plasma membrane, activate PKC, and inhibit store-operated Ca^{2+} entry (SOCE) (indicated in *red*), as described in more detail in Fig. 30.4. Exendin-4 is proposed to bind to GLP-1 receptors on the hepatocyte plasma membrane which, in turn, activate adenylate cyclase, increase the concentration of cyclic AMP, and activate protein kinase A (PKA). This enhances lipolysis by increasing fatty acid oxidation (indicated in *green*). The decrease in lipid droplets leads to a decrease in diacylglycerol (DAG), decreased activation of PKC, and decreased PKC-mediated inhibition of Orai1 and hence SOCE (from Ali et al. 2016)

(Ali et al. 2016). However, further experiments are needed to test components of this proposed pathway and to determine whether exendin-4 also increases $[Ca^{2+}]_{ER}$ and reduces ER stress.

As suggested above, the mechanism by which exendin-4 reverses lipid-induced inhibition of SOCE does not appear to involve a direct action of exendin-4, cyclic AMP, or PKA on STIM1 or Orai1. Another potential strategy to increase $[Ca^{2+}]_{ER}$ and decrease ER stress and hence ultimately to decrease liver lipid in steatotic hepatocytes involves direct activation of the SOCE pathway in steatotic hepatocytes. However, how this could be achieved depends on the mechanism by which PKC inhibits SOCE in steatotic hepatocytes. Moreover, the success of such an approach would depend on whether the activation of lipid-induced inhibition of SOCE does actually lead to an increase in $[Ca^{2+}]_{ER}$ under conditions where SERCA2b is inhibited. A strategy involving the activation of lipid-induced inhibition of SOCE could be used in conjunction with use of an allosteric activator of SERCA2b such as CDN1163 (Kang et al. 2016).

At present, there are, to our knowledge, no known small molecule activators of SOCE. Such an "allosteric" activator could potentially bind to the Orai1 polypeptide and activate the channel pore or bind to STIM1 or the ER membrane to enhance the oligomerization of STIM1 and its interaction with Orai1 at the ER-PM junction. In addition, the selective inhibition of a PKC isoform to inhibit the hypothesized phosphorylation of Orai, or the use of taurodeoxycholic acid, which has been shown to activate SOCE in hepatocytes (Aromataris et al. 2008; Castro et al. 2009), offer possible strategies.

30.9 Cytoplasmic-Free Ca^{2+}, Calmodulin-Dependent Protein Kinase II, and Lipid Autophagy in Steatotic Hepatocytes

As described above, the increase in lipid in steatotic hepatocytes inhibits SOCE, alters the lipid composition of the ER membrane, and decreases the expression and activity of SERCA2b, leading to a decrease in $[Ca^{2+}]_{ER}$ (Fu et al. 2011; Park et al. 2010; Wilson et al. 2015). Ozcan et al. (2012) have also shown that in obese mice, the increased blood glucagon concentration activates PKA in hepatocytes, increases the activity of IP_3R, and hence increases the outflow of Ca^{2+} from the ER. Other studies have demonstrated that activated CaMKII is, in part, responsible for increased hepatic glucose production and decreased sensitivity to insulin in steatotic hepatocytes (Ozcan et al. 2013). It has been proposed that a "pseudo steady state" involving increased activity of IP_3R and decreased activity of SERCA2b causes an increase in $[Ca^{2+}]_{cyt}$ which is subsequently responsible for the increased activities of CaMKII and other Ca^{2+}-regulated enzymes (Ozcan et al. 2012, 2013; Park et al. 2014). However, there have been few direct measurements of basal (resting) $[Ca^{2+}]_{cyt}$ in steatotic hepatocytes compared to that in normal hepatocytes. Park et al. (2014) reported an increase in $[Ca^{2+}]_{cyt}$ in the HepG2 human liver cells incubated with the unsaturated fatty acid palmitate. However, two other groups found no change in basal (resting) $[Ca^{2+}]_{cyt}$ in steatotic compared to normal rat hepatocytes and liver cells (Puljak et al. 2005; Wilson et al. 2015).

Comparison of values of resting $[Ca^{2+}]_{cyt}$ between steatotic and normal hepatocytes poses some technical challenges. While it is relatively easy to measure agonist-induced changes in $[Ca^{2+}]_{cyt}$ in a given hepatocyte preparation cultured on a single coverslip using a Ca^{2+} reporter such as fura-2 and fluorescence imaging, it is much more difficult to compare the value of basal $[Ca^{2+}]_{cyt}$ between coverslips of steatotic and normal hepatocytes. Among other things, very careful calibration of the fluorescence signals is required (Thomas and Delaville 1991). In addition, the isolation of hepatocytes from the liver removes the cells from their local in vivo environment created by the blood, including the presence or absence of hormones, and replaces this with an artificial in vitro environment.

It is possible that in steatotic hepatocytes SOCE contributes to an elevation in $[Ca^{2+}]_{cyt}$ even though it is observed, as described above, that maximal activity of SOCE is inhibited in these cells. The proposed mechanism for the increase in $[Ca^{2+}]_{cyt}$ involves the consequences of the persistent decrease in $[Ca^{2+}]_{ER}$ and low activity of SERCA2b. The persistent decrease in $[Ca^{2+}]_{ER}$ may lead to a re-localization of some STIM1 to ER-PM junctions and the activation of Orai1 resulting in a low persistent increase in SOCE. This low level of SOCE in steatotic hepatocytes may be maintained in part due to the high ratio of Orai1 to STIM1 and consequent reduction in Ca^{2+}-dependent inhibition of SOCE (Scrimgeour et al. 2009, 2014). Since in these steatotic cells, SERCA2b is inhibited and IP_3Rs are activated, less Ca^{2+} entering the cell via SOCE may be transported to the ER than in normal cells. It can be predicted that this would result in an elevated basal $[Ca^{2+}]_{cyt}$. As suggested above, further experiments are required to accurately determine whether baseline (resting) $[Ca^{2+}]_{cyt}$ is increased in steatotic compared to normal hepatocytes and to test this hypothesis.

In a separate investigation, Park et al. (2014) studied the role of $[Ca^{2+}]_{cyt}$ in lipid autophagy in steatotic liver cells. As mentioned above, the autophagic pathway can remove lipid from lipid-loaded hepatocytes (Dong and Czaja 2011; Liu and Czaja 2013; Martinez-Lopez and Singh 2015). In the autophagy of lipid droplets, cytoplasm containing a small lipid droplet or part of a large lipid droplet is engulfed by an autophagosome, a vesicle comprised of a double membrane. The autophagosome containing the lipid droplet then fuses with a lysosome to form an autolysosome in which lipids are then degraded by lysosomal enzymes. Park et al. argued that the lipid-induced inhibition of SERCA2b in steatotic hepatocytes would lead to an increase in $[Ca^{2+}]_{cyt}$. They showed that preincubation of HepG2 liver cells with the unsaturated fatty acid palmitate increased $[Ca^{2+}]_{cyt}$, as indicated by increased fluorescence of X-Rhod-1 and fura-2 (Park et al. 2014). This increase was inhibited by verapamil (50 μM) and nifedipine (100 μM), blockers of VOCCs (Park and Lee 2014; Park et al. 2014). The authors also detected a low expression of mRNA encoding VOCCs in RNA extracts of whole liver (Park et al. 2014). However, as mentioned above, VOCCs are expressed in a number of liver cell types other than hepatocytes (Bataller et al. 2001; Hijioka et al. 1992; Lobeck et al.

2016; Pinzani et al. 1992), and the possibility that the mRNA encoding VOCC was derived from one or more of these other cell types was not excluded.

The administration of verapamil to obese mice was found to increase autophagosome-lysosome fusion in hepatocytes and reduce lipid droplets, inflammation, and insulin resistance (Park et al. 2014). These effects were attributed to the action of verapamil on hepatocytes leading to inhibition of Ca^{2+} entry through VOCCs and a decrease in $[Ca^{2+}]_{cyt}$. However, it is possible that the actions of verapamil on the livers of steatotic mice are exerted indirectly by the binding of the drug to VOCCs expressed in cell types other than hepatocytes (Bataller et al. 2001; Hijioka et al. 1992; Lobeck et al. 2016; Pinzani et al. 1992). Moreover, several previous studies have found no evidence for the expression of VOCCs in rat or mouse primary hepatocytes but have detected mRNA encoding variants of VOCCs in a liver cell line (Auld et al. 2000; Brereton et al. 1997; Graf and Häussinger 1996; Sawanobori et al. 1989). In addition, at relatively high concentrations, verapamil and nifedipine can inhibit SOCs (Auld et al. 2000; Hughes et al. 1986; Striggow and Bohnensack 1993; Wu et al. 1995).

In the recent study of Maus et al. (2017) described above in which the authors investigated the effect of inhibition of SOCE on lipid accumulation in fibroblasts and in some other cell types, the authors also found that inhibition of SOCE and the subsequent increase in intracellular lipid accumulation were associated with an increase in lipid autophagy. It was suggested that the activation of lipid autophagy is a compensatory mechanism directed toward the reduction of cellular lipid. As described above, Park et al. have provided some evidence to indicate that in steatotic liver cells $[Ca^{2+}]_{cyt}$ is increased compared to normal and this inhibits lipid autophagy. They suggested that inhibition of Ca^{2+} entry across the plasma membrane leads to a decrease in $[Ca^{2+}]_{cyt}$ and increase in lipid autophagy (Park et al. 2014). Further experiments, including accurate comparison of $[Ca^{2+}]_{cyt}$ under the different conditions employed, are warranted in order to gain a better understanding of the roles of SOCE, SERCA2b, IP_3R, and $[Ca^{2+}]_{cyt}$ in the regulation of lipid autophagy.

30.10 Conclusions

Liver steatosis is associated with a substantial PKC-mediated inhibition of SOCE in hepatocytes. This lipid-induced inhibition of SOCE enhances (amplifies) further lipid accumulation. In steatotic hepatocytes, the GLP-1 receptor agonist exendin-4 reverses the lipid-induced inhibition of SOCE and decreases hepatocyte lipid with rapid onset. Together with lipid-induced inhibition of SERCA2b, the inhibition of SOCE in steatotic hepatocytes and the subsequent decrease in $[Ca^{2+}]_{ER}$ are thought to make a major contribution to the development of ER stress and the ER stress response. This can mediate the progression of NAFLD to NASH, insulin resistance, and hepatocellular carcinoma. Both SOCE and SERCA2b offer potential targets for pharmacological intervention directed toward the restoration of normal $[Ca^{2+}]_{ER}$ concentrations, decreasing ER stress and thus inhibiting this progression. Many questions, though, remain unresolved. These include details of the mechanisms by which PKC inhibits SOCE and by which exendin-4 reverses this inhibition, and by which inhibited SOCE leads to enhanced lipid synthesis. Moreover, further experiments would be needed to expand knowledge of the concentrations of Ca^{2+} in the cytoplasmic space, lumen of the ER, and mitochondrial matrix in steatotic compared to normal hepatocytes.

Acknowledgements This research was supported by the Diabetes Australia Research Trust (grant number Y11-BARG); the Rebecca L. Cooper Medical Research Foundation and the Flinders Medical Centre Foundation.

References

Ali ES, Hua J, Wilson CH, Tallis GA, Zhou FH, Rychkov GY, Barritt GJ (2016) The glucagon-like peptide-1 analogue exendin-4 reverses impaired intracellular Ca2+ signalling in steatotic hepatocytes. Biochim Biophys Acta 1863(9):2135–2146

Aromataris EC, Roberts ML, Barritt GJ, Rychkov GY (2006) Glucagon activates Ca2+ and Cl- channels in rat hepatocytes. J Physiol (Lond) 573(Pt 3):611–625

Aromataris EC, Castro J, Rychkov G, Barritt GJ (2008) Store-operated Ca2+ channels and Stromal Interaction Molecule 1 (STIM1) are targets for the actions of bile acids on liver cells. Biochim Biophys Acta 1783:874–885

Arruda AP, Hotamisligil GS (2015) Calcium homeostasis and organelle function in the pathogenesis of obesity and diabetes. Cell Metab 22(3):381–397

Arruda AP, Pers BM, Parlakgul G, Guney E, Inouye K, Hotamisligil GS (2014) Chronic enrichment of hepatic endoplasmic reticulum-mitochondria contact leads to mitochondrial dysfunction in obesity. Nat Med 20(12):1427–1435

Auld A, Chen J, Brereton HM, Wang YJ, Gregory RB, Barritt GJ (2000) Store-operated Ca(2+) inflow in Reuber hepatoma cells is inhibited by voltage-operated Ca(2+) channel antagonists and, in contrast to freshly isolated hepatocytes, does not require a pertussis toxin-sensitive trimeric GTP-binding protein. Biochim Biophys Acta 1497(1):11–26

Barritt GJ (1999) Does a decrease in subplasmalemmal Ca2+ explain how store-operated Ca2+ channels are opened? Cell Calcium 23(1):65–75

Barritt GJ, Chen J, Rychkov GY (2008) Ca(2+)-permeable channels in the hepatocyte plasma membrane and their roles in hepatocyte physiology. Biochim Biophys Acta 1783(5):651–672

Bartlett PJ, Gaspers LD, Pierobon N, Thomas AP (2014) Calcium-dependent regulation of glucose homeostasis in the liver. Cell Calcium 55(6):306–316

Bataller R, Gasull X, Gines P, Hellemans K, Görbig MN, Nicolas JM, Sancho-Bru P, De Las Heras D, Gual A, Geerts A, Arroyo V, Rodes J (2001) In vitro and in vivo activation of rat hepatic stellate cells results in de novo expression of L-type voltage-operated calcium channels. Hepatology 33(4):956–962

Baumbach J, Hummel P, Bickmeyer I, Kowalczyk KM, Frank M, Knorr K, Hildebrandt A, Riedel D, Jäckle H, Kühnlein RP (2014) A Drosophila in vivo screen identifies store-operated calcium entry as a key regulator of adiposity. Cell Metab 19(2):331–343

Beebe SJ, Koch SR, Chu DT, Corbin JD, Granner DK (1987) Regulation of phosphoenolpyruvate carboxykinase gene transcription in H4IIE hepatoma cells: evidence for a primary role of the catalytic subunit of 3′,5′-cyclic adenosine monophosphate-dependent protein kinase. Mol Endocrinol 1(9):639–647

Boyer JL (2002) Bile formation and cholestasis Schiffs diseases of the liver, vol 1, 9th edn. Lippincott, Williams & Wilkins, Philadelphia, pp 135–165

Bozaykut P, Sahin A, Karademir B, Ozer NK (2016) Endoplasmic reticulum stress related molecular mechanisms in nonalcoholic steatohepatitis. Mech Ageing Dev 157:17–29

Brereton HM, Harland ML, Froscio M, Petronijevic T, Barritt GJ (1997) Novel variants of voltage-operated calcium channel alpha 1-subunit transcripts in a rat liver-derived cell line: deletion in the IVS4 voltage sensing region. Cell Calcium 22(1):39–52

Burgess GM, Godfrey PP, McKinney JS, Berridge MJ, Irvine RF, Putney JW Jr (1984) The second messenger linking receptor activation to internal Ca release in liver. Nature 309(5963):63–66

Cabre N, Camps J, Joven J (2016) Inflammation, mitochondrial metabolism and nutrition: the multi-faceted progression of non-alcoholic fatty liver disease to hepatocellular carcinoma. Hepatobiliary Surg Nutr 5(5):438–443

Campbell JE, Drucker DJ (2013) Pharmacology, physiology, and mechanisms of incretin hormone action. Cell Metab 17(6):819–837

Cantley JL, Yoshimura T, Camporez JP, Zhang D, Jornayvaz FR, Kumashiro N, Guebre-Egziabher F, Jurczak MJ, Kahn M, Guigni BA, Serr J, Hankin J, Murphy RC, Cline GW, Bhanot S, Manchem VP, Brown JM, Samuel VT, Shulman GI (2013) CGI-58 knockdown sequesters diacylglycerols in lipid droplets/ER-preventing diacylglycerol-mediated hepatic insulin resistance. Proc Natl Acad Sci U S A 110(5):1869–1874

Carrasco S, Meyer T (2011) STIM proteins and the endoplasmic reticulum-plasma membrane junctions. Annu Rev Biochem 80:973–1000

Castro J, Aromataris EC, Rychkov GY, Barritt GJ (2009) A small component of the endoplasmic reticulum is required for store-operated Ca2+ channel activation in liver cells: evidence from studies using TRPV1 and taurodeoxycholic acid. Biochem J 418(3):553–566

Chan C, Harland ML, Webb SE, Chen J, Miller AL, Barritt GJ (2004) Evaluation, using targeted aequorins, of the roles of the endoplasmic reticulum and its (Ca2++Mg2+)ATP-ases in the activation of store-operated Ca2+ channels in liver cells. Cell Calcium 35(4):317–331

Chiono M, Mahey R, Tate G, Cooper DM (1995) Capacitative Ca2+ entry exclusively inhibits cAMP synthesis in C6-2B glioma cells. Evidence that physiologically evoked Ca2+ entry regulates Ca(2+)-inhibitable adenylyl cyclase in non-excitable cells. J Biol Chem 270 (3):1149–1155

Cooper DM (2015) Store-operated Ca(2)(+)-entry and adenylyl cyclase. Cell Calcium 58 (4):368–375

Courjaret R, Machaca K (2014) Mid-range Ca2+ signalling mediated by functional coupling between store-operated Ca2+ entry and IP3-dependent Ca2+ release. Nat Commun 5:3916

Crane J, McGowan B (2016) The GLP-1 agonist, liraglutide, as a pharmacotherapy for obesity. Ther Adv Chronic Dis 7(2):92–107

Dagli ML, Guerra JL, Sinhorini IL, Wu TS, Rizzi MB, Penteado MV, Moreno FS (1998) Beta-carotene reduces the ductular (oval) cell reaction in the liver of Wistar rats submitted to the resistant hepatocyte model of carcinogenesis. Pathology 30(3):259–266

Dara L, Ji C, Kaplowitz N (2011) The contribution of endoplasmic reticulum stress to liver diseases. Hepatology 53(5):1752–1763

Derler I, Jardin I, Romanin C (2016a) Molecular mechanisms of STIM/Orai communication. Am J Physiol Cell Physiol 310(8):C643–C662

Derler I, Jardin I, Stathopulos PB, Muik M, Fahrner M, Zayats V, Pandey SK, Poteser M, Lackner B, Absolonova M, Schindl R, Groschner K, Ettrich R, Ikura M, Romanin C (2016b) Cholesterol modulates Orai1 channel function. Sci Signal 9(412):ra10

Dong H, Czaja MJ (2011) Regulation of lipid droplets by autophagy. Trends Endocrinol Metab 22 (6):234–240

Egnatchik RA, Leamy AK, Jacobson DA, Shiota M, Young JD (2014) ER calcium release promotes mitochondrial dysfunction and hepatic cell lipotoxicity in response to palmitate overload. Mol Metab 3(5):544–553

El Boustany C, Bidaux G, Enfissi A, Delcourt P, Prevarskaya N, Capiod T (2008) Capacitative calcium entry and transient receptor potential canonical 6 expression control human hepatoma cell proliferation. Hepatology 47(6):2068–2077

Engin F, Hotamisligil GS (2010) Restoring endoplasmic reticulum function by chemical chaperones: an emerging therapeutic approach for metabolic diseases. Diabetes Obes Metab 12(Suppl 2):108–115

Exton JH, Blackmore PF, El-Refai MF, Dehaye JP, Strickland WG, Cherrington AD, Chan TM, Assimacopoulos-Jeannet FD, Chrisman TD (1981) Mechanisms of hormonal regulation of liver metabolism. Adv Cyclic Nucleotide Res 14:491–505

Fahrner M, Derler I, Jardin I, Romanin C (2013) The STIM1/Orai signaling machinery. Channels 7 (5):330–343

Fernando KC, Gregory RB, Barritt GJ (1998) Protein kinase A regulates the disposition of Ca2+ which enters the cytoplasmic space through store-activated Ca2+ channels in rat hepatocytes by diverting inflowing Ca2+ to mitochondria. Biochem J 330(Pt 3):1179–1187

Font-Burgada J, Sun B, Karin M (2016) Obesity and cancer: the oil that feeds the flame. Cell Metab 23(1):48–62

Fu S, Yang L, Li P, Hofmann O, Dicker L, Hide W, Lin X, Watkins SM, Ivanov AR, Hotamisligil GS (2011) Aberrant lipid metabolism disrupts calcium homeostasis causing liver endoplasmic reticulum stress in obesity. Nature 473(7348):528–531

Fu S, Watkins SM, Hotamisligil GS (2012) The role of endoplasmic reticulum in hepatic lipid homeostasis and stress signaling. Cell Metab 15(5):623–634

Gao J, Zhang Y, Yu C, Tan F, Wang L (2016) Spontaneous nonalcoholic fatty liver disease and ER stress in Sidt2 deficiency mice. Biochem Biophys Res Commun 475(4):326–332

Gaspers LD, Thomas AP (2005) Calcium signaling in liver. Cell Calcium 38(3–4):329–342

Golovina VA (2005) Visualization of localized store-operated calcium entry in mouse astrocytes. Close proximity to the endoplasmic reticulum. J Physiol (Lond) 564(Pt 3):737–749

Graf J, Häussinger D (1996) Ion transport in hepatocytes: mechanisms and correlations to cell volume, hormone actions and metabolism. J Hepatol 24(Suppl 1):53–77

Greene MW, Burrington CM, Ruhoff MS, Johnson AK, Chongkrairatanakul T, Kangwanpornsiri A (2010) PKC{delta} is activated in a dietary model of steatohepatitis and regulates endoplasmic reticulum stress and cell death. J Biol Chem 285(53):42115–42129

Gregory RB, Wilcox RA, Berven LA, van Straten NC, van der Marel GA, van Boom JH, Barritt GJ (1999) Evidence for the involvement of a small subregion of the endoplasmic reticulum in the inositol trisphosphate receptor-induced activation of Ca2+ inflow in rat hepatocytes. Biochem J 341(Pt 2):401–408

Han J, Kaufman RJ (2016) The role of ER stress in lipid metabolism and lipotoxicity. J Lipid Res 57(8):329–338

Hijioka T, Rosenberg RL, Lemasters JJ, Thurman RG (1992) Kupffer cells contain voltage-dependent calcium channels. Mol Pharmacol 41(3):435–440

Holowka D, Korzeniowski MK, Bryant KL, Baird B (2014) Polyunsaturated fatty acids inhibit stimulated coupling between the ER Ca(2+) sensor STIM1 and the Ca(2+) channel protein Orai1 in a process that correlates with inhibition of stimulated STIM1 oligomerization. Biochim Biophys Acta 1841(8):1210–1216

Huang W, Bansode R, Mehta M, Mehta KD (2009) Loss of protein kinase Cbeta function protects mice against diet-induced obesity and development of hepatic steatosis and insulin resistance. Hepatology 49(5):1525–1536

Hughes BP, Milton SE, Barritt GJ, Auld AM (1986) Studies with verapamil and nifedipine provide evidence for the presence in the liver cell plasma membrane of two types of Ca2+ inflow transporter which are dissimilar to potential-operated Ca2+ channels. Biochem Pharmacol 35 (18):3045–3052

Jones BF, Boyles RR, Hwang SY, Bird GS, Putney JW Jr (2008) Calcium influx mechanisms underlying calcium oscillations in rat hepatocytes. Hepatology 48(4):1273–1281

Jornayvaz FR, Shulman GI (2012) Diacylglycerol activation of protein kinase Ce and hepatic insulin resistance. Cell Metab 15:574–584

Jornayvaz FR, Birkenfeld AL, Jurczak MJ, Kanda S, Guigni BA, Jiang DC, Zhang D, Lee HY, Samuel VT, Shulman GI (2011) Hepatic insulin resistance in mice with hepatic overexpression of diacylglycerol acyltransferase 2. Proc Natl Acad Sci U S A 108(14):5748–5752

Jousset H, Frieden M, Demaurex N (2007) STIM1 knockdown reveals that store-operated Ca2+ channels located close to sarco/endoplasmic Ca2+ ATPases (SERCA) pumps silently refill the endoplasmic reticulum. J Biol Chem 282(15):11456–11464

Kammoun HL, Chabanon H, Hainault I, Luquet S, Magnan C, Koike T, Ferré P, Foufelle F (2009) GRP78 expression inhibits insulin and ER stress-induced SREBP-1c activation and reduces hepatic steatosis in mice. J Clin Invest 119(5):1201–1215

Kang S, Dahl R, Hsieh W, Shin A, Zsebo KM, Buettner C, Hajjar RJ, Lebeche D (2016) Small molecular allosteric activator of the sarco/endoplasmic reticulum Ca2+-ATPase (SERCA) attenuates diabetes and metabolic disorders. J Biol Chem 291(10):5185–5198

Karin M, Dhar D (2016) Liver carcinogenesis: from naughty chemicals to soothing fat and the surprising role of NRF2. Carcinogenesis 37(6):541–546

Kars M, Yang L, Gregor MF, Mohammed BS, Pietka TA, Finck BN, Patterson BW, Horton JD, Mittendorfer B, Hotamisligil GS, Klein S (2010) Tauroursodeoxycholic Acid may improve liver and muscle but not adipose tissue insulin sensitivity in obese men and women. Diabetes 59(8):1899–1905

Kawasaki T, Ueyama T, Lange I, Feske S, Saito N (2010) Protein kinase C-induced phosphorylation of Orai1 regulates the intracellular Ca2+ level via the store-operated Ca2+ channel. J Biol Chem 285(33):25720–25730

Khera R, Murad MH, Chandar AK, Dulai PS, Wang Z, Prokop LJ, Loomba R, Camilleri M, Singh S (2016) Association of pharmacological treatments for obesity with weight loss and adverse events: a systematic review and meta-analysis. JAMA 315(22):2424–2434

Kheradpezhouh E, Ma L, Morphett A, Barritt GJ, Rychkov GY (2014) TRPM2 channels mediate acetaminophen-induced liver damage. Proc Natl Acad Sci U S A 111(8):3176–3181

Kim S (2016) Drugs to treat obesity: do they work? Postgrad Med J 92(1089):401–406

Lau JK, Zhang X, Yu J (2016) Animal models of non-alcoholic fatty liver disease: current perspectives and recent advances. J Pathol 241(1):36–44

Lee J, Hong SW, Chae SW, Kim DH, Choi JH, Bae JC, Park SE, Rhee EJ, Park CY, Oh KW, Park SW, Kim SW, Lee WY (2012) Exendin-4 improves steatohepatitis by increasing Sirt1 expression in high-fat diet-induced obese C57BL/6J mice. PLoS One 7(2):e31394

Lee J, Hong SW, Park SE, Rhee EJ, Park CY, Oh KW, Park SW, Lee WY (2014) Exendin-4 regulates lipid metabolism and fibroblast growth factor 21 in hepatic steatosis. Metabolism 63 (8):1041–1048

Leite MF, Nathanson MH (2001) The liver biology and pathobiology. Lippincott, Williams and Wilkins, Philadelphia, pp 537–554

Litjens T, Nguyen T, Castro J, Aromataris EC, Jones L, Barritt GJ, Rychkov GY (2007) Phospholipase C-gamma1 is required for the activation of store-operated Ca2+ channels in liver cells. Biochem J 405(2):269–276

Liu K, Czaja MJ (2013) Regulation of lipid stores and metabolism by lipophagy. Cell Death Differ 20(1):3–11

Liu W, Baker RD, Bhatia T, Zhu L, Baker SS (2016) Pathogenesis of nonalcoholic steatohepatitis. Cell Mol Life Sci 73(10):1969–1987

Lobeck I, Donnelly B, Dupree P, Mahe MM, McNeal M, Mohanty SK, Tiao G (2016) Rhesus rotavirus VP6 regulates ERK-dependent calcium influx in cholangiocytes. Virology 499:185–195

Ma C, Kesarwala AH, Eggert T, Medina-Echeverz J, Kleiner DE, Jin P, Stroncek DF, Terabe M, Kapoor V, ElGindi M, Han M, Thornton AM, Zhang H, Egger M, Luo J, Felsher DW, McVicar DW, Weber A, Heikenwalder M, Greten TF (2016) NAFLD causes selective CD4(+) T lymphocyte loss and promotes hepatocarcinogenesis. Nat Med 531(7593):253–257

Manjarres IM, Alonso MT, Garcia-Sancho J (2011) Calcium entry-calcium refilling (CECR) coupling between store-operated Ca(2+) entry and sarco/endoplasmic reticulum Ca(2+)-ATPase. Cell Calcium 49(3):153–161

Martinez-Lopez N, Singh R (2015) Autophagy and lipid droplets in the liver. Annu Rev Nutr 35:215–237

Mashek DG, Khan SA, Sathyanarayan A, Ploeger JM, Franklin MP (2015) Hepatic lipid droplet biology: getting to the root of fatty liver. Hepatology 62(3):964–967

Maus M, Cuk M, Patel B, Lian J, Ouimet M, Kaufmann U, Yang J, Horvath R, Hornig-Do HT, Chrzanowska-Lightowlers ZM, Moore KJ, Cuervo AM, Feske S (2017) Store-operated Ca2+ entry controls induction of lipolysis and the transcriptional reprogramming to lipid metabolism. Cell Metab 25(3):698–712

Mekahli D, Bultynck G, Parys JB, De Smedt H, Missiaen L (2011) Endoplasmic-reticulum calcium depletion and disease. Cold Spring Harb Perspect Biol 3(6):a004317

Musso G, Cassader M, Gambino R (2016) Non-alcoholic steatohepatitis: emerging molecular targets and therapeutic strategies. Nat Rev Drug Discov 15(4):249–274

Nakamura S, Takamura T, Matsuzawa-Nagata N, Takayama H, Misu H, Noda H, Nabemoto S, Kurita S, Ota T, Ando H, Miyamoto K, Kaneko S (2009) Palmitate induces insulin resistance in H4IIEC3 hepatocytes through reactive oxygen species produced by mitochondria. J Biol Chem 284(22):14809–14818

Nyenwe EA, Jerkins TW, Umpierrez GE, Kitabchi AE (2011) Management of type 2 diabetes: evolving strategies for the treatment of patients with type 2 diabetes. Metabolism 60(1):1–23

Ozcan L, Tabas I (2016) Calcium signalling and ER stress in insulin resistance and atherosclerosis. J Intern Med 280(5):457–464

Ozcan U, Yilmaz E, Ozcan L, Furuhashi M, Vaillancourt E, Smith RO, Görgün CZ, Hotamisligil GS (2006) Chemical chaperones reduce ER stress and restore glucose homeostasis in a mouse model of type 2 diabetes. Science 313(5790):1137–1140

Ozcan L, Wong CC, Li G, Xu T, Pajvani U, Park SK, Wronska A, Chen BX, Marks AR, Fukamizu A, Backs J, Singer HA, Yates JR 3rd, Accili D, Tabas I (2012) Calcium signaling through CaMKII regulates hepatic glucose production in fasting and obesity. Cell Metab 15:739–751

Ozcan L, Cristina de Souza J, Harari AA, Backs J, Olson EN, Tabas I (2013) Activation of calcium/calmodulin-dependent protein kinase II in obesity mediates suppression of hepatic insulin signaling. Cell Metab 18(6):803–815

Pagliassotti MJ, Kim PY, Estrada AL, Stewart CM, Gentile CL (2016) Endoplasmic reticulum stress in obesity and obesity-related disorders: an expanded view. Metabolism 65:1238–1246

Parekh AB (2011) Decoding cytosolic Ca2+ oscillations. Trends Biochem Sci 36(2):78–87

Park HW, Lee JH (2014) Calcium channel blockers as potential therapeutics for obesity-associated autophagy defects and fatty liver pathologies. Autophagy 10(12):2385–2386

Park SW, Zhou Y, Lee J, Ozcan U (2010) Sarco(endo)plasmic reticulum Ca2+-ATPase 2b is a major regulator of endoplasmic reticulum stress and glucose homeostasis in obesity. Proc Natl Acad Sci U S A 107(45):19320–19325

Park HW, Park H, Semple IA, Jang I, Ro SH, Kim M, Cazares VA, Stuenkel EL, Kim JJ, Kim JS, Lee JH (2014) Pharmacological correction of obesity-induced autophagy arrest using calcium channel blockers. Nat Commun 5:4834

Perry RJ, Samuel VT, Petersen KF, Shulman GI (2014) The role of hepatic lipids in hepatic insulin resistance and type 2 diabetes. Nature 510(7503):84–91

Pinzani M, Failli P, Ruocco C, Casini A, Milani S, Baldi E, Giotti A, Gentilini P (1992) Fat-storing cells as liver-specific pericytes. Spatial dynamics of agonist-stimulated intracellular calcium transients. J Clin Invest 90(2):642–646

Prakriya M, Lewis RS (2015) Store-operated calcium channels. Physiol Rev 95(4):1383–1436

Puljak L, Pagliassotti MJ, Wei Y, Qadri I, Parameswara V, Esser V, Fitz JG, Kilic G (2005) Inhibition of cellular responses to insulin in a rat liver cell line. A role for PKC in insulin resistance. J Physiol 563(Pt 2):471–482

Rieusset J (2017) Endoplasmic reticulum-mitochondria calcium signalling in hepatic metabolic diseases. Biochim Biophys Acta pii:S0167-4889(17)30003-4. doi:10.1016/j.bbamcr.2017.01.001

Rieusset J, Fauconnier J, Paillard M, Belaidi E, Tubbs E, Chauvin MA, Durand A, Bravard A, Teixeira G, Bartosch B, Michelet M, Theurey P, Vial G, Demion M, Blond E, Zoulim F, Gomez L, Vidal H, Lacampagne A, Ovize M (2016) Disruption of calcium transfer from ER to mitochondria links alterations of mitochondria-associated ER membrane integrity to hepatic insulin resistance. Diabetologia 59(3):614–623

Röhrig F, Schulze A (2016) The multifaceted roles of fatty acid synthesis in cancer. Nat Rev Cancer 16(11):732–749

Rychkov GY, Barritt GJ (2011) Expression and function of TRP channels in liver cells. Adv Exp Med Biol 704:667–686

Rychkov G, Brereton HM, Harland ML, Barritt GJ (2001) Plasma membrane Ca2+ release-activated Ca2+ channels with a high selectivity for Ca2+ identified by patch-clamp recording in rat liver cells. Hepatology 33(4):938–947

Samuel VT, Shulman GI (2012) Mechanisms for insulin resistance: common threads and missing links. Cell 148(5):852–871

Samuel VT, Shulman GI (2016) The pathogenesis of insulin resistance: integrating signaling pathways and substrate flux. J Clin Invest 126(1):12–22

Sawanobori T, Takanashi H, Hiraoka M, Iida Y, Kamisaka K, Maezawa H (1989) Electrophysiological properties of isolated rat liver cells. J Cell Physiol 139(3):580–585

Scherer A, Dufour JF (2016) Treatment of non-alcoholic fatty liver disease. Dig Dis 34(Suppl 1):27–31

Scrimgeour N, Litjens T, Ma L, Barritt GJ, Rychkov GY (2009) Properties of Orai1 mediated store-operated current depend on the expression levels of STIM1 and Orai1 proteins. J Physiol (Lond) 587(Pt 12):2903–2918

Scrimgeour NR, Wilson DP, Barritt GJ, Rychkov GY (2014) Structural and stoichiometric determinants of Ca2+ release-activated Ca2+ (CRAC) channel Ca2+-dependent inactivation. Biochim Biophys Acta 1838(5):1281–1287

Sharma A, Umar S, Kar P, Singh K, Sachdev M, Goel A (2016) A new type of biocompatible fluorescent probe AFN for fixed and live cell imaging of intracellular lipid droplets. Analyst 141(1):137–143

Srikanth S, Ribalet B, Gwack Y (2013) Regulation of CRAC channels by protein interactions and post-translational modification. Channels (Austin) 7(5):354–363

Srivats S, Balasuriya D, Pasche M, Vistal G, Edwardson JM, Taylor CW, Murrell-Lagnado RD (2016) Sigma1 receptors inhibit store-operated Ca2+ entry by attenuating coupling of STIM1 to Orai1. J Cell Biol 213(1):65–79

Steenks M, van Baal MC, Nieuwenhuijs VB, de Bruijn MT, Schiesser M, Teo MH, Callahan T, Padbury RT, Barritt GJ (2010) Intermittent ischaemia maintains function after ischaemia reperfusion in steatotic livers. HPB (Oxford) 12(4):250–261

Striggow F, Bohnensack R (1993) Verapamil and diltiazem inhibit receptor-operated calcium channels and intracellular calcium oscillations in rat hepatocytes. FEBS Lett 318(3):341–344

Sun Z, Lazar MA (2013) Dissociating fatty liver and diabetes. Trends Endocrinol Metab 24 (1):4–12

Thomas A, Delaville F (1991) The use of fluorescent indicators for measurements of cytosolic-free calcium concentration in cell populations and single cells. In: McCormack JG, Cobbold PH (eds) Cellular calcium: a practical approach. IRL, Oxford, pp 1–54

Tilg H, Moschen AR, Roden M (2017) NAFLD and diabetes mellitus. Nat Rev Gastroenterol Hepatol 14(1):32–42

To MS, Aromataris EC, Castro J, Roberts ML, Barritt GJ, Rychkov GY (2010) Mitochondrial uncoupler FCCP activates proton conductance but does not block store-operated Ca(2+) current in liver cells. Arch Biochem Biophys 495(2):152–158

Vaca L (2010) SOCIC: the store-operated calcium influx complex. Cell Calcium 47(3):199–209

Walther TC, Farese RV Jr (2012) Lipid droplets and cellular lipid metabolism. Annu Rev Biochem 81:687–714
Wang S, Kaufman RJ (2014) How does protein misfolding in the endoplasmic reticulum affect lipid metabolism in the liver? Curr Opin Lipidol 25(2):125–132
Wei Y, Wang D, Gentile CL, Pagliassotti MJ (2009) Reduced endoplasmic reticulum luminal calcium links saturated fatty acid-mediated endoplasmic reticulum stress and cell death in liver cells. Mol Cell Biochem 331(1–2):31–40
Wilson CH, Ali ES, Scrimgeour N, Martin AM, Hua J, Tallis GA, Rychkov GY, Barritt GJ (2015) Steatosis inhibits liver cell store-operated Ca(2)(+) entry and reduces ER Ca(2)(+) through a protein kinase C-dependent mechanism. Biochem J 466(2):379–390
Wong CR, Nguyen MH, Lim JK (2016) Hepatocellular carcinoma in patients with non-alcoholic fatty liver disease. World J Gastroenterol 22(37):8294–8303
Woods NM, Cuthbertson KS, Cobbold PH (1986) Repetitive transient rises in cytoplasmic free calcium in hormone-stimulated hepatocytes. Nature 319(6054):600–602
Woods NM, Cuthbertson KS, Cobbold PH (1987) Agonist-induced oscillations in cytoplasmic free calcium concentration in single rat hepatocytes. Cell Calcium 8(1):79–100
Wu J, Danielsson A, Lindstrom P, Karlsson K, Sehlin J (1995) Protective effects of calcium channel blockers on acute bromobenzene toxicity to isolated rat hepatocytes. Inhibition of phenylephrine-induced calcium oscillations. Scand J Gastroenterol 30(6):590–600
Xiao C, Giacca A, Lewis GF (2011) Sodium phenylbutyrate, a drug with known capacity to reduce endoplasmic reticulum stress, partially alleviates lipid-induced insulin resistance and beta-cell dysfunction in humans. Diabetes 60(3):918–924
Young B, Heath JW (2000) Liver and pancreas Wheater's functional histology, 4th edn. Churchill Livingstone, Edinburgh, pp 274–275
Zhang C, Thomas DW (2016) Stromal Interaction Molecule 1 rescues store-operated calcium entry and protects NG115-401L cells against cell death induced by endoplasmic reticulum and mitochondrial oxidative stress. Neurochem Int 97:137–145
Zhang J, Li Y, Jiang S, Yu H, An W (2014) Enhanced endoplasmic reticulum SERCA activity by overexpression of hepatic stimulator substance gene prevents hepatic cells from ER stress-induced apoptosis. Am J Physiol Cell Physiol 306(3):C279–C290

Metabolic Disorders and Cancer: Store-Operated Ca^{2+} Entry in Cancer – Focus on IP_3R-Mediated Ca^{2+} Release from Intracellular Stores and Its Role in Migration and Invasion

31

Abigaël Ritaine, George Shapovalov, and Natalia Prevarskaya

Abstract

Store-operated calcium entry (SOCE) plays important roles in a multitude of cellular processes, from muscle contraction to cellular proliferation and migration. Dysregulation of SOCE is responsible for the advancement of multiple diseases, ranging from immune diseases, myopathies, to terminal ones like cancer. Naturally, SOCE has been a focus of many studies and review papers which, however, primarily concentrated on the principal players localized to the plasma membrane and responsible for Ca^{2+} entry into the cell. Much less has been said about other players participating in the entire SOCE event. This review aims to address this shortcoming by discussing the accumulated scientific knowledge focused on the inositol trisphosphate receptors (IP_3Rs), principal player responsible for emptying intracellular Ca^{2+} stores in a majority of cells, and their involvement in regulation of cell migration and invasion in cancer.

George Shapovalov and Natalia Prevarskaya shared senior co-authorship.

A. Ritaine
Inserm U1003, Equipe Labellisée par la Ligue Nationale Contre le Cancer, Université de Sciences et Technologies de Lille (USTL), 59655 Villeneuve d'Ascq, France

Laboratory of Excellence, Ion Channels Science and Therapeutics, Université Lille I Sciences et Technologies, Villeneuve d'Ascq, France

Laboratory of Molecular and Cellular Signaling, Department of Cellular and Molecular Medicine, Katholieke Universiteit Leuven, Leuven, Belgium

G. Shapovalov • N. Prevarskaya (✉)
Inserm U1003, Equipe Labellisée par la Ligue Nationale Contre le Cancer, Université de Sciences et Technologies de Lille (USTL), 59655 Villeneuve d'Ascq, France

Laboratory of Excellence, Ion Channels Science and Therapeutics, Université Lille I Sciences et Technologies, Villeneuve d'Ascq, France
e-mail: natacha.prevarskaya@univ-lille1.fr

K. Groschner et al. (eds.), *Store-Operated Ca^{2+} Entry (SOCE) Pathways*, Advances in Experimental Medicine and Biology 993, DOI 10.1007/978-3-319-57732-6_31

Keywords
Calcium • IP_3R • SOCE • SOC channels • Migration • Invasion • Cancer

31.1 Introduction

Cancer is one of the most common terminal diseases in developed countries. It is among the main causes of death in the world with more than 100 different types of cancer identified at present, each represented by a specific genotype. Cancer progression development follows three typical steps: initiation, promotion, and progression. The initiation is the step when one or more errors appear in the original DNA sequence, so-called driver mutations, by lesions due to endogenous or exogenous factors. These driver mutations lead to the transformation of the cells which thus become abnormal. The promotion is the identical multiplication of transformed cells. These cells form an abnormal mass called neoplasm or tumor. We talk about cancer once transformed cells have acquired the capacity to trigger the formation of blood vessels (neo-angiogenesis) allowing, in particular, their feeding in nutrients and oxygen ensuring the tumor growth. The last stage is the progression during which cancer cells cross the blood vessel and lymphatic basal lamina in order to migrate to other tissues. Once the cells have arrived at the new tissue, they cross the vessels' basal lamina for the second time to insert themselves into the tissues. This process is called metastasis, and it is the main cause of death by cancer (Hanahan and Weinberg 2011; Jin et al. 2017).

Among the proteins of particular importance that can be targeted by genetic alterations, we find ion channels which are implicated in various carcinogenesis processes such as proliferation, migration, or cancer cell survival. This importance of ion channels stems from their role in maintaining the intracellular ion equilibrium which is crucial for most physiological functions. An imbalance in any of these processes, caused by modification of expression and/or functionality of an ion channel, is called a channelopathy and can be a root cause of a severe disease, such as cancer (Prevarskaya et al. 2010).

One of the more important ions in cellular homeostasis is Ca^{2+} which is a key second messenger, regulating numerous physiological and pathological processes ranging from muscle contraction to much more fatally important ones such as cell motility and death. While, classically, ion channels have been considered to be localized to the plasma membrane (PM), many ion channels have recently been found in the membranes of various intracellular organelles, such as the endoplasmic reticulum (ER), mitochondria, or the nuclear membrane (Leanza et al. 2013). Interestingly, while plasma membrane-localized ion channels are best known to participate in fast-paced cellular processes like muscle contraction or action potential propagation, the majority of intracellular ion channels are known to play important roles in regulating cellular function and fate. For instance, the IP_3Rs and the ryanodine receptors (RyRs), acting as Ca^{2+} release channels in the ER, were shown to regulate various important cellular functions, such as apoptosis, contraction, cell motility, proliferation and migration (Prevarskaya et al. 2011; Vervloessem et al. 2014).

In general, increase of intracellular Ca^{2+} concentration happens by Ca^{2+} entering the cell via the Ca^{2+}-permeable ion channels of the plasma membrane or by the release of Ca^{2+} from internal stores, typically the endoplasmic or sarcoplasmic reticulum. The release of Ca^{2+} from the ER correspondingly requires the presence of Ca^{2+}-permeable ion channels with the two principal and best studied such protein classes being IP_3Rs and RyRs. The basal cytoplasmic Ca^{2+} concentration is regulated also by specialized transport proteins, such as PMCA (plasma membrane Ca^{2+} ATPase), responsible for the transport of excessive Ca^{2+} out of the cell through the plasma membrane, and SERCA, present in the ER membrane and responsible for Ca^{2+} reuptake by the internal stores. Additionally, store-operated channels (SOC) are known to be present in the plasma membrane, where they are responsible for Ca^{2+} entry into the cells when internal stores are exhausted, and implicate Orai and STIM (stromal interaction molecule) proteins (Yang et al. 2012). The mechanism of such replenishment of Ca^{2+} upon store depletion is called store-operated Ca^{2+} entry (SOCE). In the majority of studies, SOC response is stimulated by blocking SERCA chemically (Michelangeli and East 2011). A classical irreversible SERCA inhibitor is thapsigargin, which blocks Ca^{2+} reuptake into the ER and leads to gradual decrease of its Ca^{2+} content and eventual opening of SOCs.

Since their discovery at the beginning of the twenty-first century, a lot of studies have involved STIM and Orai in various cellular processes from immune cell activation, muscle contraction, and sperm chemotaxis to regulation of gene transcription and cell fate (Kahl and Means 2003; Yoshida et al. 2003). In these studies, Orai1 and STIM1 have been found to be important for cellular migration. For example, suppression of these proteins by siRNA or inhibition of SOC currents by SKF96365 inhibited the migration of the human breast adenocarcinoma cell line MDA-MB-231 and decreased the formation of metastases in mice (Yang et al. 2009). Thus, the importance of SOCE in various cellular processes, including such critical ones as migration and invasion, is well known. The role of STIM and ORAI proteins in hallmarks of cancer and in particular in tissue invasion or metastasis processes has also been extensively studied (Fiorio Pla et al. 2016). However, the majority of associated research focused on immediate players responsible for Ca^{2+} entry into the cells via the PM, with less evidence put forward on how other players, such as IP_3Rs, participate in this big picture. This review aims to close this gap by focusing on the recent findings of how IP_3R activity is regulated by different players and how changes of this activity affect the complex processes of migration and invasion of cancer cells.

31.2 Cellular Migration and Invasion

During metastasis formation, one of the six initial hallmarks of cancer, the migration of epithelial and endothelial cells is a key step leading to the propagation of the primary tumor and to the invasion of neighboring tissues (Hanahan and Weinberg

2000, 2011). However, cell migration is also a central mechanism in homeostasis of healthy tissues. Important physiological processes such as embryogenesis, immune defense, angiogenesis, and wound healing are closely linked to the capability of cells to migrate.

In general, cell migration is conceptualized as a cyclic process (Lauffenburger and Horwitz 1996). It consists of multiple steps, such as the formation of organized adhesive structures, or focal adhesion sites, containing receptors and integrins linked to the extremities of actin fibers (Westhoff et al. 2004) and membrane protrusions, such as lamellipodes or invadopodes, controlled by small GTPases, Rac and Cdc42, with a finely controlled balance between detachment and adhesion at the focal adhesion sites (Ridley et al. 2003). The lamellipodes are surface-attached structures that are found at the leading edge of migrating cells. In order to become invasive, tumor cells need to acquire abilities that allow increased migration through the extracellular matrix (ECM). In order to do this, tumor cells have specialized membrane protrusions, called invadopodia, that provide important functions for invasion. It should be noted that the term invadopodia is primarily used in conjunction with cancer cells, while the term podosome can often be found in the literature when talking about normal cells. Invadopodia are actin-rich plasma membrane protrusions at the ventral surface of the cell, often situated under the nucleus, that are proteolytically active and responsible for the focal degradation of ECM components through matrix metalloproteinases (MMP) (Nabeshima et al. 2002; Yoon et al. 2003).

Both, Ca^{2+} release and Ca^{2+} influx, have been linked to cell migration depending on cell types and stimuli (Agle et al. 2010; Tsai et al. 2014; Yang and Huang 2005). Thus, migrating cells are polarized and exhibit a transient and steady gradient of intracellular Ca^{2+}. That gradient increases from the front of the cell to the rear and is thought to be responsible for rear-end retraction (Brundage et al. 1991; Hahn et al. 1992). The increased cytosolic Ca^{2+} concentration ($[Ca^{2+}]_i$) at the rear end is related among others to myosin II contraction and calpain-mediated release of focal adhesions leading to the retraction of the uropod (Eddy et al. 2000; Palecek et al. 1998). The retraction is accomplished via myosin II contraction, which is regulated by myosin light chain (MLC) phosphorylation through Ca^{2+}-dependent MLC kinase (MLCK) and by disassembly of adhesions at the rear of the cell owing to calpain-mediated cleavage of focal adhesion proteins such as integrins, talin, vinculin, and focal adhesion kinase (FAK) (Ridley et al. 2003). However, the cyclic morphological and adherence changes that are observed during cell migration are also supported by repetitive Ca^{2+} signals, which take the form of Ca^{2+} spikes or oscillations (Pettit and Fay 1998).

It is therefore natural that migration of the tumor, being strongly dependent on these Ca^{2+} signals, is accompanied by changes in expression and activity of ion channels. Below we discuss how IP_3Rs participate in these processes, regulate intracellular Ca^{2+}, and work in ensemble with SOCs.

31.3 Intracellular Ca^{2+} Release Channels: The IP_3Rs

31.3.1 The Discovery of the IP_3Rs

Involvement of IP_3R in intracellular Ca^{2+} regulation was observed for the first time over 30 years ago, when Streb et al. (1983) found a second messenger that was able to increase $[Ca^{2+}]_i$ from a non-mitochondrial store: inositol 1,4,5-trisphosphate (IP_3). This discovery was followed by concentrated efforts of many researchers to discover the target of this second messenger. This goal was achieved first by K. Mikoshiba's team, which was working on a P400 protein which was absent in Purkinje-neuron-degenerating mutant mice and found that this protein binds IP_3 (Furuichi et al. 1989a; Maeda et al. 1988), and established its complete primary sequence (Furuichi et al. 1989b; Maeda et al. 1990).

One year later the P400 protein was found to form a channel by incorporating it into a planar lipid bilayer and recording its activity and was renamed IP3R. Moreover, in the same paper, it was shown that this channel exhibited a Ca^{2+} conductance (Maeda et al. 1991). These findings were confirmed by the work of other teams that achieved similar results, establishing the identity of the IP_3R at the same time (Sudhof et al. 1991; Supattapone et al. 1988).

The IP_3R is one of the first ion channels that were discovered to be localized in the ER (Ross et al. 1989). The channel is a tetramer, formed from monomers of 260 kDa in size. Each monomer contains binding sites for IP_3 and Ca^{2+} in the N-terminal region, six transmembrane domains forming a pore region and the determinants for tetramer formation in the C-terminus (Fan et al. 2015; Taylor and Konieczny 2016).

The IP_3R is a principal Ca^{2+} release channel from the internal stores in the majority of cells, participating in the regulation of intracellular Ca^{2+} in the cytosol and in Ca^{2+} transfer between ER and mitochondria (Rizzuto et al. 2009). Over the years, a lot of different studies have implicated the IP_3R in plenty of physiological and pathological mechanisms. Indeed, an overexpression of the IP_3R is observed in numerous forms of cancer, typically in correlation with an increase in proliferation or migration (Ivanova et al. 2014).

As mentioned earlier, the first identified physiological trigger for the activation of these channels is known to be IP_3. However, further studies elucidated the specific mechanism of IP_3R activation by IP_3 in a Ca^{2+}-sensitive manner. The IP_3R is regulated in a bimodal manner by Ca^{2+} itself—its activity increases with the increase of $[Ca^{2+}]_i$ to a certain level, a phenomenon that is known as Ca^{2+}-induced Ca^{2+} release (CICR). However, IP_3R activity is inhibited at higher $[Ca^{2+}]_i$, with IP_3 removing this inhibition of IP_3R activity at high $[Ca^{2+}]_i$ (Foskett et al. 2007). IP_3 itself is produced in the cells in response to stimulation of surface receptors such as receptor tyrosine kinases (RTK) and G-protein-coupled receptors (GPCR), leading to the activation of phospholipase C (PLC). The PLC is an enzyme that metabolizes phosphatidylinositol 4,5-biphosphate (PiP2) into IP_3 and diacylglycerol (DAG), both of which are second messengers, thus triggering various signaling cascades that can involve IP_3Rs and lead, for example, to cellular growth.

31.3.2 Role of the IP_3Rs in Regulating Migration and Invasion

While the identity of IP_3R channels and their role in Ca^{2+} release from internal stores have been known for a long time, the question of their involvement in regulating cellular migration and invasion has only recently became a topic of active studies. Publications covering this subject are few, but their number is steadily increasing in the last few years (see Table 31.1).

Treatment with caffeine was found to inhibit the IP_3R-mediated Ca^{2+} release by selectively targeting IP_3R3 in glioblastoma cell lines. Interestingly, the mRNA

Table 31.1 Implication of IP_3R in migration and invasion

Cell type	Mechanism and function	References
Swiss 3T3 fibroblasts	GPCR were stimulated (bombesin, bradykinin, vasopressin) resulting in IP_3-mediated Ca^{2+} release in fibroblasts. Released Ca^{2+} associates with calmodulin (CaM), modulating CaMKII (CaM-dependent protein kinase II). CaMKII, in turn, phosphorylates FAK at Ser-843. Once FAK is phosphorylated, it leads to increased formation of focal adhesion sites	Fan et al. (2005)
WI-38 (human embryonic lung fibroblasts)	Alterations in the membrane potential via transient receptor potential melastatin 7 (TRPM7) and chemoattractant signal transduction via IP_3R2 are responsible for Ca^{2+} flickers at the front of a migrating cell, controlling the direction of cell migration	Wei et al. (2009); for review Wei et al. (2010)
Melanoma cells (Mel-2 and SK-Mel-24)	The PLCε/IP_3/IP_3R1 pathway mediates EPAC (exchange protein activated by cyclic AMP)-induced Ca^{2+} elevation which increases melanoma cell migration through the interaction between S100A4 and MHCIIA (myosin heavy chain IIA) and the resultant actin assembly	Baljinnyam et al. (2010)
Glioblastoma cell lines (U178MG, U87MG, and T98G); mouse xenograft model of glioblastoma	Caffeine inhibits invasion and increases survival rate of subject animal (in mouse xenograft model of glioblastoma) by selectively targeting the IP_3R3	Kang et al. (2010)
Pancreatic ductal adenocarcinoma cells (PDAC; PANC-1 cell line)	The migration of PDAC cells was strongly suppressed by selective inhibition of IP_3Rs and SOCE (using xestospongin B and siRNA)	Okeke et al. (2016)

level of IP_3R3 was significantly increased in glioblastoma cells compared to normal cells (Kang et al. 2010). The inhibition of IP_3R3 by caffeine suppresses the migration and invasion of glioblastoma cell lines when assessed by scrape motility, matrigel invasion, soft agar, and mouse brain slice implantation. To show this, U87MG cells, a glioblastoma cell line known for its high tumorigenicity, were injected into the skin of nude mice, some of which were supplied with caffeine in drinking water. The supplementation with caffeine considerably increased the mean survival rate of mice, proposing IP_3R3 as a promising novel therapeutic target for glioblastoma treatment and showing that caffeine could be a useful adjunct therapy (Kang et al. 2010). Noteworthy, the final concentration of caffeine measured in the mice brain corresponded to levels of caffeine produced by drinking two to five cups of coffee per day by humans. Caffeine is a well-known activator of RyRs but has also been shown to inhibit IP_3R activity (Brown et al. 1992), although ryanodine, a known inhibitor of RyR activity, did not show any effect on migration and invasion of the transformed cells. The type 3 IP_3R was also implicated in peritoneal dissemination of gastric cancer (Sakakura et al. 2003).

IP_3R-mediated Ca^{2+} release has also been linked to metastatic melanoma. Interestingly, Ca^{2+} release from the stores promoted the migration of metastatic epithelial breast adenocarcinoma (MDA-MB-231) and cervical carcinoma (HeLa) cell lines via the activation of S100A4 protein by Ca^{2+} (Mueller et al. 1999). Indeed, the PLC/IP_3/IP_3R1 pathway has been shown to mediate EPAC (exchange protein activated by cyclic AMP)-induced Ca^{2+} elevation that in turn increases melanoma cell migration through the interaction between S100A4 and MHCIIA and the resultant actin assembly (Baljinnyam et al. 2010). This suggests that IP_3R can also be a potential target for the suppression of melanoma cell migration opening a perspective of treatments against the metastasis development (see Fig. 31.1).

In addition, the use of the PLC inhibitor U73122 significantly decreased the Ca^{2+} release from the ER and decreased the migration of melanoma cell through monolayers of human umbilical vein endothelial cells (HUVECs) (Peng et al. 2009).

A key constituent involved in the regulation of focal adhesion turnover is FAK. It is a non-receptor kinase that promotes cell migration by transferring signals between growth factor receptors and integrins. FAK is regulated by phosphorylation at tyrosine and serine residues. By using GPCR agonists such as bombesin, bradykinin, and vasopressin stimulates the PLC/IP_3/IP_3R pathway leading to Ca^{2+} mobilization from ER stores in Swiss 3T3 fibroblasts. This released Ca^{2+} associates with CaM, and, subsequentially, the Ca^{2+}/CaM complex regulates CaMKII (calmodulin-dependent protein kinase II) which then rapidly phosphorylates FAK at Ser-843. Once FAK is phosphorylated at this and three other locations (Ser-722, Ser-732 and Ser-910), it facilitates the formation of focal adhesion points (Fan et al. 2005) (see Fig. 31.1).

Localized Ca^{2+} flickers at the leading lamella permit the decision-making for turning direction of the migrating cells: the cell turns in the direction of higher Ca^{2+} flicker activity. These flickering Ca^{2+} microdomains result in membrane tension via TRPM7 and chemoattractant signal transduction via IP_3R2. However, by

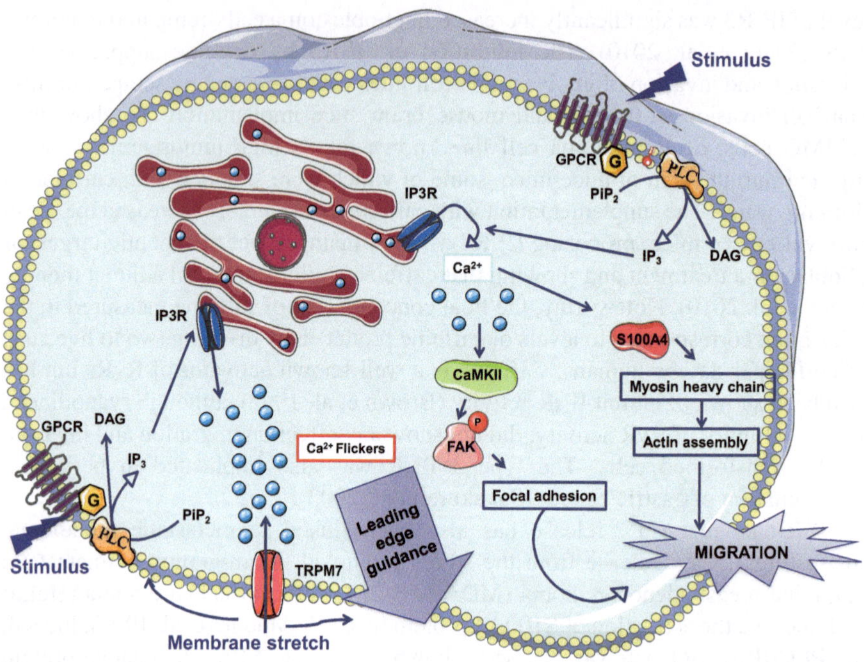

Fig. 31.1 This figure represents the cross section of a migrating cell at the leading edge. It demonstrates the known pathways by which the IP_3Rs participate in cell migration/invasion through controlling the release of Ca^{2+} from the ER. Activation of GPCRs by external stimuli is represented via lightning bolts and the activation of further pathways, as well as Ca^{2+} release and production of secondary messengers are indicated by the *arrows*. The membrane stretch-activated, transient receptor potential cation channel subfamily M member 7 (TRPM7)-mediated Ca^{2+} influx coupled to IP_3R-mediated Ca^{2+} release participates in the guidance of the leading edge toward a chemoattractant (Wei et al. 2010). Formation of focal adhesions is facilitated by IP_3R-mediated Ca^{2+} release that is stimulated through surface GPCRs or EPAC leading to CaMKII-dependent FAK phosphorylation (Fan et al. 2005). IP_3R-mediated Ca^{2+}mobilization from the endoplasmic reticulum (ER) stores also promotes cell migration in an S100A4-dependent manner (Baljinnyam et al. 2010)

themselves, these flickers only determine the turning direction but not the migration speed. Indeed, the use of EGTA to chelate intracellular Ca^{2+} slows the turning but not the migration speed in a dose-dependent manner, showing that spatiotemporal organization of Ca^{2+} microdomains is important in such convoluted cellular processes as cell migration (Wei et al. 2009, 2010) (see Fig. 31.1).

Recently, IP_3Rs have been shown to translocate from cell-cell contacts to the leading edge of migrating pancreatic ductal adenocarcinoma cells (PDAC) which was accompanied by the formation of ER-PM junctions and STIM1 puncta. Xestospongin B, a selective IP_3R inhibitor, and the STIM-Orai inhibitor GSK-79751 suppressed the migration of the cell line PANC-1. Taken together, these data emphasize that both, Ca^{2+} release and Ca^{2+} influx, are important for cell migration (Okeke et al. 2016).

31.3.3 Role of IP_3Rs in SOCE

The principal pathways by which ion channels regulate cellular migration are associated with their primary functions, such as regulation of cellular adhesion, volume or migration speed, maintenance of cellular potential, and regulation of intra- and extracellular pH. Unfortunately, the link between the expression and function of IP_3Rs and the mechanisms of their involvement in cancer cell migration, invasion, and metastasis is not yet well understood. Nonetheless, there is general consensus that Ca^{2+} signaling alterations play a major role in carcinogenesis (Parkash and Asotra 2011; Wei et al. 2012).

Ion channel activity can essentially be described as a series of transitions between open and closed states, with a particular probability of being open (P_{open}). The opening/closing of the IP_3Rs can be modulated by different factors such as Ca^{2+}, IP_3, and Mg^{2+}. For example, the IP_3R activity is regulated by Ca^{2+} in a bimodal manner. P_{open} of IP_3R, and thus Ca^{2+} release from ER, is very low at low Ca^{2+} levels, gradually increasing with the rise of $[Ca^{2+}]_i$, reaching a plateau at μM concentrations, and finally being again inhibited at higher $[Ca^{2+}]_i$, thus also leading to a decrease in Ca^{2+} release. In the majority of publications, involvement of the IP_3Rs in migration and invasion was found through the use of its major agonist, IP_3. Once the IP_3R is activated, Ca^{2+} is released from the ER and can either stimulate Ca^{2+}-dependent proteins such as calpain or Ca^{2+}-activated ion channels (Fig. 31.2). In that way, the mobilization of Ca^{2+} from the intracellular stores via IP_3Rs can magnify the Ca^{2+} entry signal via SOCs and result in a strong response capable of inducing cell migration and invasion or even stimulate metastasis formation. The release of Ca^{2+} from the ER via IP_3Rs has also been shown to induce the SOC opening. The Ca^{2+} release induces Ca^{2+} entry via STIM1/ORAI1 resulting in Ras and Rac protein (small GTPases) activation (Yang et al. 2009) which in turn accelerates the turnover of focal adhesions and further increases migration. In support of these observations, there is now considerable evidence that STIM1 activates both, Orai1 and TRPC1, via distinct domains at its C-terminus (Ong et al. 2016).

Concerning the invasion mechanism, there is no evidence that the IP_3Rs can directly affect the activation of metalloproteinases and, thus, the digestion of the extracellular matrix necessary for cell invasion. Otherwise, it has been shown that the induction of MMP2 and MMP9 can happen via Ca^{2+} flux through TRPV2 (Monet et al. 2010). Moreover, S100A4 has been found to control the invasive potential of human prostate cancer cells through the regulation of MMP9 expression, contributing to the proteolysis of ECM by invadopodia (Saleem et al. 2006). It has been suggested that the link between IP_3Rs and ECM degradation can be S100A4 (also known as mts1) which is a Ca^{2+}-binding protein associated with invasion and metastasis of cancer cells and can, therefore, be activated by Ca^{2+} release via the IP_3Rs (Baljinnyam et al. 2010). Further studies on this interesting topic are needed to confirm these hypotheses.

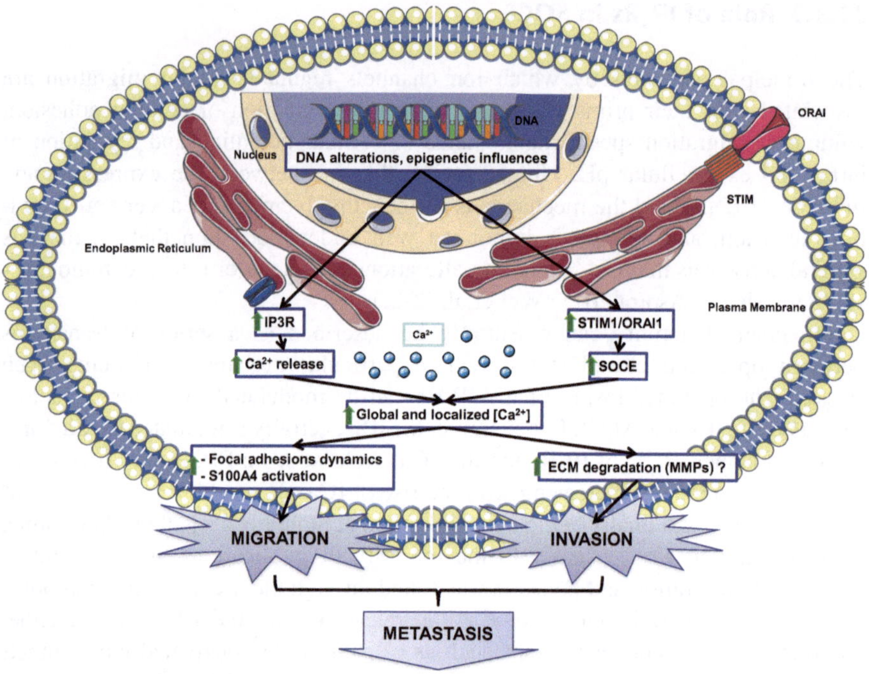

Fig. 31.2 This scheme illustrates the roles IP3Rs play in SOCE regulation for the promotion of cancer cell metastasis. The *black arrows* show the sequence of events; the *green upward arrows* indicate an increase in the expression and/or function of the corresponding protein(s) or augmentation of the corresponding process. The "?" indicates that the associated step has been recently hypothesized in literature (Baljinnyam et al. 2010) but requires further evidence

31.3.4 Role of IP_3Rs Protein Partners in SOCE

It is well known that IP_3R activity is heavily modulated by various partner proteins. Among them are in particular the proteins of the Bcl-2 family that modulate the activity of IP_3Rs directly or indirectly. Multiple studies, such as Rong et al. (2008), have shown that Bcl-2 is present in the ER where it can interact directly with all three IP_3R isoforms, inhibiting their activity via multiple binding sites proposed to explain this interaction (Fouque et al. 2016; Ivanova et al. 2016). Other proteins of the Bcl-2 family such as Bcl-xL or Mcl-1 are also known to interact with the IP_3Rs and modulate its activity (Eckenrode et al. 2010; Monaco et al. 2012).

Interestingly, Bcl-2 and Bcl-xL are found to be overexpressed in many cancers. Overexpression of Bcl-2 is observed in 30–60% of androgen-dependent and nearly 100% of androgen-independent prostate cancers, with similar results observed for Bcl-xL. Unfortunately, the studies carried out so far reported primarily on the changes in the expression levels, with little to no data on the possible impact of the modulation of IP_3R activity by Bcl-2/Bcl-xL on cancer cell migration/invasion. To find this likely link, it could be very interesting to test the effects of different drugs already known to specifically disrupt the interaction between Bcl-2 protein family members and the IP_3Rs, such as ABT-199, a BH3 mimetic that specifically binds Bcl-2's hydrophobic cleft (Vervloessem et al. 2016), on cancer cell migration and invasion properties.

While the interaction between Bcl-2 family members and IP_3Rs has been known for a while, in the past few years scientists have shown that the IP_3Rs also interact with many other proteins, including the Sigma1 receptor which has also been found to modulate SOCE (Rosado 2016). Additionally, IP_3Rs have been suggested to interact with some of the channels belonging to the family of the TRP channels (Fiorio Pla and Gkika 2013). Thus, for example, TRPC3 was successfully co-immunoprecipitated with the IP_3Rs by glutathione S-transferase (GST) pulldown experiments (Boulay et al. 1999). Moreover, TRPC3 has been shown to be activated by IP_3, whereby the N-terminal IP_3-binding domain of IP_3R was essential (Kiselyov et al. 1999). Similar experiment permits allowed the identification of biochemical interactions between IP_3Rs and TRPC1 (Mehta et al. 2003). It is also interesting to note that there is now considerable evidence that STIM1 activates both, Orai1 and TRPC1, via distinct domains in its C-terminus (Ong et al. 2016).

31.4 Conclusion and Perspectives

Ca^{2+} plays many important roles in regulating cellular health and migration properties with SOCE being one of the primary mechanisms regulating intracellular Ca^{2+} homeostasis. It is therefore natural that SOCE is at the core of many diseases, including such important ones as cancer, with many publications focusing on the mechanisms and dysregulations of SOCE. This review has focused in particular on a less well-covered area of store-operated Ca^{2+}—the IP_3R channels responsible for the depletion of internal stores in a majority of cells and their various roles in modifying SOCE and affecting cellular motility and invasion.

The role of IP_3Rs in SOCE and associated diseases is a relatively novel topic. Recent publications only start to address the question of how IP3Rs modulate cancer progression in particular concerning the questions of migration and invasion. However, IP3R activity is known to be modulated by numerous protein partners. While multitude of studies have been devoted of the topics of this regulation itself, its involvement in SOCE regulation where cancer progression is concerned has not yet been properly addressed in scientific literature. Considering the increasing number of IP_3R protein partners discovered regularly in recent years, the importance of this angle should not be underestimated. However, as one can easily see, there is still a lot of research to be carried out in order to comprehensively understand the mechanisms by which these channels and SOCE regulate migration and invasion of cancer cells. Further research on any of these topics provides a great promise to develop treatments that could counteract or at least arrest progression of various forms of cancer toward metastatic stages.

Acknowledgments This work is supported by grant from LabEx ICST (Laboratory of Excellence Ion Channel Science and Therapeutics).

Illustrations were produced in part using the content of Servier Medical Art image bank, made available under the terms of Creative Commons Attribution Licence 3.0 by Servier (France).

References

Agle KA, Vongsa RA, Dwinell MB (2010) Calcium mobilization triggered by the chemokine CXCL12 regulates migration in wounded intestinal epithelial monolayers. J Biol Chem 285:16066–16075

Baljinnyam E, De Lorenzo MS, Xie L, Iwatsubo M, Chen S, Goydos JS, Nowycky MC, Iwatsubo K (2010) Exchange protein directly activated by cyclic AMP increases melanoma cell migration by a Ca2+-dependent mechanism. Cancer Res 70:5607–5617

Boulay G, Brown DM, Qin N, Jiang M, Dietrich A, Zhu MX, Chen Z, Birnbaumer M, Mikoshiba K, Birnbaumer L (1999) Modulation of Ca(2+) entry by polypeptides of the inositol 1,4, 5-trisphosphate receptor (IP3R) that bind transient receptor potential (TRP): evidence for roles of TRP and IP3R in store depletion-activated Ca(2+) entry. Proc Natl Acad Sci USA 96:14955–14960

Brown GR, Sayers LG, Kirk CJ, Michell RH, Michelangeli F (1992) The opening of the inositol 1, 4, 5-trisphosphate-sensitive Ca2+ channel in rat cerebellum is inhibited by caffeine. Biochem J 282:309–312

Brundage RA, Fogarty KE, Tuft RA, Fay FS (1991) Calcium gradients underlying polarization and chemotaxis of eosinophils. Science 254:703–706

Eckenrode EF, Yang J, Velmurugan GV, Foskett JK, White C (2010) Apoptosis protection by Mcl-1 and Bcl-2 modulation of inositol 1,4,5-trisphosphate receptor-dependent Ca2+ signaling. J Biol Chem 285:13678–13684

Eddy RJ, Pierini LM, Matsumura F, Maxfield FR (2000) Ca2+-dependent myosin II activation is required for uropod retraction during neutrophil migration. J Cell Sci 1298:1287–1298

Fan RS, Jacoma RO, Jiang X, Sinnett-Smith J, Rozengurt E (2005) G protein-coupled receptor activation rapidly stimulates focal adhesion kinase phosphorylation at Ser-843. Mediation by Ca2+, calmodulin, and Ca2+/calmodulin-dependent kinase II. J Biol Chem 280:24212–24220

Fan G, Baker ML, Wang Z, Baker MR, Sinyagovskiy PA, Chiu W, Ludtke SJ, Serysheva II (2015) Gating machinery of InsP3R channels revealed by electron cryomicroscopy. Nature 527:336–341

Fiorio Pla A, Gkika D (2013) Emerging role of TRP channels in cell migration: from tumor vascularization to metastasis. Front Physiol 4:311

Fiorio Pla A, Kondratska K, Prevarskaya N (2016) STIM and ORAI proteins: crucial roles in hallmarks of cancer. Am J Physiol Cell Physiol 310:C509–C519

Foskett JK, White C, Cheung KH, Mak DO (2007) Inositol trisphosphate receptor Ca2+ release channels. Physiol Rev 87:593–658

Fouque A, Lepvrier E, Debure L, Gouriou Y, Malleter M, Delcroix V, Ovize M, Ducret T, Li C, Hammadi M, Vacher P, Legembre P (2016) The apoptotic members CD95, BclxL, and Bcl-2 cooperate to promote cell migration by inducing Ca(2+) flux from the endoplasmic reticulum to mitochondria. Cell Death Differ 23:1702–1716

Furuichi T, Yoshikawa S, Miyawaki A, Wada K, Maeda N, Mikoshiba K (1989a) Primary structure and functional expression of the inositol 1,4,5-trisphosphate-binding protein P400. Nature 342:32–38

Furuichi T, Yoshikawa S, Mikoshiba K (1989b) Nucleotide sequence of cDNA encoding P400 protein in the mouse cerebellum. Nucleic Acids Res 17:5385–5386

Hahn K, DeBiasio R, Taylor DL (1992) Patterns of elevated free calcium and calmodulin activation in living cells. Nature 359:736–738

Hanahan D, Weinberg RA (2000) The hallmarks of cancer. Cell 100:57–70

Hanahan D, Weinberg RA (2011) Hallmarks of cancer: the next generation. Cell 144:646–674

Ivanova H, Vervliet T, Missiaen L, Parys JB, De Smedt H, Bultynck G (2014) IP3R-isoform diversity in cell death and survival. Biochim Biophys Acta 1843:2164–2183

Ivanova H, Ritaine A, Wagner L, Luyten T, Shapovalov G, Welkenhuyzen K, Seitaj B, Monaco G, De Smedt H, Prevarskaya N, Yule DI, Parys JB, Bultynck G (2016) The trans-membrane domain of Bcl-2α, but not its hydrophobic cleft, is a critical determinant for efficient IP3 receptor inhibition. Oncotarget 7:55704–55720

Jin K, Li T, van Dam H, Zhou F, Zhang L (2017) Molecular insights into tumour metastasis: tracing the dominant events. J Pathol 241(5):567–577

Kahl CR, Means AR (2003) Regulation of cell cycle progression by calcium/calmodulin-dependent pathways. Endocr Rev 24:719–736

Kang SS, Han KS, Ku BM, Lee YK, Hong J, Shin HY, Almonte AG, Woo DH, Brat DJ, Hwang EM, Yoo SH, Chung CK, Park SH, Paek SH, Roh EJ, Lee SJ, Park JY, Traynelis SF, Lee CJ (2010) Caffeine-mediated inhibition of calcium release channel inositol 1,4,5-trisphosphate receptor subtype 3 blocks glioblastoma invasion and extends survival. Cancer Res 70:1173–1183

Kiselyov K, Mignery GA, Zhu MX, Muallem S (1999) The N-terminal domain of the IP3 receptor gates store-operated hTrp3 channels. Mol Cell 4:423–429

Lauffenburger DA, Horwitz AF (1996) Cell migration: a physically integrated molecular process. Cell 84:359–369

Leanza L, Biasutto L, Manago A, Gulbins E, Zoratti M, Szabo I (2013) Intracellular ion channels and cancer. Front Physiol 4:227

Maeda N, Niinobe M, Nakahira K (1988) Purification and characterization of P400 protein, a glycoprotein characteristic of Purkinje cell, from mouse cerebellum. J Neurochem 51:1724–1730

Maeda N, Niinobe M, Mikoshiba K (1990) A cerebellar Purkinje cell marker P400 is an inositol 1,4,5-trisphosphate (InsP3) receptor protein. Purification and characterization of InsP3 receptor complex. EMBO J 9:61–67

Maeda N, Kawasaki T, Nakade S, Yokota N, Taguchi T, Kasaiq M, Mikoshiba K (1991) Structural and functional characterization of inositol 1,4,5-trisphosphate receptor channel from mouse cerebellum. J Biol Chem 266:1109–1116

Mehta D, Ahmmed GU, Paria BC, Holinstat M, Voyno-Yasenetskaya T, Tiruppathi C, Minshall RD, Malik AB (2003) RhoA interaction with inositol 1,4,5-trisphosphate receptor and transient receptor potential channel-1 regulates Ca2+ entry. Role in signaling increased endothelial permeability. J Biol Chem 278:33492–33500

Michelangeli F, East JM (2011) A diversity of SERCA Ca2+ pump inhibitors. Biochem Soc Trans 39:789–797

Monaco G, Beckers M, Ivanova H, Missiaen L, Parys JB, De Smedt H, Bultynck G (2012) Profiling of the Bcl-2/Bcl-X(L)-binding sites on type 1 IP(3) receptor. Biochem Biophys Res Commun 428:31–35

Monet M, Lehen'kyi V, Gackiere F, Firlej V, Vandenberghe M, Roudbaraki M, Gkika D, Pourtier A, Bidaux G, Slomianny C, Delcourt P, Rassendren F, Bergerat JP, Ceraline J, Cabon F, Humez S, Prevarskaya N (2010) Role of cationic channel TRPV2 in promoting prostate cancer migration and progression to androgen resistance. Cancer Res 70:1225–1235

Mueller A, Bächi T, Höchli M, Schäfer BW, Heizmann CW (1999) Subcellular distribution of S100 proteins in tumor cells and their relocation in response to calcium activation. Histochem Cell Biol 111:453–459

Nabeshima K, Inoue T, Shimao Y, Sameshima T (2002) Matrix metalloproteinases in tumor invasion: role for cell migration. Pathol Int 52:255–264

Okeke E, Parker T, Dingsdale H, Concannon M, Awais M, Voronina S, Molgo J, Begg M, Metcalf D, Knight AE, Sutton R, Haynes L, Tepikin AV (2016) Epithelial-mesenchymal transition, IP3 receptors and ER-PM junctions: translocation of Ca2+ signalling complexes and regulation of migration. Biochem J 473:757–767

Ong HL, de Souza LB, Ambudkar IS (2016) Role of TRPC channels in store-operated calcium entry. In: Rosado JA (ed) Calcium entry pathways in non-excitable cells, Advances in experimental medicine and biology. Springer, Cham, pp 87–109
Palecek SP, Huttenlocher A, Horwitz AF, Lauffenburger DA (1998) Physical and biochemical regulation of integrin release during rear detachment of migrating cells. J Cell Sci 940:929–940
Parkash J, Asotra K (2011) Calcium wave signaling in cancer cells. Life Sci 87:587–595
Peng H, Hodgson L, Henderson AJ, Dong C (2009) Involvement of phospholipase C signaling in melanoma cell-induced endothelial junction disassembly. Front Biosci 1:1597–1606
Pettit EJ, Fay FS (1998) Cytosolic free calcium and the cytoskeleton in the control of leukocyte chemotaxis. Physiol Rev 78:949–967
Prevarskaya N, Skryma R, Shuba Y (2010) Ion channels and the hallmarks of cancer. Trends Mol Med 16:107–121
Prevarskaya N, Skryma R, Shuba Y (2011) Calcium in tumour metastasis: new roles for known actors. Nat Rev Cancer 11:609–618
Ridley A, Schwartz M, Burridge K, Firtel R, Ginsberg M, Borisy G, Parsons J, Horwitz A (2003) Cell migration: integrating signals from front to back. Science 302:1704–1710
Rizzuto R, Marchi S, Bonora M, Aguiari P, Bononi A, De Stefani D, Giorgi C, Leo S, Rimessi A, Siviero R, Zecchini E, Pinton P (2009) Ca(2+) transfer from the ER to mitochondria: when, how and why. Biochim Biophys Acta 1787:1342–1351
Rong YP, Aromolaran AS, Bultynck G, Zhong F, Li X, McColl K, Matsuyama S, Herlitze S, Roderick HL, Bootman MD, Mignery GA, Parys JB, De Smedt H, Distelhorst CW (2008) Targeting Bcl-2-IP3 receptor interaction to reverse Bcl-2's inhibition of apoptotic calcium signals. Mol Cell 31:255–265
Rosado JA (2016) Sigma-1 receptors: a new pathway for the modulation of store-operated calcium entry. Biochem J 473:e9–e10
Ross CA, Meldolesi J, Milner TA, Satoh T, Supattapone S, Snyder SH (1989) Inositol 1,4,5-trisphosphate receptor localized to endoplasmic reticulum in cerebellar Purkinje neurons. Nature 339:468–470
Sakakura C, Hagiwara A, Fukuda K, Shimomura K, Takagi T, Kin S, Nakase Y, Fujiyama J, Mikoshiba K, Okazaki Y, Yamagishi H (2003) Possible involvement of inositol 1,4,5-trisphosphate receptor type 3 (IP3R3) in the peritoneal dissemination of gastric cancers. Anticancer Res 23:3691–3697
Saleem M, Kweon MH, Johnson JJ, Adhami VM, Elcheva I, Khan N, Bin Hafeez B, Bhat KM, Sarfaraz S, Reagan-Shaw S, Spiegelman VS, Setaluri V, Mukhtar H (2006) S100A4 accelerates tumorigenesis and invasion of human prostate cancer through the transcriptional regulation of matrix metalloproteinase 9. Proc Natl Acad Sci USA 103:14825–14830
Streb H, Irvine RF, Berridge MJ, Schulz I (1983) Release of Ca2+ from a nonmitochondrial intracellular store in pancreatic acinar cells by inositol-1,4,5-trisphosphate. Nature 306:67–69
Sudhof TC, Newton CL, Archer BT, Ushkaryov YA, Mignery GA (1991) Structure of novel InsP3 receptor. EMBO J 1:3199–3206
Supattapone S, Worley PF, Baraban JM, Snyder SH (1988) Solubilization, purification, and characterization of an inositol trisphosphate receptor. J Biol Chem 263:1530–1534
Taylor CW, Konieczny V (2016) IP 3 receptors: take four IP 3 to open. Sci Signal 9:1–9
Tsai F, Seki A, Yang HW, Hayer A, Carrasco S, Meyer T (2014) A polarized Ca2+, diacylglycerol, and STIM1 signaling system regulates directed cell migration. Nat Cell Biol 16:133–144
Vervloessem T, Yule DI, Bultynck G, Parys JB (2014) The type 2 inositol 1,4,5-trisphosphate receptor, emerging functions for an intriguing Ca^{2+}-release channel. Biochim Biophys Acta 1853:1992–2005
Vervloessem T, Ivanova H, Luyten T, Parys JB, Bultynck G (2016) The selective Bcl-2 inhibitor venetoclax, a BH3 mimetic, does not dysregulate intracellular Ca2+ signaling. Biochim Biophys Acta 1864:968–976
Wei C, Wang X, Chen M, Ouyang K, Song LS, Cheng H (2009) Calcium flickers steer cell migration. Nature 457:901–905

Wei C, Wang X, Chen M, Ouyang K, Zheng M, Cheng H (2010) Flickering calcium microdomains signal turning of migrating cells. Can J Physiol Pharmacol 88:105–110

Wei C, Wang X, Zheng M, Cheng H (2012) Calcium gradients underlying cell migration. Curr Opin Cell Biol 24:254–261

Westhoff MA, Serrels B, Fincham VJ, Frame MC, Carragher NO (2004) SRC-mediated phosphorylation of focal adhesion kinase couples actin and adhesion dynamics to survival signaling. Mol Cell Biol 24:8113–8133

Yang S, Huang XY (2005) Ca2+ influx through L-type Ca2+ channels controls the trailing tail contraction in growth factor-induced fibroblast cell migration. J Biol Chem 280:27130–27137

Yang S, Zhang JJ, Huang XY (2009) Orai1 and STIM1 are critical for breast tumor cell migration and metastasis. Cancer Cell 15:124–134

Yang X, Jin H, Cai X, Li S, Shen Y (2012) Structural and mechanistic insights into the activation of Stromal interaction molecule 1 (STIM1). Proc Natl Acad Sci USA 109:5657–5662

Yoon S, Park S, Yun C, Chung A (2003) Roles of matrix metalloproteinases in tumor metastasis and angiogenesis. J Biochem Mol Biol 36:128–137

Yoshida M, Ishikawa M, Izumi H, De Santis R, Morisawa M (2003) Store-operated calcium channel regulates the chemotactic behavior of ascidian sperm. Proc Natl Acad Sci USA 100:149–154

Index

K. Groschner et al. (eds.), *Store-Operated Ca^{2+} Entry (SOCE) Pathways*, Advances in Experimental Medicine and Biology 993, DOI 10.1007/978-3-319-57732-6

D